# BNi Building News

# REMODELING 1999 COSTBOOK

NINTH EDITION

BNi Building News

Los Angeles • Anaheim

Boston • Washington, D.C.

BNi Building News

EDITOR-IN-CHIEF
*William D. Mahoney, P.E.*

TECHNICAL SERVICES
*Edward B. Wetherill, CSI*

GRAPHIC DESIGN
*Robert O. Wright*

COVER DESIGN
*Hannus Design Associates*

**BNI Publications, Inc.**

**LOS ANGELES**
10801 NATIONAL BLVD. SUITE 100
LOS ANGELES, CA 90064

**ANAHEIM**
1612 S. CLEMENTINE STREET
ANAHEIM, CA 92802

**BOSTON**
629 HIGHLAND AVENUE
NEEDHAM HEIGHTS, MA 02494

**WASHINGTON, D.C.**
502 MAPLE AVENUE WEST
VIENNA, VA 22180

**1-888-BNI-BOOK**

**ISBN 1-55701-250-4**

# PREFACE

For the past 50 years, Building News has been dedicated to providing construction professionals with timely and reliable information. Based on this experience, our staff has researched and compiled thousands of up-to-the-minute costs for the **Building News 1999 Costbooks**. This book is an essential reference for contractors, engineers, architects, facilities managers — any construction professional who must provide an estimate or any type of building project.

Whether working up a preliminary estimate or submitting a formal bid, the costs listed here can quickly and easily be tailored to your needs. All costs are based on national averages, while a table of modifiers is provided for regional adjustments. Overhead and profit are included in all costs.

Complete man-hour tables follow the unit costs to provide data on typical durations of specific tasks. This information can be used to schedule projects as well as to determine specific labor costs based on local labor rates.

All data is categorized according to the MASTERFORMAT of the Construction Specifications Institute (CSI). This industry standard provides an all-inclusive checklist to ensure that no element of a project is overlooked. In addition, to make specific items even easier to locate, there is a complete alphabetical index.

This costbook contains an appendix with reference charts and tables taken from an array of sources. Text explains the costs in certain categories and provides helpful pointers that should be taken into account with every estimate.

A section on square foot costs provides an overview of project costs for different building types — commercial, residential, etc. — with summaries of the actual projects. Square foot costs are invaluable for making budget estimates and checking prices when time is a factor.

The "Features in this Book" section presents a clear overview of the many features of this book. Included is an explanation of the data, sample page layout and discussion of how to best use the information in the book.

Of course, all buildings and construction projects are unique. The costs provided in this book are based on averages from well-managed projects with good labor productivity under normal working conditions (eight hours a day). Other circumstances affecting costs such as overtime, unusual working conditions, savings from buying bulk quantities for large projects, and unusual or hidden costs must be factored in as they arise.

The data provided in this book is for estimating purposes only. Check all applicable federal, state and local codes and regulations for specific requirements.

BNi Building News

# TABLE OF CONTENTS


CSI MASTERFORMAT ........................................ vi
Features In This Book ........................................ ix
Construction Dictionary ........................................ xi
Sample Costbook Page ........................................ xxxvii

Costbook Pages ........................................ 1
Man-Hour Tables ........................................ 271
Supporting Construction Reference Data ......... 399
Geographic Cost Modifiers ........................................ 475
Square Foot Tables ........................................ 479

Index ........................................ 491

Instructions for Software ........................................ 505

# CSI MASTERFORMAT

All data in the Costbook pages and Man-Hour tables is organized according to the CSI MASTERFORMAT — the industry standard numbering/classification system. The data is divided into the 16 divisions as shown below. The five digit numbers within each division correspond to the MASTERFORMAT Broadscope designations. Each section is further broken down into MASTERFORMAT Mediumscope designations. These numbers, in most cases, are the same as those used in architectural and engineering specifications.

The construction estimating information in this book is divided into two main sections: Costbook Pages and Man-Hour Tables. Each is organized to the 16 divisions of the CSI MASTERFORMAT as shown below. In addition, there are extensive Construction Reference Tables, Geographic Cost Modifiers, Square Foot Tables and a detailed Index. Sample pages with graphic explanations are included before the Costbook pages, Man-Hour Tables and Square Foot Tables. These explanations, along with the discussions below, will provide a good understanding of what is included in this book and how it can best be used for construction estimating.

**GENERAL REQUIREMENTS** — *Division 1*

- Summary of Work .......... 01010
- Allowances .......... 01020
- Measurement and Payment .......... 01025
- Alternates/Alternatives .......... 01030
- Modification Procedures .......... 01035
- Coordination .......... 01040
- Field Engineering .......... 01050
- Regulatory Requirements .......... 01060
- Identification Systems .......... 01070
- References .......... 01090
- Special Project Procedures .......... 01100
- Project Meetings .......... 01200
- Submittals .......... 01300
- Quality Control .......... 01400
- Construction Facilities & Temporary Controls .......... 01500
- Material and Equipment .......... 01600
- Facility Startup/Commissioning .......... 01650
- Contract Closeout .......... 01700
- Maintenance .......... 01800

**SITEWORK** — *Division 2*

- Subsurface Investigation .......... 02010
- Demolition .......... 02050
- Site Preparation .......... 02100
- Dewatering .......... 02140
- Shoring and Underpinning .......... 02150
- Excavation Support Systems .......... 02160
- Cofferdams .......... 02170
- Earthwork .......... 02200
- Tunneling .......... 02300
- Piles and Caissons .......... 02350
- Railroad Work .......... 02450
- Marine Work .......... 02480
- Paving and Surfacing .......... 02500
- Utility Piping Materials .......... 02600
- Water Distribution .......... 02660
- Fuel and Steam Distribution .......... 02680
- Sewerage and Drainage .......... 02700
- Restoration and Underground Pipe .......... 02760
- Ponds and Reservoirs .......... 02770
- Power and Communications .......... 02780
- Site Improvements .......... 02800
- Landscaping .......... 02900

**CONCRETE** — *Division 3*

- Concrete Formwork .......... 03100
- Concrete Reinforcement .......... 03200
- Concrete Accessories .......... 03250
- Cast-In-Place Concrete .......... 03300
- Concrete Curing .......... 03370
- Precast Concrete .......... 03400
- Cementitious Decks and Toppings .......... 03500
- Grout .......... 03600
- Concrete Restoration and Cleaning .......... 03700
- Mass Concrete .......... 03800

**MASONRY** — *Division 4*

- Mortar and Masonry Grout .......... 04100
- Masonry Accessories .......... 04150
- Unit Masonry .......... 04200
- Stone .......... 04400
- Masonry Restoration and Cleaning .......... 04500
- Refractories .......... 04550
- Corrosion Resistant Masonry .......... 04600
- Simulated Masonry .......... 04700

**METALS** — *Division 5*

- Metal Materials .......... 05010
- Metal Coatings .......... 05030
- Metal Fastening .......... 05050
- Structural Metal Framing .......... 05100
- Metal Joists .......... 05200
- Metal Decking .......... 05300
- Cold Formed Metal Framing .......... 05400

Metal Fabrication .....05500
Sheet Metal Fabrications .....05580
Ornamental Metal .....05700
Expansion Control .....05800
Hydraulic Structures .....05900

**WOOD AND PLASTICS** *Division 6*

Fasteners and Adhesives .....06050
Rough Carpentry .....06100
Heavy Timber Construction .....06130
Wood and Metal Systems .....06150
Prefabricated Structural Wood .....06170
Finish Carpentry .....06200
Wood Treatment .....06300
Architectural Woodwork .....06400
Structural Plastics .....06500
Plastic Fabrications .....06600
Solid Polymer Fabrications .....06650

**THERMAL AND MOISTURE PROTECTION** *Division 7*

Waterproofing .....07100
Dampproofing .....07150
Water Repellents .....07180
Vapor Retarders .....07190
Air Barriers .....07195
Insulation .....07200
Exterior Insulation and Finish Systems .....07240
Fireproofing .....07250
Firestopping .....07270
Shingles and Roofing Tiles .....07280
Manufactured Roofing and Siding .....07400
Exterior Wall Assemblies .....07480
Membrane Roofing .....07500
Traffic Coatings .....07570
Flashing and Sheet Metal .....07600
Roof Specialties and Accessories .....07700
Skylights .....07800
Joint Sealers .....07900

**DOORS AND WINDOWS** *Division 8*

Metal Doors and Frames .....08100
Wood and Plastic Doors .....08200
Door Opening Assemblies .....08250
Special Doors .....08300
Entrances and Storefronts .....08400
Metal Windows .....08500
Wood and Plastic Windows .....08600
Special Windows .....08650
Hardware .....08700
Glazing .....08800
Glazed Curtain Walls .....08900

**FINISHES** *Division 9*

Metal Support Systems .....09100
Lath and Plaster .....09200
Gypsum Board .....09250
Tile .....09300
Terrazzo .....09400
Stone Facing .....09450
Acoustical Treatment .....09500
Special Wall Surfaces .....09540
Special Ceiling Surfaces .....09545
Wood Flooring .....09550
Stone Flooring .....09600
Unit Masonry Flooring .....09630
Resilient Flooring .....09650
Carpet .....09680
Special Flooring .....09700
Floor Treatment .....09780
Special Coatings .....09800
Painting .....09900
Wall Coverings .....09950

**SPECIALTIES** *Division 10*

Visual Display Boards .....10100
Compartments and Cubicles .....10150
Louvers and Vents .....10200
Grilles and Screens .....10240
Service Wall Systems .....10250
Wall and Corner Guards .....10260
Access Flooring .....10270
Pest Control .....10290
Fireplaces and Stoves .....10300
Manufactured Exterior Specialties .....10340
Flagpoles .....10350
Identifying Devices .....10400
Pedestrian Control Devices .....10450
Lockers .....10500
Fire Protection Specialties .....10520
Protective Covers .....10530
Postal Specialties .....10550
Partitions .....10600
Operable Partitions .....10650
Storage Shelving .....10670
Exterior Protection Devices for Openings .....10700
Telephone Specialties .....10750
Toilet and Bath Accessories .....10800
Scales .....10880
Wardrobe and Closet Specialties .....10900

**EQUIPMENT** *Division 11*

Maintenance Equipment .....11010
Security and Vault Equipment .....11020
Teller and Service Equipment .....11030
Ecclesiastical Equipment .....11040
Library Equipment .....11050
Theater and Stage Equipment .....11060
Instrumental Equipment .....11070
Registration Equipment .....11080
Checkroom Equipment .....11090
Mercantile Equipment .....11100
Commercial Laundry and Dry Cleaning Equip. .....11110
Vending Equipment .....11120
Audio-Visual Equipment .....11130
Vehicle Service Equipment .....11140
Parking Control Equipment .....11150
Loading Dock Equipment .....11160
Solid Waste Handling Equipment .....11170

Detention Equipment ... 11190
Water Supply and Treatment Equipment ... 11200
Hydraulic Gates and Valves ... 11280
Fluid Waste Treatment and Disposal Equipment .. 11300
Food Service Equipment ... 11400
Residential Equipment ... 11450
Unit Kitchens ... 11460
Darkroom Equipment ... 11470
Athletic, Recreational & Therapeutic Equipment .. 11480
Industrial and Process Equipment ... 11500
Laboratory Equipment ... 11600
Planetarium Equipment ... 11650
Observatory Equipment ... 11660
Office Equipment ... 11680
Medical Equipment ... 11700
Mortuary Equipment ... 11780
Navigation Equipment ... 11850
Agricultural Equipment ... 11870

**FURNISHINGS** — *Division 12*

Fabrics ... 12060
Artwork ... 12100
Manufactured Casework ... 12300
Window Treatment ... 12500
Furniture and Accessories ... 12600
Rugs and Mats ... 12670
Multiple Seating ... 12700
Interior Parts and Planters ... 12800

**SPECIAL CONSTRUCTION** — *Division 13*

Air Supported Structures ... 13010
Integrated Assemblies ... 13020
Special Purposes Rooms ... 13030
Sound, Vibration and Seismic Control ... 13080
Radiation Protection ... 13090
Nuclear Reactors ... 13100
Pre-Engineered Structures ... 13120
Aquatic Facilities ... 13150
Ice Rinks ... 13175
Site Constructed Incinerators ... 13180
Kennels and Animal Shelters ... 13185
Liquid and Gas Storage Tanks ... 13200
Filter Underdrains and Media ... 13220
Digester Covers and Appurtenances ... 13230
Oxygenation Systems ... 13240
Sludge Conditioning Systems ... 13260
Utility Control Systems ... 13330
Industrial and Process Control Systems ... 13400
Recording Instrumentation ... 13500
Transportation Control Instrumentation ... 13550
Solar Energy Systems ... 13600
Wind Energy Systems ... 13700
Cogeneration Systems ... 13700
Building Automation Systems ... 13800
Fire Suppression and Supervisory Systems ... 13900
Special Security Construction ... 13950

**CONVEYING SYSTEMS** — *Division 14*

Dumbwaiters ... 14100
Elevators ... 14200
Escalators and Moving Walks ... 14300
Lifts ... 14440
Material Handling Systems ... 14500
Hoists and Cranes ... 14600
Turntables ... 14700
Scaffolding ... 14800
Transportation Systems ... 14900

**MECHANICAL** — *Division 15*

Basic Mechanical Material and Methods ... 15050
Mechanical Insulation ... 15250
Fire Protection ... 15300
Plumbing ... 15400
Heating, Ventilating and Air Conditioning ... 15500
Heat Generation ... 15550
Refrigeration ... 15650
Heat Transfer ... 15750
Air Handling ... 15850
Air Distribution ... 15880
Controls ... 15950
Testing, Adjusting and Balancing ... 15990

**ELECTRICAL** — *Division 16*

Basic Electrical Materials and Methods ... 16050
Power Generation — Built-Up Systems ... 16200
Medium Voltage Distribution ... 16300
Service and Distribution ... 16400
Lighting ... 16500
Special Systems ... 16600
Communications ... 16700
Electric Resistance Heating ... 16850
Controls ... 16900
Testing ... 16950

# FEATURES IN THIS BOOK

The construction estimating information in this book is divided into two main sections: Costbook Pages and Man-Hour Tables. Each section is organized according to the 16 divisions of the MASTERFORMAT as shown on the previous pages. In addition, there are extensive Supporting Construction Reference tables, Geographic Costs Modifiers, Square Foot tables and a detailed Index.

Sample pages with graphic explanations are included before the Costbook pages and Man-Hour tables. These explanations, along with the discussions below, will provide a good understanding of what is included in this book and how it can best be used in construction estimating.

## Material Costs

The material costs used in this book represent national averages for prices that a contractor would expect to pay plus an allowance for freight (if applicable), handling and storage. These costs reflect neither the lowest or highest prices, but rather a typical average cost over time. Periodic fluctuations in availability and in certain commodities (e.g. copper, lumber) can significantly affect local material pricing. In the final estimating and bidding stages of a project when the highest degree of accuracy is required, it is best to check local, current prices.

## Labor Costs

Labor costs include the basic wage, plus commonly applicable taxes, insurance and markups for overhead and profit. The labor rates used here to develop the costs are typical average prevailing wage rates. Rates for different trades are used where appropriate for each type of work.

Taxes and insurance which are most often applied to labor rates include employer-paid Social Security/Medicare taxes (FICA), Worker's Compensation insurance, state and federal unemployment taxes, and business insurance. Fixed government rates as well as average allowances are included in the labor costs.

However, most of these items vary significantly allowances are included in the labor costs. from state to state and within states. For more specific data, local agencies and sources should be consulted.

## Equipment Costs

Costs for various types and pieces of equipment are included in Division 1 - General Requirements and can be included in an estimate when required either as a total "Equipment" category or with specific appropriate trades. Costs for equipment are included when appropriate in the installation of costs in the Costbook pages.

## Overhead And Profit

Included in the labor costs are allowances for overhead and profit for the contractor/employer whose workers are performing the specific tasks. No cost allowances or fees are included for management of subcontractors by the general contractor or construction manager. These costs, where appropriate, must be added to the costs as listed in the book.

The allowance for overhead is included to account for office overhead, the contractors' typical costs of doing business. These costs normally include in-house office staff salaries and benefits, office rent and operating expenses, professional fees, vehicle costs and other operating costs which are not directly applicable to specific jobs. It should be noted for this book that office overhead as included should be distinguished from project overhead, the General Requirements (CSI Division 1) which are specific to particular projects. Project overhead should be included on an item by item basis for each job.

Depending on the trade, an allowance of 10-15 percent is incorporated into the labor/installation costs to account for typical profit of the installing contractor. See Division 1, General Requirements, for a more detailed review of typical profit allowances.

## Adjustments to Costs

The costs as presented in this book attempt to represent national averages. Costs, however, vary among regions, states and even between adjacent localities.

In order to more closely approximate the probable costs for specific locations throughout the U.S., a table of Geographic Cost Modifiers is provided. These adjustment factors are used to modify costs obtained from this book to help account for regional variations of construction costs. Whenever local current costs are known, whether material or equipment prices or labor rates, they should be used if more accuracy is required.

## Man-Hour Tables

The man-hour data used to develop the labor costs are listed in the second main section of this book, the "Man-Hour Tables". These productivties represent typical installation labor for thousands of construction items. The data takes into account all activities involved in normal construction under commonly experienced working conditions such as site movement, material handling, start-up, etc. As with the Costbook pages, these items are listed according to the CSI MASTERFORMAT.

## Square Foot Tables

Included as an additional reference are Square Foot Tables which list hundreds of actual projects for dozens of building types, each with associated building size and total square foot building cost. This data provides an overview of construction costs by building type. These costs are for actual projects. The variations within similar building types may be due, among other factors, to size, location, quality and specified components, material and processes. Depending upon all such factors, specific building costs can vary significantly and may not necessarily fall within the range of costs as presented.

**Editor's Note:** The **Building News 1999 Costbooks** are intended to provide accurate, reliable, average costs and typical productivities for thousands of common construction components. The data is developed and compiled from various industry sources, including government, manufacturers, suppliers and working professionals. The intent of the information is to provide assistance and guidelines to construction professionals in estimating. The user should be aware that local conditions, material and labor availability and cost variations, economic considerations, weather, local codes and regulations, etc., all affect the actual cost of construction.These and other such factors must be considered and incorporated into any and all construction estimates.

**BNi® Building News**

# CONSTRUCTION DICTIONARY

# COSTBOOK EDITION

**An Abridged Edition Of
"Building News
Illustrated Construction Dictionary"**

**Published By**

**BNi® Building News**

**Division Of BNI Publications, Inc.**

## A

**Abandonment.** The failure of both parties to a contract to abide by its terms.

**Above Ground Tank.** A large above ground vessel used for the storage of liquids.

**Abrasion.** Wearing away by friction.

**Abrasive.** A substance used for wearing, grinding, or rubbing away by friction.

**Abrasive Surface.** A surface that has been roughened for safety or for warning.

**Abrasive Surface Tile.** Floor tile that has been roughened to be slip-resistant.

**ABS.** Acrylonitrile-butadiene-styrene, hard plastic used because of its resistance to impact, heat and chemicals.

**Absorption.** The relationship of the weight of the water absorbed by a material specimen subjected to prescribed immersion procedure, to the weight of the dry specimen, expressed in percent.

**AC.** Initials for alternating current. Alternating current flows alternately in one direction, then the opposite, completing 60 cycles every second.

**Acceleration.** Rate of change of velocity.

**Accelerator.** A substance which, when added to concrete, mortar, or grout, increases the rate of hydration of the hydraulic cement, shortens the time of setting, or increases the rate of hardening of strength development, or both.

**Access Door.** A door or panel creating a means of access for the inspection or repair of concealed equipment.

**Access Floor.** A raised floor platform with removable panels to provide access to the area below.

**Access Stair.** A stair system to provide specific access to roofs, mechanical equipment rooms or as a means of exit in an emergency.

**Accessory.** An object or device aiding or contributing in a secondary way.

**Accessory, Concrete.** An implement or device used in the formwork, pouring, spreading, finishing, or facilitate in the installation of masonry or concrete reinforcing.

**Accordion Folding Door.** A folding, hinged or creased door with rollers which run along a track.

**Accordion Partition.** A folded, creased, or hinged interior dividing wall.

**Acid Etch.** The use of acid to cut lines into metal or glass. The use of acid to remove the surface of concrete.

**Acid-Proof Counter.** A horizontal work surface resistant to acid spills.

**Acoustical Treatment.** The act or process of applying acoustical materials to walls and ceilings.

**Acrylic.** A general class of resinous polymers derived from esters, amides or other acrylic aid derivatives. A transparent plastic material used in sheet form for window glass and skylights.

**AD Plywood.** A designation or gradation of plywood. The "A" and the "D" designate quality of surface layers.

**Adapter.** A device for connecting two different parts.

**Additive.** A term frequently used as a synonym for addition or admixture.

**Adhesive.** A substance that dries or cures and binds two surfaces together. A substance capable of holding materials together by surface attachment.

**Adjustable Bar Hanger.** A metal hanger that can be made to fit the varying distances between floor and ceiling joists or rafters to securely hold electrical outlet boxes and devices.

**Adjustable Shelf.** A shelf that can be adjusted to different heights.

**Admixture.** A material other than water, aggregates, and hydraulic cement used as an ingredient of concrete or mortar, and added immediately before or during its mixing. A chemical additive used to alter the normal properties of concrete.

**Aggregate.** Inert particles such as sand, gravel, crushed stone, or expanded materials, in a concrete or plaster mixture.

**Air Compressor.** A tool which takes air and forces it at a high pressure into a storage tank. The air is released through a regulator and a hose to power small tools.

**Air Conditioning.** Equipment with an apparatus for controlling the humidity and temperature of air.

**Air Distribution.** To force air to desired locations in a building or facility.

**Air Eliminator.** A mechanical device that expels excess air.

**Air Entraining Agent.** A substance added to concrete, mortar or cement that produces air bubbles during mixing, making it easier to work with and increasing its resistance to frost and freezing.

**Air Handling.** Single or variable-speed fans pushing air over hot or cold coils, through dampers and ducts to heat or cool a building or structure.

**Air Plenum.** Any space usually used to convey return air in a building or structure.

**Air Powered Hoist.** A hoist that is operated by compressed air.

**Air Tool.** Attachments using compressed air to saw, spray-paint, sand, drill or nail, etc.

**Air Vent.** An opening in a building or structure for the passage of air.

**Aluminum Faced Tile.** A thin, rectangular unit that has a sheet of aluminum bonded to it.

**Amperes (Amp).** A unit of measure of electrical flow in a conductor.

**Anchor.** Irons or metals of special form and shapes used to fasten together and secure timbers or masonry.

**Anchor Bolt.** A bolt embedded in concrete for the purpose of fastening a building frame to a concrete or masonry foundation.

**Anchor, Rafter.** A bolt or fastening device which attaches the rafters used to support a roof to the walls or rafter plate.

**Anchor Slot.** A groove in an object into which a fastener or connector is inserted to attach objects together.

**Anchorage.** An attachment for resistance to movement. The movement can be a result of overturning, sliding or uplift. The most common anchorage for these movements are tie-downs (hold-downs) for overturning and uplift, and anchor bolts for sliding.

**Anchoring Cement.** Grout used in sleeves to anchor tubing in place.

**Angle.** A structural section of steel which resembles an "L" in cross section.

**Angle Valve.** A valve in which the shut-off in the pipe openings are set at right angles to each other.

**Annunciator Panel.** A panel mounted on a surface which indicates by lights which circuits have been activated.

**Anodized.** A metal that has been subjected to electrolytic action in order to coat with a protective or decorative film.

**Anti-Siphon.** A device to prevent the removal of fluid caused by suction produced by fluid flow.

**Apron.** The lower trim member under the sill of the interior casing of a window. An upward or downward vertical extension of a sink or lavatory. A paved area immediately adjacent to a building, structure or facility.

**Arch.** A curved structure in which compression is the essential cause of internal stresses.

**Arch Culvert.** A curved shaped drain under a roadway, canal or embankment.

**Architectural Equipment.** The implements, apparatus, or equipment used in the construction and initial outfitting of a building.

**Architectural Fee.** The amount of money charged for the preparation of plans with specifications for the construction of a building or facility.

**Architectural Woodwork.** Finish work using wood or composites for ornamental designs or casework.

**Armor Plate.** A kick-plate made of metal installed on the bottom of a door to protect it from denting and scratching.

**Asbestos.** A fibrous mineral that was used for its noncombustible, non-conducting or chemically resistant properties.



*For More Terms and Definitions See*
*Building News Illustrated Construction Dictionary*

**Ashlar Veneer.** An ornamental or protective facing of masonry composed of squared stones.

**Asphalt.** A residue in petroleum or coal-tar refining that is used for pavements and as a waterproofing cement.

**Asphalt Shingle.** Saturated roofing felt either in large rolls or cut into composition shingles, impregnated with aggregate particles applied to a roof surfaces.

**Astragal Weatherstripping.** Fabric, rubber or plastic strips attached to the molding that is attached to one of a pair doors or casement windows to cover up the joint between the two stiles.

**Attenuation Blanket.** Material utilized to help in the reduction of the energy or intensity of sound.

**Auger.** An instrument or device used for boring or forcing through materials or soil.

**Auxiliary Switch.** A standby device for switching.

**Awning Window.** A window hinged at the top.

## B

**Back Splash.** A protective panel installed behind a counter, sink or lavatory.

**Back-Up Block.** Block which is used for the inner load bearing, structural portion of a masonry wall.

**Back-Up Brick.** A load bearing or structural portion of a masonry wall constructed of brick against which a veneer is attached.

**Backfill.** Earth or earthen material used to fill the excavation around a foundation; the act of filling around a foundation.

**Backflow.** The unintentional flow of water or other substances into the distribution pipes of a water drainage system from a source other than the intended source.

**Backhoe.** An excavating machine with a bucket rigidly attached to a hinged stick on a boom that is drawn toward the machine in operation.

**Backhoe/Loader.** An excavation machine combining a bucket on a hinged stick on a boom on one end, and a bucket or scoop at the other.

**Ball Valve.** A valve in which a ball regulates the opening by its rise and fall due to fluid pressure, a spring, or its own weight.

**Ballast.** A device used with a fluorescent and high intensity lamp, to provide the necessary circuit condition for starting and operation. Any material used as non-structural fill or dead weight.

**Ballast, Roof.** Crushed rock or gravel which is spread on a roof surface to form its final surface.

**Baluster.** Vertical members that extend from a stair, or floor, to a handrail. A small pillar or column used to support a rail.

**Balustrade.** A series of balusters connected by a rail, generally used for porches, balconies and the like.

**Band.** A low, flat molding.

**Bank Run Gravel.** Excavated material that is generally 1/4 inch minimum to 6 inches maximum.

**Bar, Reinforcing.** A manufactured usually deformed steel bar, used in concrete and masonry construction to provide additional strength.

**Barbed Wire.** Wire that is twisted with barbs or sharp points.

**Bare Solid Wire.** Un-insulated single wire used as an electric conductor.

**Bare Stranded Wire.** Un-insulated group of fine wires used as a single electric conductor.

**Barge.** A floating platform or vessel from which construction activities may be performed. Often used in rivers to install bridge piers and also used extensively in waterfront construction.

**Barge Board.** The installation of ornamental boards to conceal roof timbers projecting over gables.

**Barricade.** An obstruction to prevent passage or to prevent access.

**Barrier, Vapor.** A type of plastic sheeting that both eliminates drafts and keeps moisture from damaging a building or structure.

**Base Cabinet.** Case, box, or piece of furniture which sets on floor with sets of drawers or shelves with doors, primarily used for storage.

**Baseboard Heater.** Heating strips that are installed at the juncture of the wall and floor and may be either recessed or surface-mounted; generally along the outside walls of rooms.

**Baseplate.** A steel plate inserted between a column and the foundation, used to level the column and to spread the load of the column to a larger area of the foundation.

**Batt Insulation.** Loosely matted plant or glass fibers with one or both sides faced with kraft paper or aluminum foil available in specifically sized sections.

**Batten Siding.** Vertical siding which has narrow strips of metal or wood covering the joints.

**Batter Pile.** Pile driven at an angle to brace a structure against lateral or horizontal thrust.

**Bay.** A rectangular area of a building defined by four adjacent columns; a portion of a building that projects from a facade.

**Bead.** A narrow line of weld metal or sealant; a strip of metal or wood used to hold a sheet of glass in place; a narrow, convex molding profile; a metal edge or corner accessory for plaster.

**Beam.** A straight structural member that acts primarily to resist transverse loads. A structural element which sustains transverse loading and develops internal forces of bending and shear in resisting the loads. An inclusive term for joists, girders, rafters, and purlins.

**Beam, Grade.** An end-supported horizontal load-bearing foundation member that supports an exterior wall.

**Beam, Reinforcing.** A horizontal member installed to strengthen and support the load of a structure.

**Bearing Wall.** A wall which supports any vertical loads in addition to its own weight.

**Bedding.** A filling of mortar, putty, or other substance in order to secure a firm bearing.

**Bell and Spigot.** Cast iron pipe joints formed with sections that have a wide opening (bell) at one end and a narrow end (spigot) at the other. They are then fitted by caulking with oakum and lead.

**Bend, Soil.** A piece of short, curved pipe, like an elbow, used to connect two straight links of conduit in a sewage system.

**Bevel.** An end or edge cut at an angle other than a right angle.

**Bevel Board (Pitch Board).** A board used in framing roof or stairway to lay out bevels.

**Bevel Siding.** Wood siding that tapers in cross section.

**Bibb, Hose.** A water spigot or faucet with its nozzle threaded or coupling attached to accept a hose.

**Bi-Fold Door.** A door with two leaves, hinged together to close on itself. One edge of each leaf is hinged at the jamb and the other edge is connected and guided by an overhead track.

**Birch Veneer.** Thin sheets of strong fine-grained hardwood used in furniture, flooring, etc.

**Bitumen.** A tar based mixture, such as asphalt or coal tar.

**Bituminous.** Resembling, containing, or impregnated with various mixtures of hydrocarbons (like tar) together with their non-metallic derivatives.

**Bituminous Sidewalk.** A walkway constructed with an impregnated mixture of hydrocarbons together with aggregate such as sand or stone. Commonly called "blacktop".

**Blanket Insulation.** Thermal insulating material made of fibrous glass or mineral wool, sometimes offered with paper or foil surfacing, formed in batts or rolls.

**Block, Concrete.** A hollow concrete masonry unit constructed a composite material consisting of sand, coarse aggregate, cement and water.

**Block, Glass.** A hollow masonry unit made of glass.

**Block, Granite.** A masonry unit consisting of a very hard natural igneous rock used for its firmness and endurance.

**Block, Grout.** Mortar mixes used in block walls to fill voids and joints.

**Block, Splash.** A small masonry block placed in the ground beneath a downspout to receive roof drainage and prevent standing water or soil erosion.

**Block Vent.** An opening serving as an outlet or inlet for air in a block structure.

**Blocking.** Pieces of wood inserted tightly between joists, studs, or rafters in a building frame to stabilize the structure, inhibit the passage of fire, provide a nailing surface for finish materials, or retain insulation.

**Blown-In Insulation.** Insulation made of cellulose (paper treated with a fire retardant) that is blown into an attic, crawl space or walls by a blowing machine.

**Bluestone.** A sandstone of a dark-greenish to bluish-gray color that splits into thin slabs, commonly used to pave surfaces for pedestrian traffic.

**Board.** Lumber less than 2 inches thick.

**Board Foot.** A unit of lumber volume, a rectangular solid nominally 12" x 12" x 1". The equivalent of a board 1 foot square and 1 inch thick.

**Board Siding.** A type of lumber installed on the exterior walls of a building or structure to act as the finish sheathing.

**Boiler.** A closed vessel in which heat is produced by combustion or electricity and transferred to water. A heating system in which water is used as the distribution medium.

**Bollard.** Short steel post (usually filled with concrete) set to prevent vehicular access to or to protect property from damage by vehicular encroachment. Steel or cast iron post to which ships are tied.

**Bolster.** A long wire type "chair" used to support reinforcing bars in a concrete slab while the concrete is being placed.

**Bolt.** A fastener consisting of a cylindrical metal body with a head at one end and a helical thread at the other, intended to be inserted through holes in adjoining pieces of material and closed with a threaded nut.

**Bolted Steel.** Steel members bolted with a metallic pin or rod having a head at one end and threads on the other for attaching the nut.

**Bond Beam.** A horizontal grouted element within masonry in which reinforcement is embedded.

**Bond Breaker.** A material used to prevent adhesion of newly placed concrete to other surfaces.

**Bond Coat.** A material used between the back of the tile and the prepared surface. Suitable bond coats include pure portland cement. Dry-Set portland cement mortar, latex-type portland cement mortar, organic adhesive, and the like.

**Bonding Agent.** A substance applied to a suitable substrate to create a bond between it and a succeeding layer as between a subsurface and a terrazzo topping or a succeeding plaster application.

**Borrow.** Excavated material that has been taken from one area to be used as fill at another location.

**Bottom Pivot Hinge.** A flexible pair of plates joined by a pin to allow swinging of a door or gate installed at the bottom.

**Bowl, Toilet.** The oval part of a toilet which receives the waste and fills with water after flushing the toilet tank.

**Box Culvert.** A concrete drainage structure rectangular shaped, reinforced and cast in place or made of precast sections.

**Box, Distribution.** A box which contains the circuit breakers, connects to the service wires and delivers current to the various outlets throughout a building or structure.

**Box, Floor.** An electrical, metal outlet box providing outlets from conduits in or under a floor.

**Box, Gang.** Electrical boxes constructed of metal or hard plastic, manufactured with knockout holes to pull wire through to connect outlets or switches.

**Bracing.** Diagonal members, either temporary or permanent, installed to stabilize a structure against lateral loads. Structural member used to prevent buckling or rotation of wood studs.

**Bracket.** A projecting support for a shelf or other structure.

**Brass Fitting.** Threaded pipe connector constructed of brass, used to join two pieces of pipe together.

**Breaker, Circuit.** A switch which stops the flow of current by opening the circuit automatically when more electricity flows through the circuit than the circuit is capable of carrying. Resetting may be either automatic or manual.

**Breaker, Main.** A switch in a main electrical service panel where the two hot service wires attach.

**Brick.** A solid masonry unit having the shape of a rectangular prism. Usually made from clay, shale, fire clay, or a mixture of these.

**Bridge Crane.** A hoisting device spanning two overhead rails. The hoisting device moves laterally along the bridge with the bridge moving longitudinally along the rails.

**Bridge Deck.** The slab or other structure forming the travel surface of a bridge.

**Bridging.** Pieces fitted in pairs from the bottom of one floor joist to the top of adjacent joists, and crossed to distribute the floor load; sometimes pieces of width equal to the joists and fitted neatly between them. Diagonal or longitudinal members used to keep horizontal members properly spaced, in lateral position, vertically plumb, and to distribute load.

**Bronze.** An alloy of copper and tin and sometimes other elements.

**Broom Finish.** A finish applied to an uncured concrete surface, to provide skid or slip resistance, made by dragging a broom across the freshly placed concrete surface.

**Brown Coat.** The second of three coats of a plaster or stucco application.

**BTU.** British thermal unit; measurement of the heat energy required to raise one pound of water one degree Fahrenheit.

**Bucket Trap.** A mechanical steam trap operating on buoyancy that prevents the passage of steam through the mechanical system it protects.

**Building, Metal.** A building or structure constructed of a structural steel frame covered by metal roof and wall panels. Commonly prefabricated in a factory and assembled at the site.

**Building Paper.** Water repellent paper used to assist in shedding incidental moisture what may penetrate exterior finishes of exterior wall construction. Thick paper, used to insulate a building before the siding or cladding is installed. Occasionally used in flooring between double floors.

**Built-Up Roof.** A roof covering made of continuous rolls or sheets of saturated or coated felt. They are then cemented together with bitumen and may have a final coating of gravel or slag.

**Bulb Tee.** Rolled steel in the form of a "T" with a formed bulb on the edge of the web.

**Bulk Excavation.** The digging out of large amounts of dirt and debris.

**Bulkhead Formwork.** The temporary formwork that blocks fresh concrete from a section of forms or closes the end of a form at a construction joint.

**Bulldozer.** A tractor driven machine with a horizontal blade for clearing land, road building, or similar work.

**Bullnose.** A trim tile with a convex radius on one edge. This tile is used for finishing the top of a wainscot or for turning an outside corner.

**Bumper, Dock.** Thick rubber units placed under loading dock openings to absorb the shock and prevent damage when trucks back in for loading or unloading.

**Bumper, Door.** Rubber tip devices mounted on walls or baseboards that prevent door knobs from marring walls.



*For More Terms and Definitions See*
*Building News Illustrated Construction Dictionary*

**Burlap Rub.** A finish obtained by rubbing burlap to remove surface irregularities from concrete.

**Bus Duct.** A prefabricated unit containing one or more electric conductors, often a metal bar, that serves as a common connection for two or more circuits.

**Bush Hammer.** A hammer used to dress concrete or stone with its serrated face and may pyramidal points. A hammer that has a rectangular head with serrated or jagged faces. Used for roughing concrete.

**Bushed Nipple.** A pipe threaded at both ends to connect two pipes of different dimensions.

**Busway.** A rigid assembly consisting of one or more busbars.

**Butt Hinge.** A hinge consisting of two plates with a removable connecting pin.

**Butt, Pile.** The large end of a pile. The small end of the pile is called the tip. The pile butt is the end of the pile which the pile driver impacts or "hits" and after completion, the part which the structure is connected to.

**Butterfly Valve.** A valve constructed with a disc that rotates 90 degrees within the valve body.

**Butyl Caulk.** Caulking that is made from various synthetic rubbers derived from butanes.

**Butyl Membrane.** Pliable thin sheets or layers made from synthetic rubber.

**BX Cable.** Type of indoor wiring consisting of two or more insulated wires protected by a wound, galvanized steel strip cover. The metal winding forms a flexible tube.

## C

**Cabinet.** Case, box, or piece of furniture with sets of drawers or shelves, with doors, primarily used for storage.

**Cabinet, Base.** The floor-mounted cabinets in a room that a counter, sink or appliance is installed upon.

**Cable.** A thin, flexible line which carries only tensile forces. A bundle of two or more electrical conductors.

**Cable Bus.** An assembly of insulated cables.

**Cable, Coaxial.** A cable to transmit telephone, television and computer signals. A cable that has a tube of conducting material surrounding a central conductor.

**Cable Tray.** Open track for support of insulated cables.

**Cage Ladder.** A vertical device with rungs for climbing that has, for safety, a surrounding structure to prevent the climber from falling off.

**Caisson.** A cylindrical, sitecast concrete foundation that penetrates through unsatisfactory soil to rest upon an underlying stratum of rock or satisfactory soil. A type of drilled or augured piling.

**Cant Strip.** A strip of material, usually treated wood or fiber, with a sloping face used to ease the transition from a horizontal to a vertical surface at the edge of a flat roof. This prevents the roofing material from abruptly stopping at the parapet wall and also helps prevent leakage at that juncture.

**Cap.** A trim tile with a convex radius on one edge. Used for finishing the top of a wainscot or for turning an outside corner, a bullnose.

**Cap, Pile.** A structural member usually fastened to, and placed on the top of a slender timber, concrete or structural element. Used to transmit loads into the pile or group of piles and to connect them.

**Capacitor.** A device which introduces capacitance into an electric circuit.

**Carbon Steel.** Low carbon or mild steel.

**Carpentry, Rough.** The preliminary framing, boxing and sheeting of a wood frame building.

**Carpet Tile.** Carpet that comes in sheets or small squares and is installed by the use of adhesives.

**Carrier Channel.** The main supporting metal members used in the construction of suspended ceilings.

**Casement.** A window in which the sash opens with hinges and pivots on an axis along the vertical line of the frame.

**Casework.** Assembled cabinetry or millwork.

**Casing.** The wood finish pieces surrounding the frame of a window or door, or the finished lumber around a post or beam. A cylindrical steel tube used to line a drilled or driven hole such as a well or caisson.

**Cast Iron.** Iron with a high carbon content, which cannot, because of the percentage of carbon, be classified as steel.

**Cast-In-Place.** Mortar or concrete which is deposited in the place where it is required to harden as part of the structure, as opposed to precast concrete.

**Catch Basin.** Formed pan, usually constructed of masonry, which collects run-off and debris. The basin usually includes a drain connected to plumbing or stormwater system.

**Caulking.** Waterproof material used to stop up and fill seams to make watertight.

**Cavity Wall.** A masonry wall that includes a continuous airspace between its outermost wythe and the remainder of the wall.

**CDX.** A grading system mark for plywood which means: grade C and D, exterior glue.

**Ceiling Diffuser.** A mechanical device through which warm or cold air is blown into an enclosure. Its purpose is to distribute conditioned air.

**Cement.** Usually refers to portland cement which when mixed with sand, gravel, and water forms concrete. Generally, cement is an adhesive; specifically, it is that type of adhesive which sets by virtue of a chemical reaction.

**Cement Grout.** A cementitious mixture of portland cement, sand or other ingredients and water which produces a uniform paste used to fill joints and cavities between masonry units.

**Cement, Mortar.** A mixture of cement, lime, sand, or other aggregates, and water, used for plastering over masonry or to lay block, brick or tile.

**Cement, Plaster.** Plaster used on exterior surfaces or in damp areas. Plaster having Portland Cement as a binder.

**Cementitious Topping.** A compound that is capable of setting like concrete and is applied on a concrete base to form a floor surface.

**Centering.** Temporary formwork for an arch, dome, vault, or other overhead surface.

**Centigrade.** A scale of temperature which features 0 degrees and 100 degrees as the freezing and boiling points of water respectively. To convert centigrade to Fahrenheit multiply by 1.8 and add 32, e.g. (100 degrees x 1.8)+32 = 212 degrees F.

**Centrifugal Pump.** A pump which draws water into the center of a high speed impeller and forces the fluid outward with velocity and pressure.

**Ceramic Tile.** A ceramic surfacing unit, usually relatively thin in relation to facial area, made from clay or a mixture of clay; and other ceramic material, called the body of the tile, having either a "glazed" or "unglazed" face, and fired above red heat in the course of manufacture to a temperature sufficiently high to produce specific physical properties and characteristics.

**CFM.** Cubic feet per minute.

**Chain Trencher.** A self-propelled machine with blades attached to a continuous chain, used to excavate trenches.

**Chair, Reinforcing.** Metal supports made of fabricated wire, made to hold reinforcing steel in place until concrete is poured.

**Chamfer.** A beveled surface cut on the edge of a piece of wood. A strip of wood or other substance cut on an angle and placed in concrete formwork to create a beveled finished surface on the corners of the final concrete shape (beam, column, etc.).

**Channel.** A U-shaped steel or aluminum section shaped like a rectangular box with one side removed.

**Chicken Wire.** Thin, woven wire mounted on an exterior wall as a base for stucco plaster to cling to, often backed by paper.

**Chilled Water System.** A cooling system in which the entire refrigeration cycle occurs within a single piece of equipment. Water is used to bring the heat from the space to the evaporator section of the chiller, and water is also used to carry the heat from the condenser to the outside.

**Chimney.** A vertical, noncombustible, structure with one or more flues to carry smoke and other gases of combustion into the atmosphere.

**Chimney Brick.** Brick, chosen for the specific use of the construction of chimneys because of its ability to withstand high temperatures without cracking.

**Chimney Flue.** A channel or shaft in a chimney for conveying smoke and exhaust gases.

**Chord.** One of the main members of a truss braced by web members of the truss. Perimeter member of a building or structure which resists lateral forces.

**Circuit Breaker.** An overcurrent protection device.

**Cladding Panel.** A panel applied to a structure to provide durability, weathering, corrosion and impact resistance.

**Clapboards.** A special form of outside covering of a house; siding. Usually made of wood, clapboards are tapered with the thicker side of the board exposed and the thinner side lapped under the next course of boards.

**Clay Pipe.** Pipe used for drainage systems and sanitary sewers made of earthenware and glazed to eliminate porosity.

**Clay Tile.** Earthenware tile that is fired and is used on roofs. Known as quarry tile when used for flooring.

**Cleaning Masonry.** The final removal of excess grout, excess concrete, etc. from an exterior masonry structure.

**Cleanout.** An opening to the bottom of a space of sufficient size and spacing to allow the removal of debris. In plumbing, a fitting in a pipeline which can be easily accessed to remove foreign objects or provide an opening to insert cleaning type devices.

**Cleft, Natural.** A natural V-shaped channel, space, opening or fissure in a material.

**Closer.** The last masonry unit laid in a course; a partial masonry unit used at the corner of a course to adjust the joint spacing. A hydraulic device used to close doors.

**Closet Flange.** The fitting attached to a subfloor onto which the toilet bowl is attached.

**CMU.** Concrete Masonry Unit (concrete block).

**Cofferdam.** A watertight enclosure from which water is pumped to expose the bottom of a body of water and permit construction.

**Cold Formed Steel.** Process of shaping steel without using heat.

**Collar.** A compression ring around a small circular opening.

**Column.** A structural member used primarily to support axial compression loads and with a height of at least three times its least lateral dimension. An upright structural member acting primarily in compression. A square, rectangular, or cylindrical support for roofs, ceilings, and so forth, composed of base, shaft, and capital.

**Column Capital.** The uppermost member of a column crowning the shaft and taking the weight of the beam or girder.

**Column Footing.** Commonly known as individual footing, generally square or rectangular in shape.

**Combined Footing.** A concrete footing which supports two or more columns.

**Common Bond.** Brickwork laid with each five courses of alternating stretchers followed by one course of headers.

**Compaction.** The process whereby the volume of freshly placed material is reduced or flattened by vibration or tamping, or some combination of these.

**Composite Beam.** A beam that is composed of two different materials; for example, a wood and a steel beam, or a steel beam and concrete slab, in which the two act as one.

**Compression.** Force which tends to crush adjacent particles of a material together and cause overall shortening in the direction of its action. Stress which tends to shorten a member.

**Compression Fitting.** Bends, couplings, crosses, elbows, tees, unions, etc., which use a force when connecting that pushes together and squeezes a metal or rubber gasket.

**Compressor.** A machine that compresses gases or air.

**Concave Joint.** A mortar joint tooled into a curved, indented profile.

**Concealed Grid.** A suspended ceiling framework that is completely hidden by the tiles or panels it supports.

**Concealed Z Bar.** A hidden z-shaped bar that is used as a wall tie.

**Concrete.** A composite material which consists essentially of a binding medium within which is embedded particles or fragments of aggregate; in portland cement concrete, the binder is a mixture of portland cement and water. A mixture of cement, water, fine aggregate (sand), and coarse aggregate (gravel). An artificial building material made by mixing cement and sand with gravel, broken stone, or other aggregate, and sufficient water to cause the cement to set and bind the entire mass.

**Concrete Masonry Unit (CMU).** A block of hardened concrete, with or without hollow cores, designed to be laid in the same manner as a brick; a concrete block.

**Concrete Pipe.** Pipe manufactured from concrete. The manufacturing is done in a plant under controlled conditions. Concrete pipe is usually used for drainage but may be used for sanitary sewers also.

**Concrete Placement.** The placing and finishing of concrete during a continuous operation. Also known as pouring.

**Concrete Plank.** A solid or hollow-core, flat-beam used for floor or roof decking. Usually precast and prestressed.

**Concrete, Precast.** Cast and cured concrete manufactured in a plant under controlled conditions. Examples are; precast concrete slabs, precast reinforced lintels, beams, columns, piles, parts of walls and floors, etc.

**Concrete, Prestressed.** Concrete in which internal stresses of such magnitude and distribution are introduced so that the tensile stresses resulting from the structure's loads are counteracted to a desired degree. The stresses are usually developed by inserting tendons (cables) through pre-formed tubes in the concrete member, at which time the tendon in stressed (tightened) and then grouted in place.

**Concrete Pump.** An apparatus which forces concrete to the placing position through a pipeline or hose.

**Concrete Reinforcement.** Steel rods that are embedded in wet concrete to give additional strength.

**Concrete Testing.** Testing to determine the plasticity or strength of concrete.

**Concrete Topping.** A rich mixture of fine aggregate concrete used to top concrete floor surfaces for durability, safety and appearance.

**Condenser.** A heat exchanger in a refrigeration cycle used to discharge heat to the outside. Three commonly used types of condensers can be classified as water cooled, air cooled, and evaporative. Condenser water is normally circulated through a cooling tower through which heat is distributed to the atmosphere. A heat exchanger in a refrigeration cycle used to discharge heat to the outside.

**Conductive Floor.** Flooring material specifically designed to prevent electrostatic buildup and sparking.

**Conductor.** A substance or body capable of transmitting electricity, heat or sound.

**Conductors.** Pipes for conducting water from a roof to the ground or to a receptacle or drain; downspout. Any electrical wire used to convey electricity.

**Conduit.** A protective sleeve or pipe commonly used for individual electrical conductors.

**Connector, Compression.** A connecting device which when attaching uses a force that pushes together and squeezes.

**Connector Set Screw.** A screw on a connector fitting that when tightened connects two components together.

**Continuous Footing.** A concrete footing that supports a wall or two or more columns. The footing can vary in width and depth. Sometimes called a strip footing.

**Continuous High Chair.** A rigid wire device used to hold steel reinforcements off the bottom of the slab. As opposed to a single chair, the continuous chair is manufactured in strips.



*For More Terms and Definitions See*
*Building News Illustrated Construction Dictionary*

**Control Joint.** An intentional linear discontinuity in a structure or component, designed to form a plane of weakness where cracking or movement can occur in response to various forces so as to minimize or eliminate cracking elsewhere in the structure.

**Controls.** A mechanism used to regulate or guide the operation of a machine, apparatus, or HVAC system.

**Convector.** A heat exchange device that uses the heat in steam, hot water, or an electric resistance element to warm the air in a room; often called, inaccurately, a radiator.

**Conveying System.** A device used to move material from one place to another. A broad term which includes elevators, escalators, moving walks, dumbwaiters, etc..

**Coping.** The material or units used to form a cap or finish on top of a wall, pier, pilaster, or chimney. A protective cap at the top of a masonry wall.

**Corbel.** A projection from the face of a beam, girder, column, or wall used as a beam seat or a decoration. A spanning device in which masonry units in successive courses are cantilevered slightly over one another; a projecting bracket of masonry or concrete. A masonry unit such as brick or stone which projects beyond the unit below.

**Core Drilling.** The process of drilling which extracts a cylindrical sample of concrete, rock or soil. Sometimes used to install pipe or conduit in or through an existing concrete or masonry wall.

**Corner Bead.** A metal or plastic strip used to form a neat, durable edge at an outside corner of two walls of plaster or gypsum board.

**Corner Guard.** Type of molding that is mounted on outside corners in a room or space for finishing and for the protection of the corner from damage.

**Cornice.** The exterior detail at the meeting of a wall and a roof overhang; a decorative molding at the intersection of a wall and a ceiling. The molded projection which finishes the top of the wall of a building.

**Cost Estimate.** Determined by one of the following methods: a. Area and volume method: The estimate is based on a cost per square foot of the building and/or a cost per cubic foot of the building. b. Unit use: The estimate is based on the cost of one unit multiplied by the number of units in the project. For example, in a hospital, the cost of one patient unit multiplied by the number of patient units in the project. c. In-place unit: The estimate is based on the cost in-place of a unit, such as doors, cubic yards of concrete etc..

**Counterflashing.** An inverted L-shaped metal strip built into a wall to overlap flashing and make a roof or wall watertight. A flashing turned down from above to overlap another flashing turned up from below so as to shed water.

**Countersunk Plug.** A wooden peg used to fill a drilled hole in a wooden surface.

**Course.** A horizontal layer of masonry units, one unit high; a horizontal line of shingles.

**Coursed.** Laid in courses with straight bed joints.

**Cove.** A trim tile unit having one edge with a concave radius. A cove is used to form a junction between the bottom wall course and the floor or to form an inside corner.

**Cove Base.** A flexible strip of plastic or synthetic rubber used to finish the junction between floor and wall.

**Cove Molding.** Molding that is concave-shaped. Used to cover interior angles, such as that between the ceiling and a wall.

**Cover, Manhole.** A heavy, usually round, steel or iron cover used to gain access to underground work through a manhole.

**Crane.** A machine for raising, shifting and lowering heavy weights by means of a projecting swinging arm or with the hoisting apparatus supported on an overhead track.

**Crane, Traveling.** A crane which is connected to tires or tracks and can be readily moved.

**Creosote.** A brownish oily liquid obtained by distillation of coal tar, used as a wood preservative.

**Crew Trailer.** A trailer provided on a job site for use by employees.

**Cross Brace.** Bracing with two intersecting diagonals. Slender diagonal member within framed wall or partition, to support wall or partition and to withstand structural loads imposed by wind and suction loads, building loads, movement and deflection of structure.

**Crown Molding.** A decorative type of molding used to make ceiling to wall transitions.

**Crusher Run.** Gravel, rock, boulders, or blasted rock that has been reduced in size by a machine, but has not been sorted for size.

**Culvert.** A drain pipe or masonry crossing made of concrete, galvanized corrugated metal, aluminum or steel constructed under a road or embankment to provide a waterway.

**Curb Inlet.** The opening in a curb through which water flows and drains.

**Curb Roof.** A roof with a slope that is divided into two pitches of each side. Also known as a gambrel roof or mansard roof.

**Curing.** Maintenance of humidity and temperature of freshly placed concrete during some definite period following placing, casting, or finishing to assure satisfactory hydration of the cementitious materials and proper hardening of the concrete. The hardening of concrete or plaster.

**Curtain Wall.** An exterior building wall that is supported, entirely by the frame of the building, rather than being self-supporting or loadbearing.

**Cut and Fill.** Excavated material removed from one location and used as fill material in another location.

**Cylinder Test.** A test to determine the compressive strength of concrete.

## D

**Damper.** A flap to control or obstruct the flow of air or other gasses; specifically, a metal control flap in the throat of a fireplace, or in an air duct. Controls that vary airflow through an air outlet, inlet, or duct. A damper position may be immovable, manually adjustable, or part of an automated control system.

**Dampproofing.** The treatment of concrete or mortar to help prevent the passage or absorption of water. Applied with a suitable coating to exposed surfaces or applied with a suitable admixture of treated cement.

**Deck.** The form on which concrete for a slab is placed, also the floor or roof slab itself.

**Decorative Tile.** Tile with a ceramic decoration on the surface.

**Deformed Reinforcement.** Deformed reinforcing bars, bar and rod mats, and deformed wire.

**Demolish.** To raze or tear down a building or structure.

**Dewatering.** To remove water from the ground or excavations with pumps, wellpoints or drainage systems.

**Diaphragm Pump.** A water pump used to continuously remove water from excavations containing mud and small stones.

**Diffuser.** A device for distributing air in a forced air heating/cooling system. Often flush-mounted on a ceiling, it has slats to direct the conditioned air evenly into the room or space. Components of the ventilation system that distribute and diffuse air to promote air circulation in the occupied space. A lens in a light fixture to diffuse light.

**Dimension Lumber.** The most common type of framing lumber. For example, lumber which is 2 inches (nominal) thick and commonly is called "2 by 4", "2 by 6", etc.

**Direct Burial Conduit.** Electric conduit suitable for burial in exterior applications with an outer surface that resists moisture, fungus, and corrosion.

**Direct Current (DC).** An electrical current that has a constant flow rate induced by a constant voltage.

**Domestic Marble.** Marble which comes from the country or area where the structure in which it used is built.

**Door Bumper.** Rubber tip devices mounted on walls or baseboards that prevent door knobs from marring walls.

**Door Closer.** A mechanical device attached to a door to make it close automatically.

**Door Coordinator.** A device installed on a pair of doors with an overlapping astragal, that permits the proper closing sequence. The door without the overlapping astragal would be closed first.

**Door, Flush.** Doors that have two flat surfaces. They can come in 3 basic types - solid (wood or metal), solid-core, or hollow-core.

**Door, Folding.** Two or more hinged door leaves that fold upon each other when opened.

**Door Frame.** The area which completely surrounds a door, made of wood or metal, where the hinges are attached.

**Door Guard.** Guard fabricated of steel components to provide protection over interior face of door or prevent damage to glass and/or to prevent intrusion.

**Door, Roll-up.** A door which raises and rolls into a coiled configuration and lowers on tracks on either side.

**Door Trim.** Wood or metal finishing work, often ornamental, used for covering the joints between door jambs and plaster walls. The locks, knobs and hinges on a door.

**Dormitory Wardrobe.** A closet in student housing where clothes are kept. Commonly a free-standing cabinet made to match the cabinets of the dormitory room.

**Double Glazing.** Two parallel sheets of glass with an airspace between.

**Double Hung Window.** The most common style of operable window. It has two sashes that slide vertically in parallel tracks. A window with two overlapping sashes that slide vertically in tracks.

**Douglas Fir.** A species of framing lumber. A tree which supplies framing lumber because of its straight growth.

**Dovetail Anchor Slot.** A matching interlocking strip or slot which is used with a dovetail fastener.

**Dowel.** A steel bar, which extends into two adjoining portions of a concrete construction, as at a joint in a pavement slab, so as to connect the portions and transfer shear loads.

**Downspout.** The vertical part of a drainage system from a roof which forms a channel or pipe to remove water from the gutters.

**Dozer.** A term used in the trade for a bulldozer.

**Dragline.** A bucket attachment for a crane commonly used in a marsh or marine area, that digs soft materials that must be excavated at some distance from the crane, and draws the bucket towards itself using a cable.

**Drain.** A trench, ditch, or pipe designed to carry away waste water.

**Drain Field.** A system of trenches filled with sand, gravel or crushed stone and a series of pipes to distribute septic tank effluent into the surrounding soil.

**Drip Edge.** A discontinuity or strip installed at roof eaves or over a window or wall component to force adhering drops of water to fall free of the face of the building rather than move farther toward the interior. The projection of a window header or water table to allow the water to drip clear of the side of the building below it.

**Drop Panel.** A thickening of a two-way concrete structure at the head of a column.

**Drum Trap.** A plumbing trap shaped like a cylinder.

**Dry Stone Wall.** A wall of stone that has been constructed without the use of mortar or concrete in its joints.

**Drywall.** An interior facing panel consisting of a gypsum core sandwiched between paper faces, also called gypsum board or plasterboard. Different types are available for standard, fire-resistant, water-resistant, and other applications.

**Duct, Bus.** A prefabricated unit containing one or more electric conductors, often a metal bar, that serves as a common connection for two or more circuits.

**Duct Heater.** A heating element in a duct of an air-handling system.

**Ductile Iron Pipe.** Iron pipe that is manufactured so as to render it able to be flexed. Ductile iron pipe has the non-corrosive qualities of cast iron but is not brittle and has the handling characteristics of steel.

**Ductwork, Metal.** The rigid sheet material out of which the ducts of an HVAC system are manufactured, commonly, galvanized sheet steel, aluminum, or stainless steel.

**Dump Truck.** A truck used for transporting and dumping loose materials.

**Dumpster.** A large, heavy metal container used to hold and haul rubbish.

**Duplex Box.** An electrical box for wiring switches or duplex receptacles.

**DWV Fitting.** A fitting that is either soldered or compressed on lengths of pipe to connect them in a drain waste vent plumbing system.

## E

**Earthwork.** An embankment or other construction made of earth. Any work involving movement or use of soil and other earthen material.

**Eaves.** The horizontal edge at the low side of a sloping roof.

**Edge Form.** A forming member used to limit the horizontal spread of fresh concrete on flat surfaces.

**Elastomeric/Plastomeric Membrane.** A rubber-like sheet material used as a roof covering.

**Elbow.** A pipe fitting which joins two pipes at 90 degree angles.

**Electric Heating.** A system that imparts heat or holds something to be heated, controlled by electric current.

**Electrical.** Relating to, or operated by electric current.

**Elevated Floor.** A floor system not supported by a subgrade.

**Elevated Slab.** A roof slab or floor supported by structural members.

**Elevated Slab Formwork.** The system of support for a freshly poured or placed concrete elevated slab.

**Elevated Stairs.** A stair system not supported by the subgrade.

**Elevator.** A cable or chain hoist conveying system used for raising material or passengers in a cab, cage, or platform.

**EMT Conduit.** "Electrical Metallic Tubing" or "thin wall". For running of electrical wires, that is not threaded, easier to handle than "rigid", and installed more rapidly because of the type of non-threaded fittings used with it.

**Enamel Paint.** A paint which dries to a hard gloss or semi-gloss smooth finish.

**Encased Burial Conduit.** Metal or plastic conduit EB for outdoor wiring with type TW wires encased, or type UF (underground feeder) cable.

**End Bell.** Pipe with a wide opening at one end, commonly cast iron.

**Engineered Fill.** Earth compacted into place in such a way that it has predictable physical properties, based on laboratory tests and specified, supervised, installation procedures.

**Epoxy Adhesive.** A two-part adhesive system employing epoxy resin and epoxy hardener used for bonding of ceramic tile to back-up materials.

**Equipment, Architectural.** The implements or apparatus installed during the construction or alteration of a structure. Architectural equipment is usually connected to the structure and either plumbed or wired to the building's services.

**Equipment Mobilization.** The assembly and movement of equipment to a jobsite.

**Equipment Pad.** A thick slab-type stone or precast concrete block placed under mechanical devices to spread the weight and load of the machinery evenly and to prevent excessive vibration.

**Escalator.** A power-driven set of stairs arranged like an endless belt that ascend or descend continuously.

**Estimate.** A prediction of the cost of performing work. A value judgement based on experience. An approximation of construction costs.

**Etch.** The art of producing designs on metal or glass by the use of the corrosive action of an acid. The use of acid to cut lines into metal or remove the surface of concrete.



For More Terms and Definitions See
Building News Illustrated Construction Dictionary

**Excavation.** A cavity formed by cutting, digging or scooping.

**Exhaust Fan.** An electrical powered device to withdraw fumes, dusts, or odors from an enclosure.

**Exhaust Hood.** Usually a square or rectangular hood housing an exhaust fan to withdraw fumes, dust, or odors from an enclosure.

**Expanded Metal Lath.** Open mesh cut and drawn from solid sheet of ferrous or nonferrous metal. Made in various patterns and metal thicknesses with uneven or flattened surface.

**Expansion Joint.** A joint through tile, mortar, concrete or masonry down to the substrate, intended to allow for gross movement due to thermal stress or material shrinkage.

**Exposed Aggregate Finish.** A concrete surface in which the coarse aggregate is revealed. A decorative finish for concrete work achieved by removing, generally before the concrete has fully hardened, the outer skin of mortar and exposing the coarse aggregate.

**Extrusion.** The process of squeezing a material through a shaped orifice to produce a linear element with the desired cross section; an element produced by this process.

## F

**Facade.** An exterior face of a building.

**Face Brick.** A brick selected on the basis of appearance and durability for use in the exposed surface of a wall.

**Fahrenheit.** A scale for registering temperature where freezing is 32 degrees above zero and boiling is 212 degrees.

**Fan Coil Unit (FCU).** A packaged unit consisting of a heating/cooling coil, fan, and filter, without ductwork, used to serve a space or a group of small spaces.

**Fan, Exhaust.** An electrical powered device used to withdraw fumes, dusts, or odors from an enclosure.

**Fascia.** A flat, vertical face member or band at the surface of a building or the edge beam of a bridge, or exposed eaves of a building; often inappropriately called facia. A flat member of a cornice or other finish; generally the board of the cornice to which the gutter is fastened.

**Fastener.** Generic term for welds, bolts, rivets, screws, and other connecting devices.

**Fastener, Insulation.** A device used to fasten different types of insulation to floors, walls, roofs, or ceiling surfaces.

**Feeder Duct.** An enclosure for a group of electrical conductors that runs power from a central, large source to one or more secondary distribution centers.

**Fence.** A barrier used to prevent escape or intrusion or to mark a boundary, usually made of posts, boards or wire.

**Fertilizing.** The act or process of adding a substance (as manure or a chemical mixture) in order to make soil more fertile.

**Fiberboard.** A prefabricated sheet of compressed wood or plant fibers used for building. A homogeneous panel made from wood or cane fibers.

**Fiberglass.** The name for products made of or with glass fibers ranging from 5 to 600 hundred-thousandths inch in diameter. Used for making textile fabrics, and for heat or sound insulation.

**Fibrous Concrete.** A light concrete made from a fibrous aggregate, like sawdust or asbestos, for increased tensile strength and making it easy to nail.

**Field Engineer.** An engineer who works primarily at the jobsite as opposed to the home office.

**Field Welded Truss.** A truss fabricated and welded at a jobsite.

**Filler, Joint.** Compressible material used to fill a joint to prevent the infiltration of debris and to provide support for sealants.

**Finger Joint.** A glued end connection between two pieces of wood, using an interlocking pattern of deeply cut "fingers". A finger joint creates a large surface for the glue bond, to allow it to develop the full tensile strength of the wood it connects.

**Finish Carpentry.** Installation of wood finish and materials to a building, such as doors, moldings, and window trim.

**Finish Plaster.** The final layer of plaster coating.

**Finishing Sheetrock.** The taping and sanding of sheetrock seams to make ready for painting or finish.

**Fink Truss.** A three triangle symmetrical truss, commonly used in supporting large, sloping roofs.

**Fire Damper.** A damper that automatically closes a duct or opening upon detection of fire.

**Fire Hydrant.** A discharge fitting or apparatus with a valve and spout at which water may be drawn from the mains of waterworks. Used primarily in fire-fighting, also used to service water mains and systems.

**Fire Rated Door.** A door which has been given a rating of how long in time it can withstand fire before failure.

**Fire Retardant.** Denotes a reduction in flammability and in spread of fire. Applied or pressurized chemical treatment to retard combustion.

**Firebrick.** A brick made from special clays to withstand very high temperatures, as in a fireplace, furnace, or industrial chimney.

**Fireproofing.** Material applied to an element to insulate it against excessive temperatures in case of fire.

**Firestop.** A wood baffle used to close an opening between studs or joists in a balloon or platform framing in order to retard the spread of fire through the opening.

**Firewall.** A wall constructed to prevent or slow down the spread of fire.

**Fixed Window.** Glass that is immovably mounted in a wall.

**Fixture Carrier.** A wall mounted frame for mounting and support of a plumbing fixture.

**Fixture, Lighting.** An electrical device installed to illuminated an area with light.

**Fixture, Plumbing.** Plumbing devices such as sinks, water closets, shower/bath units, etc..

**Flange.** A rib or rim on an object for strength, for guiding, or for attachment to another object. A projecting hard ring, ridge, or collar placed on a pipe or shaft to strengthen, prevent sliding, or accommodate attachments.

**Flashing.** A thin, continuous sheet of metal, plastic, rubber or waterproof paper used to prevent the passage of water through a joint in a wall, roof or at a chimney. The material used and the process of making watertight the roof intersections and other exposed places on the outside of a structure.

**Flat Seam.** A sheet metal roofing seam that is formed flat against the surface of the roof.

**Flat Slab.** A reinforced concrete slab that is designed to span, without any beams or girders, in two directions to supporting columns.

**Flat Washer.** A washer which goes under a bolt head or a nut to spread the load, prevent loosening, and protect the surface.

**Flexible Conduit.** Metallic conduit that is flexible and is put up in coils instead of lengths. May be used in lieu of the rigid type in many dry locations.

**Flexible Ductwork.** Flexible ductwork manufactured in various diameters, made from spiral wire covered in plastic and commonly insulated, for use in the transfer of air in heating, cooling and ventilating systems.

**Flexible Wiring.** Electrical wiring that permits movement from expansion, contraction, vibration, and/or rotation.

**Float Trap.** A floating device in a plumbing fixture which opens or closes a valve to prevent sewer air gases from escaping back through the fixture.

**Floor Box.** A flush or raised electrical outlet box providing outlets from conduits in or under a floor.

**Floor Joist.** A support beam, commonly installed in parallel with other beams to create a structural floor system, after which floor sheathing is fastened.

**Floor Pedestal.** A member, such as a short pier, used as a base for a floor system.

**Floor Slab.** A reinforced concrete slab on grade or elevated, used as a floor.

**Flooring.** Any type of material used to create a final floor surface.

**Flue Liner.** Heat-resistant firebrick or other fire clay materials that make up the lining of a chimney.

**Fluid Applied Roof.** A roof coated with an asphalt-based liquid.

**Flush.** Adjacent surfaces even, or in same plane (with reference to two structural or finish pieces).

**Flush Door.** A door with a smooth surface and no protrusions.

**Fluted Column.** Columns with vertical decorative, semi-circular channeled shafts.

**Foamglass Insulation.** A thermal insulation made by foaming glass with hydrogen sulfide. It is manufactured in the form of block or board and has a low fire hazard rating.

**Folding Door.** A door with hinged leaves, often provided with a ceiling or floor track.

**Folding Stair.** A hinged stair, attached to and concealed within the ceiling, which can be raised and lowered.

**Food Service Equipment.** Appliances, materials, etc., used in the cooking and serving of food in large quantities.

**Footing.** The widened base of a foundation that spreads a load from the building across a broader area of soil. An enlargement at the lower end of a wall, pier, or column, to distribute the load.

**Foreman.** An specially trained workman/manager who works with and usually leads a crew or gang.

**Form Deck.** Thin, corrugated steel decking that serves as permanent formwork for a reinforced concrete deck.

**Form Fabric.** Welded-wire fabric used to reinforce concrete while it is setting and gaining sufficient strength to be self-supporting.

**Form Oil.** Oil applied to interior surface of formwork to promote easy release from the concrete when forms are removed.

**Form Tie.** A steel rod with fasteners on each end, used to hold together the formwork for a concrete wall.

**Formica.** Brand name and trademark for any of the various laminated plastic products, usually used for surface finish on cabinets or millwork.

**Formwork.** Temporary structures of wood, steel or plastic that serve to give shape to poured concrete.

**Foundation.** The portion of a building that has the sole purpose of transmitting structural loads from the building into the earth. That part of a building or wall which supports the superstructure.

**Foundation Vent.** Opening in foundation wall to provide natural ventilation to foundation crawl spaces.

**Framing.** The rough wooden structural skeleton of a building, including interior and exterior walls, floor, roof, and ceilings.

**Framing Lumber.** Wood members of framing systems which are manufactured by sawing, resawing, passing lengthwise through standard planing machine, crosscutting to length, and matching, but without further manufacturing.

**Friction Connection.** Two or more structural steel members clamped together by high-strength bolts with sufficient force that the loads on the members are transmitted between them by friction along their mating surfaces.

**Front End Loader.** A tractor or bulldozer with a bucket which operates from the front of the vehicle.

**Furnace.** A unit which draws in cool air from an occupied space and passes the air through a heating chamber, combustion or electric, and then is returned to the occupied space; a heat system using air as the distribution fluid.

**Furnishings.** Articles, especially furniture, found in the interior of a structure, generally to increase comfort or utility.

**Furring.** Wood or metal strips used to build out a surface such as a studded wall. Narrow strips fastened to the walls and ceilings to form a straight surface upon which to lay the lath or other finish.

**Fuse.** An overcurrent protection device.

**Fused Reducer.** A pipe coupling with a larger size at one end that the other and is attached to a length of pipe by welding.

## G

**Gabion.** A metal or wire cage filled with ballast or stone, used in large scale retaining walls.

**Gable.** The triangular wall beneath the end of a gable roof. The vertical triangular end of a building from the eaves to the apex of the roof.

**Galvanized.** Zinc plated for corrosion protection achieved by hot dipping into molten zinc or by electrolysis.

**Galvanized Mesh.** Mesh screening that has been galvanized. Commonly used as wire lath, reinforcing, or fencing.

**Gang Box.** Electrical boxes constructed of metal or hard plastic, manufactured with knockout holes to pull wire through to connect outlets, switches and other devices.

**Gas Furnace.** A heating system that burns gas to produce heat.

**Gas Meter.** A mechanical device measuring and recording the volume of gas passing a given point.

**Gate, Swing.** The operable member of a fence system that is hinged for opening and closing.

**Gate Valve.** A valve utilizing a wedge-shaped gate, which allows fluid flow when the gate is lifted from the seat.

**Gauging Plaster.** A gypsum plaster formulated for use in combination with finish lime in finish coat plaster.

**General Conditions.** A written document, supplementing the specifications, which indicates and defines areas of the project relating to other than specific building trades.

**Geotextile.** Synthetic fabrics used to separate backfill materials for proper drainage. Used in high retaining walls and landscape design.

**GFI Breaker.** Supplies power, as any breaker does, but also monitors the amount of incoming and outgoing current. Whenever the entering current does not equal the leaving current, indicating current leakage, the GFI instantly opens the circuit. A faster overcurrent protection device than either a fuse or circuit breaker.

**Girder.** A beam that supports other beams; a very large beam, especially one that is built up from smaller elements. A timber used to support wall beams or joists.

**Glass.** An amorphous inorganic usually transparent or translucent substance consisting of a mixture of silicates or sometimes borates or phosphates formed by fusion of silica or of oxides of boron or phosphorous with a flux and a stabilizer into a mass that cools to a rigid condition without crystallization.

**Glass Bead.** A narrow strip of plastic, metal, or wood used to hold glass in a sash. Removable trim that holds glass in place.

**Glass Block.** A hollow masonry unit made of glass.

**Glass Pipe.** Glass and glass-lined pipe used in process piping and in laboratories.

**Glass, Wire.** Glass in which a wire mesh was embedded during manufacture.

**Glass-Fiber-Reinforced Concrete (GFRC).** Concrete with a strengthening admixture of short alkali-resistant glass fiber.

**Glazed Block.** Concrete blocks with a surface produced by fusing it with a glazing material.

**Glazed Brick.** Brick or tile with a surface produced by fusing it with a glazing material.

**Glazing.** The act or process of furnishing or fitting with glass. A transparent or translucent color applied to modify the effect of a painted surface.

**Glazing Compound.** Any of a number of types of mastic or putty used to bed small lights of glass in a frame.

**Globe Valve.** A valve with a rounded disc that shuts off the flow when closed, and seats to prevent fluid flow.

**Glue Laminated Timber.** A timber made up of a large number of small strips of wood glued together.

**Gradall.** A hydraulic, wheel-mounted backhoe often used with a wide bucket for dressing earth slopes.

**Grade Beam.** A reinforced concrete beam that transmits the load from a bearing wall into spaced foundations such as pile caps or caissons.

**Grading.** The act or process of leveling earth.

**Granite Block.** A masonry unit consisting of a very hard natural igneous rock used for its firmness and endurance.

**Granolithic Topping.** A covering layer consisting of an artificial stone of crushed granite and cement.

**Grass Cloth.** A wall covering manufactured from vegetable fibers, woven and laminated to paper for backing.

**Grating.** Open grid of metal bars structurally formed.

**Grating Frame.** Frame, usually metal, to contain floor grating and provide means to anchor to floor construction.

**Gravel.** Small rock particles resulting from natural disintegration and weathering such as river gravel. Or a loose term for mechanically crushed stone.

**Gravel Roof.** A roof composed of layers of roofing felt for waterproofing, then sealed with tar or pitch and covered with a layer of gravel to assist in protection from wear and the sun.

**Gravel Stop.** A metal flange or strip with a vertical lip placed around a built-up roof to prevent loose gravel from falling off the roof.

**Gravity Ventilator.** A device installed in an opening in a room or building which is activated by air passing through to remove stale air and replace it with fresh air.

**Gravity Wall.** A retaining wall which depends solely on its weight to resist lateral forces of retained earth.

**Grease Trap.** A device to trap and retain the grease content of wastewater and sewage.

**Grille.** Component of the ventilation system that promotes air circulation in the occupied space by providing a means to return air. A metal screen or grating that allows for the circulation of air.

**Ground.** A strip of wood assisting the plasterer in making a straight wall and in giving a place to which the finish trim of the room may be nailed. A strip attached to a wall or ceiling to establish the level to which plaster should be applied.

**Ground Fault Circuit Interrupter (GFCI).** Supplies power, as any receptacle does, but also monitors the amount of incoming and outgoing current. Whenever the entering current does not equal the leaving current, indicating current leakage, the GFCI instantly opens the circuit. A faster overcurrent protection device than either a fuse or circuit breaker.

**Ground Hydrant.** A water hydrant for the use in fighting fires, installed in the ground.

**Grounding.** The act or process of making an electrical connection with the earth. A large conduction body (as the earth) used as a common return for an electric circuit.

**Grout.** A rich or strong cementitious or chemically setting mix used for filling masonry or tile joints and/or voids. A mixture of portland cement, aggregates, and water, which can be poured or pumped into cavities in concrete or masonry.

**Grout Lift.** An increment of height that grout is poured.

**Grubbing.** The act or process of clearing and digging up roots and stumps.

**Guardrail.** A safety device used as a barrier to prevent encroachment. In street or highway construction, a barrier to keep vehicles in their lanes. A device for protecting a machine part or the operator of a machine.

**Gunite.** A term used to describe a concrete material applied by pumping through a hose.

**Gusset Plate.** A flat steel plate to which the chords of a truss are connected at a joint; a stiffener plate.

**Gutter.** A channel to collect rainwater and snow melt at the eaves of a roof.

**Guy Cable.** Cable anchored at one end and supporting or stabilizing an object at other end.

**Guy Rod.** A metal rod with a cable or rope attached, leading to an object to support and stabilize it.

**Gypsum Board.** An interior facing panel consisting of a gypsum core sandwiched between paper faces. Also called drywall or plasterboard.

**Gypsum Lath.** Small sheets of gypsum board manufactured specifically for use as a plaster base.

## H

**H Beam.** A steel beam which in cross section resembles the letter "H". Commonly used in earthwork as a retaining structure or piling.

**Half Round Molding.** An ornamental strip having one flat side and one rounded side.

**Halide Fixture.** An electric-discharge light fixture that makes its light from metal vapor such as sodium or mercury.

**Halon System.** A system using halon gas for the fire protection of water sensitive equipment.

**Hand Excavation.** The act or process of digging out earth using hand tools.

**Handling, Air.** Single or variable speed fans pushing air over hot or cold coils, through dampers and ducts to heat or cool a building or structure.

**Handrail.** Member which is normally grasped by hand for support.

**Handsplit Shingle.** A shingle made by splitting a block of wood, usually cedar or redwood, along its grain and thereby creating a shingle which may be used for roofing or siding. Another term used may be "shake shingles".

**Hanger, Joist.** A metal stirrup that supports the ends of joists so that they are flush with the girder.

**Hanger Rod.** A rod for connecting pipe, gutters, or ceiling framework to a support.

**Hardboard.** A very dense panel product, usually with at least one smooth face, made of highly compressed wood fibers. Flat sheet material of fibers consolidated under heat and pressure in hot press.

**Hardware.** A general term for metal or plastic fittings used in or on a building and its parts.

**Hasp Assembly.** A fastener for a door or lid consisting of a hinged metal strap that fits over a loop and is secured by a pin or padlock.

**Hatch.** A hinged or removable cover in a floor or roof which permits ventilation or the passage of persons or objects.

**Head, Pop-Up.** A watering device in an irrigation or sprinkling system that pops up when the system is charged with water.

**Headwall.** A wall, usually constructed of concrete or masonry, that is placed at the outlet side of a drain or culvert to protect fill from scouring, undermining or to divert flow.

**Heat Pump.** A unit that supplies heating or cooling to an interior space by either absorbing heat from or rejecting heat to the outside.

**Heater, Unit.** A device for heating a space without the use of ductwork.

**Heavy Timber.** Construction requiring noncombustible exterior walls with a minimal fire-resistance rating of two-hours, laminated or solid interior members, heavy plank or laminated wood floors and roofs.

**Hinge.** A joint fixing the relative position of the ends of two or more structural members, but permitting their relative rotation. A piece of door hardware that permits the opening and closing of a door by joining the door to the jamb with a flexible device.

**Hip Roof.** A roof consisting of four sloping planes that intersect to form a pyramidal or elongated pyramid shape. A roof which slopes up toward the center from all sides, necessitating a hip rafter at each corner.

**Hollow Block.** Concrete blocks that can be filled with insulation or reinforced.

**Hood, Exhaust.** Usually square or rectangular hood housing an exhaust fan to withdraw fumes, dusts, or odors from an enclosure.

**Horizontal Damper.** A damper that disperses forces in a horizontal plane.

**Horizontal Siding.** Linear horizontal material, usually overlapping, used as exterior surface or cladding for exterior framed wall to provide protection from exterior elements.

**Hose Bibb.** An outdoor water faucet, usually at sill height, used as a hose connection.

**HP Sodium Lamp.** A high-pressure (HP) sodium vapor lamp that produces a wide-spectrum yellow light.

**Hub Union.** A pipe fitting used to join two pipes without turning either pipe.

**HVAC.** Heating, ventilation, and air-conditioning system.

**Hydrant.** A connection with a valve to a water main, used for the dispensation and delivery of water.

**Hydraulic Elevator.** A lifting device powered by pressured fluid.

**Hydronic Heating System.** A system that circulates heated water through convectors to heat a building or space.

## I

**I Beam.** An obsolete term; an American Standard designation for a particular section of hot-rolled steel which in cross section is shaped like a capital "I".

**IMC (Conduit).** Intermediate Metal Conduit.

**Incandescent Fixture.** An electric light fixture in which the lamp or bulb has a filament that gives off light when heated to incandescence by an electric current.

**Indirect Luminaire.** A lighting fixture that distributes most of its light upward.

**Induced Draft (in a Boiler).** Combustion air drawn through the burner or fuel bed by a power driven fan in the flue.

**Industrial Equipment.** Mechanical or non-mechanical devices used in an industrial setting.

**Industrial Wood Floor.** A heavy duty wood floor made of decking (2 inches thick) or of wooden blocks laid on end. Also called a "factory" floor. Very resistant to heavy loads and traffic.

**Inlet.** An opening for intake.

**Insulated Block.** Hollow masonry block filled with insulation.

**Insulated Glass.** A type of glass constructed in a manner to protect against sound, heat, heat loss or moisture. Double or triple glazed glass.

**Insulation.** Any material used to reduce the effects of heat, cold or sound transmission and to reduce fire hazard. Any material used in the prevention of the transfer of electricity, heat, cold, moisture and sound.

**Insulation, Rigid.** Sheet or board form of rigid foam.

**Insulation, Sprayed.** A plastic foam or substance sprayed on a surface to insulate.

**Insulation, Sill Sealer.** Insulation place between sill plate and supporting concrete or masonry to prevent air leaks.

**Interceptor.** A device that collects foreign matter and prevents it from reaching the sewer system.

**Invert, Manhole.** The lowest inside surface of a manhole. A channel in the manhole through which wastewater or stormwater flows.

**Inverted Roof.** A membrane roof assembly in which the thermal insulation lies above the membrane.

**Irregular Stone.** Stone cut to or quarried in different shapes and/or sizes.

**Irrigation, Lawn.** To supply water to grassy areas by artificial means.

## J

**Jack Rafter.** A shortened rafter that joins a hip or valley rafter to the top of the wall plate.

**Jacking Pipe.** Forcing pipe through the ground in a tunnel created by the pipe itself. The pipe is generally jacked horizontally in short lengths.

**Jamb Anchor.** Steel anchor to fasten steel door frame to wall or partition construction.

**Jamb.** The vertical side of a door or window. The side piece or post of an opening; commonly applied to the door frame.

**Jib Crane.** A crane which has a projecting arm of its derrick boom.

**Job Built Form.** A temporary structure or mold constructed on a jobsite, for the support of concrete while it is setting and gaining sufficient strength to be self-supporting.

**Job Requirements.** A list of specific, necessary and essential tasks to bring to completion a building or structure. Sometimes called "General Requirements".

**Joint, Contraction.** Formed, sawed, or tooled groove in a structure to create a weakened plane and regulate the location or cracking resulting from the dimensional change of different parts of the structure.

**Joint, Expansion.** A joint through tile, mortar, concrete or masonry down to the substrate, intended to allow for gross movement due to thermal stress or material shrinkage. A discontinuity or break intended to allow movement.

**Joist.** One of a group of light, closely spaced beams used to support a floor deck or flat roof. Timbers supporting the floorboards.

**Joist Girder.** A light steel truss used to support open-web steel joists.

**Joist, Sister.** The reinforcement of a joist by nailing, or attaching alongside the existing joist, another joist or reinforcing member.

**Joist, Wood.** A piece of lumber used horizontally as a support for a floor, measuring two to four inches thick and six or more inches wide.

**Junction Box.** A metal box in which runs of cable meet and are protectively enclosed.

**Jute Padding.** A padding made of a durable yarn that comes from plant fiber used as an underlayment for carpet..

## K

**Keene's Cement.** A cement composed of finely ground, anhydrous, calcined gypsum, the set of which is accelerated by the addition of other materials, used in areas subjected to moisture. A hard, strong finishing plaster that is made from gypsum and maintains a high polish.

**Kerf.** The cut made by a saw.

**Key.** A slot formed into a concrete surface for the purpose of interlocking with a subsequent pour of concrete; a slot at the edge of a precast member into which grout will be poured to lock it to an adjacent member.

**Kick Plate.** A metal plate or strip that runs along the bottom edge of a door to protect against the marring of the finished surface.

**Kiln Dried.** Lumber that has been heated in a kiln to dry and control the amounts of moisture.

**King Post.** In a roof system, the member placed vertically between the center of the horizontal tie beam at the lower end of the rafters and the ridge, or apex of the inclined rafters. A term often used to refer to a certain type of fabricated truss.

**Knee Wall.** A short wall under the slope of a roof.

**Knotty Pine.** Pine wood whose knots are exposed, often used for cabinets and interior paneling.

## L

**Ladder.** Frame consisting of two parallel side pieces connected by rungs at suitable distances to form steps on which persons may climb up or down.

**Lag Rod.** A large diameter rod with a square or hexagonal head.

**Laminated Glass.** A glazing material consisting of outer layers of glass laminated to, and encasing, an inner layer of transparent plastic.

**Laminated Plastic.** Sheets of synthetic resin-soaked material bonded by heat and pressure into a stiff board. Often referred to as "Formica."

**Lamp, Incandescent.** An electric lamp in which a filament gives off light when heated to incandescence by an electric current.



*For More Terms and Definitions See*
*Building News Illustrated Construction Dictionary*

**Landing.** A platform part of a staircase system at the top, bottom, or between two flights of stairs.

**Latex.** A water based emulsion of a synthetic rubber or plastic obtained by polymerization and used commonly in coatings and adhesives.

**Lath.** A wood strip, metal mesh, or gypsum board which acts as a backing and/or reinforcing agent for the plaster scratch coat or initial mortar coat.

**Lath, Expanded.** Metal lath that has been stamped with a pattern and then expanded to create an open grid to accept wet plaster.

**Lath, Stucco.** Wood strips or metal mesh which form the base for the application of a cement plaster on an exterior wall surface. Sometimes made of gypsum which is manufactured in a board form.

**Lattice.** Crossed wood, or steel bars to create a decorative screen..

**Lauan Veneer.** Very thin sheets of wood, in this case Philippine Mahogany.

**Lavatory.** A basin with drainage and running water primarily used for washing the face and hands. A room with a toilet and wash basin.

**Lawn Irrigation.** A system where water is transported and distributed to water grass.

**Lb.** Symbol for pound or pounds.

**Leaching Pit.** An excavated hole (pit) that can hold solids but allows liquids to pass through and leach into the ground.

**Let-In Bracing.** Diagonal bracing nailed into notches cut in the face of the studs so as to avoid an increase in the thickness of the wall.

**LH Joist.** A type of long-span high strength bar joist.

**Lift-Slab Construction.** A method of building multi-story sitecast concrete buildings by casting all the slabs in a stack on the ground, then lifting them up the columns with jacks and welding them in place.

**Lighting.** An artificial supply of light or the apparatus providing it.

**Lightning Arrester.** A device that protects electrical units from lightning and power surges.

**Lightweight Aggregate.** Aggregate of low specific gravity, such as expanded or sintered clay, shale, slate, diatomaceous shale, perlite, vermiculite, or slag; natural pumice, scoria, volcanic cinders, tuff, and diatomite, sintered fly ash or industrial cinders; used to produce lightweight concrete. Aggregate with a dry, loose weight of 70 pounds per cubic foot or less.

**Lightweight Block.** A concrete unit constructed of lightweight materials and used to reduce the weight of walls.

**Lightweight Concrete.** Concrete that achieves a significant reduction in weight by the substitution of lighter materials for the concrete aggregate.

**Limestone Panel.** A limestone slab, relatively thin with respect to other dimensions, and rectangular in shape.

**Liner, Flue.** Heat-resistant firebrick or other fireclay materials that make up the lining of a chimney.

**Liner Panel.** A panel used for interior finish.

**Lintel.** The horizontal beam placed over an opening.

**Load Bearing.** Supporting a superimposed weight or force.

**Loader.** An excavating machine with a movable bucket or scoop, used to transport earth, crushed stone, or other construction materials.

**Lockset.** A device installed in a door that has both a deadbolt and doorknob assembly.

**Lockwasher.** A flat, split ring of metal or steel that when tightened with a nut is used prevent loosening.

**Louver.** A construction of numerous sloping, closely spaced slats used to prevent the entry of rainwater into a ventilating opening. A kind of window, generally in peaks of gables and the tops of towers, provided with horizontal slots which exclude rain and snow and allow ventilation.

**Low-Lift Grouting.** A method of constructing a reinforced masonry wall in which the wall is grouted in increments not higher than 4 feet.

**Lugs.** Spacers, or protuberances on the sides of ceramic tiles. These devices automatically space the tile for the grout joints.

**Lumber.** Sawed parts of a log such as boards, planks, and timber. Wood members which are manufactured by sawing, resawing, passing lengthwise through standard planing machine, crosscutting to length, and matching, but without further manufacturing.

**Lumen.** A quantitative unit for the measurement of the flow of light energy.

**Luminaire.** A complete lighting unit consisting of a light source, switch, globe, reflector, housing and wiring.

## M

**Magnetic Door Holder.** A door holder using a magnet to hold it in an open position.

**Mahogany Veneer.** A thin layer of straight-grained medium density wood for an outer finish or decoration.

**Main Breaker.** The main electrical service protective device where the power enters a building.

**Malleable Iron.** Iron that can be hammered or bent without breaking.

**Manhole.** A hole through which a person may go to gain access to an underground or enclosed structure.

**Manufactured Roof.** A factory-finished roof system.

**Manufactured Wall.** A factory-finished wall system.

**Maple Floor.** A floor constructed from a dense, durable wood found from Eastern Canada down to the Southern U.S. Commonly used in gymnasium floors.

**Marble.** Limestone that is more or less crystallized by metamorphism, that ranges from granular to compact in texture, that is capable of taking a high polish, and that is used in architecture and sculpture.

**Masonry.** Brickwork, blockwork, and stonework.

**Masterformat.** The copyrighted title of a uniform indexing system for construction specifications, as created by the Construction Specifications Institute and Construction Specifications Canada, commonly (but inaccurately) called the CSI format or numbering system.

**Mastic.** Organic tile adhesive. A viscous, dough-like, adhesive substance; can be any of a large number of formulations for different purposes such as sealants, adhesives, glazing compounds, or roofing membranes.

**Mat, Concrete.** A grid of metal reinforcement for concrete foundations, slabs, or mats.

**Mat Foundation.** A concrete slab used as a building or equipment foundation.

**Matching.** The method used in cutting the edges of a board to make a tongue on one edge and a groove on the other.

**Material Handling.** The act or process of transporting materials on or to a jobsite.

**Mechanical.** Of or relating to machinery or tools. Relating to, governed by, or in accordance the principles of mechanics. Loosely used as a term for anything in the plumbing, heating, air-conditioning, or fire sprinkler trades.

**Membrane.** A sheet material that is impervious to water or water vapor.

**Membrane Roof.** A roof structure with a covering of a sheet material that is impervious to water or water vapor. Commonly a single sheet of material.

**Membrane Waterproofing.** A membrane, usually made of built-up roofing or sheet material, to provide a positive waterproof floor over the substrate, which is to receive a tile installation using a wire reinforced mortar bed.

**Mercury Fixture.** A light fixture that has an electric discharge lamp that produces a blue-white light by creating an arc in mercury vapor enclosed in a tube or globe.

**Mesh, Slab.** Welded-wire fabric in sheets or rolls used to reinforce concrete slabs.

**Mesh Wire.** A series of longitudinal and transverse wires arranged at right angles to each other in sheets or rolls, used to reinforce mortar and concrete. Welded-wire fabric.

*For More Terms and Definitions See*
*Building News Illustrated Construction Dictionary*

**Metal Building.** A building or structure constructed of metal. A term used to denote buildings that are prefabricated in a factory and installed on the jobsite.

**Metal Chimney.** Vertical metal structure with one or more flues to carry smoke and other gases or combustion into atmosphere.

**Metal Clad Cable.** Electrical conduit of a flexible steel jacket wrapped around insulated wires.

**Metal Clad Door.** A flush wooden door covered in sheet metal.

**Metal Fabrication.** The building, construction or manufacture of metal structures or metal devices.

**Metal Framing.** The construction of a building or structure by using steel. Loosely used term to denote the construction of frame houses and partitions by using light gauge metal studs and members.

**Metal Halide Lamp.** A lamp that uses an electric-discharge to produce light from a metal vapor such as sodium or mercury.

**Metal Joist.** Horizontal cold formed metal framing member of floor, ceiling or flat roof to transmit loads to bearing points. Often refers to a bar joist.

**Metal Lath.** A steel mesh used primarily as a base for the application of plaster.

**Metal Window.** Any window manufactured from metal.

**Mild Steel.** Steel containing less than three-tenths of one percent carbon, not used as structural steel because of its low strength.

**Millwork.** Interior components such as trim work, cabinets, doors and windows, etc. but not including floors, siding and ceilings. Often made of wood or plastic laminates and produced in a shop or factory.

**Mineral Wool.** Any of various lightweight fibrous materials used in heat and sound insulation.

**Mirror, Plate.** Thick mirror glass manufactured with a high-quality standard.

**Mixer.** A machine used for blending the constituents of concrete, grout, mortar, cement paste, or other mixtures.

**Mobilization.** The act of putting into movement or circulation. The assembly and movement of equipment to a jobsite.

**Mogul Base.** A screw-in style base for an incandescent lamp of generally 300 watts or more.

**Moisture Barrier.** A membrane used to prevent the migration of liquid water through a floor or wall.

**Moisture Protection.** The act or process of retarding the seepage of moisture.

**Molding.** A strip of wood, plastic or plaster with an ornamental profile.

**Molding, Lip.** A molding with a lip which overlaps the piece against which the back of the molding rests.

**Monolithic Concrete.** Concrete cast with no joints other than construction joints or as one piece, generally, the term is used on larger structures.

**Mop Sink.** A deep well plumbing fixture with a faucet and a drain used for collecting and dispensing water for mopping/janitorial purposes.

**Mortar.** A mixture of cement paste and fine aggregate; in fresh concrete, the material occupying the interstices among particles of coarse aggregate; in masonry construction, mortar may contain masonry cement, or may contain hydraulic cement with lime (and possibly other admixtures) to afford greater plasticity and workability than are attainable with standard hydraulic cement mortar. A substance used to join masonry units consisting of cementitious materials, fine aggregate, and water.

**Mortise-and-Tenon.** A joint in which a tongue-like protrusion (tenon) on the end of one piece is tightly fitted into a rectangular slot (mortise) in the side of the other piece. A joint made by cutting a hole or mortise, in one piece, and a tenon, or piece to fit the hole, upon the other.

**Moss, Peat.** Moss containing partially carbonized vegetable tissue formed by the partial decomposition of water in that moss.

**Movable Partition.** A dividing wall that can be moved and arranged to form different walled spaces.

**Mud.** A term used in the trade for mortar.

**Mulch.** A mixture, as of leaves and compost, that covers or is mixed with the earth, often to help enrich the soil. Bark, crushed stone or other material used to cover planting beds, retain moisture, reduce weeds, and improve appearance.

**Mullion.** A vertical or horizontal bar between adjacent window or door units. The member between the openings of a window frame to accommodate two or more windows.

**Muntin.** A small vertical or horizontal bar between small lights of glass in a sash. The vertical member between two panels of the same piece of panel work. The vertical sash-bars separating the different panels of glass.

## N

**Nailable Concrete.** Concrete, usually made with suitable lightweight aggregate, with or without the addition of sawdust, into which nails can be driven.

**Natural Cleft Slate.** Slate which has been split into thinner pieces along its natural cleft or seam. It is rougher in appearance than machined slate.

**Natural Draft.** A natural stream of air up a chimney, caused by the stack effect, which draws combustion air through the burner or fuel bed.

**Neat Cement.** Hydraulic cement in the unhydrated state.

**Needle Beam.** A steel or wood beam threaded through a hole in a bearing wall and used to support the wall and its superimposed loads during underpinning of its foundation.

**NEMA Switch.** A electrical switch approved by the National Electrical Manufacturer's Association (NEMA).

**Neoprene.** A type of synthetic rubber with outstanding oil resistance. Can be used for quick-setting, high strength adhesives.

**Net, Safety.** A meshed fabric that is spread below activity, to protect materials or people that may fall from dangerous heights.

**Newel.** The principal post of the foot of a staircase; also the central support of a winding flight of stairs.

**Nipple, Bushed.** A pipe threaded at both ends to connect two pipes of different dimensions.

**Nipple, Offset.** A fitting, threaded at both ends, that is a combination of elbows or bends which brings one section of pipe out of line with, but into a line parallel with, another section.

**No-Hub Pipe.** Pipe usually manufactured of cast iron, which is fabricated without hubs and coupled together by a stainless steel and rubber fastener.

**Nominal Size (Lumber).** The commercial size designation of width and depth, in standard sawn lumber and glued-laminated lumber grades; larger than the standard actual net size of the finished, dressed lumber.

**Non-Metallic Tubing.** A round sheath product, of round cross-section, fabricated from a moisture-resistant, flame-retardant material.

**Nonbearing.** Not carrying a load.

**Non-shrink Grout.** Cementitious or epoxy based mix used to fill gap created between bearing components or base plates and foundation or other supporting element.

**Norman Brick.** A brick with 2-3/4 by 4 by 12 inch dimensions.

**Nosing.** The projecting forward edge of a stair tread. The part of a stair tread which projects over the riser, or any similar projection; a term applied to the rounded edge of a board.

**Nut, Hex.** A six-sided, short metal nut with a threaded hole for receiving a rod or threaded bolt.

**Nylon Carpet.** A carpet made from synthetic nylon fibers.

## O

**Oak.** A strong, hard, heavy wood.

**Oak Floor.** A common type of flooring, usually installed in small tongue and groove slats made of oak.



*For More Terms and Definitions See*
*Building News Illustrated Construction Dictionary*

**Oak Veneer.** A thin layer of oak glued to an interior wood or bonded together to form a protective or ornamental facing. Types available are white oak and red oak.

**Office Partition.** Interior walls, which sometimes move, to partition-off an office area, assembled together to create cubicles for employees.

**Office Trailer.** A highway vehicle parked on a job site, designed to serve wherever needed, as an office and a place to carry out business.

**Offset Connector.** A fitting that is a combination of elbows or bends which brings one section of pipe out of line with, but into a line parallel with, another section.

**Ogee Molding.** A molding with a modified S-shaped profile.

**Oil Fired Boiler.** A boiler that is heated by a unit that sends oil under high pressure to a nozzle, where it is sprayed as a mist and ignited by an electric spark.

**Oil Furnace.** A furnace that burns heating oil.

**Open Type Decking.** A deck in which the joists on the underside are exposed.

**Open-Web Steel Joist.** A prefabricated, welded steel truss used at closely spaced intervals to support floor or roof decking.

**Opposed Blade.** Two sets of blades in a damper, linked so that the adjacent blades can open and turn in opposite directions.

**Oriented Strand Board (OSB).** A building panel composed of long shreds of wood fiber oriented in specific directions and bonded together under pressure. Commonly called "strand board".

**Ornamental Metal.** A detail that is added to a building constructed of metal, with the purpose of embellishment or decorating the structure.

**OS&Y Valve.** A type of valve, with external exposed threads supported by a yoke, indicating the open or closed position of the valve.

**Overhead Door.** A door, commonly used in garages and warehouse, that opens upward from the ground.

## P

**P-Trap.** A P-shaped drain connection that prevents sewer gas from escaping from a plumbing fixture.

**Packaged Heater.** A factory assembled piece of equipment that supplies heat producing units for circulation to a desired location or use.

**Packaged Terminal Air Conditioner (PTAC).** A self contained unit, operated in the direct expansion method (DX), located in the space served.

**Pad, Bearing.** A thick slab-type stone or precast concrete block placed under a structural member of a building to spread the load or weight evenly.

**Pad, Equipment.** Typically, a cast-in-place or precast concrete block placed under mechanical devices to spread the weight and load of the machinery evenly and to prevent excessive vibration.

**Pad Eyes.** Metal rings mounted vertically on a plate for tying small vessels.

**Pad, Transformer.** A precast or cast-in-place concrete block placed under a transformer to spread and support its weight.

**Pan.** A form used to produce a cavity between joists in a one-way concrete joist system.

**Panel Cladding.** Metal sheathing panels used to provide durability, weathering and corrosion, or impact resistance.

**Panel, Limestone.** A limestone cladding member, usually machined before installation, relatively thin with respect to other dimensions, and rectangular in shape.

**Panelboard.** A collection of service disconnect switches along with switches to major branch lines (feeders) mounted together in a panel.

**Panic Device.** A horizontal bar mounted across the full length of a door or a sort of large push-plate, acting as a latching system and operated by inside pressure on the bar or plate. A mechanical device that opens a door automatically if pressure is exerted against the device from the interior of the building. Often called "panic hardware".

**Paper Backed Insulation.** Insulation that comes with paper facing on one side which serves as a vapor retarder.

**Paper Backed Lath.** Rectangular sheets of gypsum plaster applied to interior walls, to provide a smooth finished surface with a thin layer of paper acting as a vapor retarder.

**Paper, Building.** A generic term for felts, papers and sheet materials used in building applications.

**Paper, Curing.** Waterproof paper placed over freshly finished concrete, to help control the humidity and temperature, aiding in the proper curing of concrete.

**Parapet.** The region of an exterior wall that projects above the level of the roof.

**Parging.** Portland cement plaster applied over masonry.

**Parking Barrier.** A structure either temporary or permanent, that is placed to prevent the encroachment of vehicles.

**Parquet Floor.** A floor covering laid out in a geometric design composed of small pieces of wood.

**Particleboard.** A building panel composed of small particles of wood and resins bonded together under pressure. Flat sheet material producing durable and dimensionally stable product which is often used in dry conditions in place of plywood.

**Parting Bead.** A narrow vertical strip in a double-hung window frame separating the upper and lower sashes.

**Partition.** A permanent interior wall which serves to divide a building into rooms.

**Partition Toilet.** Pre-finished panels used in a toilet enclosure.

**Passenger Elevator.** An elevator used for people.

**Patio Block.** Lightweight concrete paving slabs installed in lightly used foot traffic areas.

**Pavement Cutting.** The process of scoring or cutting through pavement surfaces with a power saw with a specific blade for that purpose.

**Pavement Marking.** The act or process of applying painted lines or necessary instructional signage on pavement surfaces for pedestrians or vehicle drivers.

**Paver, Masonry.** Shaped or molded units, composed of stone, ceramic brick or tile, concrete, or cast-in place concrete used for driveways and patios.

**Paver, Stone.** Blocks of rock processed by shaping, cutting or sizing, used for driveways and patios.

**Paving, Concrete.** The use of a mixture of portland cement, fine aggregate (sand) coarse aggregate (gravel or crushed stone) and water, to make a hard surface in areas such as walks, roadways, ramps, parking areas, etc..

**Pea Gravel.** Screened gravel particles most of which would pass through a 3/8 inch (9.5mm) sieve.

**Peat Moss.** Moss containing a partially carbonized vegetable tissue formed by the partial decomposition in that moss.

**Pecan Veneer.** A thin layer of hardwood glued over a core of sturdier less valuable solid wood, or plywood, used in flooring and furniture.

**Pedestal.** A short compression member of reinforced concrete that is placed between a column and the footing to distribute the load to the footing.

**Pedestal Floor.** A flooring system which has short piers or legs used as a base and the flooring laid over those piers to provide a floor system. Special flooring designed to prevent electrostatic buildup and sparking in a computer room. Usually elevated over the existing floor to facilitate the installation of wires between the components in the room.

**Pegboard.** A board with holes in which pegs/fittings may be inserted to hang tools or objects.

**Penthouse Louver.** A louvered wall around the mechanical penthouse area of a structure, with fixed or movable flaps. This protects it from the elements and provides a visual screen around equipment.

**Perforated Pipe.** Pipe with one or more rows of uniform holes along the length. Buried in the ground alongside building foundations or structures, to aid in drainage of ground-water and moisture.

**Performance Bond.** A bond, secured by the general contractor, which guarantees that the contract will be performed. An undertaking by a surety that a contractor will perform a contract.

**Perlite.** Expanded volcanic rock, used as a lightweight aggregate in concrete and plaster, and as an insulating fill.

**Personnel Lift.** An elevator for use by persons at a job site, a building, or structure.

**Pest Control.** The act or process of the placement of devices or spraying of chemicals or powders to control the spread of insects and pests.

**Photographs.** Used as a loose term for the pictures or photographs taken before a job commences to provide an accurate representation of what the site was like before construction.

**Pickup Truck.** A light truck having an open body with low sides and tailboard.

**Picture Window.** A large, often fixed, window, usually of plate or insulating glass.

**Pier.** Timber, concrete, or masonry supports for girders, posts, or arches. Intermediate supports for a bridge span. Structure extending outward from shore into water used as a dock for ships.

**Pilaster.** A vertical, integral stiffening rib in a masonry or concrete wall. A portion of a square column, usually set within or against a wall.

**Pile.** A long, slender, piece of material driven into the ground to act as a foundation. A member embedded into the ground that supports vertical loads, can be made of wood, steel or concrete.

**Pipe.** A long tube or hollow body for conducting a liquid, gas, or finely divided solid. May be used for structural elements as well.

**Pipe Cleanout.** A pipe fitting placed in a drain waste (DWV) system for access to remove blockages.

**Pipe Flange.** Projecting ring, ridge or collar placed on pipe to strengthen, prevent sliding, or accommodate attachments.

**Pipe Insulation.** Insulation that covers pipes to help in the reduction of heat loss or gain.

**Pipe Jacking.** Forcing pipe through the ground in a tunnel created by the pipe itself. The pipe is generally jacked horizontally in short lengths.

**Pipe, No-Hub.** Pipe manufactured in cast iron, which is fabricated without hubs for coupling.

**Pipe Sleeve.** Cylindrical insert cast into concrete wall or floor to provide for later passage or anchorage of pipe.

**Pipe, Structural.** Pipe used in a structure to transfer imposed loads to the ground.

**Pitch Board.** A board sawed to the exact shape formed by the stair tread, riser, and slope of the stairs and used to lay out the carriage and stringers.

**Pitch Pocket.** An opening between growth rings of a tree which usually contains resin, bark, or both. In roof construction, a metal container placed around a roof penetration at roof level to receive hot bitumen or caulking and provide a roof seal. Commonly found at columns or plumbing stacks.

**Plank.** A wide piece of sawed timber, usually 1-1/2 to 4-1/2 inches thick and 6 inches or more wide.

**Plaster.** A cementitious material, usually based on gypsum or portland cement, which is applied to lath or masonry in paste form, to harden into a finished surface. A mixture of lime, hair, and sand, or of lime, cement, and sand, used to cover exterior or interior wall surfaces.

**Plaster Ring.** A guide with a metal collar attached to a base to apply plaster to a certain thickness or provide a fastener for trim.

**Plasterboard.** A board used in large sheets as a backing or as a substitute for plaster in walls and consisting of fiberboard, paper, or felt, bonded to a hardened gypsum plaster core. Loose term for drywall or sheetrock.

**Plastic Coated Conduit.** A type of conduit for electrical wiring that is used around moist areas and highly corrosive fumes.

**Plastic Laminate.** Sheet material manufactured of multiple layers of paper with top layer of plastic usually 1/16 inch (1.59 mm) thick with decorative finish. Plastic laminate may be used in flat sheets or heat formed bent, and adhered to single curved base material. Commonly referred to by the brand name of "Formica".

**Plastic Pipe.** Pipe manufactured from hard plastic to resist corrosion and rust.

**Plate.** The top horizontal piece of the wall of a frame building upon which the roof or other structural elements rests.

**Plate Glass.** Glass of high optical quality produced by grinding and polishing both faces of a glass sheet.

**Plate, Toe.** A metal bar fastened to the outer edge of a grating; rear of a tread; the bottom rail of a door.

**Platform.** Horizontal landing in stair either at the end of a flight or between flights, either at floor level or between floors.

**Plenum.** Air compartment connected to a duct or ducts.

**Plexiglass.** A plastic resilient material comparable to glass in use, usually manufactured in sheets.

**Plumbing.** The act or process of installing in a building or structure the pipes, fixtures, or other apparatus for supplying potable water and removing liquid and water-borne wastes. The installed fixtures and piping of a building or structure.

**Plumbing Fixture.** Plumbing equipment, usually installed last, such as sinks, water closets, shower/bath units, etc..

**Ply.** A term used to denote a layer or thickness of building or roofing paper as two-ply, three-ply, etc.

**Plywood.** A wood panel composed of a number of layers of wood veneer bonded and glued together under pressure. Flat sheet material built up of sheets of veneer called plies, united under pressure by bonding agent to create a panel with adhesive bond between plies as strong as or stronger than the wood alone.

**Plywood, Finish.** The finest grade of plywood.

**Plywood Sheathing.** A flat panel made up of a number of thin sheets, or veneers, of wood used to close up side walls, or roofs preparatory to installation of finish materials on the surface.

**Pneumatic Concrete.** Concrete that is delivered by equipment powered by compressed air.

**Pneumatic Tool.** A tool powered by compressed air.

**Pointing.** The process of applying and compacting mortar to the surface of a mortar joint after the masonry has been laid, either as a means of finishing the joint or to repair a defective joint.

**Pole, Utility.** A vertical pole usually with cross arms, where utilities install their service or supply lines.

**Polyethylene Pipe.** Pipe manufactured from a thermoplastic, high-molecular-weight organic compound.

**Polyethylene Vapor Barrier.** A sheet form thermoplastic membrane, high molecular weight, organic compound, used as a protective cover to prevent the passage of air and/or moisture.

**Polypropylene Pipe.** A tough plastic pipe with resistance to chemicals and heat.

**Pop-Up Head.** The part of a sink drain assembly that is operated by a linkage to open or close the drain. In lawn irrigation systems, a watering head that retracts when not in use, flush with the ground.

**Portland Cement.** The element used as the binder in concrete, mortar, and stucco.

**Post.** A timber set on end to support a wall, girder, or other member of the structure.

**Post-Tensioning.** A method by which concrete is compressed after it has been cast by stressing the steel reinforcing. The compressing of the concrete in a structural member by means of tensioning high-strength steel tendons against it after the concrete has cured.

**Pour.** To place concrete; a continuous increment of concrete casting carried out without interruption.

**Power Tool.** An apparatus or device used in construction, powered by electric current.



*For More Terms and Definitions See*
*Building News Illustrated Construction Dictionary*

**Precast.** A concrete component or member cast and cured in other than its final position.

**Precast Specialty.** Special shapes or ornamental objects of precast masonry or concrete. A loose term which may refer to any manufacturing of precast members.

**Prefinished Gypsum Board.** Gypsum board finished at the factory with a decorative layer of paint, paper, or plastic.

**Prehung Door.** A door that is hinged to its frame in a factory or shop.

**Preservative Treated.** Applied or pressurized chemical treatment of wood or plywood to make it resistant to deterioration from moisture and insects.

**Pressure Reducing Valve.** A valve which maintains fluid pressure uniformly on its outlet side as long as pressure on the inlet side is at or above a design pressure.

**Pressure-Treated Lumber.** Lumber that has been impregnated with chemicals under pressure, for the purpose of retarding either decay or fire.

**Prestressed Concrete.** Concrete that has the reinforcing pretensioned prior to placement.

**Prestressing.** Applying compressive stress to a concrete structural member to increase its strength.

**Pretensioning.** A method by which the design tensile force is applied to the steel reinforcing before the concrete is set.

**Processor, Photo.** A piece of equipment used to develop photographs from negatives.

**Professional Fee.** The amount of money charged by a person hired to perform a service.

**Profit.** The excess of returns over expenditures in transactions or series of transactions.

**Projecting Sash.** A window with sashes that swing either inward or outward.

**Protective Board.** A board or sheet of material that is installed next to a waterproofing membrane and then backfilled against thus protecting the membrane from puncture or damage.

**PSF.** Pounds per square foot.

**PSI.** Pounds per square inch.

**Pull Box.** A box, in electrical work, placed in a length of conduit, through which cables can be pulled.

**Pump.** A device that raises, transfers, or compresses fluids or attenuates gases by suction or pressure or both.

**Pump, Sump.** A small capacity pump that empties pits receiving water, sewage, or liquid waste.

**Pumped Concrete.** Concrete that is pumped through a hose or pipe.

**Purger, Air.** A mechanical device that removes unwanted air.

**Purlin.** A timber supporting several rafters at one or more points. Beams or struts that span across a roof to support the roof framing system.

**Push Plate.** A metal plate on a door installed at hand level to serve as an opening device, used for protection from damage and easier cleaning.

**PVC.** Polyvinyl Chloride.

**PVC Waterstop.** A nonmetallic synthetic resin prepared by polymerization of vinyl chloride inserted across a joint to obstruct the seepage of water through the joint.

## Q

**Quadruplex Cable.** Four wire cable.

**Quarry Tile.** A large clay floor tile, usually unglazed.

**Quarter Round.** Wood molding in the shape of a quarter circle.

## R

**R-Value.** The number of minutes (seconds) required for 1 Btu (joule) to penetrate one square foot (square meter) of a material for each degree of temperature difference between the two sides of the material. The resistance of a material to the passage of heat. The reciprocal of conductors (1/c).

**Rabbet.** A corner cut out along edge of a piece of wood.

**Raceway.** Any channel courses supporting and protecting electrical conductors, including conduits, wireways, surface metal raceway, cable trays, floor and ceiling raceways, busways and cable bus.

**Radiant Heat.** Heat transmitted by radiation.

**Raft.** A footing or foundation, usually a large thick concrete mat.

**Rafter.** A framing member that runs up and down the slope of a pitched roof. The beams that slope from the ridge of a roof to the eaves and make up the main body of the roof's framework.

**Rafter Anchor.** A bolt or fastening device which attaches the parallel beams used to support a roof covering to the rafter plate.

**Rafter, Hip.** The diagonal outside intersecting rafter at the meeting planes in a hip roof. A structural member of a roof forming the junction of an external roof angle or, where the planes of a hip roof meet.

**Rafter, Jack.** A rafter that fills the space between the hip rafter and the top of the wall plate.

**Rafter, Valley.** The rafter extending from an inside angle of the plates toward the ridge or center line of the house.

**Rafters, Common.** Those which run square with the plate and extend to the ridge.

**Rail.** The horizontal members of the balustrade or panel work.

**Railing.** An open fence or guard for safety, made of rails and posts. A banding in cabinetwork. On plywood, the solid wood band around one or more edges.

**Railing, Pipe.** A metal railing made of pipe.

**Rake.** The sloping edge of a pitched roof. The trim of a building extending in an oblique line, as rake dado or molding.

**Ranch Molding.** An architectural style of molding that is installed as finish work or ornamentation. A type or style of trim which is gently curved and devoid of ornate design.

**Random Width Flooring.** Flooring materials of wood with varying widths, often called a "plank floor".

**Receptacle, Cable.** An interrupting outlet box device installed in an electric supply circuit for the connection of electric cables.

**Receptor.** A metallic or nonmetallic waterproof base for a shower stall.

**Recessed Light Fixture.** A light fixture installed in a suspended ceiling or a recess in a plaster or gypsum board ceiling.

**Reducer.** A pipe fitting which connects pipes of different sizes. A tile trim unit used to reduce the radius of a bullnose or a cove to another radius or to a square.

**Redwood Siding.** Wood species that is resistant to decay used for the final covering of an exterior wall surface of a building or structure.

**Reflective Coated Glass.** Glass onto which a thin layer of metal or metal oxide has been deposited to reflect light and/or heat.

**Reflectorized Sign.** A sign made with highly reflective material.

**Refractories.** Heat-resistant non-metallic ceramic materials.

**Refrigeration.** The cooling of a material or space.

**Register.** An opening to a room or space with a grille and a damper at the end of ductwork for the distribution of conditioned air. May be mounted in a ceiling, wall, or floor.

**Reglet, Roof.** A thin narrow groove cut or cast into a building to accept pieces of flashing.

**Reinforced Concrete.** Concrete containing adequate reinforcement (pre-stressed or not prestressed) and designed on the assumption that concrete and steel act together in resisting forces. Concrete work into which steel bars have been embedded to impart tensile strength to the member.

**Reinforced Grouted Masonry.** Wall construction consisting of brick or block that is grouted solid throughout its entire height and has both vertical and horizontal reinforcing.

**Reinforcement.** The action or state of strengthening by additional assistance, material, or support. The action or state of strengthening or increasing by fresh additions. A term used for reinforcing steel.

**Reinforcement, Mesh.** A series of longitudinal and transverse wires arranged at right angles to each other in sheets or rolls, used to reinforce mortar and concrete. Welded-wire fabric.

*For More Terms and Definitions See*
*Building News Illustrated Construction Dictionary*

**Reinforcing Accessory.** The items used to install reinforcing in concrete. These include but are not limited to; chairs, couplings, tie wire etc.

**Release Agent.** Material used to prevent bonding of concrete to a form surface.

**Reline Pipe.** To install new linings in pipes. Commonly includes the cleaning of built-up scale or debris from the existing pipe and relining with a compatible material.

**Remove.** The act or process of demolishing, dismantling, and carrying away an old fitting or component.

**Restoration.** The act or process of bringing a structure back to its former position or condition. The act of restoring to an original condition.

**Retaining Wall.** A wall that is designed to resist the lateral pressures of retained soil. A wall that holds back a hillside or is backfilled to create a level surface.

**Retardant, Fire.** A material or treatment which effects a reduction in flammability and in spread of fire.

**Return.** The ending of a small splash wall or a wainscot at right angle to the major wall. The continuation of a molding or finish of any kind in a different direction. In HVAC, a term for the return-air duct of a forced air heating/cooling system.

**Revolving Door.** Typically a four panel door attached at 90 degrees to each other that turns on a center axis.

**Ribband.** The support for the second-floor joists of a balloon-frame house. The horizontal member of a wood frame wall which supports floor joists or roof rafters.

**Ridge.** The top edge or corner formed by the intersection of two roof surfaces. The board against which the tips of rafters are fastened.

**Ridge Cut.** Any cut made in a vertical plane; the vertical cut at the top end of a rafter.

**Ridge Vent.** A construction element mounted along the ridge of a roof to aid in ventilating an attic space.

**Rigid Insulation.** Thermal insulating material made of polystyrene, polyurethane, polyisocyanurate, cellular glass, or glass fiber, sometimes offered with a skin surfacing, formed to flat board shape of constant thickness.

**Riprap.** A foundation or sustaining wall of stones placed together without order, as in deep water or on an embankment, to prevent erosion.

**Rise.** The vertical distance through which anything rises, as the rise of a roof or stair.

**Riser.** A single vertical increment of a stair; the vertical face between two treads in a stair; a vertical run of plumbing, wiring or ductwork.

**Rivet.** A structural fastener on which a second head is formed after the fastener is in place.

**Rock Anchor.** A post-tensioned rod or cable inserted into a rock formation for the purpose of anchoring a structure to it.

**Rock Drilling.** The act or process of boring holes into rock.

**Rodding.** A method of using a straightedge to align mortar with the float strips or screeds. This technique also is called dragging, pulling, floating, or rodding off.

**Roll Roof.** A roof that has been covered with an asphaltic material that comes in rolls.

**Roll-Up Door.** A door which raises and rolls into a coiled configuration and lowers on tracks on either side.

**Roof.** The top cover of a building or structure.

**Roof Ballast.** Crushed rock or gravel which is spread on a roof surface to form its final surface.

**Roof, Built-Up.** A roof covering made of continuous rolls or sheets of saturated or coated felt embedded in a bituminous coating. The roofing may have a final ballast coating of gravel or slag.

**Roof Cricket.** Relatively small elevated area of wood roof constructed to divert water around chimney, curb, or other projection.

**Roof, Curb.** A roof with a slope that is divided into two pitches of each side. Also known as a gambrel or mansard roof.

**Roof, Fluid-Applied.** A roof coated with an asphalt-based liquid.

**Roof Insulation.** Materials used in between rafters or roof supports for the protection from heat and cold. Solid sheets of insulating material installed on a flat roof.

**Roof Scupper.** A fitting cut through a roof system, acting as a waterway so water can be discharged off the roof.

**Roof Sheathing.** The first layer of covering on a roof, fastened to the rafter boards, used to support the roofing material.

**Roof Shingle.** Roof covering pieces of asphalt, asbestos, wood, tile, slate, etc. Pieces of wedge shaped wood or other material used in overlapping courses to cover a roof.

**Roof Slab.** The flat section of a reinforced concrete roof, supported by beams, columns, or other framework.

**Roof Specialties.** A fitting or piece of trim used in the installation of a roof, such as gravel stop, flashing, vent strips, etc..

**Roof Walkway.** A permanent aisle for safe access across a roof. Also serves as a protection for the roofing material when maintenance is being done.

**Roofing.** The material put on a roof to make it weatherproof.

**Rooftop AC.** Air-conditioning system or equipment installed or placed on a roof.

**Rooftop Heater.** A heating unit installed or placed on a roof.

**Rooftop Unit.** A system of conditioning air in a self contained unit, installed directly above the conditioned space, in a weatherproof enclosure.

**Rosewood Veneer.** A thin layer of cabinet wood with a dark red or purplish color streaked and variegated with black, applied to give a superior and decorative surface.

**Rough Opening.** The clear dimensions of the framed opening that must be provided in a wall to accept a given door or window unit.

**Roughing In.** The act of preparing a surface by applying tar paper and metal lath (or wire mesh). Sometimes called "wiring". In plumbing and electrical, the pre-wiring or pre-piping of a building before the finished walls, fixtures, and devices are installed.

**Round, Quarter.** Molding in the shape of a quarter circle.

**Rowlock.** A brick laid on its long edge, with its end exposed in the face of the wall.

**Rubble Masonry.** Uncut stone, used for rough work, foundations, backfilling, and the like.

**Run.** Horizontal dimension in a stair. The length of the horizontal projection of a piece such as a rafter when in position.

**Runner Channel.** A steel member from which furring channels and lath are supported in a suspended plaster ceiling.

**Running Bond.** Brickwork consisting entirely of stretchers.

**Rusticated Concrete.** Beveled edges of concrete making the joints conspicuous.

## S

**Saddle Board.** The finish of the ridge of a pitch-roof house. Sometimes called comb board.

**Safety Chain.** Chain installed horizontally in railing assembly to provide for ease in providing temporary opening in railing.

**Safety Glass.** Specific type of glass having the ability to withstand breaking into large jagged pieces.

**Safety Net.** A woven meshed fabric that is spread below activity to protect materials and people that may fall from dangerous heights.

**Safety Nosing.** Stair nosing with abrasive non-slip strip surface flush with tread surface.

**Salamander.** A portable source of heat, customarily kerosene or oil-burning, used to temporarily heat an enclosure. Commonly used around newly placed concrete to prevent freezing.

**Sample Soil.** A representative specimen of soil from a site.

**Samples.** Material samples requested by the architect of the general contractor or materials specified for the project.

**Sampling.** The method of obtaining small amounts of material for testing from an agreed-upon lot.

**Sandblast.** A system of cutting or abrading a surface such as concrete or metal by a stream of sand ejected from a nozzle at high speed.

**Sandstone.** A sedimentary rock formed from sand.

**Sanitary Piping.** Drain, waste, and vent plumbing systems.

**Sash.** A frame that holds glass. The framework which holds the glass in a window.

**Sash Bead.** A strip with one edge molded, against which a sash slides.

**Saw Cut, Concrete.** A cut in hardened concrete utilizing diamond or silicone-carbide blades or discs.

**Scab.** A short piece of lumber used to splice, or to prevent movement of two other pieces.

**Scaffold or Staging.** A temporary structure or platform enabling workmen to reach high places.

**Scale.** A short measurement used as a proportionate part of a larger, actual dimension. The scale of a drawing is expressed as 1/4 inch = 1 foot.

**Scarifier.** A piece of thin metal with teeth or serrations cut in the edge. It is used to roughen fresh mortar surfaces to achieve a good bond for the tile. A scarifier also can be used to roughen the surface of concrete.

**Scotia.** A hollow molding used as a part of a cornice, and often under the nosing of a stair tread.

**Scratch Coat.** The first coat in a three-coat application of plaster.

**Scratcher.** Any serrated or sharply tined object that is used to roughen the surface of one coat of mortar to provide a mechanical key or bond for the next coat.

**Screed.** A strip of wood, metal, or plaster that establishes the level to which concrete or plaster will be placed. To strike or plane off wet mortar or concrete which is above the desired plane or shape.

**Scupper.** An opening through which water can drain from the edge of a flat roof. An opening cut through a roof system, acting as a waterway so water can be discharged off the roof.

**Sealant.** An elastomeric material that is used to fill and seal cracks and joints. At expansion joints, this material prevents the passage of moisture and allows horizontal and lateral movement.

**Seat Cut.** The cut at the bottom end of a rafter to allow it to fit upon the plate.

**Seated Connection.** A connection in which a steel beam rests on top of a steel angle fastened to a column or girder.

**Security Grille.** Steel grille to prevent intrusion or entry through openings.

**Self Furring Lath.** Metal lath with dimples that space the lath away from the sheathing behind it to allow plaster to penetrate the lath.

**Self Tapping.** Creates its own screw threads on the inside of a hole.

**Septic Tank.** A private sewage system holding tank, installed together with a leaching field, that collects sewage and allows the solid waste to settle to the bottom of an area while the liquid particles of the sewage drain into the leaching field area.

**Service Disconnect (or, Main Switch).** A switch for, or a means of disconnecting an entire building from electrical service.

**Service Sink.** A deep basin with a faucet used in janitorial applications.

**Service Wye.** A pipe fitting which joins three pipes at a 45 degree angle. A drainage fitting in a plumbing system.

**Set Screw.** A headless screw used to secure two separate parts in a relative position to one another, preventing the independent motion of either part. A screw to adjust the tension of a spring.

**Shaft.** An unbroken vertical passage through a multistory building, used for elevators, wiring, plumbing, ductwork, etc.

**Shaft Wall.** A wall surrounding a shaft, also used loosely as a term for a certain type of drywall system (firewall) which is used to enclose a shaft.

**Shake Roof.** A roof covering system made up of thicker, hand-cut cedar wood shingles (shakes).

**Shear Plate.** In heavy timber construction, a round steel plate used for connecting wood to non-wood materials.

**Sheathing.** The rough covering applied to the outside of the roof, wall, or floor framing of a structure. In clay tile or wood shingle/shake roofs, roofing boards generally installed as narrow boards laid with a space between them, according to the length of a shingle or tile exposed to weather.

**Shed.** A building or dormer with a single pitched roof.

**Sheet Floor.** A term denoting any type of resilient flooring material that is manufactured and installed in sheets or rolls.

**Sheet Glass.** Window glass.

**Sheet Metal.** Flat rolled metal less than 1/4 inches in thickness.

**Sheet Metal Roof.** A roof covering of aluminum, copper, stainless steel, or galvanized metal sheets. In high-finish applications, loosely referred to as an "architectural roof".

**Sheet Piling.** Planking or sheeting made of concrete, timber, or steel that is driven in, interlocked or tongue and grooved together to provide a tight wall to resist the lateral pressure of water, adjacent earth or other materials.

**Sheetrock.** Plasterboard sheets, an interior facing panel consisting of a gypsum core sandwiched between paper faces, also called gypsum board or plasterboard. Different types are available for standard, fire-resistant, water-resistant, and other applications.

**Sheetrock Finishing.** The final sanding and coating of sheetrock (gypsum board) seams to make ready for painting.

**Shield, Expansion.** Device inserted in pre-drilled holes (usually in concrete or masonry) which expands as screw or bolt is tightened within it, used to fasten items to concrete or masonry.

**Shingle.** A small thin piece of building material often with one end thicker than the other for laying in overlapping rows as a covering for the roof or sides of a building or structure.

**Shiplap.** A board with edges rabbeted so as to overlap flush from one board to the next.

**Shop Painted.** Coating(s) of paint applied in shop, usually a primer coat to protect metal from corrosion, which may or may not receive additional coats in the field.

**Short Circuiting.** In electrical, an interrupted circuit. In HVAC, a situation that occurs when the supply air flows to exhaust registers before entering the breathing zone. To avoid short-circuiting, the supply air must be delivered at a temperature and velocity that results in mixing throughout the space.

**Shotcrete.** A low-slump concrete mixture deposited by being blown from a nozzle at high speed with a stream of compressed air.

**Shower Compartment.** An enclosure in which water is showered on a person.

**Shower Pan.** A prefabricated assembly to provide a bottom for a shower, may be made of sheet metal, plastic, or masonry. Terminology used in some areas for a waterproof membrane.

**Shower Receptor.** The floor and side walls of the shower up to and including the curb of the shower.

**Sidelight.** A tall, narrow window alongside a door.

**Sidewalk.** A walk for pedestrians at the side of a street.

**Siding.** The outside finish of an exterior wall.

**Signage.** Any of a group of posted commands, warnings or directions.

**Sills.** The horizontal timbers of a house which either rest upon the masonry foundations or, in the absence of such, form the foundation.

**Sink.** A basin with a drainage system and water supply, used for washing and drainage. A pit or pool used for the deposit of waste.

**Sister Joist.** The reinforcement of a joist by nailing, or attaching alongside the existing joist, another joist or reinforcing member.

**Sitecast Concrete.** Concrete that is cast-in-place.

**Sizing.** Working material to the desired size; a coating of glue, shellac, or other substance applied to a surface to prepare it for painting or other method of finish.

**Skim Coat.** A thin coat of plaster over any base system. May be the final or finish coat on plaster base or a certain type of drywall.

**Slab Bolster.** Continuous, individual support used to hold reinforcing bars in the proper position.

**Slab Form.** The formwork used for the pouring or placing of a concrete slab. A type of manufactured metal decking which is made expressly to receive a final layer of poured concrete.

**Slate.** A form of geologically hardened clay, easily split into thin sheets.

**Sleeper.** A timber laid on the ground to support a floor joist. A framing system, usually for a wood floor system, which is fastened directly to a concrete floor thus facilitating the installation of the finished floor.

**Sliding Glass Door.** An exterior glass door mounted above and below on tracks for ease in movement.

**Slip Coating.** A ceramic material or mixture other than a glaze, applied to a ceramic body and fired to the maturity required to develop specified characteristics.

**Slot.** An opening in a member to receive a connection with another part.

**Slot, Anchor.** A groove in an object into which a fastener or connector is inserted to attach objects together.

**Slump Test.** The procedure for measuring slump with a slump cone.

**Slurry.** A mixture of water and any finely divided insoluble material, such as portland cement, slag, or clay in suspension. A watery mixture of insoluble materials.

**Smoke Detector.** A device that detects the presence of smoke, usually resulting in the sounding of an alarm.

**Sodium Fixture.** An electric lamp that contains sodium vapor and electrodes between which a luminous discharge takes place, commonly used outdoors.

**Soffit.** The underside part of a member of a structure, such as a beam, stairway, roof, or arch. The undersurface of a horizontal element of a building, especially the underside of a stair or a roof overhang.

**Soil.** A generic term for unconsolidated natural surface material above bedrock.

**Soil and Waste Pipe.** Plastic, copper, cast iron or DWV drainage, water and vent.

**Soldier Course.** Oblong tile or brick laid with the long side vertical and all joints in alignment.

**Sole Plate.** The horizontal piece of dimension lumber to which the bottom of the studs are attached in a wall of a light frame building.

**Solid Block.** A block with small or no internal cavities.

**Sound Absorption.** The process of dissipating or removing sound energy. The property possessed by materials, objects, and structures such as rooms, of absorbing sound energy.

**Sound Attenuation.** A process in which sound is reduced as its energy is converted to motion or heat.

**Spall.** A fragment, usually in the shape of a flake, detached from a larger mass by a blow, by the action of weather, by pressure, or by expansion within the larger mass.

**Span.** The distance between supports for a beam, girder, truss, vault, arch or other horizontal structural device; to carry a load between supports. The distance between the bearings of a timber or arch.

**Spandrel.** The wall area between the head of a window on one story and the sill of a window on the floor above; the area of a wall between adjacent arches.

**Spandrel Beam.** A beam that runs along the outside edge of a floor or roof.

**Special Door.** A door that has a unique use, such as a bank vault door.

**Specialties.** A designation of construction materials or components which commonly typifies items that furnish and finish off the structure. A loosely used term to denote any item which aids in the installation of a structure's parts ( i.e. roof specialties).

**Specifications.** The written or printed directions regarding the details of a building or other construction.

**Spiral Reinforcing.** Continuously wound reinforcing in the form of a cylindrical helix.

**Splash Block.** A small precast block of concrete or plastic used to divert water at the bottom of a downspout.

**Splice.** Joining of two similar members in a straight line.

**Split Jamb.** A door frame fabricated in two interlocking halves, to be installed from the opposite sides of an opening.

**Split Ring.** A connector for timbers consisting of a metal ring set in circular grooves in two pieces. The assembly is held together by bolts.

**Spotlight.** A light or lamp which directs a narrow intense beam of light on a small area.

**Spray Booth.** A room or enclosed space that provides a ventilated, dust-free environment for the application of paints.

**Sprinkler Head.** The outlets from which water is sprayed from a sprinkler or irrigation system.

**Stability.** The ability to remain unchanged. Ability to restore to original condition after being disturbed by some force.

**Staging.** Temporary scaffolding with platforms placed in and around a building to create elevated work areas.

**Staining.** Discoloration caused by a foreign matter chemically affecting the material itself. A paint which lets the grain of wood show through.

**Stainless Tubing.** Material in the form of a tube constructed of stainless steel.

**Stair, Access.** A stair system to provide specific access to roofs, mechanical equipment rooms etc.

**Stair, Concrete.** A stair system constructed solely from concrete.

**Stairs, Box.** Stairs built between walls, and usually with no support except the wall.

**Standing Seam Roof.** A sheet metal roof system that has seams that project at right angles to the plane of the roof.

**Standpipe.** A pipe that extends the full height of a building, with hose connections, used to provide water exclusively for the fighting of fires. A dry standpipe does not have water in the pipe until a valve is activated. A wet standpipe is directly connected to a water supply and usually located in a stairway.

**Staple.** Double pointed, U-shaped metal fastener used for same purposes as nails, but providing additional head holding power.

**Starter.** A device that insures that a motor does not receive too high a current when starting up.

**Static.** State of being at rest, having no motion.

**Steam Boiler.** A boiler for producing steam.

**Steam Trap.** A device that allows the passage of a condensate and/or air, but prevents the passage of steam.

**Steel.** Iron compounded with other metals to increase strength and wearing or rust resistance.

**Steel Angle.** An L-shaped member constructed of steel, often used as a lintel or carrying shelf for masonry.

**Steel Joist.** Open web, parallel chord, load-carrying members suitable for direct support of floors and roof decks, utilizing hot rolled or cold formed steel.

**Steel Plate.** Sheet steel of a heavier thickness.

**Steel, Structural.** Steel that is rolled in a variety of shapes and manufactured for use as structural load-bearing members.

**Steel Stud.** Steel pin or rod having head at one end for driving into material used for holding members or parts of members together. In light gage construction, a vertical framing member such as a "2 by 4".

**Stile.** A vertical framing member in a panel door.

**Stirrup.** A vertical loop of steel bar used to reinforce a concrete beam against diagonal tension forces.

**Stone.** Earthy or mineral matter of indeterminate size or shape such as rock, etc.

**Stone Paver.** Blocks of rock processed by shaping, cutting or sizing, used for driveways, patios and walkways.



*For More Terms and Definitions See*
*Building News Illustrated Construction Dictionary*

**Stop, Gravel.** A metal flange or strip with a vertical lip placed around the edge of a built-up roof to prevent loose gravel from falling off the roof.

**Stop Notice.** A charge against construction funds in the hands of a property owner or a construction lender for the value of work or materials incorporated into a construction project.

**Storefront.** The facade which is constructed on the street side of a building or structure into which persons can enter and transact business. A loosely used term to denote a steel or aluminum tube frame and glass wall.

**Strap Tie.** A metal plate that fastens two parts together as a post, rod or beam.

**Stressed-Skin Panel.** A panel consisting of two face sheets of wood or metal bonded to perpendicular spacer strips.

**Stretcher.** A masonry unit laid with its length horizontal and parallel with the face of a wall or other masonry member.

**Striking Joints.** A process of removing excess grout from the joints by wiping with a sponge or cloth or scraping, compacting or rubbing with a curved instrument.

**Stringer.** The sloping wood or steel member supporting the treads of a stair. A long horizontal timber in a structure supporting a floor.

**Structural Lumber.** Wood members of a structural system which are manufactured by sawing, resawing, passing lengthwise through standard planing machine, crosscutting to length, but without further manufacturing.

**Structural Pipe.** Pipe used in a structure to transfer imposed loads to the ground.

**Structural Plywood.** The highest grade of plywood. Plywood of exterior structural grade, secured to top side of floor joists used to create rigidity in building superstructure and also to provide smooth and even surface to receive finish floor covering.

**Structural Steel.** Steel hot-rolled into variety of shapes for use as load-bearing structural members.

**Structural Tube.** A hollow metal product used to carry imposed loads in a building or structure.

**Strut.** A compression member, a column, usually implying that it can be placed at any angle, not just vertically.

**Stucco.** A cement plaster used for coating exterior walls and other exterior surfaces of buildings. Portland cement plaster used as an exterior cladding or siding material. A plaster used for interior decoration and finish work; also for rough outside wall coverings.

**Stucco Lath.** Wood or metal lath strips which form the base for the application of a cement plaster on an exterior wall surface.

**Stud.** One of a set of small vertical elements, usually wood or steel, used to produce a framed wall. An upright beam in the framework of a building. Vertical member of appropriate size (2x4 to 4x10 in.) (50x100 to 100x250 mm) and spacing (16 to 30in.) (400 to 750 mm) to support sheathing of concrete forms; also a headed steel device used to anchor steel plates or shapes to concrete members.

**Sub-Flooring.** Certain material, like plywood, that is installed on the floor joists of a building or structure, onto which the walls and finished flooring is attached.

**Subfloor.** Typically a wood floor which is laid over the floor joists and on which the finished floor is laid.

**Subpurlin.** A small roof framing member that spans between joists and purlins.

**Sump Pump.** A small capacity pump that empties pits receiving groundwater, sewage, or liquid waste.

**Superintendent.** An individual who is at the top level of a construction team in the field. A person with executive oversight, often of a board or building complex.

**Surcharge.** An increase in the lateral earth pressure of a retaining wall, caused by a vertical load behind the wall. A load placed over an area to compact it or change its characteristics.

**Suspended Ceiling.** A finish ceiling that is hung on wires from the structure above.

**Swale Excavation.** The digging up of low-lying land such as a small meadow, swamp, or marshy depression.

**Swing Gate.** The operable member of a fence system that is hinged for opening and closing.

**Switch, Fusible.** An electric switch that has a fusible link in the wiring for an electric circuit in the event of an overload.

**Switchboard.** An apparatus consisting of a panel on which measuring, controlling, etc., devices are arranged to that a number of circuits may be measured, controlled, etc.

**Switchgear.** A freestanding assembly including primary (disconnect) switches, secondary (feeder) switches, and overcurrent protection device (fuses and circuit breakers).

## T

**T&G Siding.** Tongue and groove exterior siding.

**T-Beam.** A reinforced concrete beam that contains of a portion of the slab above and which the two act together.

**Tack Coat.** Application of material made from asphalt on old surface to insure its bond to new construction.

**Tapping Valve.** A device to open or close a duct, pipe, or other passage, or to regulate flow. It is inserted into an existing pipeline by piercing the wall of the pipe and thus tapping into the flow.

**Tarpaulin.** A waterproofed canvas or other material used for protecting construction projects, athletic fields, goods or other exposed objects or areas.

**Teak Veneer.** Thin sheets of teak, a dark wood, used for plywood or other finishes.

**Tee.** A metal or precast concrete member with a cross section resembling the letter "T". A pipe fitting which joins three pipes at 90 degree angles.

**Tee, Bulb.** Rolled steel in the form of a "T" with a formed bulb at the end of the web.

**Tee Weld.** Weld in joint between two members located approximately at right angles to each other in form of "T".

**Tempered Glass.** Glass that has been treated to increase it toughness and its resistance to breakage.

**Temporary Facility.** A structure erected for temporary use.

**Tendon.** A steel strand used for prestressing a concrete member.

**Tension Ring.** A structural element, forming a closed curve in plan, which is in tension because of the action of the rest of the structure. A concrete or masonry dome commonly has a tension ring.

**Terne.** An alloy of lead and tin, used to coat sheets of carbon steel, stainless steel, or copper for use as metal roofing sheet.

**Terra Cotta.** Hard baked clayware, or tile, of variable color, averaging reddish red-yellow in hue and of high saturation.

**Terrazzo.** A finish floor material consisting of concrete with an aggregate of marble chips selected for size and color, which is ground and polished smooth after curing.

**Test Pile.** A pile which is driven before the final design is done to see what bearing is developed and what length of pile is actually needed. A pile which is tested by placing a predetermined load on it, commonly done by erecting a crib on the pile and then filling it with ballast.

**Thermostat.** A device which measures temperature (typically of a space) and activates an HVAC system to heat or cool the air.

**Thin Coat.** A loosely used term for a one coat plaster system over gypsum board.

**Thin Set.** A term used to describe the bonding of tile with suitable materials, applied approximately 1/8" thick.

**Threshold.** The wood or metal beveled floor piece at door openings which commonly separates non-continuous floor types.

**Threshold, Door.** A beveled piece of floor trim over which a door swings.

**Tie.** A device for holding components together, a structural device that acts in tension.

**Tie Rod.** A steel rod that acts in tension and commonly holds together wall forms while concrete is being poured.

**Tile.** A fired clay product that is thin in cross section as compared to a brick; either a thin/flat element (ceramic tile or quarry tile), a thin/curved element (roofing tile), a hollow element with thin walls (masonry tile), or a pipe-like shape (drainage tile).

**Tile, Acoustical.** Finished ceiling panels in board form used for its sound absorbing properties. Sometimes used on walls.

**Tile, Quarry.** Clay tile that is fired and used for flooring.

**Tile Resilient.** Tile manufactured from rubber, vinyl, or other resilient materials.

**Tilt-Up Construction.** A method of constructing concrete walls in which panels are cast and cured flat on the floor slab, then tilted up into their final positions.

**Timber.** Lumber with cross-section over 4 by 6 inches, such as posts, sills, and girders.

**Timekeeper.** A clerk who keeps records of the time worked by employees.

**Tinted Glass.** Glass that has been treated to reduce transmitted glare.

**To the Weather.** A term applied to any part of the structure which faces the elements. A shingled roof is "to the weather" the framing system is not.

**Toggle Bolt.** A bolt and nut assembly to fasten objects to hollow construction assembly from only one side. Nut has pivoted wings that close against spring when nut end of assembly is pushed through hole and spring open on other side in void of construction assembly.

**Toilet Partition.** Privacy panels in a toilet enclosure.

**Tongue and Groove.** A type of lumber, metal or precast concrete having matching or mated edges to provide a tight fit, abbreviated "T & G".

**Toothed Ring.** A timber connector used in the manufacturing of large member wood trusses.

**Top Bars.** Steel reinforcing bars near the top of reinforced concrete.

**Top Plate.** The horizontal member at the top of a stud wall, usually supporting rafters.

**Topping, Concrete.** Concrete layer placed to form a floor surface on a concrete base.

**Topsoil.** Surface soil at and including the average plow depth, soil which is used as a planting or growing medium.

**Torsion.** The rotation of a diaphragm caused by lateral forces and whose center of mass does not coincide with the center of rigidity.

**Tort.** A negligent or intentional wrongful act that damages the person or property of another, the wrongful nature of which is independent of any contractual relationship.

**Total Cost.** A method of computing damages sustained by a contractor because of breaches of contract causing the contractor to operate in an inefficient or unproductive manner.

**Towel Dispenser.** A container that holds paper towels for future use, and which has an opening allowing one towel to be removed at a time.

**Tower.** A tall structure, constructed of frames, braces, and accessories rising to a greater height than the surrounding area.

**Trailer.** A vehicle designed to be hauled or to serve parked as dwelling or place of business.

**Transformer.** A device that changes, or transforms, alternating current from one voltage to another.

**Transformer Pad.** A precast concrete block or stone placed under a transformer to spread and support its weight.

**Translucent Panel.** A building panel that permits the passage of light but not vision.

**Transom.** A transverse piece in a structure, lintel. A horizontal crossbar in a window, over a door, or between a door and a window or a fanlight above it. A window above a door or other window built on and commonly hinged to a transom.

**Trap.** Located at a plumbing fixture, designed to hold a quantity of water that prevents gasses in the sewer system from entering a room.

**Trash Chute.** A device either constructed on the inside of a building or structure or hung outside for the removal of waste materials from upper floors.

**Traveling Crane.** A tower crane mounted on tires, rails or crawlers.

**Tray, Cable.** Open track support for insulated cables.

**Tread.** One of the horizontal planes of a stair. The horizontal part of a step.

**Treated Lumber.** Lumber infused or coated with stain or chemicals to retard fire, decay, insect damage or deterioration due to weather.

**Tremie.** A large funnel with a tube attached, used to deliver concrete into deep forms or beneath water or slurry. A tremie slows down the concrete and resists segregation of the aggregates.

**Trench.** A long, thin excavation.

**Trencher.** A mechanical device used to dig narrow channels in the ground.

**Trenching.** The act or process of digging narrow channels in the ground.

**Trespass.** Unauthorized entry upon the real property of another.

**Trim.** A term sometimes applied to outside or interior finish woodwork and the finish around openings.

**Troffer.** A channel-like enclosure for light sources.

**Trowel Finish.** The final finish of concrete, plaster, stucco, etc., by the use of a hand trowel.

**Truck.** A wheeled vehicle for moving heavy articles. A strong automotive vehicle for hauling. A small wheelbarrow consisting of a rectangular frame having at one end a pair of handles and at the other end a pair of small, heavy wheels and a projecting edge to slide under a load.

**Truck Crane.** A mechanical device for hoisting or lifting materials mounted on the bed of a truck.

**Truss.** A triangular arrangement of structural members that reduces nonaxial forces on the truss to a set of axial forces in the members. Structural framework of triangular units for supporting loads over long spans.

**TSI.** Thermal System Insulation, insulative material applied to pipes, fittings, boilers, tanks, ducts or other interior structural components to prevent heat loss or gain or water condensation.

**Tube, Structural.** A hollow metal or plastic product used to carry imposed loads in a building or structure.

**Tuck Pointing.** The process of removing deteriorated mortar from the surface of an existing brick wall, and inserting fresh mortar.

**Tunneling.** The act or process of digging a horizontal passageway through or under an obstruction.

**Turnout.** A widened space in a highway for vehicles to pass or park. A railroad siding.

**Two-Way Flat Slab.** A reinforced concrete framing system in which columns with mushroom capitals and/or drop panels directly support a two-way slab that is planar on both its surfaces.

**Type X Gypsum Board.** A gypsum board used where increased fire resistance is required.

## U

**U Bolt.** A U-shaped, bent iron bar that has bolts and threads at both ends.

**U Stirrup.** An open-top, U-shaped loop of steel bar used as reinforcing against diagonal tension in a beam.

**UHF Cable.** Cable that is designed for ultra-high frequency.

**Undercourse.** A course of shingles laid beneath an exposed course of shingles at the lower edge of a wall or roof, in order to provide a waterproof layer behind the joints in the exposed course.

**Underdrain.** Perforated pipe drain installed in crushed stone under a slab to intercept ground water and drain it away from the structure.

**Underfloor Duct.** A round or rectangular metal pipe placed in a concrete floor to distribute warm air from a heating or air conditioning system.

**Underground Tank.** A container that holds various liquid or solid matter and is found underground.

**Underlayment.** A layer of hardboard, particleboard, plywood, mastic with latex binders, mastic with asphalt binders, mastic with polyvinyl-acetate, etc., placed to cover subfloor irregularities, to absorb the movement of wood subfloors and to provide a smooth surface for the finish flooring material.

**Underpinning.** The process of placing new foundations beneath an existing structure.

**Underslab Drainage.** The process of continuous interception and removal of ground water from under a concrete slab with the installation of perforated pipe.

**Ungrounded Cable.** Two-wire nonmetallic sheathed cable that contains one neutral wire and one hot wire.

**Union.** A type of pipe fitting used to join two pipes in line without turning either pipe.

**Union T.** A pipe tee with a fitting on one end that joins two pipes without turning either pipe.

**Unit Heater.** A device for heating a space without the use of ductwork.

**Unit Masonry.** Manufactured or natural building units of concrete, burned clay, glass, stone, gypsum, etc..

**Unit Substation.** A freestanding assembly including a transformer, switchgear and meter(s).

**Unit Ventilator.** A fan coil unit (FCU) with an opening through an exterior wall to admit outside air.

**Unit-and-Mullion System.** A curtain wall system consisting of prefabricated panel units secured with site-applied mullions.

**Unitary Package Unit.** A system of conditioning air in an outdoor package unit, similar to a rooftop unit, except that the supply and return ductwork passes horizontally through an outside wall.

**Unlined Stack.** Chimney or vent fabricated of single pipe.

**Unrated Door.** A door which has not been rated for any ability to withstand the spread of fire.

**Unreinforced.** Constructed without steel reinforcing bars or welded wire fabric.

**Urethane Board.** Rigid form of plastic foam of polyurethane.

**Urinal.** A plumbing fixture used to collect urine and which can be flushed.

**Urinal Screen.** A privacy panel that separates urinals from each other. The screens are either floor, wall or ceiling mounted.

**Utility Excavation.** The act or process of either digging up existing cable buried in the ground, or trenching to lay new cable.

**Utility Pole.** A vertical pole with cross arms, where utility company lines are carried.

## V

**V Beam Roof.** A roof with corrugated sheeting with flat, V-angled surfaces.

**Vacuum Breaker.** An electrical breaker with a space that contains reduced air pressure.

**Valley.** A trough or internal angle formed by the intersection of two roof slopes. The internal angle formed by the two slopes of a roof.

**Valley Flashing.** Thin sheet metal used to line the valley of a roof.

**Valley Rafter.** A diagonal rafter that supports a roof valley.

**Valve.** Numerous mechanical devices by which the flow of liquid, gas, or loose material in bulk may be started, stopped, or regulated by a movable part that opens, shuts, or partially obstructs one or more passageways.

**Vanity Cabinet.** Case, box, or piece of furniture which sets on floor and receives a lavatory, commonly has shelves and doors and is primarily used as storage for below lavatory.

**Vapor Barrier.** Waterproof membrane placed under concrete floor slabs that are on grade.

**Varnish.** A colorless, clear, resinous product dissolved in oil, alcohol, or a number of volatile liquids, applied to wood to provide a hard, glossy, protective film.

**Vault.** An arched surface. An arch translated along an axis normal to the plane of its centerline curve. A room to store valuable items.

**VAV.** Variable air volume system.

**Veneer.** A thin layer, sheet or facing.

**Veneer, Ashlar.** An ornamental or protective facing of masonry composed of squared stones.

**Veneer Plaster.** A wall finish system in which a thin layer of plaster is applied over a special gypsum board base.

**Vent.** A vertical pipe connected to a waste or soil distribution system that prevents a vacuum that might suck the water out of a trap. Vertical pipe to provide passageway for expulsion of vent gases from gas-burning equipment to outside air.

**Vent, Foundation.** Opening in foundation wall to provide natural ventilation to foundation crawl spaces.

**Vent Stack.** A plumbing vent (pipe) in a multistory building, a separate pipe used for venting, that either connects with a stack vent above the highest fixture, or extends through the roof.

**Ventilator, Gravity.** A device installed in an opening in a room or building which is activated by air passing through to remove stale air and replace it with fresh air.

**Verge Boards.** The boards which serve as the eaves finish on the gable end of a building.

**Vermiculite.** Expanded mica, used as an insulating fill or a lightweight aggregate.

**Vermont Slate.** A fine grained thin-layered rock used for roofing, paving, etc..

**Vertical Bar.** An upright reinforcing bar in a reinforced concrete shape.

**Vertical Siding.** Exterior wall covering attached vertically to the wood frame of a building or structure.

**Vestibule.** An entrance to a house; usually enclosed.

**Vibration.** A periodic motion which repeats itself after a definite interval of time.

**Vinyl Sheet.** The rolled form of vinyl.

**Vinyl Sheetrock.** Plasterboard with a thin layer of vinyl as the finished surface.

**Vinyl Siding.** Exterior wall coverings made from a thermoplastic compound.

**Vinyl Tile.** A semi-flexible, resilient floor tile made from polymerized vinyl chloride, vinylide chloride, or vinyl acetate.

**Vision Panel.** Glass placed in an opening of a door.

**Vitrified Clay Pipe.** Pipes used especially for underground drainage, that are made of clay baked hard.

**Voltage.** Electricity is caused by creating a higher electric charge at one point in a conductor than at another. This potential difference is called voltage.

## W

**Waferboard.** A building panel made by bonding together large, flat flakes of wood.

**Waffle Slab.** A two-way concrete joist system. Two-way slab or flat slab made up of a double system of narrow ribs or joists, usually at right angles to each other, forming a pattern of waffle-like coffers.

**Wainscoting.** A wall facing, usually of wood, cut stone, or ceramic tile, that is carried only part way up a wall. Matched boarding or panel work covering the lower portion of a wall.

**Wale.** A horizontal beam.

**Walk.** A path specially arranged or paved for walking.

**Walkway, Roof.** A permanent aisle with handrails for safe access across a roof.

**Wall.** A member, usually vertical, used to enclose or separate spaces.

**Wall Blocking.** Framing lumber cut in short lengths and installed horizontally between wall studs as filler pieces to stabilize the framing or to provide a backing for fastening a finish item.

**Wall Flange.** A ridge on a wall that prevents movement. A supporting rim on a wall for attachments.

**Wall Footing.** A continuous spread footing that supports a uniform load from a wall.

*For More Terms and Definitions See*
*Building News Illustrated Construction Dictionary*

**Wall Heater.** A heating unit installed in a wall.

**Wall Hydrant.** A connection to a water main cut through and mounted on a wall.

**Wall, Retaining.** A wall that is designed to resist the lateral pressures of retained soil.

**Wall Sheathing.** The first layer of covering on an exterior wall, fastened to the wall studs.

**Wall Tie.** A mechanical metal fastener which connects wythes of masonry to each other or to other materials.

**Walnut Veneer.** An overlay of a thin layer of walnut wood for outer finish or decoration.

**Wardrobe.** A room or freestanding closet where clothes are kept.

**Wardrobe Locker.** A cabinet with a locking door where clothes are kept.

**Wash.** The slant upon a sill, capping, etc., to allow the water to run off easily.

**Wash Fountain.** A waist high sink which supplies a steady stream of water to cleanse the hands.

**Washer.** A flat thin ring or a perforated plate used in joints or assemblies to ensure tightness or relieve friction.

**Washer, Flat.** A washer which goes under a bolt head or a nut to spread the load and protect the surface.

**Washroom Faucet.** A device that dispenses hot and cold water, mounted above a sink.

**Waste Stack.** A vertical pipe that carries liquids other than human waste.

**Watchman.** A guard who keeps watch over a certain area.

**Water Closet.** A plumbing fixture for the disposal of human wastes, a toilet.

**Water Cooler.** An apparatus that holds and dispenses cold water.

**Water Heater.** An apparatus for storing and heating water.

**Water Meter.** A device for measuring the flow of water.

**Water Reducing Admixture.** Material added to cement or a concrete mix to cut down on its water content.

**Water Slide.** A sloping trough down which water is carried by gravity.

**Water Softener.** ' device attached to a water system to remove unwanted minerals and substances.

**Water Valve.** A device to regulate the flow of water in a pipe or other passage.

**Water Resistant Gypsum Board.** A gypsum board designed for use in locations where it may be exposed to occasional dampness. Plasterboard that has had a chemical treatment to make it resistant to moisture, but not necessarily waterproof.

**Waterproof Membrane.** A membrane, which can be made of built-up roofing or an elastomeric sheet, to provide positive waterproofing for a floor wall.

**Waterproofing.** The act or process of making something waterproof. A coating capable of stopping penetration of water or moisture.

**Waterstop.** A synthetic rubber strip used to seal joints in concrete foundation walls.

**Watertight Manhole.** A cover for a vertical access shaft that prevents the elements from coming in.

**Wearing Course.** A topping or surface treatment to increase the resistance of a pavement or slab to abrasion.

**Weather Seal.** A flanged channel installed on the edges of an exterior door.

**Weathering Steel.** A steel alloy that forms a tenacious, self-protecting rust layer when exposed to the atmosphere.

**Weatherproof Box.** An electrical outlet or switch box which has been manufactured to withstand the outside elements.

**Weatherstripping.** The process of reducing air or rain infiltration by covering joints of doors or windows with strips of resilient material.

**Weed Control.** The act or process of spraying chemicals or placing powders to control the spread of weeds.

**Weep Hole.** A small opening, the purpose of which is to permit drainage of water that accumulates inside a building component.

**Weld.** A joint between two pieces of metal formed by fusing the pieces together, usually with the aid of additional metal melted from a rod or electrode. Join two pieces of metal together by heating until fusion of material either with or without filler metal.

**Welded Pipe.** Piping where connections and fittings are welded.

**Welded Railing.** Railing sections with the components fastened with welds.

**Welded Truss.** Trusses with components fastened together with welds.

**Welded Wire Fabric.** A series of longitudinal and transverse wires arranged substantially at right angles to each other in sheets or rolls, used to reinforce mortar and concrete.

**Well.** A pit or hole sunk into the earth to reach a supply of water. An open space extending vertically through floors of a structure.

**Wellpoint.** A perforated pipe surrounded by sand to permit the pumping of ground water.

**Wellpoint System.** A series of vertical pipes in the ground connected to a header and pump to drain marshy areas or to control ground seepage.

**Wet Location Fluorescent.** A watertight fluorescent fixture that is sealed to protect against moisture.

**Wet Sprinkler System.** A sprinkler system that is filled with water at design pressure for immediate use upon activation.

**Wheel Barrow.** A small vehicle with handles and one or more wheels used for carrying small loads.

**Wheel Chair Partition.** A dividing wall in a bathroom or bathing room which forms the perimeter of a private area that has been made accessible to the disabled.

**Wheeled Extinguisher.** A fire extinguisher mounted on a wheeled cart that can be pushed or pulled by a person.

**White Cedar Shingle.** A light colored weather-resistant cedar wood used for roofing and siding.

**Wide Stile Door.** Wider than normal vertical members forming the outside framework of a door.

**Wide Flange Section.** Any of a wide range of steel sections rolled in the shape of a letter "T" or "H".

**Window.** An opening in the wall of a building or structure for the admission of light and air, closed by casements or sashes containing glass panes.

**Window, Drive-Up.** An opening in a wall through which transactions can be made with persons in motor vehicles.

**Window Frame.** The structure which holds a window assembly in place.

**Window Guard, Steel Bar Grille.** Guard fabricated from a steel bar grille.

**Window Guard, Woven Wire.** Guard fabricated of woven wire..

**Window Header.** A horizontal construction member placed across the top of a window opening to support the load above.

**Window, Service.** An opening in a wall or partition through which business is transacted.

**Window Sill.** The horizontal member at the bottom of a window.

**Window Sill, Marble.** Marble installed on the horizontal member at the bottom of a window.

**Window Stool.** Wood, ceramic tile, or masonry installed on the plate at a window sill on the inside of the window, fitted against the bottom rail of the lower sash.

**Window Sub-Sill.** Component anchored to wall construction and located just below window sill to receive window sill.

**Window Treatment.** The addition of hanging fabrics, curtains, or blinds to the interior of a window.

**Window Well.** Recess located at or below grade to allow for natural light to reach a ground level or basement window, often created by use of corrugated metal in half-round shape.

**Windowwall.** The opening in a wall surface which contains a window assembly or wall of assemblies. Often referred to loosely as tube-framing or storefront.

**Wire, Aluminum.** Electrical conductors and cable manufactured from aluminum.

**Wire, Chicken.** Thin, woven wire mounted on an exterior wall as a base for stucco plaster.



*For More Terms and Definitions See*
*Building News Illustrated Construction Dictionary*

**Wire Glass.** Glass in which a wire mesh was embedded during manufacture.

**Wire Hanger.** A wire that supports or connects material.

**Wire Mesh.** A series of longitudinal and transverse wires arranged substantially at right angles to each other sheets or rolls, used to reinforce mortar and concrete.

**Wire Mesh Partition.** Dividing wall constructed of metal framing and wire mesh.

**Wire, Stranded.** Fine wires twisted together in a group to create a larger stronger cable or wire.

**Wireway.** A sheet metal trough with hinge or removable cover to carry several electrical cables.

**Wiring, Flexible.** Electrical wiring that permits ease of installation and movement from expansion, contraction, vibration, and/or rotation.

**Wood Anchor.** A bolt or fastening device which attaches wood to wood or wood to another material.

**Wood and Plastics.** A category of the CSI Masterformat which is represented in Division 6 of the format. Commonly called just "Wood" or "Carpentry".

**Wood Batten.** Wood strips covering vertical joints on boards used as exterior siding.

**Wood Beam.** Horizontal wood structural member that supports uniform and/or concentrated loads.

**Wood Buck.** Wood frame typically built into concrete or masonry wall to accommodate finish door frame.

**Wood Bumper.** Wood component used to absorb impact and prevent damage to other surfaces.

**Wood Cap.** Wood member used on top of an assembly to provide termination and/or finish.

**Wood Carriage.** Sloping beam installed between stringers to support steps of wood stair.

**Wood Casing.** Wood exposed millwork or trim molding around doors, windows, beams, etc.

**Wood Chip Mulch.** Wood chips, spread on the ground to prevent erosion, control weeds, minimize evaporation and improve the soil.

**Wood Decking.** Plywood, lumber, or glued laminated member placed over roof or floor structural members for structural rigidity of building frame and to provide a surface for traffic or substrate for roofing or flooring system.

**Wood Fascia.** Flat vertical wood member of cornice, eaves, or gable or other finish, generally that part of the assembly to which the gutter is secured.

**Wood Fiber Insulation.** Wood particles and fiber used to reduce heat transfer.

**Wood Fiber Panel.** Form sheathing manufactured from glued and bonded wood particles.

**Wood Finish Concrete.** The act or process of using a wood float to smooth irregularities left in curing concrete, work the surface or compact the concrete.

**Wood Frame.** Floors, roofs, exterior and bearing walls of a building or structure constructed with wood.

**Wood Molding.** Wood strip with curved or projecting surface, used for decorative purposes.

**Wood Nailer.** Strip of wood attached to steel or concrete to facilitate making nailed connections.

**Wood Roof Curb.** Wood member elevated above plane of roof surface used for mounting of equipment or other elements.

**Wood Roof Nailer.** Wood strip (usually in plane of roof insulation) secured to structural roof deck used for securing roofing membrane.

**Wood Saddle.** Short horizontal wood member set on top of wood column to serve as seat for a girder.

**Wood Screw.** A screw for fastening objects in wood.

**Wood Shake.** A hand-split shingle.

**Wood Shingle.** Factory cut and shaped roof covering of wood (usually of cedar), cut into modular lengths, widths, and rectangular profile.

**Wood Siding.** Wood used as exterior surface or cladding for exterior framed wall to provide protection from the elements.

**Wood Sleeper.** Wood member laid on concrete floor to support and receive fastening of wood subfloor or finish flooring.

**Wood Soffit.** The wooden underside of a part or member in a building, such as the under surface of an arch, beam, cornice, or stairway.

**Wood Treatment.** The act or process of applying a variety of stains or chemicals to retard fire, decay, insect damage or deterioration, due to the elements.

**Wood Trim.** Wood millwork, primarily moldings, used to finish off and cover joints and openings.

**Wood Truss.** A structural component formed of wood members in a triangular arrangement, often used to support a roof.

**Woodwork, Architectural.** A higher than average feature of finish work using wood for ornamental design.

**Wye.** A pipe fitting which joins three pipes at a 45 degree angle.

**Wythe.** A vertical layer of masonry one masonry unit thick.

## X,Y,Z

**X Bracing.** That form of bracing wherein a pair of diagonal braces cross near mid-length of the bracing members.

**X-Ray.** Electromagnetic radiations of a short wavelength that can penetrate various thicknesses of all solids.

**X-Ray, Weld.** To examine, treat, or photograph the connection of surfaces that have been welded together.

**Y Strainer.** A device in the shape of a "Y" for withholding foreign matter from a flowing liquid or gas.

**Z Tie, Wall.** A Z-shaped reinforcing strip used as a support bracket from the structural wall to the masonry veneer.

**Zero Slump Concrete.** A concrete mixed with so little water that it has a slump of zero when tested.

BNi Building News

# Sample Costbook Page

In order to best use the information in this book, please review this sample page and read the "Features In This Book" section.

**CSI MASTERFORMAT Division**

**CSI Broadscope Category**

**CSI Mediumscope Category (First 5 Digits)**

**03 CONCRETE**

| CAST-IN-PLACE CONCRETE | UNIT | MAT. | INST. | TOTAL |
|---|---|---|---|---|
| **03350.10 CONCRETE FINISHES** | | | | |
| Wet rub | S.F. | 0.00 | 1.20 | 1.20 |
| Floor hardeners | | | | |
| Metallic | | | | |
| Light service | S.F. | 0.20 | 0.30 | 0.50 |
| Heavy service | " | 0.65 | 0.40 | 1.05 |
| Non-metallic | | | | |
| Light service | S.F. | 0.10 | 0.30 | 0.40 |
| Heavy service | " | 0.46 | 0.40 | 0.86 |
| Rusticated concrete finish | | | | |
| Beveled edge | L.F. | 0.23 | 1.35 | 1.58 |
| Square edge | " | 0.29 | 1.75 | 2.04 |
| Solid board concrete finish | | | | |
| Standard | S.F. | 0.60 | 2.00 | 2.60 |
| Rustic | " | 0.53 | 2.40 | 2.93 |
| **03360.10 PNEUMATIC CONCRETE** | | | | |
| Pneumatic applied concrete (gunite) | | | | |
| 2" thick | S.F. | 2.95 | 1.65 | 4.60 |
| 3" thick | " | 3.80 | 2.20 | 6.00 |
| 4" thick | " | 4.50 | 2.60 | 7.10 |
| Finish surface | | | | |
| Minimum | S.F. | 0.00 | 1.55 | 1.55 |
| Maximum | " | 0.00 | 3.10 | 3.10 |
| **03370.10 CURING CONCRETE** | | | | |
| Sprayed membrane | | | | |
| Slabs | S.F. | 0.04 | 0.05 | 0.09 |
| Walls | " | 0.07 | 0.06 | 0.13 |
| Curing paper | | | | |
| Slabs | S.F. | 0.07 | 0.06 | 0.13 |
| Walls | " | 0.07 | 0.07 | 0.14 |
| Burlap | | | | |
| 7.5 oz. | S.F. | 0.06 | 0.08 | 0.14 |
| 12 oz. | " | 0.08 | 0.09 | 0.17 |

| PLACING CONCRETE | UNIT | MAT. | INST. | TOTAL |
|---|---|---|---|---|
| **03380.05 BEAM CONCRETE** | | | | |
| Beams and girders | | | | |
| 2500# or 3000# concrete | | | | |
| By crane | C.Y. | 64.50 | 46.20 | 110.70 |
| By pump | " | 64.50 | 42.00 | 106.50 |

**Material Cost**
Material costs represent average contractor prices plus an allowance for freight, handling and storage.

**Installation Cost**
Installation cost includes basic wage rates, markups for taxes, insurance overhead and profit and also includes equipment costs where appropriate.

**Detailed Descriptions**
Complete descriptions of items may include information listed above a particular line. Review of the whole category is recommended for a complete description.

**Total Cost**
The total cost is the sum of material and installation costs. This total represents typical contractors' costs including overhead and profit, but does not include markups for the general contractor or construction management fees.

**Unit of Measurement**
Each item (and cost) is defined in terms of the common estimating unit. All costs are listed in dollars per unit.

BNi Building News

# 01 GENERAL

## REQUIREMENTS

| 01020.10 ALLOWANCES | UNIT | MAT. | INST. | TOTAL |
|---|---|---|---|---|
| Overhead | | | | |
| $20,000 project | | | | |
| Minimum | PCT. | | | 15.00 |
| Average | " | | | 20.00 |
| Maximum | " | | | 40.00 |
| $100,000 project | | | | |
| Minimum | PCT. | | | 12.00 |
| Average | " | | | 15.00 |
| Maximum | " | | | 25.00 |
| $500,000 project | | | | |
| Minimum | PCT. | | | 10.00 |
| Average | " | | | 12.00 |
| Maximum | " | | | 20.00 |
| $1,000,000 project | | | | |
| Minimum | PCT. | | | 6.00 |
| Average | " | | | 10.00 |
| Maximum | " | | | 12.00 |
| Profit | | | | |
| $20,000 project | | | | |
| Minimum | PCT. | | | 10.00 |
| Average | " | | | 15.00 |
| Maximum | " | | | 25.00 |
| $100,000 project | | | | |
| Minimum | PCT. | | | 10.00 |
| Average | " | | | 12.00 |
| Maximum | " | | | 20.00 |
| $500,000 project | | | | |
| Minimum | PCT. | | | 5.00 |
| Average | " | | | 10.00 |
| Maximum | " | | | 15.00 |
| $1,000,000 project | | | | |
| Minimum | PCT. | | | 3.00 |
| Average | " | | | 8.00 |
| Maximum | " | | | 15.00 |
| Professional fees | | | | |
| Architectural | | | | |
| $100,000 project | | | | |
| Minimum | PCT. | | | 5.00 |
| Average | " | | | 10.00 |
| Maximum | " | | | 20.00 |
| $500,000 project | | | | |
| Minimum | PCT. | | | 5.00 |
| Average | " | | | 8.00 |
| Maximum | " | | | 12.00 |
| $1,000,000 project | | | | |
| Minimum | PCT. | | | 3.50 |
| Average | " | | | 7.00 |
| Maximum | " | | | 10.00 |
| Structural engineering | | | | |
| Minimum | PCT. | | | 2.00 |
| Average | " | | | 3.00 |
| Maximum | " | | | 5.00 |

# 01 GENERAL

## REQUIREMENTS

### 01020.10 ALLOWANCES

| | UNIT | MAT. | INST. | TOTAL |
|---|---|---|---|---|
| Mechanical engineering | | | | |
| Minimum | PCT. | | | 4.00 |
| Average | " | | | 5.00 |
| Maximum | " | | | 15.00 |
| Electrical engineering | | | | |
| Minimum | PCT. | | | 3.00 |
| Average | " | | | 5.00 |
| Maximum | " | | | 12.00 |
| Taxes | | | | |
| Sales tax | | | | |
| Minimum | PCT. | | | 4.00 |
| Average | " | | | 5.00 |
| Maximum | " | | | 8.25 |
| Unemployment | | | | |
| Minimum | PCT. | | | 2.00 |
| Average | " | | | 5.00 |
| Maximum | " | | | 7.00 |
| Social security (FICA) | " | | | 7.85 |

### 01050.10 FIELD STAFF

| | UNIT | MAT. | INST. | TOTAL |
|---|---|---|---|---|
| Superintendent | | | | |
| Minimum | YEAR | | | 60,750 |
| Average | " | | | 87,750 |
| Maximum | " | | | 128,250 |
| Field engineer | | | | |
| Minimum | YEAR | | | 47,250 |
| Average | " | | | 70,200 |
| Maximum | " | | | 101,250 |
| Foreman | | | | |
| Minimum | YEAR | | | 33,750 |
| Average | " | | | 54,000 |
| Maximum | " | | | 81,000 |
| Bookkeeper/timekeeper | | | | |
| Minimum | YEAR | | | 20,250 |
| Average | " | | | 25,650 |
| Maximum | " | | | 40,500 |
| Watchman | | | | |
| Minimum | YEAR | | | 13,500 |
| Average | " | | | 16,875 |
| Maximum | " | | | 27,000 |

### 01310.10 SCHEDULING

| | UNIT | MAT. | INST. | TOTAL |
|---|---|---|---|---|
| Scheduling for | | | | |
| $100,000 project | | | | |
| Minimum | PCT. | | | 1.00 |
| Average | " | | | 2.00 |
| Maximum | " | | | 4.00 |
| $500,000 project | | | | |
| Minimum | PCT. | | | 0.50 |
| Average | " | | | 1.00 |
| Maximum | " | | | 2.00 |
| $1,000,000 project | | | | |

# 01 GENERAL

| REQUIREMENTS | UNIT | MAT. | INST. | TOTAL |
|---|---|---|---|---|
| **01310.10 SCHEDULING** | | | | |
| Minimum | PCT. | | | 0.30 |
| Average | " | | | 0.80 |
| Maximum | " | | | 1.50 |
| Scheduling software, not including computer | | | | |
| Minimum | EA. | | | 613.64 |
| Average | " | | | 3,068 |
| Maximum | " | | | 49,091 |
| **01330.10 SURVEYING** | | | | |
| Surveying | | | | |
| Small crew | DAY | 0.00 | 560.00 | 560.00 |
| Average crew | " | 0.00 | 840.00 | 840.00 |
| Large crew | " | 0.00 | 1,110 | 1,110 |
| **01380.10 JOB REQUIREMENTS** | | | | |
| Job photographs, small jobs | | | | |
| Minimum | EA. | | | 67.50 |
| Average | " | | | 135.00 |
| Maximum | " | | | 297.00 |
| Large projects | | | | |
| Minimum | EA. | | | 472.50 |
| Average | " | | | 675.00 |
| Maximum | " | | | 1,620 |
| **01410.10 TESTING** | | | | |
| Testing concrete, per test | | | | |
| Minimum | EA. | | | 14.85 |
| Average | " | | | 27.00 |
| Maximum | " | | | 54.00 |
| Soil, per test | | | | |
| Minimum | EA. | | | 27.00 |
| Average | " | | | 81.00 |
| Maximum | " | | | 216.00 |
| Welding, per test | | | | |
| Minimum | EA. | | | 13.50 |
| Average | " | | | 27.00 |
| Maximum | " | | | 101.25 |
| **01500.10 TEMPORARY FACILITIES** | | | | |
| Barricades, temporary | | | | |
| Highway | | | | |
| Concrete | L.F. | 9.45 | 2.80 | 12.25 |
| Wood | " | 2.70 | 1.10 | 3.80 |
| Steel | " | 2.95 | 0.93 | 3.88 |
| Pedestrian barricades | | | | |
| Plywood | S.F. | 1.90 | 0.93 | 2.83 |
| Chain link fence | " | 2.30 | 0.93 | 3.23 |
| Trailers, general office type, per month | | | | |
| Minimum | EA. | | | 168.75 |
| Average | " | | | 270.00 |

# 01 GENERAL

## REQUIREMENTS

| REQUIREMENTS | UNIT | MAT. | INST. | TOTAL |
|---|---|---|---|---|
| **01500.10 TEMPORARY FACILITIES** | | | | |
| Maximum | EA. | | | 540.00 |
| Crew change trailers, per month | | | | |
| Minimum | EA. | | | 94.50 |
| Average | " | | | 108.00 |
| Maximum | " | | | 162.00 |
| **01505.10 MOBILIZATION** | | | | |
| Equipment mobilization | | | | |
| Bulldozer | | | | |
| Minimum | EA. | | | 135.00 |
| Average | " | | | 283.50 |
| Maximum | " | | | 472.50 |
| Backhoe/front-end loader | | | | |
| Minimum | EA. | | | 81.00 |
| Average | " | | | 141.75 |
| Maximum | " | | | 310.50 |
| Crane, crawler type | | | | |
| Minimum | EA. | | | 1,485 |
| Average | " | | | 3,645 |
| Maximum | " | | | 7,830 |
| Truck crane | | | | |
| Minimum | EA. | | | 351.00 |
| Average | " | | | 526.50 |
| Maximum | " | | | 904.50 |
| Pile driving rig | | | | |
| Minimum | EA. | | | 6,750 |
| Average | " | | | 13,500 |
| Maximum | " | | | 24,300 |
| **01525.10 CONSTRUCTION AIDS** | | | | |
| Scaffolding/staging, rent per month | | | | |
| Measured by lineal feet of base | | | | |
| 10' high | L.F. | | | 9.72 |
| 20' high | " | | | 17.55 |
| 30' high | " | | | 24.57 |
| 40' high | " | | | 28.35 |
| 50' high | " | | | 33.75 |
| Measured by square foot of surface | | | | |
| Minimum | S.F. | | | 0.43 |
| Average | " | | | 0.73 |
| Maximum | " | | | 1.32 |
| Safety nets, heavy duty, per job | | | | |
| Minimum | S.F. | | | 0.27 |
| Average | " | | | 0.34 |
| Maximum | " | | | 0.74 |
| Tarpaulins, fabric, per job | | | | |
| Minimum | S.F. | | | 0.20 |
| Average | " | | | 0.34 |
| Maximum | " | | | 0.88 |

# 01 GENERAL

| REQUIREMENTS | UNIT | MAT. | INST. | TOTAL |
|---|---|---|---|---|
| **01570.10 SIGNS** | | | | |
| Construction signs, temporary | | | | |
| Signs, 2' x 4' | | | | |
| Minimum | EA. | | | 27.00 |
| Average | " | | | 64.80 |
| Maximum | " | | | 229.50 |
| Signs, 4' x 8' | | | | |
| Minimum | EA. | | | 56.70 |
| Average | " | | | 148.50 |
| Maximum | " | | | 634.50 |
| Signs, 8' x 8' | | | | |
| Minimum | EA. | | | 72.90 |
| Average | " | | | 229.50 |
| Maximum | " | | | 2,363 |
| **01600.10 EQUIPMENT** | | | | |
| Air compressor | | | | |
| 60 cfm | | | | |
| By day | EA. | | | 70.20 |
| By week | " | | | 210.60 |
| By month | " | | | 634.50 |
| 300 cfm | | | | |
| By day | EA. | | | 148.50 |
| By week | " | | | 445.50 |
| By month | " | | | 1,337 |
| 600 cfm | | | | |
| By day | EA. | | | 405.00 |
| By week | " | | | 1,215 |
| By month | " | | | 3,645 |
| Air tools, per compressor, per day | | | | |
| Minimum | EA. | | | 27.00 |
| Average | " | | | 33.75 |
| Maximum | " | | | 47.25 |
| Generators, 5 kw | | | | |
| By day | EA. | | | 67.50 |
| By week | " | | | 202.50 |
| By month | " | | | 621.00 |
| Heaters, salamander type, per week | | | | |
| Minimum | EA. | | | 81.00 |
| Average | " | | | 114.75 |
| Maximum | " | | | 243.00 |
| Pumps, submersible | | | | |
| 50 gpm | | | | |
| By day | EA. | | | 54.00 |
| By week | " | | | 162.00 |
| By month | " | | | 486.00 |
| 100 gpm | | | | |
| By day | EA. | | | 67.50 |
| By week | " | | | 202.50 |
| By month | " | | | 607.50 |
| 500 gpm | | | | |
| By day | EA. | | | 108.00 |

# 01 GENERAL

| REQUIREMENTS | UNIT | MAT. | INST. | TOTAL |
|---|---|---|---|---|
| **01600.10 EQUIPMENT** | | | | |
| By week | EA. | | | 324.00 |
| By month | " | | | 972.00 |
| Diaphragm pump, by week | | | | |
| Minimum | EA. | | | 94.50 |
| Average | " | | | 162.00 |
| Maximum | " | | | 337.50 |
| Pickup truck | | | | |
| By day | EA. | | | 101.25 |
| By week | " | | | 297.00 |
| By month | " | | | 918.00 |
| Dump truck | | | | |
| 6 cy truck | | | | |
| By day | EA. | | | 270.00 |
| By week | " | | | 810.00 |
| By month | " | | | 2,430 |
| 10 cy truck | | | | |
| By day | EA. | | | 337.50 |
| By week | " | | | 1,013 |
| By month | " | | | 3,038 |
| 16 cy truck | | | | |
| By day | EA. | | | 540.00 |
| By week | " | | | 1,620 |
| By month | " | | | 4,860 |
| Backhoe, track mounted | | | | |
| 1/2 cy capacity | | | | |
| By day | EA. | | | 553.50 |
| By week | " | | | 1,688 |
| By month | " | | | 4,995 |
| 1 cy capacity | | | | |
| By day | EA. | | | 877.50 |
| By week | " | | | 2,633 |
| By month | " | | | 7,898 |
| 2 cy capacity | | | | |
| By day | EA. | | | 1,485 |
| By week | " | | | 4,455 |
| By month | " | | | 13,365 |
| 3 cy capacity | | | | |
| By day | EA. | | | 2,835 |
| By week | " | | | 8,505 |
| By month | " | | | 25,515 |
| Backhoe/loader, rubber tired | | | | |
| 1/2 cy capacity | | | | |
| By day | EA. | | | 337.50 |
| By week | " | | | 1,013 |
| By month | " | | | 3,038 |
| 3/4 cy capacity | | | | |
| By day | EA. | | | 405.00 |
| By week | " | | | 1,215 |
| By month | " | | | 3,645 |
| Bulldozer | | | | |
| 75 hp | | | | |
| By day | EA. | | | 472.50 |

# 01 GENERAL

| REQUIREMENTS | UNIT | MAT. | INST. | TOTAL |
|---|---|---|---|---|
| **01600.10 EQUIPMENT** | | | | |
| By week | EA. | | | 1,418 |
| By month | " | | | 4,253 |
| Cranes, crawler type | | | | |
| 15 ton capacity | | | | |
| By day | EA. | | | 607.50 |
| By week | " | | | 1,823 |
| By month | " | | | 5,468 |
| 25 ton capacity | | | | |
| By day | EA. | | | 742.50 |
| By week | " | | | 2,228 |
| By month | " | | | 6,683 |
| 50 ton capacity | | | | |
| By day | EA. | | | 1,350 |
| By week | " | | | 4,050 |
| By month | " | | | 12,150 |
| 100 ton capacity | | | | |
| By day | EA. | | | 2,025 |
| By week | " | | | 6,075 |
| By month | " | | | 18,225 |
| Truck mounted, hydraulic | | | | |
| 15 ton capacity | | | | |
| By day | EA. | | | 573.75 |
| By week | " | | | 1,721 |
| By month | " | | | 5,164 |
| Loader, rubber tired | | | | |
| 1 cy capacity | | | | |
| By day | EA. | | | 405.00 |
| By week | " | | | 1,215 |
| By month | " | | | 3,645 |
| **01740.10 BONDS** | | | | |
| Performance bonds | | | | |
| Minimum | PCT. | | | 0.60 |
| Average | " | | | 1.90 |
| Maximum | " | | | 3.00 |

# 02 SITEWORK

| DEMOLITION | UNIT | MAT. | INST. | TOTAL |
|---|---|---|---|---|
| **02060.10 BUILDING DEMOLITION** | | | | |
| Building, complete with disposal | | | | |
| Wood frame | C.F. | 0.00 | 0.20 | 0.20 |
| Concrete | " | 0.00 | 0.30 | 0.30 |
| Steel frame | " | 0.00 | 0.40 | 0.40 |
| Partition removal | | | | |
| Concrete block partitions | | | | |
| 4" thick | S.F. | 0.00 | 1.40 | 1.40 |
| 8" thick | " | 0.00 | 1.85 | 1.85 |
| 12" thick | " | 0.00 | 2.50 | 2.50 |
| Brick masonry partitions | | | | |
| 4" thick | S.F. | 0.00 | 1.40 | 1.40 |
| 8" thick | " | 0.00 | 1.75 | 1.75 |
| 12" thick | " | 0.00 | 2.30 | 2.30 |
| 16" thick | " | 0.00 | 3.45 | 3.45 |
| Cast in place concrete partitions | | | | |
| Unreinforced | | | | |
| 6" thick | S.F. | 0.00 | 9.60 | 9.60 |
| 8" thick | " | 0.00 | 10.30 | 10.30 |
| 10" thick | " | 0.00 | 12.00 | 12.00 |
| 12" thick | " | 0.00 | 14.40 | 14.40 |
| Reinforced | | | | |
| 6" thick | S.F. | 0.00 | 11.10 | 11.10 |
| 8" thick | " | 0.00 | 14.40 | 14.40 |
| 10" thick | " | 0.00 | 16.00 | 16.00 |
| 12" thick | " | 0.00 | 19.20 | 19.20 |
| Terra cotta | | | | |
| To 6" thick | S.F. | 0.00 | 1.40 | 1.40 |
| Stud partitions | | | | |
| Metal or wood, with drywall both sides | S.F. | 0.00 | 1.40 | 1.40 |
| Metal studs, both sides, lath and plaster | " | 0.00 | 1.85 | 1.85 |
| Door and frame removal | | | | |
| Hollow metal in masonry wall | | | | |
| Single | | | | |
| 2'6"x6'8" | EA. | 0.00 | 34.70 | 34.70 |
| 3'x7' | " | 0.00 | 46.30 | 46.30 |
| Double | | | | |
| 3'x7' | EA. | 0.00 | 55.50 | 55.50 |
| 4'x8' | " | 0.00 | 55.50 | 55.50 |
| Wood in framed wall | | | | |
| Single | | | | |
| 2'6"x6'8" | EA. | 0.00 | 19.85 | 19.85 |
| 3'x6'8" | " | 0.00 | 23.15 | 23.15 |
| Double | | | | |
| 2'6"x6'8" | EA. | 0.00 | 27.80 | 27.80 |
| 3'x6'8" | " | 0.00 | 30.80 | 30.80 |
| Remove for re-use | | | | |
| Hollow metal | EA. | 0.00 | 69.50 | 69.50 |
| Wood | " | 0.00 | 46.30 | 46.30 |
| Floor removal | | | | |
| Brick flooring | S.F. | 0.00 | 1.10 | 1.10 |
| Ceramic or quarry tile | " | 0.00 | 0.62 | 0.62 |
| Terrazzo | " | 0.00 | 1.25 | 1.25 |

## 02 SITEWORK

| DEMOLITION | UNIT | MAT. | INST. | TOTAL |
|---|---|---|---|---|
| **02060.10 BUILDING DEMOLITION** | | | | |
| Heavy wood | S.F. | 0.00 | 0.74 | 0.74 |
| Residential wood | " | 0.00 | 0.79 | 0.79 |
| Resilient tile or linoleum | " | 0.00 | 0.28 | 0.28 |
| Ceiling removal | | | | |
| Acoustical tile ceiling | | | | |
| Adhesive fastened | S.F. | 0.00 | 0.28 | 0.28 |
| Furred and glued | " | 0.00 | 0.23 | 0.23 |
| Suspended grid | " | 0.00 | 0.17 | 0.17 |
| Drywall ceiling | | | | |
| Furred and nailed | S.F. | 0.00 | 0.31 | 0.31 |
| Nailed to framing | " | 0.00 | 0.28 | 0.28 |
| Plastered ceiling | | | | |
| Furred on framing | S.F. | 0.00 | 0.69 | 0.69 |
| Suspended system | " | 0.00 | 0.93 | 0.93 |
| Roofing removal | | | | |
| Steel frame | | | | |
| Corrugated metal roofing | S.F. | 0.00 | 0.56 | 0.56 |
| Built-up roof on metal deck | " | 0.00 | 0.93 | 0.93 |
| Wood frame | | | | |
| Built up roof on wood deck | S.F. | 0.00 | 0.85 | 0.85 |
| Roof shingles | " | 0.00 | 0.46 | 0.46 |
| Roof tiles | " | 0.00 | 0.93 | 0.93 |
| Concrete frame | C.F. | 0.00 | 1.85 | 1.85 |
| Concrete plank | S.F. | 0.00 | 1.40 | 1.40 |
| Built-up roof on concrete | " | 0.00 | 0.79 | 0.79 |
| Cut-outs | | | | |
| Concrete, elevated slabs, mesh reinforcing | | | | |
| Under 5 cf | C.F. | 0.00 | 27.80 | 27.80 |
| Over 5 cf | " | 0.00 | 23.15 | 23.15 |
| Bar reinforcing | | | | |
| Under 5 cf | C.F. | 0.00 | 46.30 | 46.30 |
| Over 5 cf | " | 0.00 | 34.70 | 34.70 |
| Window removal | | | | |
| Metal windows, trim included | | | | |
| 2'x3' | EA. | 0.00 | 27.80 | 27.80 |
| 2'x4' | " | 0.00 | 30.80 | 30.80 |
| 2'x6' | " | 0.00 | 34.70 | 34.70 |
| 3'x4' | " | 0.00 | 34.70 | 34.70 |
| 3'x6' | " | 0.00 | 39.70 | 39.70 |
| 3'x8' | " | 0.00 | 46.30 | 46.30 |
| 4'x4' | " | 0.00 | 46.30 | 46.30 |
| 4'x6' | " | 0.00 | 55.50 | 55.50 |
| 4'x8' | " | 0.00 | 69.50 | 69.50 |
| Wood windows, trim included | | | | |
| 2'x3' | EA. | 0.00 | 15.40 | 15.40 |
| 2'x4' | " | 0.00 | 16.35 | 16.35 |
| 2'x6' | " | 0.00 | 17.35 | 17.35 |
| 3'x4' | " | 0.00 | 18.50 | 18.50 |
| 3'x6' | " | 0.00 | 19.85 | 19.85 |
| 3'x8' | " | 0.00 | 21.35 | 21.35 |
| 6'x4' | " | 0.00 | 23.15 | 23.15 |
| 6'x6' | " | 0.00 | 25.20 | 25.20 |

# 02 SITEWORK

## DEMOLITION

### 02060.10 BUILDING DEMOLITION

| | UNIT | MAT. | INST. | TOTAL |
|---|---|---|---|---|
| 6'x8' | EA. | 0.00 | 27.80 | 27.80 |
| Walls, concrete, bar reinforcing | | | | |
| Small jobs | C.F. | 0.00 | 18.50 | 18.50 |
| Large jobs | " | 0.00 | 15.40 | 15.40 |
| Brick walls, not including toothing | | | | |
| 4" thick | S.F. | 0.00 | 1.40 | 1.40 |
| 8" thick | " | 0.00 | 1.75 | 1.75 |
| 12" thick | " | 0.00 | 2.30 | 2.30 |
| 16" thick | " | 0.00 | 3.45 | 3.45 |
| Concrete block walls, not including toothing | | | | |
| 4" thick | S.F. | 0.00 | 1.55 | 1.55 |
| 6" thick | " | 0.00 | 1.65 | 1.65 |
| 8" thick | " | 0.00 | 1.75 | 1.75 |
| 10" thick | " | 0.00 | 2.00 | 2.00 |
| 12" thick | " | 0.00 | 2.30 | 2.30 |
| Rubbish handling | | | | |
| Load in dumpster or truck | | | | |
| Minimum | C.F. | 0.00 | 0.62 | 0.62 |
| Maximum | " | 0.00 | 0.93 | 0.93 |
| For use of elevators, add | | | | |
| Minimum | C.F. | 0.00 | 0.14 | 0.14 |
| Maximum | " | 0.00 | 0.28 | 0.28 |
| Rubbish hauling | | | | |
| Hand loaded on trucks, 2 mile trip | C.Y. | 0.00 | 21.35 | 21.35 |
| Machine loaded on trucks, 2 mile trip | " | 0.00 | 14.40 | 14.40 |

## HIGHWAY DEMOLITION

### 02065.10 PAVEMENT DEMOLITION

| | UNIT | MAT. | INST. | TOTAL |
|---|---|---|---|---|
| Bituminous pavement, up to 3" thick | | | | |
| On streets | S.Y. | 0.00 | 5.75 | 5.75 |
| On pipe trench | " | 0.00 | 7.20 | 7.20 |
| Concrete pavement, 6" thick | | | | |
| No reinforcement | S.Y. | 0.00 | 9.60 | 9.60 |
| With wire mesh | " | 0.00 | 14.40 | 14.40 |
| With rebars | " | 0.00 | 18.00 | 18.00 |
| 9" thick | | | | |
| No reinforcement | S.Y. | 0.00 | 12.00 | 12.00 |
| With wire mesh | " | 0.00 | 18.00 | 18.00 |
| With rebars | " | 0.00 | 24.00 | 24.00 |
| 12" thick | | | | |
| No reinforcement | S.Y. | 0.00 | 14.40 | 14.40 |
| With wire mesh | " | 0.00 | 20.60 | 20.60 |

# 02 SITEWORK

## HIGHWAY DEMOLITION

| HIGHWAY DEMOLITION | UNIT | MAT. | INST. | TOTAL |
|---|---|---|---|---|
| **02065.10 PAVEMENT DEMOLITION** | | | | |
| With rebars | S.Y. | 0.00 | 28.80 | 28.80 |
| Sidewalk, 4" thick, with disposal | " | 0.00 | 4.80 | 4.80 |
| Removal of pavement markings by waterblasting | S.F. | 0.00 | 0.14 | 0.14 |
| **02065.15 SAW CUTTING PAVEMENT** | | | | |
| Pavement, bituminous | | | | |
| 2" thick | L.F. | 0.00 | 1.05 | 1.05 |
| 3" thick | " | 0.00 | 1.35 | 1.35 |
| Concrete pavement, with wire mesh | | | | |
| 4" thick | L.F. | 0.00 | 2.05 | 2.05 |
| 5" thick | " | 0.00 | 2.20 | 2.20 |
| Plain concrete, unreinforced | | | | |
| 4" thick | L.F. | 0.00 | 1.80 | 1.80 |
| 5" thick | " | 0.00 | 2.05 | 2.05 |
| **02065.80 CURB & GUTTER** | | | | |
| Removal, plain concrete curb | L.F. | 0.00 | 3.60 | 3.60 |
| Plain concrete curb and 2' gutter | " | 0.00 | 4.95 | 4.95 |
| **02065.85 GUARDRAILS** | | | | |
| Remove standard guardrail | | | | |
| Steel | L.F. | 0.00 | 4.80 | 4.80 |
| Wood | " | 0.00 | 3.70 | 3.70 |
| **02075.80 CORE DRILLING** | | | | |
| Concrete | | | | |
| 6" thick | | | | |
| 3" dia. | EA. | 0.00 | 25.20 | 25.20 |
| 4" dia. | " | 0.00 | 29.40 | 29.40 |
| 6" dia. | " | 0.00 | 35.30 | 35.30 |
| 8" dia. | " | 0.00 | 59.00 | 59.00 |
| 8" thick | | | | |
| 3" dia. | EA. | 0.00 | 35.30 | 35.30 |
| 4" dia. | " | 0.00 | 44.10 | 44.10 |
| 6" dia. | " | 0.00 | 50.50 | 50.50 |
| 8" dia. | " | 0.00 | 70.50 | 70.50 |
| 10" thick | | | | |
| 3" dia. | EA. | 0.00 | 44.10 | 44.10 |
| 4" dia. | " | 0.00 | 50.50 | 50.50 |
| 6" dia. | " | 0.00 | 59.00 | 59.00 |
| 8" dia. | " | 0.00 | 88.00 | 88.00 |
| 12" thick | | | | |
| 3" dia. | EA. | 0.00 | 59.00 | 59.00 |
| 4" dia. | " | 0.00 | 70.50 | 70.50 |
| 6" dia. | " | 0.00 | 88.00 | 88.00 |
| 8" dia. | " | 0.00 | 120.00 | 120.00 |

# 02 SITEWORK

## HAZARDOUS WASTE

| HAZARDOUS WASTE | UNIT | MAT. | INST. | TOTAL |
|---|---|---|---|---|
| **02080.10 ASBESTOS REMOVAL** | | | | |
| Enclosure using wood studs & poly, install & remove | S.F. | 0.90 | 0.69 | 1.59 |
| Trailer (change room) | DAY | | | 83.86 |
| Disposal suits (4 suits per man day) | " | | | 34.70 |
| Type C respirator mask, includes hose & filters, per man | " | | | 17.35 |
| Respirator mask & filter, light contamination | " | | | 6.79 |
| Air monitoring test, 12 tests per day | | | | |
| Off job testing | DAY | | | 896.40 |
| On the job testing | " | | | 1,186 |
| Asbestos vacuum with attachments | EA. | | | 520.56 |
| Hydraspray piston pump | " | | | 665.14 |
| Negative air pressure system | " | | | 607.23 |
| Grade D breathing air equipment | " | | | 1,519 |
| Glove bag, 44" x 60" x 6 mil plastic | " | | | 4.62 |
| 40 CY asbestos dumpster | | | | |
| Weekly rental | EA. | | | 577.80 |
| Pick up/delivery | " | | | 274.73 |
| Asbestos dump fee | " | | | 144.45 |
| **02080.12 DUCT INSULATION REMOVAL** | | | | |
| Remove duct insulation, duct size | | | | |
| 6" x 12" | L.F. | 0.00 | 1.55 | 1.55 |
| x 18" | " | 0.00 | 2.15 | 2.15 |
| x 24" | " | 0.00 | 3.10 | 3.10 |
| 8" x 12" | " | 0.00 | 2.30 | 2.30 |
| x 18" | " | 0.00 | 2.50 | 2.50 |
| x 24" | " | 0.00 | 3.45 | 3.45 |
| 12" x 12" | " | 0.00 | 2.30 | 2.30 |
| x 18" | " | 0.00 | 3.10 | 3.10 |
| x 24" | " | 0.00 | 3.95 | 3.95 |
| **02080.15 PIPE INSULATION REMOVAL** | | | | |
| Removal, asbestos insulation | | | | |
| 2" thick, pipe | | | | |
| 1" to 3" dia. | L.F. | 0.00 | 2.30 | 2.30 |
| 4" to 6" dia. | " | 0.00 | 2.65 | 2.65 |
| 3" thick | | | | |
| 7" to 8" dia. | L.F. | 0.00 | 2.80 | 2.80 |
| 9" to 10" dia. | " | 0.00 | 2.90 | 2.90 |
| 11" to 12" dia. | " | 0.00 | 3.10 | 3.10 |
| 13" to 14" dia. | " | 0.00 | 3.25 | 3.25 |
| 15" to 18" dia. | " | 0.00 | 3.45 | 3.45 |

# 02 SITEWORK

## SITE DEMOLITION

| SITE DEMOLITION | UNIT | MAT. | INST. | TOTAL |
|---|---|---|---|---|
| **02105.10 CATCH BASINS/MANHOLES** | | | | |
| Abandon catch basin or manhole (fill with sand) | EA. | 0.00 | 290.00 | 290.00 |
| Remove and reset frame and cover | " | 0.00 | 140.00 | 140.00 |
| Remove catch basin, to 10' deep | | | | |
| Masonry | EA. | 0.00 | 360.00 | 360.00 |
| Concrete | " | 0.00 | 480.00 | 480.00 |
| **02105.20 FENCES** | | | | |
| Remove fencing | | | | |
| Chain link, 8' high | | | | |
| For disposal | L.F. | 0.00 | 1.40 | 1.40 |
| For reuse | " | 0.00 | 3.45 | 3.45 |
| Wood | | | | |
| 4' high | S.F. | 0.00 | 0.93 | 0.93 |
| 6' high | " | 0.00 | 1.10 | 1.10 |
| 8' high | " | 0.00 | 1.40 | 1.40 |
| Masonry | | | | |
| 8" thick | | | | |
| 4' high | S.F. | 0.00 | 2.80 | 2.80 |
| 6' high | " | 0.00 | 3.45 | 3.45 |
| 8' high | " | 0.00 | 3.95 | 3.95 |
| 12" thick | | | | |
| 4' high | S.F. | 0.00 | 4.65 | 4.65 |
| 6' high | " | 0.00 | 5.55 | 5.55 |
| 8' high | " | 0.00 | 6.95 | 6.95 |
| 12' high | " | 0.00 | 9.25 | 9.25 |
| **02105.42 DRAINAGE PIPING** | | | | |
| Remove drainage pipe, not including excavation | | | | |
| 12" dia. | L.F. | 0.00 | 6.00 | 6.00 |
| 18" dia. | " | 0.00 | 7.60 | 7.60 |
| 24" dia. | " | 0.00 | 9.60 | 9.60 |
| 36" dia. | " | 0.00 | 12.00 | 12.00 |
| **02105.43 GAS PIPING** | | | | |
| Remove welded steel pipe, not including excavation | | | | |
| 4" dia. | L.F. | 0.00 | 9.00 | 9.00 |
| 5" dia. | " | 0.00 | 14.40 | 14.40 |
| 6" dia. | " | 0.00 | 18.00 | 18.00 |
| 8" dia. | " | 0.00 | 28.80 | 28.80 |
| 10" dia. | " | 0.00 | 36.00 | 36.00 |
| **02105.45 SANITARY PIPING** | | | | |
| Remove sewer pipe, not including excavation | | | | |
| 4" dia. | L.F. | 0.00 | 5.75 | 5.75 |
| 6" dia. | " | 0.00 | 6.55 | 6.55 |
| 8" dia. | " | 0.00 | 7.20 | 7.20 |
| 10" dia. | " | 0.00 | 7.60 | 7.60 |
| 12" dia. | " | 0.00 | 8.00 | 8.00 |
| 15" dia. | " | 0.00 | 8.50 | 8.50 |

# 02 SITEWORK

## SITE DEMOLITION

| | UNIT | MAT. | INST. | TOTAL |
|---|---|---|---|---|
| **02105.45 SANITARY PIPING** | | | | |
| 18" dia. | L.F. | 0.00 | 9.60 | 9.60 |
| 24" dia. | " | 0.00 | 12.00 | 12.00 |
| 30" dia. | " | 0.00 | 14.40 | 14.40 |
| 36" dia. | " | 0.00 | 18.00 | 18.00 |
| **02105.48 WATER PIPING** | | | | |
| Remove water pipe, not including excavation | | | | |
| 4" dia. | L.F. | 0.00 | 6.55 | 6.55 |
| 6" dia. | " | 0.00 | 6.85 | 6.85 |
| 8" dia. | " | 0.00 | 7.60 | 7.60 |
| 10" dia. | " | 0.00 | 8.00 | 8.00 |
| 12" dia. | " | 0.00 | 8.50 | 8.50 |
| 14" dia. | " | 0.00 | 9.00 | 9.00 |
| 16" dia. | " | 0.00 | 9.60 | 9.60 |
| 18" dia. | " | 0.00 | 10.30 | 10.30 |
| 20" dia. | " | 0.00 | 11.10 | 11.10 |
| Remove valves | | | | |
| 6" | EA. | 0.00 | 72.00 | 72.00 |
| 10" | " | 0.00 | 80.00 | 80.00 |
| 14" | " | 0.00 | 90.00 | 90.00 |
| 18" | " | 0.00 | 120.00 | 120.00 |
| **02105.60 UNDERGROUND TANKS** | | | | |
| Remove underground storage tank, and backfill | | | | |
| 50 to 250 gals | EA. | 0.00 | 480.00 | 480.00 |
| 600 gals | " | 0.00 | 480.00 | 480.00 |
| 1000 gals | " | 0.00 | 720.00 | 720.00 |
| 4000 gals | " | 0.00 | 1,150 | 1,150 |
| 5000 gals | " | 0.00 | 1,150 | 1,150 |
| 10,000 gals | " | 0.00 | 1,920 | 1,920 |
| 12,000 gals | " | 0.00 | 2,400 | 2,400 |
| 15,000 gals | " | 0.00 | 2,880 | 2,880 |
| 20,000 gals | " | 0.00 | 3,600 | 3,600 |
| **02105.66 SEPTIC TANKS** | | | | |
| Remove septic tank | | | | |
| 1000 gals | EA. | 0.00 | 120.00 | 120.00 |
| 2000 gals | " | 0.00 | 140.00 | 140.00 |
| 5000 gals | " | 0.00 | 180.00 | 180.00 |
| 15,000 gals | " | 0.00 | 1,440 | 1,440 |
| 25,000 gals | " | 0.00 | 1,920 | 1,920 |
| 40,000 gals | " | 0.00 | 2,880 | 2,880 |
| **02105.80 WALLS, EXTERIOR** | | | | |
| Concrete wall | | | | |
| Light reinforcing | | | | |
| 6" thick | S.F. | 0.00 | 7.20 | 7.20 |
| 8" thick | " | 0.00 | 7.60 | 7.60 |
| 10" thick | " | 0.00 | 8.00 | 8.00 |

# 02 SITEWORK

## SITE DEMOLITION

| SITE DEMOLITION | UNIT | MAT. | INST. | TOTAL |
|---|---|---|---|---|
| **02105.80 WALLS, EXTERIOR** | | | | |
| 12" thick | S.F. | 0.00 | 9.00 | 9.00 |
| Medium reinforcing | | | | |
| 6" thick | S.F. | 0.00 | 7.60 | 7.60 |
| 8" thick | " | 0.00 | 8.00 | 8.00 |
| 10" thick | " | 0.00 | 9.00 | 9.00 |
| 12" thick | " | 0.00 | 10.30 | 10.30 |
| Heavy reinforcing | | | | |
| 6" thick | S.F. | 0.00 | 8.50 | 8.50 |
| 8" thick | " | 0.00 | 9.00 | 9.00 |
| 10" thick | " | 0.00 | 10.30 | 10.30 |
| 12" thick | " | 0.00 | 12.00 | 12.00 |
| Masonry | | | | |
| No reinforcing | | | | |
| 8" thick | S.F. | 0.00 | 3.20 | 3.20 |
| 12" thick | " | 0.00 | 3.60 | 3.60 |
| 16" thick | " | 0.00 | 4.10 | 4.10 |
| Horizontal reinforcing | | | | |
| 8" thick | S.F. | 0.00 | 3.60 | 3.60 |
| 12" thick | " | 0.00 | 3.90 | 3.90 |
| 16" thick | " | 0.00 | 4.65 | 4.65 |
| Vertical reinforcing | | | | |
| 8" thick | S.F. | 0.00 | 4.65 | 4.65 |
| 12" thick | " | 0.00 | 5.35 | 5.35 |
| 16" thick | " | 0.00 | 6.55 | 6.55 |
| Remove concrete headwall | | | | |
| 15" pipe | EA. | 0.00 | 100.00 | 100.00 |
| 24" pipe | " | 0.00 | 130.00 | 130.00 |
| **02162.10 TRENCH SHEETING** | | | | |
| Closed timber, including pull and salvage, excavation | | | | |
| 8' deep | S.F. | 2.65 | 4.80 | 7.45 |
| 20' deep | " | 2.65 | 7.40 | 10.05 |

## EARTHWORK

| EARTHWORK | UNIT | MAT. | INST. | TOTAL |
|---|---|---|---|---|
| **02210.10 HAULING MATERIAL** | | | | |
| Haul material by 10 cy dump truck, round trip distance | | | | |
| 1 mile | C.Y. | 0.00 | 2.95 | 2.95 |
| 2 mile | " | 0.00 | 3.55 | 3.55 |
| 5 mile | " | 0.00 | 4.85 | 4.85 |
| 10 mile | " | 0.00 | 5.35 | 5.35 |
| 20 mile | " | 0.00 | 5.95 | 5.95 |

# 02 SITEWORK

| EARTHWORK | UNIT | MAT. | INST. | TOTAL |
|---|---|---|---|---|
| **02210.10 HAULING MATERIAL** | | | | |
| 30 mile | C.Y. | 0.00 | 7.10 | 7.10 |
| Site grading, cut & fill, sandy clay, 200' haul, 75 hp dozer | " | 0.00 | 2.15 | 2.15 |
| Spread topsoil by equipment on site | " | 0.00 | 2.35 | 2.35 |
| Site grading (cut and fill to 6") less than 1 acre | | | | |
| 75 hp dozer | C.Y. | 0.00 | 3.55 | 3.55 |
| 1.5 cy backhoe/loader | " | 0.00 | 5.35 | 5.35 |
| **02210.30 BULK EXCAVATION** | | | | |
| Hydraulic excavator | | | | |
| 1 cy capacity | | | | |
| Light material | C.Y. | 0.00 | 2.40 | 2.40 |
| Medium material | " | 0.00 | 2.90 | 2.90 |
| Wet material | " | 0.00 | 3.60 | 3.60 |
| Blasted rock | " | 0.00 | 4.10 | 4.10 |
| Wheel mounted front-end loader | | | | |
| 7/8 cy capacity | | | | |
| Light material | C.Y. | 0.00 | 1.85 | 1.85 |
| Medium material | " | 0.00 | 2.15 | 2.15 |
| Wet material | " | 0.00 | 2.50 | 2.50 |
| Blasted rock | " | 0.00 | 3.00 | 3.00 |
| Track mounted front-end loader | | | | |
| 1-1/2 cy capacity | | | | |
| Light material | C.Y. | 0.00 | 1.25 | 1.25 |
| Medium material | " | 0.00 | 1.35 | 1.35 |
| Wet material | " | 0.00 | 1.50 | 1.50 |
| Blasted rock | " | 0.00 | 1.65 | 1.65 |
| **02220.40 BUILDING EXCAVATION** | | | | |
| Structural excavation, unclassified earth | | | | |
| 3/8 cy backhoe | C.Y. | 0.00 | 9.95 | 9.95 |
| 3/4 cy backhoe | " | 0.00 | 7.50 | 7.50 |
| 1 cy backhoe | " | 0.00 | 6.25 | 6.25 |
| Foundation backfill and compaction by machine | " | 0.00 | 14.95 | 14.95 |
| **02220.50 UTILITY EXCAVATION** | | | | |
| Trencher, sandy clay, 8" wide trench | | | | |
| 18" deep | L.F. | 0.00 | 1.20 | 1.20 |
| 24" deep | " | 0.00 | 1.35 | 1.35 |
| 36" deep | " | 0.00 | 1.50 | 1.50 |
| Trench backfill, 95% compaction | | | | |
| Tamp by hand | C.Y. | 0.00 | 17.35 | 17.35 |
| Vibratory compaction | " | 0.00 | 13.90 | 13.90 |
| Trench backfilling, with borrow sand, place & compact | " | 10.10 | 13.90 | 24.00 |
| **02220.60 TRENCHING** | | | | |
| Trenching and continuous footing excavation | | | | |
| By gradall | | | | |
| 1 cy capacity | | | | |
| Light soil | C.Y. | 0.00 | 2.15 | 2.15 |
| Medium soil | " | 0.00 | 2.30 | 2.30 |

# 02 SITEWORK

## EARTHWORK

### 02220.60 TRENCHING

| | UNIT | MAT. | INST. | TOTAL |
|---|---|---|---|---|
| Heavy/wet soil | C.Y. | 0.00 | 2.50 | 2.50 |
| Loose rock | " | 0.00 | 2.70 | 2.70 |
| Blasted rock | " | 0.00 | 2.90 | 2.90 |
| By hydraulic excavator | | | | |
| 1/2 cy capacity | | | | |
| Light soil | C.Y. | 0.00 | 2.50 | 2.50 |
| Medium soil | " | 0.00 | 2.70 | 2.70 |
| Heavy/wet soil | " | 0.00 | 3.00 | 3.00 |
| Loose rock | " | 0.00 | 3.30 | 3.30 |
| Blasted rock | " | 0.00 | 3.75 | 3.75 |
| Hand excavation | | | | |
| Bulk, wheeled 100' | | | | |
| Normal soil | C.Y. | 0.00 | 30.80 | 30.80 |
| Sand or gravel | " | 0.00 | 27.80 | 27.80 |
| Medium clay | " | 0.00 | 39.70 | 39.70 |
| Heavy clay | " | 0.00 | 55.50 | 55.50 |
| Loose rock | " | 0.00 | 69.50 | 69.50 |
| Trenches, up to 2' deep | | | | |
| Normal soil | C.Y. | 0.00 | 34.70 | 34.70 |
| Sand or gravel | " | 0.00 | 30.80 | 30.80 |
| Medium clay | " | 0.00 | 46.30 | 46.30 |
| Heavy clay | " | 0.00 | 69.50 | 69.50 |
| Loose rock | " | 0.00 | 92.50 | 92.50 |
| Trenches, to 6' deep | | | | |
| Normal soil | C.Y. | 0.00 | 39.70 | 39.70 |
| Sand or gravel | " | 0.00 | 34.70 | 34.70 |
| Medium clay | " | 0.00 | 55.50 | 55.50 |
| Heavy clay | " | 0.00 | 92.50 | 92.50 |
| Loose rock | " | 0.00 | 140.00 | 140.00 |
| Backfill trenches | | | | |
| With compaction | | | | |
| By hand | C.Y. | 0.00 | 23.15 | 23.15 |
| By 60 hp tracked dozer | " | 0.00 | 1.35 | 1.35 |
| By 200 hp tracked dozer | " | 0.00 | 0.83 | 0.83 |
| By small front-end loader | " | 0.00 | 1.50 | 1.50 |
| Backfill trenches, sand bedding, no compaction | | | | |
| By hand | C.Y. | 10.10 | 23.15 | 33.25 |
| By small front-end loader | " | 10.10 | 2.15 | 12.25 |

### 02220.90 HAND EXCAVATION

| | UNIT | MAT. | INST. | TOTAL |
|---|---|---|---|---|
| Excavation | | | | |
| To 2' deep | | | | |
| Normal soil | C.Y. | 0.00 | 30.80 | 30.80 |
| Sand and gravel | " | 0.00 | 27.80 | 27.80 |
| Medium clay | " | 0.00 | 34.70 | 34.70 |
| Heavy clay | " | 0.00 | 39.70 | 39.70 |
| Loose rock | " | 0.00 | 46.30 | 46.30 |
| To 6' deep | | | | |
| Normal soil | C.Y. | 0.00 | 39.70 | 39.70 |
| Sand and gravel | " | 0.00 | 34.70 | 34.70 |
| Medium clay | " | 0.00 | 46.30 | 46.30 |
| Heavy clay | " | 0.00 | 55.50 | 55.50 |

# 02 SITEWORK

| EARTHWORK | UNIT | MAT. | INST. | TOTAL |
|---|---|---|---|---|
| **02220.90 HAND EXCAVATION** | | | | |
| Loose rock | C.Y. | 0.00 | 69.50 | 69.50 |
| Backfilling foundation without compaction, 6" lifts | " | 0.00 | 17.35 | 17.35 |
| Compaction of backfill around structures or in trench | | | | |
| By hand with air tamper | C.Y. | 0.00 | 19.85 | 19.85 |
| By hand with vibrating plate tamper | " | 0.00 | 18.50 | 18.50 |
| 1 ton roller | " | 0.00 | 26.70 | 26.70 |
| Miscellaneous hand labor | | | | |
| Trim slopes, sides of excavation | S.F. | 0.00 | 0.05 | 0.05 |
| Trim bottom of excavation | " | 0.00 | 0.06 | 0.06 |
| Excavation around obstructions and services | C.Y. | 0.00 | 92.50 | 92.50 |
| **02240.05 SOIL STABILIZATION** | | | | |
| Straw bale secured with rebar | L.F. | 1.50 | 0.93 | 2.42 |
| Filter barrier, 18" high filter fabric | " | 1.50 | 2.80 | 4.30 |
| Sediment fence, 36" fabric with 6" mesh | " | 3.50 | 3.45 | 6.95 |
| **02280.20 SOIL TREATMENT** | | | | |
| Soil treatment, termite control pretreatment | | | | |
| Under slabs | S.F. | 0.09 | 0.15 | 0.24 |
| By walls | " | 0.09 | 0.19 | 0.28 |
| **02290.30 WEED CONTROL** | | | | |
| Weed control, bromicil, 15 lb./acre, wettable powder | ACRE | 220.00 | 140.00 | 360.00 |
| Vegetation control, by application of plant killer | S.Y. | 0.03 | 0.11 | 0.14 |
| Weed killer, lawns and fields | " | 0.16 | 0.06 | 0.22 |

| PILES AND CAISSONS | UNIT | MAT. | INST. | TOTAL |
|---|---|---|---|---|
| **02360.50 PRESTRESSED PILING** | | | | |
| Prestressed concrete piling, less than 60' long | | | | |
| 10" sq. | L.F. | 9.45 | 3.50 | 12.95 |
| 12" sq. | " | 13.25 | 3.70 | 16.95 |
| 14" sq. | " | 13.75 | 3.75 | 17.50 |
| 16" sq. | " | 16.95 | 3.85 | 20.80 |
| 18" sq. | " | 23.40 | 4.10 | 27.50 |
| 20" sq. | " | 31.70 | 4.25 | 35.95 |
| 24" sq. | " | 40.50 | 4.35 | 44.85 |
| **02360.60 STEEL PILES** | | | | |
| H-section piles | | | | |
| 8x8 | | | | |

# 02 SITEWORK

## PILES AND CAISSONS

| PILES AND CAISSONS | UNIT | MAT. | INST. | TOTAL |
|---|---|---|---|---|
| **02360.60 STEEL PILES** | | | | |
| 36 lb/ft | | | | |
| 30' long | L.F. | 13.75 | 7.05 | 20.80 |
| 40' long | " | 13.75 | 5.65 | 19.40 |
| 50' long | " | 13.75 | 4.70 | 18.45 |
| 10x10 | | | | |
| 42 lb/ft | | | | |
| 30' long | L.F. | 15.50 | 7.05 | 22.55 |
| 40' long | " | 15.50 | 5.65 | 21.15 |
| 50' long | " | 15.50 | 4.70 | 20.20 |
| 57 lb/ft | | | | |
| 30' long | L.F. | 19.95 | 7.05 | 27.00 |
| 40' long | " | 19.95 | 5.65 | 25.60 |
| 50' long | " | 19.95 | 4.70 | 24.65 |
| Splice | | | | |
| 8" | EA. | 71.50 | 46.30 | 117.80 |
| 10" | " | 79.00 | 55.50 | 134.50 |
| **02360.65 STEEL PIPE PILES** | | | | |
| Concrete filled, 3000# concrete, up to 40' | | | | |
| 8" dia. | L.F. | 14.85 | 6.05 | 20.90 |
| 10" dia. | " | 18.25 | 6.25 | 24.50 |
| 12" dia. | " | 22.30 | 6.50 | 28.80 |
| Pipe piles, non-filled | | | | |
| 8" dia. | L.F. | 12.30 | 4.70 | 17.00 |
| 10" dia. | " | 14.50 | 4.85 | 19.35 |
| 12" dia. | " | 17.85 | 4.95 | 22.80 |
| Splice | | | | |
| 8" dia. | EA. | 42.10 | 55.50 | 97.60 |
| 10" dia. | " | 44.60 | 55.50 | 100.10 |
| 12" dia. | " | 50.00 | 69.50 | 119.50 |
| Standard point | | | | |
| 8" dia. | EA. | 54.00 | 55.50 | 109.50 |
| 10" dia. | " | 61.00 | 55.50 | 116.50 |
| 12" dia. | " | 96.00 | 69.50 | 165.50 |
| Heavy duty point | | | | |
| 8" dia. | EA. | 67.50 | 69.50 | 137.00 |
| 10" dia. | " | 81.00 | 69.50 | 150.50 |
| 12" dia. | " | 110.00 | 92.50 | 202.50 |
| **02360.70 STEEL SHEET PILING** | | | | |
| Steel sheet piling, 12" wide | | | | |
| 20' long | S.F. | 11.25 | 8.45 | 19.70 |
| 35' long | " | 11.25 | 6.05 | 17.30 |
| **02360.80 WOOD AND TIMBER PILES** | | | | |
| Treated wood piles, 12" butt, 8" tip | | | | |
| 25' long | L.F. | 7.85 | 8.45 | 16.30 |
| 30' long | " | 8.35 | 7.05 | 15.40 |
| 35' long | " | 8.35 | 6.05 | 14.40 |
| 40' long | " | 8.35 | 5.30 | 13.65 |

# 02 SITEWORK

## PILES AND CAISSONS

| 02380.10 CAISSONS | UNIT | MAT. | INST. | TOTAL |
|---|---|---|---|---|
| Caisson, including 3000# concrete, in stable ground | | | | |
| 18" dia. | L.F. | 12.85 | 16.90 | 29.75 |
| 24" dia. | " | 21.45 | 17.60 | 39.05 |
| Wet ground, casing required but pulled | | | | |
| 18" dia. | L.F. | 12.85 | 21.15 | 34.00 |
| 24" dia. | " | 21.45 | 23.50 | 44.95 |
| Soft rock | | | | |
| 18" dia. | L.F. | 12.85 | 60.50 | 73.35 |
| 24" dia. | " | 21.45 | 110.00 | 131.45 |

## PAVING AND SURFACING

| 02510.20 ASPHALT SURFACES | UNIT | MAT. | INST. | TOTAL |
|---|---|---|---|---|
| Asphalt wearing surface, for flexible pavement | | | | |
| 1" thick | S.Y. | 2.15 | 1.40 | 3.55 |
| 1-1/2" thick | " | 3.30 | 1.70 | 5.00 |
| 2" thick | " | 4.40 | 2.10 | 6.50 |
| 3" thick | " | 6.50 | 2.80 | 9.30 |
| Binder course | | | | |
| 1-1/2" thick | S.Y. | 2.85 | 1.55 | 4.40 |
| 2" thick | " | 3.80 | 1.90 | 5.70 |
| 3" thick | " | 5.65 | 2.55 | 8.20 |
| 4" thick | " | 7.55 | 2.80 | 10.35 |
| 5" thick | " | 9.70 | 3.15 | 12.85 |
| 6" thick | " | 11.60 | 3.50 | 15.10 |
| Bituminous sidewalk, no base | | | | |
| 2" thick | S.Y. | 4.75 | 1.70 | 6.45 |
| 3" thick | " | 7.00 | 1.80 | 8.80 |
| **02520.10 CONCRETE PAVING** | | | | |
| Concrete paving, reinforced, 5000 psi concrete | | | | |
| 6" thick | S.Y. | 20.30 | 13.20 | 33.50 |
| 7" thick | " | 22.80 | 14.10 | 36.90 |
| 8" thick | " | 25.70 | 15.10 | 40.80 |
| Concrete paving, for pipe trench, reinforced | | | | |
| 7" thick | S.Y. | 42.50 | 14.40 | 56.90 |
| 8" thick | " | 48.50 | 16.00 | 64.50 |
| **02545.10 ASPHALT REPAIR** | | | | |
| Coal tar emulsion seal coat, rubber additive, fuel resistant | S.Y. | 1.05 | 0.40 | 1.45 |
| Bituminous surface treatment, single | " | 0.95 | 0.28 | 1.23 |

## PAVING AND SURFACING

| 02545.10 ASPHALT REPAIR | UNIT | MAT. | INST. | TOTAL |
|---|---|---|---|---|
| Double | S.Y. | 1.25 | 0.03 | 1.28 |
| Bituminous prime coat | " | 0.63 | 0.04 | 0.67 |
| Tack coat | " | 0.31 | 0.03 | 0.34 |
| Crack sealing, concrete paving | L.F. | 0.53 | 0.19 | 0.71 |
| Bituminous paving for pipe trench, 4" thick | S.Y. | 6.05 | 9.60 | 15.65 |
| Polypropylene, nonwoven paving fabric | " | 0.80 | 0.14 | 0.94 |
| Rubberized asphalt | " | 1.25 | 2.50 | 3.75 |
| Asphalt slurry seal | " | 3.10 | 1.65 | 4.75 |

| 02580.10 PAVEMENT MARKINGS | UNIT | MAT. | INST. | TOTAL |
|---|---|---|---|---|
| Pavement line marking, paint | | | | |
| 4" wide | L.F. | 0.07 | 0.07 | 0.14 |
| 6" wide | " | 0.09 | 0.15 | 0.24 |
| 8" wide | " | 0.11 | 0.23 | 0.34 |
| Reflective paint, 4" wide | " | 0.27 | 0.23 | 0.50 |
| Preformed tape, 4" wide | | | | |
| Inlaid reflective | L.F. | 1.25 | 0.04 | 1.29 |
| Reflective paint | " | 0.85 | 0.07 | 0.92 |
| Thermoplastic | | | | |
| White | L.F. | 0.47 | 0.14 | 0.61 |
| Yellow | " | 0.47 | 0.14 | 0.61 |
| 12" wide, thermoplastic, white | " | 1.00 | 0.40 | 1.40 |
| Directional arrows, reflective preformed tape | EA. | 89.00 | 27.80 | 116.80 |
| Messages, reflective preformed tape (per letter) | " | 44.60 | 13.90 | 58.50 |
| Handicap symbol, preformed tape | " | 74.50 | 27.80 | 102.30 |
| Parking stall painting | " | 1.15 | 5.55 | 6.70 |

## SITE IMPROVEMENTS

| 02810.40 LAWN IRRIGATION | UNIT | MAT. | INST. | TOTAL |
|---|---|---|---|---|
| Residential system, complete | | | | |
| Minimum | ACRE | | | 13,500 |
| Maximum | " | | | 27,000 |
| Commercial system, complete | | | | |
| Minimum | ACRE | | | 20,250 |
| Maximum | " | | | 33,750 |
| Components | | | | |
| Pop-up head | EA. | 11.35 | 17.35 | 28.70 |
| Impact head | " | 11.35 | 17.35 | 28.70 |
| Rotary impact head | " | 31.20 | 34.70 | 65.90 |
| Shrub head | " | 2.95 | 17.35 | 20.30 |
| Hose bibb | " | 5.55 | 34.70 | 40.25 |

# 02 SITEWORK

## SITE IMPROVEMENTS

### 02830.10 CHAIN LINK FENCE

| | UNIT | MAT. | INST. | TOTAL |
|---|---|---|---|---|
| Chain link fence, 9 ga., galvanized, with posts 10' o.c. | | | | |
| 4' high | L.F. | 5.20 | 2.00 | 7.20 |
| 5' high | " | 7.00 | 2.50 | 9.50 |
| 6' high | " | 7.90 | 3.45 | 11.35 |
| 7' high | " | 8.95 | 4.25 | 13.20 |
| 8' high | " | 10.35 | 5.55 | 15.90 |
| For barbed wire with hangers, add | | | | |
| 3 strand | L.F. | 1.90 | 1.40 | 3.30 |
| 6 strand | " | 3.20 | 2.30 | 5.50 |
| Corner or gate post, 3" post | | | | |
| 4' high | EA. | 61.00 | 9.25 | 70.25 |
| 5' high | " | 67.00 | 10.30 | 77.30 |
| 6' high | " | 74.50 | 12.05 | 86.55 |
| 7' high | " | 89.50 | 13.90 | 103.40 |
| 8' high | " | 94.00 | 15.40 | 109.40 |
| 4" post | | | | |
| 4' high | EA. | 100.00 | 10.30 | 110.30 |
| 5' high | " | 120.00 | 12.05 | 132.05 |
| 6' high | " | 140.00 | 13.90 | 153.90 |
| 7' high | " | 150.00 | 15.40 | 165.40 |
| 8' high | " | 160.00 | 17.35 | 177.35 |
| Gate with gate posts, galvanized, 3' wide | | | | |
| 4' high | EA. | 67.50 | 69.50 | 137.00 |
| 5' high | " | 86.50 | 92.50 | 179.00 |
| 6' high | " | 100.00 | 92.50 | 192.50 |
| 7' high | " | 120.00 | 140.00 | 260.00 |
| 8' high | " | 130.00 | 140.00 | 270.00 |
| Fabric, galvanized chain link, 2" mesh, 9 ga. | | | | |
| 4' high | L.F. | 2.40 | 0.93 | 3.33 |
| 5' high | " | 2.90 | 1.10 | 4.00 |
| 6' high | " | 3.05 | 1.40 | 4.45 |
| 8' high | " | 4.65 | 1.85 | 6.50 |
| Line post, no rail fitting, galvanized, 2-1/2" dia. | | | | |
| 4' high | EA. | 17.95 | 7.95 | 25.90 |
| 5' high | " | 19.55 | 8.70 | 28.25 |
| 6' high | " | 21.40 | 9.25 | 30.65 |
| 7' high | " | 24.30 | 11.10 | 35.40 |
| 8' high | " | 27.10 | 13.90 | 41.00 |
| 1-7/8" H beam | | | | |
| 4' high | EA. | 24.60 | 7.95 | 32.55 |
| 5' high | " | 27.50 | 8.70 | 36.20 |
| 6' high | " | 32.90 | 9.25 | 42.15 |
| 7' high | " | 37.30 | 11.10 | 48.40 |
| 8' high | " | 40.40 | 13.90 | 54.30 |
| 2-1/4" H beam | | | | |
| 4' high | EA. | 17.95 | 7.95 | 25.90 |
| 5' high | " | 22.35 | 8.70 | 31.05 |
| 6' high | " | 25.70 | 9.25 | 34.95 |
| 7' high | " | 29.80 | 11.10 | 40.90 |
| 8' high | " | 34.60 | 13.90 | 48.50 |
| Vinyl coated, 9 ga., with posts 10' o.c. | | | | |
| 4' high | L.F. | 5.65 | 2.00 | 7.65 |

## SITE IMPROVEMENTS

### 02830.10 CHAIN LINK FENCE

| SITE IMPROVEMENTS | UNIT | MAT. | INST. | TOTAL |
|---|---|---|---|---|
| 5' high | L.F. | 6.70 | 2.50 | 9.20 |
| 6' high | " | 8.00 | 3.45 | 11.45 |
| 7' high | " | 8.75 | 4.25 | 13.00 |
| 8' high | " | 10.00 | 5.55 | 15.55 |
| For barbed wire w/hangers, add | | | | |
| 3 strand | L.F. | 2.05 | 1.40 | 3.45 |
| 6 Strand | " | 3.30 | 2.30 | 5.60 |
| Corner, or gate post, 4' high | | | | |
| 3" dia. | EA. | 74.50 | 9.25 | 83.75 |
| 4" dia. | " | 110.00 | 9.25 | 119.25 |
| 6" dia. | " | 130.00 | 11.10 | 141.10 |
| Gate, with posts, 3' wide | | | | |
| 4' high | EA. | 81.50 | 69.50 | 151.00 |
| 5' high | " | 96.50 | 92.50 | 189.00 |
| 6' high | " | 110.00 | 92.50 | 202.50 |
| 7' high | " | 130.00 | 140.00 | 270.00 |
| 8' high | " | 150.00 | 140.00 | 290.00 |
| Line post, no rail fitting, 2-1/2" dia. | | | | |
| 4' high | EA. | 31.40 | 7.95 | 39.35 |
| 5' high | " | 42.70 | 8.70 | 51.40 |
| 6' high | " | 51.50 | 9.25 | 60.75 |
| 7' high | " | 60.50 | 11.10 | 71.60 |
| 8' high | " | 66.00 | 13.90 | 79.90 |
| Corner post, no top rail fitting, 4" dia. | | | | |
| 4' high | EA. | 120.00 | 9.25 | 129.25 |
| 5' high | " | 140.00 | 10.30 | 150.30 |
| 6' high | " | 160.00 | 12.05 | 172.05 |
| 7' high | " | 180.00 | 13.90 | 193.90 |
| 8' high | " | 190.00 | 15.40 | 205.40 |
| Fabric, vinyl, chain link, 2" mesh, 9 ga. | | | | |
| 4' high | L.F. | 4.30 | 0.93 | 5.22 |
| 5' high | " | 5.20 | 1.10 | 6.30 |
| 6' high | " | 6.10 | 1.40 | 7.50 |
| 8' high | " | 8.20 | 1.85 | 10.05 |
| Swing gates, galvanized, 4' high | | | | |
| Single gate | | | | |
| 3' wide | EA. | 160.00 | 69.50 | 229.50 |
| 4' wide | " | 170.00 | 69.50 | 239.50 |
| Double gate | | | | |
| 10' wide | EA. | 400.00 | 110.00 | 510.00 |
| 12' wide | " | 430.00 | 110.00 | 540.00 |
| 14' wide | " | 450.00 | 110.00 | 560.00 |
| 16' wide | " | 500.00 | 110.00 | 610.00 |
| 18' wide | " | 540.00 | 160.00 | 700.00 |
| 20' wide | " | 560.00 | 160.00 | 720.00 |
| 22' wide | " | 630.00 | 160.00 | 790.00 |
| 24' wide | " | 640.00 | 190.00 | 830.00 |
| 26' wide | " | 670.00 | 190.00 | 860.00 |
| 28' wide | " | 720.00 | 220.00 | 940.00 |
| 30' wide | " | 760.00 | 220.00 | 980.00 |
| 5' high | | | | |
| Single gate | | | | |

| SITE IMPROVEMENTS | UNIT | MAT. | INST. | TOTAL |
|---|---|---|---|---|
| **02830.10 CHAIN LINK FENCE** | | | | |
| 3' wide | EA. | 160.00 | 92.50 | 252.50 |
| 4' wide | " | 190.00 | 92.50 | 282.50 |
| Double gate | | | | |
| 10' wide | EA. | 430.00 | 140.00 | 570.00 |
| 12' wide | " | 460.00 | 140.00 | 600.00 |
| 14' wide | " | 800.00 | 140.00 | 940.00 |
| 16' wide | " | 520.00 | 140.00 | 660.00 |
| 18' wide | " | 540.00 | 160.00 | 700.00 |
| 20' wide | " | 610.00 | 160.00 | 770.00 |
| 22' wide | " | 640.00 | 160.00 | 800.00 |
| 24' wide | " | 670.00 | 190.00 | 860.00 |
| 26' wide | " | 690.00 | 190.00 | 880.00 |
| 28' wide | " | 770.00 | 220.00 | 990.00 |
| 30' wide | " | 800.00 | 220.00 | 1,020 |
| 6' high | | | | |
| Single gate | | | | |
| 3' wide | EA. | 190.00 | 92.50 | 282.50 |
| 4' wide | " | 200.00 | 92.50 | 292.50 |
| Double gate | | | | |
| 10' wide | EA. | 450.00 | 140.00 | 590.00 |
| 12' wide | " | 510.00 | 140.00 | 650.00 |
| 14' wide | " | 540.00 | 140.00 | 680.00 |
| 16' wide | " | 580.00 | 140.00 | 720.00 |
| 18' wide | " | 620.00 | 160.00 | 780.00 |
| 20' wide | " | 640.00 | 160.00 | 800.00 |
| 22' wide | " | 690.00 | 160.00 | 850.00 |
| 24' wide | " | 740.00 | 190.00 | 930.00 |
| 26' wide | " | 760.00 | 190.00 | 950.00 |
| 28' wide | " | 820.00 | 220.00 | 1,040 |
| 30' wide | " | 870.00 | 220.00 | 1,090 |
| 7' high | | | | |
| Single gate | | | | |
| 3' wide | EA. | 210.00 | 140.00 | 350.00 |
| 4' wide | " | 230.00 | 140.00 | 370.00 |
| Double gate | | | | |
| 10' wide | EA. | 540.00 | 190.00 | 730.00 |
| 12' wide | " | 590.00 | 190.00 | 780.00 |
| 14' wide | " | 630.00 | 190.00 | 820.00 |
| 16' wide | " | 670.00 | 190.00 | 860.00 |
| 18' wide | " | 720.00 | 220.00 | 940.00 |
| 20' wide | " | 760.00 | 220.00 | 980.00 |
| 22' wide | " | 800.00 | 220.00 | 1,020 |
| 24' wide | " | 860.00 | 280.00 | 1,140 |
| 26' wide | " | 920.00 | 280.00 | 1,200 |
| 28' wide | " | 970.00 | 350.00 | 1,320 |
| 30' wide | " | 1,110 | 350.00 | 1,460 |
| 8' high | | | | |
| Single gate | | | | |
| 3' wide | EA. | 220.00 | 140.00 | 360.00 |
| 4' wide | " | 240.00 | 140.00 | 380.00 |
| Double gate | | | | |
| 10' wide | EA. | 590.00 | 190.00 | 780.00 |

| SITE IMPROVEMENTS | UNIT | MAT. | INST. | TOTAL |
|---|---|---|---|---|
| **02830.10 CHAIN LINK FENCE** | | | | |
| 12' wide | EA. | 620.00 | 190.00 | 810.00 |
| 14' wide | " | 670.00 | 190.00 | 860.00 |
| 16' wide | " | 730.00 | 190.00 | 920.00 |
| 18' wide | " | 760.00 | 220.00 | 980.00 |
| 20' wide | " | 790.00 | 220.00 | 1,010 |
| 22' wide | " | 830.00 | 220.00 | 1,050 |
| 24' wide | " | 920.00 | 280.00 | 1,200 |
| 26' wide | " | 950.00 | 280.00 | 1,230 |
| 28' wide | " | 1,020 | 350.00 | 1,370 |
| 30' wide | " | 1,140 | 350.00 | 1,490 |
| Vinyl coated swing gates, 4' high | | | | |
| Single gate | | | | |
| 3' wide | EA. | 240.00 | 69.50 | 309.50 |
| 4' wide | " | 260.00 | 69.50 | 329.50 |
| Double gate | | | | |
| 10' wide | EA. | 560.00 | 110.00 | 670.00 |
| 12' wide | " | 720.00 | 110.00 | 830.00 |
| 14' wide | " | 760.00 | 110.00 | 870.00 |
| 16' wide | " | 850.00 | 110.00 | 960.00 |
| 18' wide | " | 950.00 | 160.00 | 1,110 |
| 20' wide | " | 1,130 | 160.00 | 1,290 |
| 22' wide | " | 1,250 | 160.00 | 1,410 |
| 24' wide | " | 1,350 | 190.00 | 1,540 |
| 26' wide | " | 1,410 | 190.00 | 1,600 |
| 28' wide | " | 1,640 | 220.00 | 1,860 |
| 30' wide | " | 1,720 | 220.00 | 1,940 |
| 5' high | | | | |
| Single gate | | | | |
| 3' wide | EA. | 260.00 | 92.50 | 352.50 |
| 4' wide | " | 290.00 | 92.50 | 382.50 |
| Double gate | | | | |
| 10' wide | EA. | 770.00 | 140.00 | 910.00 |
| 12' wide | " | 820.00 | 140.00 | 960.00 |
| 14' wide | " | 920.00 | 140.00 | 1,060 |
| 16' wide | " | 950.00 | 140.00 | 1,090 |
| 18' wide | " | 1,150 | 160.00 | 1,310 |
| 20' wide | " | 1,250 | 160.00 | 1,410 |
| 22' wide | " | 1,370 | 160.00 | 1,530 |
| 24' wide | " | 1,560 | 190.00 | 1,750 |
| 26' wide | " | 1,630 | 190.00 | 1,820 |
| 28' wide | " | 1,820 | 220.00 | 2,040 |
| 30' wide | " | 1,960 | 220.00 | 2,180 |
| 6' high | | | | |
| Single gate | | | | |
| 3' wide | EA. | 280.00 | 92.50 | 372.50 |
| 4' wide | " | 290.00 | 92.50 | 382.50 |
| Double gate | | | | |
| 10' wide | EA. | 680.00 | 140.00 | 820.00 |
| 12' wide | " | 780.00 | 140.00 | 920.00 |
| 14' wide | " | 900.00 | 140.00 | 1,040 |
| 16' wide | " | 1,040 | 140.00 | 1,180 |
| 18' wide | " | 1,160 | 160.00 | 1,320 |

# 02 SITEWORK

| SITE IMPROVEMENTS | UNIT | MAT. | INST. | TOTAL |
|---|---|---|---|---|
| **02830.10 CHAIN LINK FENCE** | | | | |
| 20' wide | EA. | 1,250 | 160.00 | 1,410 |
| 22' wide | " | 1,360 | 160.00 | 1,520 |
| 24' wide | " | 1,520 | 190.00 | 1,710 |
| 26' wide | " | 1,610 | 190.00 | 1,800 |
| 28' wide | " | 1,790 | 220.00 | 2,010 |
| 30' wide | " | 1,970 | 220.00 | 2,190 |
| 7' high | | | | |
| Single gate | | | | |
| 3' wide | EA. | 310.00 | 140.00 | 450.00 |
| 4' wide | " | 390.00 | 140.00 | 530.00 |
| Double gate | | | | |
| 10' wide | EA. | 770.00 | 190.00 | 960.00 |
| 12' wide | " | 910.00 | 190.00 | 1,100 |
| 14' wide | " | 1,030 | 190.00 | 1,220 |
| 16' wide | " | 1,150 | 190.00 | 1,340 |
| 18' wide | " | 1,300 | 220.00 | 1,520 |
| 20' wide | " | 1,490 | 220.00 | 1,710 |
| 22' wide | " | 1,650 | 220.00 | 1,870 |
| 24' wide | " | 1,810 | 280.00 | 2,090 |
| 26' wide | " | 1,960 | 280.00 | 2,240 |
| 28' wide | " | 2,110 | 350.00 | 2,460 |
| 30' wide | " | 2,270 | 350.00 | 2,620 |
| 8' high | | | | |
| Single gate | | | | |
| 3' wide | EA. | 320.00 | 140.00 | 460.00 |
| 4' wide | " | 390.00 | 140.00 | 530.00 |
| Double gate | | | | |
| 10' wide | EA. | 780.00 | 190.00 | 970.00 |
| 12' wide | " | 910.00 | 190.00 | 1,100 |
| 14' wide | " | 1,040 | 190.00 | 1,230 |
| 16' wide | " | 740.00 | 190.00 | 930.00 |
| 18' wide | " | 1,310 | 220.00 | 1,530 |
| 20' wide | " | 1,530 | 220.00 | 1,750 |
| 22' wide | " | 1,750 | 220.00 | 1,970 |
| 24' wide | " | 1,870 | 280.00 | 2,150 |
| 28' wide | " | 2,170 | 280.00 | 2,450 |
| 30' wide | " | 2,330 | 350.00 | 2,680 |
| Motor operator for gates, no wiring | " | | | 4,484 |
| Drilling fence post holes | | | | |
| In soil | | | | |
| By hand | EA. | 0.00 | 13.90 | 13.90 |
| By machine auger | " | 0.00 | 8.80 | 8.80 |
| In rock | | | | |
| By jackhammer | EA. | 0.00 | 120.00 | 120.00 |
| By rock drill | " | 0.00 | 35.30 | 35.30 |
| Aluminum privacy slats, installed vertically | S.F. | 0.81 | 0.69 | 1.50 |
| Post hole, dig by hand | EA. | 0.00 | 18.50 | 18.50 |
| Set fence post in concrete | " | 7.95 | 13.90 | 21.85 |
| **02840.30 GUARDRAILS** | | | | |
| Pipe bollard, steel pipe, concrete filled, painted | | | | |
| 6" dia. | EA. | 140.00 | 23.15 | 163.15 |

# 02 SITEWORK

| SITE IMPROVEMENTS | UNIT | MAT. | INST. | TOTAL |
|---|---|---|---|---|
| **02840.30 GUARDRAILS** | | | | |
| 8" dia. | EA. | 200.00 | 34.70 | 234.70 |
| 12" dia. | " | 320.00 | 92.50 | 412.50 |
| Corrugated steel, guardrail, galvanized | L.F. | 18.65 | 2.40 | 21.05 |
| End section, wrap around or flared | EA. | 53.00 | 27.80 | 80.80 |
| Timber guardrail, 4" x 8" | L.F. | 22.30 | 1.80 | 24.10 |
| Guard rail, 3 cables, 3/4" dia. | | | | |
| Steel posts | L.F. | 10.10 | 7.20 | 17.30 |
| Wood posts | " | 10.80 | 5.75 | 16.55 |
| Steel box beam | | | | |
| 6" x 6" | L.F. | 41.90 | 8.00 | 49.90 |
| 6" x 8" | " | | | |
| Concrete posts | EA. | 27.00 | 13.90 | 40.90 |
| Barrel type impact barrier | " | 350.00 | 27.80 | 377.80 |
| Light shield, 6' high | L.F. | 22.75 | 5.55 | 28.30 |
| **02840.40 PARKING BARRIERS** | | | | |
| Timber, treated, 4' long | | | | |
| 4" x 4" | EA. | 14.85 | 23.15 | 38.00 |
| 6" x 6" | " | 27.00 | 27.80 | 54.80 |
| Precast concrete, 6' long, with dowels | | | | |
| 12" x 6" | EA. | 28.40 | 13.90 | 42.30 |
| 12" x 8" | " | 31.10 | 15.40 | 46.50 |
| **02870.10 PREFABRICATED PLANTERS** | | | | |
| Concrete precast, circular | | | | |
| 24" dia., 18" high | EA. | 240.00 | 27.80 | 267.80 |
| 42" dia., 30" high | " | 320.00 | 34.70 | 354.70 |
| Fiberglass, circular | | | | |
| 36" dia., 27" high | EA. | 450.00 | 13.90 | 463.90 |
| 60" dia., 39" high | " | 1,030 | 15.40 | 1,045 |
| Tapered, circular | | | | |
| 24" dia., 36" high | EA. | 350.00 | 12.60 | 362.60 |
| 40" dia., 36" high | " | 580.00 | 13.90 | 593.90 |
| Square | | | | |
| 2' by 2', 17" high | EA. | 300.00 | 12.60 | 312.60 |
| 4' by 4', 39" high | " | 1,030 | 15.40 | 1,045 |
| Rectangular | | | | |
| 4' by 1', 18" high | EA. | 330.00 | 13.90 | 343.90 |

| FORMWORK | UNIT | MAT. | INST. | TOTAL |
|---|---|---|---|---|
| **03110.05 BEAM FORMWORK** | | | | |
| Beam forms, job built | | | | |
| Beam bottoms | | | | |
| 1 use | S.F. | 3.75 | 5.90 | 9.65 |
| 5 uses | " | 1.30 | 5.05 | 6.35 |
| Beam sides | | | | |
| 1 use | S.F. | 2.70 | 3.95 | 6.65 |
| 5 uses | " | 1.15 | 3.20 | 4.35 |
| **03110.15 COLUMN FORMWORK** | | | | |
| Column, square forms, job built | | | | |
| 8" x 8" columns | | | | |
| 1 use | S.F. | 2.95 | 7.10 | 10.05 |
| 5 uses | " | 1.05 | 6.10 | 7.15 |
| 12" x 12" columns | | | | |
| 1 use | S.F. | 2.70 | 6.45 | 9.15 |
| 5 uses | " | 0.88 | 5.65 | 6.53 |
| 16" x 16" columns | | | | |
| 1 use | S.F. | 2.55 | 5.90 | 8.45 |
| 5 uses | " | 0.81 | 5.20 | 6.01 |
| Round fiber forms, 1 use | | | | |
| 10" dia. | L.F. | 3.75 | 7.10 | 10.85 |
| 12" dia. | " | 4.60 | 7.25 | 11.85 |
| 18" dia. | " | 13.00 | 8.45 | 21.45 |
| 24" dia. | " | 15.90 | 9.10 | 25.00 |
| 36" dia. | " | 29.60 | 10.75 | 40.35 |
| **03110.18 CURB FORMWORK** | | | | |
| Curb forms | | | | |
| Straight, 6" high | | | | |
| 1 use | L.F. | 1.75 | 3.55 | 5.30 |
| 5 uses | " | 0.63 | 2.95 | 3.58 |
| Curved, 6" high | | | | |
| 1 use | L.F. | 1.90 | 4.45 | 6.35 |
| 5 uses | " | 0.78 | 3.60 | 4.38 |
| **03110.20 ELEVATED SLAB FORMWORK** | | | | |
| Elevated slab formwork | | | | |
| Slab, with drop panels | | | | |
| 1 use | S.F. | 2.95 | 2.85 | 5.80 |
| 5 uses | " | 1.05 | 2.45 | 3.50 |
| Floor slab, hung from steel beams | | | | |
| 1 use | S.F. | 2.40 | 2.75 | 5.15 |
| 5 uses | " | 0.88 | 2.35 | 3.23 |
| Floor slab, with pans or domes | | | | |
| 1 use | S.F. | 4.25 | 3.20 | 7.45 |
| 5 uses | " | 2.15 | 2.75 | 4.90 |
| Equipment curbs, 12" high | | | | |
| 1 use | L.F. | 2.25 | 3.55 | 5.80 |
| 5 uses | " | 0.95 | 2.95 | 3.90 |

# 03 CONCRETE

| FORMWORK | UNIT | MAT. | INST. | TOTAL |
|---|---|---|---|---|
| **03110.25 EQUIPMENT PAD FORMWORK** | | | | |
| Equipment pad, job built | | | | |
| 1 use | S.F. | 2.80 | 4.45 | 7.25 |
| 2 uses | " | 1.70 | 4.15 | 5.85 |
| 3 uses | " | 1.35 | 3.95 | 5.30 |
| 4 uses | " | 1.05 | 3.75 | 4.80 |
| 5 uses | " | 0.86 | 3.55 | 4.41 |
| **03110.35 FOOTING FORMWORK** | | | | |
| Wall footings, job built, continuous | | | | |
| 1 use | S.F. | 1.35 | 3.55 | 4.90 |
| 5 uses | " | 0.61 | 2.95 | 3.56 |
| Column footings, spread | | | | |
| 1 use | S.F. | 1.40 | 4.45 | 5.85 |
| 5 uses | " | 0.57 | 3.55 | 4.12 |
| **03110.50 GRADE BEAM FORMWORK** | | | | |
| Grade beams, job built | | | | |
| 1 use | S.F. | 2.20 | 3.55 | 5.75 |
| 5 uses | " | 0.68 | 2.95 | 3.63 |
| **03110.53 PILE CAP FORMWORK** | | | | |
| Pile cap forms, job built | | | | |
| Square | | | | |
| 1 use | S.F. | 2.40 | 4.45 | 6.85 |
| 5 uses | " | 0.81 | 3.55 | 4.36 |
| Triangular | | | | |
| 1 use | S.F. | 2.55 | 5.05 | 7.60 |
| 5 uses | " | 0.88 | 3.95 | 4.83 |
| **03110.55 SLAB/MAT FORMWORK** | | | | |
| Mat foundations, job built | | | | |
| 1 use | S.F. | 2.10 | 4.45 | 6.55 |
| 5 uses | " | 0.59 | 3.55 | 4.14 |
| Edge forms | | | | |
| 6" high | | | | |
| 1 use | L.F. | 2.10 | 3.20 | 5.30 |
| 5 uses | " | 0.59 | 2.75 | 3.34 |
| 12" high | | | | |
| 1 use | L.F. | 1.95 | 3.55 | 5.50 |
| 5 uses | " | 0.55 | 2.95 | 3.50 |
| Formwork for openings | | | | |
| 1 use | S.F. | 2.80 | 7.10 | 9.90 |
| 5 uses | " | 0.88 | 5.05 | 5.93 |
| **03110.60 STAIR FORMWORK** | | | | |
| Stairway forms, job built | | | | |
| 1 use | S.F. | 3.70 | 7.10 | 10.80 |
| 5 uses | " | 1.25 | 5.05 | 6.30 |

# 03 CONCRETE

## FORMWORK

| | UNIT | MAT. | INST. | TOTAL |
|---|---|---|---|---|
| **03110.60 STAIR FORMWORK** | | | | |
| Stairs, elevated | | | | |
| 1 use | S.F. | 4.45 | 7.10 | 11.55 |
| 5 uses | " | 1.50 | 4.45 | 5.95 |
| **03110.65 WALL FORMWORK** | | | | |
| Wall forms, exterior, job built | | | | |
| Up to 8' high wall | | | | |
| 1 use | S.F. | 2.45 | 3.55 | 6.00 |
| 5 uses | " | 0.89 | 2.95 | 3.84 |
| Over 8' high wall | | | | |
| 1 use | S.F. | 2.70 | 4.45 | 7.15 |
| 5 uses | " | 1.10 | 3.55 | 4.65 |
| Column pier and pilaster | | | | |
| 1 use | S.F. | 2.70 | 7.10 | 9.80 |
| 5 uses | " | 1.20 | 5.05 | 6.25 |
| Interior wall forms | | | | |
| Up to 8' high | | | | |
| 1 use | S.F. | 2.45 | 3.20 | 5.65 |
| 5 uses | " | 0.88 | 2.75 | 3.63 |
| Over 8' high | | | | |
| 1 use | S.F. | 2.70 | 3.95 | 6.65 |
| 5 uses | " | 1.10 | 3.20 | 4.30 |
| PVC form liner, per side, smooth finish | | | | |
| 1 use | S.F. | 4.75 | 2.95 | 7.70 |
| 5 uses | " | 1.35 | 2.35 | 3.70 |
| **03110.90 MISCELLANEOUS FORMWORK** | | | | |
| Keyway forms (5 uses) | | | | |
| 2 x 4 | L.F. | 0.15 | 1.75 | 1.90 |
| 2 x 6 | " | 0.23 | 1.95 | 2.18 |
| Bulkheads | | | | |
| Walls, with keyways | | | | |
| 2 piece | L.F. | 2.65 | 3.20 | 5.85 |
| 3 piece | " | 3.25 | 3.55 | 6.80 |
| Elevated slab, with keyway | | | | |
| 2 piece | L.F. | 2.95 | 2.95 | 5.90 |
| 3 piece | " | 4.30 | 3.20 | 7.50 |
| Ground slab, with keyway | | | | |
| 2 piece | L.F. | 3.25 | 2.55 | 5.80 |
| 3 piece | " | 4.30 | 2.75 | 7.05 |
| Chamfer strips | | | | |
| Wood | | | | |
| 1/2" wide | L.F. | 0.20 | 0.79 | 0.99 |
| 3/4" wide | " | 0.26 | 0.79 | 1.05 |
| 1" wide | " | 0.32 | 0.79 | 1.11 |
| PVC | | | | |
| 1/2" wide | L.F. | 0.69 | 0.79 | 1.48 |
| 3/4" wide | " | 0.76 | 0.79 | 1.55 |
| 1" wide | " | 0.97 | 0.79 | 1.76 |
| Radius | | | | |
| 1" | L.F. | 0.95 | 0.84 | 1.79 |

# 03 CONCRETE

## FORMWORK

| 03110.90 MISCELLANEOUS FORMWORK | UNIT | MAT. | INST. | TOTAL |
|---|---|---|---|---|
| 1-1/2" | L.F. | 1.65 | 0.84 | 2.49 |
| Reglets | | | | |
| Galvanized steel, 24 ga. | L.F. | 1.10 | 1.40 | 2.50 |
| Metal formwork | | | | |
| Straight edge forms | | | | |
| 4" high | L.F. | 0.15 | 2.20 | 2.35 |
| 6" high | " | 0.16 | 2.35 | 2.51 |
| 8" high | " | 0.19 | 2.55 | 2.74 |
| 12" high | " | 0.24 | 2.75 | 2.99 |
| 16" high | " | 0.30 | 2.95 | 3.25 |
| Curb form, S-shape | | | | |
| 12" x | | | | |
| 1'-6" | L.F. | 0.30 | 5.05 | 5.35 |
| 2' | " | 0.36 | 4.75 | 5.11 |
| 2'-6" | " | 0.41 | 4.45 | 4.86 |
| 3' | " | 0.46 | 3.95 | 4.41 |

## REINFORCEMENT

| 03210.05 BEAM REINFORCING | UNIT | MAT. | INST. | TOTAL |
|---|---|---|---|---|
| Beam-girders | | | | |
| #3 - #4 | TON | 890.00 | 990.00 | 1,880 |
| #5 - #6 | " | 780.00 | 790.00 | 1,570 |
| #7 - #8 | " | 740.00 | 660.00 | 1,400 |
| Galvanized | | | | |
| #3 - #4 | TON | 1,490 | 990.00 | 2,480 |
| #5 - #6 | " | 1,400 | 790.00 | 2,190 |
| #7 - #8 | " | 1,350 | 660.00 | 2,010 |
| Bond Beams | | | | |
| #3 - #4 | TON | 890.00 | 1,310 | 2,200 |
| #5 - #6 | " | 780.00 | 990.00 | 1,770 |
| #7 - #8 | " | 740.00 | 880.00 | 1,620 |
| Galvanized | | | | |
| #3 - #4 | TON | 1,490 | 1,310 | 2,800 |
| #5 - #6 | " | 1,400 | 990.00 | 2,390 |
| #7 - #8 | " | 1,350 | 880.00 | 2,230 |

| 03210.15 COLUMN REINFORCING | UNIT | MAT. | INST. | TOTAL |
|---|---|---|---|---|
| Columns | | | | |
| #3 - #4 | TON | 890.00 | 1,130 | 2,020 |
| #5 - #6 | " | 780.00 | 880.00 | 1,660 |
| #7 - #8 | " | 740.00 | 790.00 | 1,530 |
| Galvanized | | | | |
| #3 - #4 | TON | 1,490 | 1,130 | 2,620 |

# 03 CONCRETE

## REINFORCEMENT

| REINFORCEMENT | UNIT | MAT. | INST. | TOTAL |
|---|---|---|---|---|
| **03210.15 COLUMN REINFORCING** | | | | |
| #5 - #6 | TON | 1,400 | 880.00 | 2,280 |
| #7 - #8 | " | 1,350 | 790.00 | 2,140 |
| **03210.20 ELEVATED SLAB REINFORCING** | | | | |
| Elevated slab | | | | |
| #3 - #4 | TON | 890.00 | 490.00 | 1,380 |
| #5 - #6 | " | 780.00 | 440.00 | 1,220 |
| #7 - #8 | " | 740.00 | 390.00 | 1,130 |
| Galvanized | | | | |
| #3 - #4 | TON | 1,490 | 490.00 | 1,980 |
| #5 - #6 | " | 1,400 | 440.00 | 1,840 |
| **03210.25 EQUIP. PAD REINFORCING** | | | | |
| Equipment pad | | | | |
| #3 - #4 | TON | 890.00 | 790.00 | 1,680 |
| #5 - #6 | " | 780.00 | 720.00 | 1,500 |
| #7 - #8 | " | 740.00 | 660.00 | 1,400 |
| #9 - #10 | " | 740.00 | 610.00 | 1,350 |
| #11 - #12 | " | 740.00 | 560.00 | 1,300 |
| **03210.35 FOOTING REINFORCING** | | | | |
| Footings | | | | |
| Grade 50 | | | | |
| #3 - #4 | TON | 890.00 | 660.00 | 1,550 |
| #5 - #6 | " | 780.00 | 560.00 | 1,340 |
| #7 - #8 | " | 740.00 | 490.00 | 1,230 |
| #9 - #10 | " | 740.00 | 440.00 | 1,180 |
| Grade 60 | | | | |
| #3 - #4 | TON | 890.00 | 660.00 | 1,550 |
| #5 - #6 | " | 780.00 | 560.00 | 1,340 |
| #7 - #8 | " | 740.00 | 490.00 | 1,230 |
| Straight dowels, 24" long | | | | |
| 1" dia. (#8) | EA. | 2.75 | 3.95 | 6.70 |
| 3/4" dia. (#6) | " | 2.50 | 3.95 | 6.45 |
| 5/8" dia. (#5) | " | 2.15 | 3.30 | 5.45 |
| 1/2" dia. (#4) | " | 1.60 | 2.80 | 4.40 |
| **03210.45 FOUNDATION REINFORCING** | | | | |
| Foundations | | | | |
| #3 - #4 | TON | 890.00 | 660.00 | 1,550 |
| #5 - #6 | " | 780.00 | 560.00 | 1,340 |
| #7 - #8 | " | 740.00 | 490.00 | 1,230 |
| Galvanized | | | | |
| #3 - #4 | TON | 1,490 | 660.00 | 2,150 |
| #5 - #6 | " | 1,400 | 560.00 | 1,960 |
| #7 - #8 | " | 1,350 | 490.00 | 1,840 |

# 03 CONCRETE

| REINFORCEMENT | UNIT | MAT. | INST. | TOTAL |
|---|---|---|---|---|
| **03210.50 GRADE BEAM REINFORCING** | | | | |
| Grade beams | | | | |
| #3 - #4 | TON | 890.00 | 610.00 | 1,500 |
| #5 - #6 | " | 780.00 | 530.00 | 1,310 |
| #7 - #8 | " | 740.00 | 460.00 | 1,200 |
| Galvanized | | | | |
| #3 - #4 | TON | 1,490 | 610.00 | 2,100 |
| #5 - #6 | " | 1,400 | 530.00 | 1,930 |
| #7 - #8 | " | 1,350 | 460.00 | 1,810 |
| **03210.53 PILE CAP REINFORCING** | | | | |
| Pile caps | | | | |
| #3 - #4 | TON | 890.00 | 990.00 | 1,880 |
| #5 - #6 | " | 780.00 | 880.00 | 1,660 |
| #7 - #8 | " | 740.00 | 790.00 | 1,530 |
| Galvanized | | | | |
| #3 - #4 | TON | 1,490 | 990.00 | 2,480 |
| #5 - #6 | " | 1,400 | 880.00 | 2,280 |
| #7 - #8 | " | 1,350 | 790.00 | 2,140 |
| **03210.55 SLAB/MAT REINFORCING** | | | | |
| Bars, slabs | | | | |
| #3 - #4 | TON | 890.00 | 660.00 | 1,550 |
| #5 - #6 | " | 780.00 | 560.00 | 1,340 |
| #7 - #8 | " | 740.00 | 490.00 | 1,230 |
| Galvanized | | | | |
| #3 - #4 | TON | 1,490 | 660.00 | 2,150 |
| #5 - #6 | " | 1,400 | 560.00 | 1,960 |
| #7 - #8 | " | 1,350 | 490.00 | 1,840 |
| Wire mesh, slabs | | | | |
| Galvanized | | | | |
| 4x4 | | | | |
| W1.4xW1.4 | S.F. | 0.20 | 0.26 | 0.46 |
| W2.0xW2.0 | " | 0.27 | 0.28 | 0.55 |
| W2.9xW2.9 | " | 0.39 | 0.30 | 0.69 |
| W4.0xW4.0 | " | 0.58 | 0.33 | 0.91 |
| 6x6 | | | | |
| W1.4xW1.4 | S.F. | 0.19 | 0.20 | 0.39 |
| W2.0xW2.0 | " | 0.27 | 0.22 | 0.49 |
| W2.9xW2.9 | " | 0.38 | 0.23 | 0.61 |
| W4.0xW4.0 | " | 0.41 | 0.26 | 0.67 |
| Standard | | | | |
| 2x2 | | | | |
| W.9xW.9 | S.F. | 0.22 | 0.26 | 0.48 |
| 4x4 | | | | |
| W1.4xW1.4 | S.F. | 0.15 | 0.26 | 0.41 |
| W4.0xW4.0 | " | 0.41 | 0.33 | 0.74 |
| 6x6 | | | | |
| W1.4xW1.4 | S.F. | 0.09 | 0.20 | 0.29 |
| W4.0xW4.0 | " | 0.27 | 0.26 | 0.53 |

# 03 CONCRETE

## REINFORCEMENT

| REINFORCEMENT | UNIT | MAT. | INST. | TOTAL |
|---|---|---|---|---|
| **03210.60 STAIR REINFORCING** | | | | |
| Stairs | | | | |
| #3 - #4 | TON | 890.00 | 790.00 | 1,680 |
| #5 - #6 | " | 780.00 | 660.00 | 1,440 |
| Galvanized | | | | |
| #3 - #4 | TON | 1,490 | 790.00 | 2,280 |
| #5 - #6 | " | 1,400 | 660.00 | 2,060 |
| **03210.65 WALL REINFORCING** | | | | |
| Walls | | | | |
| #3 - #4 | TON | 890.00 | 560.00 | 1,450 |
| #5 - #6 | " | 780.00 | 490.00 | 1,270 |
| #7 - #8 | " | 740.00 | 440.00 | 1,180 |
| Galvanized | | | | |
| #3 - #4 | TON | 1,490 | 560.00 | 2,050 |
| #5 - #6 | " | 1,400 | 490.00 | 1,890 |
| #7 - #8 | " | 1,350 | 440.00 | 1,790 |
| Masonry wall (horizontal) | | | | |
| #3 - #4 | TON | 890.00 | 1,580 | 2,470 |
| #5 - #6 | " | 780.00 | 1,310 | 2,090 |
| Galvanized | | | | |
| #3 - #4 | TON | 1,490 | 1,580 | 3,070 |
| #5 - #6 | " | 1,400 | 1,310 | 2,710 |
| Masonry wall (vertical) | | | | |
| #3 - #4 | TON | 890.00 | 1,970 | 2,860 |
| #5 - #6 | " | 780.00 | 1,580 | 2,360 |
| Galvanized | | | | |
| #3 - #4 | TON | 1,490 | 1,970 | 3,460 |
| #5 - #6 | " | 1,400 | 1,580 | 2,980 |

## ACCESSORIES

| ACCESSORIES | UNIT | MAT. | INST. | TOTAL |
|---|---|---|---|---|
| **03250.40 CONCRETE ACCESSORIES** | | | | |
| Expansion joint, poured | | | | |
| Asphalt | | | | |
| 1/2" x 1" | L.F. | 0.46 | 0.56 | 1.01 |
| 1" x 2" | " | 1.40 | 0.60 | 2.00 |
| Liquid neoprene, cold applied | | | | |
| 1/2" x 1" | L.F. | 1.70 | 0.57 | 2.27 |
| 1" x 2" | " | 6.70 | 0.62 | 7.32 |
| Polyurethane, 2 parts | | | | |
| 1/2" x 1" | L.F. | 1.55 | 0.93 | 2.48 |
| 1" x 2" | " | 6.45 | 1.00 | 7.45 |
| Rubberized asphalt, cold | | | | |
| 1/2" x 1" | L.F. | 0.38 | 0.56 | 0.94 |

## ACCESSORIES

### 03250.40 CONCRETE ACCESSORIES

| ACCESSORIES | UNIT | MAT. | INST. | TOTAL |
|---|---|---|---|---|
| 1" x 2" | L.F. | 1.20 | 0.60 | 1.80 |
| Hot, fuel resistant | | | | |
| 1/2" x 1" | L.F. | 0.72 | 0.56 | 1.27 |
| 1" x 2" | " | 3.60 | 0.60 | 4.20 |
| Expansion joint, premolded, in slabs | | | | |
| Asphalt | | | | |
| 1/2" x 6" | L.F. | 0.51 | 0.69 | 1.20 |
| 1" x 12" | " | 0.86 | 0.93 | 1.78 |
| Cork | | | | |
| 1/2" x 6" | L.F. | 1.15 | 0.69 | 1.84 |
| 1" x 12" | " | 4.60 | 0.93 | 5.53 |
| Neoprene sponge | | | | |
| 1/2" x 6" | L.F. | 1.55 | 0.69 | 2.24 |
| 1" x 12" | " | 5.70 | 0.93 | 6.63 |
| Polyethylene foam | | | | |
| 1/2" x 6" | L.F. | 0.58 | 0.69 | 1.27 |
| 1" x 12" | " | 2.85 | 0.93 | 3.77 |
| Polyurethane foam | | | | |
| 1/2" x 6" | L.F. | 0.77 | 0.69 | 1.46 |
| 1" x 12" | " | 1.75 | 0.93 | 2.67 |
| Polyvinyl chloride foam | | | | |
| 1/2" x 6" | L.F. | 1.65 | 0.69 | 2.34 |
| 1" x 12" | " | 6.10 | 0.93 | 7.03 |
| Rubber, gray sponge | | | | |
| 1/2" x 6" | L.F. | 2.70 | 0.69 | 3.39 |
| 1" x 12" | " | 11.50 | 0.93 | 12.43 |
| Asphalt felt control joints or bond breaker, screed joints | | | | |
| 4" slab | L.F. | 0.72 | 0.56 | 1.27 |
| 6" slab | " | 0.93 | 0.62 | 1.55 |
| 8" slab | " | 1.25 | 0.69 | 1.94 |
| 10" slab | " | 1.75 | 0.79 | 2.54 |
| Keyed cold expansion and control joints, 24 ga. | | | | |
| 4" slab | L.F. | 0.65 | 1.75 | 2.40 |
| 5" slab | " | 0.80 | 1.75 | 2.55 |
| 6" slab | " | 0.93 | 1.85 | 2.78 |
| 8" slab | " | 1.15 | 2.00 | 3.15 |
| 10" slab | " | 1.30 | 2.15 | 3.45 |
| Waterstops | | | | |
| Polyvinyl chloride | | | | |
| Ribbed | | | | |
| 3/16" thick x | | | | |
| 4" wide | L.F. | 1.10 | 1.40 | 2.50 |
| 6" wide | " | 1.30 | 1.55 | 2.85 |
| 1/2" thick x | | | | |
| 9" wide | L.F. | 3.65 | 1.75 | 5.40 |
| Ribbed with center bulb | | | | |
| 3/16" thick x 9" wide | L.F. | 3.10 | 1.75 | 4.85 |
| 3/8" thick x 9" wide | " | 3.80 | 1.75 | 5.55 |
| Dumbbell type, 3/8" thick x 6" wide | " | 3.80 | 1.55 | 5.35 |
| Plain, 3/8" thick x 9" wide | " | 4.65 | 1.75 | 6.40 |
| Center bulb, 3/8" thick x 9" wide | " | 6.30 | 1.75 | 8.05 |
| Rubber | | | | |

# 03 CONCRETE

## ACCESSORIES

| 03250.40 CONCRETE ACCESSORIES | UNIT | MAT. | INST. | TOTAL |
|---|---|---|---|---|
| Vapor barrier | | | | |
| 4 mil polyethylene | S.F. | 0.03 | 0.09 | 0.12 |
| 6 mil polyethylene | " | 0.04 | 0.09 | 0.13 |
| Gravel porous fill, under floor slabs, 3/4" stone | C.Y. | 17.55 | 46.30 | 63.85 |

## CAST-IN-PLACE CONCRETE

| 03300.10 CONCRETE ADMIXTURES | UNIT | MAT. | INST. | TOTAL |
|---|---|---|---|---|
| Concrete admixtures | | | | |
| Water reducing admixture | GAL | | | 12.42 |
| Set retarder | " | | | 22.27 |
| Air entraining agent | " | | | 7.56 |

| 03350.10 CONCRETE FINISHES | UNIT | MAT. | INST. | TOTAL |
|---|---|---|---|---|
| Floor finishes | | | | |
| Broom | S.F. | 0.00 | 0.40 | 0.40 |
| Screed | " | 0.00 | 0.35 | 0.35 |
| Darby | " | 0.00 | 0.35 | 0.35 |
| Steel float | " | 0.00 | 0.46 | 0.46 |
| Wall finishes | | | | |
| Burlap rub, with cement paste | S.F. | 0.07 | 0.46 | 0.53 |
| Break ties and patch holes | " | 0.00 | 0.56 | 0.56 |
| Carborundum | | | | |
| Dry rub | S.F. | 0.00 | 0.93 | 0.93 |
| Wet rub | " | 0.00 | 1.40 | 1.40 |
| Floor hardeners | | | | |
| Metallic | | | | |
| Light service | S.F. | 0.24 | 0.35 | 0.59 |
| Heavy service | " | 0.80 | 0.46 | 1.26 |
| Non-metallic | | | | |
| Light service | S.F. | 0.12 | 0.35 | 0.47 |
| Heavy service | " | 0.55 | 0.46 | 1.01 |

| 03360.10 PNEUMATIC CONCRETE | UNIT | MAT. | INST. | TOTAL |
|---|---|---|---|---|
| Pneumatic applied concrete (gunite) | | | | |
| 2" thick | S.F. | 3.90 | 1.80 | 5.70 |
| 3" thick | " | 4.85 | 2.40 | 7.25 |
| 4" thick | " | 5.80 | 2.90 | 8.70 |
| Finish surface | | | | |
| Minimum | S.F. | 0.00 | 1.75 | 1.75 |
| Maximum | " | 0.00 | 3.55 | 3.55 |

# 03 CONCRETE

## CAST-IN-PLACE CONCRETE

### 03370.10 CURING CONCRETE

| | UNIT | MAT. | INST. | TOTAL |
|---|---|---|---|---|
| Sprayed membrane | | | | |
| Slabs | S.F. | 0.05 | 0.06 | 0.11 |
| Walls | " | 0.08 | 0.07 | 0.15 |
| Curing paper | | | | |
| Slabs | S.F. | 0.08 | 0.07 | 0.15 |
| Walls | " | 0.08 | 0.08 | 0.16 |
| Burlap | | | | |
| 7.5 oz. | S.F. | 0.07 | 0.09 | 0.16 |
| 12 oz. | " | 0.09 | 0.10 | 0.19 |

## PLACING CONCRETE

### 03380.05 BEAM CONCRETE

| | UNIT | MAT. | INST. | TOTAL |
|---|---|---|---|---|
| Beams and girders | | | | |
| 2500# or 3000# concrete | | | | |
| By pump | C.Y. | 86.50 | 46.60 | 133.10 |
| By hand buggy | " | 86.50 | 27.80 | 114.30 |
| 5000# concrete | | | | |
| By pump | C.Y. | 97.00 | 46.60 | 143.60 |
| By hand buggy | " | 97.00 | 27.80 | 124.80 |
| Bond beam, 3000# concrete | | | | |
| By pump | | | | |
| 8" high | | | | |
| 4" wide | L.F. | 0.23 | 1.05 | 1.28 |
| 6" wide | " | 0.55 | 1.15 | 1.70 |
| 8" wide | " | 0.72 | 1.30 | 2.02 |
| 10" wide | " | 0.96 | 1.45 | 2.41 |
| 12" wide | " | 1.30 | 1.60 | 2.90 |

### 03380.15 COLUMN CONCRETE

| | UNIT | MAT. | INST. | TOTAL |
|---|---|---|---|---|
| Columns | | | | |
| 2500# or 3000# concrete | | | | |
| By pump | C.Y. | 86.50 | 42.80 | 129.30 |
| 5000# concrete | | | | |
| By pump | C.Y. | 97.00 | 42.80 | 139.80 |
| 5000# concrete | | | | |

### 03380.25 EQUIPMENT PAD CONCRETE

| | UNIT | MAT. | INST. | TOTAL |
|---|---|---|---|---|
| Equipment pad | | | | |
| 2500# or 3000# concrete | | | | |
| By chute | C.Y. | 86.50 | 9.25 | 95.75 |
| By pump | " | 86.50 | 36.70 | 123.20 |

# 03 CONCRETE

| PLACING CONCRETE | UNIT | MAT. | INST. | TOTAL |
|---|---|---|---|---|
| **03380.25 EQUIPMENT PAD CONCRETE** | | | | |
| 3500# or 4000# concrete | | | | |
| By chute | C.Y. | 92.00 | 9.25 | 101.25 |
| By pump | " | 92.00 | 36.70 | 128.70 |
| 5000# concrete | | | | |
| By chute | C.Y. | 97.00 | 9.25 | 106.25 |
| By pump | " | 97.00 | 36.70 | 133.70 |
| **03380.35 FOOTING CONCRETE** | | | | |
| Continuous footing | | | | |
| 2500# or 3000# concrete | | | | |
| By chute | C.Y. | 86.50 | 9.25 | 95.75 |
| By pump | " | 86.50 | 32.10 | 118.60 |
| 5000# concrete | | | | |
| By chute | C.Y. | 97.00 | 9.25 | 106.25 |
| By pump | " | 97.00 | 32.10 | 129.10 |
| Spread footing | | | | |
| 2500# or 3000# concrete | | | | |
| Under 5 cy | | | | |
| By chute | C.Y. | 86.50 | 9.25 | 95.75 |
| By pump | " | 86.50 | 34.20 | 120.70 |
| 5000# concrete | | | | |
| Under 5 c.y. | | | | |
| By chute | C.Y. | 97.00 | 9.25 | 106.25 |
| By pump | " | 97.00 | 34.20 | 131.20 |
| **03380.50 GRADE BEAM CONCRETE** | | | | |
| Grade beam | | | | |
| 2500# or 3000# concrete | | | | |
| By chute | C.Y. | 86.50 | 9.25 | 95.75 |
| By pump | " | 86.50 | 32.10 | 118.60 |
| By hand buggy | " | 86.50 | 27.80 | 114.30 |
| 5000# concrete | | | | |
| By chute | C.Y. | 97.00 | 9.25 | 106.25 |
| By pump | " | 97.00 | 32.10 | 129.10 |
| By hand buggy | " | 97.00 | 27.80 | 124.80 |
| **03380.53 PILE CAP CONCRETE** | | | | |
| Pile cap | | | | |
| 2500# or 3000 concrete | | | | |
| By chute | C.Y. | 86.50 | 9.25 | 95.75 |
| By pump | " | 86.50 | 36.70 | 123.20 |
| By hand buggy | " | 86.50 | 27.80 | 114.30 |
| 5000# concrete | | | | |
| By chute | C.Y. | 97.00 | 9.25 | 106.25 |
| By pump | " | 97.00 | 36.70 | 133.70 |
| By hand buggy | " | 97.00 | 27.80 | 124.80 |

# 03 CONCRETE

## PLACING CONCRETE

| PLACING CONCRETE | UNIT | MAT. | INST. | TOTAL |
|---|---|---|---|---|
| **03380.55 SLAB/MAT CONCRETE** | | | | |
| Slab on grade | | | | |
| 2500# or 3000# concrete | | | | |
| By chute | C.Y. | 86.50 | 6.95 | 93.45 |
| By pump | " | 86.50 | 18.35 | 104.85 |
| By hand buggy | " | 86.50 | 18.50 | 105.00 |
| 5000# concrete | | | | |
| By chute | C.Y. | 97.00 | 6.95 | 103.95 |
| By pump | " | 97.00 | 18.35 | 115.35 |
| By hand buggy | " | 97.00 | 18.50 | 115.50 |
| **03380.58 SIDEWALKS** | | | | |
| Walks, cast in place with wire mesh, base not incl. | | | | |
| 4" thick | S.F. | 1.15 | 0.93 | 2.08 |
| 5" thick | " | 1.60 | 1.10 | 2.70 |
| 6" thick | " | 1.90 | 1.40 | 3.30 |
| **03380.60 STAIR CONCRETE** | | | | |
| Stairs | | | | |
| 2500# or 3000# concrete | | | | |
| By chute | C.Y. | 86.50 | 9.25 | 95.75 |
| By pump | " | 86.50 | 36.70 | 123.20 |
| By hand buggy | " | 86.50 | 27.80 | 114.30 |
| 3500# or 4000# concrete | | | | |
| By chute | C.Y. | 92.00 | 9.25 | 101.25 |
| By pump | " | 92.00 | 36.70 | 128.70 |
| By hand buggy | " | 92.00 | 27.80 | 119.80 |
| 5000# concrete | | | | |
| By chute | C.Y. | 97.00 | 9.25 | 106.25 |
| By pump | " | 97.00 | 36.70 | 133.70 |
| By hand buggy | " | 97.00 | 27.80 | 124.80 |
| **03380.65 WALL CONCRETE** | | | | |
| Walls | | | | |
| 2500# or 3000# concrete | | | | |
| To 4' | | | | |
| By chute | C.Y. | 86.50 | 7.95 | 94.45 |
| By pump | " | 86.50 | 39.50 | 126.00 |
| To 8' | | | | |
| By pump | C.Y. | 86.50 | 42.80 | 129.30 |
| 3500# or 4000# concrete | | | | |
| To 4' | | | | |
| By chute | C.Y. | 92.00 | 7.95 | 99.95 |
| By pump | " | 92.00 | 39.50 | 131.50 |
| To 8' | | | | |
| By pump | C.Y. | 92.00 | 42.80 | 134.80 |
| Filled block (CMU) | | | | |
| 3000# concrete, by pump | | | | |
| 4" wide | S.F. | 0.31 | 1.85 | 2.16 |
| 6" wide | " | 0.72 | 2.15 | 2.87 |
| 8" wide | " | 1.10 | 2.55 | 3.65 |

# 03 CONCRETE

## PLACING CONCRETE

| 03380.65 WALL CONCRETE | UNIT | MAT. | INST. | TOTAL |
|---|---|---|---|---|
| 10" wide | S.F. | 1.50 | 3.00 | 4.50 |
| 12" wide | " | 1.90 | 3.65 | 5.55 |
| Pilasters, 3000# concrete | C.F. | 4.40 | 51.50 | 55.90 |
| Wall cavity, 2" thick, 3000# concrete | S.F. | 0.81 | 1.70 | 2.51 |

## CEMENTITOUS TOPPINGS

| 03550.10 CONCRETE TOPPINGS | UNIT | MAT. | INST. | TOTAL |
|---|---|---|---|---|
| Gypsum fill | | | | |
| 2" thick | S.F. | 1.25 | 0.26 | 1.51 |
| 2-1/2" thick | " | 1.40 | 0.27 | 1.67 |
| 3" thick | " | 1.70 | 0.28 | 1.98 |
| 3-1/2" thick | " | 1.95 | 0.28 | 2.23 |
| 4" thick | " | 2.30 | 0.32 | 2.62 |
| Formboard | | | | |
| Mineral fiber board | | | | |
| 1" thick | S.F. | 1.15 | 0.69 | 1.84 |
| 1-1/2" thick | " | 3.10 | 0.79 | 3.89 |
| Cement fiber board | | | | |
| 1" thick | S.F. | 0.90 | 0.93 | 1.83 |
| 1-1/2" thick | " | 1.15 | 1.05 | 2.20 |
| Glass fiber board | | | | |
| 1" thick | S.F. | 1.40 | 0.69 | 2.09 |
| 1-1/2" thick | " | 1.90 | 0.79 | 2.69 |
| Poured deck | | | | |
| Vermiculite or perlite | | | | |
| 1 to 4 mix | C.Y. | 130.00 | 42.80 | 172.80 |
| 1 to 6 mix | " | 110.00 | 39.50 | 149.50 |
| Vermiculite or perlite | | | | |
| 2" thick | | | | |
| 1 to 4 mix | S.F. | 1.20 | 0.27 | 1.47 |
| 1 to 6 mix | " | 0.86 | 0.24 | 1.10 |
| 3" thick | | | | |
| 1 to 4 mix | S.F. | 1.60 | 0.40 | 2.00 |
| 1 to 6 mix | " | 1.30 | 0.37 | 1.67 |
| Concrete plank, lightweight | | | | |
| 2" thick | S.F. | 6.20 | 2.10 | 8.30 |
| 2-1/2" thick | " | 6.35 | 2.10 | 8.45 |
| 3-1/2" thick | " | 6.60 | 2.35 | 8.95 |
| 4" thick | " | 6.90 | 2.35 | 9.25 |
| Channel slab, lightweight, straight | | | | |
| 2-3/4" thick | S.F. | 5.00 | 2.10 | 7.10 |
| 3-1/2" thick | " | 5.20 | 2.10 | 7.30 |
| 3-3/4" thick | " | 5.55 | 2.10 | 7.65 |

# 03 CONCRETE

## CEMENTITOUS TOPPINGS

### 03550.10 CONCRETE TOPPINGS

| | UNIT | MAT. | INST. | TOTAL |
|---|---|---|---|---|
| 4-3/4" thick | S.F. | 7.05 | 2.35 | 9.40 |
| Gypsum plank | | | | |
| 2" thick | S.F. | 2.35 | 2.10 | 4.45 |
| 3" thick | " | 2.45 | 2.10 | 4.55 |
| Cement fiber, T and G planks | | | | |
| 1" thick | S.F. | 1.30 | 1.90 | 3.20 |
| 1-1/2" thick | " | 1.35 | 1.90 | 3.25 |
| 2" thick | " | 1.55 | 2.10 | 3.65 |
| 2-1/2" thick | " | 1.65 | 2.10 | 3.75 |
| 3" thick | " | 2.20 | 2.10 | 4.30 |
| 3-1/2" thick | " | 2.50 | 2.35 | 4.85 |
| 4" thick | " | 2.80 | 2.35 | 5.15 |

## CONCRETE RESTORATION

### 03730.10 CONCRETE REPAIR

| | UNIT | MAT. | INST. | TOTAL |
|---|---|---|---|---|
| Epoxy grout floor patch, 1/4" thick | S.F. | 4.60 | 2.80 | 7.40 |
| Grout, epoxy, 2 component system | C.F. | | | 231.39 |
| Epoxy sand | BAG | | | 15.55 |
| Epoxy modifier | GAL | | | 99.77 |
| Epoxy gel grout | S.F. | 2.30 | 27.80 | 30.10 |
| Injection valve, 1 way, threaded plastic | EA. | 6.35 | 5.55 | 11.90 |
| Grout crack seal, 2 component | C.F. | 540.00 | 27.80 | 567.80 |
| Grout, non shrink | " | 55.00 | 27.80 | 82.80 |
| Concrete, epoxy modified | | | | |
| Sand mix | C.F. | 87.00 | 11.10 | 98.10 |
| Gravel mix | " | 66.50 | 10.30 | 76.80 |
| Concrete repair | | | | |
| Soffit repair | | | | |
| 16" wide | L.F. | 2.75 | 5.55 | 8.30 |
| 18" wide | " | 2.90 | 5.80 | 8.70 |
| 24" wide | " | 3.40 | 6.15 | 9.55 |
| 30" wide | " | 3.90 | 6.60 | 10.50 |
| 32" wide | " | 4.20 | 6.95 | 11.15 |
| Edge repair | | | | |
| 2" spall | L.F. | 1.30 | 6.95 | 8.25 |
| 3" spall | " | 1.30 | 7.30 | 8.60 |
| 4" spall | " | 1.40 | 7.50 | 8.90 |
| 6" spall | " | 1.45 | 7.70 | 9.15 |
| 8" spall | " | 1.50 | 8.15 | 9.65 |
| 9" spall | " | 1.55 | 9.25 | 10.80 |
| Crack repair, 1/8" crack | " | 2.55 | 2.80 | 5.35 |
| Reinforcing steel repair | | | | |
| 1 bar, 4 ft | | | | |

# 03 CONCRETE

| CONCRETE RESTORATION | UNIT | MAT. | INST. | TOTAL |
|---|---|---|---|---|
| **03730.10 CONCRETE REPAIR** | | | | |
| #4 bar | L.F. | 0.36 | 4.90 | 5.26 |
| #5 bar | " | 0.51 | 4.90 | 5.41 |
| #6 bar | " | 0.65 | 5.25 | 5.90 |
| #8 bar | " | 1.15 | 5.25 | 6.40 |
| #9 bar | " | 1.45 | 5.65 | 7.10 |
| #11 bar | " | 2.25 | 5.65 | 7.90 |
| Form fabric, nylon | | | | |
| 18" diameter | L.F. | | | 9.68 |
| 20" diameter | " | | | 9.84 |
| 24" diameter | " | | | 16.19 |
| 30" diameter | " | | | 16.63 |
| 36" diameter | " | | | 19.08 |
| Pile repairs | | | | |
| Polyethylene wrap | | | | |
| 30 mil thick | | | | |
| 60" wide | S.F. | 10.40 | 9.25 | 19.65 |
| 72" wide | " | 11.30 | 11.10 | 22.40 |
| 60 mil thick | | | | |
| 60" wide | S.F. | 12.50 | 9.25 | 21.75 |
| 80" wide | " | 14.45 | 12.60 | 27.05 |
| Pile spall, average repair 3' | | | | |
| 18" x 18" | EA. | 32.50 | 23.15 | 55.65 |
| 20" x 20" | " | 43.30 | 27.80 | 71.10 |

# 04 MASONRY

## MORTAR AND GROUT

| 04100.10 MASONRY GROUT | UNIT | MAT. | INST. | TOTAL |
|---|---|---|---|---|
| Grout, non shrink, non-metallic, trowelable | C.F. | 10.35 | 0.96 | 11.31 |
| Grout door frame, hollow metal | | | | |
| Single | EA. | 15.05 | 36.00 | 51.05 |
| Double | " | 22.50 | 37.90 | 60.40 |
| Grout-filled concrete block (CMU) | | | | |
| 4" wide | S.F. | 0.43 | 1.20 | 1.63 |
| 6" wide | " | 0.90 | 1.30 | 2.20 |
| 8" wide | " | 1.55 | 1.45 | 3.00 |
| 12" wide | " | 2.25 | 1.50 | 3.75 |
| Grout-filled individual CMU cells | | | | |
| 4" wide | L.F. | 0.22 | 0.72 | 0.94 |
| 6" wide | " | 0.47 | 0.72 | 1.19 |
| 8" wide | " | 0.62 | 0.72 | 1.34 |
| 10" wide | " | 0.82 | 0.82 | 1.64 |
| 12" wide | " | 0.93 | 0.82 | 1.75 |
| Bond beams or lintels, 8" deep | | | | |
| 6" thick | L.F. | 0.95 | 1.15 | 2.10 |
| 8" thick | " | 1.20 | 1.30 | 2.50 |
| 10" thick | " | 1.50 | 1.45 | 2.95 |
| 12" thick | " | 1.75 | 1.60 | 3.35 |
| Cavity walls | | | | |
| 2" thick | S.F. | 1.05 | 1.70 | 2.75 |
| 3" thick | " | 1.55 | 1.70 | 3.25 |
| 4" thick | " | 2.05 | 1.85 | 3.90 |
| 6" thick | " | 3.10 | 2.15 | 5.25 |

| 04150.10 MASONRY ACCESSORIES | UNIT | MAT. | INST. | TOTAL |
|---|---|---|---|---|
| Foundation vents | EA. | 22.50 | 13.75 | 36.25 |
| Bar reinforcing | | | | |
| Horizontal | | | | |
| #3 - #4 | Lb. | 0.46 | 1.35 | 1.81 |
| #5 - #6 | " | 0.43 | 1.15 | 1.58 |
| Vertical | | | | |
| #3 - #4 | Lb. | 0.46 | 1.70 | 2.16 |
| #5 - #6 | " | 0.43 | 1.35 | 1.78 |
| Horizontal joint reinforcing | | | | |
| Truss type | | | | |
| 4" wide, 6" wall | L.F. | 0.18 | 0.14 | 0.32 |
| 6" wide, 8" wall | " | 0.18 | 0.14 | 0.32 |
| 8" wide, 10" wall | " | 0.22 | 0.15 | 0.37 |
| 10" wide, 12" wall | " | 0.22 | 0.16 | 0.38 |
| 12" wide, 14" wall | " | 0.26 | 0.16 | 0.42 |
| Ladder type | | | | |
| 4" wide, 6" wall | L.F. | 0.14 | 0.14 | 0.28 |
| 6" wide, 8" wall | " | 0.15 | 0.14 | 0.29 |
| 8" wide, 10" wall | " | 0.16 | 0.15 | 0.31 |
| 10" wide, 12" wall | " | 0.18 | 0.15 | 0.33 |
| Rectangular wall ties | | | | |
| 3/16" dia., galvanized | | | | |
| 2" x 6" | EA. | 0.19 | 0.57 | 0.76 |
| 2" x 8" | " | 0.20 | 0.57 | 0.77 |

## MORTAR AND GROUT

| 04150.10 MASONRY ACCESSORIES | UNIT | MAT. | INST. | TOTAL |
|---|---|---|---|---|
| 2" x 10" | EA. | 0.23 | 0.57 | 0.80 |
| 2" x 12" | " | 0.27 | 0.57 | 0.84 |
| 4" x 6" | " | 0.22 | 0.69 | 0.91 |
| 4" x 8" | " | 0.24 | 0.69 | 0.93 |
| 4" x 10" | " | 0.32 | 0.69 | 1.01 |
| 4" x 12" | " | 0.36 | 0.69 | 1.05 |
| 1/4" dia., galvanized | | | | |
| 2" x 6" | EA. | 0.36 | 0.57 | 0.93 |
| 2" x 8" | " | 0.39 | 0.57 | 0.96 |
| 2" x 10" | " | 0.45 | 0.57 | 1.02 |
| 2" x 12" | " | 0.50 | 0.57 | 1.07 |
| 4" x 6" | " | 0.41 | 0.69 | 1.10 |
| 4" x 8" | " | 0.45 | 0.69 | 1.14 |
| 4" x 10" | " | 0.50 | 0.69 | 1.19 |
| 4" x 12" | " | 0.53 | 0.69 | 1.22 |
| "Z" type wall ties, galvanized | | | | |
| 6" long | | | | |
| 1/8" dia. | EA. | 0.19 | 0.57 | 0.76 |
| 3/16" dia. | " | 0.20 | 0.57 | 0.77 |
| 1/4" dia. | " | 0.22 | 0.57 | 0.79 |
| 8" long | | | | |
| 1/8" dia. | EA. | 0.20 | 0.57 | 0.77 |
| 3/16" dia. | " | 0.22 | 0.57 | 0.79 |
| 1/4" dia. | " | 0.23 | 0.57 | 0.80 |
| 10" long | | | | |
| 1/8" dia. | EA. | 0.22 | 0.57 | 0.79 |
| 3/16" dia. | " | 0.24 | 0.57 | 0.81 |
| 1/4" dia. | " | 0.28 | 0.57 | 0.85 |
| Dovetail anchor slots | | | | |
| Galvanized steel, filled | | | | |
| 24 ga. | L.F. | 0.50 | 0.86 | 1.36 |
| 20 ga. | " | 0.63 | 0.86 | 1.49 |
| 16 oz. copper, foam filled | " | 1.25 | 0.86 | 2.11 |
| Dovetail anchors | | | | |
| 16 ga. | | | | |
| 3-1/2" long | EA. | 0.15 | 0.57 | 0.72 |
| 5-1/2" long | " | 0.18 | 0.57 | 0.75 |
| 12 ga. | | | | |
| 3-1/2" long | EA. | 0.19 | 0.57 | 0.76 |
| 5-1/2" long | " | 0.39 | 0.57 | 0.96 |
| Dovetail, triangular galvanized ties, 12 ga. | | | | |
| 3" x 3" | EA. | 0.36 | 0.57 | 0.93 |
| 5" x 5" | " | 0.39 | 0.57 | 0.96 |
| 7" x 7" | " | 0.43 | 0.57 | 1.00 |
| 7" x 9" | " | 0.46 | 0.57 | 1.03 |
| Brick anchors | | | | |
| Corrugated, 3-1/2" long | | | | |
| 16 ga. | EA. | 0.15 | 0.57 | 0.72 |
| 12 ga. | " | 0.24 | 0.57 | 0.81 |
| Non-corrugated, 3-1/2" long | | | | |
| 16 ga. | EA. | 0.19 | 0.57 | 0.76 |
| 12 ga. | " | 0.35 | 0.57 | 0.92 |

## MORTAR AND GROUT

### 04150.10 MASONRY ACCESSORIES

| | UNIT | MAT. | INST. | TOTAL |
|---|---|---|---|---|
| Cavity wall anchors, corrugated, galvanized | | | | |
| 5" long | | | | |
| 16 ga. | EA. | 0.43 | 0.57 | 1.00 |
| 12 ga. | " | 0.63 | 0.57 | 1.20 |
| 7" long | | | | |
| 28 ga. | EA. | 0.47 | 0.57 | 1.04 |
| 24 ga. | " | 0.59 | 0.57 | 1.16 |
| 22 ga. | " | 0.61 | 0.57 | 1.18 |
| 16 ga. | " | 0.69 | 0.57 | 1.26 |
| Mesh ties, 16 ga., 3" wide | | | | |
| 8" long | EA. | 0.57 | 0.57 | 1.14 |
| 12" long | " | 0.63 | 0.57 | 1.20 |
| 20" long | " | 0.88 | 0.57 | 1.45 |
| 24" long | " | 0.97 | 0.57 | 1.54 |

### 04150.20 MASONRY CONTROL JOINTS

| | UNIT | MAT. | INST. | TOTAL |
|---|---|---|---|---|
| Control joint, cross shaped PVC | L.F. | 3.35 | 0.86 | 4.21 |
| Closed cell joint filler | | | | |
| 1/2" | L.F. | 0.50 | 0.86 | 1.36 |
| 3/4" | " | 0.92 | 0.86 | 1.78 |
| Rubber, for | | | | |
| 4" wall | L.F. | 3.65 | 0.86 | 4.51 |
| 6" wall | " | 5.75 | 0.90 | 6.65 |
| 8" wall | " | 6.75 | 0.95 | 7.70 |
| PVC, for | | | | |
| 4" wall | L.F. | 2.80 | 0.86 | 3.66 |
| 6" wall | " | 3.50 | 0.90 | 4.40 |
| 8" wall | " | 4.00 | 0.95 | 4.95 |

### 04150.50 MASONRY FLASHING

| | UNIT | MAT. | INST. | TOTAL |
|---|---|---|---|---|
| Through-wall flashing | | | | |
| 5 oz. coated copper | S.F. | 3.25 | 2.85 | 6.10 |
| 0.030" elastomeric | " | 0.72 | 2.30 | 3.02 |

## UNIT MASONRY

### 04210.10 BRICK MASONRY

| | UNIT | MAT. | INST. | TOTAL |
|---|---|---|---|---|
| Standard size brick, running bond | | | | |
| Face brick, red (6.4/sf) | | | | |
| Veneer | S.F. | 3.25 | 5.70 | 8.95 |
| Cavity wall | " | 3.25 | 4.90 | 8.15 |
| 9" solid wall | " | 6.20 | 9.80 | 16.00 |
| Common brick (6.4/sf) | | | | |

# 04 MASONRY

## UNIT MASONRY

| | UNIT | MAT. | INST. | TOTAL |
|---|---|---|---|---|
| **04210.10 BRICK MASONRY** | | | | |
| Select common for veneers | S.F. | 3.65 | 5.70 | 9.35 |
| Back-up | | | | |
| 4" thick | S.F. | 3.25 | 4.30 | 7.55 |
| 8" thick | " | 6.35 | 6.85 | 13.20 |
| Firewall | | | | |
| 12" thick | S.F. | 9.70 | 11.45 | 21.15 |
| 16" thick | " | 12.85 | 15.60 | 28.45 |
| Glazed brick (7.4/sf) | | | | |
| Veneer | S.F. | 10.35 | 6.25 | 16.60 |
| Buff or gray face brick (6.4/sf) | | | | |
| Veneer | S.F. | 3.50 | 5.70 | 9.20 |
| Cavity wall | " | 3.50 | 4.90 | 8.40 |
| Jumbo or oversize brick (3/sf) | | | | |
| 4" veneer | S.F. | 4.05 | 3.45 | 7.50 |
| 4" back-up | " | 4.05 | 2.85 | 6.90 |
| 8" back-up | " | 9.45 | 4.90 | 14.35 |
| 12" firewall | " | 13.50 | 8.60 | 22.10 |
| 16" firewall | " | 19.95 | 11.45 | 31.40 |
| Norman brick, red face, (4.5/sf) | | | | |
| 4" veneer | S.F. | 4.75 | 4.30 | 9.05 |
| Cavity wall | " | 4.75 | 3.80 | 8.55 |
| Chimney, standard brick, including flue | | | | |
| 16" x 16" | L.F. | 18.85 | 34.30 | 53.15 |
| 16" x 20" | " | 22.30 | 34.30 | 56.60 |
| 16" x 24" | " | 26.40 | 34.30 | 60.70 |
| 20" x 20" | " | 28.90 | 42.90 | 71.80 |
| 20" x 24" | " | 37.20 | 42.90 | 80.10 |
| 20" x 32" | " | 46.80 | 49.00 | 95.80 |
| Window sill, face brick on edge | " | 2.80 | 8.60 | 11.40 |
| **04210.20 STRUCTURAL TILE** | | | | |
| Structural glazed tile | | | | |
| 6T series, 5-1/2" x 12" | | | | |
| Glazed on one side | | | | |
| 2" thick | S.F. | 7.00 | 3.45 | 10.45 |
| 4" thick | " | 8.80 | 3.45 | 12.25 |
| 6" thick | " | 12.85 | 3.80 | 16.65 |
| 8" thick | " | 16.85 | 4.30 | 21.15 |
| Glazed on two sides | | | | |
| 4" thick | S.F. | 13.50 | 4.30 | 17.80 |
| 6" thick | " | 20.25 | 4.90 | 25.15 |
| Special shapes | | | | |
| Group 1 | S.F. | 10.10 | 6.85 | 16.95 |
| Group 2 | " | 12.85 | 6.85 | 19.70 |
| Group 3 | " | 14.20 | 6.85 | 21.05 |
| Group 4 | " | 25.70 | 6.85 | 32.55 |
| Group 5 | " | 29.70 | 6.85 | 36.55 |
| Fire rated | | | | |
| 4" thick, 1 hr rating | S.F. | 18.90 | 3.45 | 22.35 |
| 6" thick, 2 hr rating | " | 22.95 | 3.80 | 26.75 |
| 8W series, 8" x 16" | | | | |

# 04 MASONRY

## UNIT MASONRY

### 04210.20 STRUCTURAL TILE

| | UNIT | MAT. | INST. | TOTAL |
|---|---|---|---|---|
| Glazed on one side | | | | |
| 2" thick | S.F. | 7.45 | 2.30 | 9.75 |
| 4" thick | " | 8.10 | 2.30 | 10.40 |
| 6" thick | " | 12.15 | 2.65 | 14.80 |
| 8" thick | " | 13.50 | 2.65 | 16.15 |
| Glazed on two sides | | | | |
| 4" thick | S.F. | 13.50 | 2.85 | 16.35 |
| 6" thick | " | 16.20 | 3.45 | 19.65 |
| 8" thick | " | 22.95 | 3.45 | 26.40 |
| Special shapes | | | | |
| Group 1 | S.F. | 12.15 | 4.90 | 17.05 |
| Group 2 | " | 14.85 | 4.90 | 19.75 |
| Group 3 | " | 16.20 | 4.90 | 21.10 |
| Group 4 | " | 27.00 | 4.90 | 31.90 |
| Group 5 | " | 67.50 | 4.90 | 72.40 |
| Fire rated | | | | |
| 4" thick, 1 hr rating | S.F. | 20.25 | 4.90 | 25.15 |
| 6" thick, 2 hr rating | " | 24.30 | 4.90 | 29.20 |

### 04210.60 PAVERS, MASONRY

| | UNIT | MAT. | INST. | TOTAL |
|---|---|---|---|---|
| Brick walk laid on sand, sand joints | | | | |
| Laid flat, (4.5 per sf) | S.F. | 3.15 | 3.80 | 6.95 |
| Laid on edge, (7.2 per sf) | " | 4.95 | 5.70 | 10.65 |
| Precast concrete patio blocks | | | | |
| 2" thick | | | | |
| Natural | S.F. | 2.45 | 1.15 | 3.60 |
| Colors | " | 3.35 | 1.15 | 4.50 |
| Exposed aggregates, local aggregate | | | | |
| Natural | S.F. | 2.85 | 1.15 | 4.00 |
| Colors | " | 3.80 | 1.15 | 4.95 |
| Granite or limestone aggregate | " | 5.20 | 1.15 | 6.35 |
| White tumblestone aggregate | " | 4.05 | 1.15 | 5.20 |
| Stone pavers, set in mortar | | | | |
| Bluestone | | | | |
| 1" thick | | | | |
| Irregular | S.F. | 2.55 | 8.60 | 11.15 |
| Snapped rectangular | " | 3.90 | 6.85 | 10.75 |
| 1-1/2" thick, random rectangular | " | 4.45 | 8.60 | 13.05 |
| 2" thick, random rectangular | " | 5.25 | 9.80 | 15.05 |
| Slate | | | | |
| Natural cleft | | | | |
| Irregular, 3/4" thick | S.F. | 2.45 | 9.80 | 12.25 |
| Random rectangular | | | | |
| 1-1/4" thick | S.F. | 5.25 | 8.60 | 13.85 |
| 1-1/2" thick | " | 5.95 | 9.55 | 15.50 |
| Granite blocks | | | | |
| 3" thick, 3" to 6" wide | | | | |
| 4" to 12" long | S.F. | 5.65 | 11.45 | 17.10 |
| 6" to 15" long | " | 3.30 | 9.80 | 13.10 |
| Crushed stone, white marble, 3" thick | " | 1.35 | 0.56 | 1.91 |

# 04 MASONRY

| UNIT MASONRY | UNIT | MAT. | INST. | TOTAL |
|---|---|---|---|---|
| **04220.10 CONCRETE MASONRY UNITS** | | | | |
| Hollow, load bearing | | | | |
| 4" | S.F. | 1.45 | 2.55 | 4.00 |
| 6" | " | 1.85 | 2.65 | 4.50 |
| 8" | " | 2.40 | 2.85 | 5.25 |
| 10" | " | 3.15 | 3.10 | 6.25 |
| 12" | " | 3.35 | 3.45 | 6.80 |
| Solid, load bearing | | | | |
| 4" | S.F. | 1.95 | 2.55 | 4.50 |
| 6" | " | 2.35 | 2.65 | 5.00 |
| 8" | " | 2.70 | 2.85 | 5.55 |
| 10" | " | 3.70 | 3.10 | 6.80 |
| 12" | " | 4.10 | 3.45 | 7.55 |
| Back-up block, 8" x 16" | | | | |
| 2" | S.F. | 0.96 | 1.95 | 2.91 |
| 4" | " | 1.40 | 2.00 | 3.40 |
| 6" | " | 1.75 | 2.15 | 3.90 |
| 8" | " | 2.05 | 2.30 | 4.35 |
| 10" | " | 2.75 | 2.45 | 5.20 |
| 12" | " | 2.85 | 2.65 | 5.50 |
| Foundation wall, 8" x 16" | | | | |
| 6" | S.F. | 1.90 | 2.45 | 4.35 |
| 8" | " | 2.10 | 2.65 | 4.75 |
| 10" | " | 3.15 | 2.85 | 6.00 |
| 12" | " | 3.50 | 3.10 | 6.60 |
| Solid | | | | |
| 6" | S.F. | 2.60 | 2.65 | 5.25 |
| 8" | " | 3.25 | 2.85 | 6.10 |
| 10" | " | 3.95 | 3.10 | 7.05 |
| 12" | " | 4.80 | 3.45 | 8.25 |
| Exterior, styrofoam inserts, standard weight, 8" x 16" | | | | |
| 6" | S.F. | 2.65 | 2.65 | 5.30 |
| 8" | " | 2.95 | 2.85 | 5.80 |
| 10" | " | 4.20 | 3.10 | 7.30 |
| 12" | " | 4.40 | 3.45 | 7.85 |
| Lightweight | | | | |
| 6" | S.F. | 2.85 | 2.65 | 5.50 |
| 8" | " | 3.45 | 2.85 | 6.30 |
| 10" | " | 4.40 | 3.10 | 7.50 |
| 12" | " | 4.65 | 3.45 | 8.10 |
| Acoustical slotted block | | | | |
| 4" | S.F. | 3.10 | 3.10 | 6.20 |
| 6" | " | 3.80 | 3.10 | 6.90 |
| 8" | " | 5.65 | 3.45 | 9.10 |
| Filled cavities | | | | |
| 4" | S.F. | 3.90 | 3.80 | 7.70 |
| 6" | " | 4.85 | 4.05 | 8.90 |
| 8" | " | 6.50 | 4.30 | 10.80 |
| Hollow, split face | | | | |
| 4" | S.F. | 2.95 | 2.55 | 5.50 |
| 6" | " | 3.45 | 2.65 | 6.10 |
| 8" | " | 3.10 | 2.85 | 5.95 |
| 10" | " | 4.05 | 3.10 | 7.15 |

## UNIT MASONRY

### 04220.10 CONCRETE MASONRY UNITS

| | UNIT | MAT. | INST. | TOTAL |
|---|---|---|---|---|
| 12" | S.F. | 4.30 | 3.45 | 7.75 |
| Split rib profile | | | | |
| 4" | S.F. | 2.15 | 3.10 | 5.25 |
| 6" | " | 2.50 | 3.10 | 5.60 |
| 8" | " | 2.70 | 3.45 | 6.15 |
| 10" | " | 2.95 | 3.45 | 6.40 |
| 12" | " | 3.25 | 3.45 | 6.70 |
| High strength block, 3500 psi | | | | |
| 2" | S.F. | 1.40 | 2.55 | 3.95 |
| 4" | " | 1.75 | 2.65 | 4.40 |
| 6" | " | 2.10 | 2.65 | 4.75 |
| 8" | " | 2.40 | 2.85 | 5.25 |
| 10" | " | 2.75 | 3.10 | 5.85 |
| 12" | " | 3.25 | 3.45 | 6.70 |
| Solar screen concrete block | | | | |
| 4" thick | | | | |
| 6" x 6" | S.F. | 4.20 | 7.65 | 11.85 |
| 8" x 8" | " | 5.00 | 6.85 | 11.85 |
| 12" x 12" | " | 2.90 | 5.30 | 8.20 |
| 8" thick | | | | |
| 8" x 16" | S.F. | 3.10 | 4.90 | 8.00 |
| Glazed block | | | | |
| Cove base, glazed 1 side, 2" | L.F. | 9.45 | 3.80 | 13.25 |
| 4" | " | 9.45 | 3.80 | 13.25 |
| 6" | " | 10.00 | 4.30 | 14.30 |
| 8" | " | 10.65 | 4.30 | 14.95 |
| Single face | | | | |
| 2" | S.F. | 7.00 | 2.85 | 9.85 |
| 4" | " | 7.60 | 2.85 | 10.45 |
| 6" | " | 7.70 | 3.10 | 10.80 |
| 8" | " | 7.95 | 3.45 | 11.40 |
| 10" | " | 8.25 | 3.80 | 12.05 |
| 12" | " | 8.50 | 4.05 | 12.55 |
| Double face | | | | |
| 4" | S.F. | 11.20 | 3.60 | 14.80 |
| 6" | " | 12.15 | 3.80 | 15.95 |
| 8" | " | 12.50 | 4.30 | 16.80 |
| Corner or bullnose | | | | |
| 2" | EA. | 9.45 | 4.30 | 13.75 |
| 4" | " | 12.15 | 4.90 | 17.05 |
| 6" | " | 14.85 | 4.90 | 19.75 |
| 8" | " | 16.20 | 5.70 | 21.90 |
| 10" | " | 17.55 | 6.25 | 23.80 |
| 12" | " | 18.90 | 6.85 | 25.75 |
| Gypsum unit masonry | | | | |
| Partition blocks (12"x30") | | | | |
| Solid | | | | |
| 2" | S.F. | 0.84 | 1.35 | 2.19 |
| Hollow | | | | |
| 3" | S.F. | 0.85 | 1.35 | 2.20 |
| 4" | " | 0.96 | 1.45 | 2.41 |
| 6" | " | 1.05 | 1.55 | 2.60 |

# 04 MASONRY

| UNIT MASONRY | UNIT | MAT. | INST. | TOTAL |
|---|---|---|---|---|
| **04220.10 CONCRETE MASONRY UNITS** | | | | |
| Vertical reinforcing | | | | |
| 4' o.c., add 5% to labor | | | | |
| 2'8" o.c., add 15% to labor | | | | |
| Interior partitions, add 10% to labor | | | | |
| **04220.90 BOND BEAMS & LINTELS** | | | | |
| Bond beam, no grout or reinforcement | | | | |
| 8" x 16" x | | | | |
| 4" thick | L.F. | 1.85 | 2.65 | 4.50 |
| 6" thick | " | 2.25 | 2.75 | 5.00 |
| 8" thick | " | 2.75 | 2.85 | 5.60 |
| 10" thick | " | 2.90 | 3.00 | 5.90 |
| 12" thick | " | 2.90 | 3.10 | 6.00 |
| Beam lintel, no grout or reinforcement | | | | |
| 8" x 16" x | | | | |
| 10" thick | L.F. | 5.10 | 3.45 | 8.55 |
| 12" thick | " | 6.50 | 3.80 | 10.30 |
| Precast masonry lintel | | | | |
| 6 lf, 8" high x | | | | |
| 4" thick | L.F. | 4.85 | 5.70 | 10.55 |
| 6" thick | " | 6.20 | 5.70 | 11.90 |
| 8" thick | " | 7.00 | 6.25 | 13.25 |
| 10" thick | " | 8.35 | 6.25 | 14.60 |
| 10 lf, 8" high x | | | | |
| 4" thick | L.F. | 6.10 | 3.45 | 9.55 |
| 6" thick | " | 7.50 | 3.45 | 10.95 |
| 8" thick | " | 8.35 | 3.80 | 12.15 |
| 10" thick | " | 11.35 | 3.80 | 15.15 |
| Steel angles and plates | | | | |
| Minimum | Lb. | 0.45 | 0.49 | 0.94 |
| Maximum | " | 0.61 | 0.86 | 1.47 |
| Various size angle lintels | | | | |
| 1/4" stock | | | | |
| 3" x 3" | L.F. | 3.10 | 2.15 | 5.25 |
| 3" x 3-1/2" | " | 3.25 | 2.15 | 5.40 |
| 3/8" stock | | | | |
| 3" x 4" | L.F. | 5.00 | 2.15 | 7.15 |
| 3-1/2" x 4" | " | 5.25 | 2.15 | 7.40 |
| 4" x 4" | " | 5.65 | 2.15 | 7.80 |
| 5" x 3-1/2" | " | 5.95 | 2.15 | 8.10 |
| 6" x 3-1/2" | " | 6.10 | 2.15 | 8.25 |
| 1/2" stock | | | | |
| 6" x 4" | L.F. | 7.55 | 2.15 | 9.70 |
| **04240.10 CLAY TILE** | | | | |
| Hollow clay tile, for back-up, 12" x 12" | | | | |
| Scored face | | | | |
| Load bearing | | | | |
| 4" thick | S.F. | 4.20 | 2.45 | 6.65 |
| 6" thick | " | 5.15 | 2.55 | 7.70 |
| 8" thick | " | 6.20 | 2.65 | 8.85 |

| UNIT MASONRY | UNIT | MAT. | INST. | TOTAL |
|---|---|---|---|---|
| **04240.10 CLAY TILE** | | | | |
| 10" thick | S.F. | 6.90 | 2.75 | 9.65 |
| 12" thick | " | 8.10 | 2.85 | 10.95 |
| Non-load bearing | | | | |
| 3" thick | S.F. | 3.90 | 2.35 | 6.25 |
| 4" thick | " | 4.05 | 2.45 | 6.50 |
| 6" thick | " | 4.20 | 2.55 | 6.75 |
| 8" thick | " | 5.40 | 2.65 | 8.05 |
| 12" thick | " | 5.95 | 2.85 | 8.80 |
| Partition, 12" x 12" | | | | |
| In walls | | | | |
| 3" thick | S.F. | 3.35 | 2.85 | 6.20 |
| 4" thick | " | 3.90 | 2.85 | 6.75 |
| 6" thick | " | 4.30 | 3.00 | 7.30 |
| 8" thick | " | 5.65 | 3.10 | 8.75 |
| 10" thick | " | 6.75 | 3.25 | 10.00 |
| 12" thick | " | 7.15 | 3.45 | 10.60 |
| Clay tile floors | | | | |
| 4" thick | S.F. | 4.20 | 1.90 | 6.10 |
| 6" thick | " | 5.15 | 2.00 | 7.15 |
| 8" thick | " | 6.20 | 2.15 | 8.35 |
| 10" thick | " | 6.90 | 2.30 | 9.20 |
| 12" thick | " | 8.10 | 2.45 | 10.55 |
| Terra cotta | | | | |
| Coping, 10" or 12" wide, 3" thick | L.F. | 13.50 | 6.85 | 20.35 |
| **04270.10 GLASS BLOCK** | | | | |
| Glass block, 4" thick | | | | |
| 6" x 6" | S.F. | 29.70 | 11.45 | 41.15 |
| 8" x 8" | " | 25.70 | 8.60 | 34.30 |
| 12" x 12" | " | 25.70 | 6.85 | 32.55 |
| Replacement glass blocks, 4" x 8" x 8" | | | | |
| Minimum | S.F. | 31.20 | 34.30 | 65.50 |
| Maximum | " | 36.90 | 68.50 | 105.40 |
| **04295.10 PARGING/MASONRY PLASTER** | | | | |
| Parging | | | | |
| 1/2" thick | S.F. | 0.54 | 2.30 | 2.84 |
| 3/4" thick | " | 0.68 | 2.85 | 3.53 |
| 1" thick | " | 0.95 | 3.45 | 4.40 |

# 04 MASONRY

## STONE

| 04400.10 STONE | UNIT | MAT. | INST. | TOTAL |
|---|---|---|---|---|
| Rubble stone | | | | |
| Walls set in mortar | | | | |
| 8" thick | S.F. | 11.05 | 8.60 | 19.65 |
| 12" thick | " | 13.35 | 13.75 | 27.10 |
| 18" thick | " | 17.70 | 17.15 | 34.85 |
| 24" thick | " | 22.15 | 22.90 | 45.05 |
| Dry set wall | | | | |
| 8" thick | S.F. | 12.45 | 5.70 | 18.15 |
| 12" thick | " | 14.05 | 8.60 | 22.65 |
| 18" thick | " | 19.40 | 11.45 | 30.85 |
| 24" thick | " | 23.65 | 13.75 | 37.40 |
| Cut stone | | | | |
| Imported marble | | | | |
| Facing panels | | | | |
| 3/4" thick | S.F. | 29.50 | 13.75 | 43.25 |
| 1-1/2" thick | " | 44.20 | 15.60 | 59.80 |
| 2-1/4" thick | " | 51.50 | 19.05 | 70.55 |
| Base | | | | |
| 1" thick | | | | |
| 4" high | L.F. | 13.35 | 17.15 | 30.50 |
| 6" high | " | 16.35 | 17.15 | 33.50 |
| Columns, solid | | | | |
| Plain faced | C.F. | 88.00 | 230.00 | 318.00 |
| Fluted | " | 240.00 | 230.00 | 470.00 |
| Flooring, travertine, minimum | S.F. | 12.15 | 5.30 | 17.45 |
| Average | " | 16.20 | 6.85 | 23.05 |
| Maximum | " | 24.30 | 7.65 | 31.95 |
| Domestic marble | | | | |
| Facing panels | | | | |
| 7/8" thick | S.F. | 26.70 | 13.75 | 40.45 |
| 1-1/2" thick | " | 40.00 | 15.60 | 55.60 |
| 2-1/4" thick | " | 48.50 | 19.05 | 67.55 |
| Stairs | | | | |
| 12" treads | L.F. | 23.90 | 17.15 | 41.05 |
| 6" risers | " | 17.70 | 11.45 | 29.15 |
| Thresholds, 7/8" thick, 3' long, 4" to 6" wide | | | | |
| Plain | EA. | 22.75 | 28.60 | 51.35 |
| Beveled | " | 25.20 | 28.60 | 53.80 |
| Window sill | | | | |
| 6" wide, 2" thick | L.F. | 11.45 | 13.75 | 25.20 |
| Stools | | | | |
| 5" wide, 7/8" thick | L.F. | 16.85 | 13.75 | 30.60 |
| Limestone panels up to 12' x 5', smooth finish | | | | |
| 2" thick | S.F. | 16.20 | 5.75 | 21.95 |
| 3" thick | " | 18.90 | 5.75 | 24.65 |
| 4" thick | " | 27.00 | 5.75 | 32.75 |
| Miscellaneous limestone items | | | | |
| Steps, 14" wide, 6" deep | L.F. | 57.00 | 22.90 | 79.90 |
| Coping, smooth finish | C.F. | 78.50 | 11.45 | 89.95 |
| Sills, lintels, jambs, smooth finish | " | 78.50 | 13.75 | 92.25 |
| Granite veneer facing panels, polished | | | | |
| 7/8" thick | | | | |

# 04 MASONRY

| STONE | UNIT | MAT. | INST. | TOTAL |
|---|---|---|---|---|
| **04400.10 STONE** | | | | |
| Black | S.F. | 31.60 | 13.75 | 45.35 |
| Gray | " | 27.50 | 13.75 | 41.25 |
| Base | | | | |
| 4" high | L.F. | 14.75 | 6.85 | 21.60 |
| 6" high | " | 17.70 | 7.65 | 25.35 |
| Curbing, straight, 6" x 16" | " | 16.30 | 24.00 | 40.30 |
| Radius curbs, radius over 5' | " | 19.95 | 32.00 | 51.95 |
| Ashlar veneer | | | | |
| 4" thick, random | S.F. | 27.00 | 13.75 | 40.75 |
| Pavers, 4" x 4" split | | | | |
| Gray | S.F. | 26.60 | 6.85 | 33.45 |
| Pink | " | 26.00 | 6.85 | 32.85 |
| Black | " | 25.70 | 6.85 | 32.55 |
| Slate, panels | | | | |
| 1" thick | S.F. | 18.90 | 13.75 | 32.65 |
| 2" thick | " | 25.70 | 15.60 | 41.30 |
| Sills or stools | | | | |
| 1" thick | | | | |
| 6" wide | L.F. | 8.80 | 13.75 | 22.55 |
| 10" wide | " | 14.30 | 14.90 | 29.20 |
| 2" thick | | | | |
| 6" wide | L.F. | 14.40 | 15.60 | 30.00 |
| 10" wide | " | 23.90 | 17.15 | 41.05 |

| MASONRY RESTORATION | UNIT | MAT. | INST. | TOTAL |
|---|---|---|---|---|
| **04520.10 RESTORATION AND CLEANING** | | | | |
| Masonry cleaning | | | | |
| Washing brick | | | | |
| Smooth surface | S.F. | 0.35 | 0.57 | 0.92 |
| Rough surface | " | 0.47 | 0.76 | 1.23 |
| Steam clean masonry | | | | |
| Smooth face | | | | |
| Minimum | S.F. | 0.00 | 0.44 | 0.44 |
| Maximum | " | 0.00 | 0.64 | 0.64 |
| Rough face | | | | |
| Minimum | S.F. | 0.00 | 0.59 | 0.59 |
| Maximum | " | 0.00 | 0.88 | 0.88 |
| Sandblast masonry | | | | |
| Minimum | S.F. | 0.42 | 0.70 | 1.13 |
| Maximum | " | 0.63 | 1.20 | 1.83 |
| Pointing masonry | | | | |
| Brick | S.F. | 1.10 | 1.35 | 2.45 |

# 04 MASONRY

## MASONRY RESTORATION

| 04520.10 RESTORATION AND CLEANING | UNIT | MAT. | INST. | TOTAL |
|---|---|---|---|---|
| Concrete block | S.F. | 0.49 | 0.98 | 1.47 |
| Cut and repoint | | | | |
| Brick | | | | |
| Minimum | S.F. | 0.30 | 1.70 | 2.00 |
| Maximum | " | 0.51 | 3.45 | 3.96 |
| Stone work | L.F. | 1.40 | 2.65 | 4.05 |
| Cut and recaulk | | | | |
| Oil base caulks | L.F. | 1.05 | 2.30 | 3.35 |
| Butyl caulks | " | 0.93 | 2.30 | 3.23 |
| Polysulfides and acrylics | " | 1.80 | 2.30 | 4.10 |
| Silicones | " | 2.10 | 2.30 | 4.40 |
| Cement and sand grout on walls, to 1/8" thick | | | | |
| Minimum | S.F. | 0.53 | 1.35 | 1.88 |
| Maximum | " | 1.30 | 1.70 | 3.00 |
| Brick removal and replacement | | | | |
| Minimum | EA. | 0.54 | 4.30 | 4.84 |
| Average | " | 0.68 | 5.70 | 6.38 |
| Maximum | " | 1.35 | 17.15 | 18.50 |

| 04550.10 REFRACTORIES | UNIT | MAT. | INST. | TOTAL |
|---|---|---|---|---|
| Flue liners | | | | |
| Rectangular | | | | |
| 8" x 12" | L.F. | 5.55 | 5.70 | 11.25 |
| 12" x 12" | " | 6.95 | 6.25 | 13.20 |
| 12" x 18" | " | 12.50 | 6.85 | 19.35 |
| 16" x 16" | " | 13.90 | 7.65 | 21.55 |
| 18" x 18" | " | 16.70 | 8.15 | 24.85 |
| 20" x 20" | " | 27.80 | 8.60 | 36.40 |
| 24" x 24" | " | 33.40 | 9.80 | 43.20 |
| Round | | | | |
| 18" dia. | L.F. | 27.80 | 8.15 | 35.95 |
| 24" dia. | " | 38.90 | 9.80 | 48.70 |

## METAL FASTENING

| | UNIT | MAT. | INST. | TOTAL |
|---|---|---|---|---|
| **05050.10 STRUCTURAL WELDING** | | | | |
| Welding | | | | |
| Single pass | | | | |
| 1/8" | L.F. | 0.49 | 1.95 | 2.44 |
| 3/16" | " | 0.92 | 2.55 | 3.47 |
| 1/4" | " | 1.25 | 3.20 | 4.45 |
| Miscellaneous steel shapes | | | | |
| Plain | Lb. | 0.92 | 0.08 | 1.00 |
| Galvanized | " | 1.40 | 0.13 | 1.53 |
| Plates | | | | |
| Plain | Lb. | 0.85 | 0.10 | 0.95 |
| Galvanized | " | 1.30 | 0.15 | 1.45 |
| **05050.90 METAL ANCHORS** | | | | |
| Anchor bolts | | | | |
| 3/8" x | | | | |
| 8" long | EA. | | | 2.70 |
| 10" long | " | | | 2.97 |
| 12" long | " | | | 3.24 |
| 1/2" x | | | | |
| 8" long | EA. | | | 4.05 |
| 10" long | " | | | 4.32 |
| 12" long | " | | | 4.72 |
| 18" long | " | | | 5.13 |
| 5/8" x | | | | |
| 8" long | EA. | | | 3.78 |
| 10" long | " | | | 4.18 |
| 12" long | " | | | 4.46 |
| 18" long | " | | | 4.72 |
| 24" long | " | | | 5.13 |
| 3/4" x | | | | |
| 8" long | EA. | | | 5.40 |
| 12" long | " | | | 6.08 |
| 18" long | " | | | 8.37 |
| 24" long | " | | | 11.07 |
| 7/8" x | | | | |
| 8" long | EA. | | | 5.40 |
| 12" long | " | | | 6.08 |
| 18" long | " | | | 8.37 |
| 24" long | " | | | 12.15 |
| 1" x | | | | |
| 12" long | EA. | | | 10.80 |
| 18" long | " | | | 13.50 |
| 24" long | " | | | 16.20 |
| 36" long | " | | | 24.30 |
| Expansion shield | | | | |
| 1/4" | EA. | | | 0.81 |
| 3/8" | " | | | 1.35 |
| 1/2" | " | | | 2.70 |
| 5/8" | " | | | 3.92 |
| 3/4" | " | | | 4.72 |
| 1" | " | | | 6.48 |

# 05 METALS

## METAL FASTENING

| METAL FASTENING | UNIT | MAT. | INST. | TOTAL |
|---|---|---|---|---|
| **05050.90 METAL ANCHORS** | | | | |
| Non-drilling anchor | | | | |
| 1/4" | EA. | | | 0.50 |
| 3/8" | " | | | 0.64 |
| 1/2" | " | | | 0.97 |
| 5/8" | " | | | 1.61 |
| 3/4" | " | | | 2.74 |
| Self-drilling anchor | | | | |
| 1/4" | EA. | | | 1.35 |
| 5/16" | " | | | 1.69 |
| 3/8" | " | | | 2.02 |
| 1/2" | " | | | 2.70 |
| 5/8" | " | | | 5.13 |
| 3/4" | " | | | 6.75 |
| 7/8" | " | | | 9.45 |
| Add 25% for galvanized anchor bolts | | | | |
| Channel door frame, with anchors | Lb. | 1.35 | 0.43 | 1.78 |
| Corner guard angle, with anchors | " | 1.20 | 0.64 | 1.84 |
| **05050.95 METAL LINTELS** | | | | |
| Lintels, steel | | | | |
| Plain | Lb. | 0.90 | 0.96 | 1.86 |
| Galvanized | " | 1.30 | 0.96 | 2.26 |
| **05300.10 METAL DECKING** | | | | |
| Roof, 1-1/2" deep, non-composite | | | | |
| 16 ga. | | | | |
| Primed | S.F. | 2.45 | 0.70 | 3.15 |
| Galvanized | " | 2.55 | 0.70 | 3.25 |
| Open type decking, galvanized | | | | |
| 1-1/2" deep | | | | |
| 18 ga. | S.F. | 1.95 | 0.70 | 2.65 |
| 20 ga. | " | 1.50 | 0.70 | 2.20 |
| 3" deep | | | | |
| 16 ga. | S.F. | 2.95 | 0.77 | 3.72 |
| 18 ga. | " | 2.85 | 0.77 | 3.62 |
| 20 ga. | " | 2.15 | 0.77 | 2.92 |
| Cellular type | | | | |
| 1-1/2" deep, galvanized | | | | |
| 18-18 ga. | S.F. | 4.85 | 0.84 | 5.70 |
| 22-18 ga. | " | 4.45 | 0.84 | 5.29 |
| 3" deep, galvanized | | | | |
| 16-16 ga. | S.F. | 6.95 | 0.94 | 7.89 |
| 18-16 ga. | " | 6.55 | 0.94 | 7.49 |
| 18-18 ga. | " | 6.00 | 0.94 | 6.94 |
| 20-18 ga. | " | 5.30 | 0.94 | 6.24 |
| Composite deck, non-cellular, galvanized | | | | |
| 1-1/2" deep | | | | |
| 18 ga. | S.F. | 1.60 | 0.77 | 2.37 |
| 20 ga. | " | 1.50 | 0.77 | 2.27 |
| 3" deep | | | | |

# 05 METALS

## METAL FASTENING

| 05300.10 METAL DECKING | UNIT | MAT. | INST. | TOTAL |
|---|---|---|---|---|
| 18 ga. | S.F. | 2.30 | 0.81 | 3.11 |
| 20 ga. | " | 1.50 | 0.81 | 2.31 |

## COLD FORMED FRAMING

| 05410.10 METAL FRAMING | UNIT | MAT. | INST. | TOTAL |
|---|---|---|---|---|
| Furring channel, galvanized | | | | |
| Beams and columns, 3/4" | | | | |
| 12" o.c. | S.F. | 0.41 | 3.85 | 4.26 |
| 16" o.c. | " | 0.34 | 3.50 | 3.84 |
| Walls, 3/4" | | | | |
| 12" o.c. | S.F. | 0.39 | 1.95 | 2.34 |
| 16" o.c. | " | 0.32 | 1.60 | 1.92 |
| 24" o.c. | " | 0.23 | 1.30 | 1.53 |
| 1-1/2" | | | | |
| 12" o.c. | S.F. | 0.51 | 1.95 | 2.46 |
| 16" o.c. | " | 0.41 | 1.60 | 2.01 |
| 24" o.c. | " | 0.32 | 1.30 | 1.62 |
| Stud, load bearing | | | | |
| 16" o.c. | | | | |
| 16 ga. | | | | |
| 2-1/2" | S.F. | 1.10 | 1.70 | 2.80 |
| 3-5/8" | " | 1.40 | 1.70 | 3.10 |
| 4" | " | 1.55 | 1.70 | 3.25 |
| 6" | " | 2.20 | 1.95 | 4.15 |
| 18 ga. | | | | |
| 2-1/2" | S.F. | 0.95 | 1.70 | 2.65 |
| 3-5/8" | " | 1.25 | 1.70 | 2.95 |
| 4" | " | 1.40 | 1.70 | 3.10 |
| 6" | " | 1.75 | 1.95 | 3.70 |
| 8" | " | 1.95 | 1.95 | 3.90 |
| 20 ga. | | | | |
| 2-1/2" | S.F. | 0.80 | 1.70 | 2.50 |
| 3 5/8" | " | 0.95 | 1.70 | 2.65 |
| 4" | " | 1.10 | 1.70 | 2.80 |
| 6" | " | 1.25 | 1.95 | 3.20 |
| 8" | " | 1.40 | 1.95 | 3.35 |
| 24" o.c. | | | | |
| 16 ga. | | | | |
| 2-1/2" | S.F. | 0.88 | 1.50 | 2.38 |
| 3-5/8" | " | 1.15 | 1.50 | 2.65 |
| 4" | " | 1.30 | 1.50 | 2.80 |
| 6" | " | 1.90 | 1.60 | 3.50 |
| 8" | " | 2.05 | 1.60 | 3.65 |

## COLD FORMED FRAMING

| 05410.10 METAL FRAMING | UNIT | MAT. | INST. | TOTAL |
|---|---|---|---|---|
| 18 ga. | | | | |
| 2-1/2" | S.F. | 0.73 | 1.50 | 2.23 |
| 3-5/8" | " | 1.05 | 1.50 | 2.55 |
| 4" | " | 1.15 | 1.50 | 2.65 |
| 6" | " | 1.30 | 1.60 | 2.90 |
| 8" | " | 1.45 | 1.60 | 3.05 |
| 20 ga. | | | | |
| 2-1/2" | S.F. | 0.66 | 1.50 | 2.16 |
| 3-5/8" | " | 0.85 | 1.50 | 2.35 |
| 4" | " | 0.99 | 1.50 | 2.49 |
| 6" | " | 1.10 | 1.60 | 2.70 |
| 8" | " | 1.30 | 1.60 | 2.90 |

## METAL FABRICATIONS

| 05510.10 STAIRS | UNIT | MAT. | INST. | TOTAL |
|---|---|---|---|---|
| Stock unit, steel, complete, per riser | | | | |
| Tread | | | | |
| 3'-6" wide | EA. | 98.50 | 48.20 | 146.70 |
| 4' wide | " | 110.00 | 55.00 | 165.00 |
| 5' wide | " | 130.00 | 64.50 | 194.50 |
| Metal pan stair, cement filled, per riser | | | | |
| 3'-6" wide | EA. | 110.00 | 38.60 | 148.60 |
| 4' wide | " | 120.00 | 42.80 | 162.80 |
| 5' wide | " | 140.00 | 48.20 | 188.20 |
| Landing, steel pan | S.F. | 42.10 | 9.65 | 51.75 |
| Cast iron tread, steel stringers, stock units, per riser | | | | |
| Tread | | | | |
| 3'-6" wide | EA. | 190.00 | 48.20 | 238.20 |
| 4' wide | " | 220.00 | 55.00 | 275.00 |
| 5' wide | " | 260.00 | 64.50 | 324.50 |
| Stair treads, abrasive, 12" x 3'-6" | | | | |
| Cast iron | | | | |
| 3/8" | EA. | 100.00 | 19.30 | 119.30 |
| 1/2" | " | 130.00 | 19.30 | 149.30 |
| Cast aluminum | | | | |
| 5/16" | EA. | 120.00 | 19.30 | 139.30 |
| 3/8" | " | 130.00 | 19.30 | 149.30 |
| 1/2" | " | 160.00 | 19.30 | 179.30 |

| 05515.10 LADDERS | UNIT | MAT. | INST. | TOTAL |
|---|---|---|---|---|
| Ladder, 18" wide | | | | |
| With cage | L.F. | 74.50 | 25.70 | 100.20 |

# METAL FABRICATIONS

| METAL FABRICATIONS | UNIT | MAT. | INST. | TOTAL |
|---|---|---|---|---|
| **05515.10 LADDERS** | | | | |
| Without cage | L.F. | 47.30 | 19.30 | 66.60 |
| **05520.10 RAILINGS** | | | | |
| Railing, pipe | | | | |
| 1-1/4" diameter, welded steel | | | | |
| 2-rail | | | | |
| Primed | L.F. | 14.85 | 7.70 | 22.55 |
| Galvanized | " | 18.90 | 7.70 | 26.60 |
| 3-rail | | | | |
| Primed | L.F. | 18.90 | 9.65 | 28.55 |
| Galvanized | " | 24.30 | 9.65 | 33.95 |
| Wall mounted, single rail, welded steel | | | | |
| Primed | L.F. | 6.30 | 5.95 | 12.25 |
| Galvanized | " | 9.45 | 5.95 | 15.40 |
| 1-1/2" diameter, welded steel | | | | |
| 2-rail | | | | |
| Primed | L.F. | 16.20 | 7.70 | 23.90 |
| Galvanized | " | 20.25 | 7.70 | 27.95 |
| 3-rail | | | | |
| Primed | L.F. | 20.25 | 9.65 | 29.90 |
| Galvanized | " | 25.70 | 9.65 | 35.35 |
| Wall mounted, single rail, welded steel | | | | |
| Primed | L.F. | 7.40 | 5.95 | 13.35 |
| Galvanized | " | 10.10 | 5.95 | 16.05 |
| 2" diameter, welded steel | | | | |
| 2-rail | | | | |
| Primed | L.F. | 20.25 | 8.55 | 28.80 |
| Galvanized | " | 25.70 | 8.55 | 34.25 |
| 3-rall | | | | |
| Primed | L.F. | 25.70 | 11.00 | 36.70 |
| Galvanized | " | 31.10 | 11.00 | 42.10 |
| Wall mounted, single rail, welded steel | | | | |
| Primed | L.F. | 8.10 | 6.45 | 14.55 |
| Galvanized | " | 10.10 | 6.45 | 16.55 |
| **05530.10 METAL GRATING** | | | | |
| Floor plate, checkered, steel | | | | |
| 1/4" | | | | |
| Primed | S.F. | 6.30 | 0.55 | 6.85 |
| Galvanized | " | 8.40 | 0.55 | 8.95 |
| 3/8" | | | | |
| Primed | S.F. | 8.40 | 0.59 | 8.99 |
| Galvanized | " | 11.25 | 0.59 | 11.84 |
| Aluminum grating, pressure-locked bearing bars | | | | |
| 3/4" x 1/8" | S.F. | 13.50 | 0.96 | 14.46 |
| 1" x 1/8" | " | 14.85 | 0.96 | 15.81 |
| 1-1/4" x 1/8" | " | 18.90 | 0.96 | 19.86 |
| 1-1/4" x 3/16" | " | 20.25 | 0.96 | 21.21 |
| 1-1/2" x 1/8" | " | 21.60 | 0.96 | 22.56 |

# 05 METALS

## METAL FABRICATIONS

| 05530.10 METAL GRATING | UNIT | MAT. | INST. | TOTAL |
|---|---|---|---|---|
| 1-3/4" x 3/16" | S.F. | 24.30 | 0.96 | 25.26 |
| Miscellaneous expenses | | | | |
| Cutting | | | | |
| Minimum | L.F. | 0.00 | 2.55 | 2.55 |
| Maximum | " | 0.00 | 3.85 | 3.85 |
| Banding | | | | |
| Minimum | L.F. | 0.00 | 6.45 | 6.45 |
| Maximum | " | 0.00 | 7.70 | 7.70 |
| Toe plates | | | | |
| Minimum | L.F. | 0.00 | 7.70 | 7.70 |
| Maximum | " | 0.00 | 9.65 | 9.65 |
| Steel grating, primed | | | | |
| 3/4" x 1/8" | S.F. | 6.00 | 1.30 | 7.30 |
| 1" x 1/8" | " | 6.20 | 1.30 | 7.50 |
| 1-1/4" x 1/8" | " | 6.60 | 1.30 | 7.90 |
| 1-1/4" x 3/16" | " | 8.10 | 1.30 | 9.40 |
| 1-1/2" x 1/8" | " | 7.40 | 1.30 | 8.70 |
| 1-3/4" x 3/16" | " | 10.55 | 1.30 | 11.85 |
| Galvanized | | | | |
| 3/4" x 1/8" | S.F. | 7.40 | 1.30 | 8.70 |
| 1" x 1/8" | " | 7.85 | 1.30 | 9.15 |
| 1-1/4" x 1/8" | " | 8.55 | 1.30 | 9.85 |
| 1-1/4" x 3/16" | " | 10.20 | 1.30 | 11.50 |
| 1-1/2" x 1/8" | " | 9.15 | 1.30 | 10.45 |
| 1-3/4" x 3/16" | " | 13.35 | 1.30 | 14.65 |
| Miscellaneous expenses | | | | |
| Cutting | | | | |
| Minimum | L.F. | 0.00 | 2.75 | 2.75 |
| Maximum | " | 0.00 | 4.30 | 4.30 |
| Banding | | | | |
| Minimum | L.F. | 0.00 | 7.00 | 7.00 |
| Maximum | " | 0.00 | 8.55 | 8.55 |
| Toe plates | | | | |
| Minimum | L.F. | 0.00 | 8.55 | 8.55 |
| Maximum | " | 0.00 | 11.00 | 11.00 |
| **05540.10 CASTINGS** | | | | |
| Miscellaneous castings | | | | |
| Light sections | Lb. | 1.30 | 0.77 | 2.07 |
| Heavy sections | " | 1.10 | 0.55 | 1.65 |

## MISC. FABRICATIONS

| MISC. FABRICATIONS | UNIT | MAT. | INST. | TOTAL |
|---|---|---|---|---|
| **05580.10 METAL SPECIALTIES** | | | | |
| Kick plate | | | | |
| 4" high x 1/4" thick | | | | |
| Primed | L.F. | 5.65 | 7.70 | 13.35 |
| Galvanized | " | 6.40 | 7.70 | 14.10 |
| 6" high x 1/4" thick | | | | |
| Primed | L.F. | 6.15 | 8.55 | 14.70 |
| Galvanized | " | 7.30 | 8.55 | 15.85 |
| **05700.10 ORNAMENTAL METAL** | | | | |
| Railings, vertical square bars, 6" o.c., with shaped top rails | | | | |
| Steel | L.F. | 61.00 | 19.30 | 80.30 |
| Aluminum | " | 83.00 | 19.30 | 102.30 |
| Bronze | " | 140.00 | 25.70 | 165.70 |
| Stainless steel | " | 120.00 | 25.70 | 145.70 |
| Laminated metal or wood handrails with metal supports | | | | |
| 2-1/2" round or oval shape | L.F. | 110.00 | 19.30 | 129.30 |
| Grilles and louvers | | | | |
| Fixed type louvers | | | | |
| 4 through 10 sf | S.F. | 21.60 | 6.45 | 28.05 |
| Over 10 sf | " | 25.70 | 4.80 | 30.50 |
| Movable type louvers | | | | |
| 4 through 10 sf | S.F. | 25.70 | 6.45 | 32.15 |
| Over 10 sf | " | 28.40 | 4.80 | 33.20 |
| Aluminum louvers | | | | |
| Residential use, fixed type, with screen | | | | |
| 8" x 8" | EA. | 13.50 | 19.30 | 32.80 |
| 12" x 12" | " | 14.90 | 19.30 | 34.20 |
| 12" x 18" | " | 17.85 | 19.30 | 37.15 |
| 14" x 24" | " | 25.60 | 19.30 | 44.90 |
| 18" x 24" | " | 28.80 | 19.30 | 48.10 |
| 30" x 24" | " | 39.30 | 21.40 | 60.70 |
| **05800.10 EXPANSION CONTROL** | | | | |
| Expansion joints with covers, floor assembly type | | | | |
| With 1" space | | | | |
| Aluminum | L.F. | 19.50 | 6.45 | 25.95 |
| Bronze | " | 40.70 | 6.45 | 47.15 |
| Stainless steel | " | 31.60 | 6.45 | 38.05 |
| Aluminum | " | 24.15 | 6.45 | 30.60 |
| Ceiling and wall assembly type | | | | |
| With 1" space | | | | |
| Aluminum | L.F. | 12.50 | 7.70 | 20.20 |
| Bronze | " | 42.80 | 7.70 | 50.50 |
| Stainless steel | " | 38.60 | 7.70 | 46.30 |
| Exterior roof and wall, aluminum | | | | |
| Roof to roof | | | | |
| With 1" space | L.F. | 36.50 | 6.45 | 42.95 |
| With 2" space | " | 39.50 | 6.45 | 45.95 |
| Roof to wall | | | | |
| With 1" space | L.F. | 28.40 | 7.00 | 35.40 |
| With 2" space | " | 33.70 | 7.00 | 40.70 |

# 05 METALS

## MISC. FABRICATIONS

| 05800.10 EXPANSION CONTROL | UNIT | MAT. | INST. | TOTAL |
|---|---|---|---|---|
| Flat wall to wall | | | | |
| With 1" space | L.F. | 13.75 | 6.45 | 20.20 |
| With 2" space | " | 15.05 | 6.45 | 21.50 |
| Corner to flat wall | | | | |
| With 1" space | L.F. | 16.60 | 7.70 | 24.30 |
| With 2" in space | " | 16.85 | 7.70 | 24.55 |

# 06 WOOD AND PLASTICS

## FASTENERS AND ADHESIVES

| 06050.10 ACCESSORIES | UNIT | MAT. | INST. | TOTAL |
|---|---|---|---|---|
| Column/post base, cast aluminum | | | | |
| 4" x 4" | EA. | 2.40 | 8.85 | 11.25 |
| 6" x 6" | " | 5.25 | 8.85 | 14.10 |
| Bridging, metal, per pair | | | | |
| 12" o.c. | EA. | 0.88 | 3.55 | 4.43 |
| 16" o.c. | " | 0.88 | 3.20 | 4.08 |
| Anchors | | | | |
| Bolts, threaded two ends, with nuts and washers | | | | |
| 1/2" dia. | | | | |
| 4" long | EA. | 1.15 | 2.20 | 3.35 |
| 7-1/2" long | " | 1.45 | 2.20 | 3.65 |
| 3/4" dia. | | | | |
| 7-1/2" long | EA. | 2.25 | 2.20 | 4.45 |
| 15" long | " | 2.95 | 2.20 | 5.15 |
| Framing anchors | | | | |
| 10 gauge | EA. | 0.63 | 2.95 | 3.58 |
| Bolts, carriage | | | | |
| 1/4 x 4 | EA. | 1.20 | 3.55 | 4.75 |
| 5/16 x 6 | " | 1.25 | 3.75 | 5.00 |
| 3/8 x 6 | " | 1.45 | 3.75 | 5.20 |
| 1/2 x 6 | " | 1.60 | 3.75 | 5.35 |
| Joist and beam hangers | | | | |
| 18 ga. | | | | |
| 2 x 4 | EA. | 0.58 | 3.55 | 4.13 |
| 2 x 6 | " | 0.74 | 3.55 | 4.29 |
| 2 x 8 | " | 0.88 | 3.55 | 4.43 |
| 2 x 10 | " | 1.00 | 3.95 | 4.95 |
| 2 x 12 | " | 1.15 | 4.45 | 5.60 |
| 16 ga. | | | | |
| 3 x 6 | EA. | 2.10 | 3.95 | 6.05 |
| 3 x 8 | " | 2.15 | 3.95 | 6.10 |
| 3 x 10 | " | 2.50 | 4.15 | 6.65 |
| 3 x 12 | " | 2.70 | 4.75 | 7.45 |
| 3 x 14 | " | 2.90 | 5.05 | 7.95 |
| 4 x 6 | " | 2.35 | 3.95 | 6.30 |
| 4 x 8 | " | 2.70 | 3.95 | 6.65 |
| 4 x 10 | " | 3.40 | 4.15 | 7.55 |
| 4 x 12 | " | 3.80 | 4.75 | 8.55 |
| 4 x 14 | " | 4.15 | 5.05 | 9.20 |
| Rafter anchors, 18 ga., 1-1/2" wide | | | | |
| 5-1/4" long | EA. | 0.65 | 2.95 | 3.60 |
| 10-3/4" long | " | 0.99 | 2.95 | 3.94 |
| Shear plates | | | | |
| 2-5/8" dia. | EA. | 1.80 | 2.75 | 4.55 |
| 4" dia. | " | 3.80 | 2.95 | 6.75 |
| Sill anchors | | | | |
| Embedded in concrete | EA. | 1.60 | 3.55 | 5.15 |
| Split rings | | | | |
| 2-1/2" dia. | EA. | 1.25 | 3.95 | 5.20 |
| 4" dia. | " | 2.25 | 4.45 | 6.70 |
| Strap ties, 14 ga., 1-3/8" wide | | | | |
| 12" long | EA. | 1.30 | 2.95 | 4.25 |

## FASTENERS AND ADHESIVES

| 06050.10 ACCESSORIES | UNIT | MAT. | INST. | TOTAL |
|---|---|---|---|---|
| 18" long | EA. | 1.45 | 3.20 | 4.65 |
| 24" long | " | 1.85 | 3.55 | 5.40 |
| 36" long | " | 2.80 | 3.95 | 6.75 |
| Toothed rings | | | | |
| 2-5/8" dia. | EA. | 1.10 | 5.90 | 7.00 |
| 4" dia. | " | 1.25 | 7.10 | 8.35 |

## ROUGH CARPENTRY

| 06110.10 BLOCKING | UNIT | MAT. | INST. | TOTAL |
|---|---|---|---|---|
| Steel construction | | | | |
| Walls | | | | |
| 2x4 | L.F. | 0.53 | 2.35 | 2.88 |
| 2x6 | " | 0.74 | 2.75 | 3.49 |
| 2x8 | " | 1.05 | 2.95 | 4.00 |
| 2x10 | " | 1.45 | 3.20 | 4.65 |
| 2x12 | " | 1.95 | 3.55 | 5.50 |
| Ceilings | | | | |
| 2x4 | L.F. | 0.53 | 2.75 | 3.28 |
| 2x6 | " | 0.74 | 3.20 | 3.94 |
| 2x8 | " | 1.05 | 3.55 | 4.60 |
| 2x10 | " | 1.45 | 3.95 | 5.40 |
| 2x12 | " | 1.95 | 4.45 | 6.40 |
| Wood construction | | | | |
| Walls | | | | |
| 2x4 | L.F. | 0.53 | 1.95 | 2.48 |
| 2x6 | " | 0.74 | 2.20 | 2.94 |
| 2x8 | " | 1.05 | 2.35 | 3.40 |
| 2x10 | " | 1.45 | 2.55 | 4.00 |
| 2x12 | " | 1.95 | 2.75 | 4.70 |
| Ceilings | | | | |
| 2x4 | L.F. | 0.53 | 2.20 | 2.73 |
| 2x6 | " | 0.74 | 2.55 | 3.29 |
| 2x8 | " | 1.05 | 2.75 | 3.80 |
| 2x10 | " | 1.45 | 2.95 | 4.40 |
| 2x12 | " | 1.95 | 3.20 | 5.15 |

| 06110.20 CEILING FRAMING | UNIT | MAT. | INST. | TOTAL |
|---|---|---|---|---|
| Ceiling joists | | | | |
| 12" o.c. | | | | |
| 2x4 | S.F. | 0.63 | 0.84 | 1.47 |
| 2x6 | " | 0.93 | 0.89 | 1.82 |
| 2x8 | " | 1.30 | 0.93 | 2.23 |
| 2x10 | " | 2.00 | 0.98 | 2.98 |

## ROUGH CARPENTRY

### 06110.20 CEILING FRAMING

| | UNIT | MAT. | INST. | TOTAL |
|---|---|---|---|---|
| 2x12 | S.F. | 2.65 | 1.05 | 3.70 |
| 16" o.c. | | | | |
| 2x4 | S.F. | 0.53 | 0.68 | 1.21 |
| 2x6 | " | 0.80 | 0.71 | 1.51 |
| 2x8 | " | 1.10 | 0.74 | 1.84 |
| 2x10 | " | 1.65 | 0.77 | 2.42 |
| 2x12 | " | 2.15 | 0.81 | 2.96 |
| 24" o.c. | | | | |
| 2x4 | S.F. | 0.42 | 0.56 | 0.98 |
| 2x6 | " | 0.63 | 0.59 | 1.22 |
| 2x8 | " | 0.86 | 0.62 | 1.48 |
| 2x10 | " | 1.30 | 0.66 | 1.96 |
| 2x12 | " | 1.75 | 0.69 | 2.44 |
| Headers and nailers | | | | |
| 2x4 | L.F. | 0.53 | 1.15 | 1.68 |
| 2x6 | " | 0.74 | 1.20 | 1.94 |
| 2x8 | " | 1.05 | 1.25 | 2.30 |
| 2x10 | " | 1.45 | 1.35 | 2.80 |
| 2x12 | " | 1.95 | 1.50 | 3.45 |
| Sister joists for ceilings | | | | |
| 2x4 | L.F. | 0.53 | 2.55 | 3.08 |
| 2x6 | " | 0.74 | 2.95 | 3.69 |
| 2x8 | " | 1.05 | 3.55 | 4.60 |
| 2x10 | " | 1.45 | 4.45 | 5.90 |
| 2x12 | " | 1.95 | 5.90 | 7.85 |

### 06110.30 FLOOR FRAMING

| | UNIT | MAT. | INST. | TOTAL |
|---|---|---|---|---|
| Floor joists | | | | |
| 12" o.c. | | | | |
| 2x6 | S.F. | 0.93 | 0.71 | 1.64 |
| 2x8 | " | 1.30 | 0.72 | 2.02 |
| 2x10 | " | 2.00 | 0.74 | 2.74 |
| 2x12 | " | 2.65 | 0.77 | 3.42 |
| 2x14 | " | 2.90 | 0.74 | 3.64 |
| 3x6 | " | 2.10 | 0.75 | 2.85 |
| 3x8 | " | 2.80 | 0.77 | 3.57 |
| 3x10 | " | 3.50 | 0.81 | 4.30 |
| 3x12 | " | 4.15 | 0.84 | 4.99 |
| 3x14 | " | 4.80 | 0.89 | 5.69 |
| 4x6 | " | 2.80 | 0.74 | 3.54 |
| 4x8 | " | 3.60 | 0.77 | 4.37 |
| 4x10 | " | 4.60 | 0.81 | 5.41 |
| 4x12 | " | 5.60 | 0.84 | 6.44 |
| 4x14 | " | 6.50 | 0.89 | 7.39 |
| 16" o.c. | | | | |
| 2x6 | S.F. | 0.80 | 0.59 | 1.39 |
| 2x8 | " | 1.10 | 0.60 | 1.70 |
| 2x10 | " | 1.65 | 0.61 | 2.26 |
| 2x12 | " | 2.15 | 0.63 | 2.78 |
| 2x14 | " | 2.65 | 0.66 | 3.31 |
| 3x6 | " | 1.80 | 0.61 | 2.41 |

# 06 WOOD AND PLASTICS

## ROUGH CARPENTRY

| ROUGH CARPENTRY | UNIT | MAT. | INST. | TOTAL |
|---|---|---|---|---|
| **06110.30 FLOOR FRAMING** | | | | |
| 3x8 | S.F. | 2.35 | 0.63 | 2.98 |
| 3x10 | " | 2.90 | 0.66 | 3.56 |
| 3x12 | " | 3.45 | 0.68 | 4.13 |
| 3x14 | " | 4.10 | 0.71 | 4.81 |
| 4x6 | " | 2.30 | 0.61 | 2.91 |
| 4x8 | " | 3.10 | 0.63 | 3.73 |
| 4x10 | " | 3.85 | 0.66 | 4.51 |
| 4x12 | " | 4.60 | 0.68 | 5.28 |
| 4x14 | " | 5.45 | 0.71 | 6.16 |
| Sister joists for floors | | | | |
| 2x4 | L.F. | 0.53 | 2.20 | 2.73 |
| 2x6 | " | 0.74 | 2.55 | 3.29 |
| 2x8 | " | 1.05 | 2.95 | 4.00 |
| 2x10 | " | 1.45 | 3.55 | 5.00 |
| 2x12 | " | 1.95 | 4.45 | 6.40 |
| 3x6 | " | 1.80 | 3.55 | 5.35 |
| 3x8 | " | 2.35 | 3.95 | 6.30 |
| 3x10 | " | 2.90 | 4.45 | 7.35 |
| 3x12 | " | 3.45 | 5.05 | 8.50 |
| 4x6 | " | 2.25 | 3.55 | 5.80 |
| 4x8 | " | 3.10 | 3.95 | 7.05 |
| 4x10 | " | 3.85 | 4.45 | 8.30 |
| 4x12 | " | 4.60 | 5.05 | 9.65 |
| **06110.40 FURRING** | | | | |
| Furring, wood strips | | | | |
| Walls | | | | |
| On masonry or concrete walls | | | | |
| 1x2 furring | | | | |
| 12" o.c. | S.F. | 0.24 | 1.10 | 1.34 |
| 16" o.c. | " | 0.20 | 1.00 | 1.20 |
| 24" o.c. | " | 0.18 | 0.93 | 1.11 |
| 1x3 furring | | | | |
| 12" o.c. | S.F. | 0.35 | 1.10 | 1.45 |
| 16" o.c. | " | 0.31 | 1.00 | 1.31 |
| 24" o.c. | " | 0.24 | 0.93 | 1.17 |
| On wood walls | | | | |
| 1x2 furring | | | | |
| 12" o.c. | S.F. | 0.24 | 0.79 | 1.03 |
| 16" o.c. | " | 0.20 | 0.71 | 0.91 |
| 24" o.c. | " | 0.18 | 0.64 | 0.82 |
| 1x3 furring | | | | |
| 12" o.c. | S.F. | 0.35 | 0.79 | 1.14 |
| 16" o.c. | " | 0.31 | 0.71 | 1.02 |
| 24" o.c. | " | 0.24 | 0.64 | 0.88 |
| Ceilings | | | | |
| On masonry or concrete ceilings | | | | |
| 1x2 furring | | | | |
| 12" o.c. | S.F. | 0.24 | 1.95 | 2.19 |
| 16" o.c. | " | 0.20 | 1.75 | 1.95 |
| 24" o.c. | " | 0.18 | 1.60 | 1.78 |

## ROUGH CARPENTRY

| ROUGH CARPENTRY | UNIT | MAT. | INST. | TOTAL |
|---|---|---|---|---|
| **06110.40 FURRING** | | | | |
| 1x3 furring | | | | |
| 12" o.c. | S.F. | 0.35 | 1.95 | 2.30 |
| 16" o.c. | " | 0.31 | 1.75 | 2.06 |
| 24" o.c. | " | 0.24 | 1.60 | 1.84 |
| On wood ceilings | | | | |
| 1x2 furring | | | | |
| 12" o.c. | S.F. | 0.24 | 1.30 | 1.54 |
| 16" o.c. | " | 0.20 | 1.20 | 1.40 |
| 24" o.c. | " | 0.18 | 1.05 | 1.23 |
| 1x3 | | | | |
| 12" o.c. | S.F. | 0.35 | 1.30 | 1.65 |
| 16" o.c. | " | 0.31 | 1.20 | 1.51 |
| 24" o.c. | " | 0.24 | 1.05 | 1.29 |
| **06110.50 ROOF FRAMING** | | | | |
| Roof framing | | | | |
| Rafters, gable end | | | | |
| 0-2 pitch (flat to 2-in-12) | | | | |
| 12" o.c. | | | | |
| 2x4 | S.F. | 0.63 | 0.74 | 1.37 |
| 2x6 | " | 0.95 | 0.77 | 1.72 |
| 2x8 | " | 1.30 | 0.81 | 2.11 |
| 2x10 | " | 2.00 | 0.84 | 2.84 |
| 2x12 | " | 2.65 | 0.89 | 3.54 |
| 16" o.c. | | | | |
| 2x6 | S.F. | 0.80 | 0.63 | 1.43 |
| 2x8 | " | 1.10 | 0.66 | 1.76 |
| 2x10 | " | 1.65 | 0.68 | 2.33 |
| 2x12 | " | 2.15 | 0.71 | 2.86 |
| 24" o.c. | | | | |
| 2x6 | S.F. | 0.63 | 0.54 | 1.17 |
| 2x8 | " | 0.86 | 0.55 | 1.41 |
| 2x10 | " | 1.30 | 0.57 | 1.87 |
| 2x12 | " | 1.75 | 0.59 | 2.34 |
| 4-6 pitch (4-in-12 to 6-in-12) | | | | |
| 12" o.c. | | | | |
| 2x4 | S.F. | 0.68 | 0.77 | 1.45 |
| 2x6 | " | 1.00 | 0.81 | 1.80 |
| 2x8 | " | 1.40 | 0.84 | 2.24 |
| 2x10 | " | 2.15 | 0.89 | 3.04 |
| 2x12 | " | 2.85 | 0.93 | 3.78 |
| 16" o.c. | | | | |
| 2x6 | S.F. | 0.84 | 0.66 | 1.50 |
| 2x8 | " | 1.20 | 0.68 | 1.88 |
| 2x10 | " | 1.75 | 0.71 | 2.46 |
| 2x12 | " | 2.35 | 0.74 | 3.09 |
| 24" o.c. | | | | |
| 2x6 | S.F. | 0.68 | 0.55 | 1.23 |
| 2x8 | " | 0.93 | 0.57 | 1.50 |
| 2x10 | " | 1.40 | 0.61 | 2.01 |
| 2x12 | " | 1.90 | 0.68 | 2.58 |

## ROUGH CARPENTRY

| 06110.50 ROOF FRAMING | UNIT | MAT. | INST. | TOTAL |
|---|---|---|---|---|
| 8-12 pitch (8-in-12 to 12-in-12) | | | | |
| 12" o.c. | | | | |
| 2x4 | S.F. | 0.70 | 0.81 | 1.50 |
| 2x6 | " | 1.10 | 0.84 | 1.94 |
| 2x8 | " | 1.50 | 0.89 | 2.39 |
| 2x10 | " | 2.25 | 0.93 | 3.18 |
| 2x12 | " | 3.00 | 0.98 | 3.98 |
| 16" o.c. | | | | |
| 2x6 | S.F. | 0.90 | 0.68 | 1.58 |
| 2x8 | " | 1.25 | 0.71 | 1.96 |
| 2x10 | " | 1.90 | 0.74 | 2.64 |
| 2x12 | " | 2.50 | 0.77 | 3.27 |
| 24" o.c. | | | | |
| 2x6 | S.F. | 0.70 | 0.57 | 1.27 |
| 2x8 | " | 0.99 | 0.59 | 1.58 |
| 2x10 | " | 1.50 | 0.61 | 2.11 |
| 2x12 | " | 2.00 | 0.63 | 2.63 |
| Ridge boards | | | | |
| 2x6 | L.F. | 0.74 | 1.75 | 2.49 |
| 2x8 | " | 1.05 | 1.95 | 3.00 |
| 2x10 | " | 1.45 | 2.20 | 3.65 |
| 2x12 | " | 1.95 | 2.55 | 4.50 |
| Hip rafters | | | | |
| 2x6 | L.F. | 0.74 | 1.25 | 1.99 |
| 2x8 | " | 1.05 | 1.30 | 2.35 |
| 2x10 | " | 1.45 | 1.35 | 2.80 |
| 2x12 | " | 1.95 | 1.40 | 3.35 |
| Jack rafters | | | | |
| 4-6 pitch (4-in-12 to 6-in-12) | | | | |
| 16" o.c. | | | | |
| 2x6 | S.F. | 0.85 | 1.05 | 1.90 |
| 2x8 | " | 1.20 | 1.05 | 2.25 |
| 2x10 | " | 1.75 | 1.15 | 2.90 |
| 2x12 | " | 2.40 | 1.20 | 3.60 |
| 24" o.c. | | | | |
| 2x6 | S.F. | 0.68 | 0.81 | 1.49 |
| 2x8 | " | 0.93 | 0.82 | 1.75 |
| 2x10 | " | 1.40 | 0.86 | 2.26 |
| 2x12 | " | 1.90 | 0.89 | 2.79 |
| 8-12 pitch (8-in-12 to 12-in-12) | | | | |
| 16" o.c. | | | | |
| 2x6 | S.F. | 1.30 | 1.10 | 2.40 |
| 2x8 | " | 1.85 | 1.15 | 3.00 |
| 2x10 | " | 2.80 | 1.20 | 4.00 |
| 2x12 | " | 3.70 | 1.20 | 4.90 |
| 24" o.c. | | | | |
| 2x6 | S.F. | 1.10 | 0.84 | 1.94 |
| 2x8 | " | 1.50 | 0.86 | 2.36 |
| 2x10 | " | 2.25 | 0.89 | 3.14 |
| 2x12 | " | 3.00 | 0.91 | 3.91 |
| Sister rafters | | | | |
| 2x4 | L.F. | 0.53 | 2.55 | 3.08 |

# 06 WOOD AND PLASTICS

## ROUGH CARPENTRY

| | UNIT | MAT. | INST. | TOTAL |
|---|---|---|---|---|
| **06110.50 ROOF FRAMING** | | | | |
| 2x6 | L.F. | 0.74 | 2.95 | 3.69 |
| 2x8 | " | 1.05 | 3.55 | 4.60 |
| 2x10 | " | 1.45 | 4.45 | 5.90 |
| 2x12 | " | 1.95 | 5.90 | 7.85 |
| Fascia boards | | | | |
| 2x4 | L.F. | 0.53 | 1.75 | 2.28 |
| 2x6 | " | 0.74 | 1.75 | 2.49 |
| 2x8 | " | 1.05 | 1.95 | 3.00 |
| 2x10 | " | 1.45 | 1.95 | 3.40 |
| 2x12 | " | 1.95 | 2.20 | 4.15 |
| Cant strips | | | | |
| Fiber | | | | |
| 3x3 | L.F. | 0.26 | 1.00 | 1.26 |
| 4x4 | " | 0.34 | 1.05 | 1.39 |
| Wood | | | | |
| 3x3 | L.F. | 1.40 | 1.05 | 2.45 |
| **06110.60 SLEEPERS** | | | | |
| Sleepers, over concrete | | | | |
| 12" o.c. | | | | |
| 1x2 | S.F. | 0.22 | 0.81 | 1.02 |
| 1x3 | " | 0.31 | 0.84 | 1.15 |
| 2x4 | " | 0.63 | 0.98 | 1.61 |
| 2x6 | " | 0.93 | 1.05 | 1.98 |
| 16" o.c. | | | | |
| 1x2 | S.F. | 0.19 | 0.71 | 0.90 |
| 1x3 | " | 0.26 | 0.71 | 0.97 |
| 2x4 | " | 0.53 | 0.84 | 1.37 |
| 2x6 | " | 0.80 | 0.89 | 1.69 |
| **06110.65 SOFFITS** | | | | |
| Soffit framing | | | | |
| 2x3 | L.F. | 0.42 | 2.55 | 2.97 |
| 2x4 | " | 0.53 | 2.75 | 3.28 |
| 2x6 | " | 0.74 | 2.95 | 3.69 |
| 2x8 | " | 1.05 | 3.20 | 4.25 |
| **06110.70 WALL FRAMING** | | | | |
| Framing wall, studs | | | | |
| 12" o.c. | | | | |
| 2x3 | S.F. | 0.46 | 0.66 | 1.12 |
| 2x4 | " | 0.63 | 0.66 | 1.29 |
| 2x6 | " | 0.93 | 0.71 | 1.64 |
| 2x8 | " | 1.30 | 0.74 | 2.04 |
| 16" o.c. | | | | |
| 2x3 | S.F. | 0.39 | 0.55 | 0.94 |
| 2x4 | " | 0.53 | 0.55 | 1.08 |
| 2x6 | " | 0.74 | 0.59 | 1.33 |
| 2x8 | " | 1.10 | 0.61 | 1.71 |
| 24" o.c. | | | | |

# 06 WOOD AND PLASTICS

## ROUGH CARPENTRY

| ROUGH CARPENTRY | UNIT | MAT. | INST. | TOTAL |
|---|---|---|---|---|
| **06110.70 WALL FRAMING** | | | | |
| 2x3 | S.F. | 0.30 | 0.48 | 0.78 |
| 2x4 | " | 0.43 | 0.48 | 0.91 |
| 2x6 | " | 0.63 | 0.51 | 1.14 |
| 2x8 | " | 0.86 | 0.52 | 1.38 |
| Plates, top or bottom | | | | |
| 2x3 | L.F. | 0.42 | 1.05 | 1.47 |
| 2x4 | " | 0.53 | 1.10 | 1.63 |
| 2x6 | " | 0.74 | 1.20 | 1.94 |
| 2x8 | " | 1.05 | 1.25 | 2.30 |
| Headers, door or window | | | | |
| 2x6 | | | | |
| Single | | | | |
| 3' long | EA. | 2.40 | 17.70 | 20.10 |
| 6' long | " | 4.80 | 22.15 | 26.95 |
| Double | | | | |
| 3' long | EA. | 4.80 | 19.70 | 24.50 |
| 6' long | " | 9.55 | 25.30 | 34.85 |
| 2x8 | | | | |
| Single | | | | |
| 4' long | EA. | 4.45 | 22.15 | 26.60 |
| 8' long | " | 8.90 | 27.30 | 36.20 |
| Double | | | | |
| 4' long | EA. | 8.90 | 25.30 | 34.20 |
| 8' long | " | 17.80 | 32.20 | 50.00 |
| 2x10 | | | | |
| Single | | | | |
| 5' long | EA. | 8.30 | 27.30 | 35.60 |
| 10' long | " | 16.50 | 35.40 | 51.90 |
| Double | | | | |
| 5' long | EA. | 16.50 | 29.50 | 46.00 |
| 10' long | " | 33.10 | 35.40 | 68.50 |
| 2x12 | | | | |
| Single | | | | |
| 6' long | EA. | 12.85 | 27.30 | 40.15 |
| 12' long | " | 25.70 | 35.40 | 61.10 |
| Double | | | | |
| 6' long | EA. | 25.70 | 32.20 | 57.90 |
| 12' long | " | 51.50 | 39.40 | 90.90 |
| **06115.10 FLOOR SHEATHING** | | | | |
| Sub-flooring, plywood, CDX | | | | |
| 1/2" thick | S.F. | 0.55 | 0.44 | 0.99 |
| 5/8" thick | " | 0.66 | 0.51 | 1.17 |
| 3/4" thick | " | 0.78 | 0.59 | 1.37 |
| Structural plywood | | | | |
| 1/2" thick | S.F. | 0.61 | 0.44 | 1.05 |
| 5/8" thick | " | 0.73 | 0.51 | 1.24 |
| 3/4" thick | " | 0.86 | 0.55 | 1.41 |
| Board type subflooring | | | | |
| 1x6 | | | | |
| Minimum | S.F. | 0.90 | 0.79 | 1.69 |
| Maximum | " | 1.25 | 0.89 | 2.14 |

# 06 WOOD AND PLASTICS

| ROUGH CARPENTRY | UNIT | MAT. | INST. | TOTAL |
|---|---|---|---|---|
| **06115.10 FLOOR SHEATHING** | | | | |
| 1x8 | | | | |
| Minimum | S.F. | 1.10 | 0.75 | 1.85 |
| Maximum | " | 1.25 | 0.83 | 2.08 |
| 1x10 | | | | |
| Minimum | S.F. | 1.30 | 0.71 | 2.01 |
| Maximum | " | 1.35 | 0.79 | 2.14 |
| Underlayment | | | | |
| Hardboard, 1/4" tempered | S.F. | 0.45 | 0.44 | 0.89 |
| Plywood, CDX | | | | |
| 3/8" thick | S.F. | 0.49 | 0.44 | 0.93 |
| 1/2" thick | " | 0.55 | 0.47 | 1.02 |
| 5/8" thick | " | 0.66 | 0.51 | 1.17 |
| 3/4" thick | " | 0.78 | 0.55 | 1.33 |
| **06115.20 ROOF SHEATHING** | | | | |
| Sheathing | | | | |
| Plywood, CDX | | | | |
| 3/8" thick | S.F. | 0.49 | 0.46 | 0.95 |
| 1/2" thick | " | 0.55 | 0.47 | 1.02 |
| 5/8" thick | " | 0.66 | 0.51 | 1.17 |
| 3/4" thick | " | 0.78 | 0.55 | 1.33 |
| Structural plywood | | | | |
| 3/8" thick | S.F. | 0.53 | 0.46 | 0.99 |
| 1/2" thick | " | 0.61 | 0.47 | 1.08 |
| 5/8" thick | " | 0.73 | 0.51 | 1.24 |
| 3/4" thick | " | 0.86 | 0.55 | 1.41 |
| **06115.30 WALL SHEATHING** | | | | |
| Sheathing | | | | |
| Plywood, CDX | | | | |
| 3/8" thick | S.F. | 0.49 | 0.53 | 1.01 |
| 1/2" thick | " | 0.55 | 0.55 | 1.10 |
| 5/8" thick | " | 0.66 | 0.59 | 1.25 |
| 3/4" thick | " | 0.78 | 0.64 | 1.42 |
| Waferboard | | | | |
| 3/8" thick | S.F. | 0.42 | 0.53 | 0.94 |
| 1/2" thick | " | 0.51 | 0.55 | 1.05 |
| 5/8" thick | " | 0.62 | 0.59 | 1.21 |
| 3/4" thick | " | 0.69 | 0.64 | 1.33 |
| Structural plywood | | | | |
| 3/8" thick | S.F. | 0.53 | 0.53 | 1.05 |
| 1/2" thick | " | 0.61 | 0.55 | 1.16 |
| 5/8" thick | " | 0.73 | 0.59 | 1.32 |
| 3/4" thick | " | 0.86 | 0.64 | 1.50 |
| Gypsum, 1/2" thick | " | 0.45 | 0.55 | 1.00 |
| Asphalt impregnated fiberboard, 1/2" thick | " | 0.47 | 0.55 | 1.01 |

# ROUGH CARPENTRY

| ROUGH CARPENTRY | UNIT | MAT. | INST. | TOTAL |
|---|---|---|---|---|
| **06125.10 WOOD DECKING** | | | | |
| Decking, T&G solid | | | | |
| Cedar | | | | |
| 3" thick | S.F. | 6.55 | 0.89 | 7.44 |
| 4" thick | " | 7.90 | 0.94 | 8.85 |
| Fir | | | | |
| 3" thick | S.F. | 2.75 | 0.89 | 3.64 |
| 4" thick | " | 3.35 | 0.94 | 4.29 |
| Southern yellow pine | | | | |
| 3" thick | S.F. | 2.70 | 1.00 | 3.70 |
| 4" thick | " | 2.90 | 1.10 | 4.00 |
| White pine | | | | |
| 3" thick | S.F. | 3.40 | 0.89 | 4.29 |
| 4" thick | " | 4.50 | 0.94 | 5.45 |
| **06130.10 HEAVY TIMBER** | | | | |
| Mill framed structures | | | | |
| Beams to 20' long | | | | |
| Douglas fir | | | | |
| 6x8 | L.F. | 4.85 | 4.30 | 9.15 |
| 6x10 | " | 5.70 | 4.40 | 10.10 |
| 6x12 | " | 6.80 | 4.75 | 11.55 |
| 6x14 | " | 8.20 | 4.95 | 13.15 |
| 6x16 | " | 8.90 | 5.15 | 14.05 |
| 8x10 | " | 7.55 | 4.40 | 11.95 |
| 8x12 | " | 8.90 | 4.75 | 13.65 |
| 8x14 | " | 10.30 | 4.95 | 15.25 |
| 8x16 | " | 11.70 | 5.15 | 16.85 |
| Southern yellow pine | | | | |
| 6x8 | L.F. | 3.85 | 4.30 | 8.15 |
| 6x10 | " | 4.65 | 4.40 | 9.05 |
| 6x12 | " | 5.95 | 4.75 | 10.70 |
| 6x14 | " | 6.80 | 4.95 | 11.75 |
| 6x16 | " | 7.55 | 5.15 | 12.70 |
| 8x10 | " | 6.30 | 4.40 | 10.70 |
| 8x12 | " | 7.70 | 4.75 | 12.45 |
| 8x14 | " | 8.80 | 4.95 | 13.75 |
| 8x16 | " | 10.15 | 5.15 | 15.30 |
| Columns to 12' high | | | | |
| Douglas fir | | | | |
| 6x6 | L.F. | 3.45 | 6.40 | 9.85 |
| 8x8 | " | 5.95 | 6.40 | 12.35 |
| 10x10 | " | 10.40 | 7.15 | 17.55 |
| 12x12 | " | 12.90 | 7.15 | 20.05 |
| Southern yellow pine | | | | |
| 6x6 | L.F. | 3.00 | 6.40 | 9.40 |
| 8x8 | " | 5.00 | 6.40 | 11.40 |
| 10x10 | " | 7.80 | 7.15 | 14.95 |
| 12x12 | " | 10.90 | 7.15 | 18.05 |
| Posts, treated | | | | |
| 4x4 | L.F. | 1.60 | 1.40 | 3.00 |
| 6x6 | " | 3.80 | 1.75 | 5.55 |

## ROUGH CARPENTRY

### 06190.20 WOOD TRUSSES

| | UNIT | MAT. | INST. | TOTAL |
|---|---|---|---|---|
| Truss, fink, 2x4 members | | | | |
| 3-in-12 slope | | | | |
| 24' span | EA. | 64.00 | 36.70 | 100.70 |
| 26' span | " | 74.50 | 36.70 | 111.20 |
| 28' span | " | 77.00 | 38.90 | 115.90 |
| 30' span | " | 82.00 | 38.90 | 120.90 |
| 34' span | " | 90.00 | 41.40 | 131.40 |
| 38' span | " | 110.00 | 41.40 | 151.40 |
| 5-in-12 slope | | | | |
| 24' span | EA. | 65.50 | 37.70 | 103.20 |
| 28' span | " | 77.00 | 38.90 | 115.90 |
| 30' span | " | 90.00 | 40.10 | 130.10 |
| 32' span | " | 97.50 | 40.10 | 137.60 |
| 40' span | " | 130.00 | 42.80 | 172.80 |
| Gable, 2x4 members | | | | |
| 5-in-12 slope | | | | |
| 24' span | EA. | 90.00 | 37.70 | 127.70 |
| 26' span | " | 96.00 | 37.70 | 133.70 |
| 28' span | " | 110.00 | 38.90 | 148.90 |
| 30' span | " | 120.00 | 40.10 | 160.10 |
| 32' span | " | 120.00 | 40.10 | 160.10 |
| 36' span | " | 130.00 | 41.40 | 171.40 |
| 40' span | " | 140.00 | 42.80 | 182.80 |
| King post type, 2x4 members | | | | |
| 4-in-12 slope | | | | |
| 16' span | EA. | 54.00 | 34.70 | 88.70 |
| 18' span | " | 57.50 | 35.60 | 93.10 |
| 24' span | " | 61.50 | 37.70 | 99.20 |
| 26' span | " | 66.50 | 37.70 | 104.20 |
| 30' span | " | 84.50 | 40.10 | 124.60 |
| 34' span | " | 90.00 | 40.10 | 130.10 |
| 38' span | " | 110.00 | 41.40 | 151.40 |
| 42' span | " | 130.00 | 44.20 | 174.20 |

## FINISH CARPENTRY

### 06200.10 FINISH CARPENTRY

| | UNIT | MAT. | INST. | TOTAL |
|---|---|---|---|---|
| Mouldings and trim | | | | |
| Apron, flat | | | | |
| 9/16 x 2 | L.F. | 1.00 | 1.75 | 2.75 |
| 9/16 x 3-1/2 | " | 1.80 | 1.85 | 3.65 |
| Base | | | | |
| Colonial | | | | |

# 06 WOOD AND PLASTICS

| FINISH CARPENTRY | UNIT | MAT. | INST. | TOTAL |
|---|---|---|---|---|
| **06200.10 FINISH CARPENTRY** | | | | |
| 7/16 x 2-1/4 | L.F. | 1.15 | 1.75 | 2.90 |
| 7/16 x 3 | " | 1.40 | 1.75 | 3.15 |
| 7/16 x 3-1/4 | " | 1.50 | 1.75 | 3.25 |
| 9/16 x 3 | " | 1.50 | 1.85 | 3.35 |
| 9/16 x 3-1/4 | " | 1.60 | 1.85 | 3.45 |
| 11/16 x 2-1/4 | " | 1.40 | 1.95 | 3.35 |
| Ranch | | | | |
| 7/16 x 2-1/4 | L.F. | 1.20 | 1.75 | 2.95 |
| 7/16 x 3-1/4 | " | 1.60 | 1.75 | 3.35 |
| 9/16 x 2-1/4 | " | 1.30 | 1.85 | 3.15 |
| 9/16 x 3 | " | 1.55 | 1.85 | 3.40 |
| 9/16 x 3-1/4 | " | 1.60 | 1.85 | 3.45 |
| Casing | | | | |
| 11/16 x 2-1/2 | L.F. | 1.20 | 1.60 | 2.80 |
| 11/16 x 3-1/2 | " | 1.55 | 1.70 | 3.25 |
| Chair rail | | | | |
| 9/16 x 2-1/2 | L.F. | 1.25 | 1.75 | 3.00 |
| 9/16 x 3-1/2 | " | 1.90 | 1.75 | 3.65 |
| Closet pole | | | | |
| 1-1/8" dia. | L.F. | 1.45 | 2.35 | 3.80 |
| 1-5/8" dia. | " | 1.65 | 2.35 | 4.00 |
| Cove | | | | |
| 9/16 x 1-3/4 | L.F. | 0.97 | 1.75 | 2.72 |
| 11/16 x 2-3/4 | " | 1.35 | 1.75 | 3.10 |
| Crown | | | | |
| 9/16 x 1-5/8 | L.F. | 1.45 | 2.35 | 3.80 |
| 9/16 x 2-5/8 | " | 1.60 | 2.75 | 4.35 |
| 11/16 x 3-5/8 | " | 1.90 | 2.95 | 4.85 |
| 11/16 x 4-1/4 | " | 2.70 | 3.20 | 5.90 |
| 11/16 x 5-1/4 | " | 3.45 | 3.55 | 7.00 |
| Drip cap | | | | |
| 1-1/16 x 1-5/8 | L.F. | 1.40 | 1.75 | 3.15 |
| Glass bead | | | | |
| 3/8 x 3/8 | L.F. | 0.39 | 2.20 | 2.59 |
| 1/2 x 9/16 | " | 0.49 | 2.20 | 2.69 |
| 5/8 x 5/8 | " | 0.51 | 2.20 | 2.71 |
| 3/4 x 3/4 | " | 0.63 | 2.20 | 2.83 |
| Half round | | | | |
| 1/2 | L.F. | 0.47 | 1.40 | 1.87 |
| 5/8 | " | 0.63 | 1.40 | 2.03 |
| 3/4 | " | 0.80 | 1.40 | 2.20 |
| Lattice | | | | |
| 1/4 x 7/8 | L.F. | 0.38 | 1.40 | 1.78 |
| 1/4 x 1-1/8 | " | 0.41 | 1.40 | 1.81 |
| 1/4 x 1-3/8 | " | 0.43 | 1.40 | 1.83 |
| 1/4 x 1-3/4 | " | 0.49 | 1.40 | 1.89 |
| 1/4 x 2 | " | 0.57 | 1.40 | 1.97 |
| Ogee molding | | | | |
| 5/8 x 3/4 | L.F. | 0.80 | 1.75 | 2.55 |
| 11/16 x 1-1/8 | " | 1.20 | 1.75 | 2.95 |
| 11/16 x 1-3/8 | " | 1.45 | 1.75 | 3.20 |
| Parting bead | | | | |

## FINISH CARPENTRY

### 06200.10 FINISH CARPENTRY

| FINISH CARPENTRY | UNIT | MAT. | INST. | TOTAL |
|---|---|---|---|---|
| 3/8 x 7/8 | L.F. | 0.61 | 2.20 | 2.81 |
| Quarter round | | | | |
| 1/4 x 1/4 | L.F. | 0.19 | 1.40 | 1.59 |
| 3/8 x 3/8 | " | 0.28 | 1.40 | 1.68 |
| 1/2 x 1/2 | " | 0.39 | 1.40 | 1.79 |
| 11/16 x 11/16 | " | 0.42 | 1.55 | 1.97 |
| 3/4 x 3/4 | " | 0.46 | 1.55 | 2.01 |
| 1-1/16 x 1-1/16 | " | 0.58 | 1.60 | 2.18 |
| Railings, balusters | | | | |
| 1-1/8 x 1-1/8 | L.F. | 2.10 | 3.55 | 5.65 |
| 1-1/2 x 1-1/2 | " | 2.45 | 3.20 | 5.65 |
| Screen moldings | | | | |
| 1/4 x 3/4 | L.F. | 0.50 | 2.95 | 3.45 |
| 5/8 x 5/16 | " | 0.62 | 2.95 | 3.57 |
| Shoe | | | | |
| 7/16 x 11/16 | L.F. | 0.62 | 1.40 | 2.02 |
| Sash beads | | | | |
| 1/2 x 3/4 | L.F. | 0.78 | 2.95 | 3.73 |
| 1/2 x 7/8 | " | 0.92 | 2.95 | 3.87 |
| 1/2 x 1-1/8 | " | 0.99 | 3.20 | 4.19 |
| 5/8 x 7/8 | " | 0.99 | 3.20 | 4.19 |
| Stop | | | | |
| 5/8 x 1-5/8 | | | | |
| Colonial | L.F. | 0.49 | 2.20 | 2.69 |
| Ranch | " | 0.47 | 2.20 | 2.67 |
| Stools | | | | |
| 11/16 x 2-1/4 | L.F. | 2.25 | 3.95 | 6.20 |
| 11/16 x 2-1/2 | " | 2.40 | 3.95 | 6.35 |
| 11/16 x 5-1/4 | " | 2.55 | 4.45 | 7.00 |
| Exterior trim, casing, select pine, 1x3 | " | 1.65 | 1.75 | 3.40 |
| Douglas fir | | | | |
| 1x3 | L.F. | 0.88 | 1.75 | 2.63 |
| 1x4 | " | 1.20 | 1.75 | 2.95 |
| 1x6 | " | 1.45 | 1.95 | 3.40 |
| 1x8 | " | 2.10 | 2.20 | 4.30 |
| Cornices, white pine, #2 or better | | | | |
| 1x2 | L.F. | 0.54 | 1.75 | 2.29 |
| 1x4 | " | 0.82 | 1.75 | 2.57 |
| 1x6 | " | 1.30 | 1.95 | 3.25 |
| 1x8 | " | 1.45 | 2.10 | 3.55 |
| 1x10 | " | 1.95 | 2.20 | 4.15 |
| 1x12 | " | 2.45 | 2.35 | 4.80 |
| Shelving, pine | | | | |
| 1x8 | L.F. | 1.75 | 2.75 | 4.50 |
| 1x10 | " | 2.10 | 2.85 | 4.95 |
| 1x12 | " | 2.75 | 2.95 | 5.70 |
| Plywood shelf, 3/4", with edge band, 12" wide | " | 1.90 | 3.55 | 5.45 |
| Adjustable shelf, and rod, 12" wide | | | | |
| 3' to 4' long | EA. | 12.75 | 8.85 | 21.60 |
| 5' to 8' long | " | 24.10 | 11.80 | 35.90 |
| Prefinished wood shelves with brackets and supports | | | | |
| 8" wide | | | | |

# 06 WOOD AND PLASTICS

| FINISH CARPENTRY | UNIT | MAT. | INST. | TOTAL |
|---|---|---|---|---|
| **06200.10 FINISH CARPENTRY** | | | | |
| 3' long | EA. | 37.80 | 8.85 | 46.65 |
| 4' long | " | 43.20 | 8.85 | 52.05 |
| 6' long | " | 63.50 | 8.85 | 72.35 |
| 10" wide | | | | |
| 3' long | EA. | 41.90 | 8.85 | 50.75 |
| 4' long | " | 59.50 | 8.85 | 68.35 |
| 6' long | " | 70.00 | 8.85 | 78.85 |
| **06220.10 MILLWORK** | | | | |
| Countertop, laminated plastic | | | | |
| 25" x 7/8" thick | | | | |
| Minimum | L.F. | 13.75 | 8.85 | 22.60 |
| Average | " | 24.80 | 11.80 | 36.60 |
| Maximum | " | 41.30 | 14.20 | 55.50 |
| 25" x 1-1/4" thick | | | | |
| Minimum | L.F. | 16.50 | 11.80 | 28.30 |
| Average | " | 27.50 | 14.20 | 41.70 |
| Maximum | " | 44.10 | 17.70 | 61.80 |
| Add for cutouts | EA. | 0.00 | 22.15 | 22.15 |
| Backsplash, 4" high, 7/8" thick | L.F. | 13.75 | 7.10 | 20.85 |
| Plywood, sanded, A-C | | | | |
| 1/4" thick | S.F. | 0.82 | 1.20 | 2.02 |
| 3/8" thick | " | 0.89 | 1.25 | 2.14 |
| 1/2" thick | " | 1.00 | 1.35 | 2.35 |
| A-D | | | | |
| 1/4" thick | S.F. | 0.74 | 1.20 | 1.94 |
| 3/8" thick | " | 0.85 | 1.25 | 2.10 |
| 1/2" thick | " | 0.96 | 1.35 | 2.31 |
| Base cabinets, 34-1/2" high, 24" deep, hardwood, no tops | | | | |
| Minimum | L.F. | 48.20 | 14.20 | 62.40 |
| Average | " | 82.50 | 17.70 | 100.20 |
| Maximum | " | 130.00 | 23.65 | 153.65 |
| Wall cabinets | | | | |
| Minimum | L.F. | 38.60 | 11.80 | 50.40 |
| Average | " | 62.00 | 14.20 | 76.20 |
| Maximum | " | 120.00 | 17.70 | 137.70 |

| WOOD TREATMENT | UNIT | MAT. | INST. | TOTAL |
|---|---|---|---|---|
| **06300.10 WOOD TREATMENT** | | | | |
| Creosote preservative treatment | | | | |
| 8 lb/cf | B.F. | | | 0.31 |

# 06 WOOD AND PLASTICS

## WOOD TREATMENT

### 06300.10 WOOD TREATMENT

| WOOD TREATMENT | UNIT | MAT. | INST. | TOTAL |
|---|---|---|---|---|
| 10 lb/cf | B.F. | | | 0.41 |
| Salt preservative treatment | | | | |
| Oil borne | | | | |
| Minimum | B.F. | | | 0.23 |
| Maximum | " | | | 0.35 |
| Water borne | | | | |
| Minimum | B.F. | | | 0.19 |
| Maximum | " | | | 0.28 |
| Fire retardant treatment | | | | |
| Minimum | B.F. | | | 0.39 |
| Maximum | " | | | 0.47 |
| Kiln dried, softwood, add to framing costs | | | | |
| 1" thick | B.F. | | | 0.14 |
| 2" thick | " | | | 0.20 |
| 3" thick | " | | | 0.27 |
| 4" thick | " | | | 0.32 |

## ARCHITECTURAL WOODWORK

### 06420.10 PANEL WORK

| ARCHITECTURAL WOODWORK | UNIT | MAT. | INST. | TOTAL |
|---|---|---|---|---|
| Hardboard, tempered, 1/4" thick | | | | |
| Natural faced | S.F. | 0.53 | 0.89 | 1.42 |
| Plastic faced | " | 0.88 | 1.00 | 1.88 |
| Pegboard, natural | " | 0.61 | 0.89 | 1.50 |
| Plastic faced | " | 0.80 | 1.00 | 1.80 |
| Untempered, 1/4" thick | | | | |
| Natural faced | S.F. | 0.47 | 0.89 | 1.36 |
| Plastic faced | " | 0.85 | 1.00 | 1.85 |
| Pegboard, natural | " | 0.50 | 0.89 | 1.39 |
| Plastic faced | " | 0.78 | 1.00 | 1.78 |
| Plywood unfinished, 1/4" thick | | | | |
| Birch | | | | |
| Natural | S.F. | 1.20 | 1.20 | 2.40 |
| Select | " | 1.55 | 1.20 | 2.75 |
| Knotty pine | " | 1.55 | 1.20 | 2.75 |
| Cedar (closet lining) | | | | |
| Standard boards T&G | S.F. | 1.50 | 1.20 | 2.70 |
| Particle board | " | 1.15 | 1.20 | 2.35 |
| Plywood, prefinished, 1/4" thick, premium grade | | | | |
| Birch veneer | S.F. | 2.10 | 1.40 | 3.50 |
| Cherry veneer | " | 2.35 | 1.40 | 3.75 |
| Chestnut veneer | " | 4.20 | 1.40 | 5.60 |
| Lauan veneer | " | 0.76 | 1.40 | 2.16 |
| Mahogany veneer | " | 2.40 | 1.40 | 3.80 |

# 06 WOOD AND PLASTICS

## ARCHITECTURAL WOODWORK

| | UNIT | MAT. | INST. | TOTAL |
|---|---|---|---|---|
| **06420.10 PANEL WORK** | | | | |
| Oak veneer (red) | S.F. | 2.40 | 1.40 | 3.80 |
| Pecan veneer | " | 2.85 | 1.40 | 4.25 |
| Rosewood veneer | " | 4.00 | 1.40 | 5.40 |
| Teak veneer | " | 5.15 | 1.40 | 6.55 |
| Walnut veneer | " | 3.50 | 1.40 | 4.90 |
| **06430.10 STAIRWORK** | | | | |
| Risers, 1x8, 42" wide | | | | |
| White oak | EA. | 17.20 | 17.70 | 34.90 |
| Pine | " | 12.20 | 17.70 | 29.90 |
| Treads, 1-1/16" x 9-1/2" x 42" | | | | |
| White oak | EA. | 24.35 | 22.15 | 46.50 |
| **06440.10 COLUMNS** | | | | |
| Column, hollow, round wood | | | | |
| 12" diameter | | | | |
| 10' high | EA. | 490.00 | 48.00 | 538.00 |
| 12' high | " | 590.00 | 51.50 | 641.50 |
| 14' high | " | 720.00 | 57.50 | 777.50 |
| 16' high | " | 890.00 | 72.00 | 962.00 |
| 24" diameter | | | | |
| 16' high | EA. | 2,030 | 72.00 | 2,102 |
| 18' high | " | 2,300 | 76.00 | 2,376 |
| 20' high | " | 2,840 | 76.00 | 2,916 |
| 22' high | " | 2,970 | 80.00 | 3,050 |
| 24' high | " | 3,240 | 80.00 | 3,320 |

## MOISTURE PROTECTION

### 07100.10 WATERPROOFING

| MOISTURE PROTECTION | UNIT | MAT. | INST. | TOTAL |
|---|---|---|---|---|
| Membrane waterproofing, elastomeric | | | | |
| Butyl | | | | |
| 1/32" thick | S.F. | 0.70 | 1.10 | 1.80 |
| 1/16" thick | " | 0.92 | 1.15 | 2.07 |
| Butyl with nylon | | | | |
| 1/32" thick | S.F. | 0.84 | 1.10 | 1.94 |
| 1/16" thick | " | 0.99 | 1.15 | 2.14 |
| Neoprene | | | | |
| 1/32" thick | S.F. | 1.15 | 1.10 | 2.25 |
| 1/16" thick | " | 1.90 | 1.15 | 3.05 |
| Neoprene with nylon | | | | |
| 1/32" thick | S.F. | 1.25 | 1.10 | 2.35 |
| 1/16" thick | " | 2.05 | 1.15 | 3.20 |
| Bituminous membrane waterproofing, asphalt felt, 15 lb. | | | | |
| One ply | S.F. | 0.42 | 0.69 | 1.11 |
| Two ply | " | 0.49 | 0.84 | 1.33 |
| Three ply | " | 0.62 | 0.99 | 1.61 |
| Four ply | " | 0.69 | 1.15 | 1.84 |
| Five ply | " | 0.76 | 1.45 | 2.21 |
| Modified asphalt membrane waterproofing, fibrous asphalt | | | | |
| One ply | S.F. | 0.96 | 1.15 | 2.11 |
| Two ply | " | 1.45 | 1.40 | 2.85 |
| Three ply | " | 1.65 | 1.55 | 3.20 |
| Four ply | " | 1.95 | 1.85 | 3.80 |
| Five ply | " | 2.30 | 2.20 | 4.50 |
| Asphalt coated protective board | | | | |
| 1/8" thick | S.F. | 0.54 | 0.69 | 1.23 |
| 1/4" thick | " | 0.76 | 0.69 | 1.45 |
| 3/8" thick | " | 0.84 | 0.69 | 1.53 |
| 1/2" thick | " | 1.00 | 0.73 | 1.73 |
| Cement protective board | | | | |
| 3/8" thick | S.F. | 1.90 | 0.93 | 2.83 |
| 1/2" thick | " | 2.60 | 0.93 | 3.52 |
| Fluid applied, neoprene | | | | |
| 50 mil | S.F. | 2.25 | 0.93 | 3.17 |
| 90 mil | " | 3.10 | 0.93 | 4.03 |
| Tab extended polyurethane | | | | |
| .050" thick | S.F. | 1.20 | 0.69 | 1.89 |
| Fluid applied rubber based polyurethane | | | | |
| 6 mil | S.F. | 0.65 | 0.87 | 1.52 |
| 15 mil | " | 1.20 | 0.69 | 1.89 |
| Bentonite waterproofing, panels | | | | |
| 3/16" thick | S.F. | 1.10 | 0.69 | 1.79 |
| 1/4" thick | " | 1.25 | 0.69 | 1.94 |
| 5/8" thick | " | 1.85 | 0.73 | 2.58 |
| Granular admixtures, trowel on, 3/8" thick | " | 1.50 | 0.69 | 2.19 |
| Metallic oxide waterproofing, iron compound, troweled | | | | |
| 5/8" thick | S.F. | 1.10 | 0.69 | 1.79 |
| 3/4" thick | " | 1.35 | 0.79 | 2.14 |

# 07 THERMAL AND MOISTURE

## MOISTURE PROTECTION

| MOISTURE PROTECTION | UNIT | MAT. | INST. | TOTAL |
|---|---|---|---|---|
| **07150.10 DAMPPROOFING** | | | | |
| Silicone dampproofing, sprayed on | | | | |
| Concrete surface | | | | |
| 1 coat | S.F. | 0.45 | 0.15 | 0.60 |
| 2 coats | " | 0.73 | 0.21 | 0.94 |
| Concrete block | | | | |
| 1 coat | S.F. | 0.45 | 0.19 | 0.64 |
| 2 coats | " | 0.73 | 0.25 | 0.98 |
| Brick | | | | |
| 1 coat | S.F. | 0.51 | 0.21 | 0.72 |
| 2 coats | " | 0.80 | 0.28 | 1.08 |
| **07160.10 BITUMINOUS DAMPPROOFING** | | | | |
| Building paper, asphalt felt | | | | |
| 15 lb | S.F. | 0.09 | 1.10 | 1.19 |
| 30 lb | " | 0.18 | 1.15 | 1.33 |
| Asphalt dampproofing, troweled, cold, primer plus | | | | |
| 1 coat | S.F. | 0.61 | 0.93 | 1.53 |
| 2 coats | " | 0.90 | 1.40 | 2.30 |
| 3 coats | " | 1.35 | 1.75 | 3.10 |
| Fibrous asphalt dampproofing, hot troweled, primer plus | | | | |
| 1 coat | S.F. | 0.95 | 1.10 | 2.05 |
| 2 coats | " | 1.15 | 1.55 | 2.70 |
| 3 coats | " | 1.25 | 2.00 | 3.25 |
| Asphaltic paint dampproofing, per coat | | | | |
| Brush on | S.F. | 0.18 | 0.40 | 0.58 |
| Spray on | " | 0.30 | 0.31 | 0.61 |
| **07190.10 VAPOR BARRIERS** | | | | |
| Vapor barrier, polyethylene | | | | |
| 2 mil | S.F. | 0.01 | 0.14 | 0.15 |
| 6 mil | " | 0.03 | 0.14 | 0.17 |
| 8 mil | " | 0.04 | 0.15 | 0.19 |
| 10 mil | " | 0.05 | 0.15 | 0.20 |

## INSULATION

| INSULATION | UNIT | MAT. | INST. | TOTAL |
|---|---|---|---|---|
| **07210.10 BATT INSULATION** | | | | |
| Ceiling, fiberglass, unfaced | | | | |
| 3-1/2" thick, R11 | S.F. | 0.31 | 0.33 | 0.64 |
| 6" thick, R19 | " | 0.41 | 0.37 | 0.78 |
| 9" thick, R30 | " | 0.81 | 0.43 | 1.24 |
| Suspended ceiling, unfaced | | | | |
| 3-1/2" thick, R11 | S.F. | 0.31 | 0.31 | 0.62 |

# 07 THERMAL AND MOISTURE

| INSULATION | UNIT | MAT. | INST. | TOTAL |
|---|---|---|---|---|
| **07210.10 BATT INSULATION** | | | | |
| 6" thick, R19 | S.F. | 0.41 | 0.35 | 0.76 |
| 9" thick, R30 | " | 0.81 | 0.40 | 1.21 |
| Crawl space, unfaced | | | | |
| 3-1/2" thick, R11 | S.F. | 0.31 | 0.43 | 0.74 |
| 6" thick, R19 | " | 0.41 | 0.46 | 0.87 |
| 9" thick, R30 | " | 0.81 | 0.51 | 1.31 |
| Wall, fiberglass | | | | |
| Paper backed | | | | |
| 2" thick, R7 | S.F. | 0.20 | 0.29 | 0.49 |
| 3" thick, R8 | " | 0.22 | 0.31 | 0.53 |
| 4" thick, R11 | " | 0.36 | 0.33 | 0.69 |
| 6" thick, R19 | " | 0.54 | 0.35 | 0.89 |
| Foil backed, 1 side | | | | |
| 2" thick, R7 | S.F. | 0.41 | 0.29 | 0.70 |
| 3" thick, R11 | " | 0.43 | 0.31 | 0.74 |
| 4" thick, R14 | " | 0.46 | 0.33 | 0.79 |
| 6" thick, R21 | " | 0.49 | 0.35 | 0.84 |
| Foil backed, 2 sides | | | | |
| 2" thick, R7 | S.F. | 0.47 | 0.33 | 0.80 |
| 3" thick, R11 | " | 0.59 | 0.35 | 0.94 |
| 4" thick, R14 | " | 0.72 | 0.37 | 1.09 |
| 6" thick, R21 | " | 0.77 | 0.40 | 1.17 |
| Unfaced | | | | |
| 2" thick, R7 | S.F. | 0.27 | 0.29 | 0.56 |
| 3" thick, R9 | " | 0.31 | 0.31 | 0.62 |
| 4" thick, R11 | " | 0.32 | 0.33 | 0.65 |
| 6" thick, R19 | " | 0.41 | 0.35 | 0.76 |
| Mineral wool batts | | | | |
| Paper backed | | | | |
| 2" thick, R6 | S.F. | 0.45 | 0.29 | 0.74 |
| 4" thick, R12 | " | 0.54 | 0.31 | 0.85 |
| 6" thick, R19 | " | 0.62 | 0.35 | 0.97 |
| Fasteners, self adhering, attached to ceiling deck | | | | |
| 2-1/2" long | EA. | 0.14 | 0.46 | 0.60 |
| 4-1/2" long | " | 0.16 | 0.51 | 0.67 |
| Capped, self-locking washers for fastening insulation | " | 0.14 | 0.28 | 0.42 |
| **07210.20 BOARD INSULATION** | | | | |
| Insulation, rigid | | | | |
| Fiberglass, roof | | | | |
| 0.75" thick, R2.78 | S.F. | 0.27 | 0.25 | 0.52 |
| 1.06" thick, R4.17 | " | 0.41 | 0.26 | 0.67 |
| 1.31" thick, R5.26 | " | 0.54 | 0.28 | 0.82 |
| 1.63" thick, R6.67 | " | 0.68 | 0.29 | 0.97 |
| 2.25" thick, R8.33 | " | 0.74 | 0.31 | 1.05 |
| Composite board, roof | | | | |
| 1-1/2" thick, R6.67 | S.F. | 0.88 | 0.28 | 1.16 |
| 1-5/8" thick, R7.69 | " | 0.95 | 0.29 | 1.24 |
| 2" thick, R10.0 | " | 1.70 | 0.31 | 2.01 |
| 2-1/4" thick, R12.50 | " | 1.90 | 0.33 | 2.23 |
| 2-1/2" thick, R14.29 | " | 2.05 | 0.35 | 2.40 |
| 2-3/4" thick, R16.67 | " | 2.25 | 0.37 | 2.62 |

## INSULATION

### 07210.20 BOARD INSULATION

| INSULATION | UNIT | MAT. | INST. | TOTAL |
|---|---|---|---|---|
| 3-1/4" thick, R20.00 | S.F. | 2.90 | 0.40 | 3.30 |
| Perlite board, roof | | | | |
| 1.00" thick, R2.78 | S.F. | 0.57 | 0.23 | 0.80 |
| 1.50" thick, R4.17 | " | 0.88 | 0.24 | 1.12 |
| 2.00" thick, R5.92 | " | 1.10 | 0.25 | 1.35 |
| 2.50" thick, R6.67 | " | 1.30 | 0.26 | 1.56 |
| 3.00" thick, R8.33 | " | 1.65 | 0.28 | 1.93 |
| 4.00" thick, R10.00 | " | 1.85 | 0.29 | 2.14 |
| 5.25" thick, R14.29 | " | 2.05 | 0.31 | 2.36 |
| Rigid urethane | | | | |
| Roof | | | | |
| 1" thick, R6.67 | S.F. | 0.77 | 0.23 | 1.00 |
| 1.20" thick, R8.33 | " | 0.86 | 0.24 | 1.10 |
| 1.50" thick, R11.11 | " | 1.05 | 0.24 | 1.29 |
| 2" thick, R14.29 | " | 1.40 | 0.25 | 1.65 |
| 2.25" thick, R16.67 | " | 1.75 | 0.26 | 2.01 |
| Wall | | | | |
| 1" thick, R6.67 | S.F. | 0.77 | 0.29 | 1.06 |
| 1.5" thick, R11.11 | " | 1.05 | 0.31 | 1.36 |
| 2" thick, R14.29 | " | 1.40 | 0.33 | 1.73 |
| Polystyrene | | | | |
| Roof | | | | |
| 1.0" thick, R4.17 | S.F. | 0.73 | 0.23 | 0.96 |
| 1.5" thick, R6.26 | " | 1.20 | 0.24 | 1.44 |
| 2.0" thick, R8.33 | " | 1.35 | 0.25 | 1.60 |
| Wall | | | | |
| 1.0" thick, R4.17 | S.F. | 0.73 | 0.29 | 1.02 |
| 1.5" thick, R6.26 | " | 1.20 | 0.31 | 1.51 |
| 2.0" thick, R8.33 | " | 1.35 | 0.33 | 1.68 |
| Rigid board insulation, deck | | | | |
| Mineral fiberboard | | | | |
| 1" thick, R3.0 | S.F. | 0.43 | 0.23 | 0.66 |
| 2" thick, R5.26 | " | 0.97 | 0.25 | 1.22 |
| Fiberglass | | | | |
| 1" thick, R4.3 | S.F. | 0.76 | 0.23 | 0.99 |
| 2" thick, R8.5 | " | 1.10 | 0.25 | 1.35 |
| Polystyrene | | | | |
| 1" thick, R5.4 | S.F. | 0.73 | 0.23 | 0.96 |
| 2" thick, R10.8 | " | 1.35 | 0.25 | 1.60 |
| Urethane | | | | |
| .75" thick, R5.4 | S.F. | 0.70 | 0.23 | 0.93 |
| 1" thick, R6.4 | " | 0.84 | 0.23 | 1.07 |
| 1.5" thick, R10.7 | " | 1.00 | 0.24 | 1.24 |
| 2" thick, R14.3 | " | 1.15 | 0.25 | 1.40 |
| Foamglass | | | | |
| 1" thick, R1.8 | S.F. | 2.30 | 0.23 | 2.53 |
| 2" thick, R5.26 | " | 2.90 | 0.25 | 3.15 |
| Wood fiber | | | | |
| 1" thick, R3.85 | S.F. | 2.55 | 0.23 | 2.78 |
| 2" thick, R7.7 | " | 3.10 | 0.25 | 3.35 |
| Particle board | | | | |
| 3/4" thick, R2.08 | S.F. | 0.68 | 0.23 | 0.91 |

| INSULATION | UNIT | MAT. | INST. | TOTAL |
|---|---|---|---|---|
| **07210.20 BOARD INSULATION** | | | | |
| 1" thick, R2.77 | S.F. | 0.76 | 0.23 | 0.99 |
| 2" thick, R5.50 | " | 0.78 | 0.25 | 1.03 |
| **07210.60 LOOSE FILL INSULATION** | | | | |
| Blown-in type | | | | |
| Fiberglass | | | | |
| 5" thick, R11 | S.F. | 0.26 | 0.23 | 0.49 |
| 6" thick, R13 | " | 0.31 | 0.28 | 0.59 |
| 9" thick, R19 | " | 0.36 | 0.40 | 0.76 |
| Rockwool, attic application | | | | |
| 6" thick, R13 | S.F. | 0.46 | 0.28 | 0.74 |
| 8" thick, R19 | " | 0.57 | 0.35 | 0.92 |
| 10" thick, R22 | " | 0.68 | 0.43 | 1.11 |
| 12" thick, R26 | " | 0.86 | 0.46 | 1.32 |
| 15" thick, R30 | " | 1.05 | 0.56 | 1.61 |
| Poured type | | | | |
| Fiberglass | | | | |
| 1" thick, R4 | S.F. | 0.23 | 0.17 | 0.40 |
| 2" thick, R8 | " | 0.43 | 0.20 | 0.63 |
| 3" thick, R12 | " | 0.65 | 0.23 | 0.88 |
| 4" thick, R16 | " | 0.85 | 0.28 | 1.13 |
| Mineral wool | | | | |
| 1" thick, R3 | S.F. | 0.26 | 0.17 | 0.43 |
| 2" thick, R6 | " | 0.49 | 0.20 | 0.69 |
| 3" thick, R9 | " | 0.74 | 0.23 | 0.97 |
| 4" thick, R12 | " | 0.85 | 0.28 | 1.13 |
| Vermiculite or perlite | | | | |
| 2" thick, R4.8 | S.F. | 0.58 | 0.20 | 0.78 |
| 3" thick, R7.2 | " | 0.82 | 0.23 | 1.05 |
| 4" thick, R9.6 | " | 1.10 | 0.28 | 1.38 |
| Masonry, poured vermiculite or perlite | | | | |
| 4" block | S.F. | 0.43 | 0.14 | 0.57 |
| 6" block | " | 0.55 | 0.17 | 0.72 |
| 8" block | " | 0.84 | 0.20 | 1.04 |
| 10" block | " | 1.00 | 0.21 | 1.21 |
| 12" block | " | 1.10 | 0.23 | 1.33 |
| **07210.70 SPRAYED INSULATION** | | | | |
| Foam, sprayed on | | | | |
| Polystyrene | | | | |
| 1" thick, R4 | S.F. | 0.54 | 0.28 | 0.82 |
| 2" thick, R8 | " | 1.10 | 0.37 | 1.47 |
| Urethane | | | | |
| 1" thick, R4 | S.F. | 0.49 | 0.28 | 0.77 |
| 2" thick, R8 | " | 1.00 | 0.37 | 1.37 |
| **07250.10 FIREPROOFING** | | | | |
| Sprayed on | | | | |
| 1" thick | | | | |
| On beams | S.F. | 0.53 | 0.62 | 1.15 |
| On columns | " | 0.54 | 0.56 | 1.10 |

# 07 THERMAL AND MOISTURE

## INSULATION

| 07250.10 FIREPROOFING | UNIT | MAT. | INST. | TOTAL |
|---|---|---|---|---|
| On decks | | | | |
| Flat surface | S.F. | 0.54 | 0.28 | 0.82 |
| Fluted surface | " | 0.68 | 0.35 | 1.03 |
| 1-1/2" thick | | | | |
| On beams | S.F. | 0.95 | 0.79 | 1.74 |
| On columns | " | 1.10 | 0.69 | 1.79 |
| On decks | | | | |
| Flat surface | S.F. | 0.81 | 0.35 | 1.16 |
| Fluted surface | " | 0.95 | 0.46 | 1.41 |

## SHINGLES AND TILES

| 07310.10 ASPHALT SHINGLES | UNIT | MAT. | INST. | TOTAL |
|---|---|---|---|---|
| Standard asphalt shingles, strip shingles | | | | |
| 210 lb/square | SQ. | 44.10 | 34.10 | 78.20 |
| 235 lb/square | " | 46.80 | 37.90 | 84.70 |
| 240 lb/square | " | 48.20 | 42.60 | 90.80 |
| 260 lb/square | " | 69.00 | 48.70 | 117.70 |
| 300 lb/square | " | 75.50 | 57.00 | 132.50 |
| 385 lb/square | " | 100.00 | 68.00 | 168.00 |
| Roll roofing, mineral surface | | | | |
| 90 lb | SQ. | 24.80 | 24.35 | 49.15 |
| 110 lb | " | 41.30 | 28.40 | 69.70 |
| 140 lb | " | 42.70 | 34.10 | 76.80 |

| 07310.30 METAL SHINGLES | UNIT | MAT. | INST. | TOTAL |
|---|---|---|---|---|
| Aluminum, .020" thick | | | | |
| Plain | SQ. | 210.00 | 68.00 | 278.00 |
| Colors | " | 230.00 | 68.00 | 298.00 |
| Steel, galvanized | | | | |
| 26 ga. | | | | |
| Plain | SQ. | 210.00 | 68.00 | 278.00 |
| Colors | " | 260.00 | 68.00 | 328.00 |
| 24 ga. | | | | |
| Plain | SQ. | 220.00 | 68.00 | 288.00 |
| Colors | " | 280.00 | 68.00 | 348.00 |
| Porcelain enamel, 22 ga. | | | | |
| Minimum | SQ. | 630.00 | 85.00 | 715.00 |
| Average | " | 720.00 | 85.00 | 805.00 |
| Maximum | " | 810.00 | 85.00 | 895.00 |

## SHINGLES AND TILES

| | UNIT | MAT. | INST. | TOTAL |
|---|---|---|---|---|
| **07310.60 SLATE SHINGLES** | | | | |
| Slate shingles | | | | |
| Pennsylvania | | | | |
| Ribbon | SQ. | 510.00 | 170.00 | 680.00 |
| Clear | " | 650.00 | 170.00 | 820.00 |
| Vermont | | | | |
| Black | SQ. | 550.00 | 170.00 | 720.00 |
| Gray | " | 610.00 | 170.00 | 780.00 |
| Green | " | 620.00 | 170.00 | 790.00 |
| Red | " | 1,110 | 170.00 | 1,280 |
| Replacement shingles | | | | |
| Small jobs | EA. | 9.75 | 11.35 | 21.10 |
| Large jobs | S.F. | 7.65 | 5.70 | 13.35 |
| **07310.70 WOOD SHINGLES** | | | | |
| Wood shingles, on roofs | | | | |
| White cedar, #1 shingles | | | | |
| 4" exposure | SQ. | 190.00 | 110.00 | 300.00 |
| 5" exposure | " | 170.00 | 85.00 | 255.00 |
| #2 shingles | | | | |
| 4" exposure | SQ. | 130.00 | 110.00 | 240.00 |
| 5" exposure | " | 110.00 | 85.00 | 195.00 |
| Resquared and rebutted | | | | |
| 4" exposure | SQ. | 170.00 | 110.00 | 280.00 |
| 5" exposure | " | 140.00 | 85.00 | 225.00 |
| On walls | | | | |
| White cedar, #1 shingles | | | | |
| 4" exposure | SQ. | 190.00 | 170.00 | 360.00 |
| 5" exposure | " | 170.00 | 140.00 | 310.00 |
| 6" exposure | " | 140.00 | 110.00 | 250.00 |
| #2 shingles | | | | |
| 4" exposure | SQ. | 130.00 | 170.00 | 300.00 |
| 5" exposure | " | 110.00 | 140.00 | 250.00 |
| 6" exposure | " | 94.50 | 110.00 | 204.50 |
| Add for fire retarding | " | | | 80.86 |
| **07310.80 WOOD SHAKES** | | | | |
| Shakes, hand split, 24" red cedar, on roofs | | | | |
| 5" exposure | SQ. | 190.00 | 170.00 | 360.00 |
| 7" exposure | " | 180.00 | 140.00 | 320.00 |
| 9" exposure | " | 170.00 | 110.00 | 280.00 |
| On walls | | | | |
| 6" exposure | SQ. | 180.00 | 170.00 | 350.00 |
| 8" exposure | " | 170.00 | 140.00 | 310.00 |
| 10" exposure | " | 170.00 | 110.00 | 280.00 |
| Add for fire retarding | " | | | 58.50 |

## ROOFING AND SIDING

### 07410.10 MANUFACTURED ROOFS

| | UNIT | MAT. | INST. | TOTAL |
|---|---|---|---|---|
| Aluminum roof panels, for structural steel framing | | | | |
| Corrugated | | | | |
| Unpainted finish | | | | |
| .024" | S.F. | 1.50 | 0.85 | 2.35 |
| .030" | " | 1.75 | 0.85 | 2.60 |
| Painted finish | | | | |
| .024" | S.F. | 1.90 | 0.85 | 2.75 |
| .030" | " | 2.30 | 0.85 | 3.15 |
| V-beam | | | | |
| Unpainted finish | | | | |
| .032" | S.F. | 1.95 | 0.85 | 2.80 |
| .040" | " | 2.35 | 0.85 | 3.20 |
| .050" | " | 2.95 | 0.85 | 3.80 |
| Painted finish | | | | |
| .032" | S.F. | 2.55 | 0.85 | 3.40 |
| .040" | " | 3.05 | 0.85 | 3.90 |
| .050" | " | 3.65 | 0.85 | 4.50 |
| Steel roof panels, for structural steel framing | | | | |
| Corrugated, painted | | | | |
| 18 ga. | S.F. | 4.40 | 0.85 | 5.25 |
| 20 ga. | " | 4.10 | 0.85 | 4.95 |
| 22 ga. | " | 3.65 | 0.85 | 4.50 |
| Box rib, painted | | | | |
| 18 ga. | S.F. | 5.15 | 0.90 | 6.05 |
| 20 ga. | " | 4.25 | 0.90 | 5.15 |
| 22 ga. | " | 3.80 | 0.90 | 4.70 |
| 4" rib, painted | | | | |
| 18 ga. | S.F. | 5.95 | 0.95 | 6.90 |
| 20 ga. | " | 5.00 | 0.95 | 5.95 |
| 22 ga. | " | 4.50 | 0.95 | 5.45 |
| Standing seam roof | | | | |
| 2" high seam, painted | | | | |
| 22 ga. | S.F. | 5.75 | 1.35 | 7.10 |
| 24 ga. | " | 5.65 | 1.35 | 7.00 |
| 26 ga. | " | 5.50 | 1.35 | 6.85 |

### 07410.30 MANUFACTURED WALLS

| | UNIT | MAT. | INST. | TOTAL |
|---|---|---|---|---|
| Sandwich panels with 1-1/2" fiberglass insulation | | | | |
| Galvanized 18 ga. steel interior panels | | | | |
| Exterior panels | | | | |
| 16 ga. aluminum | S.F. | 5.50 | 5.15 | 10.65 |
| 18 ga. galvanized steel | " | 7.75 | 5.15 | 12.90 |
| 20 ga. painted steel | " | 7.90 | 5.15 | 13.05 |
| 20 ga. stainless steel | " | 8.05 | 5.15 | 13.20 |
| Metal liner panels, 1-3/8" thick, 24" wide | | | | |
| Galvanized | | | | |
| 22 ga. | S.F. | 2.90 | 1.30 | 4.20 |
| 20 ga. | " | 3.15 | 1.30 | 4.45 |
| 18 ga. | " | 3.90 | 1.30 | 5.20 |
| Primed | | | | |
| 22 ga. | S.F. | 2.25 | 1.30 | 3.55 |

# ROOFING AND SIDING

| ROOFING AND SIDING | UNIT | MAT. | INST. | TOTAL |
|---|---|---|---|---|
| **07410.30 MANUFACTURED WALLS** | | | | |
| 20 ga. | S.F. | 2.70 | 1.30 | 4.00 |
| 18 ga. | " | 3.15 | 1.30 | 4.45 |
| **07440.10 AGGREGATE COATED PANELS** | | | | |
| Dryvit type system | | | | |
| 1" thick | S.F. | 2.05 | 1.30 | 3.35 |
| 1-1/2" thick | " | 2.10 | 1.40 | 3.50 |
| 2" thick | " | 2.25 | 1.60 | 3.85 |
| **07460.10 METAL SIDING PANELS** | | | | |
| Aluminum siding panels | | | | |
| Corrugated | | | | |
| Plain finish | | | | |
| .024" | S.F. | 1.30 | 1.55 | 2.85 |
| .032" | " | 1.60 | 1.55 | 3.15 |
| Painted finish | | | | |
| .024" | S.F. | 1.65 | 1.55 | 3.20 |
| .032" | " | 1.95 | 1.55 | 3.50 |
| V. beam | | | | |
| Plain finish | | | | |
| .032" | S.F. | 1.80 | 1.55 | 3.35 |
| .040" | " | 2.10 | 1.55 | 3.65 |
| .050" | " | 2.65 | 1.55 | 4.20 |
| Painted finish | | | | |
| .032" | S.F. | 2.25 | 1.55 | 3.80 |
| .040" | " | 2.65 | 1.55 | 4.20 |
| .050" | " | 3.40 | 1.55 | 4.95 |
| 4" rib | | | | |
| Plain finish | | | | |
| .032" | S.F. | 1.75 | 1.75 | 3.50 |
| .040" | " | 1.90 | 1.75 | 3.65 |
| .050" | " | 2.30 | 1.75 | 4.05 |
| Painted finish | | | | |
| .032" | S.F. | 2.15 | 1.75 | 3.90 |
| .040" | " | 2.30 | 1.75 | 4.05 |
| .050" | " | 2.65 | 1.75 | 4.40 |
| Steel siding panels | | | | |
| Corrugated | | | | |
| 22 ga. | S.F. | 1.55 | 2.55 | 4.10 |
| 24 ga. | " | 1.40 | 2.55 | 3.95 |
| 26 ga. | " | 1.25 | 2.55 | 3.80 |
| Box rib | | | | |
| 20 ga. | S.F. | 2.75 | 2.55 | 5.30 |
| 22 ga. | " | 2.30 | 2.55 | 4.85 |
| 24 ga. | " | 2.05 | 2.55 | 4.60 |
| 26 ga. | " | 1.45 | 2.55 | 4.00 |
| **07460.50 PLASTIC SIDING** | | | | |
| Horizontal vinyl siding, solid | | | | |
| 8" wide | | | | |
| Standard | S.F. | 0.90 | 1.35 | 2.25 |
| Insulated | " | 1.10 | 1.35 | 2.45 |

# 07 THERMAL AND MOISTURE

| ROOFING AND SIDING | UNIT | MAT. | INST. | TOTAL |
|---|---|---|---|---|
| **07460.50 PLASTIC SIDING** | | | | |
| 10" wide | | | | |
| Standard | S.F. | 0.93 | 1.25 | 2.18 |
| Insulated | " | 1.15 | 1.25 | 2.40 |
| Vinyl moldings for doors and windows | L.F. | 0.58 | 1.40 | 1.98 |
| **07460.60 PLYWOOD SIDING** | | | | |
| Rough sawn cedar, 3/8" thick | S.F. | 1.35 | 1.20 | 2.55 |
| Fir, 3/8" thick | " | 0.74 | 1.20 | 1.94 |
| Texture 1-11, 5/8" thick | | | | |
| Cedar | S.F. | 1.80 | 1.25 | 3.05 |
| Fir | " | 1.30 | 1.25 | 2.55 |
| Redwood | " | 1.95 | 1.20 | 3.15 |
| Southern Yellow Pine | " | 1.00 | 1.25 | 2.25 |
| **07460.70 STEEL SIDING** | | | | |
| Ribbed, sheets, galvanized | | | | |
| 22 ga. | S.F. | 1.55 | 1.55 | 3.10 |
| 24 ga. | " | 1.40 | 1.55 | 2.95 |
| 26 ga. | " | 1.10 | 1.55 | 2.65 |
| 28 ga. | " | 0.92 | 1.55 | 2.47 |
| Primed | | | | |
| 24 ga. | S.F. | 1.80 | 1.55 | 3.35 |
| 26 ga. | " | 1.30 | 1.55 | 2.85 |
| 28 ga. | " | 1.10 | 1.55 | 2.65 |
| **07460.80 WOOD SIDING** | | | | |
| Beveled siding, cedar | | | | |
| A grade | | | | |
| 1/2 x 6 | S.F. | 2.50 | 1.75 | 4.25 |
| 1/2 x 8 | " | 2.55 | 1.40 | 3.95 |
| 3/4 x 10 | " | 3.30 | 1.20 | 4.50 |
| Clear | | | | |
| 1/2 x 6 | S.F. | 2.80 | 1.75 | 4.55 |
| 1/2 x 8 | " | 2.85 | 1.40 | 4.25 |
| 3/4 x 10 | " | 3.85 | 1.20 | 5.05 |
| B grade | | | | |
| 1/2 x 6 | S.F. | 2.15 | 1.75 | 3.90 |
| 1/2 x 8 | " | 2.45 | 14.20 | 16.65 |
| 3/4 x 10 | " | 2.30 | 1.20 | 3.50 |
| Board and batten | | | | |
| Cedar | | | | |
| 1x6 | S.F. | 3.10 | 1.75 | 4.85 |
| 1x8 | " | 2.85 | 1.40 | 4.25 |
| 1x10 | " | 2.55 | 1.25 | 3.80 |
| 1x12 | " | 2.30 | 1.15 | 3.45 |
| Pine | | | | |
| 1x6 | S.F. | 0.95 | 1.75 | 2.70 |
| 1x8 | " | 0.92 | 1.40 | 2.32 |
| 1x10 | " | 0.88 | 1.25 | 2.13 |
| 1x12 | " | 0.81 | 1.15 | 1.96 |

## ROOFING AND SIDING

| 07460.80 WOOD SIDING | UNIT | MAT. | INST. | TOTAL |
|---|---|---|---|---|
| Redwood | | | | |
| 1x6 | S.F. | 4.05 | 1.75 | 5.80 |
| 1x8 | " | 3.80 | 1.40 | 5.20 |
| 1x10 | " | 3.50 | 1.25 | 4.75 |
| 1x12 | " | 3.25 | 1.15 | 4.40 |
| Tongue and groove | | | | |
| Cedar | | | | |
| 1x4 | S.F. | 3.50 | 1.95 | 5.45 |
| 1x6 | " | 3.35 | 1.85 | 5.20 |
| 1x8 | " | 3.15 | 1.75 | 4.90 |
| 1x10 | " | 3.10 | 1.70 | 4.80 |
| Pine | | | | |
| 1x4 | S.F. | 1.05 | 1.95 | 3.00 |
| 1x6 | " | 1.00 | 1.85 | 2.85 |
| 1x8 | " | 0.93 | 1.75 | 2.68 |
| 1x10 | " | 0.88 | 1.70 | 2.58 |
| Redwood | | | | |
| 1x4 | S.F. | 3.70 | 1.95 | 5.65 |
| 1x6 | " | 3.60 | 1.85 | 5.45 |
| 1x8 | " | 3.45 | 1.75 | 5.20 |
| 1x10 | " | 3.30 | 1.70 | 5.00 |

## MEMBRANE ROOFING

| 07510.10 BUILT-UP ASPHALT ROOFING | UNIT | MAT. | INST. | TOTAL |
|---|---|---|---|---|
| Built-up roofing, asphalt felt, including gravel | | | | |
| 2 ply | SQ. | 32.10 | 85.00 | 117.10 |
| 3 ply | " | 43.90 | 110.00 | 153.90 |
| 4 ply | " | 62.50 | 140.00 | 202.50 |
| Walkway, for built-up roofs | | | | |
| 3' x 3' x | | | | |
| 1/2" thick | S.F. | 3.05 | 1.15 | 4.20 |
| 3/4" thick | " | 4.05 | 1.15 | 5.20 |
| 1" thick | " | 4.45 | 1.15 | 5.60 |
| Roof bonds | | | | |
| Asphalt felt | | | | |
| 10 yrs | SQ. | | | 24.33 |
| 20 yrs | " | | | 25.72 |
| Cant strip, 4" x 4" | | | | |
| Treated wood | L.F. | 1.40 | 0.97 | 2.37 |
| Foamglass | " | 0.73 | 0.85 | 1.58 |
| Mineral fiber | " | 0.31 | 0.85 | 1.16 |
| New gravel for built-up roofing, 400 lb/sq | SQ. | 22.60 | 68.00 | 90.60 |

## MEMBRANE ROOFING

| MEMBRANE ROOFING | UNIT | MAT. | INST. | TOTAL |
|---|---|---|---|---|
| **07510.10 BUILT-UP ASPHALT ROOFING** | | | | |
| Roof gravel (ballast) | C.Y. | 15.15 | 170.00 | 185.15 |
| Aluminum coating, top surfacing, for built-up roofing | SQ. | 23.90 | 57.00 | 80.90 |
| Remove 4-ply built-up roof (includes gravel) | " | 0.00 | 170.00 | 170.00 |
| Remove & replace gravel, includes flood coat | " | 32.00 | 110.00 | 142.00 |
| **07530.10 SINGLE-PLY ROOFING** | | | | |
| Elastic sheet roofing | | | | |
| Neoprene, 1/16" thick | S.F. | 1.80 | 0.43 | 2.23 |
| EPDM rubber | | | | |
| 45 mil | S.F. | 0.97 | 0.43 | 1.40 |
| 60 mil | " | 1.35 | 0.43 | 1.78 |
| PVC | | | | |
| 45 mil | S.F. | 1.35 | 0.43 | 1.78 |
| 60 mil | " | 1.60 | 0.43 | 2.03 |
| Flashing | | | | |
| Pipe flashing, 90 mil thick | | | | |
| 1" pipe | EA. | 19.15 | 8.50 | 27.65 |
| 2" pipe | " | 20.55 | 8.50 | 29.05 |
| 3" pipe | " | 20.70 | 8.95 | 29.65 |
| 4" pipe | " | 22.40 | 8.95 | 31.35 |
| 5" pipe | " | 24.00 | 9.45 | 33.45 |
| 6" pipe | " | 26.20 | 9.45 | 35.65 |
| 8" pipe | " | 29.80 | 10.00 | 39.80 |
| 10" pipe | " | 34.00 | 11.35 | 45.35 |
| 12" pipe | " | 41.80 | 11.35 | 53.15 |
| Neoprene flashing, 60 mil thick strip | | | | |
| 6" wide | L.F. | 1.15 | 2.85 | 4.00 |
| 12" wide | " | 2.25 | 4.25 | 6.50 |
| 18" wide | " | 3.35 | 5.70 | 9.05 |
| 24" wide | " | 4.40 | 8.50 | 12.90 |
| Adhesives | | | | |
| Mastic sealer, applied at joints only | | | | |
| 1/4" bead | L.F. | 0.07 | 0.17 | 0.24 |
| Fluid applied roofing | | | | |
| Urethane, 2 components, elastomeric top membrane | | | | |
| 1" thick | S.F. | 2.00 | 0.57 | 2.57 |
| Vinyl liquid roofing, 2 coats, 2 mils per coat | " | 3.10 | 0.49 | 3.59 |
| Silicone roofing, 2 coats sprayed, 16 mil per coat | " | 2.35 | 0.57 | 2.92 |
| Inverted roof system | | | | |
| Insulated membrane with coarse gravel ballast | | | | |
| 3 ply with 2" polystyrene | S.F. | 4.25 | 0.57 | 4.82 |
| Ballast, 3/4" through 1-1/2" dia. river gravel, 100lb/sf | " | 0.28 | 34.10 | 34.38 |
| Walkway for membrane roofs, 1/2" thick | " | 1.60 | 1.15 | 2.75 |

## FLASHING AND SHEET METAL

### 07610.10 METAL ROOFING

| | UNIT | MAT. | INST. | TOTAL |
|---|---|---|---|---|
| Sheet metal roofing, copper, 16 oz, batten seam | SQ. | 570.00 | 230.00 | 800.00 |
| Standing seam | " | 560.00 | 210.00 | 770.00 |
| Aluminum roofing, natural finish | | | | |
| Corrugated, on steel frame | | | | |
| .0175" thick | SQ. | 100.00 | 97.50 | 197.50 |
| .0215" thick | " | 140.00 | 97.50 | 237.50 |
| .024" thick | " | 170.00 | 97.50 | 267.50 |
| .032" thick | " | 200.00 | 97.50 | 297.50 |
| V-beam, on steel frame | | | | |
| .032" thick | SQ. | 190.00 | 97.50 | 287.50 |
| .040" thick | " | 210.00 | 97.50 | 307.50 |
| .050" thick | " | 260.00 | 97.50 | 357.50 |
| Ridge cap | | | | |
| .019" thick | L.F. | 3.10 | 1.15 | 4.25 |
| Corrugated galvanized steel roofing, on steel frame | | | | |
| 28 ga. | SQ. | 100.00 | 97.50 | 197.50 |
| 26 ga. | " | 110.00 | 97.50 | 207.50 |
| 24 ga. | " | 130.00 | 97.50 | 227.50 |
| 22 ga. | " | 140.00 | 97.50 | 237.50 |
| 26 ga., factory insulated with 1" polystyrene | " | 290.00 | 140.00 | 430.00 |
| Ridge roll | | | | |
| 10" wide | L.F. | 1.55 | 1.15 | 2.70 |
| 20" wide | " | 3.15 | 1.35 | 4.50 |

### 07620.10 FLASHING AND TRIM

| | UNIT | MAT. | INST. | TOTAL |
|---|---|---|---|---|
| Counter flashing | | | | |
| Aluminum, .032" | S.F. | 1.15 | 3.40 | 4.55 |
| Stainless steel, .015" | " | 3.35 | 3.40 | 6.75 |
| Copper | | | | |
| 16 oz. | S.F. | 3.60 | 3.40 | 7.00 |
| 20 oz. | " | 4.60 | 3.40 | 8.00 |
| 24 oz. | " | 5.60 | 3.40 | 9.00 |
| 32 oz. | " | 7.20 | 3.40 | 10.60 |
| Valley flashing | | | | |
| Aluminum, .032" | S.F. | 1.15 | 2.15 | 3.30 |
| Stainless steel, .015 | " | 3.35 | 2.15 | 5.50 |
| Copper | | | | |
| 16 oz. | S.F. | 3.60 | 2.15 | 5.75 |
| 20 oz. | " | 4.60 | 2.85 | 7.45 |
| 24 oz. | " | 5.60 | 2.15 | 7.75 |
| 32 oz. | " | 7.20 | 2.15 | 9.35 |
| Base flashing | | | | |
| Aluminum, .040" | S.F. | 2.05 | 2.85 | 4.90 |
| Stainless steel, .018" | " | 4.10 | 2.85 | 6.95 |
| Copper | | | | |
| 16 oz. | S.F. | 3.60 | 2.85 | 6.45 |
| 20 oz. | " | 4.60 | 2.15 | 6.75 |
| 24 oz. | " | 5.60 | 2.85 | 8.45 |
| 32 oz. | " | 7.20 | 2.85 | 10.05 |
| Waterstop, "T" section, 22 ga. | | | | |
| 1-1/2" x 3" | L.F. | 1.90 | 1.70 | 3.60 |

## FLASHING AND SHEET METAL

### 07620.10 FLASHING AND TRIM

| Description | UNIT | MAT. | INST. | TOTAL |
|---|---|---|---|---|
| 2" x 2" | L.F. | 2.05 | 1.70 | 3.75 |
| 4" x 3" | " | 2.35 | 1.70 | 4.05 |
| 6" x 4" | " | 2.50 | 1.70 | 4.20 |
| 8" x 4" | " | 3.10 | 1.70 | 4.80 |
| Scupper outlets | | | | |
| 10" x 10" x 4" | EA. | 20.00 | 8.50 | 28.50 |
| 22" x 4" x 4" | " | 24.70 | 8.50 | 33.20 |
| 8" x 8" x 5" | " | 20.00 | 8.50 | 28.50 |
| Flashing and trim, aluminum | | | | |
| .019" thick | S.F. | 0.96 | 2.45 | 3.41 |
| .032" thick | " | 1.15 | 2.45 | 3.60 |
| .040" thick | " | 2.05 | 2.60 | 4.65 |
| Neoprene sheet flashing, .060" thick | " | 1.55 | 2.15 | 3.70 |
| Copper, paper backed | | | | |
| 2 oz. | S.F. | 1.15 | 3.40 | 4.55 |
| Drainage boots, roof, cast iron | | | | |
| 2 x 3 | L.F. | 37.10 | 4.25 | 41.35 |
| 3 x 4 | " | 46.30 | 4.25 | 50.55 |
| 4 x 5 | " | 66.50 | 4.55 | 71.05 |
| 4 x 6 | " | 63.50 | 4.55 | 68.05 |
| 5 x 7 | " | 75.50 | 4.85 | 80.35 |
| Pitch pocket, copper, 16 oz. | | | | |
| 4 x 4 | EA. | 62.00 | 8.50 | 70.50 |
| 6 x 6 | " | 67.50 | 8.50 | 76.00 |
| 8 x 8 | " | 73.00 | 8.50 | 81.50 |
| 8 x 10 | " | 78.50 | 8.50 | 87.00 |
| 8 x 12 | " | 94.50 | 8.50 | 103.00 |
| Reglets, copper 10 oz. | L.F. | 2.50 | 2.25 | 4.75 |
| Stainless steel, .020" | " | 1.90 | 2.25 | 4.15 |
| Gravel stop | | | | |
| Aluminum, .032" | | | | |
| 4" | L.F. | 3.85 | 1.15 | 5.00 |
| 6" | " | 4.60 | 1.15 | 5.75 |
| 8" | " | 6.95 | 1.30 | 8.25 |
| 10" | " | 8.15 | 1.30 | 9.45 |
| Copper, 16 oz. | | | | |
| 4" | L.F. | 5.65 | 1.15 | 6.80 |
| 6" | " | 7.30 | 1.15 | 8.45 |
| 8" | " | 9.20 | 1.30 | 10.50 |
| 10" | " | 10.25 | 1.30 | 11.55 |

### 07620.20 GUTTERS AND DOWNSPOUTS

| Description | UNIT | MAT. | INST. | TOTAL |
|---|---|---|---|---|
| Copper gutter and downspout | | | | |
| Downspouts, 16 oz. copper | | | | |
| Round | | | | |
| 3" dia. | L.F. | 5.15 | 2.25 | 7.40 |
| 4" dia. | " | 6.40 | 2.25 | 8.65 |
| Rectangular, corrugated | | | | |
| 2" x 3" | L.F. | 5.00 | 2.15 | 7.15 |
| 3" x 4" | " | 6.15 | 2.15 | 8.30 |
| Rectangular, flat surface | | | | |
| 2" x 3" | L.F. | 5.70 | 2.25 | 7.95 |

## FLASHING AND SHEET METAL

| 07620.20 GUTTERS AND DOWNSPOUTS | UNIT | MAT. | INST. | TOTAL |
|---|---|---|---|---|
| 3" x 4" | L.F. | 8.05 | 2.25 | 10.30 |
| Lead-coated copper downspouts | | | | |
| Round | | | | |
| 3" dia. | L.F. | 6.40 | 2.15 | 8.55 |
| 4" dia. | " | 7.70 | 2.45 | 10.15 |
| Rectangular, corrugated | | | | |
| 2" x 3" | L.F. | 4.50 | 2.25 | 6.75 |
| 3" x 4" | " | 5.40 | 2.25 | 7.65 |
| Rectangular, plain | | | | |
| 2" x 3" | L.F. | 4.25 | 2.25 | 6.50 |
| 3" x 4" | " | 5.10 | 2.25 | 7.35 |
| Gutters, 16 oz. copper | | | | |
| Half round | | | | |
| 4" wide | L.F. | 4.60 | 3.40 | 8.00 |
| 5" wide | " | 5.65 | 3.80 | 9.45 |
| Type K | | | | |
| 4" wide | L.F. | 5.15 | 3.40 | 8.55 |
| 5" wide | " | 5.40 | 3.80 | 9.20 |
| Lead-coated copper gutters | | | | |
| Half round | | | | |
| 4" wide | L.F. | 5.40 | 3.40 | 8.80 |
| 6" wide | " | 7.35 | 3.80 | 11.15 |
| Type K | | | | |
| 4" wide | L.F. | 6.15 | 3.40 | 9.55 |
| 5" wide | " | 6.65 | 3.80 | 10.45 |
| Aluminum gutter and downspout | | | | |
| Downspouts | | | | |
| 2" x 3" | L.F. | 0.81 | 2.25 | 3.06 |
| 3" x 4" | " | 1.10 | 2.45 | 3.55 |
| 4" x 5" | " | 1.25 | 2.60 | 3.85 |
| Round | | | | |
| 3" dia. | L.F. | 1.30 | 2.25 | 3.55 |
| 4" dia. | " | 1.70 | 2.45 | 4.15 |
| Gutters, stock units | | | | |
| 4" wide | L.F. | 1.35 | 3.60 | 4.95 |
| 5" wide | " | 1.45 | 3.80 | 5.25 |
| Galvanized steel gutter and downspout | | | | |
| Downspouts, round corrugated | | | | |
| 3" dia. | L.F. | 1.15 | 2.25 | 3.40 |
| 4" dia. | " | 1.30 | 2.25 | 3.55 |
| 5" dia. | " | 1.55 | 2.45 | 4.00 |
| 6" dia. | " | 2.00 | 2.45 | 4.45 |
| Rectangular | | | | |
| 2" x 3" | L.F. | 1.15 | 2.25 | 3.40 |
| 3" x 4" | " | 1.85 | 2.15 | 4.00 |
| 4" x 4" | " | 2.30 | 2.15 | 4.45 |
| Gutters, stock units | | | | |
| 5" wide | | | | |
| Plain | L.F. | 1.30 | 3.80 | 5.10 |
| Painted | " | 1.40 | 3.80 | 5.20 |
| 6" wide | | | | |
| Plain | L.F. | 1.70 | 4.00 | 5.70 |

## FLASHING AND SHEET METAL

| 07620.20 GUTTERS AND DOWNSPOUTS | UNIT | MAT. | INST. | TOTAL |
|---|---|---|---|---|
| Painted | L.F. | 2.00 | 4.00 | 6.00 |

## ROOFING SPECIALTIES

| 07700.10 MANUFACTURED SPECIALTIES | UNIT | MAT. | INST. | TOTAL |
|---|---|---|---|---|
| Moisture relief vent | | | | |
| Aluminum | EA. | 15.75 | 4.85 | 20.60 |
| Copper | " | 27.90 | 4.85 | 32.75 |
| Expansion joint | | | | |
| Aluminum | | | | |
| Opening to 2.5" | L.F. | 8.25 | 2.45 | 10.70 |
| Opening to 3.5" | " | 9.05 | 2.60 | 11.65 |
| Copper, 16 oz. | | | | |
| Opening to 2.5" | L.F. | 14.20 | 2.45 | 16.65 |
| Opening to 3.5" | " | 18.20 | 2.60 | 20.80 |
| Butyl or neoprene | | | | |
| 4" wide | | | | |
| 16 oz. copper bellows | L.F. | 14.60 | 2.85 | 17.45 |
| 28 ga. stainless steel bellows | " | 13.30 | 2.85 | 16.15 |
| 6" wide | | | | |
| Copper bellows | L.F. | 14.20 | 3.10 | 17.30 |
| Stainless steel | | | | |
| Opening to 2.5" | L.F. | 14.05 | 2.45 | 16.50 |
| Opening to 3.5" | " | 17.20 | 2.60 | 19.80 |
| Smoke vent, 48" x 48" | | | | |
| Aluminum | EA. | 740.00 | 85.00 | 825.00 |
| Galvanized steel | " | 720.00 | 85.00 | 805.00 |
| Heat/smoke vent, 48" x 96" | | | | |
| Aluminum | EA. | 2,290 | 110.00 | 2,400 |
| Galvanized steel | " | 2,150 | 110.00 | 2,260 |
| Ridge vent strips | | | | |
| Mill finish | L.F. | 3.05 | 2.25 | 5.30 |
| Connectors | EA. | 14.30 | 8.50 | 22.80 |
| End cap | " | 29.40 | 9.75 | 39.15 |
| Soffit vents | | | | |
| Mill finish | | | | |
| 2-1/2" wide | L.F. | 0.38 | 1.35 | 1.73 |
| 3" wide | " | 0.45 | 1.35 | 1.80 |
| 6" wide | " | 0.78 | 1.35 | 2.13 |
| Roof hatches | | | | |
| Steel, plain, primed | | | | |
| 2'6" x 3'0" | EA. | 500.00 | 85.00 | 585.00 |

## ROOFING SPECIALTIES

### 07700.10 MANUFACTURED SPECIALTIES

| ROOFING SPECIALTIES | UNIT | MAT. | INST. | TOTAL |
|---|---|---|---|---|
| 2'6" x 4'6" | EA. | 730.00 | 110.00 | 840.00 |
| 2'6" x 8'0" | " | 1,130 | 170.00 | 1,300 |
| Galvanized steel | | | | |
| 2'6" x 3'0" | EA. | 510.00 | 85.00 | 595.00 |
| 2'6" x 4'6" | " | 770.00 | 110.00 | 880.00 |
| 2'6" x 8'0" | " | 1,200 | 170.00 | 1,370 |
| Aluminum | | | | |
| 2'6" x 3'0" | EA. | 540.00 | 85.00 | 625.00 |
| 2'6" x 4'6" | " | 810.00 | 110.00 | 920.00 |
| 2'6" x 8'0" | " | 1,500 | 170.00 | 1,670 |
| Ceiling access doors | | | | |
| Swing up model, metal frame | | | | |
| Steel door | | | | |
| 2'6" x 2'6" | EA. | 370.00 | 34.10 | 404.10 |
| 2'6" x 3'0" | " | 400.00 | 34.10 | 434.10 |
| Aluminum door | | | | |
| 2'6" x 2'6" | EA. | 400.00 | 34.10 | 434.10 |
| 2'6" x 3'0" | " | 430.00 | 34.10 | 464.10 |
| Swing down model, metal frame | | | | |
| Steel door | | | | |
| 2'6" x 2'6" | EA. | 360.00 | 34.10 | 394.10 |
| 2'6" x 3'0" | " | 390.00 | 34.10 | 424.10 |
| Aluminum door | | | | |
| 2'6" x 2'6" | EA. | 370.00 | 34.10 | 404.10 |
| 2'6" x 3'0" | " | 400.00 | 34.10 | 434.10 |
| Gravity ventilators, with curb, base, damper and screen | | | | |
| Stationary siphon | | | | |
| 6" dia. | EA. | 28.90 | 22.70 | 51.60 |
| 12" dia. | " | 44.10 | 22.70 | 66.80 |
| 24" dia. | " | 190.00 | 34.10 | 224.10 |
| 36" dia. | " | 340.00 | 34.10 | 374.10 |
| Wind driven spinner | | | | |
| 6" dia. | EA. | 44.10 | 22.70 | 66.80 |
| 12" dia. | " | 59.00 | 22.70 | 81.70 |
| 24" dia. | " | 220.00 | 34.10 | 254.10 |
| 36" dia. | " | 450.00 | 34.10 | 484.10 |
| Stationary mushroom | | | | |
| 16" dia. | EA. | 280.00 | 34.10 | 314.10 |
| 30" dia. | " | 620.00 | 42.60 | 662.60 |
| 36" dia. | " | 800.00 | 57.00 | 857.00 |
| 42" dia. | " | 1,180 | 68.00 | 1,248 |

## SKYLIGHTS

### 07810.10 PLASTIC SKYLIGHTS

| | UNIT | MAT. | INST. | TOTAL |
|---|---|---|---|---|
| Single thickness, not including mounting curb | | | | |
| 2' x 4' | EA. | 270.00 | 42.60 | 312.60 |
| 4' x 4' | " | 360.00 | 57.00 | 417.00 |
| 5' x 5' | " | 490.00 | 85.00 | 575.00 |
| 6' x 8' | " | 1,030 | 110.00 | 1,140 |
| Double thickness, not including mounting curb | | | | |
| 2' x 4' | EA. | 350.00 | 42.60 | 392.60 |
| 4' x 4' | " | 440.00 | 57.00 | 497.00 |
| 5' x 5' | " | 650.00 | 85.00 | 735.00 |
| 6' x 8' | " | 1,150 | 110.00 | 1,260 |
| Metal framed skylights | | | | |
| Translucent panels, 2-1/2" thick | S.F. | 32.00 | 3.40 | 35.40 |
| Continuous vaults, 8' wide | | | | |
| Single glazed | S.F. | 43.10 | 4.25 | 47.35 |
| Double glazed | " | 69.50 | 4.85 | 74.35 |

## JOINT SEALERS

### 07920.10 CAULKING

| | UNIT | MAT. | INST. | TOTAL |
|---|---|---|---|---|
| Caulk exterior, two component | | | | |
| 1/4 x 1/2 | L.F. | 0.28 | 1.75 | 2.03 |
| 3/8 x 1/2 | " | 0.45 | 1.95 | 2.40 |
| 1/2 x 1/2 | " | 0.59 | 2.20 | 2.79 |
| Caulk interior, single component | | | | |
| 1/4 x 1/2 | L.F. | 0.18 | 1.70 | 1.88 |
| 3/8 x 1/2 | " | 0.27 | 1.85 | 2.12 |
| 1/2 x 1/2 | " | 0.36 | 2.10 | 2.46 |
| Butyl rubber fillers | | | | |
| 1/4" x 1/4" | L.F. | 0.54 | 0.71 | 1.25 |
| 1/2" x 1/2" | " | 0.81 | 1.20 | 2.01 |
| 1/2" x 3/4" | " | 1.20 | 1.40 | 2.60 |
| 3/4" x 3/4" | " | 1.60 | 1.40 | 3.00 |
| 1" x 1" | " | 1.90 | 1.60 | 3.50 |
| Seals, "O" ring type cord | | | | |
| 1/4" dia. | L.F. | 0.61 | 0.89 | 1.50 |
| 1/2" dia. | " | 1.80 | 0.93 | 2.73 |
| 1" dia. | " | 6.40 | 0.98 | 7.38 |
| 1-1/4" dia. | " | 8.10 | 1.05 | 9.15 |
| 1-1/2" dia. | " | 10.85 | 1.10 | 11.95 |
| 1-3/4" dia. | " | 15.95 | 1.15 | 17.10 |
| 2" dia. | " | 20.60 | 1.20 | 21.80 |
| Polyvinyl chloride, closed cell | | | | |
| 1/4" x 2" | L.F. | 0.42 | 1.25 | 1.67 |

# 07 THERMAL AND MOISTURE

## JOINT SEALERS

| 07920.10 CAULKING | UNIT | MAT. | INST. | TOTAL |
|---|---|---|---|---|
| 1/4" x 6" | L.F. | 1.30 | 1.60 | 2.90 |
| Silicon foam penetration seal | | | | |
| 1/4" x 1/2" | L.F. | 0.15 | 0.44 | 0.59 |
| 1/2" x 1/2" | " | 0.26 | 0.59 | 0.85 |
| 1/2" x 3/4" | " | 0.41 | 0.71 | 1.12 |
| 3/4" x 3/4" | " | 0.61 | 0.89 | 1.50 |
| 1/8" x 1" | " | 0.15 | 0.44 | 0.59 |
| 1/8" x 3" | " | 0.41 | 0.71 | 1.12 |
| 1/4" x 3" | " | 0.81 | 0.89 | 1.70 |
| 1/4" x 6" | " | 1.65 | 1.20 | 2.85 |
| 1/2" x 6" | " | 3.25 | 2.75 | 6.00 |
| 1/2" x 9" | " | 4.90 | 4.45 | 9.35 |
| 1/2" x 12" | " | 6.45 | 6.45 | 12.90 |
| Oil base sealants and caulking | | | | |
| 1/4" x 1/4" | L.F. | 0.04 | 0.89 | 0.93 |
| 1/4" x 3/8" | " | 0.07 | 0.92 | 0.99 |
| 1/4" x 1/2" | " | 0.08 | 0.96 | 1.04 |
| 3/8" x 3/8" | " | 0.09 | 1.00 | 1.09 |
| 3/8" x 1/2" | " | 0.12 | 1.05 | 1.17 |
| 3/8" x 5/8" | " | 0.19 | 1.15 | 1.34 |
| 3/8" x 3/4" | " | 0.22 | 1.20 | 1.42 |
| 1/2" x 1/2" | " | 0.19 | 1.35 | 1.54 |
| 1/2" x 5/8" | " | 0.24 | 1.55 | 1.79 |
| 1/2" x 3/4" | " | 0.28 | 1.75 | 2.03 |
| 1/2" x 7/8" | " | 0.34 | 1.80 | 2.14 |
| 1/2" x 1" | " | 0.38 | 1.85 | 2.23 |
| 3/4" x 3/4" | " | 0.43 | 1.90 | 2.33 |
| 1" x 1" | " | 0.76 | 1.95 | 2.71 |
| Polyurethane compounds | | | | |
| 1/4" x 1/4" | L.F. | 0.14 | 0.89 | 1.03 |
| 1/4" x 3/8" | " | 0.26 | 0.92 | 1.18 |
| 1/4" x 1/2" | " | 0.34 | 0.96 | 1.30 |
| 3/8" x 3/8" | " | 0.38 | 1.00 | 1.38 |
| 3/8" x 1/2" | " | 0.46 | 1.05 | 1.51 |
| 3/8" x 5/8" | " | 0.58 | 1.15 | 1.73 |
| 3/8" x 3/4" | " | 0.72 | 1.20 | 1.92 |
| 1/2" x 1/2" | " | 0.69 | 1.35 | 2.04 |
| 1/2" x 5/8" | " | 0.74 | 1.55 | 2.29 |
| 1/2" x 3/4" | " | 0.85 | 1.75 | 2.60 |
| 1/2" x 7/8" | " | 1.10 | 1.80 | 2.90 |
| 1/2" x 1" | " | 1.35 | 1.95 | 3.30 |
| 3/4" x 3/4" | " | 1.45 | 1.90 | 3.35 |
| 3/4" x 1" | " | 1.50 | 1.95 | 3.45 |
| Backer rod, polyethylene | | | | |
| 1/4" | L.F. | 0.04 | 0.89 | 0.93 |
| 1/2" | " | 0.07 | 0.93 | 1.00 |
| 3/4" | " | 0.08 | 0.98 | 1.06 |
| 1" | " | 0.11 | 1.05 | 1.16 |

## METAL

| 08110.10 METAL DOORS | UNIT | MAT. | INST. | TOTAL |
|---|---|---|---|---|
| Flush hollow metal, standard duty, 20 ga., 1-3/8" thick | | | | |
| 2-6 x 6-8 | EA. | 180.00 | 39.40 | 219.40 |
| 2-8 x 6-8 | " | 210.00 | 39.40 | 249.40 |
| 3-0 x 6-8 | " | 230.00 | 39.40 | 269.40 |
| 1-3/4" thick | | | | |
| 2-6 x 6-8 | EA. | 210.00 | 39.40 | 249.40 |
| 2-8 x 6-8 | " | 230.00 | 39.40 | 269.40 |
| 3-0 x 6-8 | " | 240.00 | 39.40 | 279.40 |
| 2-6 x 7-0 | " | 230.00 | 39.40 | 269.40 |
| 2-8 x 7-0 | " | 240.00 | 39.40 | 279.40 |
| 3-0 x 7-0 | " | 260.00 | 39.40 | 299.40 |
| Heavy duty, 20 ga., unrated, 1-3/4" | | | | |
| 2-8 x 6-8 | EA. | 230.00 | 39.40 | 269.40 |
| 3-0 x 6-8 | " | 240.00 | 39.40 | 279.40 |
| 2-8 x 7-0 | " | 250.00 | 39.40 | 289.40 |
| 3-0 x 7-0 | " | 260.00 | 39.40 | 299.40 |
| 3-4 x 7-0 | " | 320.00 | 39.40 | 359.40 |
| 18 ga., 1-3/4", unrated door | | | | |
| 2-0 x 7-0 | EA. | 250.00 | 39.40 | 289.40 |
| 2-4 x 7-0 | " | 260.00 | 39.40 | 299.40 |
| 2-6 x 7-0 | " | 270.00 | 39.40 | 309.40 |
| 2-8 x 7-0 | " | 280.00 | 39.40 | 319.40 |
| 3-0 x 7-0 | " | 290.00 | 39.40 | 329.40 |
| 3-4 x 7-0 | " | 340.00 | 39.40 | 379.40 |
| 2", unrated door | | | | |
| 2-0 x 7-0 | EA. | 260.00 | 44.30 | 304.30 |
| 2-4 x 7-0 | " | 280.00 | 44.30 | 324.30 |
| 2-6 x 7-0 | " | 280.00 | 44.30 | 324.30 |
| 2-8 x 7-0 | " | 300.00 | 44.30 | 344.30 |
| 3-0 x 7-0 | " | 310.00 | 44.30 | 354.30 |
| 3-4 x 7-0 | " | 380.00 | 44.30 | 424.30 |
| Galvanized metal door | | | | |
| 3-0 x 7-0 | EA. | 390.00 | 44.30 | 434.30 |
| For lead lining in doors | " | | | 788.83 |
| For sound attenuation | " | | | 36.57 |
| Vision glass | | | | |
| 8" x 8" | EA. | 83.50 | 44.30 | 127.80 |
| 8" x 48" | " | 130.00 | 44.30 | 174.30 |
| Fixed metal louver | " | 130.00 | 35.40 | 165.40 |
| For fire rating, add | | | | |
| 3 hr door | EA. | | | 146.00 |
| 1-1/2 hr door | " | | | 61.36 |
| 3/4 hr door | " | | | 30.67 |
| 1' extra height, add to material, 20% | | | | |
| 1'6" extra height, add to material, 60% | | | | |
| For dutch doors with shelf, add to material, 100% | | | | |

| 08110.40 METAL DOOR FRAMES | UNIT | MAT. | INST. | TOTAL |
|---|---|---|---|---|
| Hollow metal, stock, 18 ga., 4-3/4" x 1-3/4" | | | | |
| 2-0 x 7-0 | EA. | 110.00 | 44.30 | 154.30 |
| 2-4 x 7-0 | " | 110.00 | 44.30 | 154.30 |
| 2-6 x 7-0 | " | 120.00 | 44.30 | 164.30 |

## METAL

### 08110.40 METAL DOOR FRAMES

| | UNIT | MAT. | INST. | TOTAL |
|---|---|---|---|---|
| 2-8 x 7-0 | EA. | 120.00 | 44.30 | 164.30 |
| 3-0 x 7-0 | " | 130.00 | 44.30 | 174.30 |
| 4-0 x 7-0 | " | 130.00 | 59.00 | 189.00 |
| 5-0 x 7-0 | " | 140.00 | 59.00 | 199.00 |
| 6-0 x 7-0 | " | 140.00 | 59.00 | 199.00 |
| 16 ga., 6-3/4" x 1-3/4" | | | | |
| 2-0 x 7-0 | EA. | 110.00 | 48.20 | 158.20 |
| 2-4 x 7-0 | " | 120.00 | 48.20 | 168.20 |
| 2-6 x 7-0 | " | 120.00 | 48.20 | 168.20 |
| 2-8 x 7-0 | " | 120.00 | 48.20 | 168.20 |
| 3-0 x 7-0 | " | 130.00 | 48.20 | 178.20 |
| 4-0 x 7-0 | " | 140.00 | 64.50 | 204.50 |
| 6-0 x 7-0 | " | 150.00 | 64.50 | 214.50 |
| Transom frame | | | | |
| 3-4 x 1-6 | EA. | 74.00 | 48.20 | 122.20 |
| 3-8 x 1-6 | " | 74.00 | 48.20 | 122.20 |
| 6-4 x 1-6 | " | 100.00 | 48.20 | 148.20 |
| Transom sash | | | | |
| 3-0 x 1-4 | EA. | 500.00 | 48.20 | 548.20 |
| 3-4 x 1-4 | " | 500.00 | 48.20 | 548.20 |
| 6-0 x 1-4 | " | 560.00 | 48.20 | 608.20 |
| 1' extension of frame, add | " | | | 15.61 |
| 14 ga. frame, add | " | | | 15.61 |
| For fire rating, add | | | | |
| 3 hour | EA. | | | 42.55 |
| 1-1/2 hour | " | | | 34.03 |
| 3/4 hour | " | | | 29.78 |
| Lead lining in frame, add | " | | | 92.19 |
| Sidelights, complete | | | | |
| 1-0 x 7-2 | EA. | 300.00 | 48.20 | 348.20 |
| 1-4 x 7-2 | " | 330.00 | 48.20 | 378.20 |
| 1-0 x 8-8 | " | 340.00 | 48.20 | 388.20 |
| 1-6 x 8-8 | " | 360.00 | 48.20 | 408.20 |
| 16 ga., 4-3/4" x 1-3/4" | | | | |
| 2-0 x 7-0 | EA. | 91.00 | 48.20 | 139.20 |
| 2-4 x 7-0 | " | 92.00 | 48.20 | 140.20 |
| 2-6 x 7-0 | " | 93.50 | 48.20 | 141.70 |
| 2-8 x 7-0 | " | 95.00 | 48.20 | 143.20 |
| 3-0 x 7-0 | " | 96.50 | 48.20 | 144.70 |
| 4-0 x 7-0 | " | 100.00 | 64.50 | 164.50 |
| 6-0 x 7-0 | " | 110.00 | 64.50 | 174.50 |
| Transom frame | | | | |
| 3-4 x 1-6 | EA. | 74.00 | 48.20 | 122.20 |
| 3-8 x 1-6 | " | 74.00 | 48.20 | 122.20 |
| 6-4 x 1-6 | " | 100.00 | 48.20 | 148.20 |
| Transom sash | | | | |
| 3-0 x 1-4 | EA. | 480.00 | 48.20 | 528.20 |
| 3-4 x 1-4 | " | 480.00 | 48.20 | 528.20 |
| 6-0 x 1-4 | " | 550.00 | 48.20 | 598.20 |
| 1' extension of door frame, add | " | | | 15.61 |
| 14 ga., metal frame, add | " | | | 15.61 |
| For fire rating, add | | | | |

# 08 DOORS AND WINDOWS

| METAL | UNIT | MAT. | INST. | TOTAL |
|---|---|---|---|---|
| **08110.40 METAL DOOR FRAMES** | | | | |
| 3 hour | EA. | | | 42.52 |
| 1-1/2 hour | " | | | 34.03 |
| 3/4 hour | " | | | 29.78 |
| Lead lining in frame, add | " | | | 89.36 |
| Sidelights, complete | | | | |
| 1-0 x 7-2 | EA. | 280.00 | 48.20 | 328.20 |
| 1-4 x 7-2 | " | 300.00 | 48.20 | 348.20 |
| 1-0 x 8-8 | " | 330.00 | 48.20 | 378.20 |
| 1-4 x 8-8 | " | 330.00 | 48.20 | 378.20 |
| 16 ga., 5-3/4" x 1-3/4" | | | | |
| 2-0 x 7-0 | EA. | 92.00 | 44.30 | 136.30 |
| 2-4 x 7-0 | " | 93.50 | 44.30 | 137.80 |
| 2-6 x 7-0 | " | 96.50 | 44.30 | 140.80 |
| 2-8 x 7-0 | " | 100.00 | 44.30 | 144.30 |
| 3-0 x 7-0 | " | 100.00 | 44.30 | 144.30 |
| 4-0 x 7-0 | " | 110.00 | 59.00 | 169.00 |
| 5-0 x 7-0 | " | 110.00 | 59.00 | 169.00 |
| 6-0 x 7-0 | " | 120.00 | 59.00 | 179.00 |
| Mullions, vertical | | | | |
| 5-1/4" x 1-3/4" | L.F. | 10.65 | 4.45 | 15.10 |
| 5-1/4" x 2" | " | 13.20 | 4.45 | 17.65 |
| Horizontal | | | | |
| 5-1/4" x 1-3/4" | L.F. | 10.65 | 4.45 | 15.10 |
| 5-1/4" x 2" | " | 13.20 | 4.45 | 17.65 |
| **08120.10 ALUMINUM DOORS** | | | | |
| Aluminum doors, commercial | | | | |
| Narrow stile | | | | |
| 2-6 x 7-0 | EA. | 540.00 | 190.00 | 730.00 |
| 3-0 x 7-0 | " | 570.00 | 190.00 | 760.00 |
| 3-6 x 7-0 | " | 580.00 | 190.00 | 770.00 |
| Pair | | | | |
| 5-0 x 7-0 | EA. | 900.00 | 390.00 | 1,290 |
| 6-0 x 7-0 | " | 920.00 | 390.00 | 1,310 |
| 7-0 x 7-0 | " | 960.00 | 390.00 | 1,350 |
| Wide stile | | | | |
| 2-6 x 7-0 | EA. | 830.00 | 190.00 | 1,020 |
| 3-0 x 7-0 | " | 860.00 | 190.00 | 1,050 |
| 3-6 x 7-0 | " | 890.00 | 190.00 | 1,080 |
| Pair | | | | |
| 5-0 x 7-0 | EA. | 1,530 | 390.00 | 1,920 |
| 6-0 x 7-0 | " | 1,600 | 390.00 | 1,990 |
| 7-0 x 7-0 | " | 1,630 | 390.00 | 2,020 |

# 08 DOORS AND WINDOWS

## WOOD AND PLASTIC

### 08210.10 WOOD DOORS

| | UNIT | MAT. | INST. | TOTAL |
|---|---|---|---|---|
| Solid core, 1-3/8" thick | | | | |
| Birch faced | | | | |
| 2-4 x 7-0 | EA. | 130.00 | 44.30 | 174.30 |
| 2-8 x 7-0 | " | 140.00 | 44.30 | 184.30 |
| 3-0 x 7-0 | " | 150.00 | 44.30 | 194.30 |
| 3-4 x 7-0 | " | 160.00 | 44.30 | 204.30 |
| 2-4 x 6-8 | " | 130.00 | 44.30 | 174.30 |
| 2-6 x 6-8 | " | 140.00 | 44.30 | 184.30 |
| 2-8 x 6-8 | " | 140.00 | 44.30 | 184.30 |
| 3-0 x 6-8 | " | 150.00 | 44.30 | 194.30 |
| Lauan faced | | | | |
| 2-4 x 6-8 | EA. | 110.00 | 44.30 | 154.30 |
| 2-8 x 6-8 | " | 120.00 | 44.30 | 164.30 |
| 3-0 x 6-8 | " | 130.00 | 44.30 | 174.30 |
| 3-4 x 6-8 | " | 120.00 | 44.30 | 164.30 |
| Tempered hardboard faced | | | | |
| 2-4 x 7-0 | EA. | 120.00 | 44.30 | 164.30 |
| 2-8 x 7-0 | " | 130.00 | 44.30 | 174.30 |
| 3-0 x 7-0 | " | 140.00 | 44.30 | 184.30 |
| 3-4 x 7-0 | " | 150.00 | 44.30 | 194.30 |
| Hollow core, 1-3/8" thick | | | | |
| Birch faced | | | | |
| 2-4 x 7-0 | EA. | 69.50 | 44.30 | 113.80 |
| 2-8 x 7-0 | " | 74.00 | 44.30 | 118.30 |
| 3-0 x 7-0 | " | 76.50 | 44.30 | 120.80 |
| 3-4 x 7-0 | " | 83.50 | 44.30 | 127.80 |
| Lauan faced | | | | |
| 2-4 x 6-8 | EA. | 44.00 | 44.30 | 88.30 |
| 2-6 x 6-8 | " | 46.80 | 44.30 | 91.10 |
| 2-8 x 6-8 | " | 49.60 | 44.30 | 93.90 |
| 3-0 x 6-8 | " | 54.00 | 44.30 | 98.30 |
| 3-4 x 6-8 | " | 56.50 | 44.30 | 100.80 |
| Tempered hardboard faced | | | | |
| 2-4 x 7-0 | EA. | 46.80 | 44.30 | 91.10 |
| 2-6 x 7-0 | " | 49.60 | 44.30 | 93.90 |
| 2-8 x 7-0 | " | 54.00 | 44.30 | 98.30 |
| 3-0 x 7-0 | " | 56.50 | 44.30 | 100.80 |
| 3-4 x 7-0 | " | 59.50 | 44.30 | 103.80 |
| Solid core, 1-3/4" thick | | | | |
| Birch faced | | | | |
| 2-4 x 7-0 | EA. | 160.00 | 44.30 | 204.30 |
| 2-6 x 7-0 | " | 160.00 | 44.30 | 204.30 |
| 2-8 x 7-0 | " | 170.00 | 44.30 | 214.30 |
| 3-0 x 7-0 | " | 180.00 | 44.30 | 224.30 |
| 3-4 x 7-0 | " | 190.00 | 44.30 | 234.30 |
| Lauan faced | | | | |
| 2-4 x 7-0 | EA. | 140.00 | 44.30 | 184.30 |
| 2-6 x 7-0 | " | 160.00 | 44.30 | 204.30 |
| 2-8 x 7-0 | " | 160.00 | 44.30 | 204.30 |
| 3-0 x 7-0 | " | 170.00 | 44.30 | 214.30 |
| 3-0 x 7-0 | " | 180.00 | 44.30 | 224.30 |
| Tempered hardboard faced | | | | |

| WOOD AND PLASTIC | UNIT | MAT. | INST. | TOTAL |
|---|---|---|---|---|
| **08210.10 WOOD DOORS** | | | | |
| 2-4 x 7-0 | EA. | 140.00 | 44.30 | 184.30 |
| 2-6 x 7-0 | " | 150.00 | 44.30 | 194.30 |
| 2-8 x 7-0 | " | 170.00 | 44.30 | 214.30 |
| 3-0 x 7-0 | " | 170.00 | 44.30 | 214.30 |
| 3-4 x 7-0 | " | 180.00 | 44.30 | 224.30 |
| Hollow core, 1-3/4" thick | | | | |
| Birch faced | | | | |
| 2-4 x 7-0 | EA. | 71.00 | 44.30 | 115.30 |
| 2-6 x 7-0 | " | 74.00 | 44.30 | 118.30 |
| 2-8 x 7-0 | " | 78.00 | 44.30 | 122.30 |
| 3-0 x 7-0 | " | 81.00 | 44.30 | 125.30 |
| 3-4 x 7-0 | " | 85.00 | 44.30 | 129.30 |
| Lauan faced | | | | |
| 2-4 x 6-8 | EA. | 54.00 | 44.30 | 98.30 |
| 2-6 x 6-8 | " | 56.50 | 44.30 | 100.80 |
| 2-8 x 6-8 | " | 59.50 | 44.30 | 103.80 |
| 3-0 x 6-8 | " | 62.50 | 44.30 | 106.80 |
| 3-4 x 6-8 | " | 66.50 | 44.30 | 110.80 |
| Tempered hardboard | | | | |
| 2-4 x 7-0 | EA. | 54.00 | 44.30 | 98.30 |
| 2-6 x 7-0 | " | 56.50 | 44.30 | 100.80 |
| 2-8 x 7-0 | " | 59.50 | 44.30 | 103.80 |
| 3-0 x 7-0 | " | 62.50 | 44.30 | 106.80 |
| 3-4 x 7-0 | " | 66.50 | 44.30 | 110.80 |
| Add-on, louver | " | 17.00 | 35.40 | 52.40 |
| Glass | " | 62.50 | 35.40 | 97.90 |
| Exterior doors, 3-0 x 7-0 x 2-1/2", solid core | | | | |
| Carved | | | | |
| One face | EA. | 800.00 | 88.50 | 888.50 |
| Two faces | " | 1,210 | 88.50 | 1,299 |
| Closet doors, 1-3/4" thick | | | | |
| Bi-fold or bi-passing, includes frame and trim | | | | |
| Paneled | | | | |
| 4-0 x 6-8 | EA. | 290.00 | 59.00 | 349.00 |
| 6-0 x 6-8 | " | 350.00 | 59.00 | 409.00 |
| Louvered | | | | |
| 4-0 x 6-8 | EA. | 200.00 | 59.00 | 259.00 |
| 6-0 x 6-8 | " | 250.00 | 59.00 | 309.00 |
| Flush | | | | |
| 4-0 x 6-8 | EA. | 170.00 | 59.00 | 229.00 |
| 6-0 x 6-8 | " | 210.00 | 59.00 | 269.00 |
| Primed | | | | |
| 4-0 x 6-8 | EA. | 180.00 | 59.00 | 239.00 |
| 6-0 x 6-8 | " | 220.00 | 59.00 | 279.00 |
| **08210.90 WOOD FRAMES** | | | | |
| Frame, interior, pine | | | | |
| 2-6 x 6-8 | EA. | 35.10 | 50.50 | 85.60 |
| 2-8 x 6-8 | " | 36.50 | 50.50 | 87.00 |
| 3-0 x 6-8 | " | 37.80 | 50.50 | 88.30 |
| 5-0 x 6-8 | " | 39.20 | 50.50 | 89.70 |
| 6-0 x 6-8 | " | 41.90 | 50.50 | 92.40 |

# 08 DOORS AND WINDOWS

## WOOD AND PLASTIC

| WOOD AND PLASTIC | UNIT | MAT. | INST. | TOTAL |
|---|---|---|---|---|
| **08210.90 WOOD FRAMES** | | | | |
| 2-6 x 7-0 | EA. | 37.80 | 50.50 | 88.30 |
| 2-8 x 7-0 | " | 41.90 | 50.50 | 92.40 |
| 3-0 x 7-0 | " | 43.20 | 50.50 | 93.70 |
| 5-0 x 7-0 | " | 47.30 | 71.00 | 118.30 |
| 6-0 x 7-0 | " | 48.60 | 71.00 | 119.60 |
| Exterior, custom, with threshold, including trim | | | | |
| Walnut | | | | |
| 3-0 x 7-0 | EA. | 220.00 | 88.50 | 308.50 |
| 6-0 x 7-0 | " | 260.00 | 88.50 | 348.50 |
| Oak | | | | |
| 3-0 x 7-0 | EA. | 190.00 | 88.50 | 278.50 |
| 6-0 x 7-0 | " | 250.00 | 88.50 | 338.50 |
| Pine | | | | |
| 2-4 x 7-0 | EA. | 81.00 | 71.00 | 152.00 |
| 2-6 x 7-0 | " | 83.50 | 71.00 | 154.50 |
| 2-8 x 7-0 | " | 85.00 | 71.00 | 156.00 |
| 3-0 x 7-0 | " | 93.00 | 71.00 | 164.00 |
| 3-4 x 7-0 | " | 100.00 | 71.00 | 171.00 |
| 6-0 x 7-0 | " | 180.00 | 120.00 | 300.00 |
| **08300.10 SPECIAL DOORS** | | | | |
| Vault door and frame, class 5, steel | EA. | 4,050 | 350.00 | 4,400 |
| Overhead door, coiling insulated | | | | |
| Chain gear, no frame, 12' x 12' | EA. | 2,160 | 440.00 | 2,600 |
| Aluminum, bronze glass panels, 12-9 x 13-0 | " | 2,430 | 350.00 | 2,780 |
| Garage door, flush insulated metal, primed, 9-0 x 7-0 | " | 770.00 | 120.00 | 890.00 |
| Sliding metal fire doors, motorized, fusible link, 3 hr. | | | | |
| 3-0 x 6-8 | EA. | 2,770 | 710.00 | 3,480 |
| 3-8 x 6-8 | " | 2,770 | 710.00 | 3,480 |
| 4-0 x 8-0 | " | 2,980 | 710.00 | 3,690 |
| 5-0 x 8-0 | " | 3,050 | 710.00 | 3,760 |
| Metal clad doors, including electric motor | | | | |
| Light duty | | | | |
| Minimum | S.F. | 29.70 | 5.90 | 35.60 |
| Maximum | " | 48.60 | 14.20 | 62.80 |
| Heavy duty | | | | |
| Minimum | S.F. | 58.00 | 17.70 | 75.70 |
| Maximum | " | 98.50 | 22.15 | 120.65 |
| Hangar doors, based on 150' openings | | | | |
| To 20' high | S.F. | 33.80 | 5.15 | 38.95 |
| 20' to 40' high | " | 31.10 | 3.20 | 34.30 |
| 40' to 60' high | " | 33.80 | 2.15 | 35.95 |
| 60' to 80' high | " | 39.20 | 1.30 | 40.50 |
| Over 80' high | " | 52.50 | 1.05 | 53.55 |
| Counter doors, (roll-up shutters), standard, manual | | | | |
| Opening, 4' high | | | | |
| 4' wide | EA. | 990.00 | 300.00 | 1,290 |
| 6' wide | " | 1,350 | 300.00 | 1,650 |
| 8' wide | " | 1,560 | 320.00 | 1,880 |
| 10' wide | " | 1,700 | 440.00 | 2,140 |
| 14' wide | " | 2,130 | 440.00 | 2,570 |

| WOOD AND PLASTIC | UNIT | MAT. | INST. | TOTAL |
|---|---|---|---|---|
| **08300.10 SPECIAL DOORS** | | | | |
| 6' high | | | | |
| 4' wide | EA. | 1,220 | 300.00 | 1,520 |
| 6' wide | " | 1,560 | 320.00 | 1,880 |
| 8' wide | " | 1,690 | 350.00 | 2,040 |
| 10' wide | " | 1,890 | 440.00 | 2,330 |
| 14' wide | " | 2,030 | 510.00 | 2,540 |
| For stainless steel, add to material, 40% | | | | |
| For motor operator, add | EA. | | | 1,080 |
| Service doors, (roll up shutters), standard, manual | | | | |
| Opening | | | | |
| 8' high x 8' wide | EA. | 1,260 | 200.00 | 1,460 |
| 10' high x 10' wide | " | 1,490 | 300.00 | 1,790 |
| 12' high x 12' wide | " | 1,760 | 440.00 | 2,200 |
| 14' high x 14' wide | " | 2,300 | 590.00 | 2,890 |
| 16' high x 14' wide | " | 3,110 | 590.00 | 3,700 |
| 20' high x 14' wide | " | 4,400 | 890.00 | 5,290 |
| 24' high x 16' wide | " | 5,960 | 790.00 | 6,750 |
| For motor operator | | | | |
| Up to 12-0 x 12-0, add | EA. | | | 1,220 |
| Over 12-0 x 12-0, add | " | | | 1,561 |
| Roll-up doors | | | | |
| 13-0 high x 14-0 wide | EA. | 1,080 | 510.00 | 1,590 |
| 12-0 high x 14-0 wide | " | 1,350 | 510.00 | 1,860 |
| Top coiling grilles, manually operated, steel or aluminum | | | | |
| Opening, 4' high x | | | | |
| 4' wide | EA. | 1,280 | 140.00 | 1,420 |
| 6' wide | " | 1,300 | 140.00 | 1,440 |
| 8' wide | " | 1,490 | 200.00 | 1,690 |
| 12' wide | " | 1,760 | 200.00 | 1,960 |
| 16' wide | " | 2,030 | 300.00 | 2,330 |
| 6' high x | | | | |
| 4' wide | EA. | 1,350 | 300.00 | 1,650 |
| 6' wide | " | 1,490 | 320.00 | 1,810 |
| 8' wide | " | 1,550 | 350.00 | 1,900 |
| 12' wide | " | 1,890 | 390.00 | 2,280 |
| 16' wide | " | 2,430 | 510.00 | 2,940 |
| Side coiling grilles, manually operated, aluminum | | | | |
| Opening, 8' high x | | | | |
| 18' wide | EA. | 9,450 | 3,210 | 12,660 |
| 24' wide | " | 12,150 | 3,670 | 15,820 |
| 12' high x | | | | |
| 12' wide | EA. | 9,450 | 3,210 | 12,660 |
| 18' wide | " | 12,150 | 3,670 | 15,820 |
| 24' wide | " | 16,200 | 4,280 | 20,480 |
| Accordion folding doors, tracks and fittings included | | | | |
| Vinyl covered, 2 layers | S.F. | 9.90 | 14.20 | 24.10 |
| Woven mahogany and vinyl | " | 12.75 | 14.20 | 26.95 |
| Economy vinyl | " | 8.50 | 14.20 | 22.70 |
| Rigid polyvinyl chloride | " | 14.20 | 14.20 | 28.40 |
| Sectional wood overhead doors, frames not included | | | | |
| Commercial grade, heavy duty, 1-3/4" thick, manual | | | | |
| 8' x 8' | EA. | 540.00 | 300.00 | 840.00 |

| WOOD AND PLASTIC | UNIT | MAT. | INST. | TOTAL |
|---|---|---|---|---|
| **08300.10 SPECIAL DOORS** | | | | |
| 10' x 10' | EA. | 950.00 | 320.00 | 1,270 |
| 12' x 12' | " | 1,220 | 350.00 | 1,570 |
| Chain hoist | | | | |
| 12' x 16' high | EA. | 1,990 | 590.00 | 2,580 |
| 14' x 14' high | " | 2,130 | 440.00 | 2,570 |
| 20' x 8' high | " | 1,840 | 710.00 | 2,550 |
| 16' high | " | 4,260 | 890.00 | 5,150 |
| Sectional metal overhead doors, complete | | | | |
| Residential grade, manual | | | | |
| 9' x 7' | EA. | 440.00 | 140.00 | 580.00 |
| 16' x 7' | " | 840.00 | 180.00 | 1,020 |
| Commercial grade | | | | |
| 8' x 8' | EA. | 570.00 | 300.00 | 870.00 |
| 10' x 10' | " | 720.00 | 320.00 | 1,040 |
| 12' x 12' | " | 1,270 | 350.00 | 1,620 |
| 20' x 14', with chain hoist | " | 2,700 | 710.00 | 3,410 |
| Sliding glass doors | | | | |
| Tempered plate glass, 1/4" thick | | | | |
| 6' wide | | | | |
| Economy grade | EA. | 800.00 | 120.00 | 920.00 |
| Premium grade | " | 920.00 | 120.00 | 1,040 |
| 12' wide | | | | |
| Economy grade | EA. | 1,130 | 180.00 | 1,310 |
| Premium grade | " | 1,700 | 180.00 | 1,880 |
| Insulating glass, 5/8" thick | | | | |
| 6' wide | | | | |
| Economy grade | EA. | 990.00 | 120.00 | 1,110 |
| Premium grade | " | 1,280 | 120.00 | 1,400 |
| 12' wide | | | | |
| Economy grade | EA. | 1,280 | 180.00 | 1,460 |
| Premium grade | " | 1,980 | 180.00 | 2,160 |
| 1" thick | | | | |
| 6' wide | | | | |
| Economy grade | EA. | 1,250 | 120.00 | 1,370 |
| Premium grade | " | 1,440 | 120.00 | 1,560 |
| 12' wide | | | | |
| Economy grade | EA. | 1,940 | 180.00 | 2,120 |
| Premium grade | " | 2,840 | 180.00 | 3,020 |
| Added costs | | | | |
| Custom quality, add to material, 30% | | | | |
| Tempered glass, 6' wide, add | S.F. | | | 3.17 |
| Vertical lift doors, channel frame construction | | | | |
| 20' high x | | | | |
| 10' wide | EA. | 21,600 | 580.00 | 22,180 |
| 15' wide | " | 27,000 | 580.00 | 27,580 |
| 20' wide | " | 29,700 | 1,030 | 30,730 |
| 25' wide | " | 32,400 | 1,030 | 33,430 |
| 25' high | | | | |
| 20' wide | EA. | 33,750 | 1,030 | 34,780 |
| 25' wide | " | 37,800 | 1,200 | 39,000 |
| 30' high | | | | |
| 25' wide | EA. | 40,500 | 1,200 | 41,700 |

# 08 DOORS AND WINDOWS

## WOOD AND PLASTIC

| 08300.10 SPECIAL DOORS | UNIT | MAT. | INST. | TOTAL |
|---|---|---|---|---|
| 30' wide | EA. | 44,550 | 1,200 | 45,750 |
| 35' wide | " | 52,650 | 1,200 | 53,850 |
| Residential storm door | | | | |
| Minimum | EA. | 120.00 | 59.00 | 179.00 |
| Average | " | 180.00 | 59.00 | 239.00 |
| Maximum | " | 380.00 | 88.50 | 468.50 |

## STOREFRONTS

| 08410.10 STOREFRONTS | UNIT | MAT. | INST. | TOTAL |
|---|---|---|---|---|
| Storefront, aluminum and glass | | | | |
| Minimum | S.F. | 15.05 | 4.80 | 19.85 |
| Average | " | 21.50 | 5.50 | 27.00 |
| Maximum | " | 35.80 | 6.45 | 42.25 |
| Entrance doors, premium, incl. glass, closers, panic dev.,etc. | | | | |
| 1/2" thick glass | | | | |
| 3' x 7' | EA. | 2,430 | 320.00 | 2,750 |
| 6' x 7' | " | 4,190 | 480.00 | 4,670 |
| 3/4" thick glass | | | | |
| 3' x 7' | EA. | 2,570 | 320.00 | 2,890 |
| 6' x 7' | " | 4,320 | 480.00 | 4,800 |
| 1" thick glass | | | | |
| 3' x 7' | EA. | 2,700 | 320.00 | 3,020 |
| 6' x 7' | " | 4,860 | 480.00 | 5,340 |
| Revolving doors | | | | |
| 7'diameter, 7' high | | | | |
| Minimum | EA. | 14,850 | 3,600 | 18,450 |
| Average | " | 20,250 | 5,760 | 26,010 |
| Maximum | " | 27,000 | 7,200 | 34,200 |

## METAL WINDOWS

| 08510.10 STEEL WINDOWS | UNIT | MAT. | INST. | TOTAL |
|---|---|---|---|---|
| Steel windows, primed | | | | |
| Casements | | | | |
| Operable | | | | |

| METAL WINDOWS | UNIT | MAT. | INST. | TOTAL |
|---|---|---|---|---|
| **08510.10 STEEL WINDOWS** | | | | |
| Minimum | S.F. | 28.40 | 2.25 | 30.65 |
| Maximum | " | 42.50 | 2.55 | 45.05 |
| Fixed sash | " | 22.70 | 1.95 | 24.65 |
| Double hung | " | 42.50 | 2.15 | 44.65 |
| Industrial windows | | | | |
| Horizontally pivoted sash | S.F. | 39.70 | 2.55 | 42.25 |
| Fixed sash | " | 31.20 | 2.15 | 33.35 |
| Security sash | | | | |
| Operable | S.F. | 49.60 | 2.55 | 52.15 |
| Fixed | " | 43.90 | 2.15 | 46.05 |
| Picture window | " | 21.25 | 2.15 | 23.40 |
| Projecting sash | | | | |
| Minimum | S.F. | 36.90 | 2.40 | 39.30 |
| Maximum | " | 45.40 | 2.40 | 47.80 |
| Mullions | L.F. | 9.70 | 1.95 | 11.65 |
| **08520.10 ALUMINUM WINDOWS** | | | | |
| Jalousie | | | | |
| 3-0 x 4-0 | EA. | 250.00 | 48.20 | 298.20 |
| 3-0 x 5-0 | " | 290.00 | 48.20 | 338.20 |
| Fixed window | | | | |
| 6 sf to 8 sf | S.F. | 12.40 | 5.50 | 17.90 |
| 12 sf to 16 sf | " | 11.00 | 4.30 | 15.30 |
| Projecting window | | | | |
| 6 sf to 8 sf | S.F. | 27.50 | 9.65 | 37.15 |
| 12 sf to 16 sf | " | 24.80 | 6.45 | 31.25 |
| Horizontal sliding | | | | |
| 6 sf to 8 sf | S.F. | 17.90 | 4.80 | 22.70 |
| 12 sf to 16 sf | " | 16.50 | 3.85 | 20.35 |
| Double hung | | | | |
| 6 sf to 8 sf | S.F. | 24.80 | 7.70 | 32.50 |
| 10 sf to 12 sf | " | 22.05 | 6.45 | 28.50 |
| Storm window, 0.5 cfm, up to | | | | |
| 60 u.i. (united inches) | EA. | 58.00 | 19.30 | 77.30 |
| 70 u.i. | " | 59.00 | 19.30 | 78.30 |
| 80 u.i. | " | 66.00 | 19.30 | 85.30 |
| 90 u.i. | " | 67.50 | 21.40 | 88.90 |
| 100 u.i. | " | 69.00 | 21.40 | 90.40 |
| 2.0 cfm, up to | | | | |
| 60 u.i. | EA. | 74.50 | 19.30 | 93.80 |
| 70 u.i. | " | 75.50 | 19.30 | 94.80 |
| 80 u.i. | " | 77.00 | 19.30 | 96.30 |
| 90 u.i. | " | 82.50 | 21.40 | 103.90 |
| 100 u.i. | " | 84.00 | 21.40 | 105.40 |

## WOOD AND PLASTIC

### 08600.10 WOOD WINDOWS

| | UNIT | MAT. | INST. | TOTAL |
|---|---|---|---|---|
| Double hung | | | | |
| 24" x 36" | | | | |
| Minimum | EA. | 150.00 | 35.40 | 185.40 |
| Average | " | 220.00 | 44.30 | 264.30 |
| Maximum | " | 300.00 | 59.00 | 359.00 |
| 24" x 48" | | | | |
| Minimum | EA. | 170.00 | 35.40 | 205.40 |
| Average | " | 260.00 | 44.30 | 304.30 |
| Maximum | " | 370.00 | 59.00 | 429.00 |
| 30" x 48" | | | | |
| Minimum | EA. | 190.00 | 39.40 | 229.40 |
| Average | " | 270.00 | 50.50 | 320.50 |
| Maximum | " | 390.00 | 71.00 | 461.00 |
| 30" x 60" | | | | |
| Minimum | EA. | 200.00 | 39.40 | 239.40 |
| Average | " | 330.00 | 50.50 | 380.50 |
| Maximum | " | 410.00 | 71.00 | 481.00 |
| Casement | | | | |
| 1 leaf, 22" x 38" high | | | | |
| Minimum | EA. | 220.00 | 35.40 | 255.40 |
| Average | " | 270.00 | 44.30 | 314.30 |
| Maximum | " | 320.00 | 59.00 | 379.00 |
| 2 leaf, 50" x 50" high | | | | |
| Minimum | EA. | 620.00 | 44.30 | 664.30 |
| Average | " | 690.00 | 59.00 | 749.00 |
| Maximum | " | 800.00 | 88.50 | 888.50 |
| 3 leaf, 71" x 62" high | | | | |
| Minimum | EA. | 1,030 | 44.30 | 1,074 |
| Average | " | 1,000 | 59.00 | 1,059 |
| Maximum | " | 1,190 | 88.50 | 1,279 |
| 4 leaf, 95" x 75" high | | | | |
| Minimum | EA. | 1,310 | 50.50 | 1,361 |
| Average | " | 1,430 | 71.00 | 1,501 |
| Maximum | " | 1,860 | 120.00 | 1,980 |
| 5 leaf, 119" x 75" high | | | | |
| Minimum | EA. | 1,650 | 50.50 | 1,701 |
| Average | " | 1,860 | 71.00 | 1,931 |
| Maximum | " | 2,360 | 120.00 | 2,480 |
| Picture window, fixed glass, 54" x 54" high | | | | |
| Minimum | EA. | 340.00 | 44.30 | 384.30 |
| Average | " | 390.00 | 50.50 | 440.50 |
| Maximum | " | 720.00 | 59.00 | 779.00 |
| 68" x 55" high | | | | |
| Minimum | EA. | 660.00 | 44.30 | 704.30 |
| Average | " | 750.00 | 50.50 | 800.50 |
| Maximum | " | 970.00 | 59.00 | 1,029 |
| Sliding, 40" x 31" high | | | | |
| Minimum | EA. | 190.00 | 35.40 | 225.40 |
| Average | " | 300.00 | 44.30 | 344.30 |
| Maximum | " | 370.00 | 59.00 | 429.00 |
| 52" x 39" high | | | | |
| Minimum | EA. | 240.00 | 44.30 | 284.30 |

# 08 DOORS AND WINDOWS

## WOOD AND PLASTIC

### 08600.10 WOOD WINDOWS

| WOOD AND PLASTIC | UNIT | MAT. | INST. | TOTAL |
|---|---|---|---|---|
| Average | EA. | 380.00 | 50.50 | 430.50 |
| Maximum | " | 430.00 | 59.00 | 489.00 |
| 64" x 72" high | | | | |
| Minimum | EA. | 400.00 | 44.30 | 444.30 |
| Average | " | 650.00 | 59.00 | 709.00 |
| Maximum | " | 720.00 | 71.00 | 791.00 |
| Awning windows | | | | |
| 34" x 21" high | | | | |
| Minimum | EA. | 210.00 | 35.40 | 245.40 |
| Average | " | 250.00 | 44.30 | 294.30 |
| Maximum | " | 290.00 | 59.00 | 349.00 |
| 40" x 21" high | | | | |
| Minimum | EA. | 250.00 | 39.40 | 289.40 |
| Average | " | 280.00 | 50.50 | 330.50 |
| Maximum | " | 310.00 | 71.00 | 381.00 |
| 48" x 27" high | | | | |
| Minimum | EA. | 260.00 | 39.40 | 299.40 |
| Average | " | 320.00 | 50.50 | 370.50 |
| Maximum | " | 370.00 | 71.00 | 441.00 |
| 60" x 36" high | | | | |
| Minimum | EA. | 280.00 | 44.30 | 324.30 |
| Average | " | 510.00 | 59.00 | 569.00 |
| Maximum | " | 570.00 | 71.00 | 641.00 |
| Window frame, milled | | | | |
| Minimum | L.F. | 2.75 | 7.10 | 9.85 |
| Average | " | 4.15 | 8.85 | 13.00 |
| Maximum | " | 6.20 | 11.80 | 18.00 |

## HARDWARE

### 08710.10 HINGES

| HARDWARE | UNIT | MAT. | INST. | TOTAL |
|---|---|---|---|---|
| Hinges | | | | |
| 3 x 3 butts, steel, interior, plain bearing | PAIR | | | 14.61 |
| 4 x 4 butts, steel, standard | " | | | 21.91 |
| 5 x 4-1/2 butts, bronze/s. steel, heavy duty | " | | | 58.43 |
| Pivot hinges | | | | |
| Top pivot | EA. | | | 52.58 |
| Intermediate pivot | " | | | 56.97 |
| Bottom pivot | " | | | 116.86 |
| BHMA specifications | | | | |
| 3-1/2 x 3-1/2, full mortise butts | | | | |
| Plain bearing | PAIR | | | 17.52 |
| Ball bearing | " | | | 21.91 |

# 08 DOORS AND WINDOWS

| HARDWARE | UNIT | MAT. | INST. | TOTAL |
|---|---|---|---|---|
| **08710.10 HINGES** | | | | |
| Half surface butts | | | | |
| 4 x 4 | PAIR | | | 32.14 |
| Full mortise butts, plain bearing, standard duty | PAIR | | | 19.00 |
| Full mortise butts, ball bearing | " | | | 23.37 |
| Half surface butts | | | | |
| Standard duty | PAIR | | | 32.14 |
| Ball bearing | " | | | 32.14 |
| 4-1/2 x 4-1/2 | | | | |
| Full mortise butts, plain bearing | PAIR | | | 20.45 |
| Ball bearing, heavy duty | " | | | 51.13 |
| Half mortise and half surface butts | | | | |
| Plain bearing | PAIR | | | 32.14 |
| Full surface and half surface butts | | | | |
| Standard duty | PAIR | | | 49.67 |
| Heavy duty | " | | | 137.31 |
| Full mortise and full slide-in butts, ball bearing | " | | | 23.37 |
| Half mortise butts, ball bearing | | | | |
| Standard duty | PAIR | | | 109.55 |
| Heavy duty | " | | | 128.55 |
| 5 x 5, ball bearing | | | | |
| Full mortise butts | PAIR | | | 51.13 |
| Half mortise, full & half surface butts | " | | | 113.93 |
| Full mortise, full surface and half surface butts | " | | | 154.83 |
| 4 x 4 | | | | |
| Full mortise butts, plain bearing, standard duty | PAIR | | | 10.23 |
| 5 x 4-1/2 | | | | |
| Full mortise butts, ball bearing, heavy duty | PAIR | | | 58.43 |
| **08710.20 LOCKSETS** | | | | |
| Latchset, heavy duty | | | | |
| Cylindrical | EA. | 120.00 | 22.15 | 142.15 |
| Mortise | " | 110.00 | 35.40 | 145.40 |
| Lockset, heavy duty | | | | |
| Cylindrical | EA. | 150.00 | 22.15 | 172.15 |
| Mortise | " | 180.00 | 35.40 | 215.40 |
| Mortise locks and latchsets, chrome | | | | |
| Latchset passage or closet latch | EA. | 120.00 | 29.50 | 149.50 |
| Privacy (bath or bedroom) | " | 150.00 | 29.50 | 179.50 |
| Entry lockset | " | 180.00 | 29.50 | 209.50 |
| Classroom lockset (outside key operated) | " | 180.00 | 29.50 | 209.50 |
| Storeroom lock | " | 180.00 | 29.50 | 209.50 |
| Front door lock | " | 180.00 | 29.50 | 209.50 |
| Dormitory or exit lock | " | 180.00 | 29.50 | 209.50 |
| Preassembled locks and latches, brass | | | | |
| Latchset, passage or closet latch | EA. | 200.00 | 29.50 | 229.50 |
| Lockset | | | | |
| Privacy (bath or bathroom) | EA. | 240.00 | 29.50 | 269.50 |
| Entry lock | " | 290.00 | 29.50 | 319.50 |
| Classroom lock (outside key, operated) | " | 290.00 | 29.50 | 319.50 |
| Storeroom lock | " | 350.00 | 29.50 | 379.50 |
| Bored locks and latches, satin chrome plated | | | | |

# 08 DOORS AND WINDOWS

| HARDWARE | UNIT | MAT. | INST. | TOTAL |
|---|---|---|---|---|
| **08710.20 LOCKSETS** | | | | |
| Latchset passage or closet latch | EA. | 110.00 | 29.50 | 139.50 |
| Lockset | | | | |
| Privacy (bath or bedroom) | EA. | 140.00 | 29.50 | 169.50 |
| Entry lock | " | 160.00 | 29.50 | 189.50 |
| Classroom lock | " | 160.00 | 29.50 | 189.50 |
| Corridor lock | " | 160.00 | 29.50 | 189.50 |
| Miscellaneous locks | | | | |
| Exit lock with alarm, single door | EA. | 500.00 | 140.00 | 640.00 |
| Electric strike | | | | |
| Rim mounted wrought steel | EA. | 420.00 | 88.50 | 508.50 |
| Mortised, wrought steel with bronze plating | " | 140.00 | 140.00 | 280.00 |
| Dead bolt | | | | |
| Bored, wrought brass, keyed both sides | EA. | 83.50 | 59.00 | 142.50 |
| Mortised, cast brass | " | 91.00 | 59.00 | 150.00 |
| Lockset, cipher, mechanical | " | 1,440 | 35.40 | 1,475 |
| **08710.30 CLOSERS** | | | | |
| Door closers | | | | |
| Surface mounted, traditional type, parallel arm | | | | |
| Standard | EA. | 130.00 | 44.30 | 174.30 |
| Heavy duty | " | 150.00 | 44.30 | 194.30 |
| Modern type, parallel arm, standard duty | " | 150.00 | 44.30 | 194.30 |
| Overhead, concealed, pivot hung, single acting | | | | |
| Interior | EA. | 250.00 | 44.30 | 294.30 |
| Exterior | " | 470.00 | 44.30 | 514.30 |
| Floor concealed, single acting, offset, pivoted | | | | |
| Interior | EA. | 440.00 | 120.00 | 560.00 |
| Exterior | " | 570.00 | 120.00 | 690.00 |
| **08710.40 DOOR TRIM** | | | | |
| Door bumper, bronze, wall type | EA. | 3.00 | 7.10 | 10.10 |
| Wall type, 4" dia. with convex rubber pad, aluminum | " | 24.75 | 7.10 | 31.85 |
| Floor type | | | | |
| Aluminum | EA. | 3.85 | 7.10 | 10.95 |
| Brass | " | 4.15 | 7.10 | 11.25 |
| Door holders | | | | |
| Wall type, bronze | EA. | 3.85 | 7.10 | 10.95 |
| Overhead | " | 38.70 | 17.70 | 56.40 |
| Floor type | " | 38.70 | 17.70 | 56.40 |
| Plunger type | " | 20.90 | 17.70 | 38.60 |
| Wall type, aluminum | " | 24.75 | 17.70 | 42.45 |
| Surface bolt | " | 18.60 | 7.10 | 25.70 |
| Panic device | | | | |
| Rim type with thumb piece | EA. | 500.00 | 88.50 | 588.50 |
| Mortise | " | 630.00 | 88.50 | 718.50 |
| Vertical rod | " | 730.00 | 88.50 | 818.50 |
| Labelled, rim type | " | 650.00 | 88.50 | 738.50 |
| Mortise | " | 780.00 | 88.50 | 868.50 |
| Vertical rod | " | 880.00 | 88.50 | 968.50 |
| Silencers, rubber type | " | 0.15 | 0.71 | 0.86 |

## HARDWARE

| | UNIT | MAT. | INST. | TOTAL |
|---|---|---|---|---|
| **08710.40 DOOR TRIM** | | | | |
| Dust proof strike with plate, brass | EA. | 13.95 | 11.80 | 25.75 |
| Flush bolt, lever extension, brass, rated | " | 24.75 | 7.10 | 31.85 |
| Surface bolt with strike, brass, 6" long | " | 18.60 | 7.10 | 25.70 |
| Door coordinator, labeled, brass, satin chrome | " | 90.00 | 25.30 | 115.30 |
| Door plates | | | | |
| Kick plate, aluminum, 3 beveled edges | | | | |
| 10" x 28" | EA. | 13.95 | 17.70 | 31.65 |
| 10" x 30" | " | 15.45 | 17.70 | 33.15 |
| 10" x 34" | " | 17.05 | 17.70 | 34.75 |
| 10" x 38" | " | 19.30 | 17.70 | 37.00 |
| Push plate, 4" x 16" | | | | |
| Aluminum | EA. | 13.75 | 7.10 | 20.85 |
| Bronze | " | 69.00 | 7.10 | 76.10 |
| Stainless steel | " | 55.00 | 7.10 | 62.10 |
| Armor plate, 40" x 34" | " | 62.00 | 14.20 | 76.20 |
| Pull handle, 4" x 16" | | | | |
| Aluminum | EA. | 55.00 | 7.10 | 62.10 |
| Bronze | " | 110.00 | 7.10 | 117.10 |
| Stainless steel | " | 96.50 | 7.10 | 103.60 |
| Hasp assembly | | | | |
| 3" | EA. | 2.05 | 5.90 | 7.95 |
| 4-1/2" | " | 2.75 | 7.90 | 10.65 |
| 6" | " | 4.80 | 10.15 | 14.95 |
| Electro-magnetic door holder | | | | |
| Wall mounted | EA. | 140.00 | 120.00 | 260.00 |
| Floor mounted | " | 220.00 | 120.00 | 340.00 |
| Smoke detector door holder | | | | |
| Photo electric type | EA. | 220.00 | 120.00 | 340.00 |
| Ionization type | " | 220.00 | 120.00 | 340.00 |
| Pneumatic operators, activated by rubber mats | | | | |
| Swing | | | | |
| Single | EA. | 1,650 | 300.00 | 1,950 |
| Double | " | 2,750 | 440.00 | 3,190 |
| Sliding | | | | |
| Single | EA. | 1,790 | 300.00 | 2,090 |
| Double | " | 3,170 | 440.00 | 3,610 |
| **08710.60 WEATHERSTRIPPING** | | | | |
| Weatherstrip, head and jamb, metal strip, neoprene bulb | | | | |
| Standard duty | L.F. | 3.15 | 1.95 | 5.10 |
| Heavy duty | " | 4.00 | 2.20 | 6.20 |
| Spring type | | | | |
| Metal doors | EA. | 41.30 | 88.50 | 129.80 |
| Wood doors | " | 41.30 | 120.00 | 161.30 |
| Sponge type with adhesive backing | " | 38.60 | 35.40 | 74.00 |
| Astragal | | | | |
| 1-3/4" x 13 ga., aluminum | L.F. | 4.05 | 2.95 | 7.00 |
| 1-3/8" x 5/8", oak | " | 3.85 | 2.35 | 6.20 |
| Thresholds | | | | |
| Bronze | L.F. | 39.70 | 8.85 | 48.55 |
| Aluminum | | | | |
| Plain | L.F. | 14.90 | 8.85 | 23.75 |

## HARDWARE

| 08710.60 WEATHERSTRIPPING | UNIT | MAT. | INST. | TOTAL |
|---|---|---|---|---|
| Vinyl insert | L.F. | 22.70 | 8.85 | 31.55 |
| Aluminum with grit | " | 21.30 | 8.85 | 30.15 |
| Steel | | | | |
| Plain | L.F. | 14.90 | 8.85 | 23.75 |
| Interlocking | " | 22.70 | 29.50 | 52.20 |

## GLAZING

| 08810.10 GLAZING | UNIT | MAT. | INST. | TOTAL |
|---|---|---|---|---|
| Sheet glass, 1/8" thick | S.F. | 5.55 | 2.15 | 7.70 |
| Plate glass, bronze or grey, 1/4" thick | " | 8.25 | 3.50 | 11.75 |
| Clear | " | 6.35 | 3.50 | 9.85 |
| Polished | " | 7.00 | 3.50 | 10.50 |
| Plexiglass | | | | |
| 1/8" thick | S.F. | 2.40 | 3.50 | 5.90 |
| 1/4" thick | " | 4.40 | 2.15 | 6.55 |
| Float glass, clear | | | | |
| 3/16" thick | S.F. | 4.05 | 3.20 | 7.25 |
| 1/4" thick | " | 4.75 | 3.50 | 8.25 |
| 5/16" thick | " | 8.80 | 3.85 | 12.65 |
| 3/8" thick | " | 9.45 | 4.80 | 14.25 |
| 1/2" thick | " | 15.95 | 6.45 | 22.40 |
| 5/8" thick | " | 18.30 | 7.70 | 26.00 |
| 3/4" thick | " | 19.95 | 9.65 | 29.60 |
| 1" thick | " | 35.10 | 12.85 | 47.95 |
| Tinted glass, polished plate, twin ground | | | | |
| 3/16" thick | S.F. | 6.30 | 3.20 | 9.50 |
| 1/4" thick | " | 6.30 | 3.50 | 9.80 |
| 3/8" thick | " | 10.20 | 4.80 | 15.00 |
| 1/2" thick | " | 16.50 | 6.45 | 22.95 |
| Total, full vision, all glass window system | | | | |
| To 10' high | | | | |
| Minimum | S.F. | 25.70 | 9.65 | 35.35 |
| Average | " | 33.80 | 9.65 | 43.45 |
| Maximum | " | 40.50 | 9.65 | 50.15 |
| 10' to 20' high | | | | |
| Minimum | S.F. | 31.10 | 9.65 | 40.75 |
| Average | " | 37.80 | 9.65 | 47.45 |
| Maximum | " | 47.30 | 9.65 | 56.95 |
| Insulated glass, bronze or gray | | | | |
| 1/2" thick | S.F. | 11.80 | 6.45 | 18.25 |
| 1" thick | " | 14.00 | 9.65 | 23.65 |
| Spandrel glass, polished bronze/grey, 1 side, 1/4" thick | " | 9.45 | 3.50 | 12.95 |

## GLAZING

| 08810.10 GLAZING | UNIT | MAT. | INST. | TOTAL |
|---|---|---|---|---|
| Tempered glass (safety) | | | | |
| Clear sheet glass | | | | |
| 1/8" thick | S.F. | 6.50 | 2.15 | 8.65 |
| 3/16" thick | " | 7.85 | 2.95 | 10.80 |
| Clear float glass | | | | |
| 1/4" thick | S.F. | 6.75 | 3.20 | 9.95 |
| 5/16" thick | " | 12.15 | 3.85 | 16.00 |
| 3/8" thick | " | 14.85 | 4.80 | 19.65 |
| 1/2" thick | " | 20.25 | 6.45 | 26.70 |
| 5/8" thick | " | 22.95 | 7.70 | 30.65 |
| 3/4" thick | " | 28.40 | 12.85 | 41.25 |
| Tinted float glass | | | | |
| 3/16" thick | S.F. | 8.10 | 2.95 | 11.05 |
| 1/4" thick | " | 8.90 | 3.20 | 12.10 |
| 3/8" thick | " | 16.20 | 4.80 | 21.00 |
| 1/2" thick | " | 21.60 | 6.45 | 28.05 |
| Laminated glass | | | | |
| Float safety glass with polyvinyl plastic interlayer | | | | |
| 1/4", sheet or float | | | | |
| Two lites, 1/8" thick, clear glass | S.F. | 9.45 | 3.20 | 12.65 |
| 1/2" thick, float glass | | | | |
| Two lites, 1/4" thick, clear glass | S.F. | 14.05 | 6.45 | 20.50 |
| Tinted glass | " | 16.80 | 6.45 | 23.25 |
| Insulating glass, two lites, clear float glass | | | | |
| 1/2" thick | S.F. | 9.20 | 6.45 | 15.65 |
| 5/8" thick | " | 10.60 | 7.70 | 18.30 |
| 3/4" thick | " | 11.65 | 9.65 | 21.30 |
| 7/8" thick | " | 12.25 | 11.00 | 23.25 |
| 1" thick | " | 13.15 | 12.85 | 26.00 |
| Glass seal edge | | | | |
| 3/8" thick | S.F. | 7.75 | 6.45 | 14.20 |
| Tinted glass | | | | |
| 1/2" thick | S.F. | 15.80 | 6.45 | 22.25 |
| 1" thick | " | 16.95 | 12.85 | 29.80 |
| Tempered, clear | | | | |
| 1" thick | S.F. | 30.90 | 12.85 | 43.75 |
| Wire reinforced | " | 39.30 | 12.85 | 52.15 |
| Plate mirror glass | | | | |
| 1/4" thick | | | | |
| 15 sf | S.F. | 8.10 | 3.85 | 11.95 |
| Over 15 sf | " | 7.45 | 3.50 | 10.95 |
| Transparent, one way vision, 1/4" thick | " | 18.00 | 3.85 | 21.85 |
| Sheet mirror glass | | | | |
| 3/16" thick | S.F. | 7.85 | 3.85 | 11.70 |
| 1/4" thick | " | 8.25 | 3.20 | 11.45 |
| Wall tiles, 12" x 12" | | | | |
| Clear glass | S.F. | 2.65 | 2.15 | 4.80 |
| Veined glass | " | 3.10 | 2.15 | 5.25 |
| Wire glass, 1/4" thick | | | | |
| Clear | S.F. | 10.95 | 12.85 | 23.80 |
| Hammered | " | 11.10 | 12.85 | 23.95 |
| Obscure | " | 13.25 | 12.85 | 26.10 |

## GLAZING

### 08810.10 GLAZING

| | UNIT | MAT. | INST. | TOTAL |
|---|---|---|---|---|
| Bullet resistant glass, plate glass with inter-leaved vinyl | | | | |
| 1-3/16" thick | | | | |
| To 15 sf | S.F. | 29.50 | 19.30 | 48.80 |
| Over 15 sf | " | 33.70 | 19.30 | 53.00 |
| 2" thick | | | | |
| To 15 sf | S.F. | 56.00 | 32.10 | 88.10 |
| Over 15 sf | " | 70.00 | 32.10 | 102.10 |
| Glazing accessories | | | | |
| Neoprene glazing gaskets | | | | |
| 1/4" glass | L.F. | 1.05 | 1.55 | 2.60 |
| 3/8" glass | " | 1.30 | 1.60 | 2.90 |
| 1/2" glass | " | 1.60 | 1.70 | 3.30 |
| 3/4" glass | " | 2.20 | 1.75 | 3.95 |
| 1" glass | " | 2.50 | 1.95 | 4.45 |
| Mullion section | | | | |
| 1/4" glass | L.F. | 0.46 | 0.77 | 1.23 |
| 3/8" glass | " | 0.59 | 0.96 | 1.55 |
| 1/2" glass | " | 0.88 | 1.10 | 1.98 |
| 3/4" glass | " | 1.35 | 1.30 | 2.65 |
| 1" glass | " | 1.80 | 1.55 | 3.35 |
| Molded corners | EA. | 1.95 | 25.70 | 27.65 |

## GLAZED CURTAIN WALLS

### 08910.10 GLAZED CURTAIN WALLS

| | UNIT | MAT. | INST. | TOTAL |
|---|---|---|---|---|
| Curtain wall, aluminum system, framing sections | | | | |
| 2" x 3" | | | | |
| Jamb | L.F. | 8.20 | 3.20 | 11.40 |
| Horizontal | " | 8.35 | 3.20 | 11.55 |
| Mullion | " | 11.15 | 3.20 | 14.35 |
| 2" x 4" | | | | |
| Jamb | L.F. | 11.15 | 4.80 | 15.95 |
| Horizontal | " | 11.50 | 4.80 | 16.30 |
| Mullion | " | 11.15 | 4.80 | 15.95 |
| 3" x 5-1/2" | | | | |
| Jamb | L.F. | 14.75 | 4.80 | 19.55 |
| Horizontal | " | 16.40 | 4.80 | 21.20 |
| Mullion | " | 14.85 | 4.80 | 19.65 |
| 4" corner mullion | " | 19.65 | 6.45 | 26.10 |
| Coping sections | | | | |
| 1/8" x 8" | L.F. | 20.60 | 6.45 | 27.05 |
| 1/8" x 9" | " | 20.75 | 6.45 | 27.20 |
| 1/8" x 12-1/2" | " | 21.25 | 7.70 | 28.95 |

# 08 DOORS AND WINDOWS

## GLAZED CURTAIN WALLS

| 08910.10 GLAZED CURTAIN WALLS | UNIT | MAT. | INST. | TOTAL |
|---|---|---|---|---|
| Sill section | | | | |
| 1/8" x 6" | L.F. | 21.30 | 3.85 | 25.15 |
| 1/8" x 7" | " | 20.50 | 3.85 | 24.35 |
| 1/8" x 8-1/2" | " | 20.95 | 3.85 | 24.80 |
| Column covers, aluminum | | | | |
| 1/8" x 26" | L.F. | 20.50 | 9.65 | 30.15 |
| 1/8" x 34" | " | 20.50 | 10.15 | 30.65 |
| 1/8" x 38" | " | 20.75 | 10.15 | 30.90 |
| Doors | | | | |
| Aluminum framed, standard hardware | | | | |
| Narrow stile | | | | |
| 2-6 x 7-0 | EA. | 460.00 | 190.00 | 650.00 |
| 3-0 x 7-0 | " | 460.00 | 190.00 | 650.00 |
| 3-6 x 7-0 | " | 480.00 | 190.00 | 670.00 |
| Wide stile | | | | |
| 2-6 x 7-0 | EA. | 790.00 | 190.00 | 980.00 |
| 3-0 x 7-0 | " | 850.00 | 190.00 | 1,040 |
| 3-6 x 7-0 | " | 910.00 | 190.00 | 1,100 |
| Flush panel doors, to match adjacent wall panels | | | | |
| 2-6 x 7-0 | EA. | 670.00 | 240.00 | 910.00 |
| 3-0 x 7-0 | " | 700.00 | 240.00 | 940.00 |
| 3-6 x 7-0 | " | 730.00 | 240.00 | 970.00 |
| Wall panel, insulated | | | | |
| "U"=.08 | S.F. | 8.35 | 3.20 | 11.55 |
| "U"=.10 | " | 7.90 | 3.20 | 11.10 |
| "U"=.15 | " | 7.15 | 3.20 | 10.35 |
| Window wall system, complete | | | | |
| Minimum | S.F. | 14.75 | 3.85 | 18.60 |
| Average | " | 23.60 | 4.30 | 27.90 |
| Maximum | " | 54.50 | 5.50 | 60.00 |
| Added costs | | | | |
| For bronze, add 20% to material | | | | |
| For stainless steel, add 50% to material | | | | |

## SUPPORT SYSTEMS

### 09110.10 METAL STUDS

| Description | UNIT | MAT. | INST. | TOTAL |
|---|---|---|---|---|
| Studs, non load bearing, galvanized | | | | |
| 2-1/2", 20 ga. | | | | |
| 12" o.c. | S.F. | 0.74 | 0.74 | 1.48 |
| 16" o.c. | " | 0.63 | 0.59 | 1.22 |
| 25 ga. | | | | |
| 12" o.c. | S.F. | 0.53 | 0.74 | 1.27 |
| 16" o.c. | " | 0.38 | 0.59 | 0.97 |
| 24" o.c. | " | 0.27 | 0.49 | 0.76 |
| 3-5/8", 20 ga. | | | | |
| 12" o.c. | S.F. | 0.90 | 0.89 | 1.79 |
| 16" o.c. | " | 0.76 | 0.71 | 1.47 |
| 24" o.c. | " | 0.62 | 0.59 | 1.21 |
| 25 ga. | | | | |
| 12" o.c. | S.F. | 0.51 | 0.89 | 1.40 |
| 16" o.c. | " | 0.43 | 0.71 | 1.14 |
| 24" o.c. | " | 0.36 | 0.59 | 0.95 |
| 4", 20 ga. | | | | |
| 12" o.c. | S.F. | 1.05 | 0.89 | 1.94 |
| 16" o.c. | " | 0.84 | 0.71 | 1.55 |
| 24" o.c. | " | 0.68 | 0.59 | 1.27 |
| 25 ga. | | | | |
| 12" o.c. | S.F. | 0.62 | 0.89 | 1.51 |
| 16" o.c. | " | 0.51 | 0.71 | 1.22 |
| 24" o.c. | " | 0.42 | 0.59 | 1.01 |
| 6", 20 ga. | | | | |
| 12" o.c. | S.F. | 1.25 | 1.10 | 2.35 |
| 16" o.c. | " | 1.05 | 0.89 | 1.94 |
| 24" o.c. | " | 0.82 | 0.74 | 1.56 |
| 25 ga. | | | | |
| 12" o.c. | S.F. | 0.77 | 1.10 | 1.87 |
| 16" o.c. | " | 0.63 | 0.89 | 1.52 |
| 24" o.c. | " | 0.53 | 0.74 | 1.27 |
| Load bearing studs, galvanized | | | | |
| 3-5/8", 16 ga. | | | | |
| 12" o.c. | S.F. | 1.50 | 0.89 | 2.39 |
| 16" o.c. | " | 1.20 | 0.71 | 1.91 |
| 18 ga. | | | | |
| 12" o.c. | S.F. | 1.35 | 0.59 | 1.94 |
| 16" o.c. | " | 1.10 | 0.71 | 1.81 |
| 4", 16 ga. | | | | |
| 12" o.c. | S.F. | 1.65 | 0.89 | 2.54 |
| 16" o.c. | " | 1.35 | 0.71 | 2.06 |
| 6", 16 ga. | | | | |
| 12" o.c. | S.F. | 2.00 | 1.10 | 3.10 |
| 16" o.c. | " | 1.60 | 0.89 | 2.49 |
| Furring | | | | |
| On beams and columns | | | | |
| 7/8" channel | L.F. | 0.39 | 2.35 | 2.74 |
| 1-1/2" channel | " | 0.46 | 2.75 | 3.21 |
| On ceilings | | | | |
| 3/4" furring channels | | | | |
| 12" o.c. | S.F. | 0.38 | 1.50 | 1.88 |

# 09 FINISHES

## SUPPORT SYSTEMS

| 09110.10 METAL STUDS | UNIT | MAT. | INST. | TOTAL |
|---|---|---|---|---|
| 16" o.c. | S.F. | 0.32 | 1.40 | 1.72 |
| 24" o.c. | " | 0.23 | 1.25 | 1.48 |
| 1-1/2" furring channels | | | | |
| 12" o.c. | S.F. | 0.54 | 1.60 | 2.14 |
| 16" o.c. | " | 0.46 | 1.50 | 1.96 |
| 24" o.c. | " | 0.39 | 1.35 | 1.74 |
| On walls | | | | |
| 3/4" furring channels | | | | |
| 12" o.c. | S.F. | 0.38 | 1.20 | 1.58 |
| 16" o.c. | " | 0.32 | 1.10 | 1.42 |
| 24" o.c. | " | 0.23 | 1.05 | 1.28 |
| 1-1/2" furring channels | | | | |
| 12" o.c. | S.F. | 0.54 | 1.25 | 1.79 |
| 16" o.c. | " | 0.46 | 1.20 | 1.66 |
| 24" o.c. | " | 0.39 | 1.10 | 1.49 |

## LATH AND PLASTER

| 09205.10 GYPSUM LATH | UNIT | MAT. | INST. | TOTAL |
|---|---|---|---|---|
| Gypsum lath, 1/2" thick | | | | |
| Clipped | S.Y. | 5.40 | 1.95 | 7.35 |
| Nailed | " | 5.25 | 2.20 | 7.45 |

| 09205.20 METAL LATH | UNIT | MAT. | INST. | TOTAL |
|---|---|---|---|---|
| Diamond expanded, galvanized | | | | |
| 2.5 lb., on walls | | | | |
| Nailed | S.Y. | 2.70 | 4.45 | 7.15 |
| Wired | " | 2.75 | 5.05 | 7.80 |
| On ceilings | | | | |
| Nailed | S.Y. | 2.70 | 5.05 | 7.75 |
| Wired | " | 2.75 | 5.90 | 8.65 |
| 3.4 lb., on walls | | | | |
| Nailed | S.Y. | 3.10 | 4.45 | 7.55 |
| Wired | " | 3.15 | 5.05 | 8.20 |
| On ceilings | | | | |
| Nailed | S.Y. | 3.10 | 5.05 | 8.15 |
| Wired | " | 3.15 | 5.90 | 9.05 |
| Flat rib | | | | |
| 2.75 lb., on walls | | | | |
| Nailed | S.Y. | 2.65 | 4.45 | 7.10 |
| Wired | " | 2.75 | 5.05 | 7.80 |
| On ceilings | | | | |

# 09 FINISHES

## LATH AND PLASTER

| LATH AND PLASTER | UNIT | MAT. | INST. | TOTAL |
|---|---|---|---|---|
| **09205.20 METAL LATH** | | | | |
| Nailed | S.Y. | 2.65 | 5.05 | 7.70 |
| Wired | " | 2.75 | 5.90 | 8.65 |
| 3.4 lb., on walls | | | | |
| Nailed | S.Y. | 3.30 | 4.45 | 7.75 |
| Wired | " | 3.40 | 5.05 | 8.45 |
| On ceilings | | | | |
| Nailed | S.Y. | 3.30 | 5.05 | 8.35 |
| Wired | " | 3.40 | 5.90 | 9.30 |
| Stucco lath | | | | |
| 1.8 lb. | S.Y. | 4.30 | 4.45 | 8.75 |
| 3.6 lb. | " | 4.85 | 4.45 | 9.30 |
| Paper backed | | | | |
| Minimum | S.Y. | 3.35 | 3.55 | 6.90 |
| Maximum | " | 5.40 | 5.05 | 10.45 |
| **09205.60 PLASTER ACCESSORIES** | | | | |
| Expansion joint, 3/4", 26 ga., galvanized, one piece | L.F. | 0.84 | 0.89 | 1.73 |
| Plaster corner beads, 3/4", galvanized | " | 0.49 | 1.00 | 1.49 |
| Casing bead, expanded flange, galvanized | " | 0.49 | 0.89 | 1.38 |
| Expanded wing, 1-1/4" wide, galvanized | " | 0.41 | 0.89 | 1.30 |
| Joint clips for lath | EA. | 0.22 | 0.18 | 0.40 |
| Metal base, galvanized, 2-1/2" high | L.F. | 0.65 | 1.20 | 1.85 |
| Stud clips for gypsum lath | EA. | 0.22 | 0.18 | 0.40 |
| Tie wire galvanized, 18 ga., 25 lb. hank | " | | | 41.30 |
| Sound deadening board, 1/4", nailed or clipped | S.F. | 0.41 | 0.59 | 1.00 |
| **09210.10 PLASTER** | | | | |
| Gypsum plaster, trowel finish, 2 coats | | | | |
| Ceilings | S.Y. | 4.55 | 10.95 | 15.50 |
| Walls | " | 4.55 | 10.30 | 14.85 |
| 3 coats | | | | |
| Ceilings | S.Y. | 6.30 | 15.20 | 21.50 |
| Walls | " | 6.30 | 13.45 | 19.75 |
| Vermiculite plaster | | | | |
| 2 coats | | | | |
| Ceilings | S.Y. | 5.20 | 16.65 | 21.85 |
| Walls | " | 5.20 | 15.20 | 20.40 |
| 3 coats | | | | |
| Ceilings | S.Y. | 8.15 | 20.60 | 28.75 |
| Walls | " | 8.15 | 18.40 | 26.55 |
| Keenes cement plaster | | | | |
| 2 coats | | | | |
| Ceilings | S.Y. | 6.90 | 13.45 | 20.35 |
| Walls | " | 6.90 | 11.65 | 18.55 |
| 3 coats | | | | |
| Ceilings | S.Y. | 8.40 | 15.20 | 23.60 |
| Walls | " | 8.40 | 13.45 | 21.85 |
| On columns, add to installation, 50% | " | | | |
| Chases, fascia, and soffits, add to installation, 50% | " | | | |
| Beams, add to installation, 50% | " | | | |
| Patch holes, average size holes | | | | |

# 09 FINISHES

## LATH AND PLASTER

| LATH AND PLASTER | UNIT | MAT. | INST. | TOTAL |
|---|---|---|---|---|
| **09210.10 PLASTER** | | | | |
| 1 sf to 5 sf | | | | |
| Minimum | S.F. | 1.80 | 5.85 | 7.65 |
| Average | " | 2.45 | 7.00 | 9.45 |
| Maximum | " | 3.30 | 8.75 | 12.05 |
| Over 5 sf | | | | |
| Minimum | S.F. | 1.60 | 3.50 | 5.10 |
| Average | " | 2.15 | 5.00 | 7.15 |
| Maximum | " | 2.70 | 5.85 | 8.55 |
| Patch cracks | | | | |
| Minimum | S.F. | 0.95 | 1.15 | 2.10 |
| average | " | 1.10 | 1.75 | 2.85 |
| Maximum | " | 1.35 | 3.50 | 4.85 |
| **09220.10 PORTLAND CEMENT PLASTER** | | | | |
| Stucco, portland, gray, 3 coat, 1" thick | | | | |
| Sand finish | S.Y. | 5.85 | 15.20 | 21.05 |
| Trowel finish | " | 5.85 | 15.90 | 21.75 |
| White cement | | | | |
| Sand finish | S.Y. | 6.70 | 15.90 | 22.60 |
| Trowel finish | " | 6.70 | 17.50 | 24.20 |
| Scratch coat | | | | |
| For ceramic tile | S.Y. | 2.15 | 3.50 | 5.65 |
| For quarry tile | " | 2.15 | 3.50 | 5.65 |
| Portland cement plaster | | | | |
| 2 coats, 1/2" | S.Y. | 4.25 | 7.00 | 11.25 |
| 3 coats, 7/8" | " | 5.05 | 8.75 | 13.80 |
| **09250.10 GYPSUM BOARD** | | | | |
| Drywall, plasterboard, 3/8" clipped to | | | | |
| Metal furred ceiling | S.F. | 0.27 | 0.39 | 0.66 |
| Columns and beams | " | 0.27 | 0.89 | 1.16 |
| Walls | " | 0.27 | 0.35 | 0.62 |
| Nailed or screwed to | | | | |
| Wood framed ceiling | S.F. | 0.27 | 0.35 | 0.62 |
| Columns and beams | " | 0.27 | 0.79 | 1.06 |
| Walls | " | 0.27 | 0.32 | 0.59 |
| 1/2", clipped to | | | | |
| Metal furred ceiling | S.F. | 0.28 | 0.39 | 0.67 |
| Columns and beams | " | 0.28 | 0.89 | 1.17 |
| Walls | " | 0.28 | 0.35 | 0.63 |
| Nailed or screwed to | | | | |
| Wood framed ceiling | S.F. | 0.28 | 0.35 | 0.63 |
| Columns and beams | " | 0.28 | 0.79 | 1.07 |
| Walls | " | 0.28 | 0.32 | 0.60 |
| 5/8", clipped to | | | | |
| Metal furred ceiling | S.F. | 0.31 | 0.44 | 0.75 |
| Columns and beams | " | 0.31 | 0.98 | 1.29 |
| Walls | " | 0.31 | 0.39 | 0.70 |
| Nailed or screwed to | | | | |
| Wood framed ceiling | S.F. | 0.31 | 0.44 | 0.75 |
| Columns and beams | " | 0.31 | 0.98 | 1.29 |
| Walls | " | 0.31 | 0.39 | 0.70 |

## LATH AND PLASTER

### 09250.10 GYPSUM BOARD

| LATH AND PLASTER | UNIT | MAT. | INST. | TOTAL |
|---|---|---|---|---|
| Vinyl faced, clipped to metal studs | | | | |
| 1/2" | S.F. | 1.40 | 0.44 | 1.84 |
| 5/8" | " | 1.45 | 0.44 | 1.89 |
| Add for | | | | |
| Fire resistant | S.F. | | | 0.08 |
| Water resistant | " | | | 0.14 |
| Water and fire resistant | " | | | 0.18 |
| Taping and finishing joints | | | | |
| Minimum | S.F. | 0.04 | 0.24 | 0.28 |
| Average | " | 0.05 | 0.29 | 0.34 |
| Maximum | " | 0.07 | 0.35 | 0.42 |
| Casing bead | | | | |
| Minimum | L.F. | 0.12 | 1.00 | 1.12 |
| Average | " | 0.14 | 1.20 | 1.34 |
| Maximum | " | 0.18 | 1.75 | 1.93 |
| Corner bead | | | | |
| Minimum | L.F. | 0.14 | 1.00 | 1.14 |
| Average | " | 0.16 | 1.20 | 1.36 |
| Maximum | " | 0.20 | 1.75 | 1.95 |

## TILE

### 09310.10 CERAMIC TILE

| TILE | UNIT | MAT. | INST. | TOTAL |
|---|---|---|---|---|
| Glazed wall tile, 4-1/4" x 4-1/4" | | | | |
| Minimum | S.F. | 1.95 | 2.45 | 4.40 |
| Average | " | 3.10 | 2.85 | 5.95 |
| Maximum | " | 8.40 | 3.45 | 11.85 |
| Base, 4-1/4" high | | | | |
| Minimum | L.F. | 3.00 | 4.30 | 7.30 |
| Average | " | 3.35 | 4.30 | 7.65 |
| Maximum | " | 5.35 | 4.30 | 9.65 |
| Unglazed floor tile | | | | |
| Portland cement bed, cushion edge, face mounted | | | | |
| 1" x 1" | S.F. | 5.85 | 3.10 | 8.95 |
| 1" x 2" | " | 9.40 | 3.00 | 12.40 |
| 2" x 2" | " | 6.20 | 2.85 | 9.05 |
| Adhesive bed, with white grout | | | | |
| 1" x 1" | S.F. | 5.20 | 3.10 | 8.30 |
| 1" x 2" | " | 5.40 | 3.00 | 8.40 |
| 2" x 2" | " | 5.50 | 2.85 | 8.35 |
| Organic adhesive bed, thin set, back mounted | | | | |
| 1" x 1" | S.F. | 5.75 | 3.10 | 8.85 |
| 1" x 2" | " | 5.95 | 3.00 | 8.95 |

## TILE

| 09310.10 CERAMIC TILE | UNIT | MAT. | INST. | TOTAL |
|---|---|---|---|---|
| 2" x 2" | S.F. | 6.20 | 2.85 | 9.05 |
| For group 2 colors, add to material, 10% | | | | |
| For group 3 colors, add to material, 20% | | | | |
| For abrasive surface, add to material, 25% | | | | |
| Unglazed wall tile | | | | |
| Organic adhesive, face mounted cushion edge | | | | |
| 1" x 1" | | | | |
| Minimum | S.F. | 2.75 | 2.85 | 5.60 |
| Average | " | 3.60 | 3.10 | 6.70 |
| Maximum | " | 5.35 | 3.45 | 8.80 |
| 1" x 2" | | | | |
| Minimum | S.F. | 3.00 | 2.75 | 5.75 |
| Average | " | 4.00 | 3.00 | 7.00 |
| Maximum | " | 5.90 | 3.25 | 9.15 |
| 2" x 2" | | | | |
| Minimum | S.F. | 3.15 | 2.65 | 5.80 |
| Average | " | 3.60 | 2.85 | 6.45 |
| Maximum | " | 4.15 | 3.10 | 7.25 |
| Back mounted | | | | |
| 1" x 1" | | | | |
| Minimum | S.F. | 2.75 | 2.85 | 5.60 |
| Average | " | 3.60 | 3.10 | 6.70 |
| Maximum | " | 5.35 | 3.45 | 8.80 |
| 1" x 2" | | | | |
| Minimum | S.F. | 3.00 | 2.75 | 5.75 |
| Average | " | 4.00 | 3.00 | 7.00 |
| Maximum | " | 5.80 | 3.25 | 9.05 |
| 2" x 2" | | | | |
| Minimum | S.F. | 3.15 | 2.65 | 5.80 |
| Average | " | 3.60 | 2.85 | 6.45 |
| Maximum | " | 5.80 | 3.10 | 8.90 |
| For glazed finish, add to material, 25% | | | | |
| For glazed mosaic, add to material, 100% | | | | |
| For metallic colors, add to material, 125% | | | | |
| For exterior wall use, add to total, 25% | | | | |
| For exterior soffit, add to total, 25% | | | | |
| For portland cement bed, add to total, 25% | | | | |
| For dry set portland cement bed, add to total, 10% | | | | |
| Conductive floor tile, unglazed square edged | | | | |
| Portland cement bed | | | | |
| 1 x 1 | S.F. | 5.60 | 4.30 | 9.90 |
| 1-9/16 x 1-9/16 | " | 5.35 | 4.30 | 9.65 |
| Dry set | | | | |
| 1 x 1 | S.F. | 5.90 | 4.30 | 10.20 |
| 1-9/16 x 1-9/16 | " | 5.55 | 4.30 | 9.85 |
| Epoxy bed with epoxy joints | | | | |
| 1 x 1 | S.F. | 4.85 | 4.30 | 9.15 |
| 1-9/16 x 1-9/16 | " | 5.90 | 4.30 | 10.20 |
| For WWF in bed add to total, 15% | | | | |
| For abrasive surface, add to material, 40% | | | | |
| Ceramic accessories | | | | |
| Towel bar, 24" long | | | | |

# 09 FINISHES

| TILE | UNIT | MAT. | INST. | TOTAL |
|---|---|---|---|---|
| **09310.10 CERAMIC TILE** | | | | |
| Minimum | EA. | 22.50 | 13.75 | 36.25 |
| Average | " | 29.50 | 17.15 | 46.65 |
| Maximum | " | 44.90 | 22.90 | 67.80 |
| Soap dish | | | | |
| Minimum | EA. | 19.65 | 22.90 | 42.55 |
| Average | " | 23.90 | 28.60 | 52.50 |
| Maximum | " | 33.70 | 34.30 | 68.00 |
| **09330.10 QUARRY TILE** | | | | |
| Floor | | | | |
| 4 x 4 x 1/2" | S.F. | 4.50 | 4.60 | 9.10 |
| 6 x 6 x 1/2" | " | 4.55 | 4.30 | 8.85 |
| 6 x 6 x 3/4" | " | 5.35 | 4.30 | 9.65 |
| Wall, applied to 3/4" portland cement bed | | | | |
| 4 x 4 x 1/2" | S.F. | 4.10 | 6.85 | 10.95 |
| 6 x 6 x 3/4" | " | 5.15 | 5.70 | 10.85 |
| Cove base | | | | |
| 5 x 6 x 1/2" straight top | L.F. | 4.00 | 5.70 | 9.70 |
| 6 x 6 x 3/4" round top | " | 4.25 | 5.70 | 9.95 |
| Stair treads 6 x 6 x 3/4" | " | 7.10 | 8.60 | 15.70 |
| Window sill 6 x 8 x 3/4" | " | 5.70 | 6.85 | 12.55 |
| For abrasive surface, add to material, 25% | | | | |
| **09410.10 TERRAZZO** | | | | |
| Floors on concrete, 1-3/4" thick, 5/8" topping | | | | |
| Gray cement | S.F. | 3.75 | 5.00 | 8.75 |
| White cement | " | 4.10 | 5.00 | 9.10 |
| Sand cushion, 3" thick, 5/8" top, 1/4" | | | | |
| Gray cement | S.F. | 4.45 | 5.85 | 10.30 |
| White cement | " | 4.95 | 5.85 | 10.80 |
| Monolithic terrazzo, 3-1/2" base slab, 5/8" topping | " | 3.15 | 4.35 | 7.50 |
| Terrazzo wainscot, cast-in-place, 1/2" thick | " | 6.60 | 8.75 | 15.35 |
| Base, cast in place, terrazzo cove type, 6" high | L.F. | 7.10 | 5.00 | 12.10 |
| Curb, cast in place, 6" wide x 6" high, polished top | " | 7.85 | 17.50 | 25.35 |
| For venetian type terrazzo, add to material, 10% | | | | |
| For abrasive heavy duty terrazzo, add to material, 15% | | | | |
| Divider strips | | | | |
| Zinc | L.F. | | | 1.19 |
| Brass | " | | | 2.23 |
| Stairs, cast-in-place, topping on concrete or metal | | | | |
| 1-1/2" thick treads, 12" wide | L.F. | 4.75 | 17.50 | 22.25 |
| Combined tread and riser | " | 7.10 | 43.80 | 50.90 |
| Precast terrazzo, thin set | | | | |
| Terrazzo tiles, non-slip surface | | | | |
| 9" x 9" x 1" thick | S.F. | 14.95 | 5.00 | 19.95 |
| 12" x 12" | | | | |
| 1" thick | S.F. | 16.15 | 4.65 | 20.80 |
| 1-1/2" thick | " | 16.85 | 5.00 | 21.85 |
| 18" x 18" x 1-1/2" thick | " | 22.05 | 5.00 | 27.05 |
| 24" x 24" x 1-1/2" thick | " | 28.40 | 4.10 | 32.50 |

## TILE

### 09410.10 TERRAZZO

| TILE | UNIT | MAT. | INST. | TOTAL |
|---|---|---|---|---|
| For white cement, add to material, 10% | | | | |
| For venetian type terrazzo, add to material, 25% | | | | |
| Terrazzo wainscot | | | | |
| 12" x 12" x 1" thick | S.F. | 16.55 | 8.75 | 25.30 |
| 18" x 18" x 1-1/2" thick | " | 26.00 | 10.00 | 36.00 |
| Base | | | | |
| 6" high | | | | |
| Straight | L.F. | 10.65 | 2.70 | 13.35 |
| Coved | " | 12.60 | 2.70 | 15.30 |
| 8" high | | | | |
| Straight | L.F. | 11.95 | 2.90 | 14.85 |
| Coved | " | 14.00 | 2.90 | 16.90 |
| Terrazzo curbs | | | | |
| 8" wide x 8" high | L.F. | 25.20 | 14.00 | 39.20 |
| 6" wide x 6" high | " | 22.85 | 11.65 | 34.50 |
| Precast terrazzo stair treads, 12" wide | | | | |
| 1-1/2" thick | | | | |
| Diamond pattern | L.F. | 26.80 | 6.35 | 33.15 |
| Non-slip surface | " | 29.90 | 6.35 | 36.25 |
| 2" thick | | | | |
| Diamond pattern | L.F. | 29.90 | 6.35 | 36.25 |
| Non-slip surface | " | 33.10 | 7.00 | 40.10 |
| Stair risers, 1" thick to 6" high | | | | |
| Straight sections | L.F. | 11.25 | 3.50 | 14.75 |
| Cove sections | " | 13.25 | 3.50 | 16.75 |
| Combined tread and riser | | | | |
| Straight sections | | | | |
| 1-1/2" tread, 3/4" riser | L.F. | 55.00 | 10.00 | 65.00 |
| 3" tread, 1" riser | " | 58.00 | 10.00 | 68.00 |
| Curved sections | | | | |
| 2" tread, 1" riser | L.F. | 61.50 | 11.65 | 73.15 |
| 3" tread, 1" riser | " | 64.00 | 11.65 | 75.65 |
| Stair stringers, notched for treads and risers | | | | |
| 1" thick | L.F. | 29.10 | 8.75 | 37.85 |
| 2" thick | " | 30.20 | 11.65 | 41.85 |
| Landings, structural, nonslip | | | | |
| 1-1/2" thick | S.F. | 22.85 | 5.85 | 28.70 |
| 3" thick | " | 24.40 | 7.00 | 31.40 |
| Conductive terrazzo, spark proof industrial floor | | | | |
| Epoxy terrazzo | | | | |
| Floor | S.F. | 10.55 | 2.20 | 12.75 |
| Base | " | 10.55 | 2.90 | 13.45 |
| Polyacrylate | | | | |
| Floor | S.F. | 8.50 | 2.20 | 10.70 |
| Base | " | 8.50 | 2.90 | 11.40 |
| Polyester | | | | |
| Floor | S.F. | 3.00 | 1.40 | 4.40 |
| Base | " | 3.00 | 1.75 | 4.75 |
| Synthetic latex mastic | | | | |
| Floor | S.F. | 4.90 | 2.20 | 7.10 |
| Base | " | 4.90 | 2.90 | 7.80 |

# 09 FINISHES

## ACOUSTICAL TREATMENT

### 09510.10 CEILINGS AND WALLS

| ACOUSTICAL TREATMENT | UNIT | MAT. | INST. | TOTAL |
|---|---|---|---|---|
| Acoustical panels, suspension system not included | | | | |
| Fiberglass panels | | | | |
| 5/8" thick | | | | |
| 2' x 2' | S.F. | 0.72 | 0.51 | 1.23 |
| 2' x 4' | " | 0.72 | 0.39 | 1.11 |
| 3/4" thick | | | | |
| 2' x 2' | S.F. | 0.86 | 0.51 | 1.37 |
| 2' x 4' | " | 0.86 | 0.39 | 1.25 |
| Glass cloth faced fiberglass panels | | | | |
| 3/4" thick | S.F. | 1.95 | 0.59 | 2.54 |
| 1" thick | " | 2.15 | 0.59 | 2.74 |
| Mineral fiber panels | | | | |
| 5/8" thick | | | | |
| 2' x 2' | S.F. | 0.86 | 0.51 | 1.37 |
| 2' x 4' | " | 0.86 | 0.39 | 1.25 |
| 3/4" thick | | | | |
| 2' x 2' | S.F. | 1.15 | 0.51 | 1.66 |
| 2' x 4' | " | 1.15 | 0.39 | 1.54 |
| For aluminum faced panels, add to material, 80% | | | | |
| For vinyl faced panels, add to total, 125% | | | | |
| For fire rated panels, add to material, 75% | | | | |
| Wood fiber panels | | | | |
| 2" thick | | | | |
| 2' x 2' | S.F. | 2.05 | 0.59 | 2.64 |
| 2' x 4' | " | 2.05 | 0.44 | 2.49 |
| For flameproofing, add to material, 10% | | | | |
| For sculptured finish, add to material, 15% | | | | |
| Air distributing panels | | | | |
| 3/4" thick | S.F. | 2.05 | 0.89 | 2.94 |
| 5/8" thick | " | 1.75 | 0.71 | 2.46 |
| Acoustical tiles, suspension system not included | | | | |
| Fiberglass tile, 12" x 12" | | | | |
| 5/8" thick | S.F. | 0.51 | 0.64 | 1.15 |
| 3/4" thick | " | 0.55 | 0.79 | 1.34 |
| Glass cloth faced fiberglass tile | | | | |
| 3/4" thick | S.F. | 1.95 | 0.79 | 2.74 |
| 3" thick | " | 2.00 | 0.89 | 2.89 |
| Mineral fiber tile, 12" x 12" | | | | |
| 5/8" thick | | | | |
| Standard | S.F. | 1.10 | 0.71 | 1.81 |
| Vinyl faced | " | 1.25 | 0.71 | 1.96 |
| 3/4" thick | | | | |
| Standard | S.F. | 1.30 | 0.71 | 2.01 |
| Vinyl faced | " | 1.45 | 0.71 | 2.16 |
| Fire rated | " | 2.05 | 0.71 | 2.76 |
| Aluminum or mylar faced | " | 2.60 | 0.71 | 3.31 |
| Wood fiber tile, 12" x 12" | | | | |
| 1/2" thick | S.F. | 0.82 | 0.71 | 1.53 |
| 3/4" thick | " | 1.15 | 0.71 | 1.86 |
| For flameproofing, add to material, 10% | | | | |
| For sculptured 3 dimensional, add to material, 50% | | | | |
| Metal pan units, 24 ga. steel | | | | |

# 09 FINISHES

## ACOUSTICAL TREATMENT

### 09510.10 CEILINGS AND WALLS

| ACOUSTICAL TREATMENT | UNIT | MAT. | INST. | TOTAL |
|---|---|---|---|---|
| 12" x 12" | S.F. | 4.15 | 1.40 | 5.55 |
| 12" x 24" | " | 4.25 | 1.20 | 5.45 |
| Aluminum, .025" thick | | | | |
| 12" x 12" | S.F. | 4.40 | 1.40 | 5.80 |
| 12" x 24" | " | 4.50 | 1.20 | 5.70 |
| Anodized aluminum, 0.25" thick | | | | |
| 12" x 12" | S.F. | 5.05 | 1.40 | 6.45 |
| 12" x 24" | " | 6.00 | 1.20 | 7.20 |
| Stainless steel, 24 ga. | | | | |
| 12" x 12" | S.F. | 9.70 | 1.40 | 11.10 |
| 12" x 24" | " | 10.00 | 1.20 | 11.20 |
| For flameproof sound absorbing pads, add to material | " | | | 1.50 |
| Metal ceiling systems | | | | |
| .020" thick panels | | | | |
| 10', 12', and 16' lengths | S.F. | 3.75 | 1.00 | 4.75 |
| Custom lengths, 3' to 20' | " | 3.75 | 1.00 | 4.75 |
| .025" thick panels | | | | |
| 32 sf, 38 sf, and 52 sf pieces | S.F. | 4.30 | 1.20 | 5.50 |
| Custom lengths, 10 sf to 65 sf | " | 4.40 | 1.20 | 5.60 |
| Carriers, black, add | " | | | 2.25 |
| Recess filler strip, add | " | | | 0.76 |
| Custom lengths, add | " | | | 1.13 |
| Sound absorption walls, with fabric cover | | | | |
| 2-6" x 9' x 3/4" | S.F. | 5.05 | 1.20 | 6.25 |
| 2' x 9' x 1" | " | 5.50 | 1.20 | 6.70 |
| Starter spline | L.F. | 1.20 | 0.89 | 2.09 |
| Internal spline | " | 1.05 | 0.89 | 1.94 |
| Acoustical treatment | | | | |
| Barriers for plenums | | | | |
| Leaded vinyl | | | | |
| 0.48 lb per sf | S.F. | 2.20 | 1.70 | 3.90 |
| 0.87 lb per sf | " | 2.20 | 1.75 | 3.95 |
| Aluminum foil, fiberglass reinforcement | | | | |
| Minimum | S.F. | 0.85 | 1.20 | 2.05 |
| Maximum | " | 0.96 | 1.75 | 2.71 |
| Aluminum mesh, paper backed | " | 0.82 | 1.20 | 2.02 |
| Fibered cement sheet, 3/16" thick | " | 1.65 | 1.25 | 2.90 |
| Sheet lead, 1/64" thick | " | 2.05 | 0.89 | 2.94 |
| Sound attenuation blanket | | | | |
| 1" thick | S.F. | 0.42 | 3.55 | 3.97 |
| 1-1/2" thick | " | 0.55 | 3.55 | 4.10 |
| 2" thick | " | 0.69 | 3.55 | 4.24 |
| 3" thick | " | 0.82 | 3.95 | 4.77 |
| Ceiling suspension systems | | | | |
| T bar system | | | | |
| 2' x 4' | S.F. | 0.86 | 0.35 | 1.21 |
| 2' x 2' | " | 0.93 | 0.39 | 1.32 |
| Concealed Z bar suspension system, 12" module | " | 0.81 | 0.59 | 1.40 |
| For 1-1/2" carrier channels, 4' o.c., add | " | | | 0.28 |
| Carrier channel for recessed light fixtures | " | | | 0.51 |

| FLOORING | UNIT | MAT. | INST. | TOTAL |
|---|---|---|---|---|
| **09550.10 WOOD FLOORING** | | | | |
| Wood strip flooring, unfinished | | | | |
| Fir floor | | | | |
| C and better | | | | |
| Vertical grain | S.F. | 2.75 | 1.20 | 3.95 |
| Flat grain | " | 2.60 | 1.20 | 3.80 |
| Oak floor | | | | |
| Minimum | S.F. | 3.10 | 1.70 | 4.80 |
| Average | " | 3.70 | 1.70 | 5.40 |
| Maximum | " | 4.65 | 1.70 | 6.35 |
| Maple floor | | | | |
| 25/32" x 2-1/4" | | | | |
| Minimum | S.F. | 3.35 | 1.70 | 5.05 |
| Maximum | " | 3.95 | 1.70 | 5.65 |
| 33/32" x 3-1/4" | | | | |
| Minimum | S.F. | 4.35 | 1.70 | 6.05 |
| Maximum | " | 4.90 | 1.70 | 6.60 |
| Added costs | | | | |
| For factory finish, add to material, 10% | | | | |
| For random width floor, add to total, 20% | | | | |
| For simulated pegs, add to total, 10% | | | | |
| Wood block industrial flooring | | | | |
| Creosoted | | | | |
| 2" thick | S.F. | 3.25 | 0.93 | 4.18 |
| 2-1/2" thick | " | 3.35 | 1.10 | 4.45 |
| 3" thick | " | 3.50 | 1.20 | 4.70 |
| Parquet, 5/16", white oak | | | | |
| Finished | S.F. | 7.15 | 1.75 | 8.90 |
| Unfinished | " | 2.55 | 1.75 | 4.30 |
| Gym floor, 2 ply felt, 25/32" maple, finished, in mastic | " | 6.05 | 1.95 | 8.00 |
| Over wood sleepers | " | 6.75 | 2.20 | 8.95 |
| Finishing, sand, fill, finish, and wax | " | 0.57 | 0.89 | 1.46 |
| Refinish sand, seal, and 2 coats of polyurethane | " | 0.95 | 1.20 | 2.15 |
| Clean and wax floors | " | 0.16 | 0.18 | 0.34 |
| **09630.10 UNIT MASONRY FLOORING** | | | | |
| Clay brick | | | | |
| 9 x 4-1/2 x 3" thick | | | | |
| Glazed | S.F. | 6.75 | 2.95 | 9.70 |
| Unglazed | " | 6.45 | 2.95 | 9.40 |
| 8 x 4 x 3/4" thick | | | | |
| Glazed | S.F. | 6.10 | 3.10 | 9.20 |
| Unglazed | " | 6.00 | 3.10 | 9.10 |
| For herringbone pattern, add to labor, 15% | | | | |
| **09660.10 RESILIENT TILE FLOORING** | | | | |
| Solid vinyl tile, 1/8" thick, 12" x 12" | | | | |
| Marble patterns | S.F. | 2.05 | 0.89 | 2.94 |
| Solid colors | " | 2.60 | 0.89 | 3.49 |
| Travertine patterns | " | 2.95 | 0.89 | 3.84 |
| Conductive resilient flooring, vinyl tile | | | | |
| 1/8" thick, 12" x 12" | S.F. | 4.60 | 1.00 | 5.60 |

## FLOORING

| FLOORING | UNIT | MAT. | INST. | TOTAL |
|---|---|---|---|---|
| **09665.10 RESILIENT SHEET FLOORING** | | | | |
| Vinyl sheet flooring | | | | |
| Minimum | S.F. | 1.75 | 0.35 | 2.10 |
| Average | " | 2.55 | 0.43 | 2.98 |
| Maximum | " | 5.35 | 0.59 | 5.94 |
| Cove, to 6" | L.F. | 1.10 | 0.71 | 1.81 |
| Fluid applied resilient flooring | | | | |
| Polyurethane, poured in place, 3/8" thick | S.F. | 7.25 | 2.95 | 10.20 |
| Vinyl sheet goods, backed | | | | |
| 0.070" thick | S.F. | 2.65 | 0.44 | 3.09 |
| 0.093" thick | " | 4.15 | 0.44 | 4.59 |
| 0.125" thick | " | 4.80 | 0.44 | 5.24 |
| 0.250" thick | " | 5.50 | 0.44 | 5.94 |
| **09678.10 RESILIENT BASE AND ACCESSORIES** | | | | |
| Wall base, vinyl | | | | |
| Group 1 | | | | |
| 4" high | L.F. | 0.72 | 1.20 | 1.92 |
| 6" high | " | 0.99 | 1.20 | 2.19 |
| Group 2 | | | | |
| 4" high | L.F. | 0.63 | 1.20 | 1.83 |
| 6" high | " | 0.74 | 1.20 | 1.94 |
| Group 3 | | | | |
| 4" high | L.F. | 0.69 | 1.20 | 1.89 |
| 6" high | " | 0.82 | 1.20 | 2.02 |
| Stair accessories | | | | |
| Treads, 1/4" x 12", rubber diamond surface | | | | |
| Marbled | L.F. | 5.70 | 2.95 | 8.65 |
| Plain | " | 5.60 | 2.95 | 8.55 |
| Grit strip safety tread, 12" wide, colors | | | | |
| 3/16" thick | L.F. | 7.85 | 2.95 | 10.80 |
| 5/16" thick | " | 11.00 | 2.95 | 13.95 |
| Risers, 7" high, 1/8" thick, colors | | | | |
| Flat | L.F. | 1.65 | 1.75 | 3.40 |
| Coved | " | 2.75 | 1.75 | 4.50 |
| Nosing, rubber | | | | |
| 3/16" thick, 3" wide | | | | |
| Black | L.F. | 2.75 | 1.75 | 4.50 |
| Colors | " | 2.90 | 1.75 | 4.65 |
| 6" wide | | | | |
| Black | L.F. | 3.45 | 2.95 | 6.40 |
| Colors | " | 3.60 | 2.95 | 6.55 |

| CARPET | UNIT | MAT. | INST. | TOTAL |
|---|---|---|---|---|
| **09680.10 FLOOR LEVELING** | | | | |
| Repair and level floors to receive new flooring | | | | |
| Minimum | S.Y. | 0.68 | 1.20 | 1.88 |
| Average | " | 2.70 | 2.95 | 5.65 |
| Maximum | " | 3.50 | 3.55 | 7.05 |
| **09682.10 CARPET PADDING** | | | | |
| Carpet padding | | | | |
| Foam rubber, waffle type, 0.3" thick | S.Y. | 4.75 | 1.75 | 6.50 |
| Jute padding | | | | |
| Minimum | S.Y. | 3.45 | 1.60 | 5.05 |
| Average | " | 4.45 | 1.75 | 6.20 |
| Maximum | " | 6.75 | 1.95 | 8.70 |
| Sponge rubber cushion | | | | |
| Minimum | S.Y. | 4.05 | 1.60 | 5.65 |
| Average | " | 5.40 | 1.75 | 7.15 |
| Maximum | " | 6.75 | 1.95 | 8.70 |
| Urethane cushion, 3/8" thick | | | | |
| Minimum | S.Y. | 4.05 | 1.60 | 5.65 |
| Average | " | 4.75 | 1.75 | 6.50 |
| Maximum | " | 5.40 | 1.95 | 7.35 |
| **09685.10 CARPET** | | | | |
| Carpet, acrylic | | | | |
| 24 oz., light traffic | S.Y. | 22.75 | 1.95 | 24.70 |
| 28 oz., medium traffic | " | 28.80 | 1.95 | 30.75 |
| Residential | | | | |
| Nylon | | | | |
| 15 oz., light traffic | S.Y. | 16.40 | 1.95 | 18.35 |
| 28 oz., medium traffic | " | 19.65 | 1.95 | 21.60 |
| Commercial | | | | |
| Nylon | | | | |
| 28 oz., medium traffic | S.Y. | 20.95 | 1.95 | 22.90 |
| 35 oz., heavy traffic | " | 25.10 | 1.95 | 27.05 |
| Wool | | | | |
| 30 oz., medium traffic | S.Y. | 34.20 | 1.95 | 36.15 |
| 36 oz., medium traffic | " | 36.10 | 1.95 | 38.05 |
| 42 oz., heavy traffic | " | 47.70 | 1.95 | 49.65 |
| Carpet tile | | | | |
| Foam backed | | | | |
| Minimum | S.F. | 2.60 | 0.35 | 2.95 |
| Average | " | 3.05 | 0.39 | 3.44 |
| Maximum | " | 4.80 | 0.44 | 5.24 |
| Tufted loop or shag | | | | |
| Minimum | S.F. | 2.80 | 0.35 | 3.15 |
| Average | " | 3.40 | 0.39 | 3.79 |
| Maximum | " | 5.50 | 0.44 | 5.94 |
| Clean and vacuum carpet | | | | |
| Minimum | S.Y. | 0.23 | 0.14 | 0.37 |
| Average | " | 0.38 | 0.24 | 0.62 |
| Maximum | " | 0.51 | 0.35 | 0.86 |

# 09 FINISHES

## CARPET

| 09700.10 SPECIAL FLOORING | UNIT | MAT. | INST. | TOTAL |
|---|---|---|---|---|
| Epoxy flooring, marble chips | | | | |
| Epoxy with colored quartz chips in 1/4" base | S.F. | 8.75 | 1.95 | 10.70 |
| Heavy duty epoxy topping, 3/16" thick | " | 7.70 | 1.95 | 9.65 |
| Epoxy terrazzo | | | | |
| 1/4" thick chemical resistant | S.F. | 8.95 | 2.20 | 11.15 |

## PAINTING

| 09910.10 EXTERIOR PAINTING | UNIT | MAT. | INST. | TOTAL |
|---|---|---|---|---|
| Exterior painting | | | | |
| Wood surfaces, 1 coat primer, two coats paint | | | | |
| Door and frame | EA. | 4.50 | 59.00 | 63.50 |
| Windows | S.F. | 0.09 | 0.89 | 0.98 |
| Wood trim | " | 0.23 | 0.89 | 1.12 |
| Wood siding | " | 0.23 | 0.44 | 0.67 |
| Hardboard surfaces | | | | |
| One coat primer, two coats paint | S.F. | 0.23 | 0.44 | 0.67 |
| Asbestos cement surfaces | | | | |
| One coat primer, two coats paint | S.F. | 0.26 | 0.44 | 0.70 |
| Galvanized surfaces, galvanized primer | | | | |
| One coat primer, two coats paint | S.F. | 0.24 | 0.42 | 0.66 |
| Stucco surfaces, acrylic primer, acrylic latex paint | | | | |
| One coat primer, two coats paint | S.F. | 0.24 | 0.59 | 0.83 |
| Concrete masonry unit surfaces, brush work | | | | |
| One coat filler, one coat paint | S.F. | 0.27 | 0.44 | 0.71 |
| Two coats epoxy | " | 0.39 | 0.59 | 0.98 |
| Texture coating | " | 0.36 | 0.35 | 0.71 |
| Concrete surfaces | | | | |
| One coat filler, one coat paint | S.F. | 0.23 | 0.44 | 0.67 |
| Two coats paint | " | 0.31 | 0.59 | 0.90 |
| Structural steel | | | | |
| One field coat paint, brush work | | | | |
| Light framing | S.F. | 0.12 | 0.29 | 0.41 |
| Heavy framing | " | 0.12 | 0.18 | 0.30 |
| One field coat paint, spray work | | | | |
| Light framing | S.F. | 0.09 | 0.09 | 0.18 |
| Heavy framing | " | 0.09 | 0.07 | 0.16 |
| Miscellaneous steel items, spray work, one coat | | | | |
| Exposed decking | S.F. | 0.12 | 0.29 | 0.41 |
| Joist | " | 0.12 | 0.44 | 0.56 |
| Columns | " | 0.12 | 0.44 | 0.56 |
| Pipes, one coat primer, one coat paint | | | | |
| 4" dia. | L.F. | 0.09 | 0.44 | 0.53 |

# 09 FINISHES

## PAINTING

| | UNIT | MAT. | INST. | TOTAL |
|---|---|---|---|---|
| **09910.10 EXTERIOR PAINTING** | | | | |
| 8" dia. | L.F. | 0.23 | 0.59 | 0.82 |
| 12" dia. | " | 0.41 | 0.89 | 1.30 |
| Paint letters on pipe with brush | " | 0.46 | 0.89 | 1.35 |
| Paint pipe insulation cloth cover | S.F. | 0.27 | 0.44 | 0.71 |
| Sprinkler system piping | L.F. | 0.28 | 0.29 | 0.57 |
| Miscellaneous surfaces | | | | |
| Stair pipe rails | | | | |
| Two rails | L.F. | 0.20 | 1.20 | 1.40 |
| One rail | " | 0.11 | 0.71 | 0.82 |
| Stair to 4' wide, including rails, per riser | EA. | 0.76 | 5.05 | 5.81 |
| Gratings and frames | S.F. | 0.20 | 1.20 | 1.40 |
| Ladders | L.F. | 0.23 | 1.00 | 1.23 |
| Miscellaneous exposed metal | S.F. | 0.20 | 0.35 | 0.55 |
| **09920.10 INTERIOR PAINTING** | | | | |
| Walls, concrete and masonry, brush, primer, acrylic | | | | |
| One coat primer, one coat paint | S.F. | 0.12 | 0.44 | 0.56 |
| Two coats paint | " | 0.20 | 0.59 | 0.79 |
| Plywood, paint | " | 0.07 | 0.20 | 0.27 |
| Natural finish | " | 0.08 | 0.22 | 0.30 |
| Wood, paint | " | 0.08 | 0.22 | 0.30 |
| Natural finish | " | 0.11 | 0.25 | 0.36 |
| Metal | | | | |
| One coat filler | S.F. | 0.20 | 0.22 | 0.42 |
| One coat primer, one coat paint | " | 0.23 | 0.44 | 0.67 |
| Two coats paint | " | 0.28 | 0.59 | 0.87 |
| Plaster or gypsum board, paint | " | 0.11 | 0.20 | 0.31 |
| Epoxy | " | 0.15 | 0.22 | 0.37 |
| Ceilings, one coat paint, wood | " | 0.09 | 0.25 | 0.34 |
| Concrete | " | 0.15 | 0.22 | 0.37 |
| Plaster | " | 0.09 | 0.20 | 0.29 |
| Miscellaneous metal, brushwork | " | 0.20 | 0.44 | 0.64 |
| **09920.30 DOORS AND MILLWORK** | | | | |
| Painting, doors | | | | |
| Minimum | S.F. | 0.16 | 1.20 | 1.36 |
| Average | " | 0.28 | 1.75 | 2.03 |
| Maximum | " | 0.47 | 2.35 | 2.82 |
| Cabinets, shelves, and millwork | | | | |
| Minimum | S.F. | 0.11 | 0.59 | 0.70 |
| Average | " | 0.23 | 1.00 | 1.23 |
| Maximum | " | 0.35 | 1.75 | 2.10 |
| **09920.60 WINDOWS** | | | | |
| Painting, windows | | | | |
| Minimum | S.F. | 0.07 | 0.71 | 0.78 |
| Average | " | 0.11 | 0.89 | 1.00 |
| Maximum | " | 0.18 | 1.40 | 1.58 |

# 09 FINISHES

## PAINTING

| PAINTING | UNIT | MAT. | INST. | TOTAL |
|---|---|---|---|---|
| **09955.10 WALL COVERING** | | | | |
| Vinyl wall covering | | | | |
| Medium duty | S.F. | 1.10 | 0.51 | 1.61 |
| Heavy duty | " | 1.60 | 0.59 | 2.19 |
| Over pipes and irregular shapes | | | | |
| Lightweight, 13 oz. | S.F. | 1.10 | 0.71 | 1.81 |
| Medium weight, 25 oz. | " | 1.25 | 0.79 | 2.04 |
| Heavy weight, 34 oz. | " | 1.60 | 0.89 | 2.49 |
| Cork wall covering | | | | |
| 1' x 1' squares | | | | |
| 1/4" thick | S.F. | 1.70 | 0.89 | 2.59 |
| 1/2" thick | " | 2.15 | 0.89 | 3.04 |
| 3/4" thick | " | 2.45 | 0.89 | 3.34 |
| Wall fabrics | | | | |
| Natural fabrics, grass cloths | | | | |
| Minimum | S.F. | 1.10 | 0.55 | 1.65 |
| Average | " | 1.20 | 0.59 | 1.79 |
| Maximum | " | 4.05 | 0.71 | 4.76 |
| Flexible gypsum coated wall fabric, fire resistant | " | 1.20 | 0.35 | 1.55 |
| Vinyl corner guards | | | | |
| 3/4" x 3/4" x 8' | EA. | 5.75 | 4.45 | 10.20 |
| 2-3/4" x 2-3/4" x 4' | " | 3.35 | 4.45 | 7.80 |
| **09980.10 PAINTING PREPARATION** | | | | |
| Cleaning, light | | | | |
| Wood | S.F. | 0.04 | 0.09 | 0.13 |
| Plaster or gypsum wallboard | " | 0.04 | 0.08 | 0.12 |
| Prepare sprinkler piping for painting | L.F. | 0.04 | 0.07 | 0.11 |
| Prepare insulated pipe for painting | S.F. | 0.04 | 0.09 | 0.13 |
| Normal painting prep, masonry and concrete | | | | |
| Unpainted | S.F. | 0.04 | 0.06 | 0.10 |
| Painted | " | 0.04 | 0.09 | 0.13 |
| Plaster or gypsum | | | | |
| Unpainted | S.F. | 0.04 | 0.06 | 0.10 |
| Painted | " | 0.04 | 0.09 | 0.13 |
| Wood | | | | |
| Unpainted | S.F. | 0.04 | 0.06 | 0.10 |
| Painted | " | 0.04 | 0.09 | 0.13 |
| Painted steel, light rusting | " | 0.05 | 0.12 | 0.17 |
| Sandblasting | | | | |
| Brush off blast | S.F. | 0.15 | 0.24 | 0.39 |
| Commercial blast | " | 0.30 | 0.59 | 0.89 |
| Near white metal blast | " | 0.43 | 1.00 | 1.43 |
| White metal blast | " | 0.58 | 1.20 | 1.78 |
| **09980.15 PAINT** | | | | |
| Paint, enamel | | | | |
| 600 sf per gal. | GAL | | | 32.94 |
| 550 sf per gal. | " | | | 29.70 |
| 500 sf per gal. | " | | | 21.60 |
| 450 sf per gal. | " | | | 20.25 |
| 350 sf per gal. | " | | | 18.90 |

## PAINTING

### 09980.15 PAINT

| PAINTING | UNIT | MAT. | INST. | TOTAL |
|---|---|---|---|---|
| Filler, 60 sf per gal. | GAL | | | 22.95 |
| Latex, 400 sf per gal. | " | | | 18.90 |
| Aluminum | | | | |
| 400 sf per gal. | GAL | | | 27.00 |
| 500 sf per gal. | " | | | 48.60 |
| Red lead, 350 sf per gal. | " | | | 40.50 |
| Primer | | | | |
| 400 sf per gal. | GAL | | | 25.20 |
| 300 sf per gal. | " | | | 25.04 |
| Latex base, interior, white | " | | | 20.95 |
| Sealer and varnish | | | | |
| 400 sf per gal. | GAL | | | 19.36 |
| 425 sf per gal. | " | | | 28.32 |
| 600 sf per gal. | " | | | 37.95 |

| SPECIALTIES | UNIT | MAT. | INST. | TOTAL |
|---|---|---|---|---|
| **10110.10 CHALKBOARDS** | | | | |
| Chalkboard, metal frame, 1/4" thick | | | | |
| 48"x60" | EA. | 240.00 | 35.40 | 275.40 |
| 48"x96" | " | 390.00 | 39.40 | 429.40 |
| 48"x144" | " | 590.00 | 44.30 | 634.30 |
| 48"x192" | " | 760.00 | 50.50 | 810.50 |
| Liquid chalkboard | | | | |
| 48"x60" | EA. | 290.00 | 35.40 | 325.40 |
| 48"x96" | " | 430.00 | 39.40 | 469.40 |
| 48"x144" | " | 670.00 | 44.30 | 714.30 |
| 48"x192" | " | 850.00 | 50.50 | 900.50 |
| Map rail, deluxe | L.F. | 7.25 | 1.75 | 9.00 |
| **10165.10 TOILET PARTITIONS** | | | | |
| Toilet partition, plastic laminate | | | | |
| Ceiling mounted | EA. | 740.00 | 120.00 | 860.00 |
| Floor mounted | " | 700.00 | 88.50 | 788.50 |
| Metal | | | | |
| Ceiling mounted | EA. | 540.00 | 120.00 | 660.00 |
| Floor mounted | " | 510.00 | 88.50 | 598.50 |
| Wheel chair partition, plastic laminate | | | | |
| Ceiling mounted | EA. | 990.00 | 120.00 | 1,110 |
| Floor mounted | " | 920.00 | 88.50 | 1,009 |
| Painted metal | | | | |
| Ceiling mounted | EA. | 780.00 | 120.00 | 900.00 |
| Floor mounted | " | 700.00 | 88.50 | 788.50 |
| Urinal screen, plastic laminate | | | | |
| Wall hung | EA. | 350.00 | 44.30 | 394.30 |
| Floor mounted | " | 300.00 | 44.30 | 344.30 |
| Porcelain enameled steel, floor mounted | " | 410.00 | 44.30 | 454.30 |
| Painted metal, floor mounted | " | 260.00 | 44.30 | 304.30 |
| Stainless steel, floor mounted | " | 540.00 | 44.30 | 584.30 |
| Metal toilet partitions | | | | |
| Toilet partitions, front door and side divider, floor mounted | | | | |
| Porcelain enameled steel | EA. | 920.00 | 88.50 | 1,009 |
| Painted steel | " | 510.00 | 88.50 | 598.50 |
| Stainless steel | " | 1,220 | 88.50 | 1,309 |
| **10185.10 SHOWER STALLS** | | | | |
| Shower receptors | | | | |
| Precast, terrazzo | | | | |
| 32" x 32" | EA. | 410.00 | 32.50 | 442.50 |
| 32" x 48" | " | 510.00 | 39.00 | 549.00 |
| Concrete | | | | |
| 32" x 32" | EA. | 200.00 | 32.50 | 232.50 |
| 48" x 48" | " | 240.00 | 43.30 | 283.30 |
| Shower door, trim and hardware | | | | |
| Economy, 24" wide, chrome frame, tempered glass | EA. | 220.00 | 39.00 | 259.00 |
| Porcelain enameled steel, flush | " | 390.00 | 39.00 | 429.00 |
| Baked enameled steel, flush | " | 230.00 | 39.00 | 269.00 |
| Aluminum frame, tempered glass, 48" wide, sliding | " | 500.00 | 48.80 | 548.80 |

| SPECIALTIES | UNIT | MAT. | INST. | TOTAL |
|---|---|---|---|---|
| **10185.10 SHOWER STALLS** | | | | |
| Folding | EA. | 470.00 | 48.80 | 518.80 |
| Aluminum frame and tempered glass, molded plastic | | | | |
| Complete with receptor and door | | | | |
| 32" x 32" | EA. | 620.00 | 97.50 | 717.50 |
| 36" x 36" | " | 700.00 | 97.50 | 797.50 |
| 40" x 40" | " | 780.00 | 110.00 | 890.00 |
| Shower compartment, precast concrete receptor | | | | |
| Single entry type | | | | |
| Porcelain enameled steel | EA. | 1,840 | 390.00 | 2,230 |
| Baked enameled steel | " | 1,060 | 390.00 | 1,450 |
| Stainless steel | " | 2,130 | 390.00 | 2,520 |
| Double entry type | | | | |
| Porcelain enameled steel | EA. | 5,250 | 490.00 | 5,740 |
| Baked enameled steel | " | 3,400 | 490.00 | 3,890 |
| Stainless steel | " | 5,320 | 490.00 | 5,810 |
| **10190.10 CUBICLES** | | | | |
| Hospital track | | | | |
| Ceiling hung | L.F. | 4.65 | 3.95 | 8.60 |
| Suspended | " | 5.20 | 5.05 | 10.25 |
| Hospital metal dividers, galvanized steel | | | | |
| Baked enamel finish | | | | |
| 54" high | | | | |
| 10" glass light | L.F. | 100.00 | 17.70 | 117.70 |
| 14" glass light | " | 100.00 | 17.70 | 117.70 |
| 24" glass light | " | 96.00 | 17.70 | 113.70 |
| 60" high | | | | |
| 10" glass light | L.F. | 110.00 | 19.70 | 129.70 |
| 14" glass light | " | 110.00 | 19.70 | 129.70 |
| 24" glass light | " | 100.00 | 19.70 | 119.70 |
| Stainless steel | | | | |
| 54" high | | | | |
| 10" glass light | L.F. | 170.00 | 19.70 | 189.70 |
| 14" glass light | " | 190.00 | 19.70 | 209.70 |
| 24" glass light | " | 220.00 | 19.70 | 239.70 |
| 60" high | | | | |
| 14" glass light | L.F. | 200.00 | 22.15 | 222.15 |
| 14" glass light | " | 200.00 | 22.15 | 222.15 |
| 24" glass light | " | 250.00 | 22.15 | 272.15 |
| **10210.10 VENTS AND WALL LOUVERS** | | | | |
| Block vent, 8"x16"x4" aluminum, w/screen, mill finish | EA. | 120.00 | 12.85 | 132.85 |
| Standard | " | 63.00 | 12.05 | 75.05 |
| Vents w/screen, 4" deep, 8" wide, 5" high | | | | |
| Modular | EA. | 73.50 | 12.05 | 85.55 |
| Aluminum gable louvers | S.F. | 13.15 | 6.45 | 19.60 |
| Vent screen aluminum, 4" wide, continuous | L.F. | 3.65 | 1.30 | 4.95 |
| Louvers, aluminum, anodized, fixed blade | | | | |
| Horizontal line | S.F. | 42.40 | 9.65 | 52.05 |
| Vertical line | " | 42.40 | 9.65 | 52.05 |
| Wall louver, aluminum mill finish | | | | |

| SPECIALTIES | UNIT | MAT. | INST. | TOTAL |
|---|---|---|---|---|
| **10210.10 VENTS AND WALL LOUVERS** | | | | |
| Under, 2 sf | S.F. | 27.80 | 4.80 | 32.60 |
| 2 to 4 sf | " | 23.35 | 4.30 | 27.65 |
| 5 to 10 sf | " | 22.65 | 4.30 | 26.95 |
| Galvanized steel | | | | |
| Under 2 sf | S.F. | 26.30 | 4.80 | 31.10 |
| 2 to 4 sf | " | 19.00 | 4.30 | 23.30 |
| 5 to 10 sf | " | 17.55 | 4.30 | 21.85 |
| **10225.10 DOOR LOUVERS** | | | | |
| Fixed, 1" thick, enameled steel | | | | |
| 8"x8" | EA. | 42.40 | 4.45 | 46.85 |
| 12"x8" | " | 48.20 | 4.45 | 52.65 |
| 12"x12" | " | 54.00 | 5.05 | 59.05 |
| 16"x12" | " | 77.50 | 5.45 | 82.95 |
| 18"x12" | " | 77.50 | 8.85 | 86.35 |
| 20"x8" | " | 86.00 | 5.05 | 91.05 |
| 20"x12" | " | 98.00 | 10.15 | 108.15 |
| 20"x16" | " | 100.00 | 11.80 | 111.80 |
| 20"x20" | " | 100.00 | 14.20 | 114.20 |
| 24"x12" | " | 87.00 | 11.80 | 98.80 |
| 24"x16" | " | 91.50 | 12.65 | 104.15 |
| 24"x18" | " | 110.00 | 13.65 | 123.65 |
| 24"x20" | " | 110.00 | 14.75 | 124.75 |
| 24"x24" | " | 110.00 | 16.10 | 126.10 |
| 26"x26" | " | 130.00 | 22.15 | 152.15 |
| **10270.40 ACCESS & PEDESTAL FLOOR** | | | | |
| Panels, no covering, 2'x2' | | | | |
| Plain | S.F. | 7.95 | 0.44 | 8.39 |
| Perforated | " | 10.85 | 17.70 | 28.55 |
| Pedestals | | | | |
| For 6" to 12" clearance | EA. | 6.10 | 3.55 | 9.65 |
| Stringers | | | | |
| 2' | L.F. | 1.05 | 1.70 | 2.75 |
| 6' | " | 1.05 | 1.20 | 2.25 |
| Accessories | | | | |
| Ramp assembly | S.F. | 37.60 | 1.40 | 39.00 |
| Elevated floor assembly | " | 58.50 | 1.30 | 59.80 |
| Handrail | L.F. | 45.60 | 17.70 | 63.30 |
| Fascia plate | " | 21.70 | 8.85 | 30.55 |
| For carpet tiles, add | S.F. | | | 5.06 |
| For vinyl flooring, add | " | | | 6.51 |
| RF shielding components, floor liner | | | | |
| Hot rolled steel sheet | | | | |
| 14 ga. | S.F. | 2.45 | 0.89 | 3.34 |
| 11 ga. | " | 2.60 | 2.75 | 5.35 |

# 10 SPECIALTIES

| SPECIALTIES | UNIT | MAT. | INST. | TOTAL |
|---|---|---|---|---|
| **10290.10 PEST CONTROL** | | | | |
| Termite control | | | | |
| Under slab spraying | | | | |
| Minimum | S.F. | 0.15 | 0.07 | 0.22 |
| Average | " | 0.19 | 0.14 | 0.33 |
| Maximum | " | 0.26 | 0.28 | 0.54 |
| **10350.10 FLAGPOLES** | | | | |
| Installed in concrete base | | | | |
| Fiberglass | | | | |
| 25' high | EA. | 1,080 | 240.00 | 1,320 |
| 50' high | " | 4,860 | 590.00 | 5,450 |
| Aluminum | | | | |
| 25' high | EA. | 1,350 | 240.00 | 1,590 |
| 50' high | " | 4,460 | 590.00 | 5,050 |
| Bonderized steel | | | | |
| 25' high | EA. | 2,030 | 270.00 | 2,300 |
| 50' high | " | 4,730 | 710.00 | 5,440 |
| Freestanding tapered, fiberglass | | | | |
| 30' high | EA. | 1,220 | 250.00 | 1,470 |
| 40' high | " | 2,160 | 320.00 | 2,480 |
| 50' high | " | 4,320 | 350.00 | 4,670 |
| 60' high | " | 5,670 | 420.00 | 6,090 |
| Wall mounted, with collar, brushed aluminum finish | | | | |
| 15' long | EA. | 950.00 | 180.00 | 1,130 |
| 18' long | " | 1,220 | 180.00 | 1,400 |
| 20' long | " | 1,350 | 190.00 | 1,540 |
| 24' long | " | 1,490 | 210.00 | 1,700 |
| Outrigger, wall, including base | | | | |
| 10' long | EA. | 810.00 | 240.00 | 1,050 |
| 20' long | " | 1,220 | 300.00 | 1,520 |
| **10400.10 IDENTIFYING DEVICES** | | | | |
| Directory and bulletin boards | | | | |
| Open face boards | | | | |
| Chrome plated steel frame | S.F. | 24.35 | 17.70 | 42.05 |
| Aluminum framed | " | 21.40 | 17.70 | 39.10 |
| Bronze framed | " | 23.60 | 17.70 | 41.30 |
| Stainless steel framed | " | 45.70 | 17.70 | 63.40 |
| Tack board, aluminum framed | " | 13.70 | 17.70 | 31.40 |
| Visual aid board, aluminum framed | " | 13.70 | 17.70 | 31.40 |
| Glass encased boards, hinged and keyed | | | | |
| Aluminum framed | S.F. | 53.00 | 44.30 | 97.30 |
| Bronze framed | " | 57.50 | 44.30 | 101.80 |
| Stainless steel framed | " | 130.00 | 44.30 | 174.30 |
| Chrome plated steel framed | " | 130.00 | 44.30 | 174.30 |
| Metal plaque | | | | |
| Cast bronze | S.F. | 350.00 | 29.50 | 379.50 |
| Aluminum | " | 320.00 | 29.50 | 349.50 |
| Metal engraved plaque | | | | |
| Porcelain steel | S.F. | 440.00 | 29.50 | 469.50 |

# 10 SPECIALTIES

| SPECIALTIES | UNIT | MAT. | INST. | TOTAL |
|---|---|---|---|---|
| **10400.10 IDENTIFYING DEVICES** | | | | |
| Stainless steel | S.F. | 400.00 | 29.50 | 429.50 |
| Brass | " | 440.00 | 29.50 | 469.50 |
| Aluminum | " | 200.00 | 29.50 | 229.50 |
| Metal built-up plaque | | | | |
| Bronze | S.F. | 340.00 | 35.40 | 375.40 |
| Copper and bronze | " | 350.00 | 35.40 | 385.40 |
| Copper and aluminum | " | 380.00 | 35.40 | 415.40 |
| Metal nameplate plaques | | | | |
| Cast bronze | S.F. | 340.00 | 22.15 | 362.15 |
| Cast aluminum | " | 310.00 | 22.15 | 332.15 |
| Engraved, 1-1/2" x 6" | | | | |
| Bronze | EA. | 110.00 | 22.15 | 132.15 |
| Aluminum | " | 110.00 | 22.15 | 132.15 |
| Letters, on masonry or concrete, aluminum, satin finish | | | | |
| 1/2" thick | | | | |
| 2" high | EA. | 12.55 | 14.20 | 26.75 |
| 4" high | " | 14.00 | 17.70 | 31.70 |
| 6" high | " | 18.45 | 19.70 | 38.15 |
| 3/4" thick | | | | |
| 8" high | EA. | 24.35 | 22.15 | 46.50 |
| 10" high | " | 31.00 | 25.30 | 56.30 |
| 1" thick | | | | |
| 12" high | EA. | 40.60 | 29.50 | 70.10 |
| 14" high | " | 45.70 | 35.40 | 81.10 |
| 16" high | " | 63.50 | 44.30 | 107.80 |
| For polished aluminum add, 15% | | | | |
| For clear anodized aluminum add, 15% | | | | |
| For colored anodic aluminum add, 30% | | | | |
| For profiled and color enameled letters add, 50% | | | | |
| Cast bronze, satin finish letters | | | | |
| 3/8" thick | | | | |
| 2" high | EA. | 22.85 | 14.20 | 37.05 |
| 4" high | " | 24.80 | 17.70 | 42.50 |
| 1/2" thick, 6" high | " | 32.20 | 19.70 | 51.90 |
| 5/8" thick, 8" high | " | 41.60 | 22.15 | 63.75 |
| 1" thick | | | | |
| 10" high | EA. | 63.00 | 25.30 | 88.30 |
| 12" high | " | 80.50 | 29.50 | 110.00 |
| 14" high | " | 90.00 | 35.40 | 125.40 |
| 16" high | " | 110.00 | 44.30 | 154.30 |
| Interior door signs, adhesive, flexible | | | | |
| 2" x 8" | EA. | 15.50 | 6.95 | 22.45 |
| 4" x 4" | " | 15.50 | 6.95 | 22.45 |
| 6" x 7" | " | 20.20 | 6.95 | 27.15 |
| 6" x 9" | " | 26.60 | 6.95 | 33.55 |
| 10" x 9" | " | 35.40 | 6.95 | 42.35 |
| 10" x 12" | " | 45.70 | 6.95 | 52.65 |
| Hard plastic type, no frame | | | | |
| 3" x 8" | EA. | 32.40 | 6.95 | 39.35 |
| 4" x 4" | " | 31.00 | 6.95 | 37.95 |
| 4" x 12" | " | 38.40 | 6.95 | 45.35 |
| Hard plastic type, with frame | | | | |

| SPECIALTIES | UNIT | MAT. | INST. | TOTAL |
|---|---|---|---|---|
| **10400.10 IDENTIFYING DEVICES** | | | | |
| 3" x 8" | EA. | 100.00 | 6.95 | 106.95 |
| 4" x 4" | " | 84.00 | 6.95 | 90.95 |
| 4" x 12" | " | 140.00 | 6.95 | 146.95 |
| **10450.10 CONTROL** | | | | |
| Access control, 7' high, indoor or outdoor impenetrability | | | | |
| Remote or card control, type B | EA. | 3,510 | 480.00 | 3,990 |
| Free passage, type B | " | 3,240 | 480.00 | 3,720 |
| Remote or card control, type AA | " | 4,590 | 480.00 | 5,070 |
| Free passage, type AA | " | 4,190 | 480.00 | 4,670 |
| **10500.10 LOCKERS** | | | | |
| Locker bench, floor mounted, laminated maple | | | | |
| 4' | EA. | 110.00 | 29.50 | 139.50 |
| 6' | " | 140.00 | 29.50 | 169.50 |
| Wardrobe locker, single tier type | | | | |
| 12" x 15" x 72" | EA. | 140.00 | 35.40 | 175.40 |
| 18" x 15" x 72" | " | 180.00 | 37.30 | 217.30 |
| 12" x 18" x 72" | " | 160.00 | 39.40 | 199.40 |
| 18" x 18" x 72" | " | 190.00 | 41.70 | 231.70 |
| Double tier type | | | | |
| 12" x 15" x 36" | EA. | 100.00 | 17.70 | 117.70 |
| 18" x 15" x 36" | " | 100.00 | 17.70 | 117.70 |
| 12" x 18" x 36" | " | 92.50 | 17.70 | 110.20 |
| 18" x 18" x 36" | " | 110.00 | 17.70 | 127.70 |
| Two person unit | | | | |
| 18" x 15" x 72" | EA. | 280.00 | 59.00 | 339.00 |
| 18" x 18" x 72" | " | 160.00 | 71.00 | 231.00 |
| Duplex unit | | | | |
| 15" x 15" x 72" | EA. | 270.00 | 35.40 | 305.40 |
| 15" x 21" x 72" | " | 280.00 | 35.40 | 315.40 |
| Basket lockers, basket sets with baskets | | | | |
| 24 basket set | SET | 580.00 | 180.00 | 760.00 |
| 30 basket set | " | 690.00 | 220.00 | 910.00 |
| 36 basket set | " | 770.00 | 300.00 | 1,070 |
| 42 basket set | " | 870.00 | 350.00 | 1,220 |
| **10520.10 FIRE PROTECTION** | | | | |
| Portable fire extinguishers | | | | |
| Water pump tank type | | | | |
| 2.5 gal. | | | | |
| Red enameled galvanized | EA. | 95.50 | 18.50 | 114.00 |
| Red enameled copper | " | 110.00 | 18.50 | 128.50 |
| Polished copper | " | 160.00 | 18.50 | 178.50 |
| Carbon dioxide type, red enamel steel | | | | |
| Squeeze grip with hose and horn | | | | |
| 2.5 lb | EA. | 55.00 | 18.50 | 73.50 |
| 5 lb | " | 120.00 | 21.35 | 141.35 |
| 10 lb | " | 190.00 | 27.80 | 217.80 |
| 15 lb | " | 220.00 | 34.70 | 254.70 |

| SPECIALTIES | UNIT | MAT. | INST. | TOTAL |
|---|---|---|---|---|
| **10520.10 FIRE PROTECTION** | | | | |
| 20 lb | EA. | 260.00 | 34.70 | 294.70 |
| Wheeled type | | | | |
| 125 lb | EA. | 1,390 | 55.50 | 1,446 |
| 250 lb | " | 1,430 | 55.50 | 1,486 |
| 500 lb | " | 2,680 | 55.50 | 2,736 |
| Dry chemical, pressurized type | | | | |
| Red enameled steel | | | | |
| 2.5 lb | EA. | 33.00 | 18.50 | 51.50 |
| 5 lb | " | 55.00 | 21.35 | 76.35 |
| 10 lb | " | 69.00 | 27.80 | 96.80 |
| 20 lb | " | 110.00 | 34.70 | 144.70 |
| 30 lb | " | 290.00 | 34.70 | 324.70 |
| Chrome plated steel, 2.5 lb | " | 69.00 | 18.50 | 87.50 |
| Other type extinguishers | | | | |
| 2.5 gal, stainless steel, pressurized water tanks | EA. | 69.00 | 18.50 | 87.50 |
| Soda and acid type | " | 120.00 | 18.50 | 138.50 |
| Cartridge operated, water type | " | 210.00 | 18.50 | 228.50 |
| Loaded stream, water type | " | 96.50 | 18.50 | 115.00 |
| Foam type | " | 80.00 | 18.50 | 98.50 |
| 40 gal, wheeled foam type | " | 4,200 | 55.50 | 4,256 |
| Fire extinguisher cabinets | | | | |
| Enameled steel | | | | |
| 8" x 12" x 27" | EA. | 150.00 | 55.50 | 205.50 |
| 8" x 16" x 38" | " | 200.00 | 55.50 | 255.50 |
| Aluminum | | | | |
| 8" x 12" x 27" | EA. | 200.00 | 55.50 | 255.50 |
| 8" x 16" x 38" | " | 240.00 | 55.50 | 295.50 |
| 8" x 12" x 27" | " | 340.00 | 55.50 | 395.50 |
| Stainless steel | | | | |
| 8" x 16" x 38" | EA. | 430.00 | 55.50 | 485.50 |
| **10550.10 POSTAL SPECIALTIES** | | | | |
| Mail chutes | | | | |
| Single mail chute | | | | |
| Finished aluminum | L.F. | 540.00 | 88.50 | 628.50 |
| Bronze | " | 800.00 | 88.50 | 888.50 |
| Single mail chute receiving box | | | | |
| Finished aluminum | EA. | 850.00 | 180.00 | 1,030 |
| Bronze | " | 1,590 | 180.00 | 1,770 |
| Twin mail chute, double parallel | | | | |
| Finished aluminum | FLR | 1,020 | 180.00 | 1,200 |
| Bronze | " | 1,590 | 180.00 | 1,770 |
| Receiving box, 36" x 20" x 12" | | | | |
| Finished aluminum | EA. | 1,770 | 300.00 | 2,070 |
| Bronze | " | 2,530 | 300.00 | 2,830 |
| Locked receiving mail box | | | | |
| Finished aluminum | EA. | 850.00 | 180.00 | 1,030 |
| Bronze | " | 1,450 | 180.00 | 1,630 |
| Commercial postal accessories for mail chutes | | | | |
| Letter slot, brass | EA. | 280.00 | 59.00 | 339.00 |
| Bulk mail slot, brass | " | 810.00 | 59.00 | 869.00 |
| Mail boxes | | | | |

| SPECIALTIES | UNIT | MAT. | INST. | TOTAL |
|---|---|---|---|---|
| **10550.10 POSTAL SPECIALTIES** | | | | |
| Residential postal accessories | | | | |
| Letter slot | EA. | 84.00 | 17.70 | 101.70 |
| Rural letter box | " | 59.50 | 44.30 | 103.80 |
| Apartment house, keyed, 3.5" x 4.5" x 16" | " | 87.00 | 11.80 | 98.80 |
| Ranch style | " | 110.00 | 17.70 | 127.70 |
| Commercial postal accessories | | | | |
| Letter box, with combination lock | EA. | 71.00 | 12.65 | 83.65 |
| Key lock | " | 64.50 | 12.65 | 77.15 |
| Mail box, aluminum w/glass front, 4x5 | | | | |
| Horizontal rear load | EA. | 88.00 | 10.15 | 98.15 |
| Vertical front load | " | 68.00 | 10.15 | 78.15 |
| **10600.10 MOVABLE PARTITIONS** | | | | |
| Partition, movable office type, 2-1/2" thick, vinyl-gypsum | S.F. | 11.35 | 1.75 | 13.10 |
| Enameled steel frame, with 1/4" thick clear glass | " | 15.60 | 1.75 | 17.35 |
| Door frame and hardware for movable partitions | EA. | 450.00 | 120.00 | 570.00 |
| Add for acoustic movable partition | S.F. | | | 0.97 |
| Accordion partition, 12' high | | | | |
| Vinyl | S.F. | 6.95 | 5.90 | 12.85 |
| Acoustical | " | 8.25 | 5.90 | 14.15 |
| Standard office cubicles, 8' high, steel framed | | | | |
| Baked enamel finish | | | | |
| 100% flush | L.F. | 110.00 | 8.85 | 118.85 |
| 75% flush and 25% glass | " | 130.00 | 9.85 | 139.85 |
| 50% flush and 50% glass | " | 190.00 | 9.85 | 199.85 |
| 100% glass | " | 230.00 | 11.80 | 241.80 |
| Natural hardwood panels | | | | |
| 100% flush | L.F. | 110.00 | 11.80 | 121.80 |
| 50% flush and 50% glass | " | 160.00 | 12.65 | 172.65 |
| Plastic laminated panels | | | | |
| 100% flush | L.F. | 140.00 | 11.80 | 151.80 |
| 75% flush and 25% glass | " | 170.00 | 12.65 | 182.65 |
| 50% flush and 50% glass | " | 200.00 | 12.65 | 212.65 |
| Vinyl covered panels | | | | |
| 100% flush | L.F. | 120.00 | 12.20 | 132.20 |
| 75% flush and 25% glass | " | 160.00 | 13.15 | 173.15 |
| 50% and 50% glass | " | 180.00 | 13.15 | 193.15 |
| Aluminum framed | | | | |
| Enameled or anodized aluminum panels | | | | |
| 100% flush | L.F. | 93.50 | 8.85 | 102.35 |
| 75% flush and 25% glass | " | 89.50 | 9.85 | 99.35 |
| 50% flush and 50% glass | " | 83.00 | 9.85 | 92.85 |
| Vinyl covered panels | | | | |
| 100% flush | L.F. | 100.00 | 12.20 | 112.20 |
| 75% flush and 25% glass | " | 120.00 | 13.15 | 133.15 |
| 50% flush and 50% glass | " | 150.00 | 13.15 | 163.15 |
| 60" high partitions, steel framed | | | | |
| Enameled panels | L.F. | 49.60 | 7.90 | 57.50 |
| Natural hardwood panels, two sides | " | 210.00 | 8.25 | 218.25 |
| Plastic laminated panels | " | 110.00 | 8.25 | 118.25 |
| Vinyl covered panels | " | 150.00 | 7.90 | 157.90 |

| SPECIALTIES | UNIT | MAT. | INST. | TOTAL |
|---|---|---|---|---|
| **10600.10 MOVABLE PARTITIONS** | | | | |
| Aluminum framed | | | | |
| Anodized or baked enamel panels | L.F. | 71.50 | 7.90 | 79.40 |
| Natural hardwood panels | " | 220.00 | 8.25 | 228.25 |
| Plastic laminated panels | " | 110.00 | 8.25 | 118.25 |
| Vinyl covered panels | " | 140.00 | 7.90 | 147.90 |
| Wire mesh partitions | | | | |
| Wall panels | | | | |
| 4' x 7' | EA. | 110.00 | 22.15 | 132.15 |
| 4' x 8' | " | 120.00 | 23.65 | 143.65 |
| 4' x 10' | " | 150.00 | 27.30 | 177.30 |
| Wall filler panels | | | | |
| 1' x 7' | EA. | 62.50 | 22.15 | 84.65 |
| 1' x 8' | " | 64.00 | 23.65 | 87.65 |
| 1' x 10' | " | 85.00 | 25.30 | 110.30 |
| 2' x 7' | " | 61.00 | 22.15 | 83.15 |
| 2' x 8' | " | 65.00 | 23.65 | 88.65 |
| 2' x 10' | " | 76.50 | 25.30 | 101.80 |
| 3' x 7' | " | 81.00 | 22.15 | 103.15 |
| 3' x 8' | " | 88.00 | 23.65 | 111.65 |
| 3' x 10' | " | 110.00 | 25.30 | 135.30 |
| Ceiling panels | | | | |
| 10' x 2' | EA. | 74.00 | 50.50 | 124.50 |
| 10' x 4' | " | 140.00 | 71.00 | 211.00 |
| Wall panel with service window | | | | |
| 5' wide | | | | |
| 7' high | EA. | 350.00 | 22.15 | 372.15 |
| 8' high | " | 360.00 | 23.65 | 383.65 |
| 10' high | " | 380.00 | 25.30 | 405.30 |
| Doors | | | | |
| Sliding | | | | |
| 3' x 7' | EA. | 430.00 | 88.50 | 518.50 |
| 3' x 8' | " | 390.00 | 100.00 | 490.00 |
| 3' x 10' | " | 400.00 | 140.00 | 540.00 |
| 4' x 7' | " | 400.00 | 100.00 | 500.00 |
| 4' x 8' | " | 410.00 | 140.00 | 550.00 |
| 4' x 10' | " | 450.00 | 180.00 | 630.00 |
| 5' x 7' | " | 440.00 | 140.00 | 580.00 |
| 5' x 8' | " | 520.00 | 180.00 | 700.00 |
| 5' x 10' | " | 610.00 | 180.00 | 790.00 |
| Swing door | | | | |
| 3' x 7' | EA. | 310.00 | 88.50 | 398.50 |
| 4' x 7' | " | 350.00 | 100.00 | 450.00 |
| Swing door, with 1' transom | | | | |
| 3' x 7' | EA. | 340.00 | 100.00 | 440.00 |
| 4' x 7' | " | 380.00 | 140.00 | 520.00 |
| Swing door, with 3' transom | | | | |
| 3' x 7' | EA. | 300.00 | 140.00 | 440.00 |
| 4' x 7' | " | 320.00 | 180.00 | 500.00 |
| **10670.10 SHELVING** | | | | |
| Shelving, enamel, closed side and back, 12" x 36" | | | | |
| 5 shelves | EA. | 140.00 | 59.00 | 199.00 |

# 10 SPECIALTIES

| SPECIALTIES | UNIT | MAT. | INST. | TOTAL |
|---|---|---|---|---|
| **10670.10 SHELVING** | | | | |
| 8 shelves | EA. | 180.00 | 79.00 | 259.00 |
| Open | | | | |
| 5 shelves | EA. | 95.50 | 59.00 | 154.50 |
| 8 shelves | " | 95.50 | 79.00 | 174.50 |
| Metal storage shelving, baked enamel | | | | |
| 7 shelf unit, 72" or 84" high | | | | |
| 10" shelf | L.F. | 39.10 | 35.40 | 74.50 |
| 12" shelf | " | 40.50 | 37.30 | 77.80 |
| 15" shelf | " | 46.30 | 39.40 | 85.70 |
| 18" shelf | " | 46.30 | 41.70 | 88.00 |
| 24" shelf | " | 53.50 | 44.30 | 97.80 |
| 30" shelf | " | 58.00 | 47.30 | 105.30 |
| 36" shelf | " | 69.50 | 50.50 | 120.00 |
| 4 shelf unit, 40" high | | | | |
| 10" shelf | L.F. | 33.30 | 29.50 | 62.80 |
| 12" shelf | " | 34.70 | 32.20 | 66.90 |
| 15" shelf | " | 43.40 | 35.40 | 78.80 |
| 18" shelf | " | 43.40 | 37.30 | 80.70 |
| 24" shelf | " | 49.20 | 39.40 | 88.60 |
| 3 shelf unit, 32" high | | | | |
| 10" shelf | L.F. | 27.50 | 17.70 | 45.20 |
| 12" shelf | " | 27.50 | 18.65 | 46.15 |
| 15" shelf | " | 30.40 | 19.70 | 50.10 |
| 18" shelf | " | 31.80 | 20.85 | 52.65 |
| 24" shelf | " | 36.20 | 22.15 | 58.35 |
| Single shelf unit, attached to masonry | | | | |
| 10" shelf | L.F. | 9.85 | 5.90 | 15.75 |
| 12" shelf | " | 10.85 | 6.45 | 17.30 |
| 15" shelf | " | 11.15 | 6.80 | 17.95 |
| 18" shelf | " | 12.60 | 7.25 | 19.85 |
| 24" shelf | " | 15.20 | 7.70 | 22.90 |
| For stainless steel, add to material, 120% | | | | |
| For attachment to gypsum board, add to labor, 50% | | | | |
| Built-in wood shelves | | | | |
| Posts and trimmed plywood | L.F. | 2.25 | 5.05 | 7.30 |
| Solid clear pine | " | 3.75 | 5.45 | 9.20 |
| Closet shelf, pine with rod | " | 2.55 | 5.45 | 8.00 |
| For lumber edge band, add to material | " | | | 1.31 |
| For prefinished shelves, add to material, 225% | | | | |
| **10750.10 TELEPHONE ENCLOSURES** | | | | |
| Telephone enclosure, wall mounted, shelf, 28" x 30" x 15" | EA. | 950.00 | 88.50 | 1,039 |
| Directory shelf, stainless steel, 3 binders | " | 740.00 | 59.00 | 799.00 |
| **10800.10 BATH ACCESSORIES** | | | | |
| Ash receiver, wall mounted, aluminum | EA. | 150.00 | 17.70 | 167.70 |
| Grab bar, 1-1/2" dia., stainless steel, wall mounted | | | | |
| 24" long | EA. | 42.00 | 17.70 | 59.70 |
| 36" long | " | 61.00 | 18.65 | 79.65 |
| 42" long | " | 69.50 | 19.70 | 89.20 |
| 48" long | " | 84.00 | 20.85 | 104.85 |

# 10 SPECIALTIES

| SPECIALTIES | UNIT | MAT. | INST. | TOTAL |
|---|---|---|---|---|
| **10800.10 BATH ACCESSORIES** | | | | |
| 52" long | EA. | 91.00 | 22.15 | 113.15 |
| 1" dia., stainless steel | | | | |
| 12" long | EA. | 33.30 | 15.40 | 48.70 |
| 18" long | " | 36.20 | 16.10 | 52.30 |
| 24" long | " | 39.10 | 17.70 | 56.80 |
| 30" long | " | 43.40 | 18.65 | 62.05 |
| 36" long | " | 49.20 | 19.70 | 68.90 |
| 48" long | " | 50.50 | 20.85 | 71.35 |
| Hand dryer, surface mounted, 110 volt | " | 470.00 | 44.30 | 514.30 |
| Medicine cabinet, 16 x 22, baked enamel, steel, lighted | " | 75.00 | 14.20 | 89.20 |
| With mirror, lighted | " | 130.00 | 23.65 | 153.65 |
| Mirror, 1/4" plate glass, up to 10 sf | S.F. | 7.50 | 3.55 | 11.05 |
| Mirror, stainless steel frame | | | | |
| 18"x24" | EA. | 120.00 | 11.80 | 131.80 |
| 18"x32" | " | 120.00 | 14.20 | 134.20 |
| 18"x36" | " | 130.00 | 17.70 | 147.70 |
| 24"x30" | " | 150.00 | 17.70 | 167.70 |
| 24"x36" | " | 170.00 | 19.70 | 189.70 |
| 24"x48" | " | 300.00 | 29.50 | 329.50 |
| 24"x60" | " | 310.00 | 35.40 | 345.40 |
| 30"x30" | " | 170.00 | 35.40 | 205.40 |
| 30"x72" | " | 470.00 | 44.30 | 514.30 |
| 48"x72" | " | 500.00 | 59.00 | 559.00 |
| With shelf, 18"x24" | " | 170.00 | 14.20 | 184.20 |
| Sanitary napkin dispenser, stainless steel, wall mounted | " | 410.00 | 23.65 | 433.65 |
| Shower rod, 1" diameter | | | | |
| Chrome finish over brass | EA. | 68.00 | 17.70 | 85.70 |
| Stainless steel | " | 65.00 | 17.70 | 82.70 |
| Soap dish, stainless steel, wall mounted | " | 82.50 | 23.65 | 106.15 |
| Toilet tissue dispenser, stainless, wall mounted | | | | |
| Single roll | EA. | 40.50 | 8.85 | 49.35 |
| Double roll | " | 63.50 | 10.15 | 73.65 |
| Towel dispenser, stainless steel | | | | |
| Flush mounted | EA. | 120.00 | 19.70 | 139.70 |
| Surface mounted | " | 110.00 | 17.70 | 127.70 |
| Combination towel dispenser and waste receptacle | " | 360.00 | 23.65 | 383.65 |
| Towel bar, stainless steel | | | | |
| 18" long | EA. | 34.70 | 14.20 | 48.90 |
| 24" long | " | 40.50 | 16.10 | 56.60 |
| 30" long | " | 43.40 | 17.70 | 61.10 |
| 36" long | " | 47.70 | 19.70 | 67.40 |
| Toothbrush and tumbler holder | " | 30.40 | 11.80 | 42.20 |
| Waste receptacle, stainless steel, wall mounted | " | 250.00 | 29.50 | 279.50 |
| **10900.10 WARDROBE SPECIALTIES** | | | | |
| Hospital wardrobe units, 24" x 24" x 76", with door | | | | |
| Baked enameled steel | EA. | 1,810 | 200.00 | 2,010 |
| Hardwood | " | 970.00 | 200.00 | 1,170 |
| Stainless steel | " | 3,040 | 200.00 | 3,240 |
| Plastic laminated | " | 1,090 | 200.00 | 1,290 |
| Dormitory wardrobe units, 24" x 76", with door | | | | |

| SPECIALTIES | UNIT | MAT. | INST. | TOTAL |
|---|---|---|---|---|
| **10900.10 WARDROBE SPECIALTIES** | | | | |
| Hardwood | EA. | 970.00 | 200.00 | 1,170 |
| Plastic laminated | " | 850.00 | 200.00 | 1,050 |
| Hat and coat rack | | | | |
| Single tier | | | | |
| Baked enameled steel | L.F. | 120.00 | 8.85 | 128.85 |
| Stainless steel | " | 130.00 | 8.85 | 138.85 |
| Aluminum | " | 130.00 | 8.85 | 138.85 |
| Double tier | | | | |
| Baked enameled steel | L.F. | 190.00 | 10.15 | 200.15 |
| Stainless steel | " | 270.00 | 10.15 | 280.15 |
| Aluminum | " | 240.00 | 10.15 | 250.15 |

# 11 EQUIPMENT

## ARCHITECTURAL EQUIPMENT

| ARCHITECTURAL EQUIPMENT | UNIT | MAT. | INST. | TOTAL |
|---|---|---|---|---|
| **11010.10 MAINTENANCE EQUIPMENT** | | | | |
| Vacuum cleaning system | | | | |
| 3 valves | | | | |
| 1.5 hp | EA. | 1,350 | 390.00 | 1,740 |
| 2.5 hp | " | 1,620 | 510.00 | 2,130 |
| 5 valves | " | 2,160 | 710.00 | 2,870 |
| 7 valves | " | 2,430 | 890.00 | 3,320 |
| **11020.10 SECURITY EQUIPMENT** | | | | |
| Bulletproof teller window | | | | |
| 4' x 4' | EA. | 9,450 | 590.00 | 10,040 |
| 5' x 4' | " | 10,800 | 710.00 | 11,510 |
| Bulletproof partitions | | | | |
| Up to 12' high, 2.5" thick | S.F. | 51.00 | 2.35 | 53.35 |
| Counter for banks | | | | |
| Minimum | L.F. | 580.00 | 71.00 | 651.00 |
| Maximum | " | 2,620 | 120.00 | 2,740 |
| Drive-up window | | | | |
| Minimum | EA. | 6,950 | 510.00 | 7,460 |
| Maximum | " | 33,750 | 1,180 | 34,930 |
| Night depository | | | | |
| Minimum | EA. | 9,450 | 510.00 | 9,960 |
| Maximum | " | 18,900 | 1,180 | 20,080 |
| Office safes, 30" x 20" x 20", 1 hr rating | " | 2,160 | 88.50 | 2,249 |
| 30" x 16" x 15", 2 hr rating | " | 2,030 | 71.00 | 2,101 |
| 30" x 28" x 20", H&G rating | " | 9,450 | 44.30 | 9,494 |
| Service windows, pass through painted steel | | | | |
| 24" x 36" | EA. | 4,050 | 350.00 | 4,400 |
| 48" x 40" | " | 6,080 | 440.00 | 6,520 |
| 72" x 40" | " | 6,750 | 710.00 | 7,460 |
| Special doors and windows | | | | |
| 3' x 7' bulletproof door with frame | EA. | 4,730 | 510.00 | 5,240 |
| 12" x 12" vision panel | " | 2,700 | 250.00 | 2,950 |
| Surveillance system | | | | |
| Minimum | EA. | 5,040 | 710.00 | 5,750 |
| Maximum | " | 39,000 | 3,540 | 42,540 |
| Vault door, 3' wide, 6'6" high | | | | |
| 3-1/2" thick | EA. | 18,440 | 4,430 | 22,870 |
| 7" thick | " | 21,980 | 5,910 | 27,890 |
| 10" thick | " | 26,240 | 7,090 | 33,330 |
| Insulated vault door | | | | |
| 2 hr rating | | | | |
| 32" wide | EA. | 2,980 | 350.00 | 3,330 |
| 40" wide | " | 3,260 | 370.00 | 3,630 |
| 4 hr rating | | | | |
| 32" wide | EA. | 3,190 | 390.00 | 3,580 |
| 40" wide | " | 3,330 | 440.00 | 3,770 |
| 6 hr rating | | | | |
| 32" wide | EA. | 3,400 | 390.00 | 3,790 |
| 40" wide | " | 3,760 | 440.00 | 4,200 |
| Insulated file room door | | | | |
| 1 hr rating | | | | |
| 32" wide | EA. | 3,400 | 350.00 | 3,750 |

# 11 EQUIPMENT

| ARCHITECTURAL EQUIPMENT | UNIT | MAT. | INST. | TOTAL |
|---|---|---|---|---|
| **11020.10 SECURITY EQUIPMENT** | | | | |
| 40" wide | EA. | 3,620 | 390.00 | 4,010 |
| **11060.10 THEATER EQUIPMENT** | | | | |
| Roll out stage, steel frame, wood floor | | | | |
| Manual | S.F. | 29.80 | 2.20 | 32.00 |
| Electric | " | 32.60 | 3.55 | 36.15 |
| Portable stages | | | | |
| 8" high | S.F. | 9.50 | 1.75 | 11.25 |
| 18" high | " | 10.90 | 1.95 | 12.85 |
| 36" high | " | 12.05 | 2.10 | 14.15 |
| 48" high | " | 12.75 | 2.20 | 14.95 |
| Band risers | | | | |
| Minimum | S.F. | 26.20 | 1.75 | 27.95 |
| Maximum | " | 42.60 | 1.75 | 44.35 |
| Chairs for risers | | | | |
| Minimum | EA. | 72.50 | 1.25 | 73.75 |
| Maximum | " | 130.00 | 1.25 | 131.25 |
| **11080.10 POLICE EQUIPMENT** | | | | |
| Firing range equipment, rifle | | | | |
| 3 position | EA. | 11,350 | 1,180 | 12,530 |
| 4 position | " | 14,890 | 1,770 | 16,660 |
| 5 position | " | 18,440 | 1,970 | 20,410 |
| 6 position | " | 22,690 | 2,080 | 24,770 |
| **11090.10 CHECKROOM EQUIPMENT** | | | | |
| Motorized checkroom equipment | | | | |
| No shelf system, 6'4" height | | | | |
| 7'6" length | EA. | 2,200 | 350.00 | 2,550 |
| 14'6" length | " | 2,480 | 350.00 | 2,830 |
| 28' length | " | 3,400 | 350.00 | 3,750 |
| One shelf, 6'8" height | | | | |
| 7'6" length | EA. | 2,690 | 350.00 | 3,040 |
| 14'6" length | " | 3,260 | 350.00 | 3,610 |
| 28' length | " | 4,820 | 350.00 | 5,170 |
| Two shelves, 7'5" height | | | | |
| 7'6" length | EA. | 3,260 | 350.00 | 3,610 |
| 14'6" length | " | 3,970 | 350.00 | 4,320 |
| 28' length | " | 5,670 | 350.00 | 6,020 |
| Three shelves, 8' height | | | | |
| 7'6" length | EA. | 3,400 | 710.00 | 4,110 |
| 14'6" length | " | 4,110 | 710.00 | 4,820 |
| 28' length | " | 5,810 | 710.00 | 6,520 |
| Four shelves, 8'7" height | | | | |
| 7'6" length | EA. | 3,400 | 710.00 | 4,110 |
| 14'6" length | " | 4,260 | 710.00 | 4,970 |
| 28' length | " | 5,810 | 710.00 | 6,520 |
| 200 lb | | | | |

# 11 EQUIPMENT

## ARCHITECTURAL EQUIPMENT

### 11110.10 LAUNDRY EQUIPMENT

| ARCHITECTURAL EQUIPMENT | UNIT | MAT. | INST. | TOTAL |
|---|---|---|---|---|
| High capacity, heavy duty | | | | |
| Washer extractors | | | | |
| 135 lb | | | | |
| Standard | EA. | 50,350 | 300.00 | 50,650 |
| Pass through | " | 58,150 | 300.00 | 58,450 |
| 200 lb | | | | |
| Standard | EA. | 56,020 | 300.00 | 56,320 |
| Pass through | " | 63,120 | 300.00 | 63,420 |
| 110 lb dryer | " | 9,930 | 300.00 | 10,230 |
| Hand operated presser | " | 6,810 | 390.00 | 7,200 |
| Mushroom press | " | 4,400 | 390.00 | 4,790 |
| Spreader feeders | | | | |
| 2 station | EA. | 65,240 | 390.00 | 65,630 |
| 4 station | " | 71,620 | 710.00 | 72,330 |
| Delivery carts | | | | |
| 12 bushel | EA. | 430.00 | 4.45 | 434.45 |
| 16 bushel | " | 510.00 | 4.75 | 514.75 |
| 18 bushel | " | 520.00 | 5.05 | 525.05 |
| 30 bushel | " | 750.00 | 5.90 | 755.90 |
| 40 bushel | " | 920.00 | 7.10 | 927.10 |
| Low capacity | | | | |
| Pressers | | | | |
| Air operated | EA. | 5,390 | 140.00 | 5,530 |
| Hand operated | " | 4,400 | 140.00 | 4,540 |
| Extractor, low capacity | " | 6,310 | 140.00 | 6,450 |
| Ironer, 48" | " | 2,980 | 71.00 | 3,051 |
| Coin washers | | | | |
| 10 lb capacity | EA. | 1,330 | 71.00 | 1,401 |
| 20 lb capacity | " | 3,400 | 71.00 | 3,471 |
| Coin dryer | " | 550.00 | 44.30 | 594.30 |
| Coin dry cleaner, 20 lb | " | 26,950 | 140.00 | 27,090 |

### 11161.10 LOADING DOCK EQUIPMENT

| ARCHITECTURAL EQUIPMENT | UNIT | MAT. | INST. | TOTAL |
|---|---|---|---|---|
| Dock leveler, 10 ton capacity | | | | |
| 6' x 8' | EA. | 4,050 | 350.00 | 4,400 |
| 7' x 8' | " | 4,190 | 350.00 | 4,540 |
| Bumpers, laminated rubber | | | | |
| 4-1/2" thick | | | | |
| 6" x 14" | EA. | 69.00 | 7.10 | 76.10 |
| 6" x 36" | " | 70.00 | 7.90 | 77.90 |
| 10" x 14" | " | 74.00 | 8.85 | 82.85 |
| 10" x 24" | " | 83.50 | 10.15 | 93.65 |
| 10" x 36" | " | 95.00 | 11.80 | 106.80 |
| 12" x 14" | " | 81.00 | 9.35 | 90.35 |
| 12" x 24" | " | 86.50 | 11.05 | 97.55 |
| 12" x 36" | " | 110.00 | 13.15 | 123.15 |
| 6" thick | | | | |
| 10" x 14" | EA. | 86.50 | 10.15 | 96.65 |
| 10" x 24" | " | 110.00 | 12.20 | 122.20 |
| 10" x 36" | " | 120.00 | 17.70 | 137.70 |
| Extruded rubber bumpers | | | | |

# 11 EQUIPMENT

| ARCHITECTURAL EQUIPMENT | UNIT | MAT. | INST. | TOTAL |
|---|---|---|---|---|
| **11161.10 LOADING DOCK EQUIPMENT** | | | | |
| T-section, 22" x 22" x 3" | EA. | 110.00 | 7.10 | 117.10 |
| Molded rubber bumpers | | | | |
| 24" x 12" x 3" thick | EA. | 66.50 | 17.70 | 84.20 |
| Door seal, 12" x 12", vinyl covered | L.F. | 38.30 | 8.85 | 47.15 |
| Dock boards, heavy duty, 5' x 5' | | | | |
| 5000 lb | | | | |
| Minimum | EA. | 1,080 | 300.00 | 1,380 |
| Maximum | " | 1,840 | 300.00 | 2,140 |
| 9000 lb | | | | |
| Minimum | EA. | 1,230 | 300.00 | 1,530 |
| Maximum | " | 2,060 | 320.00 | 2,380 |
| 15,000 lb | " | 1,620 | 320.00 | 1,940 |
| Truck shelters | | | | |
| Minimum | EA. | 1,290 | 270.00 | 1,560 |
| Maximum | " | 2,130 | 510.00 | 2,640 |
| **11170.10 WASTE HANDLING** | | | | |
| Incinerator, electric | | | | |
| 100 lb/hr | | | | |
| Minimum | EA. | 19,570 | 360.00 | 19,930 |
| Maximum | " | 33,750 | 360.00 | 34,110 |
| 400 lb/hr | | | | |
| Minimum | EA. | 34,750 | 730.00 | 35,480 |
| Maximum | " | 67,500 | 730.00 | 68,230 |
| 1000 lb/hr | | | | |
| Minimum | EA. | 67,500 | 1,100 | 68,600 |
| Maximum | " | 108,000 | 1,100 | |
| Incinerator, medical-waste | | | | |
| 25 lb/hr, 2-7 x 4-0 | EA. | 20,250 | 730.00 | 20,980 |
| 50 lb/hr, 2-11 x 4-11 | " | 21,600 | 730.00 | 22,330 |
| 75 lb/hr, 3-8 x 5-0 | " | 29,700 | 1,450 | 31,150 |
| 100 lb/hr, 3-8 x 6-0 | " | 33,750 | 1,450 | 35,200 |
| Industrial compactor | | | | |
| 1 cy | EA. | 9,860 | 400.00 | 10,260 |
| 3 cy | " | 19,860 | 520.00 | 20,380 |
| 5 cy | " | 25,650 | 730.00 | 26,380 |
| Trash chutes steel, including sprinklers | | | | |
| 18" dia. | L.F. | 66.50 | 180.00 | 246.50 |
| 24" dia. | " | 86.50 | 190.00 | 276.50 |
| 30" dia. | " | 110.00 | 200.00 | 310.00 |
| 36" dia. | " | 150.00 | 210.00 | 360.00 |
| Refuse bottom hopper | EA. | 2,130 | 200.00 | 2,330 |
| **11400.10 FOOD SERVICE EQUIPMENT** | | | | |
| Unit kitchens | | | | |
| 30" compact kitchen | | | | |
| Refrigerator, with range, sink | EA. | 3,260 | 180.00 | 3,440 |
| Sink only | " | 2,980 | 120.00 | 3,100 |
| Range only | " | 2,770 | 91.00 | 2,861 |
| Cabinet for upper wall section | " | 380.00 | 52.00 | 432.00 |

## ARCHITECTURAL EQUIPMENT

### 11400.10 FOOD SERVICE EQUIPMENT

| ARCHITECTURAL EQUIPMENT | UNIT | MAT. | INST. | TOTAL |
|---|---|---|---|---|
| Stainless shield, for rear wall | EA. | 120.00 | 14.55 | 134.55 |
| Side wall | " | 65.00 | 14.55 | 79.55 |
| 42" compact kitchen | | | | |
| Refrigerator with range, sink | EA. | 4,040 | 200.00 | 4,240 |
| Sink only | " | 3,470 | 180.00 | 3,650 |
| Cabinet for upper wall section | " | 530.00 | 60.50 | 590.50 |
| Stainless shield, for rear wall | " | 170.00 | 15.15 | 185.15 |
| Side wall | " | 65.00 | 15.15 | 80.15 |
| 54" compact kitchen | | | | |
| Refrigerator, oven, range, sink | EA. | 4,820 | 260.00 | 5,080 |
| Cabinet for upper wall section | " | 690.00 | 72.50 | 762.50 |
| Stainless shield, for | | | | |
| Rear wall | EA. | 240.00 | 16.50 | 256.50 |
| Side wall | " | 65.00 | 16.50 | 81.50 |
| 60" compact kitchen | | | | |
| Refrigerator, oven, range, sink | EA. | 5,600 | 260.00 | 5,860 |
| Cabinet for upper wall section | " | 800.00 | 72.50 | 872.50 |
| Stainless shield, for | | | | |
| Rear wall | EA. | 240.00 | 16.50 | 256.50 |
| Side wall | " | 65.00 | 16.50 | 81.50 |
| 72" compact kitchen | | | | |
| Refrigerator, oven, range, sink | EA. | 5,460 | 300.00 | 5,760 |
| Cabinet for upper wall section | " | 1,560 | 72.50 | 1,633 |
| Stainless shield for | | | | |
| Rear wall | EA. | 280.00 | 18.15 | 298.15 |
| Side wall | " | 65.00 | 18.15 | 83.15 |
| Bake oven | | | | |
| Single deck | | | | |
| Minimum | EA. | 1,840 | 45.40 | 1,885 |
| Maximum | " | 4,260 | 91.00 | 4,351 |
| Double deck | | | | |
| Minimum | EA. | 2,130 | 60.50 | 2,191 |
| Maximum | " | 8,100 | 91.00 | 8,191 |
| Triple deck | | | | |
| Minimum | EA. | 2,980 | 60.50 | 3,041 |
| Maximum | " | 10,130 | 120.00 | 10,250 |
| Convection type oven, electric, 40" x 45" x 57" | | | | |
| Minimum | EA. | 2,980 | 45.40 | 3,025 |
| Maximum | " | 5,400 | 91.00 | 5,491 |
| Broiler, without oven, 69" x 26" x 39" | | | | |
| Minimum | EA. | 2,970 | 45.40 | 3,015 |
| Maximum | " | 5,110 | 60.50 | 5,171 |
| Coffee urns, 10 gallons | | | | |
| Minimum | EA. | 5,810 | 120.00 | 5,930 |
| Maximum | " | 9,360 | 180.00 | 9,540 |
| Fryer, with submerger | | | | |
| Single | | | | |
| Minimum | EA. | 1,350 | 72.50 | 1,423 |
| Maximum | " | 33,750 | 120.00 | 33,870 |
| Double | | | | |
| Minimum | EA. | 2,570 | 91.00 | 2,661 |
| Maximum | " | 5,400 | 120.00 | 5,520 |

# 11 EQUIPMENT

## ARCHITECTURAL EQUIPMENT

### 11400.10 FOOD SERVICE EQUIPMENT

| ARCHITECTURAL EQUIPMENT | UNIT | MAT. | INST. | TOTAL |
|---|---|---|---|---|
| Griddle, counter | | | | |
| 3' long | | | | |
| Minimum | EA. | 1,760 | 60.50 | 1,821 |
| Maximum | " | 4,400 | 72.50 | 4,473 |
| 5' long | | | | |
| Minimum | EA. | 2,570 | 91.00 | 2,661 |
| Maximum | " | 5,670 | 120.00 | 5,790 |
| Kettles, steam, jacketed | | | | |
| 20 gallons | | | | |
| Minimum | EA. | 2,700 | 91.00 | 2,791 |
| Maximum | " | 5,530 | 180.00 | 5,710 |
| 40 gallons | | | | |
| Minimum | EA. | 3,240 | 91.00 | 3,331 |
| Maximum | " | 6,100 | 180.00 | 6,280 |
| 60 gallons | | | | |
| Minimum | EA. | 3,690 | 91.00 | 3,781 |
| Maximum | " | 8,160 | 180.00 | 8,340 |
| Range | | | | |
| Heavy duty, single oven, open top | | | | |
| Minimum | EA. | 2,060 | 45.40 | 2,105 |
| Maximum | " | 5,670 | 120.00 | 5,790 |
| Fry top | | | | |
| Minimum | EA. | 2,970 | 45.40 | 3,015 |
| Maximum | " | 6,210 | 120.00 | 6,330 |
| Steamers, electric | | | | |
| 27 kw | | | | |
| Minimum | EA. | 7,800 | 91.00 | 7,891 |
| Maximum | " | 15,600 | 120.00 | 15,720 |
| 18 kw | | | | |
| Minimum | EA. | 4,050 | 91.00 | 4,141 |
| Maximum | " | 9,860 | 120.00 | 9,980 |
| Dishwasher, rack type | | | | |
| Single tank, 190 racks/hr | EA. | 12,200 | 180.00 | 12,380 |
| Double tank | | | | |
| 234 racks/hr | EA. | 17,730 | 200.00 | 17,930 |
| 265 racks/hr | " | 19,570 | 240.00 | 19,810 |
| Dishwasher, automatic 100 meals/hr | " | 57,300 | 120.00 | 57,420 |
| Disposals | | | | |
| 100 gal/hr | EA. | 1,350 | 120.00 | 1,470 |
| 120 gal/hr | " | 1,910 | 130.00 | 2,040 |
| 250 gal/hr | " | 4,890 | 130.00 | 5,020 |
| Exhaust hood for dishwasher, gutter 4 sides, s-steel | | | | |
| 4'x4'x2' | EA. | 2,160 | 130.00 | 2,290 |
| 4'x7'x2' | " | 3,780 | 150.00 | 3,930 |
| Food preparation machines | | | | |
| Vertical cutter mixers | | | | |
| 25 quart | EA. | 7,590 | 120.00 | 7,710 |
| 40 quart | " | 9,150 | 120.00 | 9,270 |
| 80 quart | " | 10,420 | 180.00 | 10,600 |
| 130 quart | " | 11,490 | 300.00 | 11,790 |
| Choppers | | | | |
| 5 lb | EA. | 2,160 | 91.00 | 2,251 |

# 11 EQUIPMENT

| ARCHITECTURAL EQUIPMENT | UNIT | MAT. | INST. | TOTAL |
|---|---|---|---|---|
| **11400.10 FOOD SERVICE EQUIPMENT** | | | | |
| 16 lb | EA. | 2,570 | 120.00 | 2,690 |
| 40 lb | " | 4,590 | 180.00 | 4,770 |
| Mixers, floor models | | | | |
| 20 quart | EA. | 3,380 | 45.40 | 3,425 |
| 60 quart | " | 9,450 | 45.40 | 9,495 |
| 80 quart | " | 11,490 | 52.00 | 11,542 |
| 140 quart | " | 16,200 | 72.50 | 16,273 |
| Ice cube maker | | | | |
| 50 lb per day | | | | |
| Minimum | EA. | 1,350 | 360.00 | 1,710 |
| Maximum | " | 2,410 | 360.00 | 2,770 |
| 500 lb per day | | | | |
| Minimum | EA. | 4,610 | 610.00 | 5,220 |
| Maximum | " | 8,370 | 610.00 | 8,980 |
| Ice flakers | | | | |
| 300 lb per day | EA. | 3,780 | 360.00 | 4,140 |
| 600 lb per day | " | 5,130 | 610.00 | 5,740 |
| 1000 lb per day | " | 6,750 | 810.00 | 7,560 |
| 2000 lb per day | " | 12,150 | 910.00 | 13,060 |
| Refrigerated cases | | | | |
| Dairy products | | | | |
| Multi deck type | L.F. | 650.00 | 24.20 | 674.20 |
| For rear sliding doors, add | " | | | 73.75 |
| Delicatessen case, service deli | | | | |
| Single deck | L.F. | 610.00 | 180.00 | 790.00 |
| Multi deck | " | 670.00 | 230.00 | 900.00 |
| Meat case | | | | |
| Single deck | L.F. | 440.00 | 210.00 | 650.00 |
| Multi deck | " | 480.00 | 230.00 | 710.00 |
| Produce case | | | | |
| Single deck | L.F. | 460.00 | 210.00 | 670.00 |
| Multi deck | " | 520.00 | 230.00 | 750.00 |
| Bottle coolers | | | | |
| 6' long | | | | |
| Minimum | EA. | 1,770 | 730.00 | 2,500 |
| Maximum | " | 3,620 | 730.00 | 4,350 |
| 10' long | | | | |
| Minimum | EA. | 2,410 | 1,210 | 3,620 |
| Maximum | " | 4,680 | 1,210 | 5,890 |
| Frozen food cases | | | | |
| Chest type | L.F. | 470.00 | 210.00 | 680.00 |
| Reach-in, glass door | " | 980.00 | 230.00 | 1,210 |
| Island case, single | " | 480.00 | 210.00 | 690.00 |
| Multi deck | " | 1,060 | 230.00 | 1,290 |
| Ice storage bins | | | | |
| 500 lb capacity | EA. | 1,350 | 520.00 | 1,870 |
| 1000 lb capacity | " | 2,620 | 1,040 | 3,660 |
| **11450.10 RESIDENTIAL EQUIPMENT** | | | | |
| Compactor, 4 to 1 compaction | EA. | 1,290 | 91.00 | 1,381 |
| Dishwasher, built-in | | | | |

# 11 EQUIPMENT

## ARCHITECTURAL EQUIPMENT

| | UNIT | MAT. | INST. | TOTAL |
|---|---|---|---|---|
| **11450.10 RESIDENTIAL EQUIPMENT** | | | | |
| 2 cycles | EA. | 1,080 | 180.00 | 1,260 |
| 4 or more cycles | " | 1,700 | 180.00 | 1,880 |
| Disposal | | | | |
| Garbage disposer | EA. | 440.00 | 120.00 | 560.00 |
| Heaters, electric, built-in | | | | |
| Ceiling type | EA. | 230.00 | 120.00 | 350.00 |
| Wall type | | | | |
| Minimum | EA. | 160.00 | 91.00 | 251.00 |
| Maximum | " | 300.00 | 120.00 | 420.00 |
| Hood for range, 2-speed, vented | | | | |
| 30" wide | EA. | 160.00 | 120.00 | 280.00 |
| 42" wide | " | 610.00 | 120.00 | 730.00 |
| Ice maker, automatic | | | | |
| 30 lb per day | EA. | 750.00 | 52.00 | 802.00 |
| 50 lb per day | " | 1,700 | 180.00 | 1,880 |
| Folding access stairs, disappearing metal stair | | | | |
| 8' long | EA. | 1,770 | 52.00 | 1,822 |
| 11' long | " | 1,840 | 52.00 | 1,892 |
| 12' long | " | 1,990 | 52.00 | 2,042 |
| Wood frame, wood stair | | | | |
| 22" x 54" x 8'9" long | EA. | 95.00 | 36.30 | 131.30 |
| 25" x 54" x 10' long | " | 120.00 | 36.30 | 156.30 |
| Ranges electric | | | | |
| Built-in, 30", 1 oven | EA. | 1,060 | 120.00 | 1,180 |
| 2 oven | " | 1,630 | 120.00 | 1,750 |
| Counter top, 4 burner, standard | " | 600.00 | 91.00 | 691.00 |
| With grill | " | 1,770 | 91.00 | 1,861 |
| Free standing, 21", 1 oven | " | 680.00 | 120.00 | 800.00 |
| 30", 1 oven | " | 880.00 | 72.50 | 952.50 |
| 2 oven | " | 2,620 | 72.50 | 2,693 |
| Water softener | | | | |
| 30 grains per gallon | EA. | 890.00 | 120.00 | 1,010 |
| 70 grains per gallon | " | 1,700 | 180.00 | 1,880 |
| **11470.10 DARKROOM EQUIPMENT** | | | | |
| Dryers | | | | |
| 36" x 25" x 68" | EA. | 8,100 | 200.00 | 8,300 |
| 48" x 25" x 68" | " | 8,370 | 200.00 | 8,570 |
| Processors, film | | | | |
| Black and white | EA. | 12,840 | 200.00 | 13,040 |
| Color negatives | " | 14,180 | 200.00 | 14,380 |
| Prints | " | 15,320 | 200.00 | 15,520 |
| Transparencies | " | 17,020 | 200.00 | 17,220 |
| Sinks with cabinet and/or stand | | | | |
| 5" sink with stand | | | | |
| 24" x 48" | EA. | 1,350 | 97.50 | 1,448 |
| 32" x 64" | " | 1,490 | 130.00 | 1,620 |
| 38" x 52" | " | 1,620 | 130.00 | 1,750 |
| 42" x 132" | " | 2,700 | 200.00 | 2,900 |
| 48" x 52" | " | 2,430 | 200.00 | 2,630 |
| 5" sink with cabinet | | | | |

# 11 EQUIPMENT

| ARCHITECTURAL EQUIPMENT | UNIT | MAT. | INST. | TOTAL |
|---|---|---|---|---|
| **11470.10 DARKROOM EQUIPMENT** | | | | |
| 24" x 48" | EA. | 2,160 | 97.50 | 2,258 |
| 32" x 64" | " | 2,300 | 130.00 | 2,430 |
| 38" x 52" | " | 2,300 | 130.00 | 2,430 |
| 42" x 132" | " | 4,320 | 200.00 | 4,520 |
| 48" x 52" | " | 2,840 | 200.00 | 3,040 |
| 10" sink with stand | | | | |
| 24" x 48" | EA. | 1,350 | 97.50 | 1,448 |
| 32" x 64" | " | 1,620 | 130.00 | 1,750 |
| 38" x 52" | " | 1,890 | 130.00 | 2,020 |
| 10" sink with cabinet | | | | |
| 24" x 48" | EA. | 2,300 | 97.50 | 2,398 |
| 38" x 52" | " | 2,630 | 130.00 | 2,760 |
| **11480.10 ATHLETIC EQUIPMENT** | | | | |
| Basketball backboard | | | | |
| Fixed | EA. | 740.00 | 440.00 | 1,180 |
| Swing-up | " | 1,700 | 710.00 | 2,410 |
| Portable, hydraulic | " | 9,430 | 180.00 | 9,610 |
| Suspended type, standard | " | 3,330 | 710.00 | 4,040 |
| For glass backboard, add | " | | | 1,021 |
| For electrically operated, add | " | | | 1,333 |
| Bleacher, telescoping, manual | | | | |
| 15 tier, minimum | SEAT | 93.50 | 7.10 | 100.60 |
| Maximum | " | 110.00 | 7.10 | 117.10 |
| 20 tier, minimum | " | 95.00 | 7.90 | 102.90 |
| Maximum | " | 110.00 | 7.90 | 117.90 |
| 30 tier, minimum | " | 96.50 | 11.80 | 108.30 |
| Maximum | " | 110.00 | 11.80 | 121.80 |
| Boxing ring elevated, complete, 22' x 22' | EA. | 130.00 | 3,540 | 3,670 |
| Gym divider curtain | | | | |
| Minimum | S.F. | 3.80 | 0.47 | 4.27 |
| Maximum | " | 5.25 | 0.47 | 5.72 |
| Scoreboards, single face | | | | |
| Minimum | EA. | 3,470 | 350.00 | 3,820 |
| Maximum | " | 22,410 | 1,770 | 24,180 |
| Parallel bars | | | | |
| Minimum | EA. | 1,160 | 350.00 | 1,510 |
| Maximum | " | 3,260 | 590.00 | 3,850 |
| **11500.10 INDUSTRIAL EQUIPMENT** | | | | |
| Vehicular paint spray booth, solid back, 14'4" x 9'6" | | | | |
| 24' deep | EA. | 11,980 | 350.00 | 12,330 |
| 26'6" deep | " | 12,130 | 350.00 | 12,480 |
| 28'6" deep | " | 12,340 | 350.00 | 12,690 |
| Drive through, 14'9" x 9'6" | | | | |
| 24' deep | EA. | 12,340 | 350.00 | 12,690 |
| 26'6" deep | " | 12,640 | 350.00 | 12,990 |
| 28'6" deep | " | 12,780 | 350.00 | 13,130 |
| Water wash, paint spray booth | | | | |
| 5' x 11'2" x 10'8" | EA. | 6,210 | 350.00 | 6,560 |
| 6' x 11'2" x 10'8" | " | 6,350 | 350.00 | 6,700 |
| 8' x 11'2" x 10'8" | " | 8,030 | 350.00 | 8,380 |

# ARCHITECTURAL EQUIPMENT

## 11500.10 INDUSTRIAL EQUIPMENT

| ARCHITECTURAL EQUIPMENT | UNIT | MAT. | INST. | TOTAL |
|---|---|---|---|---|
| 10' x 11'2" x 11'2" | EA. | 8,910 | 350.00 | 9,260 |
| 12' x 12'2" x 11'2" | " | 10,880 | 350.00 | 11,230 |
| 14' x 12'2" x 11'2" | " | 12,780 | 350.00 | 13,130 |
| 16' x 12'2" x 11'2" | " | 14,390 | 350.00 | 14,740 |
| 20' x 12'2" x 11'2" | " | 17,240 | 350.00 | 17,590 |
| Dry type spray booth, with paint arrestors | | | | |
| 5'4" x 7'2" x 6'8" | EA. | 2,700 | 350.00 | 3,050 |
| 6'4" x 7'2" x 6'8" | " | 3,290 | 350.00 | 3,640 |
| 8'4" x 7'2" x 9'2" | " | 3,730 | 350.00 | 4,080 |
| 10'4" x 7'2" x 9'2" | " | 4,310 | 350.00 | 4,660 |
| 12'4" x 7'6" x 9'2" | " | 4,230 | 350.00 | 4,580 |
| 14'4" x 7'6" x 9'8" | " | 5,770 | 350.00 | 6,120 |
| 16'4" x 7'7" x 9'8" | " | 6,500 | 350.00 | 6,850 |
| 20'4" x 7'7" x 10'8" | " | 7,300 | 350.00 | 7,650 |
| Air compressor, electric | | | | |
| 1 hp | | | | |
| 115 volt | EA. | 3,290 | 240.00 | 3,530 |
| 5 hp | | | | |
| 115 volt | EA. | 6,790 | 350.00 | 7,140 |
| 230 volt | " | 5,770 | 350.00 | 6,120 |
| Hydraulic lifts | | | | |
| 8,000 lb capacity | EA. | 510.00 | 890.00 | 1,400 |
| 11,000 lb capacity | " | 7,080 | 1,420 | 8,500 |
| 24,000 lb capacity | " | 10,010 | 2,360 | 12,370 |
| Power tools | | | | |
| Band saws | | | | |
| 10" | EA. | 690.00 | 29.50 | 719.50 |
| 14" | " | 1,200 | 35.40 | 1,235 |
| Motorized shaper | " | 640.00 | 27.30 | 667.30 |
| Motorized lathe | " | 760.00 | 29.50 | 789.50 |
| Bench saws | | | | |
| 9" saw | EA. | 1,370 | 23.65 | 1,394 |
| 10" saw | " | 2,120 | 25.30 | 2,145 |
| 12" saw | " | 2,700 | 29.50 | 2,730 |
| Electric grinders | | | | |
| 1/3 hp | EA. | 260.00 | 14.20 | 274.20 |
| 1/2 hp | " | 380.00 | 15.40 | 395.40 |
| 3/4 hp | " | 450.00 | 15.40 | 465.40 |

## 11600.10 LABORATORY EQUIPMENT

| ARCHITECTURAL EQUIPMENT | UNIT | MAT. | INST. | TOTAL |
|---|---|---|---|---|
| Cabinets, base | | | | |
| Minimum | L.F. | 120.00 | 29.50 | 149.50 |
| Maximum | " | 210.00 | 29.50 | 239.50 |
| Full storage, 7' high | | | | |
| Minimum | L.F. | 180.00 | 29.50 | 209.50 |
| Maximum | " | 280.00 | 29.50 | 309.50 |
| Wall | | | | |
| Minimum | L.F. | 79.00 | 35.40 | 114.40 |
| Maximum | " | 130.00 | 35.40 | 165.40 |
| Counter tops | | | | |
| Minimum | S.F. | 16.80 | 4.45 | 21.25 |
| Average | " | 39.40 | 5.05 | 44.45 |

## ARCHITECTURAL EQUIPMENT

### 11600.10 LABORATORY EQUIPMENT

| ARCHITECTURAL EQUIPMENT | UNIT | MAT. | INST. | TOTAL |
|---|---|---|---|---|
| Maximum | S.F. | 64.50 | 5.90 | 70.40 |
| Tables | | | | |
| Open underneath | S.F. | 90.50 | 17.70 | 108.20 |
| Doors underneath | " | 170.00 | 22.15 | 192.15 |
| Medical laboratory equipment | | | | |
| Analyzer | | | | |
| Chloride | EA. | 530.00 | 18.15 | 548.15 |
| Blood | " | 23,370 | 30.30 | 23,400 |
| Bath, water, utility, countertop unit | " | 380.00 | 36.30 | 416.30 |
| Hot plate, lab, countertop | " | 120.00 | 33.00 | 153.00 |
| Stirrer | " | 210.00 | 33.00 | 243.00 |
| Incubator, anaerobic, 23x23x36" | " | 6,140 | 180.00 | 6,320 |
| Dry heat bath | " | 450.00 | 60.50 | 510.50 |
| Incinerator, for sterilizing | " | 190.00 | 3.65 | 193.65 |
| Meter, serum protein | " | 820.00 | 4.55 | 824.55 |
| Ph analog, general purpose | " | 610.00 | 5.20 | 615.20 |
| Refrigerator, blood bank, undercounter type 153 litres | " | 4,530 | 60.50 | 4,591 |
| 5.4 cf, undercounter type | " | 3,800 | 60.50 | 3,861 |
| Refrigerator/freezer, 4.4 cf, undercounter type | " | 420.00 | 60.50 | 480.50 |
| Sealer, impulse, free standing, 20x12x4" | " | 230.00 | 12.10 | 242.10 |
| Timer, electric, 1-60 minutes, bench or wall mounted | " | 61.50 | 20.20 | 81.70 |
| Glassware washer - dryer, undercounter | " | 6,940 | 450.00 | 7,390 |
| Balance, torsion suspension, tabletop, 4.5 lb capacity | " | 490.00 | 20.20 | 510.20 |
| Binocular microscope, with in base illuminator | " | 1,970 | 13.95 | 1,984 |
| Centrifuge, table model, 19x16x13" | " | 1,140 | 14.55 | 1,155 |
| Clinical model, with four place head | " | 600.00 | 8.05 | 608.05 |

### 11700.10 MEDICAL EQUIPMENT

| ARCHITECTURAL EQUIPMENT | UNIT | MAT. | INST. | TOTAL |
|---|---|---|---|---|
| Hospital equipment, lights | | | | |
| Examination, portable | EA. | 940.00 | 30.30 | 970.30 |
| Meters | | | | |
| Air flow meter | EA. | 55.00 | 20.20 | 75.20 |
| Oxygen flow meters | " | 37.60 | 15.15 | 52.75 |
| Racks | | | | |
| 40 chart, revolving open frame; mobile caddy | EA. | 1,090 | 30.30 | 1,120 |
| Scales. | | | | |
| Clinical, metric with measure rod, 350 lb | EA. | 330.00 | 33.00 | 363.00 |
| Physical therapy | | | | |
| Chair, hydrotherapy | EA. | 270.00 | 5.90 | 275.90 |
| Diathermy, shortwave, portable, on casters | " | 2,460 | 14.20 | 2,474 |
| Exercise bicycle, floor standing, 35" x 15" | " | 680.00 | 11.80 | 691.80 |
| Hydrocollator, 4 pack, portable, 129 x 90 x 160" | " | 270.00 | 5.05 | 275.05 |
| Lamp, infrared, mobile with variable heat control | " | 360.00 | 27.30 | 387.30 |
| Ultra violet, base mounted | " | 330.00 | 27.30 | 357.30 |
| Mirror, posture training, 27" wide and 72" high | " | 380.00 | 8.85 | 388.85 |
| Parallel bars, adjustable | " | 1,660 | 44.30 | 1,704 |
| Platform mat 10'x6', 1" thick | " | 450.00 | 8.85 | 458.85 |
| Pulley, duplex, wall mounted | " | 320.00 | 120.00 | 440.00 |
| Rack, crutch, wall mounted, 66 x 16 x 13" | " | 690.00 | 35.40 | 725.40 |
| Stimulator, galvanic-faradic, hand held | " | 71.50 | 2.35 | 73.85 |
| Ultrasound, muscle stimulator, portable, 13x13x8" | " | 1,810 | 2.95 | 1,813 |
| Sandbag set, velcro straps, saddle bag type | " | 98.50 | 5.05 | 103.55 |

## ARCHITECTURAL EQUIPMENT

### 11700.10 MEDICAL EQUIPMENT

| ARCHITECTURAL EQUIPMENT | UNIT | MAT. | INST. | TOTAL |
|---|---|---|---|---|
| Whirlpool, 85 gallon | EA. | 2,460 | 180.00 | 2,640 |
| 65 gallon capacity | " | 2,100 | 180.00 | 2,280 |
| Radiology | | | | |
| Radiographic table, motor driven tilting table | EA. | 26,760 | 3,540 | 30,300 |
| Fluoroscope image/tv system | " | 75,230 | 7,090 | 82,320 |
| Processor for washing and drying radiographs | | | | |
| Water filter unit, 30" x 48-1/2" x 37-1/2" | EA. | 24.15 | 610.00 | 634.15 |
| Cassette transfer cabinet | " | 750.00 | 30.30 | 780.30 |
| Base storage cabinets, sectional design | | | | |
| With back splash, 24" deep and 35" high | L.F. | 1,850 | 30.30 | 1,880 |
| Wall storage cabinets | " | 140.00 | 45.40 | 185.40 |
| Steam sterilizers | | | | |
| For heat and moisture stable materials | EA. | 2,750 | 36.30 | 2,786 |
| For fast drying after sterilization | " | 1,880 | 45.40 | 1,925 |
| Compact unit | " | 1,330 | 45.40 | 1,375 |
| Semi-automatic | " | 5,500 | 180.00 | 5,680 |
| Floor loading | | | | |
| Single door | EA. | 99,820 | 300.00 | |
| Double door | " | 125,860 | 360.00 | |
| Utensil washer, sanitizer | " | 5,640 | 280.00 | 5,920 |
| Automatic washer/sterilizer | " | 21,700 | 730.00 | 22,430 |
| 16 x 16 x 26", including generator & accessories | " | 20,980 | 1,210 | 22,190 |
| Steam generator, elec., 10 kw to 180 kw | " | 19,530 | 730.00 | 20,260 |
| Surgical scrub | | | | |
| Minimum | EA. | 2,140 | 120.00 | 2,260 |
| Maximum | " | 4,480 | 120.00 | 4,600 |
| Gas sterilizers | | | | |
| Automatic, free standing, 21x19x29" | EA. | 3,830 | 360.00 | 4,190 |
| Surgical tables | | | | |
| Minimum | EA. | 9,980 | 520.00 | 10,500 |
| Maximum | " | 44,850 | 730.00 | 45,580 |
| Surgical lights, ceiling mounted | | | | |
| Minimum | EA. | 3,760 | 610.00 | 4,370 |
| Maximum | " | 6,080 | 730.00 | 6,810 |
| Water stills | | | | |
| 4 liters/hr | EA. | 1,950 | 120.00 | 2,070 |
| 8 liters/hr | " | 2,820 | 120.00 | 2,940 |
| 19 liters/hr | " | 7,810 | 300.00 | 8,110 |
| X-ray equipment | | | | |
| Mobile unit | | | | |
| Minimum | EA. | 6,660 | 180.00 | 6,840 |
| Maximum | " | 33,270 | 360.00 | 33,630 |
| Film viewers | | | | |
| Minimum | EA. | 140.00 | 60.50 | 200.50 |
| Maximum | " | 590.00 | 120.00 | 710.00 |
| Autopsy table | | | | |
| Minimum | EA. | 3,760 | 360.00 | 4,120 |
| Maximum | " | 7,450 | 360.00 | 7,810 |
| Incubators | | | | |
| 15 cf | EA. | 3,470 | 180.00 | 3,650 |
| 29 cf | " | 4,200 | 300.00 | 4,500 |
| Infant transport, portable | " | 8,830 | 190.00 | 9,020 |

## ARCHITECTURAL EQUIPMENT

### 11700.10 MEDICAL EQUIPMENT

| ARCHITECTURAL EQUIPMENT | UNIT | MAT. | INST. | TOTAL |
|---|---|---|---|---|
| Beds | | | | |
| Stretcher, with pad, 30" x 78" | EA. | 1,350 | 91.00 | 1,441 |
| Transfer, for patient transport | " | 2,310 | 91.00 | 2,401 |
| Headwall | | | | |
| Aluminum, with back frame and console | EA. | 3,110 | 180.00 | 3,290 |
| Hospital ground detection system | | | | |
| Power ground module | EA. | 670.00 | 100.00 | 770.00 |
| Ground slave module | " | 460.00 | 79.00 | 539.00 |
| Master ground module | " | 290.00 | 68.50 | 358.50 |
| Remote indicator | " | 320.00 | 72.50 | 392.50 |
| X-ray indicator | " | 970.00 | 79.00 | 1,049 |
| Micro ammeter | " | 1,190 | 91.00 | 1,281 |
| Supervisory module | " | 920.00 | 79.00 | 999.00 |
| Ground cords | " | 76.50 | 13.45 | 89.95 |
| Hospital isolation monitors, 5 ma | | | | |
| 120v | EA. | 1,880 | 160.00 | 2,040 |
| 208v | " | 1,810 | 160.00 | 1,970 |
| 240v | " | 1,810 | 160.00 | 1,970 |
| Digital clock-timers separate display | " | 740.00 | 72.50 | 812.50 |
| One display | " | 710.00 | 72.50 | 782.50 |
| Remote control | " | 420.00 | 56.50 | 476.50 |
| Battery pack | " | 97.00 | 56.50 | 153.50 |
| Surgical chronometer clock and 3 timers | " | 1,560 | 110.00 | 1,670 |
| Auxilary control | " | 560.00 | 52.50 | 612.50 |

### 11700.20 DENTAL EQUIPMENT

| ARCHITECTURAL EQUIPMENT | UNIT | MAT. | INST. | TOTAL |
|---|---|---|---|---|
| Dental care equipment | | | | |
| Drill console with accessories | EA. | 11,000 | 610.00 | 11,610 |
| Amalgamator | " | 270.00 | 18.15 | 288.15 |
| Lathe | " | 290.00 | 12.10 | 302.10 |
| Finish polisher | " | 220.00 | 24.20 | 244.20 |
| Model trimmer | " | 350.00 | 16.50 | 366.50 |
| Motor, wall mounted | " | 380.00 | 16.50 | 396.50 |
| Cleaner, ultrasonic | " | 1,430 | 36.30 | 1,466 |
| Curing unit, bench mounted | " | 1,960 | 60.50 | 2,021 |
| Oral evacuation system, dual pump | " | 10,630 | 45.40 | 10,675 |
| Sterilizer, table top, self contained | " | 750.00 | 20.20 | 770.20 |
| Dental lights | | | | |
| Light, floor or ceiling mounted | EA. | 1,810 | 180.00 | 1,990 |
| X-ray unit | | | | |
| Portable | EA. | 2,730 | 91.00 | 2,821 |
| Wall mounted with remote control | " | 5,640 | 300.00 | 5,940 |
| Illuminator, single panel | " | 8,030 | 520.00 | 8,550 |
| X-ray film processor | " | 1,320 | 300.00 | 1,620 |
| Shield, portable x-ray, lead lined | " | 290.00 | 24.20 | 314.20 |

# 12 FURNISHINGS

| INTERIOR | UNIT | MAT. | INST. | TOTAL |
|---|---|---|---|---|
| **12302.10 CASEWORK** | | | | |
| Kitchen base cabinet, prefinished, 24" deep, 35" high | | | | |
| 12"wide | EA. | 160.00 | 35.40 | 195.40 |
| 18" wide | " | 190.00 | 35.40 | 225.40 |
| 24" wide | " | 200.00 | 39.40 | 239.40 |
| 27" wide | " | 240.00 | 39.40 | 279.40 |
| 36" wide | " | 270.00 | 44.30 | 314.30 |
| 48" wide | " | 310.00 | 44.30 | 354.30 |
| Corner cabinet, 36" wide | " | 210.00 | 44.30 | 254.30 |
| Wall cabinet, 12" deep, 12" high | | | | |
| 30" wide | EA. | 120.00 | 35.40 | 155.40 |
| 36" wide | " | 120.00 | 35.40 | 155.40 |
| 15" high | | | | |
| 30" wide | EA. | 120.00 | 39.40 | 159.40 |
| 36" wide | " | 140.00 | 39.40 | 179.40 |
| 24" high | | | | |
| 30" wide | EA. | 150.00 | 39.40 | 189.40 |
| 36" wide | " | 180.00 | 39.40 | 219.40 |
| 30" high | | | | |
| 12" wide | EA. | 110.00 | 44.30 | 154.30 |
| 18" wide | " | 130.00 | 44.30 | 174.30 |
| 24" wide | " | 140.00 | 44.30 | 184.30 |
| 27" wide | " | 170.00 | 44.30 | 214.30 |
| 30" wide | " | 190.00 | 50.50 | 240.50 |
| 36" wide | " | 190.00 | 50.50 | 240.50 |
| Corner cabinet, 30" high | | | | |
| 24" wide | EA. | 180.00 | 59.00 | 239.00 |
| 30" wide | " | 220.00 | 59.00 | 279.00 |
| 36" wide | " | 240.00 | 59.00 | 299.00 |
| Wardrobe | " | 560.00 | 88.50 | 648.50 |
| Vanity with top, laminated plastic | | | | |
| 24" wide | EA. | 350.00 | 88.50 | 438.50 |
| 30" wide | " | 430.00 | 88.50 | 518.50 |
| 36" wide | " | 450.00 | 120.00 | 570.00 |
| 48" wide | " | 500.00 | 140.00 | 640.00 |
| **12390.10 COUNTER TOPS** | | | | |
| Stainless steel, counter top, with backsplash | S.F. | 92.00 | 8.85 | 100.85 |
| Acid-proof, kemrock surface | " | 25.70 | 5.90 | 31.60 |
| **12500.10 WINDOW TREATMENT** | | | | |
| Drapery tracks, wall or ceiling mounted | | | | |
| Basic traverse rod | | | | |
| 50 to 90" | EA. | 36.20 | 17.70 | 53.90 |
| 84 to 156" | " | 54.50 | 19.70 | 74.20 |
| 136 to 250" | " | 69.50 | 19.70 | 89.20 |
| 165 to 312" | " | 100.00 | 22.15 | 122.15 |
| Traverse rod with stationary curtain rod | | | | |
| 30 to 50" | EA. | 43.40 | 17.70 | 61.10 |
| 50 to 90" | " | 53.50 | 17.70 | 71.20 |
| 84 to 156" | " | 78.00 | 19.70 | 97.70 |
| 136 to 250" | " | 89.50 | 22.15 | 111.65 |
| Double traverse rod | | | | |

# 12 FURNISHINGS

| INTERIOR | UNIT | MAT. | INST. | TOTAL |
|---|---|---|---|---|
| **12500.10 WINDOW TREATMENT** | | | | |
| 30 to 50" | EA. | 52.00 | 17.70 | 69.70 |
| 50 to 84" | " | 69.50 | 17.70 | 87.20 |
| 84 to 156" | " | 85.50 | 19.70 | 105.20 |
| 136 to 250" | " | 100.00 | 22.15 | 122.15 |
| **12510.10 BLINDS** | | | | |
| Venetian blinds | | | | |
| 2" slats | S.F. | 4.05 | 0.89 | 4.94 |
| 1" slats | " | 5.50 | 0.89 | 6.39 |
| **12690.40 FLOOR MATS** | | | | |
| Recessed entrance mat, 3/8" thick, aluminum link | S.F. | 20.65 | 17.70 | 38.35 |
| Steel, flexible | " | 9.10 | 17.70 | 26.80 |

# 13 SPECIAL CONSTRUCTION

| CONSTRUCTION | UNIT | MAT. | INST. | TOTAL |
|---|---|---|---|---|
| **13056.10 VAULTS** | | | | |
| Floor safes | | | | |
| Class C | | | | |
| 1.0 cf | EA. | 580.00 | 29.50 | 609.50 |
| 1.3 cf | " | 650.00 | 44.30 | 694.30 |
| 1.9 cf | " | 720.00 | 59.00 | 779.00 |
| 5.2 cf | " | 1,740 | 59.00 | 1,799 |
| **13121.10 PRE-ENGINEERED BUILDINGS** | | | | |
| Pre-engineered metal building, 40'x100' | | | | |
| 14' eave height | S.F. | 9.55 | 2.80 | 12.35 |
| 16' eave height | " | 10.55 | 3.25 | 13.80 |
| 20' eave height | " | 12.05 | 4.25 | 16.30 |
| 60'x100' | | | | |
| 14' eave height | S.F. | 8.65 | 2.80 | 11.45 |
| 16' eave height | " | 9.10 | 3.25 | 12.35 |
| 20' eave height | " | 10.15 | 4.25 | 14.40 |
| 80'x100' | | | | |
| 14' eave height | S.F. | 8.65 | 2.80 | 11.45 |
| 16' eave height | " | 8.95 | 3.25 | 12.20 |
| 20' eave height | " | 10.25 | 4.25 | 14.50 |
| 100'x100' | | | | |
| 14' eave height | S.F. | 12.70 | 2.80 | 15.50 |
| 16' eave height | " | 9.10 | 3.25 | 12.35 |
| 20' eave height | " | 9.95 | 4.25 | 14.20 |
| 100'x150' | | | | |
| 14' eave height | S.F. | 7.90 | 2.80 | 10.70 |
| 16' eave height | " | 8.50 | 3.25 | 11.75 |
| 20' eave height | " | 9.40 | 4.25 | 13.65 |
| 120'x150' | | | | |
| 14' eave height | S.F. | 7.90 | 2.80 | 10.70 |
| 16' eave height | " | 8.35 | 3.25 | 11.60 |
| 20' eave height | " | 9.10 | 4.25 | 13.35 |
| 140'x150' | | | | |
| 14' eave height | S.F. | 7.45 | 2.80 | 10.25 |
| 16' eave height | " | 8.05 | 3.25 | 11.30 |
| 20' eave height | " | 8.95 | 4.25 | 13.20 |
| 160'x200' | | | | |
| 14' eave height | S.F. | 7.30 | 2.80 | 10.10 |
| 16' eave height | " | 7.35 | 3.25 | 10.60 |
| 20' eave height | " | 8.35 | 4.25 | 12.60 |
| 200'x200' | | | | |
| 14' eave height | S.F. | 7.30 | 2.80 | 10.10 |
| 16' eave height | " | 7.60 | 3.25 | 10.85 |
| 20' eave height | " | 8.20 | 4.25 | 12.45 |
| Hollow metal door and frame, 6' x 7' | EA. | | | 744.19 |
| Sectional steel overhead door, manually operated | | | | |
| 8' x 8' | EA. | | | 804.15 |
| 12' x 12' | " | | | 1,206 |
| Roll-up steel door, manually operated | | | | |
| 10' x 10' | EA. | | | 1,490 |
| 12' x 12' | " | | | 1,787 |
| For gravity ridge ventilator with birdscreen | " | | | 550.99 |

# 13 SPECIAL CONSTRUCTION

| CONSTRUCTION | UNIT | MAT. | INST. | TOTAL |
|---|---|---|---|---|
| **13121.10 PRE-ENGINEERED BUILDINGS** | | | | |
| 9" throat x 10' | EA. | | | 595.63 |
| 12" throat x 10' | " | | | 625.55 |
| For 20" rotary vent with damper | " | | | 253.17 |
| For 4' x 3' fixed louver | " | | | 268.06 |
| For 4' x 3' aluminum sliding window | " | | | 275.56 |
| For 3' x 9' fiberglass panels | " | | | 110.20 |
| Liner panel, 26 ga, painted steel | S.F. | 1.70 | 0.96 | 2.66 |
| Wall panel insulated, 26 ga. steel, foam core | " | 5.50 | 0.96 | 6.46 |
| Roof panel, 26 ga. painted steel | " | 1.40 | 0.55 | 1.95 |
| Plastic (sky light) | " | 4.00 | 0.55 | 4.55 |
| Insulation, 3-1/2" thick blanket, R11 | " | 1.05 | 0.26 | 1.31 |
| **13152.10 SWIMMING POOL EQUIPMENT** | | | | |
| Diving boards | | | | |
| 14' long | | | | |
| Aluminum | EA. | 1,390 | 150.00 | 1,540 |
| Fiberglass | " | 730.00 | 150.00 | 880.00 |
| Ladders, heavy duty | | | | |
| 2 steps | | | | |
| Minimum | EA. | 470.00 | 55.50 | 525.50 |
| Maximum | " | 750.00 | 55.50 | 805.50 |
| 4 steps | | | | |
| Minimum | EA. | 530.00 | 69.50 | 599.50 |
| Maximum | " | 990.00 | 69.50 | 1,060 |
| Lifeguard chair | | | | |
| Minimum | EA. | 1,250 | 280.00 | 1,530 |
| Maximum | " | 1,750 | 280.00 | 2,030 |
| Lights, underwater | | | | |
| 12 volt, with transformer | EA. | 350.00 | 69.50 | 419.50 |
| 110 volt | | | | |
| Minimum | EA. | 320.00 | 69.50 | 389.50 |
| Maximum | " | 1,020 | 69.50 | 1,090 |
| Ground fault interrupter for 110 volt, each light | " | 130.00 | 23.15 | 153.15 |
| Pool covers | | | | |
| Reinforced polyethylene | S.F. | 0.58 | 2.15 | 2.73 |
| Vinyl water tube | | | | |
| Minimum | S.F. | 0.65 | 2.15 | 2.80 |
| Maximum | " | 1.00 | 2.15 | 3.15 |
| Slides with water tube | | | | |
| Minimum | EA. | 420.00 | 230.00 | 650.00 |
| Maximum | " | 990.00 | 230.00 | 1,220 |

# 14 CONVEYING

## ELEVATORS

| 14210.10 ELEVATORS | UNIT | MAT. | INST. | TOTAL |
|---|---|---|---|---|
| Passenger elevators, electric, geared | | | | |
| Based on a shaft of 6 stops and 6 openings | | | | |
| 50 fpm, 2000 lb | EA. | 84,250 | 1,440 | 85,690 |
| 100 fpm, 2000 lb | " | 87,370 | 1,600 | 88,970 |
| 150 fpm | | | | |
| 2000 lb | EA. | 96,730 | 1,800 | 98,530 |
| 3000 lb | " | 121,690 | 2,060 | |
| 4000 lb | " | 126,370 | 2,400 | |
| 200 fpm | | | | |
| 2500 lb | EA. | 117,790 | 2,060 | |
| 3000 lb | " | 120,910 | 2,220 | |
| 4000 lb | " | 127,930 | 2,400 | |
| 250 fpm | | | | |
| 2500 lb | EA. | 121,690 | 2,060 | |
| 3000 lb | " | 127,930 | 2,220 | |
| 4000 lb | " | 131,050 | 2,400 | |
| 300 fpm | | | | |
| 2500 lb | EA. | 121,690 | 2,060 | |
| 3000 lb | " | 126,370 | 2,220 | |
| 4000 lb | " | 131,050 | 1,440 | |
| For each additional; 50 fpm, add per stop, $3000 | | | | |
| 500 lb, add per stop, $4000 | | | | |
| Opening, add per stop, $4500 | | | | |
| Stop, add per stop, $4000 | | | | |
| Bonderized steel door, add per opening, $150 | | | | |
| Colored aluminum door, add per opening, $850 | | | | |
| Stainless steel door, add per opening, $600 | | | | |
| Cast bronze door, add per opening, $1100 | | | | |
| Two speed door, add per opening, $360 | | | | |
| Bi-parting door, add per opening, $850 | | | | |
| Custom cab interior add, $4800 | | | | |
| Based on a shaft of 8 stops and 8 openings | | | | |
| 300 fpm | | | | |
| 3000 lb | EA. | 145,090 | 2,880 | |
| 3500 lb | " | 148,210 | 2,880 | |
| 4000 lb | " | 156,010 | 3,200 | |
| 5000 lb | " | 174,740 | 3,430 | |
| 400 fpm | | | | |
| 3000 lb | EA. | 152,890 | 2,880 | |
| 3500 lb | " | 155,230 | 2,880 | |
| 4000 lb | " | 171,620 | 3,200 | |
| 5000 lb | " | 187,220 | 3,430 | |
| 600 fpm | | | | |
| 3000 lb | EA. | 218,420 | 3,200 | |
| 3500 lb | " | 221,540 | 3,430 | |
| 4000 lb | " | 223,100 | 3,510 | |
| 5000 lb | " | 233,240 | 3,600 | |
| 800 fpm | | | | |
| 3000 lb | EA. | 257,420 | 3,200 | |
| 3500 lb | " | 260,540 | 3,430 | |
| 4000 lb | " | 263,660 | 3,510 | |
| 5000 lb | " | 273,020 | 3,600 | |

| ELEVATORS | UNIT | MAT. | INST. | TOTAL |
|---|---|---|---|---|
| **14210.10 ELEVATORS** | | | | |
| For each additional; 100 fpm add per stop, $13,000 | | | | |
| 500 lb, add per stop, $6500 | | | | |
| Opening add per stop, $12,000 | | | | |
| Stop add per stop, $4800 | | | | |
| Bypass floor, add per each, $2000 | | | | |
| Bonderized steel door, add per opening, $150 | | | | |
| Colored aluminum door, add per opening, $900 | | | | |
| Stainless steel door, add per opening, $600 | | | | |
| Cast bronze door, add per opening, $600 | | | | |
| Two speed bi-parting door, add per opening, $1000 | | | | |
| Custom cab interior, add $5000 | | | | |
| Hydraulic, based on a shaft of 3 stops, 3 openings | | | | |
| 50 fpm | | | | |
| 2000 lb | EA. | 55,140 | 1,200 | 56,340 |
| 2500 lb | " | 59,010 | 1,200 | 60,210 |
| 3000 lb | " | 62,260 | 1,250 | 63,510 |
| 100 fpm | | | | |
| 2000 lb | EA. | 60,560 | 1,200 | 61,760 |
| 2500 lb | " | 63,810 | 1,250 | 65,060 |
| 3000 lb | " | 67,840 | 1,310 | 69,150 |
| 150 fpm | | | | |
| 2000 lb | EA. | 65,360 | 1,200 | 66,560 |
| 2500 lb | " | 71,550 | 1,250 | 72,800 |
| 3000 lb | " | 76,510 | 1,370 | 77,880 |
| For each additional; 50 fpm add per stop, $3500 | | | | |
| 500 lb, add per stop, $3500 | | | | |
| Opening, add, $4200 | | | | |
| Stop, add per stop, $5300 | | | | |
| Bonderized steel door, add per opening, $400 | | | | |
| Colored aluminum door, add per opening, $1500 | | | | |
| Stainless steel door, add per opening, $650 | | | | |
| Cast bronze door, add per opening, $1200 | | | | |
| Two speed door, add per opening, $400 | | | | |
| Bi-parting door, add per opening, $900 | | | | |
| Custom cab interior, add per cab, $5000 | | | | |
| Small elevators, 4 to 6 passenger capacity | | | | |
| Electric, push | | | | |
| 2 stops | EA. | 20,130 | 1,200 | 21,330 |
| 3 stops | " | 24,780 | 1,310 | 26,090 |
| 4 stops | " | 28,650 | 1,440 | 30,090 |
| Freight elevators, electric | | | | |
| Based on a shaft of 6 stops and 6 openings | | | | |
| 50 fpm | | | | |
| 3500 lb | EA. | 151,900 | 1,600 | |
| 4000 lb | " | 152,480 | 1,600 | |
| 5000 lb | " | 153,930 | 1,800 | |
| 100 fpm | | | | |
| 3500 lb | EA. | 159,860 | 1,800 | |
| 4000 lb | " | 160,580 | 1,800 | |
| 5000 lb | " | 156,680 | 2,060 | |
| 200 fpm | | | | |
| 3500 lb | EA. | 157,400 | 2,060 | |

## ELEVATORS

| ELEVATORS | UNIT | MAT. | INST. | TOTAL |
|---|---|---|---|---|
| **14210.10 ELEVATORS** | | | | |
| 4000 lb | EA. | 158,120 | 2,060 | |
| 5000 lb | " | 159,570 | 2,400 | |
| For elevator with manual door, deduct 15% | | | | |
| For variable voltage control, add 20% | | | | |
| Based on shaft of 8 stops and 8 openings | | | | |
| 100 fpm | | | | |
| 4000 lb | EA. | 160,290 | 1,800 | |
| 6000 lb | " | 154,790 | 1,850 | |
| 8000 lb | " | 157,690 | 1,950 | |
| 150 fpm | | | | |
| 4000 lb | EA. | 154,790 | 2,060 | |
| 6000 lb | " | 157,690 | 2,120 | |
| 8000 lb | " | 159,130 | 2,250 | |
| 200 fpm | | | | |
| 4000 lb | EA. | 156,240 | 2,400 | |
| 6000 lb | " | 159,130 | 2,480 | |
| 8000 lb | " | 166,370 | 2,620 | |
| For each additional; 50 fpm, add per stop, $2000 | | | | |
| 500 lb, add per stop, $600 | | | | |
| Opening, add per stop, $7000 | | | | |
| Stop, add per stop, $5500 | | | | |
| For variable voltage, add 20% | | | | |
| Hydraulic, based on 3 stops and 3 openings | | | | |
| 50 fpm | | | | |
| 3000 lb | EA. | 64,900 | 1,030 | 65,930 |
| 4000 lb | " | 70,800 | 1,070 | 71,870 |
| 6000 lb | " | 82,600 | 1,110 | 83,710 |
| 100 fpm | | | | |
| 3000 lb | EA. | 73,160 | 1,030 | 74,190 |
| 4000 lb | " | 78,180 | 1,070 | 79,250 |
| 6000 lb | " | 91,310 | 1,110 | 92,420 |
| 150 fpm | | | | |
| 3000 lb | EA. | 80,390 | 1,030 | 81,420 |
| 4000 lb | " | 86,290 | 1,070 | 87,360 |
| 6000 lb | " | 98,980 | 1,110 | |
| For each additional; 50 fpm, add per stop, $2000 | | | | |
| 500 lb, add per stop, $600 | | | | |
| Opening, add per stop, $5500 | | | | |
| Stop, add per stop, $5500 | | | | |
| For elevator with manual door deduct from total, 15% | | | | |
| **14300.10 ESCALATORS** | | | | |
| Escalators | | | | |
| 32" wide, floor to floor | | | | |
| 12' high | EA. | 110,770 | 2,400 | |
| 15' high | " | 120,130 | 2,880 | |
| 18' high | " | 129,490 | 3,600 | |
| 22' high | " | 140,410 | 4,800 | |
| 25' high | " | 156,010 | 5,760 | |
| 48" wide | | | | |
| 12' high | EA. | 123,250 | 2,480 | |
| 15' high | " | 134,170 | 3,000 | |

# 14 CONVEYING

## ELEVATORS

| 14300.10 ESCALATORS | UNIT | MAT. | INST. | TOTAL |
|---|---|---|---|---|
| 18' high | EA. | 143,530 | 3,790 | |
| 22' high | " | 160,690 | 5,150 | |
| 25' high | " | 171,620 | 5,760 | |

## LIFTS

| 14410.10 PERSONNEL LIFTS | UNIT | MAT. | INST. | TOTAL |
|---|---|---|---|---|
| Electrically operated, 1 or 2 person lift | | | | |
| With attached foot platforms | | | | |
| 3 stops | EA. | | | 17,730 |
| 5 stops | " | | | 22,694 |
| 7 stops | " | | | 26,947 |
| For each additional stop, add $1250 | | | | |
| Residential stair climber, per story | EA. | 5,250 | 300.00 | 5,550 |

| 14450.10 VEHICLE LIFTS | UNIT | MAT. | INST. | TOTAL |
|---|---|---|---|---|
| Automotive hoist, one post, semi-hydraulic, 8,000 lb | EA. | 4,320 | 1,440 | 5,760 |
| Full hydraulic, 8,000 lb | " | 4,100 | 1,440 | 5,540 |
| 2 post, semi-hydraulic, 10,000 lb | " | 4,470 | 2,060 | 6,530 |
| Full hydraulic | | | | |
| 10,000 lb | EA. | 4,770 | 2,060 | 6,830 |
| 13,000 lb | " | 5,880 | 3,600 | 9,480 |
| 18,500 lb | " | 7,670 | 3,600 | 11,270 |
| 24,000 lb | " | 10,350 | 3,600 | 13,950 |
| 26,000 lb | " | 11,990 | 3,600 | 15,590 |
| Pneumatic hoist, fully hydraulic | | | | |
| 11,000 lb | EA. | 5,730 | 4,800 | 10,530 |
| 24,000 lb | " | 10,050 | 4,800 | 14,850 |

## MATERIAL HANDLING

| 14560.10 CHUTES | UNIT | MAT. | INST. | TOTAL |
|---|---|---|---|---|
| Linen chutes, stainless steel, with supports | | | | |
| 18" dia. | L.F. | 110.00 | 2.75 | 112.75 |

# 14 CONVEYING

## MATERIAL HANDLING

| 14560.10 CHUTES | UNIT | MAT. | INST. | TOTAL |
|---|---|---|---|---|
| 24" dia. | L.F. | 140.00 | 2.95 | 142.95 |
| 30" dia. | " | 150.00 | 3.20 | 153.20 |
| Hopper | EA. | 1,710 | 25.70 | 1,736 |
| Skylight | " | 1,040 | 38.60 | 1,079 |
| Sprinkler unit at top | " | 380.00 | 42.80 | 422.80 |
| For galvanized metal, deduct from material cost, 35% | | | | |
| For aluminum, deduct from material cost, 25% | | | | |

| 14580.10 PNEUMATIC SYSTEMS | UNIT | MAT. | INST. | TOTAL |
|---|---|---|---|---|
| Pneumatic message tube system | | | | |
| Average, 20 station job | | | | |
| 3" round system | E.A. | 276,100 | 3,300 | |
| 4" round system | " | 345,130 | 3,630 | |
| 6" round system | " | 521,530 | 4,040 | |
| 4" x 7" oval system | " | 544,540 | 7,260 | |
| Trash and linen tube system | | | | |
| 10 stations | EA. | 176,400 | 7,200 | |
| 15 stations | " | 222,420 | 9,610 | |
| 20 stations | " | 298,370 | 11,080 | |
| 30 stations | " | 391,140 | 13,100 | |

## HOISTS AND CRANES

| 14600.10 INDUSTRIAL HOISTS | UNIT | MAT. | INST. | TOTAL |
|---|---|---|---|---|
| Industrial hoists, electric, light to medium duty | | | | |
| 500 lb | EA. | 5,930 | 180.00 | 6,110 |
| 1000 lb | " | 6,220 | 190.00 | 6,410 |
| 2000 lb | " | 6,380 | 200.00 | 6,580 |
| 3000 lb | " | 6,500 | 210.00 | 6,710 |
| 4000 lb | " | 7,350 | 230.00 | 7,580 |
| 5000 lb | " | 10,390 | 240.00 | 10,630 |
| 6000 lb | " | 10,730 | 250.00 | 10,980 |
| 7500 lb | " | 11,470 | 260.00 | 11,730 |
| 10,000 lb | " | 19,370 | 270.00 | 19,640 |
| 15,000 lb | " | 28,760 | 280.00 | 29,040 |
| 20,000 lb | " | 33,970 | 300.00 | 34,270 |
| 25,000 lb | " | 37,250 | 330.00 | 37,580 |
| 30,000 lb | " | 36,520 | 360.00 | 36,880 |
| Heavy duty | | | | |
| 500 lb | EA. | 8,750 | 180.00 | 8,930 |
| 1000 lb | " | 9,010 | 190.00 | 9,200 |
| 2000 lb | " | 10,030 | 200.00 | 10,230 |

## HOISTS AND CRANES

| | UNIT | MAT. | INST. | TOTAL |
|---|---|---|---|---|
| **14600.10 INDUSTRIAL HOISTS** | | | | |
| 3000 lb | EA. | 10,580 | 210.00 | 10,790 |
| 4000 lb | " | 11,180 | 230.00 | 11,410 |
| 5000 lb | " | 11,620 | 240.00 | 11,860 |
| 6000 lb | " | 11,920 | 250.00 | 12,170 |
| 7500 lb | " | 16,390 | 260.00 | 16,650 |
| 10,000 lb | " | 17,140 | 270.00 | 17,410 |
| 15,000 lb | " | 18,920 | 280.00 | 19,200 |
| 20,000 lb | " | 23,390 | 300.00 | 23,690 |
| 25,000 lb | " | 33,970 | 330.00 | 34,300 |
| 30,000 lb | " | 35,460 | 360.00 | 35,820 |
| Air powered hoists | | | | |
| 500 lb | EA. | 3,280 | 180.00 | 3,460 |
| 1000 lb | " | 3,650 | 180.00 | 3,830 |
| 2000 lb | " | 3,990 | 190.00 | 4,180 |
| 4000 lb | " | 4,660 | 210.00 | 4,870 |
| 6000 lb | " | 5,550 | 280.00 | 5,830 |
| Overhead traveling bridge crane | | | | |
| Single girder, 20' span | | | | |
| 3 ton | EA. | 18,630 | 720.00 | 19,350 |
| 5 ton | " | 20,490 | 720.00 | 21,210 |
| 7.5 ton | " | 25,930 | 720.00 | 26,650 |
| 10 ton | " | 25,930 | 900.00 | 26,830 |
| 15 ton | " | 32,040 | 900.00 | 32,940 |
| 30' span | | | | |
| 3 ton | EA. | 19,820 | 720.00 | 20,540 |
| 5 ton | " | 21,900 | 720.00 | 22,620 |
| 10 ton | " | 28,010 | 900.00 | 28,910 |
| 15 ton | " | 33,530 | 900.00 | 34,430 |
| Double girder, 40' span | | | | |
| 3 ton | EA. | 38,440 | 1,600 | 40,040 |
| 5 ton | " | 39,490 | 1,600 | 41,090 |
| 7.5 ton | " | 40,830 | 1,600 | 42,430 |
| 10 ton | " | 42,910 | 2,060 | 44,970 |
| 15 ton | " | 48,280 | 2,060 | 50,340 |
| 25 ton | " | 71,520 | 2,060 | 73,580 |
| 50' span | | | | |
| 3 ton | EA. | 41,950 | 1,600 | 43,550 |
| 5 ton | " | 43,440 | 1,600 | 45,040 |
| 7.5 ton | " | 45,000 | 1,600 | 46,600 |
| 10 ton | " | 48,870 | 2,060 | 50,930 |
| 15 ton | " | 56,620 | 2,060 | 58,680 |
| 25 ton | " | 77,480 | 2,060 | 79,540 |
| **14650.10 JIB CRANES** | | | | |
| Self supporting, swinging 8' boom, 200 deg rotation | | | | |
| 1000 lb | EA. | 1,710 | 320.00 | 2,030 |
| 2000 lb | " | 1,790 | 320.00 | 2,110 |
| 3000 lb | " | 2,090 | 640.00 | 2,730 |
| 4000 lb | " | 2,380 | 640.00 | 3,020 |
| 6000 lb | " | 2,680 | 640.00 | 3,320 |
| 10,000 lb | " | 3,530 | 640.00 | 4,170 |
| Wall mounted, 180 deg rotation | | | | |

# 14 CONVEYING

## HOISTS AND CRANES

| 14650.10 JIB CRANES | UNIT | MAT. | INST. | TOTAL |
|---|---|---|---|---|
| 2000 lb | EA. | 950.00 | 320.00 | 1,270 |
| 3000 lb | " | 1,130 | 320.00 | 1,450 |
| 4000 lb | " | 1,270 | 640.00 | 1,910 |
| 6000 lb | " | 1,430 | 640.00 | 2,070 |
| 10,000 lb | " | 2,830 | 640.00 | 3,470 |

| BASIC MATERIALS | UNIT | MAT. | INST. | TOTAL |
|---|---|---|---|---|
| **15100.10 SPECIALTIES** | | | | |
| Wall penetration | | | | |
| Concrete wall, 6" thick | | | | |
| 2" dia. | EA. | 0.00 | 9.25 | 9.25 |
| 4" dia. | " | 0.00 | 13.90 | 13.90 |
| 8" dia. | " | 0.00 | 19.85 | 19.85 |
| 12" thick | | | | |
| 2" dia. | EA. | 0.00 | 12.60 | 12.60 |
| 4" dia. | " | 0.00 | 19.85 | 19.85 |
| 8" dia. | " | 0.00 | 30.80 | 30.80 |
| Non-destructive testing, piping systems | | | | |
| X-ray of welds | | | | |
| 3" dia. pipe | EA. | 56.00 | 39.00 | 95.00 |
| 4" dia. pipe | " | 48.20 | 39.00 | 87.20 |
| 6" dia. pipe | " | 41.10 | 39.00 | 80.10 |
| 8" dia. pipe | " | 44.00 | 48.80 | 92.80 |
| 10" dia. pipe | " | 46.80 | 48.80 | 95.60 |
| Liquid penetration of welds | | | | |
| 2" dia. pipe | EA. | 28.40 | 24.35 | 52.75 |
| 3" dia. pipe | " | 29.80 | 24.35 | 54.15 |
| 4" dia. pipe | " | 32.60 | 24.35 | 56.95 |
| 6" dia. pipe | " | 32.60 | 24.35 | 56.95 |
| 8" dia. pipe | " | 39.70 | 24.35 | 64.05 |
| 10" dia. pipe | " | 39.70 | 24.35 | 64.05 |
| **15120.10 BACKFLOW PREVENTERS** | | | | |
| Backflow preventer, flanged, cast iron, with valves | | | | |
| 3" pipe | EA. | 2,970 | 200.00 | 3,170 |
| 4" pipe | " | 4,050 | 220.00 | 4,270 |
| 6" pipe | " | 6,080 | 330.00 | 6,410 |
| 8" pipe | " | 10,800 | 390.00 | 11,190 |
| Threaded | | | | |
| 3/4" pipe | EA. | 680.00 | 24.35 | 704.35 |
| 2" pipe | " | 1,220 | 39.00 | 1,259 |
| Reduced pressure assembly, bronze, threaded | | | | |
| 3/4" | EA. | 780.00 | 24.35 | 804.35 |
| 1" | " | 820.00 | 27.90 | 847.90 |
| 1-1/4" | " | 1,190 | 32.50 | 1,223 |
| 1-1/2" | " | 1,220 | 39.00 | 1,259 |
| **15140.10 PIPE HANGERS, HEAVY** | | | | |
| Hangers | | | | |
| 1/2" pipe, clevis pipe hanger | | | | |
| Black steel | EA. | 2.20 | 13.00 | 15.20 |
| Galvanized | " | 2.55 | 13.00 | 15.55 |
| U bolt | " | 1.20 | 3.90 | 5.10 |
| 3/4" pipe, clevis pipe hanger | | | | |
| Black steel | EA. | 2.20 | 13.00 | 15.20 |
| Galvanized | " | 2.60 | 13.00 | 15.60 |
| U bolt | " | 1.20 | 3.90 | 5.10 |
| 1" pipe, clevis pipe hanger | | | | |

## BASIC MATERIALS

### 15140.10 PIPE HANGERS, HEAVY

| | UNIT | MAT. | INST. | TOTAL |
|---|---|---|---|---|
| Black steel | EA. | 2.40 | 13.00 | 15.40 |
| Galvanized | " | 2.75 | 13.00 | 15.75 |
| U bolt | " | 1.25 | 3.90 | 5.15 |
| 1-1/4" pipe, clevis pipe hanger | | | | |
| Black steel | EA. | 2.65 | 13.00 | 15.65 |
| Galvanized | " | 3.10 | 13.00 | 16.10 |
| U bolt | " | 1.45 | 3.90 | 5.35 |
| 1-1/2" pipe, clevis pipe hanger | | | | |
| Black steel | EA. | 2.90 | 13.00 | 15.90 |
| Galvanized | " | 3.50 | 13.00 | 16.50 |
| U bolt | " | 1.50 | 3.90 | 5.40 |
| 2" pipe, clevis pipe hanger | | | | |
| Black steel | EA. | 3.50 | 13.00 | 16.50 |
| Galvanized | " | 3.95 | 13.00 | 16.95 |
| U bolt | " | 1.55 | 3.90 | 5.45 |
| 2-1/2" pipe, clevis pipe hanger | | | | |
| Black steel | EA. | 5.20 | 13.00 | 18.20 |
| Galvanized | " | 6.70 | 13.00 | 19.70 |
| 3" pipe, clevis pipe hanger | | | | |
| Black steel | EA. | 5.95 | 13.00 | 18.95 |
| Galvanized | " | 10.55 | 13.00 | 23.55 |
| U bolt | " | 2.60 | 3.90 | 6.50 |
| 3-1/2" pipe, clevis pipe hanger | | | | |
| Black steel | EA. | 5.20 | 13.00 | 18.20 |
| Galvanized | " | 7.85 | 13.00 | 20.85 |
| U bolt | " | 2.75 | 3.90 | 6.65 |
| 4" pipe, clevis pipe hanger | | | | |
| Black steel | EA. | 8.20 | 13.00 | 21.20 |
| Galvanized | " | 10.40 | 13.00 | 23.40 |
| U bolt | " | 2.75 | 3.90 | 6.65 |
| 5" pipe, clevis pipe hanger | | | | |
| Black steel | EA. | 9.15 | 15.60 | 24.75 |
| Galvanized | " | 11.95 | 15.60 | 27.55 |
| U bolt | " | 2.95 | 3.90 | 6.85 |
| 6" pipe, clevis pipe hanger | | | | |
| Black steel | EA. | 12.85 | 15.60 | 28.45 |
| Galvanized | " | 19.10 | 15.60 | 34.70 |
| U bolt | " | 5.30 | 4.85 | 10.15 |
| 8" pipe, clevis pipe hanger | | | | |
| Black steel | EA. | 21.00 | 15.60 | 36.60 |
| Galvanized | " | 21.20 | 15.60 | 36.80 |
| U bolt | " | 6.40 | 4.85 | 11.25 |
| 10" clevis pipe hanger | | | | |
| Black steel | EA. | 22.20 | 15.60 | 37.80 |
| Galvanized | " | 30.40 | 15.60 | 46.00 |
| 12" pipe, clevis pipe hanger | | | | |
| Black steel | EA. | 20.50 | 15.60 | 36.10 |
| Galvanized | " | 29.10 | 15.60 | 44.70 |
| Threaded rod, galvanized | | | | |
| 3/8" | L.F. | | | 0.41 |
| 1/2" | " | | | 0.68 |
| 5/8" | " | | | 0.99 |

| BASIC MATERIALS | UNIT | MAT. | INST. | TOTAL |
|---|---|---|---|---|
| **15140.10 PIPE HANGERS, HEAVY** | | | | |
| 3/4" | L.F. | | | 1.73 |
| 7/8" | " | | | 2.31 |
| 1" | " | | | 3.17 |
| Hex nuts, galvanized | | | | |
| 3/8" | EA. | | | 0.04 |
| 1/2" | " | | | 0.08 |
| 5/8" | " | | | 0.20 |
| 3/4" | " | | | 0.27 |
| 7/8" | " | | | 0.46 |
| 1" | " | | | 0.72 |
| C-clamp, steel, with lock nut | | | | |
| 3/8" | EA. | 2.05 | 4.85 | 6.90 |
| 1/2" | " | 2.45 | 4.85 | 7.30 |
| 5/8" | " | 3.80 | 4.85 | 8.65 |
| 3/4" | " | 4.85 | 4.85 | 9.70 |
| 7/8" | " | 8.00 | 4.85 | 12.85 |
| Angle support, medium, welded steel | | | | |
| 12"x18" | EA. | 85.50 | 39.00 | 124.50 |
| 18"x24" | " | 100.00 | 39.00 | 139.00 |
| 24"x30" | " | 130.00 | 39.00 | 169.00 |
| Heavy, welded steel | | | | |
| 12"x18" | EA. | 120.00 | 39.00 | 159.00 |
| 18"x24" | " | 170.00 | 39.00 | 209.00 |
| 24"x30" | " | 200.00 | 39.00 | 239.00 |
| **15140.11 PIPE HANGERS, LIGHT** | | | | |
| A band, black iron | | | | |
| 1/2" | EA. | 0.54 | 2.80 | 3.34 |
| 1" | " | 0.68 | 2.90 | 3.58 |
| 1-1/4" | " | 0.74 | 3.00 | 3.74 |
| 1-1/2" | " | 0.85 | 3.25 | 4.10 |
| 2" | " | 0.95 | 3.55 | 4.50 |
| 2-1/2" | " | 1.30 | 3.90 | 5.20 |
| 3" | " | 1.35 | 4.35 | 5.70 |
| 4" | " | 1.95 | 4.85 | 6.80 |
| 5" | " | 2.55 | 5.20 | 7.75 |
| 6" | " | 3.55 | 5.55 | 9.10 |
| 8" | " | 6.10 | 6.50 | 12.60 |
| Copper | | | | |
| 1/2" | EA. | 0.69 | 2.80 | 3.49 |
| 3/4" | " | 0.69 | 2.90 | 3.59 |
| 1" | " | 0.73 | 2.90 | 3.63 |
| 1-1/4" | " | 0.76 | 3.00 | 3.76 |
| 1-1/2" | " | 0.86 | 3.25 | 4.11 |
| 2" | " | 1.00 | 3.55 | 4.55 |
| 2-1/2" | " | 1.35 | 3.90 | 5.25 |
| 3" | " | 1.50 | 4.35 | 5.85 |
| 4" | " | 2.15 | 4.85 | 7.00 |
| Black riser friction hangers | | | | |
| 3/4" | EA. | 2.95 | 3.25 | 6.20 |
| 1" | " | 2.95 | 3.40 | 6.35 |
| 1-1/4" | " | 3.75 | 3.55 | 7.30 |

# 15 MECHANICAL

## BASIC MATERIALS

| 15140.11 PIPE HANGERS, LIGHT | UNIT | MAT. | INST. | TOTAL |
|---|---|---|---|---|
| 1-1/2" | EA. | 4.00 | 3.70 | 7.70 |
| 2" | " | 4.10 | 3.90 | 8.00 |
| 2-1/2" | " | 4.45 | 4.35 | 8.80 |
| 3" | " | 4.60 | 4.85 | 9.45 |
| 4" | " | 5.80 | 5.55 | 11.35 |
| 5" | " | 8.55 | 6.00 | 14.55 |
| 6" | " | 9.75 | 6.50 | 16.25 |
| 8" | " | 16.70 | 7.10 | 23.80 |
| 10" | " | 21.45 | 7.80 | 29.25 |
| 2 hole clips, galvanized | | | | |
| 3/4" | EA. | 0.18 | 2.60 | 2.78 |
| 1" | " | 0.20 | 2.70 | 2.90 |
| 1-1/4" | " | 0.26 | 2.80 | 3.06 |
| 1-1/2" | " | 0.32 | 2.90 | 3.22 |
| 2" | " | 0.43 | 3.00 | 3.43 |
| 2-1/2" | " | 0.78 | 3.10 | 3.88 |
| 3" | " | 1.15 | 3.25 | 4.40 |
| 4" | " | 2.45 | 3.55 | 6.00 |
| Perforated strap | | | | |
| 3/4" | | | | |
| Galvanized, 20 ga. | L.F. | 0.27 | 1.95 | 2.22 |
| Copper, 22 ga. | " | 0.32 | 1.95 | 2.27 |
| J-Hooks | | | | |
| 1/2" | EA. | 0.30 | 1.75 | 2.05 |
| 3/4" | " | 0.32 | 1.75 | 2.07 |
| 1" | " | 0.34 | 1.85 | 2.19 |
| 1-1/4" | " | 0.35 | 1.90 | 2.25 |
| 1-1/2" | " | 0.36 | 1.95 | 2.31 |
| 2" | " | 0.38 | 1.95 | 2.33 |
| 3" | " | 0.45 | 2.05 | 2.50 |
| 4" | " | 0.45 | 2.05 | 2.50 |
| PVC coated hangers, galvanized, 28 ga. | | | | |
| 1-1/2" x 12" | EA. | 0.89 | 2.60 | 3.49 |
| 2" x 12" | " | 0.99 | 2.80 | 3.79 |
| 3" x 12" | " | 1.10 | 3.00 | 4.10 |
| 4" x 12" | " | 1.20 | 3.25 | 4.45 |
| Copper, 30 ga. | | | | |
| 1-1/2" x 12" | EA. | 1.10 | 2.60 | 3.70 |
| 2" x 12" | " | 1.30 | 2.80 | 4.10 |
| 3" x 12" | " | 1.45 | 3.00 | 4.45 |
| 4" x 12" | " | 1.55 | 3.25 | 4.80 |
| 2" x 24" | " | 2.50 | 3.00 | 5.50 |
| 3" x 24" | " | 2.90 | 3.25 | 6.15 |
| 4" x 24" | " | 4.50 | 3.55 | 8.05 |
| Wire hook hangers | | | | |
| Black wire, 1/2" x | | | | |
| 4" | EA. | 0.31 | 1.95 | 2.26 |
| 6" | " | 0.36 | 2.05 | 2.41 |
| 8" | " | 0.41 | 2.15 | 2.56 |
| 10" | " | 0.50 | 2.15 | 2.65 |
| 12" | " | 0.61 | 2.30 | 2.91 |
| 3/4" x | | | | |

## BASIC MATERIALS

### 15140.11 PIPE HANGERS, LIGHT

| | UNIT | MAT. | INST. | TOTAL |
|---|---|---|---|---|
| 4" | EA. | 0.39 | 2.05 | 2.44 |
| 6" | " | 0.42 | 2.15 | 2.57 |
| 8" | " | 0.43 | 2.30 | 2.73 |
| 10" | " | 0.58 | 2.45 | 3.03 |
| 12" | " | 0.59 | 2.60 | 3.19 |
| 1" x | | | | |
| 4" | EA. | 0.39 | 2.15 | 2.54 |
| 6" | " | 0.41 | 2.30 | 2.71 |
| 8" | " | 0.43 | 2.45 | 2.88 |
| 10" | " | 0.58 | 2.60 | 3.18 |
| 12" | " | 0.63 | 2.80 | 3.43 |
| 1-1/4" x | | | | |
| 4" | EA. | 0.42 | 2.30 | 2.72 |
| 6" | " | 0.43 | 2.45 | 2.88 |
| 8" | " | 0.49 | 2.60 | 3.09 |
| 10" | " | 0.55 | 2.80 | 3.35 |
| 12" | " | 0.63 | 3.00 | 3.63 |
| 1-1/2" x | | | | |
| 6" | EA. | 0.47 | 2.60 | 3.07 |
| 8" | " | 0.49 | 2.80 | 3.29 |
| 10" | " | 0.61 | 3.00 | 3.61 |
| 12" | " | 0.66 | 3.25 | 3.91 |
| 2" x | | | | |
| 6" | EA. | 0.49 | 2.80 | 3.29 |
| 8" | " | 0.58 | 3.00 | 3.58 |
| 10" | " | 0.59 | 3.25 | 3.84 |
| 12" | " | 0.69 | 3.55 | 4.24 |
| Copper wire hooks | | | | |
| 1/2" x | | | | |
| 4" | EA. | 0.38 | 1.95 | 2.33 |
| 6" | " | 0.43 | 2.05 | 2.48 |
| 8" | " | 0.49 | 2.15 | 2.64 |
| 10" | " | 0.61 | 2.30 | 2.91 |
| 12" | " | 0.70 | 2.45 | 3.15 |
| 3/4" x | | | | |
| 4" | EA. | 0.38 | 2.05 | 2.43 |
| 6" | " | 0.47 | 2.15 | 2.62 |
| 8" | " | 0.54 | 2.30 | 2.84 |
| 10" | " | 0.62 | 2.45 | 3.07 |
| 12" | " | 0.74 | 2.60 | 3.34 |
| 1" x | | | | |
| 4" | EA. | 0.41 | 2.15 | 2.56 |
| 6" | " | 0.46 | 2.30 | 2.76 |
| 8" | " | 0.58 | 2.45 | 3.03 |
| 10" | " | 0.69 | 2.60 | 3.29 |
| 12" | " | 0.81 | 2.80 | 3.61 |
| 1-1/4" x | | | | |
| 6" | EA. | 0.53 | 2.30 | 2.83 |
| 8" | " | 0.61 | 2.45 | 3.06 |
| 10" | " | 0.76 | 2.60 | 3.36 |
| 12" | " | 0.80 | 2.80 | 3.60 |
| 1-1/2" x | | | | |

# 15 MECHANICAL

| BASIC MATERIALS | UNIT | MAT. | INST. | TOTAL |
|---|---|---|---|---|
| **15140.11 PIPE HANGERS, LIGHT** | | | | |
| 6" | EA. | 0.63 | 2.60 | 3.23 |
| 8" | " | 0.68 | 2.80 | 3.48 |
| 10" | " | 0.76 | 3.00 | 3.76 |
| 12" | " | 0.84 | 3.25 | 4.09 |
| 2" x | | | | |
| 6" | EA. | 0.66 | 2.80 | 3.46 |
| 8" | " | 0.72 | 3.00 | 3.72 |
| 10" | " | 0.77 | 3.25 | 4.02 |
| 12" | " | 0.84 | 3.55 | 4.39 |
| **15175.60 EXPANSION TANKS** | | | | |
| Expansion tank, 125 psi, steel | | | | |
| 20 gallon | EA. | 450.00 | 48.80 | 498.80 |
| 30 gallon | " | 470.00 | 65.00 | 535.00 |
| 65 gallon | " | 590.00 | 97.50 | 687.50 |
| 80 gallon | " | 670.00 | 110.00 | 780.00 |
| **15240.10 VIBRATION CONTROL** | | | | |
| Vibration isolator, in-line, stainless connector, screwed | | | | |
| 1/2" | EA. | 70.00 | 21.65 | 91.65 |
| 3/4" | " | 81.50 | 22.95 | 104.45 |
| 1" | " | 87.00 | 24.35 | 111.35 |
| 1-1/4" | " | 98.50 | 26.00 | 124.50 |
| 1-1/2" | " | 110.00 | 27.90 | 137.90 |
| 2" | " | 140.00 | 30.00 | 170.00 |
| 2-1/2" | " | 320.00 | 32.50 | 352.50 |
| 3" | " | 370.00 | 35.50 | 405.50 |
| 4" | " | 490.00 | 39.00 | 529.00 |
| 6" | " | 730.00 | 43.30 | 773.30 |
| Flanged | | | | |
| 8" | EA. | 1,130 | 48.80 | 1,179 |
| 10" | " | 1,900 | 55.50 | 1,956 |
| 12" | " | 2,390 | 65.00 | 2,455 |

| INSULATION | UNIT | MAT. | INST. | TOTAL |
|---|---|---|---|---|
| **15260.10 FIBERGLASS PIPE INSULATION** | | | | |
| Fiberglass insulation on 1/2" pipe | | | | |
| 1" thick | L.F. | 1.20 | 1.30 | 2.50 |
| 1-1/2" thick | " | 2.95 | 1.60 | 4.55 |
| 3/4" pipe | | | | |

| INSULATION | UNIT | MAT. | INST. | TOTAL |
|---|---|---|---|---|
| **15260.10 FIBERGLASS PIPE INSULATION** | | | | |
| 1" thick | L.F. | 2.60 | 1.30 | 3.90 |
| 1-1/2" thick | " | 3.15 | 1.60 | 4.75 |
| 1" pipe | | | | |
| 1" thick | L.F. | 2.70 | 1.30 | 4.00 |
| 1-1/2" thick | " | 3.40 | 1.60 | 5.00 |
| 2" pipe | | | | |
| 1" thick | L.F. | 3.80 | 1.60 | 5.40 |
| 1-1/2" thick | " | 4.45 | 1.75 | 6.20 |
| 3" pipe | | | | |
| 1" thick | L.F. | 4.30 | 1.85 | 6.15 |
| 1-1/2" thick | " | 5.25 | 1.95 | 7.20 |
| 6" pipe | | | | |
| 1" thick | L.F. | 6.65 | 2.05 | 8.70 |
| 2" thick | " | 10.45 | 2.15 | 12.60 |
| 10" pipe | | | | |
| 2" thick | L.F. | 15.55 | 2.05 | 17.60 |
| 3" thick | " | 21.60 | 2.15 | 23.75 |
| **15260.20 CALCIUM SILICATE** | | | | |
| Calcium silicate insulation, 6" pipe | | | | |
| 2" thick | L.F. | 8.45 | 2.80 | 11.25 |
| 2-1/2" thick | " | 12.90 | 3.00 | 15.90 |
| 3" thick | " | 18.70 | 3.25 | 21.95 |
| 4" thick | " | 25.40 | 3.55 | 28.95 |
| 6" thick | " | 41.00 | 3.90 | 44.90 |
| 12" pipe | | | | |
| 2" thick | L.F. | 14.10 | 3.00 | 17.10 |
| 2-1/2" thick | " | 21.05 | 3.25 | 24.30 |
| 3" thick | " | 31.10 | 3.55 | 34.65 |
| 4" thick | " | 39.30 | 3.90 | 43.20 |
| 6" thick | " | 69.00 | 4.35 | 73.35 |
| **15260.60 EXTERIOR PIPE INSULATION** | | | | |
| Fiberglass insulation, aluminum jacket | | | | |
| 1/2" pipe | | | | |
| 1" thick | L.F. | 3.45 | 3.00 | 6.45 |
| 1-1/2" thick | " | 4.05 | 3.25 | 7.30 |
| 3/4" pipe | | | | |
| 1" thick | L.F. | 3.60 | 3.00 | 6.60 |
| 1-1/2" thick | " | 4.20 | 3.25 | 7.45 |
| 1" pipe | | | | |
| 1" thick | L.F. | 3.70 | 3.00 | 6.70 |
| 1-1/2" thick | " | 4.55 | 3.25 | 7.80 |
| 2" pipe | | | | |
| 1" thick | L.F. | 4.70 | 3.55 | 8.25 |
| 1-1/2" thick | " | 5.65 | 3.70 | 9.35 |
| 3" pipe | | | | |
| 1" thick | L.F. | 5.25 | 3.90 | 9.15 |
| 1-1/2" thick | " | 6.20 | 4.10 | 10.30 |
| 6" pipe | | | | |
| 1" thick | L.F. | 6.80 | 4.35 | 11.15 |

| INSULATION | UNIT | MAT. | INST. | TOTAL |
|---|---|---|---|---|
| **15260.60 EXTERIOR PIPE INSULATION** | | | | |
| 2" thick | L.F. | 11.30 | 4.60 | 15.90 |
| 10" pipe | | | | |
| 2" thick | L.F. | 17.20 | 4.35 | 21.55 |
| 3" thick | " | 24.80 | 4.60 | 29.40 |
| **15260.90 PIPE INSULATION FITTINGS** | | | | |
| Insulation protection saddle | | | | |
| 1" thick covering | | | | |
| 1/2" pipe | EA. | 4.30 | 15.60 | 19.90 |
| 3/4" pipe | " | 4.30 | 15.60 | 19.90 |
| 1" pipe | " | 4.30 | 15.60 | 19.90 |
| 2" pipe | " | 6.20 | 15.60 | 21.80 |
| 3" pipe | " | 7.00 | 17.75 | 24.75 |
| 6" pipe | " | 8.80 | 24.35 | 33.15 |
| 1-1/2" thick covering | | | | |
| 3/4" pipe | EA. | 7.10 | 15.60 | 22.70 |
| 1" pipe | " | 7.10 | 15.60 | 22.70 |
| 2" pipe | " | 7.95 | 15.60 | 23.55 |
| 3" pipe | " | 8.95 | 15.60 | 24.55 |
| 6" pipe | " | 13.20 | 24.35 | 37.55 |
| 10" pipe | " | 16.60 | 32.50 | 49.10 |
| **15280.10 EQUIPMENT INSULATION** | | | | |
| Equipment insulation, 2" thick, cellular glass | S.F. | 2.30 | 2.45 | 4.75 |
| Urethane, rigid, field applied jacket, plastered finish | " | 2.45 | 4.85 | 7.30 |
| Fiberglass, rigid, with vapor barrier | " | 2.30 | 2.15 | 4.45 |
| **15290.10 DUCTWORK INSULATION** | | | | |
| Fiberglass duct insulation, plain blanket | | | | |
| 1-1/2" thick | S.F. | 0.69 | 0.49 | 1.18 |
| 2" thick | " | 0.82 | 0.65 | 1.47 |
| With vapor barrier | | | | |
| 1-1/2" thick | S.F. | 0.42 | 0.49 | 0.91 |
| 2" thick | " | 0.49 | 0.65 | 1.14 |
| Rigid with vapor barrier | | | | |
| 2" thick | S.F. | 1.40 | 1.30 | 2.70 |
| 3" thick | " | 2.30 | 1.55 | 3.85 |
| 4" thick | " | 3.00 | 1.95 | 4.95 |
| 6" thick | " | 4.15 | 2.60 | 6.75 |
| Weatherproof, polystyrene, 3" thick, w/vapor barrier | " | 3.05 | 3.90 | 6.95 |
| Urethane board with vapor barrier | " | 4.35 | 4.85 | 9.20 |

| FIRE PROTECTION | UNIT | MAT. | INST. | TOTAL |
|---|---|---|---|---|
| **15330.10 WET SPRINKLER SYSTEM** | | | | |
| Sprinkler head, 212 deg, brass, exposed piping | EA. | 6.45 | 15.60 | 22.05 |
| Chrome, concealed piping | " | 6.20 | 21.65 | 27.85 |
| Water motor alarm | " | 86.50 | 65.00 | 151.50 |
| Fire department inlet connection | " | 130.00 | 78.00 | 208.00 |
| Wall plate for fire dept connection | " | 63.50 | 32.50 | 96.00 |
| Swing check valve flanged iron body, 4" | " | 130.00 | 130.00 | 260.00 |
| Check valve, 6" | " | 610.00 | 200.00 | 810.00 |
| Wet pipe valve, flange to groove, 4" | " | 530.00 | 43.30 | 573.30 |
| Flange to flange | | | | |
| 6" | EA. | 680.00 | 65.00 | 745.00 |
| 8" | " | 950.00 | 130.00 | 1,080 |
| Alarm valve, flange to flange, (wet valve) | | | | |
| 4" | EA. | 630.00 | 43.30 | 673.30 |
| 8" | " | 970.00 | 330.00 | 1,300 |
| Inspector's test connection | " | 40.50 | 32.50 | 73.00 |
| Wall hydrant, polished brass, 2-1/2" x 2-1/2", single | " | 320.00 | 27.90 | 347.90 |
| 2-way | " | 730.00 | 27.90 | 757.90 |
| 3-way | " | 1,490 | 27.90 | 1,518 |
| Wet valve trim, includes retard chamber & gauges, 4"-6" | " | 510.00 | 32.50 | 542.50 |
| Retard pressure switch for wet systems | " | 890.00 | 78.00 | 968.00 |
| Air maintenance device | " | 270.00 | 32.50 | 302.50 |
| Wall hydrant non-freeze, 8" thick wall, vacuum breaker | " | 29.70 | 19.50 | 49.20 |
| 12" thick wall | " | 32.40 | 19.50 | 51.90 |
| **15330.50 DRY SPRINKLER SYSTEM** | | | | |
| Dry pipe valve, flange to flange | | | | |
| 4" | EA. | 1,030 | 78.00 | 1,108 |
| 6" | " | 1,300 | 97.50 | 1,398 |
| Trim, 4" and 6", includes gauges | " | 490.00 | 32.50 | 522.50 |
| Field testing and flushing | " | 0.00 | 330.00 | 330.00 |
| Disinfection | " | 0.00 | 330.00 | 330.00 |
| Pressure switch double circuit, open/close contacts | " | 240.00 | 97.50 | 337.50 |
| Low air | | | | |
| Supervisory unit | EA. | 800.00 | 65.00 | 865.00 |
| Pressure switch | " | 260.00 | 32.50 | 292.50 |
| **15330.70 CO2 SYSTEM** | | | | |
| CO2 system, high pressure, 75# cylinder with | | | | |
| Valve assemblies | EA. | 1,620 | 78.00 | 1,698 |
| Storage rack | " | 860.00 | 55.50 | 915.50 |
| Manifold | " | 630.00 | 280.00 | 910.00 |
| Flexible loops | " | 51.50 | 4.85 | 56.35 |
| Beam scale for cylinders | " | 430.00 | 65.00 | 495.00 |
| Mechanically control head | " | 360.00 | 26.00 | 386.00 |
| Electrically control head | " | 360.00 | 26.00 | 386.00 |
| Stop valves | " | 920.00 | 39.00 | 959.00 |
| Check valves | " | 420.00 | 48.80 | 468.80 |
| Activation station | " | 490.00 | 39.00 | 529.00 |
| Nozzles | " | 78.50 | 32.50 | 111.00 |
| Hose reel with 75' of 3/4" hose | " | 2,900 | 200.00 | 3,100 |
| Main/reserve transfer switch | " | 3,380 | 65.00 | 3,445 |

## FIRE PROTECTION

| FIRE PROTECTION | UNIT | MAT. | INST. | TOTAL |
|---|---|---|---|---|
| **15330.70 CO2 SYSTEM** | | | | |
| Pressure switch | EA. | 300.00 | 39.00 | 339.00 |
| Heat responsive device | " | 470.00 | 65.00 | 535.00 |
| Battery and charger | " | 2,700 | 200.00 | 2,900 |
| Low pressure | | | | |
| Battery and charger | EA. | 2,700 | 200.00 | 2,900 |
| Pressure switch | " | 270.00 | 43.30 | 313.30 |
| Nozzles | " | 79.50 | 35.50 | 115.00 |
| Master selector valve | " | 230.00 | 65.00 | 295.00 |
| Selector valve | " | 3,380 | 65.00 | 3,445 |
| Low pressure hose reel with 75' of 3/4" hose | " | 4,190 | 200.00 | 4,390 |
| Tank fill lines | " | 950.00 | 48.80 | 998.80 |
| Activation stations | " | 470.00 | 32.50 | 502.50 |
| Electro manual pilot panels | " | 950.00 | 48.80 | 998.80 |
| **15330.90 HALON TYPE SYSTEM** | | | | |
| Halon type fire protection system, per computer room | | | | |
| Minimum | EA. | | | 3,167 |
| Average | " | | | 4,269 |
| Maximum | " | | | 7,849 |

## PLUMBING

| PLUMBING | UNIT | MAT. | INST. | TOTAL |
|---|---|---|---|---|
| **15410.05 C.I. PIPE, ABOVE GROUND** | | | | |
| No hub pipe | | | | |
| 1-1/2" pipe | L.F. | 5.25 | 2.80 | 8.05 |
| 2" pipe | " | 6.00 | 3.25 | 9.25 |
| 3" pipe | " | 8.45 | 3.90 | 12.35 |
| 4" pipe | " | 11.55 | 6.50 | 18.05 |
| 6" pipe | " | 17.35 | 7.80 | 25.15 |
| 8" pipe | " | 26.00 | 13.00 | 39.00 |
| 10" pipe | " | 40.50 | 15.60 | 56.10 |
| No hub fittings, 1-1/2" pipe | | | | |
| 1/4 bend | EA. | 6.80 | 13.00 | 19.80 |
| 1/8 bend | " | 4.85 | 13.00 | 17.85 |
| Sanitary tee | " | 8.35 | 19.50 | 27.85 |
| Sanitary cross | " | 10.30 | 19.50 | 29.80 |
| Plug | " | | | 1.51 |
| Coupling | " | | | 9.09 |
| Wye | " | 9.10 | 19.50 | 28.60 |
| 2" pipe | | | | |
| 1/4 bend | EA. | 6.95 | 15.60 | 22.55 |
| 1/8 bend | " | 5.00 | 15.60 | 20.60 |
| Sanitary tee | " | 8.65 | 26.00 | 34.65 |

# 15 MECHANICAL

## PLUMBING

### 15410.05 C.I. PIPE, ABOVE GROUND

| | UNIT | MAT. | INST. | TOTAL |
|---|---|---|---|---|
| Sanitary cross | EA. | 10.90 | 26.00 | 36.90 |
| Plug | " | | | 1.81 |
| Coupling | " | | | 9.09 |
| Wye | " | 9.70 | 32.50 | 42.20 |
| 3" pipe | | | | |
| 1/4 bend | EA. | 8.80 | 19.50 | 28.30 |
| 1/8 bend | " | 7.85 | 19.50 | 27.35 |
| Sanitary tee | " | 10.30 | 24.35 | 34.65 |
| 3"x2" sanitary tee | " | 10.00 | 24.35 | 34.35 |
| 3"x1-1/2" sanitary tee | " | 10.15 | 24.35 | 34.50 |
| Sanitary cross | " | 15.45 | 32.50 | 47.95 |
| 3x2" sanitary cross | " | 17.05 | 32.50 | 49.55 |
| Plug | " | | | 2.73 |
| Coupling | " | | | 10.60 |
| Wye | " | 10.60 | 32.50 | 43.10 |
| 4" pipe | | | | |
| 1/4 bend | EA. | 12.10 | 19.50 | 31.60 |
| 1/8 bend | " | 10.30 | 19.50 | 29.80 |
| Sanitary tee | " | 15.15 | 32.50 | 47.65 |
| 4x3" sanitary tee | " | 13.65 | 32.50 | 46.15 |
| 4x2" sanitary tee | " | 12.10 | 32.50 | 44.60 |
| Sanitary cross | " | 33.30 | 39.00 | 72.30 |
| 4x3" sanitary cross | " | 25.70 | 39.00 | 64.70 |
| 4x2" sanitary cross | " | 22.70 | 39.00 | 61.70 |
| Plug | " | | | 4.40 |
| Coupling | " | | | 11.83 |
| Wye | " | 15.15 | 32.50 | 47.65 |
| 8" deep | " | 21.20 | 19.50 | 40.70 |
| 6" pipe | | | | |
| 1/4 bend | EA. | 28.80 | 32.50 | 61.30 |
| 1/8 bend | " | 21.20 | 32.50 | 53.70 |
| Sanitary tee | " | 45.40 | 39.00 | 84.40 |
| 6x4" sanitary tee | " | 36.40 | 39.00 | 75.40 |
| Coupling | " | | | 24.23 |
| Wye | " | 36.40 | 39.00 | 75.40 |
| 8" pipe | | | | |
| 1/4 bend | EA. | 59.00 | 32.50 | 91.50 |
| 1/8 bend | " | 53.00 | 32.50 | 85.50 |
| Sanitary tee | " | 110.00 | 48.80 | 158.80 |
| 8x6" sanitary tee | " | 75.50 | 48.80 | 124.30 |
| Plug | " | | | 16.36 |
| Coupling | " | | | 27.27 |
| Wye | " | 72.50 | 39.00 | 111.50 |
| 10" pipe | | | | |
| 1/4 bend | EA. | 110.00 | 32.50 | 142.50 |
| 1/8 bend | " | 100.00 | 32.50 | 132.50 |
| Plug | " | | | 27.27 |
| Coupling | " | | | 37.87 |
| Wye | " | 150.00 | 65.00 | 215.00 |

## PLUMBING

### 15410.06 C.I. PIPE, BELOW GROUND

| PLUMBING | UNIT | MAT. | INST. | TOTAL |
|---|---|---|---|---|
| No hub pipe | | | | |
| 1-1/2" pipe | L.F. | 5.50 | 1.95 | 7.45 |
| 2" pipe | " | 6.00 | 2.15 | 8.15 |
| 3" pipe | " | 8.45 | 2.45 | 10.90 |
| 4" pipe | " | 11.55 | 3.25 | 14.80 |
| 6" pipe | " | 17.35 | 3.55 | 20.90 |
| 8" pipe | " | 26.00 | 4.35 | 30.35 |
| 10" pipe | " | 40.50 | 4.85 | 45.35 |
| Fittings, 1-1/2" | | | | |
| 1/4 bend | EA. | 6.80 | 11.15 | 17.95 |
| 1/8 bend | " | 5.30 | 11.15 | 16.45 |
| Plug | " | | | 1.51 |
| Wye | " | 9.10 | 15.60 | 24.70 |
| Wye & 1/8 bend | " | 9.10 | 11.15 | 20.25 |
| P-trap | " | 13.20 | 11.15 | 24.35 |
| 2" | | | | |
| 1/4 bend | EA. | 6.95 | 13.00 | 19.95 |
| 1/8 bend | " | 5.00 | 13.00 | 18.00 |
| Plug | " | | | 1.81 |
| Double wye | " | 12.40 | 24.35 | 36.75 |
| Wye & 1/8 bend | " | 12.25 | 19.50 | 31.75 |
| Double wye & 1/8 bend | " | 18.15 | 24.35 | 42.50 |
| P-trap | " | 13.20 | 13.00 | 26.20 |
| 3" | | | | |
| 1/4 bend | EA. | 8.80 | 15.60 | 24.40 |
| 1/8 bend | " | 7.85 | 15.60 | 23.45 |
| Plug | " | | | 2.73 |
| Wye | " | 10.60 | 24.35 | 34.95 |
| 3x2" wye | " | 10.90 | 24.35 | 35.25 |
| Wye & 1/8 bend | " | 13.65 | 24.35 | 38.00 |
| Double wye & 1/8 bend | " | 25.00 | 24.35 | 49.35 |
| 3x2" double wye & 1/8 bend | " | 18.95 | 24.35 | 43.30 |
| 3x2" reducer | " | 5.75 | 15.60 | 21.35 |
| P-trap | " | 15.90 | 15.60 | 31.50 |
| 4" | | | | |
| 1/4 bend | EA. | 12.10 | 15.60 | 27.70 |
| 1/8 bend | " | 10.30 | 15.60 | 25.90 |
| Plug | " | | | 4.40 |
| Wye | " | 15.15 | 24.35 | 39.50 |
| 4x3" wye | " | 12.90 | 24.35 | 37.25 |
| 4x2" wye | " | 9.55 | 24.35 | 33.90 |
| Double wye | " | 34.80 | 32.50 | 67.30 |
| 4x3" double wye | " | 22.70 | 32.50 | 55.20 |
| 4x2" double wye | " | 18.15 | 32.50 | 50.65 |
| Wye & 1/8 bend | " | 19.70 | 24.35 | 44.05 |
| Double wye & 1/8 bend | " | 45.40 | 32.50 | 77.90 |
| 6" | | | | |
| 1/4 bend | EA. | 28.80 | 24.35 | 53.15 |
| 1/8 bend | " | 21.20 | 24.35 | 45.55 |
| Wye & 1/8 bend | " | 47.00 | 32.50 | 79.50 |
| P-trap | " | 57.50 | 19.50 | 77.00 |
| 8" | | | | |

| PLUMBING | UNIT | MAT. | INST. | TOTAL |
|---|---|---|---|---|
| **15410.06 C.I. PIPE, BELOW GROUND** | | | | |
| 1/4 bend | EA. | 59.00 | 24.35 | 83.35 |
| 1/8 bend | " | 53.00 | 24.35 | 77.35 |
| Plug | " | | | 16.36 |
| Wye | " | 72.50 | 32.50 | 105.00 |
| 10" | | | | |
| 1/4 bend | EA. | 110.00 | 24.35 | 134.35 |
| 1/8 bend | " | 100.00 | 24.35 | 124.35 |
| Plug | " | | | 27.27 |
| Wye | " | 150.00 | 48.80 | 198.80 |
| **15410.09 SERVICE WEIGHT PIPE** | | | | |
| Service weight pipe, single hub | | | | |
| 3" x 5' | EA. | 34.70 | 8.30 | 43.00 |
| 4" x 5' | " | 41.90 | 8.65 | 50.55 |
| 6" x 5' | " | 66.50 | 9.75 | 76.25 |
| 1/8 bend | | | | |
| 3" | EA. | 7.50 | 15.60 | 23.10 |
| 4" | " | 10.40 | 17.75 | 28.15 |
| 6" | " | 16.65 | 19.50 | 36.15 |
| 1/4 bend | | | | |
| 3" | EA. | 9.40 | 15.60 | 25.00 |
| 4" | " | 13.15 | 17.75 | 30.90 |
| 6" | " | 22.40 | 19.50 | 41.90 |
| Sweep | | | | |
| 3" | EA. | 14.45 | 15.60 | 30.05 |
| 4" | " | 20.25 | 17.75 | 38.00 |
| 6" | " | 36.20 | 19.50 | 55.70 |
| Sanitary T | | | | |
| 3" | EA. | 15.20 | 27.90 | 43.10 |
| 4" | " | 18.80 | 32.50 | 51.30 |
| 6" | " | 37.60 | 35.50 | 73.10 |
| Wye | | | | |
| 3" | EA. | 16.65 | 21.65 | 38.30 |
| 4" | " | 20.95 | 22.95 | 43.90 |
| 6" | " | 43.40 | 27.90 | 71.30 |
| **15410.10 COPPER PIPE** | | | | |
| Type "K" copper | | | | |
| 1/2" | L.F. | 2.20 | 1.20 | 3.40 |
| 3/4" | " | 4.05 | 1.30 | 5.35 |
| 1" | " | 5.10 | 1.40 | 6.50 |
| 1-1/4" | " | 6.40 | 1.50 | 7.90 |
| 1-1/2" | " | 8.10 | 1.60 | 9.70 |
| 2" | " | 11.10 | 1.75 | 12.85 |
| 2-1/2" | " | 14.40 | 1.95 | 16.35 |
| 3" | " | 19.50 | 2.05 | 21.55 |
| 4" | " | 31.40 | 2.15 | 33.55 |
| DWV, copper | | | | |
| 1-1/4" | L.F. | 4.05 | 1.60 | 5.65 |
| 1-1/2" | " | 5.25 | 1.75 | 7.00 |
| 2" | " | 5.90 | 1.95 | 7.85 |

## PLUMBING

### 15410.10 COPPER PIPE

| PLUMBING | UNIT | MAT. | INST. | TOTAL |
|---|---|---|---|---|
| 3" | L.F. | 11.10 | 2.15 | 13.25 |
| 4" | " | 18.30 | 2.45 | 20.75 |
| 6" | " | 59.00 | 2.80 | 61.80 |
| Refrigeration tubing, copper, sealed | | | | |
| 1/8" | L.F. | 0.59 | 1.55 | 2.14 |
| 3/16" | " | 0.63 | 1.60 | 2.23 |
| 1/4" | " | 0.73 | 1.70 | 2.43 |
| 5/16" | " | 0.93 | 1.75 | 2.68 |
| 3/8" | " | 1.05 | 1.85 | 2.90 |
| 1/2" | " | 1.40 | 1.95 | 3.35 |
| 7/8" | " | 2.35 | 2.25 | 4.60 |
| 1-1/8" | " | 3.80 | 2.60 | 6.40 |
| 1-3/8" | " | 4.60 | 3.00 | 7.60 |
| Type "L" copper | | | | |
| 1/4" | L.F. | 1.30 | 1.15 | 2.45 |
| 3/8" | " | 1.45 | 1.15 | 2.60 |
| 1/2" | " | 2.20 | 1.20 | 3.40 |
| 3/4" | " | 2.50 | 1.30 | 3.80 |
| 1" | " | 3.40 | 1.40 | 4.80 |
| 1-1/4" | " | 4.60 | 1.50 | 6.10 |
| 1-1/2" | " | 5.50 | 1.60 | 7.10 |
| 2" | " | 8.20 | 1.75 | 9.95 |
| Type "M" copper | | | | |
| 1/2" | L.F. | 1.55 | 1.20 | 2.75 |
| 3/4" | " | 2.20 | 1.30 | 3.50 |
| 1" | " | 2.75 | 1.40 | 4.15 |
| 1-1/4" | " | 3.95 | 1.50 | 5.45 |
| 2" | " | 7.60 | 1.75 | 9.35 |

### 15410.11 COPPER FITTINGS

| PLUMBING | UNIT | MAT. | INST. | TOTAL |
|---|---|---|---|---|
| Coupling, with stop | | | | |
| 1/4" | EA. | 0.46 | 13.00 | 13.46 |
| 3/8" | " | 0.76 | 15.60 | 16.36 |
| 1/2" | " | 0.90 | 16.95 | 17.85 |
| 5/8" | " | 1.15 | 19.50 | 20.65 |
| 3/4" | " | 1.30 | 21.65 | 22.95 |
| 1" | " | 2.60 | 22.95 | 25.55 |
| 3" | " | 23.55 | 39.00 | 62.55 |
| 4" | " | 47.10 | 48.80 | 95.90 |
| Reducing coupling | | | | |
| 1/4" x 1/8" | EA. | 1.50 | 15.60 | 17.10 |
| 3/8" x 1/4" | " | 1.65 | 16.95 | 18.60 |
| 1/2" x | | | | |
| 3/8" | EA. | 1.55 | 19.50 | 21.05 |
| 1/4" | " | 1.55 | 19.50 | 21.05 |
| 1/8" | " | 1.55 | 19.50 | 21.05 |
| 3/4" x | | | | |
| 3/8" | EA. | 2.15 | 21.65 | 23.80 |
| 1/2" | " | 2.20 | 21.65 | 23.85 |
| 1" x | | | | |
| 3/8" | EA. | 4.45 | 24.35 | 28.80 |
| 1" x 1/2" | " | 4.60 | 24.35 | 28.95 |

# 15 MECHANICAL

| PLUMBING | UNIT | MAT. | INST. | TOTAL |
|---|---|---|---|---|
| **15410.11 COPPER FITTINGS** | | | | |
| 1" x 3/4" | EA. | 4.70 | 24.35 | 29.05 |
| 1-1/4" x | | | | |
| 1/2" | EA. | 5.35 | 26.00 | 31.35 |
| 3/4" | " | 5.50 | 26.00 | 31.50 |
| 1" | " | 5.75 | 26.00 | 31.75 |
| 1-1/2" x | | | | |
| 1/2" | EA. | 9.15 | 27.90 | 37.05 |
| 3/4" | " | 9.30 | 27.90 | 37.20 |
| 1" | " | 9.40 | 27.90 | 37.30 |
| 1-1/4" | " | 9.55 | 27.90 | 37.45 |
| 2" x | | | | |
| 1/2" | EA. | 13.75 | 32.50 | 46.25 |
| 3/4" | " | 14.00 | 32.50 | 46.50 |
| 1" | " | 14.40 | 32.50 | 46.90 |
| 1-1/4" | " | 14.65 | 32.50 | 47.15 |
| 1-1/2" | " | 14.80 | 32.50 | 47.30 |
| 2-1/2" x | | | | |
| 1" | EA. | 30.10 | 39.00 | 69.10 |
| 1-1/4" | " | 30.70 | 39.00 | 69.70 |
| 1-1/2" | " | 30.70 | 39.00 | 69.70 |
| 2" | " | 31.40 | 39.00 | 70.40 |
| 3" x | | | | |
| 1-1/2" | EA. | 37.90 | 48.80 | 86.70 |
| 2" | " | 39.20 | 48.80 | 88.00 |
| 2-1/2" | " | 40.60 | 48.80 | 89.40 |
| 4" x | | | | |
| 2" | EA. | 80.00 | 55.50 | 135.50 |
| 2-1/2" | " | 81.00 | 55.50 | 136.50 |
| 3" | " | 82.50 | 55.50 | 138.00 |
| Slip coupling | | | | |
| 1/4" | EA. | 0.39 | 13.00 | 13.39 |
| 1/2" | " | 0.61 | 15.60 | 16.21 |
| 3/4" | " | 1.15 | 19.50 | 20.65 |
| 1" | " | 2.50 | 21.65 | 24.15 |
| 1-1/4" | " | 4.60 | 24.35 | 28.95 |
| 1-1/2" | " | 6.15 | 26.00 | 32.15 |
| 2" | " | 8.90 | 32.50 | 41.40 |
| 2-1/2" | " | 12.20 | 32.50 | 44.70 |
| 3" | " | 23.55 | 39.00 | 62.55 |
| 4" | " | 45.80 | 48.80 | 94.60 |
| Coupling with drain | | | | |
| 1/2" | EA. | 5.35 | 19.50 | 24.85 |
| 3/4" | " | 7.85 | 21.65 | 29.50 |
| 1" | " | 9.80 | 24.35 | 34.15 |
| Reducer | | | | |
| 3/8" x 1/4" | EA. | 1.70 | 15.60 | 17.30 |
| 1/2" x 3/8" | " | 1.85 | 15.60 | 17.45 |
| 3/4" x | | | | |
| 1/4" | EA. | 2.75 | 17.75 | 20.50 |
| 3/8" | " | 2.90 | 17.75 | 20.65 |
| 1/2" | " | 3.00 | 17.75 | 20.75 |
| 1" x | | | | |

| PLUMBING | UNIT | MAT. | INST. | TOTAL |
|---|---|---|---|---|
| **15410.11 COPPER FITTINGS** | | | | |
| 1/2" | EA. | 4.60 | 19.50 | 24.10 |
| 3/4" | " | 4.70 | 19.50 | 24.20 |
| 1-1/4" x | | | | |
| 1/2" | EA. | 6.15 | 21.65 | 27.80 |
| 3/4" | " | 7.70 | 21.65 | 29.35 |
| 1" | " | 7.85 | 21.65 | 29.50 |
| 1-1/2" x | | | | |
| 1/2" | EA. | 9.15 | 24.35 | 33.50 |
| 3/4" | " | 9.20 | 24.35 | 33.55 |
| 1" | " | 9.40 | 24.35 | 33.75 |
| 1-1/4" | " | 9.55 | 24.35 | 33.90 |
| 2" x | | | | |
| 1/2" | EA. | 14.40 | 27.90 | 42.30 |
| 3/4" | " | 14.65 | 27.90 | 42.55 |
| 1" | " | 14.65 | 27.90 | 42.55 |
| 1-1/4" | " | 15.05 | 27.90 | 42.95 |
| 1-1/2" | " | 15.05 | 27.90 | 42.95 |
| 2-1/2" x | | | | |
| 1" | EA. | 31.40 | 32.50 | 63.90 |
| 1-1/4" | " | 32.70 | 32.50 | 65.20 |
| 1-1/2" | " | 32.70 | 32.50 | 65.20 |
| 2" | " | 34.00 | 32.50 | 66.50 |
| 3" x | | | | |
| 1-1/4" | EA. | 41.90 | 39.00 | 80.90 |
| 1-1/2" | " | 41.90 | 39.00 | 80.90 |
| 2" | " | 41.90 | 39.00 | 80.90 |
| 2-1/2" | " | 41.90 | 39.00 | 80.90 |
| 4" x | | | | |
| 2" | EA. | 70.50 | 48.80 | 119.30 |
| 3" | " | 73.50 | 48.80 | 122.30 |
| Female adapters | | | | |
| 1/4" | EA. | 6.00 | 15.60 | 21.60 |
| 3/8" | " | 6.30 | 17.75 | 24.05 |
| 1/2" | " | 3.15 | 19.50 | 22.65 |
| 3/4" | " | 4.60 | 21.65 | 26.25 |
| 1" | " | 7.70 | 21.65 | 29.35 |
| 1-1/4" | " | 12.45 | 24.35 | 36.80 |
| 1-1/2" | " | 18.30 | 24.35 | 42.65 |
| 2" | " | 23.55 | 26.00 | 49.55 |
| 2-1/2" | " | 76.00 | 27.90 | 103.90 |
| 3" | " | 100.00 | 32.50 | 132.50 |
| 4" | " | 170.00 | 39.00 | 209.00 |
| Increasing female adapters | | | | |
| 1/8" x | | | | |
| 3/8" | EA. | 4.60 | 15.60 | 20.20 |
| 1/2" | " | 4.60 | 15.60 | 20.20 |
| 1/4" x 1/2" | " | 4.60 | 16.95 | 21.55 |
| 3/8" x 1/2" | " | 4.60 | 17.75 | 22.35 |
| 1/2" X | | | | |
| 3/4" | EA. | 6.15 | 19.50 | 25.65 |
| 1" | " | 11.40 | 19.50 | 30.90 |
| 3/4" X | | | | |

# 15 MECHANICAL

| PLUMBING | UNIT | MAT. | INST. | TOTAL |
|---|---|---|---|---|
| **15410.11 COPPER FITTINGS** | | | | |
| 1" | EA. | 10.45 | 21.65 | 32.10 |
| 1-1/4" | " | 18.95 | 21.65 | 40.60 |
| 1" x | | | | |
| 1-1/4" | EA. | 18.95 | 21.65 | 40.60 |
| 1-1/2" | " | 22.00 | 21.65 | 43.65 |
| 1-1/4" x | | | | |
| 1-1/2" | EA. | 22.10 | 24.35 | 46.45 |
| 2" | " | 31.40 | 24.35 | 55.75 |
| 1-1/2" x 2" | " | 32.70 | 26.00 | 58.70 |
| Reducing female adapters | | | | |
| 3/8" x 1/4" | EA. | 5.00 | 17.75 | 22.75 |
| 1/2" x | | | | |
| 1/4" | EA. | 5.00 | 19.50 | 24.50 |
| 3/8" | " | 5.25 | 19.50 | 24.75 |
| 3/4" x 1/2" | " | 6.40 | 21.65 | 28.05 |
| 1" x | | | | |
| 1/2" | EA. | 10.35 | 21.65 | 32.00 |
| 3/4" | " | 10.35 | 21.65 | 32.00 |
| 1-1/4" x | | | | |
| 1/2" | EA. | 19.65 | 24.35 | 44.00 |
| 3/4" | " | 19.65 | 24.35 | 44.00 |
| 1" | " | 19.65 | 24.35 | 44.00 |
| 1-1/2" x | | | | |
| 1" | EA. | 20.95 | 26.00 | 46.95 |
| 1-1/4" | " | 20.95 | 26.00 | 46.95 |
| 2" x | | | | |
| 1" | EA. | 30.10 | 27.90 | 58.00 |
| 1-1/4" | " | 31.10 | 27.90 | 59.00 |
| 1-1/2" | " | 32.70 | 27.90 | 60.60 |
| Female fitting adapters | | | | |
| 1/2" | EA. | 4.75 | 19.50 | 24.25 |
| 3/4" | " | 6.30 | 19.50 | 25.80 |
| 3/4" x 1/2" | " | 10.45 | 20.55 | 31.00 |
| 1" | " | 9.80 | 21.65 | 31.45 |
| 1-1/4" | " | 15.05 | 22.95 | 38.00 |
| 1-1/2" | " | 19.65 | 24.35 | 44.00 |
| 2" | " | 20.95 | 26.00 | 46.95 |
| Male adapters | | | | |
| 1/4" | EA. | 5.75 | 17.75 | 23.50 |
| 3/8" | " | 3.25 | 17.75 | 21.00 |
| 3" | " | 68.00 | 32.50 | 100.50 |
| 4" | " | 100.00 | 39.00 | 139.00 |
| Increasing male adapters | | | | |
| 3/8" x 1/2" | EA. | 3.40 | 17.75 | 21.15 |
| 1/2" x | | | | |
| 3/4" | EA. | 5.00 | 19.50 | 24.50 |
| 1" | " | 9.15 | 19.50 | 28.65 |
| 3/4" x | | | | |
| 1" | EA. | 9.15 | 20.55 | 29.70 |
| 1-1/4" | " | 13.10 | 20.55 | 33.65 |
| 1" x 1-1/4" | " | 13.10 | 21.65 | 34.75 |
| 1-1/2" x | | | | |

| PLUMBING | UNIT | MAT. | INST. | TOTAL |
|---|---|---|---|---|
| **15410.11 COPPER FITTINGS** | | | | |
| 3/4" | EA. | 17.00 | 22.95 | 39.95 |
| 1" | " | 17.25 | 22.95 | 40.20 |
| 1-1/4" | " | 17.55 | 22.95 | 40.50 |
| 2" x | | | | |
| 1" | EA. | 28.80 | 24.35 | 53.15 |
| 1-1/4" | " | 28.80 | 24.35 | 53.15 |
| 1-1/2" | " | 30.10 | 24.35 | 54.45 |
| 2" x 2-1/2" | " | 97.00 | 26.00 | 123.00 |
| Copper pipe fittings | | | | |
| 1/2" | | | | |
| 90 deg ell | EA. | 0.93 | 8.65 | 9.58 |
| 45 deg ell | " | 1.15 | 8.65 | 9.80 |
| Tee | " | 1.60 | 11.15 | 12.75 |
| Cap | " | 0.63 | 4.35 | 4.98 |
| Coupling | " | 0.68 | 8.65 | 9.33 |
| Union | " | 4.70 | 9.75 | 14.45 |
| 3/4" | | | | |
| 90 deg ell | EA. | 2.05 | 9.75 | 11.80 |
| 45 deg ell | " | 2.35 | 9.75 | 12.10 |
| Tee | " | 3.40 | 13.00 | 16.40 |
| Cap | " | 1.25 | 4.60 | 5.85 |
| Coupling | " | 1.40 | 9.75 | 11.15 |
| Union | " | 6.95 | 11.15 | 18.10 |
| 1" | | | | |
| 90 deg ell | EA. | 4.70 | 13.00 | 17.70 |
| 45 deg ell | " | 6.15 | 13.00 | 19.15 |
| Tee | " | 7.70 | 15.60 | 23.30 |
| Cap | " | 2.30 | 6.50 | 8.80 |
| Coupling | " | 3.40 | 13.00 | 16.40 |
| Union | " | 9.10 | 13.00 | 22.10 |
| 1-1/4" | | | | |
| 90 deg ell | EA. | 6.40 | 11.15 | 17.55 |
| 45 deg ell | " | 7.95 | 11.15 | 19.10 |
| Tee | " | 10.45 | 19.50 | 29.95 |
| Cap | " | 1.85 | 6.50 | 8.35 |
| Union | " | 15.05 | 13.95 | 29.00 |
| 1-1/2" | | | | |
| 90 deg ell | EA. | 8.35 | 13.95 | 22.30 |
| 45 deg ell | " | 9.95 | 13.95 | 23.90 |
| Tee | " | 13.75 | 21.65 | 35.40 |
| Cap | " | 1.85 | 6.50 | 8.35 |
| Coupling | " | 6.15 | 13.00 | 19.15 |
| Union | " | 22.90 | 17.75 | 40.65 |
| 2" | | | | |
| 90 deg ell | EA. | 16.35 | 15.60 | 31.95 |
| 45 deg ell | " | 15.05 | 24.35 | 39.40 |
| Tee | " | 23.55 | 24.35 | 47.90 |
| Cap | " | 3.80 | 7.80 | 11.60 |
| Coupling | " | 9.95 | 15.60 | 25.55 |
| Union | " | 24.85 | 19.50 | 44.35 |
| 2-1/2" | | | | |
| 90 deg ell | EA. | 31.40 | 19.50 | 50.90 |

| PLUMBING | UNIT | MAT. | INST. | TOTAL |
|---|---|---|---|---|
| **15410.11 COPPER FITTINGS** | | | | |
| 45 deg ell | EA. | 27.50 | 19.50 | 47.00 |
| Tee | " | 31.40 | 27.90 | 59.30 |
| Cap | " | 7.70 | 9.75 | 17.45 |
| Coupling | " | 15.05 | 19.50 | 34.55 |
| Union | " | 45.80 | 21.65 | 67.45 |
| **15410.15 BRASS FITTINGS** | | | | |
| Compression fittings, union | | | | |
| 3/8" | EA. | 3.15 | 6.50 | 9.65 |
| 1/2" | " | 4.70 | 6.50 | 11.20 |
| 5/8" | " | 5.80 | 6.50 | 12.30 |
| Union elbow | | | | |
| 3/8" | EA. | 5.25 | 6.50 | 11.75 |
| 1/2" | " | 8.25 | 6.50 | 14.75 |
| 5/8" | " | 10.35 | 6.50 | 16.85 |
| Union tee | | | | |
| 3/8" | EA. | 6.90 | 6.50 | 13.40 |
| 1/2" | " | 11.00 | 6.50 | 17.50 |
| 5/8" | " | 15.85 | 6.50 | 22.35 |
| Male connector | | | | |
| 3/8" | EA. | 2.20 | 6.50 | 8.70 |
| 1/2" | " | 1.50 | 6.50 | 8.00 |
| 5/8" | " | 3.45 | 6.50 | 9.95 |
| Female connector | | | | |
| 3/8" | EA. | 4.15 | 6.50 | 10.65 |
| 1/2" | " | 5.25 | 6.50 | 11.75 |
| 5/8" | " | 5.25 | 6.50 | 11.75 |
| Brass flare fittings, union | | | | |
| 3/8" | EA. | 2.35 | 6.30 | 8.65 |
| 1/2" | " | 3.50 | 6.30 | 9.80 |
| 5/8" | " | 5.15 | 6.30 | 11.45 |
| 90 deg elbow union | | | | |
| 3/8" | EA. | 5.50 | 6.30 | 11.80 |
| 1/2" | " | 7.70 | 6.30 | 14.00 |
| 5/8" | " | 11.00 | 6.30 | 17.30 |
| Three way tee | | | | |
| 3/8" | EA. | 7.30 | 10.55 | 17.85 |
| 1/2" | " | 9.50 | 10.55 | 20.05 |
| 5/8" | " | 15.85 | 10.55 | 26.40 |
| Cross | | | | |
| 3/8" | EA. | 10.20 | 13.95 | 24.15 |
| 1/2" | " | 13.75 | 13.95 | 27.70 |
| 5/8" | " | 26.20 | 13.95 | 40.15 |
| Male connector, half union | | | | |
| 3/8" | EA. | 3.15 | 6.30 | 9.45 |
| 1/2" | " | 4.05 | 6.30 | 10.35 |
| 5/8" | " | 8.55 | 6.30 | 14.85 |
| Female connector, half union | | | | |
| 3/8" | EA. | 3.15 | 6.30 | 9.45 |
| 1/2" | " | 6.35 | 6.30 | 12.65 |
| 5/8" | " | 11.00 | 6.30 | 17.30 |

# 15 MECHANICAL

| PLUMBING | UNIT | MAT. | INST. | TOTAL |
|---|---|---|---|---|
| **15410.15 BRASS FITTINGS** | | | | |
| Long forged nut | | | | |
| 3/8" | EA. | 2.15 | 6.30 | 8.45 |
| 1/2" | " | 3.85 | 6.30 | 10.15 |
| 5/8" | " | 5.80 | 6.30 | 12.10 |
| Short forged nut | | | | |
| 3/8" | EA. | 1.45 | 6.30 | 7.75 |
| 1/2" | " | 2.05 | 6.30 | 8.35 |
| 5/8" | " | 2.60 | 6.30 | 8.90 |
| **15410.18 GLASS PIPE** | | | | |
| Glass pipe | | | | |
| 1-1/2" dia. | L.F. | 10.80 | 7.80 | 18.60 |
| 2" dia. | " | 15.15 | 8.65 | 23.80 |
| 3" dia. | " | 20.25 | 9.75 | 30.00 |
| 4" dia. | " | 35.10 | 11.15 | 46.25 |
| 6" dia. | " | 67.50 | 13.00 | 80.50 |
| **15410.30 PVC/CPVC PIPE** | | | | |
| PVC schedule 40 | | | | |
| 1/2" pipe | L.F. | 1.80 | 1.60 | 3.40 |
| 3/4" pipe | " | 2.10 | 1.75 | 3.85 |
| 1" pipe | " | 2.40 | 1.95 | 4.35 |
| 1-1/4" pipe | " | 2.65 | 2.15 | 4.80 |
| 1-1/2" pipe | " | 2.80 | 2.45 | 5.25 |
| 2" pipe | " | 3.25 | 2.80 | 6.05 |
| 2-1/2" pipe | " | 4.45 | 3.25 | 7.70 |
| 3" pipe | " | 4.75 | 3.90 | 8.65 |
| 4" pipe | " | 6.95 | 4.85 | 11.80 |
| 6" pipe | " | 12.90 | 9.75 | 22.65 |
| 8" pipe | " | 19.70 | 13.00 | 32.70 |
| Fittings, 1/2" | | | | |
| 90 deg ell | EA. | 0.43 | 4.85 | 5.28 |
| 45 deg ell | " | 0.68 | 4.85 | 5.53 |
| Tee | " | 0.58 | 5.55 | 6.13 |
| 3/4" | | | | |
| 90 deg elbow | EA. | 0.45 | 6.50 | 6.95 |
| 45 deg elbow | " | 1.00 | 6.50 | 7.50 |
| Tee | " | 0.58 | 7.80 | 8.38 |
| 1" | | | | |
| 90 deg elbow | EA. | 0.76 | 7.80 | 8.56 |
| 45 deg elbow | " | 1.10 | 7.80 | 8.90 |
| Tee | " | 1.00 | 8.65 | 9.65 |
| 1-1/4" | | | | |
| 90 deg elbow | EA. | 1.30 | 11.15 | 12.45 |
| 45 deg elbow | " | 1.50 | 11.15 | 12.65 |
| Tee | " | 1.50 | 13.00 | 14.50 |
| 1-1/2" | | | | |
| 90 deg elbow | EA. | 1.45 | 11.15 | 12.60 |
| 45 deg elbow | " | 2.10 | 11.15 | 13.25 |
| Tee | " | 1.95 | 13.00 | 14.95 |
| 2" | | | | |
| 90 deg elbow | EA. | 2.25 | 13.00 | 15.25 |

# 15 MECHANICAL

| PLUMBING | UNIT | MAT. | INST. | TOTAL |
|---|---|---|---|---|
| **15410.30 PVC/CPVC PIPE** | | | | |
| 45 deg elbow | EA. | 2.90 | 13.00 | 15.90 |
| Tee | " | 3.00 | 15.60 | 18.60 |
| 2-1/2" | | | | |
| 90 deg elbow | EA. | 6.80 | 24.35 | 31.15 |
| 45 deg elbow | " | 9.85 | 24.35 | 34.20 |
| Tee | " | 8.80 | 26.00 | 34.80 |
| 3" | | | | |
| 90 deg elbow | EA. | 7.35 | 32.50 | 39.85 |
| 45 deg elbow | " | 9.50 | 32.50 | 42.00 |
| Tee | " | 11.70 | 35.50 | 47.20 |
| 4" | | | | |
| 90 deg elbow | EA. | 13.10 | 39.00 | 52.10 |
| 45 deg elbow | " | 17.10 | 39.00 | 56.10 |
| Tee | " | 19.55 | 43.30 | 62.85 |
| PVC schedule 80 pipe | | | | |
| 1-1/2" pipe | L.F. | 2.25 | 2.45 | 4.70 |
| 2" pipe | " | 2.90 | 2.80 | 5.70 |
| 3" pipe | " | 4.40 | 3.90 | 8.30 |
| 4" pipe | " | 6.05 | 4.85 | 10.90 |
| Fittings, 1-1/2" | | | | |
| 90 deg elbow | EA. | 3.95 | 13.00 | 16.95 |
| 45 deg elbow | " | 6.95 | 13.00 | 19.95 |
| Tee | " | 10.85 | 19.50 | 30.35 |
| 2" | | | | |
| 90 deg elbow | EA. | 4.70 | 15.60 | 20.30 |
| 45 deg elbow | " | 9.00 | 15.60 | 24.60 |
| Tee | " | 13.55 | 24.35 | 37.90 |
| 2-1/2" | | | | |
| 90 deg elbow | EA. | 8.90 | 24.35 | 33.25 |
| 45 deg elbow | " | 18.75 | 24.35 | 43.10 |
| Tee | " | 14.70 | 32.50 | 47.20 |
| 3" | | | | |
| 90 deg elbow | EA. | 12.10 | 32.50 | 44.60 |
| 45 deg elbow | " | 23.00 | 32.50 | 55.50 |
| Tee | " | 18.45 | 39.00 | 57.45 |
| 4" | | | | |
| 90 deg elbow | EA. | 15.30 | 39.00 | 54.30 |
| 45 deg elbow | " | 41.30 | 39.00 | 80.30 |
| Tee | " | 20.35 | 48.80 | 69.15 |
| CPVC schedule 40 | | | | |
| 1/2" pipe | L.F. | 2.65 | 1.60 | 4.25 |
| 3/4" pipe | " | 2.95 | 1.75 | 4.70 |
| 1" pipe | " | 4.15 | 1.95 | 6.10 |
| 1-1/4" pipe | " | 5.20 | 2.15 | 7.35 |
| 1-1/2" pipe | " | 5.95 | 2.45 | 8.40 |
| 2" pipe | " | 7.45 | 2.80 | 10.25 |
| Fittings, CPVC, schedule 80 | | | | |
| 1/2", 90 deg ell | EA. | 4.55 | 3.90 | 8.45 |
| Tee | " | 6.05 | 6.50 | 12.55 |
| 3/4", 90 deg ell | " | 6.20 | 3.90 | 10.10 |
| Tee | " | 6.80 | 6.50 | 13.30 |
| 1", 90 deg ell | " | 7.55 | 4.35 | 11.90 |

## PLUMBING

### 15410.30 PVC/CPVC PIPE

| PLUMBING | UNIT | MAT. | INST. | TOTAL |
|---|---|---|---|---|
| Tee | EA. | 9.55 | 7.10 | 16.65 |
| 1-1/4", 90 deg ell | " | 8.35 | 4.35 | 12.70 |
| Tee | " | 10.60 | 7.10 | 17.70 |
| 1-1/2", 90 deg ell | " | 10.45 | 7.80 | 18.25 |
| Tee | " | 13.65 | 9.75 | 23.40 |
| 2", 90 deg ell | " | 16.65 | 7.80 | 24.45 |
| Tee | " | 21.20 | 9.75 | 30.95 |
| Polypropylene, acid resistant, DWV pipe | | | | |
| Schedule 40 | | | | |
| 1-1/2" pipe | L.F. | 4.60 | 2.80 | 7.40 |
| 2" pipe | " | 6.15 | 3.25 | 9.40 |
| 3" pipe | " | 11.35 | 3.90 | 15.25 |
| 4" pipe | " | 16.05 | 4.85 | 20.90 |
| 6" pipe | " | 28.80 | 9.75 | 38.55 |
| Fittings | | | | |
| 1-1/2" | | | | |
| 1/4 bend | EA. | 16.55 | 9.75 | 26.30 |
| 1/8 bend | " | 16.30 | 9.75 | 26.05 |
| Sanitary tee | " | 20.65 | 19.50 | 40.15 |
| Cleanout with plug | " | 42.10 | 19.50 | 61.60 |
| Wye | " | 38.60 | 19.50 | 58.10 |
| 2" | | | | |
| 1/4 bend | EA. | 20.65 | 11.15 | 31.80 |
| 1/8 bend | " | 20.65 | 11.15 | 31.80 |
| Sanitary tee | " | 24.75 | 22.95 | 47.70 |
| Cleanout with plug | " | 49.30 | 22.95 | 72.25 |
| Wye | " | 52.00 | 22.95 | 74.95 |
| 3" | | | | |
| 1/4 bend | EA. | 35.00 | 13.00 | 48.00 |
| 1/8 bend | " | 36.40 | 13.00 | 49.40 |
| Sanitary tee | " | 49.50 | 26.00 | 75.50 |
| Cleanout with plug | " | 81.00 | 26.00 | 107.00 |
| Wye | " | 52.50 | 26.00 | 78.50 |
| 4" | | | | |
| 1/4 bend | EA. | 55.50 | 19.50 | 75.00 |
| 1/8 bend | " | 41.30 | 19.50 | 60.80 |
| Sanitary tee | " | 74.00 | 39.00 | 113.00 |
| Cleanout with plug | " | 110.00 | 39.00 | 149.00 |
| Wye | " | 76.50 | 39.00 | 115.50 |
| 6" | | | | |
| 1/4 bend | EA. | 140.00 | 32.50 | 172.50 |
| 1/8 bend | " | 110.00 | 32.50 | 142.50 |
| Sanitary tee | " | 210.00 | 65.00 | 275.00 |
| Cleanout with plug | " | 150.00 | 65.00 | 215.00 |
| Wye | " | 200.00 | 65.00 | 265.00 |
| Polyethylene pipe and fittings | | | | |
| SDR-21 | | | | |
| 3" pipe | L.F. | 3.00 | 4.85 | 7.85 |
| 4" pipe | " | 4.55 | 6.50 | 11.05 |
| 6" pipe | " | 7.55 | 9.75 | 17.30 |
| 8" pipe | " | 12.10 | 11.15 | 23.25 |
| 10" pipe | " | 13.65 | 13.00 | 26.65 |

# 15 MECHANICAL

| PLUMBING | UNIT | MAT. | INST. | TOTAL |
|---|---|---|---|---|
| **15410.30 PVC/CPVC PIPE** | | | | |
| 12" pipe | L.F. | 21.20 | 15.60 | 36.80 |
| 14" pipe | " | 27.30 | 19.50 | 46.80 |
| 16" pipe | " | 33.30 | 24.35 | 57.65 |
| 18" pipe | " | 36.40 | 30.00 | 66.40 |
| 20" pipe | " | 45.40 | 39.00 | 84.40 |
| 22" pipe | " | 53.00 | 43.30 | 96.30 |
| 24" pipe | " | 63.50 | 48.80 | 112.30 |
| Fittings, 3" | | | | |
| 90 deg elbow | EA. | 80.50 | 19.50 | 100.00 |
| 45 deg elbow | " | 51.00 | 19.50 | 70.50 |
| Tee | " | 46.60 | 32.50 | 79.10 |
| 4" | | | | |
| 90 deg elbow | EA. | 120.00 | 24.35 | 144.35 |
| 45 deg elbow | " | 72.50 | 24.35 | 96.85 |
| Tee | " | 100.00 | 39.00 | 139.00 |
| 8" | | | | |
| 90 deg elbow | EA. | 320.00 | 48.80 | 368.80 |
| 45 deg elbow | " | 180.00 | 48.80 | 228.80 |
| Tee | " | 300.00 | 78.00 | 378.00 |
| 10" | | | | |
| 90 deg elbow | EA. | 450.00 | 65.00 | 515.00 |
| 45 deg elbow | " | 240.00 | 65.00 | 305.00 |
| Tee | " | 400.00 | 97.50 | 497.50 |
| 12" | | | | |
| 90 deg elbow | EA. | 720.00 | 78.00 | 798.00 |
| 45 deg elbow | " | 440.00 | 78.00 | 518.00 |
| Tee | " | 570.00 | 130.00 | 700.00 |
| 14" | | | | |
| 90 deg elbow | EA. | 980.00 | 97.50 | 1,078 |
| 45 deg elbow | " | 560.00 | 97.50 | 657.50 |
| Tee | " | 720.00 | 160.00 | 880.00 |
| 16" | | | | |
| 90 deg elbow | EA. | 1,230 | 97.50 | 1,328 |
| 45 deg elbow | " | 730.00 | 97.50 | 827.50 |
| Tee | " | 890.00 | 160.00 | 1,050 |
| 18" | | | | |
| 90 deg elbow | EA. | 1,870 | 130.00 | 2,000 |
| 45 deg elbow | " | 1,210 | 130.00 | 1,340 |
| Tee | " | 1,450 | 200.00 | 1,650 |
| 20" | | | | |
| 90 deg elbow | EA. | 1,510 | 130.00 | 1,640 |
| 45 deg elbow | " | 900.00 | 130.00 | 1,030 |
| **15410.33 ABS DWV PIPE** | | | | |
| Schedule 40 ABS | | | | |
| 1-1/2" pipe | L.F. | 2.25 | 1.95 | 4.20 |
| 2" pipe | " | 2.95 | 2.15 | 5.10 |
| 3" pipe | " | 4.45 | 2.80 | 7.25 |
| 4" pipe | " | 6.70 | 3.90 | 10.60 |
| 6" pipe | " | 10.40 | 4.85 | 15.25 |
| Fittings | | | | |
| 1/8 bend | | | | |

# 15 MECHANICAL

| PLUMBING | UNIT | MAT. | INST. | TOTAL |
|---|---|---|---|---|
| **15410.33 ABS DWV PIPE** | | | | |
| 1-1/2" | EA. | 1.35 | 7.80 | 9.15 |
| 2" | " | 1.50 | 9.75 | 11.25 |
| 3" | " | 4.55 | 13.00 | 17.55 |
| 4" | " | 7.55 | 15.60 | 23.15 |
| 6" | " | 48.50 | 19.50 | 68.00 |
| Tee, sanitary | | | | |
| 1-1/2" | EA. | 2.10 | 13.00 | 15.10 |
| 2" | " | 3.15 | 15.60 | 18.75 |
| 3" | " | 7.45 | 19.50 | 26.95 |
| 4" | " | 13.65 | 24.35 | 38.00 |
| 6" | " | 79.00 | 32.50 | 111.50 |
| Tee, sanitary reducing | | | | |
| 2 x 1-1/2 x 1-1/2 | EA. | 4.10 | 15.60 | 19.70 |
| 2 x 1-1/2 x 2 | " | 5.00 | 16.25 | 21.25 |
| 2 x 2 x 1-1/2 | " | 3.00 | 17.75 | 20.75 |
| 3 x 3 x 1-1/2 | " | 6.05 | 19.50 | 25.55 |
| 3 x 3 x 2 | " | 6.05 | 21.65 | 27.70 |
| 4 x 4 x 1-1/2 | " | 16.65 | 24.35 | 41.00 |
| 4 x 4 x 2 | " | 13.65 | 27.90 | 41.55 |
| 4 x 4 x 3 | " | 18.15 | 30.00 | 48.15 |
| 6 x 6 x 4 | " | 77.00 | 32.50 | 109.50 |
| Wye | | | | |
| 1-1/2" | EA. | 3.25 | 11.15 | 14.40 |
| 2" | " | 3.80 | 15.60 | 19.40 |
| 3" | " | 7.55 | 19.50 | 27.05 |
| 4" | " | 13.65 | 24.35 | 38.00 |
| 6" | " | 79.00 | 32.50 | 111.50 |
| Reducer | | | | |
| 2 x 1-1/2 | EA. | 2.10 | 9.75 | 11.85 |
| 3 x 1-1/2 | " | 4.85 | 13.00 | 17.85 |
| 3 x 2 | " | 4.40 | 13.00 | 17.40 |
| 4 x 2 | " | 9.40 | 15.60 | 25.00 |
| 4 x 3 | " | 9.25 | 15.60 | 24.85 |
| 6 x 4 | " | 13.95 | 19.50 | 33.45 |
| P-trap | | | | |
| 1-1/2" | EA. | 4.85 | 13.00 | 17.85 |
| 2" | " | 6.35 | 14.45 | 20.80 |
| 3" | " | 28.80 | 16.95 | 45.75 |
| 4" | " | 50.50 | 19.50 | 70.00 |
| 6" | " | 82.50 | 24.35 | 106.85 |
| Double sanitary, tee | | | | |
| 1-1/2" | EA. | 6.05 | 15.60 | 21.65 |
| 2" | " | 8.20 | 19.50 | 27.70 |
| 3" | " | 18.15 | 24.35 | 42.50 |
| 4" | " | 30.30 | 32.50 | 62.80 |
| Long sweep, 1/4 bend | | | | |
| 1-1/2" | EA. | 3.00 | 7.80 | 10.80 |
| 2" | " | 3.15 | 9.75 | 12.90 |
| 3" | " | 8.50 | 13.00 | 21.50 |
| 4" | " | 17.40 | 19.50 | 36.90 |
| Wye, standard | | | | |
| 1-1/2" | EA. | 3.65 | 13.00 | 16.65 |

| PLUMBING | UNIT | MAT. | INST. | TOTAL |
|---|---|---|---|---|
| **15410.33 ABS DWV PIPE** | | | | |
| 2" | EA. | 3.30 | 15.60 | 18.90 |
| 3" | " | 8.20 | 19.50 | 27.70 |
| 4" | " | 14.40 | 24.35 | 38.75 |
| **15410.35 PLASTIC PIPE** | | | | |
| Fiberglass reinforced pipe | | | | |
| 2" pipe | L.F. | 12.10 | 3.00 | 15.10 |
| 3" pipe | " | 16.35 | 3.25 | 19.60 |
| 4" pipe | " | 20.80 | 3.55 | 24.35 |
| 6" pipe | " | 32.70 | 3.90 | 36.60 |
| 8" pipe | " | 47.50 | 6.50 | 54.00 |
| 10" pipe | " | 59.50 | 7.80 | 67.30 |
| 12" pipe | " | 74.50 | 9.75 | 84.25 |
| Fittings | | | | |
| 90 deg elbow, flanged | | | | |
| 2" | EA. | 97.00 | 39.00 | 136.00 |
| 3" | " | 120.00 | 43.30 | 163.30 |
| 4" | " | 150.00 | 48.80 | 198.80 |
| 6" | " | 270.00 | 65.00 | 335.00 |
| 8" | " | 490.00 | 78.00 | 568.00 |
| 10" | " | 650.00 | 97.50 | 747.50 |
| 12" | " | 870.00 | 130.00 | 1,000 |
| 45 deg elbow, flanged | | | | |
| 2" | EA. | 97.00 | 32.50 | 129.50 |
| 3" | " | 120.00 | 39.00 | 159.00 |
| 4" | " | 150.00 | 48.80 | 198.80 |
| 6" | " | 270.00 | 65.00 | 335.00 |
| 8" | " | 410.00 | 78.00 | 488.00 |
| 10" | " | 560.00 | 97.50 | 657.50 |
| 12" | " | 700.00 | 130.00 | 830.00 |
| Tee, flanged | | | | |
| 2" | EA. | 120.00 | 48.80 | 168.80 |
| 3" | " | 170.00 | 55.50 | 225.50 |
| 4" | " | 200.00 | 65.00 | 265.00 |
| 6" | " | 330.00 | 78.00 | 408.00 |
| 8" | " | 590.00 | 97.50 | 687.50 |
| 10" | " | 950.00 | 130.00 | 1,080 |
| 12" | " | 1,310 | 200.00 | 1,510 |
| Wye, flanged | | | | |
| 2" | EA. | 240.00 | 48.80 | 288.80 |
| 3" | " | 340.00 | 55.50 | 395.50 |
| 4" | " | 440.00 | 65.00 | 505.00 |
| 6" | " | 550.00 | 78.00 | 628.00 |
| 8" | " | 750.00 | 97.50 | 847.50 |
| 10" | " | 1,280 | 130.00 | 1,410 |
| 12" | " | 1,640 | 200.00 | 1,840 |
| **15410.70 STAINLESS STEEL PIPE** | | | | |
| Stainless steel, schedule 40, threaded | | | | |
| 1/2" pipe | L.F. | 6.10 | 5.55 | 11.65 |

## PLUMBING

### 15410.70 STAINLESS STEEL PIPE

| PLUMBING | UNIT | MAT. | INST. | TOTAL |
|---|---|---|---|---|
| 1" pipe | L.F. | 9.45 | 6.00 | 15.45 |
| 1-1/2" pipe | " | 12.85 | 6.50 | 19.35 |
| 2" pipe | " | 19.30 | 7.10 | 26.40 |
| 2-1/2" pipe | " | 27.00 | 7.80 | 34.80 |
| 3" pipe | " | 37.80 | 8.65 | 46.45 |
| 4" pipe | " | 48.60 | 9.75 | 58.35 |
| Fittings, 1/2" | | | | |
| 90 deg ell | EA. | 6.20 | 48.80 | 55.00 |
| 45 deg ell | " | 7.30 | 48.80 | 56.10 |
| Tee | " | 8.25 | 65.00 | 73.25 |
| 3/4" | | | | |
| 90 deg ell | EA. | 8.25 | 48.80 | 57.05 |
| 45 deg ell | " | 9.50 | 48.80 | 58.30 |
| Tee | " | 11.10 | 65.00 | 76.10 |
| 1" | | | | |
| 90 deg ell | EA. | 10.35 | 48.80 | 59.15 |
| 45 deg ell | " | 10.35 | 48.80 | 59.15 |
| Tee | " | 12.25 | 65.00 | 77.25 |
| 1-1/4" | | | | |
| 90 deg ell | EA. | 14.45 | 48.80 | 63.25 |
| 45 deg ell | " | 14.45 | 48.80 | 63.25 |
| Tee | " | 20.65 | 65.00 | 85.65 |
| 1-1/2" | | | | |
| 90 deg ell | EA. | 17.90 | 65.00 | 82.90 |
| 45 deg ell | " | 18.60 | 65.00 | 83.60 |
| Tee | " | 26.00 | 78.00 | 104.00 |
| 2" | | | | |
| 90 deg ell | EA. | 26.90 | 78.00 | 104.90 |
| 45 deg ell | " | 26.90 | 78.00 | 104.90 |
| Tee | " | 31.70 | 130.00 | 161.70 |
| Type 304, sch 10 pipe | | | | |
| 1" pipe | L.F. | 7.30 | 4.85 | 12.15 |
| 1-1/4" pipe | " | 9.35 | 6.50 | 15.85 |
| 1-1/2" pipe | " | 10.20 | 7.10 | 17.30 |
| 2" pipe | " | 10.90 | 8.65 | 19.55 |
| 2-1/2" pipe | " | 14.45 | 9.75 | 24.20 |
| 3" pipe | " | 17.20 | 11.15 | 28.35 |
| 4" pipe | " | 22.70 | 13.00 | 35.70 |
| 6" pipe | " | 35.80 | 15.60 | 51.40 |
| Fittings, 1" | | | | |
| 90 deg elbow | EA. | 7.15 | 78.00 | 85.15 |
| 45 deg elbow | " | 10.05 | 78.00 | 88.05 |
| Tee | " | 22.05 | 130.00 | 152.05 |
| 1-1/4" | | | | |
| 90 deg elbow | EA. | 11.70 | 78.00 | 89.70 |
| 45 deg elbow | " | 14.05 | 78.00 | 92.05 |
| Tee | " | 42.70 | 130.00 | 172.70 |
| 1-1/2" | | | | |
| 90 deg elbow | EA. | 13.10 | 65.00 | 78.10 |
| 45 deg elbow | " | 14.45 | 65.00 | 79.45 |
| Tee | " | 30.30 | 130.00 | 160.30 |
| 2" | | | | |

# 15 MECHANICAL

| PLUMBING | UNIT | MAT. | INST. | TOTAL |
|---|---|---|---|---|
| **15410.70 STAINLESS STEEL PIPE** | | | | |
| 90 deg elbow | EA. | 12.40 | 78.00 | 90.40 |
| 45 deg elbow | " | 11.30 | 78.00 | 89.30 |
| Tee | " | 30.30 | 200.00 | 230.30 |
| 2-1/2" | | | | |
| 90 deg elbow | EA. | 20.65 | 78.00 | 98.65 |
| 45 deg elbow | " | 20.65 | 78.00 | 98.65 |
| Tee | " | 66.00 | 200.00 | 266.00 |
| 3" | | | | |
| 90 deg elbow | EA. | 20.95 | 97.50 | 118.45 |
| 45 deg elbow | " | 17.35 | 97.50 | 114.85 |
| Tee | " | 52.50 | 260.00 | 312.50 |
| 4" | | | | |
| 90 deg elbow | EA. | 44.10 | 97.50 | 141.60 |
| 45 deg elbow | " | 33.00 | 97.50 | 130.50 |
| 6" | | | | |
| 90 deg elbow | EA. | 95.00 | 130.00 | 225.00 |
| 45 deg elbow | " | 62.00 | 130.00 | 192.00 |
| Type 304 tubing | | | | |
| .035 wall | | | | |
| 1/4" | L.F. | 0.90 | 2.15 | 3.05 |
| 3/8" | " | 1.10 | 2.45 | 3.55 |
| 1/2" | " | 1.25 | 2.80 | 4.05 |
| 5/8" | " | 1.80 | 3.25 | 5.05 |
| 3/4" | " | 2.10 | 3.90 | 6.00 |
| 7/8" | " | 2.20 | 4.35 | 6.55 |
| 1" | " | 2.35 | 4.85 | 7.20 |
| .049 wall | | | | |
| 1/4" | L.F. | 1.20 | 2.30 | 3.50 |
| 3/8" | " | 1.25 | 2.60 | 3.85 |
| 1/2" | " | 1.35 | 3.00 | 4.35 |
| 5/8" | " | 1.85 | 3.55 | 5.40 |
| 3/4" | " | 2.15 | 4.35 | 6.50 |
| 7/8" | " | 2.65 | 4.85 | 7.50 |
| 1" | " | 2.80 | 5.55 | 8.35 |
| .065 wall | | | | |
| 1/4" | L.F. | 1.75 | 2.60 | 4.35 |
| 3/8" | " | 2.10 | 3.25 | 5.35 |
| 1/2" | " | 2.30 | 3.55 | 5.85 |
| 5/8" | " | 2.50 | 4.35 | 6.85 |
| 3/4" | " | 3.00 | 5.55 | 8.55 |
| 7/8" | " | 3.35 | 6.50 | 9.85 |
| 1" | " | 3.70 | 7.80 | 11.50 |
| Type 316 tubing | | | | |
| .035 wall | | | | |
| 1/4" | L.F. | 0.88 | 2.15 | 3.03 |
| 3/8" | " | 1.10 | 2.45 | 3.55 |
| 1/2" | " | 1.15 | 2.80 | 3.95 |
| 5/8" | " | 2.20 | 3.25 | 5.45 |
| 3/4" | " | 2.50 | 3.90 | 6.40 |
| 7/8" | " | 3.30 | 4.35 | 7.65 |
| 1" | " | 3.80 | 4.85 | 8.65 |
| .049 wall | | | | |

## PLUMBING

### 15410.70 STAINLESS STEEL PIPE

| | UNIT | MAT. | INST. | TOTAL |
|---|---|---|---|---|
| 1/4" | L.F. | 1.50 | 2.60 | 4.10 |
| 3/8" | " | 1.65 | 3.25 | 4.90 |
| 1/2" | " | 2.05 | 3.55 | 5.60 |
| 5/8" | " | 2.60 | 4.35 | 6.95 |
| 3/4" | " | 2.75 | 5.55 | 8.30 |
| 7/8" | " | 3.85 | 6.50 | 10.35 |
| 1" | " | 2.60 | 7.80 | 10.40 |
| .065 wall | | | | |
| 1/4" | L.F. | 2.20 | 2.60 | 4.80 |
| 3/8" | " | 2.90 | 3.25 | 6.15 |
| 1/2" | " | 2.90 | 3.55 | 6.45 |
| 5/8" | " | 3.00 | 4.35 | 7.35 |
| 3/4" | " | 3.45 | 5.55 | 9.00 |
| 7/8" | " | 6.90 | 6.50 | 13.40 |
| 1" | " | 4.10 | 7.80 | 11.90 |
| Fittings, 1/4" | | | | |
| 90 deg elbow | EA. | 15.15 | 7.80 | 22.95 |
| Union tee | " | 25.50 | 13.00 | 38.50 |
| Union | " | 12.40 | 13.00 | 25.40 |
| Male connector | " | 8.25 | 9.75 | 18.00 |
| 3/8" | | | | |
| 90 deg elbow | EA. | 22.05 | 9.75 | 31.80 |
| Union tee | " | 33.00 | 15.00 | 48.00 |
| Union | " | 17.20 | 15.00 | 32.20 |
| Male connector | " | 12.40 | 9.75 | 22.15 |
| 1/2" | | | | |
| 90 deg elbow | EA. | 33.00 | 10.25 | 43.25 |
| Union tee | " | 4.15 | 16.25 | 20.40 |
| Union | " | 26.20 | 16.25 | 42.45 |
| Male connector | " | 17.90 | 9.75 | 27.65 |
| 5/8" | | | | |
| 90 deg elbow | EA. | 80.00 | 13.00 | 93.00 |
| Union tee | " | 110.00 | 19.50 | 129.50 |
| Union | " | 52.50 | 19.50 | 72.00 |
| Male connector | " | 26.20 | 13.00 | 39.20 |
| 3/4" | | | | |
| 90 deg elbow | EA. | 120.00 | 13.00 | 133.00 |
| Union tee | " | 100.00 | 19.50 | 119.50 |
| Union | " | 85.50 | 19.50 | 105.00 |
| Male connector | " | 37.20 | 13.00 | 50.20 |
| 7/8" | | | | |
| 90 deg elbow | EA. | 110.00 | 13.95 | 123.95 |
| Union tee | " | 170.00 | 21.65 | 191.65 |
| Union | " | 85.50 | 21.65 | 107.15 |
| Male connector | " | 67.50 | 13.95 | 81.45 |
| 1" | | | | |
| 90 deg elbow | EA. | 200.00 | 17.75 | 217.75 |
| Union tee | " | 250.00 | 24.35 | 274.35 |
| Union | " | 130.00 | 24.35 | 154.35 |
| Male connector | " | 89.50 | 19.50 | 109.00 |
| Type 316 valves | | | | |
| Gate valves | | | | |

## PLUMBING

| | UNIT | MAT. | INST. | TOTAL |
|---|---|---|---|---|
| **15410.70 STAINLESS STEEL PIPE** | | | | |
| 1/4" | EA. | 300.00 | 13.00 | 313.00 |
| 3/8" | " | 300.00 | 15.60 | 315.60 |
| 1/2" | " | 190.00 | 16.95 | 206.95 |
| 3/4" | " | 240.00 | 19.50 | 259.50 |
| 1" | " | 260.00 | 26.00 | 286.00 |
| Globe valves | | | | |
| 1/4" | EA. | 120.00 | 13.00 | 133.00 |
| 3/8" | " | 190.00 | 15.60 | 205.60 |
| 1/2" | " | 230.00 | 16.95 | 246.95 |
| 3/4" | " | 270.00 | 19.50 | 289.50 |
| 1" | " | 280.00 | 26.00 | 306.00 |
| Check valves | | | | |
| 1/4" | EA. | 89.50 | 13.00 | 102.50 |
| 3/8" | " | 170.00 | 15.60 | 185.60 |
| 1/2" | " | 200.00 | 16.95 | 216.95 |
| 3/4" | " | 260.00 | 19.50 | 279.50 |
| 1" | " | 280.00 | 26.00 | 306.00 |
| Test and balance | " | 0.00 | 32.50 | 32.50 |
| Pipe identification | " | 0.14 | 7.80 | 7.94 |
| Disinfect | " | 0.00 | 32.50 | 32.50 |
| **15410.80 STEEL PIPE** | | | | |
| Black steel, extra heavy pipe, threaded | | | | |
| 1/2" pipe | L.F. | 1.80 | 1.55 | 3.35 |
| 3/4" pipe | " | 2.15 | 1.55 | 3.70 |
| 1" pipe | " | 2.95 | 1.95 | 4.90 |
| 1-1/2" pipe | " | 4.45 | 2.15 | 6.60 |
| 2-1/2" pipe | " | 8.90 | 4.85 | 13.75 |
| 3" pipe | " | 11.90 | 6.50 | 18.40 |
| 4" pipe | " | 18.00 | 7.80 | 25.80 |
| 5" pipe | " | 24.25 | 9.75 | 34.00 |
| 6" pipe | " | 30.30 | 9.75 | 40.05 |
| 8" pipe | " | 44.70 | 13.00 | 57.70 |
| 10" pipe | " | 62.00 | 15.60 | 77.60 |
| 12" pipe | " | 79.00 | 19.50 | 98.50 |
| Fittings, malleable iron, threaded, 1/2" pipe | | | | |
| 90 deg ell | EA. | 1.80 | 13.00 | 14.80 |
| 45 deg ell | " | 2.90 | 13.00 | 15.90 |
| Tee | " | 2.40 | 19.50 | 21.90 |
| Reducing tee | " | 5.30 | 19.50 | 24.80 |
| Cap | " | 1.80 | 7.80 | 9.60 |
| Coupling | " | 2.20 | 15.60 | 17.80 |
| Union | " | 10.30 | 13.00 | 23.30 |
| Nipple, 4" long | " | 1.80 | 13.00 | 14.80 |
| 3/4" pipe | | | | |
| 90 deg ell | EA. | 2.40 | 13.00 | 15.40 |
| 45 deg ell | " | 3.95 | 19.50 | 23.45 |
| Tee | " | 4.00 | 19.50 | 23.50 |
| Reducing tee | " | 5.90 | 13.00 | 18.90 |
| Cap | " | 2.45 | 7.80 | 10.25 |
| Coupling | " | 2.65 | 13.00 | 15.65 |
| Union | " | 11.65 | 13.00 | 24.65 |

| PLUMBING | UNIT | MAT. | INST. | TOTAL |
|---|---|---|---|---|
| **15410.80 STEEL PIPE** | | | | |
| Nipple, 4" long | EA. | 2.65 | 13.00 | 15.65 |
| 1" pipe | | | | |
| 90 deg ell | EA. | 3.95 | 15.60 | 19.55 |
| 45 deg ell | " | 4.40 | 15.60 | 20.00 |
| Tee | " | 5.90 | 21.65 | 27.55 |
| Reducing tee | " | 7.55 | 21.65 | 29.20 |
| Cap | " | 3.30 | 7.80 | 11.10 |
| Coupling | " | 4.55 | 15.60 | 20.15 |
| Union | " | 13.95 | 15.60 | 29.55 |
| Nipple, 4" long | " | 3.15 | 15.60 | 18.75 |
| 1-1/2" pipe | | | | |
| 90 deg ell | EA. | 7.50 | 19.50 | 27.00 |
| 45 deg ell | " | 8.75 | 19.50 | 28.25 |
| Tee | " | 10.85 | 27.90 | 38.75 |
| Reducing tee | " | 16.75 | 27.90 | 44.65 |
| Cap | " | 4.85 | 9.75 | 14.60 |
| Coupling | " | 6.65 | 19.50 | 26.15 |
| Union | " | 18.15 | 19.50 | 37.65 |
| Nipple, 4" long | " | 4.70 | 19.50 | 24.20 |
| 2-1/2" pipe | | | | |
| 90 deg ell | EA. | 30.30 | 48.80 | 79.10 |
| 45 deg ell | " | 37.90 | 48.80 | 86.70 |
| Tee | " | 40.90 | 65.00 | 105.90 |
| Reducing tee | " | 53.00 | 65.00 | 118.00 |
| Cap | " | 15.90 | 24.35 | 40.25 |
| Coupling | " | 25.00 | 65.00 | 90.00 |
| Union | " | 68.00 | 65.00 | 133.00 |
| Nipple, 4" long | " | 15.90 | 65.00 | 80.90 |
| 3" pipe | | | | |
| 90 deg ell | EA. | 43.90 | 65.00 | 108.90 |
| 45 deg ell | " | 50.00 | 65.00 | 115.00 |
| Tee | " | 56.00 | 97.50 | 153.50 |
| Reducing tee | " | 89.50 | 97.50 | 187.00 |
| Cap | " | 24.25 | 32.50 | 56.75 |
| Coupling | " | 34.80 | 65.00 | 99.80 |
| Union | " | 110.00 | 65.00 | 175.00 |
| Nipple, 4" long | " | 19.70 | 65.00 | 84.70 |
| 4" pipe | | | | |
| 90 deg ell | EA. | 88.00 | 78.00 | 166.00 |
| 45 deg ell | " | 89.50 | 78.00 | 167.50 |
| Tee | " | 130.00 | 130.00 | 260.00 |
| Reducing tee | " | 140.00 | 130.00 | 270.00 |
| Cap | " | 36.40 | 130.00 | 166.40 |
| Coupling | " | 68.00 | 39.00 | 107.00 |
| Union | " | 120.00 | 130.00 | 250.00 |
| Nipple, 4" long | " | 24.25 | 130.00 | 154.25 |
| 6" pipe | | | | |
| 90 deg ell | EA. | 250.00 | 78.00 | 328.00 |
| 45 deg ell | " | 300.00 | 78.00 | 378.00 |
| Tee | " | 350.00 | 130.00 | 480.00 |
| Reducing tee | " | 340.00 | 130.00 | 470.00 |
| Cap | " | 140.00 | 39.00 | 179.00 |

| PLUMBING | UNIT | MAT. | INST. | TOTAL |
|---|---|---|---|---|
| **15410.80 STEEL PIPE** | | | | |
| 8" pipe | | | | |
| 90 deg ell | EA. | 260.00 | 160.00 | 420.00 |
| 45 deg ell | " | 320.00 | 160.00 | 480.00 |
| Tee | " | 380.00 | 240.00 | 620.00 |
| Reducing tee | " | 350.00 | 210.00 | 560.00 |
| Cap | " | 42.40 | 78.00 | 120.40 |
| 10" pipe | | | | |
| 90 deg ell | EA. | 290.00 | 200.00 | 490.00 |
| 45 deg ell | " | 340.00 | 200.00 | 540.00 |
| Tee | " | 430.00 | 240.00 | 670.00 |
| Reducing tee | " | 410.00 | 97.50 | 507.50 |
| Cap | " | 42.40 | 97.50 | 139.90 |
| 12" pipe | | | | |
| 90 deg ell | EA. | 340.00 | 240.00 | 580.00 |
| 45 deg ell | " | 390.00 | 240.00 | 630.00 |
| Tee | " | 500.00 | 330.00 | 830.00 |
| Reducing tee | " | 460.00 | 330.00 | 790.00 |
| Cap | " | 43.90 | 130.00 | 173.90 |
| **15410.82 GALVANIZED STEEL PIPE** | | | | |
| Galvanized pipe | | | | |
| 1/2" pipe | L.F. | 1.95 | 3.90 | 5.85 |
| 3/4" pipe | " | 2.50 | 4.85 | 7.35 |
| 1" pipe | " | 3.85 | 5.55 | 9.40 |
| 1-1/4" pipe | " | 4.45 | 6.50 | 10.95 |
| 1-1/2" pipe | " | 4.75 | 7.80 | 12.55 |
| 2" pipe | " | 6.85 | 9.75 | 16.60 |
| 2-1/2" pipe | " | 9.80 | 13.00 | 22.80 |
| 3" pipe | " | 13.65 | 13.95 | 27.60 |
| 4" pipe | " | 18.55 | 16.25 | 34.80 |
| 6" pipe | " | 37.10 | 32.50 | 69.60 |
| 90 degree ell, 150 lb malleable iron, galvanized | | | | |
| 1/2" | EA. | 1.50 | 7.80 | 9.30 |
| 3/4" | " | 1.80 | 9.75 | 11.55 |
| 1" | " | 3.65 | 10.25 | 13.90 |
| 1-1/4" | " | 6.05 | 11.45 | 17.50 |
| 1-1/2" | " | 6.35 | 13.00 | 19.35 |
| 2" | " | 9.10 | 15.60 | 24.70 |
| 2-1/2" | " | 15.15 | 24.35 | 39.50 |
| 3" | " | 21.20 | 30.00 | 51.20 |
| 4" | " | 33.30 | 32.50 | 65.80 |
| 5" | " | 79.00 | 39.00 | 118.00 |
| 6" | " | 95.50 | 39.00 | 134.50 |
| 45 degree ell, 150 lb m.i., galv. | | | | |
| 1/2" | EA. | 1.80 | 7.80 | 9.60 |
| 3/4" | " | 3.00 | 9.75 | 12.75 |
| 1" | " | 4.55 | 10.25 | 14.80 |
| 1-1/4" | " | 6.05 | 11.45 | 17.50 |
| 1-1/2" | " | 7.55 | 13.00 | 20.55 |
| 2" | " | 9.10 | 15.60 | 24.70 |
| 2-1/2" | " | 15.15 | 24.35 | 39.50 |
| 3" | " | 22.70 | 30.00 | 52.70 |

| PLUMBING | UNIT | MAT. | INST. | TOTAL |
|---|---|---|---|---|
| **15410.82 GALVANIZED STEEL PIPE** | | | | |
| 4" | EA. | 34.80 | 39.00 | 73.80 |
| 5" | " | 91.00 | 39.00 | 130.00 |
| 6" | " | 110.00 | 48.80 | 158.80 |
| Tees, straight, 150 lb m.i., galv. | | | | |
| 1/2" | EA. | 2.10 | 9.75 | 11.85 |
| 3/4" | " | 3.65 | 11.15 | 14.80 |
| 1" | " | 5.15 | 13.00 | 18.15 |
| 1-1/4" | " | 6.05 | 15.60 | 21.65 |
| 1-1/2" | " | 7.55 | 19.50 | 27.05 |
| 2" | " | 12.10 | 24.35 | 36.45 |
| 2-1/2" | " | 21.20 | 32.50 | 53.70 |
| 3" | " | 22.70 | 39.00 | 61.70 |
| 4" | " | 45.40 | 48.80 | 94.20 |
| 5" | " | 91.00 | 55.50 | 146.50 |
| 6" | " | 120.00 | 65.00 | 185.00 |
| Couplings, straight, 150 lb m.i., galv. | | | | |
| 1/2" | EA. | 2.40 | 7.80 | 10.20 |
| 3/4" | " | 3.00 | 8.65 | 11.65 |
| 1" | " | 4.55 | 9.75 | 14.30 |
| 1-1/4" | " | 6.05 | 11.15 | 17.20 |
| 1-1/2" | " | 6.80 | 13.00 | 19.80 |
| 2" | " | 8.35 | 15.60 | 23.95 |
| 2-1/2" | " | 10.60 | 24.35 | 34.95 |
| 3" | " | 12.90 | 32.50 | 45.40 |
| 4" | " | 28.80 | 35.50 | 64.30 |
| 5" | " | 48.50 | 39.00 | 87.50 |
| 6" | " | 66.50 | 39.00 | 105.50 |
| Caps, 150 lb m.i., galv. | | | | |
| 1/2" | EA. | 1.50 | 3.90 | 5.40 |
| 3/4" | " | 1.80 | 4.10 | 5.90 |
| 1" | " | 2.75 | 4.35 | 7.10 |
| 1-1/4" | " | 3.30 | 4.60 | 7.90 |
| 1-1/2" | " | 4.55 | 4.85 | 9.40 |
| 2" | " | 4.85 | 5.55 | 10.40 |
| 2-1/2" | " | 6.05 | 7.10 | 13.15 |
| 3" | " | 9.10 | 9.75 | 18.85 |
| 4" | " | 15.15 | 12.20 | 27.35 |
| 5" | " | 31.80 | 15.00 | 46.80 |
| 6" | " | 59.00 | 19.50 | 78.50 |
| Unions, 150 lb m.i., galv. | | | | |
| 1/2" | EA. | 4.55 | 9.75 | 14.30 |
| 3/4" | " | 4.85 | 11.15 | 16.00 |
| 1" | " | 6.05 | 13.00 | 19.05 |
| 1-1/4" | " | 7.55 | 15.60 | 23.15 |
| 1-1/2" | " | 9.10 | 19.50 | 28.60 |
| 2" | " | 12.10 | 21.65 | 33.75 |
| 2-1/2" | " | 21.20 | 26.00 | 47.20 |
| 3" | " | 33.30 | 32.50 | 65.80 |
| Nipples, galvanized steel, 4" long | | | | |
| 1/2" | EA. | 1.80 | 4.85 | 6.65 |
| 3/4" | " | 2.45 | 5.20 | 7.65 |
| 1" | " | 3.30 | 5.55 | 8.85 |

| PLUMBING | UNIT | MAT. | INST. | TOTAL |
|---|---|---|---|---|
| **15410.82 GALVANIZED STEEL PIPE** | | | | |
| 1-1/4" | EA. | 3.80 | 6.00 | 9.80 |
| 1-1/2" | " | 4.55 | 6.50 | 11.05 |
| 2" | " | 5.30 | 7.10 | 12.40 |
| 2-1/2" | " | 6.35 | 7.80 | 14.15 |
| 3" | " | 8.35 | 9.75 | 18.10 |
| 4" | " | 10.45 | 13.00 | 23.45 |
| Square head plug (C.I.) | | | | |
| 1/2" | EA. | 1.35 | 4.35 | 5.70 |
| 3/4" | " | 3.00 | 4.85 | 7.85 |
| 1" | " | 3.15 | 5.20 | 8.35 |
| 1-1/4" | " | 3.30 | 5.55 | 8.85 |
| 1-1/2" | " | 4.15 | 6.00 | 10.15 |
| 2" | " | 5.30 | 6.50 | 11.80 |
| 2-1/2" | " | 6.95 | 8.65 | 15.60 |
| 3" | " | 10.15 | 9.75 | 19.90 |
| 4" | " | 14.80 | 13.00 | 27.80 |
| 5" | " | 24.25 | 15.60 | 39.85 |
| 6" | " | 33.30 | 19.50 | 52.80 |
| Screwed flanges, galv. | | | | |
| 1" | EA. | 12.40 | 19.50 | 31.90 |
| 1-1/4" | " | 13.65 | 21.65 | 35.30 |
| 1-1/2" | " | 15.45 | 24.35 | 39.80 |
| 2" | " | 16.50 | 24.35 | 40.85 |
| 2-1/2" | " | 17.40 | 26.00 | 43.40 |
| 3" | " | 22.70 | 35.50 | 58.20 |
| 4" | " | 30.30 | 48.80 | 79.10 |
| 5" | " | 37.90 | 65.00 | 102.90 |
| 6" | " | 48.50 | 65.00 | 113.50 |
| **15430.23 CLEANOUTS** | | | | |
| Cleanout, wall | | | | |
| 2" | EA. | 67.50 | 26.00 | 93.50 |
| 3" | " | 74.50 | 26.00 | 100.50 |
| 4" | " | 97.00 | 32.50 | 129.50 |
| 6" | " | 170.00 | 39.00 | 209.00 |
| 8" | " | 200.00 | 48.80 | 248.80 |
| Floor | | | | |
| 2" | EA. | 69.50 | 32.50 | 102.00 |
| 3" | " | 88.00 | 32.50 | 120.50 |
| 4" | " | 97.00 | 39.00 | 136.00 |
| 6" | " | 130.00 | 48.80 | 178.80 |
| 8" | " | 250.00 | 55.50 | 305.50 |
| **15430.24 GREASE TRAPS** | | | | |
| Grease traps, cast iron, 3" pipe | | | | |
| 35 gpm, 70 lb capacity | EA. | 2,270 | 390.00 | 2,660 |
| 50 gpm, 100 lb capacity | " | 2,890 | 490.00 | 3,380 |

| PLUMBING | UNIT | MAT. | INST. | TOTAL |
|---|---|---|---|---|
| **15430.25 HOSE BIBBS** | | | | |
| Hose bibb | | | | |
| 1/2" | EA. | 6.35 | 13.00 | 19.35 |
| 3/4" | " | 6.75 | 13.00 | 19.75 |
| **15430.60 VALVES** | | | | |
| Gate valve, 125 lb, bronze, soldered | | | | |
| 1/2" | EA. | 15.90 | 9.75 | 25.65 |
| 3/4" | " | 18.80 | 9.75 | 28.55 |
| 1" | " | 23.15 | 13.00 | 36.15 |
| 1-1/2" | " | 40.50 | 15.60 | 56.10 |
| 2" | " | 56.50 | 19.50 | 76.00 |
| 2-1/2" | " | 120.00 | 24.35 | 144.35 |
| Threaded | | | | |
| 1/4", 125 lb | EA. | 15.60 | 15.60 | 31.20 |
| 1/2" | | | | |
| 125 lb | EA. | 17.35 | 15.60 | 32.95 |
| 150 lb | " | 23.15 | 15.60 | 38.75 |
| 300 lb | " | 40.50 | 15.60 | 56.10 |
| 3/4" | | | | |
| 125 lb | EA. | 20.25 | 15.60 | 35.85 |
| 150 lb | " | 27.50 | 15.60 | 43.10 |
| 300 lb | " | 52.00 | 15.60 | 67.60 |
| 1" | | | | |
| 125 lb | EA. | 26.00 | 15.60 | 41.60 |
| 150 lb | " | 36.20 | 15.60 | 51.80 |
| 300 lb | " | 72.50 | 19.50 | 92.00 |
| 1-1/2" | | | | |
| 125 lb | EA. | 79.50 | 19.50 | 99.00 |
| 150 lb | " | 60.50 | 19.50 | 80.00 |
| 300 lb | " | 140.00 | 21.65 | 161.65 |
| 2" | | | | |
| 125 lb | EA. | 55.00 | 27.90 | 82.90 |
| 150 lb | " | 84.00 | 27.90 | 111.90 |
| 300 lb | " | 200.00 | 32.50 | 232.50 |
| Cast iron, flanged | | | | |
| 2", 150 lb | EA. | 240.00 | 32.50 | 272.50 |
| 2-1/2" | | | | |
| 125 lb | EA. | 230.00 | 32.50 | 262.50 |
| 150 lb | " | 360.00 | 32.50 | 392.50 |
| 250 lb | " | 460.00 | 32.50 | 492.50 |
| 3" | | | | |
| 125 lb | EA. | 270.00 | 39.00 | 309.00 |
| 150 lb | " | 380.00 | 39.00 | 419.00 |
| 250 lb | " | 580.00 | 39.00 | 619.00 |
| 4" | | | | |
| 125 lb | EA. | 330.00 | 55.50 | 385.50 |
| 150 lb | " | 450.00 | 55.50 | 505.50 |
| 250 lb | " | 780.00 | 55.50 | 835.50 |
| 6" | | | | |
| 125 lb | EA. | 510.00 | 78.00 | 588.00 |
| 250 lb | " | 1,300 | 78.00 | 1,378 |

# 15 MECHANICAL

| PLUMBING | UNIT | MAT. | INST. | TOTAL |
|---|---|---|---|---|
| **15430.60 VALVES** | | | | |
| 8" | | | | |
| 125 lb | EA. | 1,010 | 97.50 | 1,108 |
| 250 lb | " | 2,460 | 97.50 | 2,558 |
| OS&Y, flanged | | | | |
| 2" | | | | |
| 125 lb | EA. | 170.00 | 32.50 | 202.50 |
| 250 lb | " | 400.00 | 32.50 | 432.50 |
| 2-1/2" | | | | |
| 125 lb | EA. | 190.00 | 32.50 | 222.50 |
| 250 lb | " | 460.00 | 39.00 | 499.00 |
| 3" | | | | |
| 125 lb | EA. | 190.00 | 39.00 | 229.00 |
| 250 lb | " | 510.00 | 39.00 | 549.00 |
| 4" | | | | |
| 125 lb | EA. | 270.00 | 65.00 | 335.00 |
| 250 lb | " | 780.00 | 65.00 | 845.00 |
| 6" | | | | |
| 125 lb | EA. | 430.00 | 78.00 | 508.00 |
| 250 lb | " | 1,260 | 78.00 | 1,338 |
| Check valve, bronze, soldered, 125 lb | | | | |
| 1/2" | EA. | 19.30 | 9.75 | 29.05 |
| 3/4" | " | 22.05 | 9.75 | 31.80 |
| 1" | " | 28.90 | 13.00 | 41.90 |
| 1-1/4" | " | 40.50 | 15.60 | 56.10 |
| 1-1/2" | " | 47.30 | 15.60 | 62.90 |
| 2" | " | 67.50 | 19.50 | 87.00 |
| Threaded | | | | |
| 1/2" | | | | |
| 125 lb | EA. | 15.90 | 13.00 | 28.90 |
| 150 lb | " | 26.00 | 13.00 | 39.00 |
| 200 lb | " | 26.60 | 13.00 | 39.60 |
| 3/4" | | | | |
| 125 lb | EA. | 20.25 | 15.60 | 35.85 |
| 150 lb | " | 31.80 | 15.60 | 47.40 |
| 200 lb | " | 31.80 | 15.60 | 47.40 |
| 1" | | | | |
| 125 lb | EA. | 26.00 | 19.50 | 45.50 |
| 150 lb | " | 41.90 | 19.50 | 61.40 |
| 200 lb | " | 41.90 | 19.50 | 61.40 |
| Flow check valve, cast iron, threaded | | | | |
| 1" | EA. | 40.60 | 15.60 | 56.20 |
| 1-1/4" | " | 43.20 | 19.50 | 62.70 |
| 1-1/2" | | | | |
| 125 lb | EA. | 39.50 | 19.50 | 59.00 |
| 150 lb | " | 65.50 | 19.50 | 85.00 |
| 200 lb | " | 65.50 | 21.65 | 87.15 |
| 2" | | | | |
| 125 lb | EA. | 58.00 | 21.65 | 79.65 |
| 150 lb | " | 96.00 | 21.65 | 117.65 |
| 200 lb | " | 96.00 | 24.35 | 120.35 |
| 2-1/2" | | | | |
| 125 lb | EA. | 140.00 | 32.50 | 172.50 |

| PLUMBING | UNIT | MAT. | INST. | TOTAL |
|---|---|---|---|---|
| **15430.60 VALVES** | | | | |
| 250 lb | EA. | 420.00 | 39.00 | 459.00 |
| 3" | | | | |
| 125 lb | EA. | 150.00 | 39.00 | 189.00 |
| 250 lb | " | 530.00 | 48.80 | 578.80 |
| 4" | | | | |
| 125 lb | EA. | 230.00 | 55.50 | 285.50 |
| 250 lb | " | 680.00 | 65.00 | 745.00 |
| 6" | | | | |
| 125 lb | EA. | 320.00 | 78.00 | 398.00 |
| 250 lb | " | 1,120 | 78.00 | 1,198 |
| Vertical check valve, bronze, 125 lb, threaded | | | | |
| 1/2" | EA. | 31.80 | 15.60 | 47.40 |
| 3/4" | " | 41.90 | 17.75 | 59.65 |
| 1" | " | 50.50 | 19.50 | 70.00 |
| 1-1/4" | " | 62.00 | 21.65 | 83.65 |
| 1-1/2" | " | 69.50 | 24.35 | 93.85 |
| 2" | " | 110.00 | 27.90 | 137.90 |
| Cast iron, flanged | | | | |
| 2-1/2" | EA. | 410.00 | 39.00 | 449.00 |
| 3" | " | 450.00 | 48.80 | 498.80 |
| 4" | " | 620.00 | 65.00 | 685.00 |
| 6 | " | 1,090 | 78.00 | 1,168 |
| 8" | " | 2,120 | 97.50 | 2,218 |
| 10" | " | 3,180 | 130.00 | 3,310 |
| 12" | " | 3,940 | 160.00 | 4,100 |
| Globe valve, bronze, soldered, 125 lb | | | | |
| 1/2" | EA. | 17.35 | 11.15 | 28.50 |
| 3/4" | " | 20.25 | 12.20 | 32.45 |
| 1" | " | 36.20 | 13.00 | 49.20 |
| 1-1/4" | " | 56.50 | 13.95 | 70.45 |
| 1-1/2" | " | 73.50 | 16.25 | 89.75 |
| 2" | " | 130.00 | 19.50 | 149.50 |
| Threaded | | | | |
| 1/2" | | | | |
| 125 lb | EA. | 17.35 | 13.00 | 30.35 |
| 150 lb | " | 49.20 | 13.00 | 62.20 |
| 300 lb | " | 60.50 | 13.00 | 73.50 |
| 3/4" | | | | |
| 125 lb | EA. | 20.25 | 15.60 | 35.85 |
| 150 lb | " | 52.00 | 15.60 | 67.60 |
| 300 lb | " | 63.50 | 15.60 | 79.10 |
| 1" | | | | |
| 125 lb | EA. | 36.20 | 19.50 | 55.70 |
| 150 lb | " | 110.00 | 19.50 | 129.50 |
| 300 lb | " | 130.00 | 19.50 | 149.50 |
| 1-1/4" | | | | |
| 125 lb | EA. | 65.00 | 19.50 | 84.50 |
| 150 lb | " | 150.00 | 19.50 | 169.50 |
| 300 lb | " | 180.00 | 19.50 | 199.50 |
| 1-1/2" | | | | |
| 125 lb | EA. | 75.00 | 21.65 | 96.65 |
| 150 lb | " | 170.00 | 21.65 | 191.65 |

# 15 MECHANICAL

## PLUMBING

### 15430.60 VALVES

| | UNIT | MAT. | INST. | TOTAL |
|---|---|---|---|---|
| 300 lb | EA. | 200.00 | 21.65 | 221.65 |
| 2" | | | | |
| 125 lb | EA. | 130.00 | 26.00 | 156.00 |
| 150 lb | " | 330.00 | 26.00 | 356.00 |
| 300 lb | " | 400.00 | 26.00 | 426.00 |
| Cast iron flanged | | | | |
| 2-1/2" | | | | |
| 125 lb | EA. | 620.00 | 39.00 | 659.00 |
| 250 lb | " | 1,130 | 39.00 | 1,169 |
| 3" | | | | |
| 125 lb | EA. | 730.00 | 48.80 | 778.80 |
| 250 lb | " | 1,220 | 48.80 | 1,269 |
| 4" | | | | |
| 125 lb | EA. | 1,040 | 65.00 | 1,105 |
| 250 lb | " | 1,700 | 65.00 | 1,765 |
| 6" | | | | |
| 125 lb | EA. | 1,860 | 78.00 | 1,938 |
| 250 lb | " | 2,290 | 78.00 | 2,368 |
| 8" | | | | |
| 125 lb | EA. | 3,060 | 97.50 | 3,158 |
| 250 lb | " | 3,580 | 97.50 | 3,678 |
| Butterfly valve, cast iron, wafer type | | | | |
| 2" | | | | |
| 150 lb | EA. | 150.00 | 27.90 | 177.90 |
| 200 lb | " | 200.00 | 32.50 | 232.50 |
| 2-1/2" | | | | |
| 150 lb | EA. | 200.00 | 32.50 | 232.50 |
| 200 lb | " | 220.00 | 35.50 | 255.50 |
| 3" | | | | |
| 150 lb | EA. | 200.00 | 39.00 | 239.00 |
| 200 lb | " | 220.00 | 43.30 | 263.30 |
| 4" | | | | |
| 150 lb | EA. | 230.00 | 55.50 | 285.50 |
| 200 lb | " | 260.00 | 65.00 | 325.00 |
| 6" | | | | |
| 150 lb | EA. | 270.00 | 78.00 | 348.00 |
| 200 lb | " | 340.00 | 78.00 | 418.00 |
| 8" | | | | |
| 150 lb | EA. | 340.00 | 86.50 | 426.50 |
| 200 lb | " | 410.00 | 97.50 | 507.50 |
| 10" | | | | |
| 150 lb | EA. | 410.00 | 97.50 | 507.50 |
| 200 lb | " | 540.00 | 130.00 | 670.00 |
| Ball valve, bronze, 250 lb, threaded | | | | |
| 1/2" | EA. | 9.65 | 15.60 | 25.25 |
| 3/4" | " | 14.45 | 15.60 | 30.05 |
| 1" | " | 16.50 | 19.50 | 36.00 |
| 1-1/4" | " | 26.90 | 21.65 | 48.55 |
| 1-1/2" | " | 42.70 | 24.35 | 67.05 |
| 2" | " | 48.20 | 27.90 | 76.10 |
| Angle valve, bronze, 150 lb, threaded | | | | |
| 1/2" | EA. | 52.50 | 13.95 | 66.45 |

## PLUMBING

### 15430.60 VALVES

| | UNIT | MAT. | INST. | TOTAL |
|---|---|---|---|---|
| 3/4" | EA. | 69.00 | 15.60 | 84.60 |
| 1" | " | 100.00 | 15.60 | 115.60 |
| 1-1/4" | " | 120.00 | 19.50 | 139.50 |
| 1-1/2" | " | 150.00 | 21.65 | 171.65 |
| Balancing valve, with meter connections, circuit setter | | | | |
| 1/2" | EA. | 50.50 | 15.60 | 66.10 |
| 3/4" | " | 53.50 | 17.75 | 71.25 |
| 1" | " | 69.50 | 19.50 | 89.00 |
| 1-1/4" | " | 95.50 | 21.65 | 117.15 |
| 1-1/2" | " | 110.00 | 26.00 | 136.00 |
| 2" | " | 170.00 | 32.50 | 202.50 |
| 2-1/2" | " | 320.00 | 39.00 | 359.00 |
| 3" | " | 470.00 | 48.80 | 518.80 |
| 4" | " | 670.00 | 65.00 | 735.00 |
| Balancing valve, straight type | | | | |
| 1/2" | EA. | 13.75 | 15.60 | 29.35 |
| 3/4" | " | 16.65 | 15.60 | 32.25 |
| Angle type | | | | |
| 1/2" | EA. | 18.50 | 15.60 | 34.10 |
| 3/4" | " | 25.70 | 15.60 | 41.30 |
| Square head cock, 125 lb, bronze body | | | | |
| 1/2" | EA. | 10.85 | 13.00 | 23.85 |
| 3/4" | " | 13.00 | 15.60 | 28.60 |
| 1" | " | 18.10 | 17.75 | 35.85 |
| 1-1/4" | " | 24.60 | 19.50 | 44.10 |
| Radiator temp control valve, with control and sensor | | | | |
| 1/2" valve | EA. | 72.50 | 24.35 | 96.85 |
| 1" valve | " | 84.00 | 24.35 | 108.35 |
| Pressure relief valve, 1/2", bronze | | | | |
| Low pressure | EA. | 17.35 | 15.60 | 32.95 |
| High pressure | " | 20.25 | 15.60 | 35.85 |
| Press and temperature relief valve | | | | |
| Bronze, 3/4" | EA. | 62.00 | 15.60 | 77.60 |
| Cast iron, 3/4" | | | | |
| High pressure | EA. | 33.30 | 15.60 | 48.90 |
| Temperature relief | " | 43.40 | 15.60 | 59.00 |
| Pressure & temp relief valve | " | 43.40 | 15.60 | 59.00 |
| Pressure reducing valve, bronze, threaded, 250 lb | | | | |
| 1/2" | EA. | 300.00 | 24.35 | 324.35 |
| 3/4" | " | 300.00 | 24.35 | 324.35 |
| 1" | " | 330.00 | 24.35 | 354.35 |
| 1-1/4" | " | 390.00 | 27.90 | 417.90 |
| 1-1/2" | " | 440.00 | 32.50 | 472.50 |
| Pressure regulating valve, bronze, class 300 | | | | |
| 1" | EA. | 630.00 | 24.35 | 654.35 |
| 1-1/2" | " | 850.00 | 30.00 | 880.00 |
| 2" | " | 960.00 | 39.00 | 999.00 |
| 3" | " | 1,090 | 55.50 | 1,146 |
| 4" | " | 1,360 | 78.00 | 1,438 |
| 5" | " | 2,070 | 97.50 | 2,168 |
| 6" | " | 2,090 | 130.00 | 2,220 |
| Solar water temperature regulating valve | | | | |

## PLUMBING

| | UNIT | MAT. | INST. | TOTAL |
|---|---|---|---|---|
| **15430.60 VALVES** | | | | |
| 3/4" | EA. | 400.00 | 32.50 | 432.50 |
| 1" | " | 400.00 | 39.00 | 439.00 |
| 1-1/4" | " | 440.00 | 43.30 | 483.30 |
| 1-1/2" | " | 480.00 | 48.80 | 528.80 |
| 2" | " | 600.00 | 55.50 | 655.50 |
| 2-1/2" | " | 1,130 | 97.50 | 1,228 |
| Tempering valve, threaded | | | | |
| 3/4" | EA. | 61.00 | 13.00 | 74.00 |
| 1" | " | 160.00 | 15.60 | 175.60 |
| 1-1/4" | " | 190.00 | 19.50 | 209.50 |
| 1-1/2" | " | 250.00 | 19.50 | 269.50 |
| 2" | " | 310.00 | 24.35 | 334.35 |
| 2-1/2" | " | 390.00 | 32.50 | 422.50 |
| 3" | " | 480.00 | 39.00 | 519.00 |
| 4" | " | 1,060 | 55.50 | 1,116 |
| Thermostatic mixing valve, threaded | | | | |
| 1/2" | EA. | 72.50 | 13.95 | 86.45 |
| 3/4" | " | 76.50 | 15.60 | 92.10 |
| 1" | " | 280.00 | 16.95 | 296.95 |
| 1-1/2" | " | 300.00 | 19.50 | 319.50 |
| 2" | " | 380.00 | 24.35 | 404.35 |
| Sweat connection | | | | |
| 1/2" | EA. | 85.00 | 13.95 | 98.95 |
| 3/4" | " | 100.00 | 15.60 | 115.60 |
| Mixing valve, sweat connection | | | | |
| 1/2" | EA. | 34.00 | 13.95 | 47.95 |
| 3/4" | " | 38.30 | 15.60 | 53.90 |
| Liquid level gauge, aluminum body | | | | |
| 3/4" | EA. | 270.00 | 15.60 | 285.60 |
| 4125 psi, pvc body | | | | |
| 3/4" | EA. | 270.00 | 15.60 | 285.60 |
| 150 psi, crs body | | | | |
| 3/4" | EA. | 210.00 | 15.60 | 225.60 |
| 1" | " | 240.00 | 15.60 | 255.60 |
| 175 psi, bronze body, 1/2" | " | 440.00 | 13.95 | 453.95 |
| **15430.65 VACUUM BREAKERS** | | | | |
| Vacuum breaker, atmospheric, threaded connection | | | | |
| 3/4" | EA. | 29.70 | 15.60 | 45.30 |
| 1" | " | 43.20 | 15.60 | 58.80 |
| Anti-siphon, brass | | | | |
| 3/4" | EA. | 33.80 | 15.60 | 49.40 |
| 1" | " | 47.30 | 15.60 | 62.90 |
| 1-1/4" | " | 74.50 | 19.50 | 94.00 |
| 1-1/2" | " | 88.00 | 21.65 | 109.65 |
| 2" | " | 140.00 | 24.35 | 164.35 |
| **15430.68 STRAINERS** | | | | |
| Strainer, Y pattern, 125 psi, cast iron body, threaded | | | | |
| 3/4" | EA. | 13.75 | 13.95 | 27.70 |

| PLUMBING | UNIT | MAT. | INST. | TOTAL |
|---|---|---|---|---|
| **15430.68 STRAINERS** | | | | |
| 1" | EA. | 16.50 | 15.60 | 32.10 |
| 1-1/4" | " | 22.05 | 19.50 | 41.55 |
| 1-1/2" | " | 27.50 | 19.50 | 47.00 |
| 2" | " | 41.30 | 24.35 | 65.65 |
| 250 psi, brass body, threaded | | | | |
| 3/4" | EA. | 31.70 | 15.60 | 47.30 |
| 1" | " | 44.10 | 15.60 | 59.70 |
| 1-1/4" | " | 55.00 | 19.50 | 74.50 |
| 1-1/2" | " | 80.00 | 19.50 | 99.50 |
| 2" | " | 130.00 | 24.35 | 154.35 |
| Cast iron body, threaded | | | | |
| 3/4" | EA. | 20.10 | 15.60 | 35.70 |
| 1" | " | 25.60 | 15.60 | 41.20 |
| 1-1/4" | " | 34.10 | 19.50 | 53.60 |
| 1-1/2" | " | 45.30 | 19.50 | 64.80 |
| 2" | " | 57.50 | 24.35 | 81.85 |
| **15430.70 DRAINS, ROOF & FLOOR** | | | | |
| Floor drain, cast iron, with cast iron top | | | | |
| 2" | EA. | 56.50 | 32.50 | 89.00 |
| 3" | " | 86.00 | 32.50 | 118.50 |
| 4" | " | 110.00 | 32.50 | 142.50 |
| 6" | " | 210.00 | 39.00 | 249.00 |
| Roof drain, cast iron | | | | |
| 2" | EA. | 140.00 | 32.50 | 172.50 |
| 3" | " | 150.00 | 32.50 | 182.50 |
| 4" | " | 170.00 | 32.50 | 202.50 |
| 5" | " | 190.00 | 39.00 | 229.00 |
| 6" | " | 220.00 | 39.00 | 259.00 |
| **15430.80 TRAPS** | | | | |
| Bucket trap, threaded | | | | |
| 3/4" | EA. | 130.00 | 24.35 | 154.35 |
| 1" | " | 370.00 | 26.00 | 396.00 |
| 1-1/4" | " | 440.00 | 30.00 | 470.00 |
| 1-1/2" | " | 650.00 | 35.50 | 685.50 |
| Inverted bucket steam trap, threaded | | | | |
| 3/4" | EA. | 160.00 | 24.35 | 184.35 |
| 1" | " | 310.00 | 24.35 | 334.35 |
| 1-1/4" | " | 470.00 | 21.65 | 491.65 |
| 1-1/2" | " | 500.00 | 32.50 | 532.50 |
| Float trap, 15 psi | | | | |
| 3/4" | EA. | 110.00 | 24.35 | 134.35 |
| 1" | " | 170.00 | 26.00 | 196.00 |
| 1-1/4" | " | 230.00 | 27.90 | 257.90 |
| 1-1/2" | " | 280.00 | 32.50 | 312.50 |
| 2" | " | 500.00 | 39.00 | 539.00 |
| Float and thermostatic trap, 15 psi | | | | |
| 3/4" | EA. | 120.00 | 24.35 | 144.35 |
| 1" | " | 130.00 | 26.00 | 156.00 |
| 1-1/4" | " | 210.00 | 27.90 | 237.90 |

## PLUMBING

### 15430.80 TRAPS

| | UNIT | MAT. | INST. | TOTAL |
|---|---|---|---|---|
| 1-1/2" | EA. | 270.00 | 32.50 | 302.50 |
| 2" | " | 500.00 | 39.00 | 539.00 |
| Steam trap, cast iron body, threaded, 125 psi | | | | |
| 3/4" | EA. | 140.00 | 24.35 | 164.35 |
| 1" | " | 170.00 | 26.00 | 196.00 |
| 1-1/4" | " | 240.00 | 27.90 | 267.90 |
| 1-1/2" | " | 390.00 | 32.50 | 422.50 |
| Thermostatic trap, low pressure, angle type, 25 psi | | | | |
| 1/2" | EA. | 43.40 | 24.35 | 67.75 |
| 3/4" | " | 72.50 | 24.35 | 96.85 |
| 1" | " | 94.00 | 26.00 | 120.00 |
| 50 psi | | | | |
| 1/2" | EA. | 69.50 | 24.35 | 93.85 |
| 3/4" | " | 91.00 | 24.35 | 115.35 |
| 1" | " | 110.00 | 26.00 | 136.00 |
| Cast iron body, threaded, 125 psi | | | | |
| 3/4" | EA. | 100.00 | 24.35 | 124.35 |
| 1" | " | 120.00 | 27.90 | 147.90 |
| 1-1/4" | " | 160.00 | 30.00 | 190.00 |
| 1-1/2" | " | 230.00 | 32.50 | 262.50 |

## PLUMBING FIXTURES

### 15440.10 BATHS

| | UNIT | MAT. | INST. | TOTAL |
|---|---|---|---|---|
| Bath tub, 5' long | | | | |
| Minimum | EA. | 430.00 | 130.00 | 560.00 |
| Average | " | 950.00 | 200.00 | 1,150 |
| Maximum | " | 1,670 | 390.00 | 2,060 |
| 6' long | | | | |
| Minimum | EA. | 490.00 | 130.00 | 620.00 |
| Average | " | 990.00 | 200.00 | 1,190 |
| Maximum | " | 1,810 | 390.00 | 2,200 |
| Square tub, whirlpool, 4'x4' | | | | |
| Minimum | EA. | 1,490 | 200.00 | 1,690 |
| Average | " | 2,100 | 390.00 | 2,490 |
| Maximum | " | 4,450 | 490.00 | 4,940 |
| 5'x5' | | | | |
| Minimum | EA. | 1,520 | 200.00 | 1,720 |
| Average | " | 2,360 | 390.00 | 2,750 |
| Maximum | " | 4,450 | 490.00 | 4,940 |
| 6'x6' | | | | |
| Minimum | EA. | 1,810 | 200.00 | 2,010 |
| Average | " | 2,640 | 390.00 | 3,030 |
| Maximum | " | 5,280 | 490.00 | 5,770 |

# PLUMBING FIXTURES

| PLUMBING FIXTURES | UNIT | MAT. | INST. | TOTAL |
|---|---|---|---|---|
| **15440.10 BATHS** | | | | |
| For trim and rough-in | | | | |
| Minimum | EA. | 140.00 | 130.00 | 270.00 |
| Average | " | 200.00 | 200.00 | 400.00 |
| Maximum | " | 290.00 | 390.00 | 680.00 |
| **15440.15 FAUCETS** | | | | |
| Kitchen | | | | |
| Minimum | EA. | 120.00 | 65.00 | 185.00 |
| Average | " | 180.00 | 78.00 | 258.00 |
| Maximum | " | 230.00 | 97.50 | 327.50 |
| Bath | | | | |
| Minimum | EA. | 180.00 | 65.00 | 245.00 |
| Average | " | 270.00 | 78.00 | 348.00 |
| Maximum | " | 310.00 | 97.50 | 407.50 |
| Lavatory, domestic | | | | |
| Minimum | EA. | 120.00 | 65.00 | 185.00 |
| Average | " | 250.00 | 78.00 | 328.00 |
| Maximum | " | 380.00 | 97.50 | 477.50 |
| Hospital, patient rooms | | | | |
| Minimum | EA. | 290.00 | 97.50 | 387.50 |
| Average | " | 410.00 | 130.00 | 540.00 |
| Maximum | " | 540.00 | 200.00 | 740.00 |
| Operating room | | | | |
| Minimum | EA. | 340.00 | 97.50 | 437.50 |
| Average | " | 440.00 | 130.00 | 570.00 |
| Maximum | " | 650.00 | 200.00 | 850.00 |
| Washroom | | | | |
| Minimum | EA. | 170.00 | 65.00 | 235.00 |
| Average | " | 280.00 | 78.00 | 358.00 |
| Maximum | " | 410.00 | 97.50 | 507.50 |
| Handicapped | | | | |
| Minimum | EA. | 210.00 | 78.00 | 288.00 |
| Average | " | 300.00 | 97.50 | 397.50 |
| Maximum | " | 450.00 | 130.00 | 580.00 |
| Shower | | | | |
| Minimum | EA. | 140.00 | 65.00 | 205.00 |
| Average | " | 260.00 | 78.00 | 338.00 |
| Maximum | " | 410.00 | 97.50 | 507.50 |
| For trim and rough-in | | | | |
| Minimum | EA. | 61.50 | 78.00 | 139.50 |
| Average | " | 92.00 | 97.50 | 189.50 |
| Maximum | " | 140.00 | 200.00 | 340.00 |
| **15440.18 HYDRANTS** | | | | |
| Wall hydrant | | | | |
| 8" thick | EA. | 130.00 | 65.00 | 195.00 |
| 12" thick | " | 150.00 | 78.00 | 228.00 |
| 18" thick | " | 170.00 | 86.50 | 256.50 |
| 24" thick | " | 190.00 | 97.50 | 287.50 |
| Ground hydrant | | | | |
| 2' deep | EA. | 250.00 | 48.80 | 298.80 |
| 4' deep | " | 260.00 | 55.50 | 315.50 |

# PLUMBING FIXTURES

| PLUMBING FIXTURES | UNIT | MAT. | INST. | TOTAL |
|---|---|---|---|---|
| **15440.18 HYDRANTS** | | | | |
| 6' deep | EA. | 290.00 | 65.00 | 355.00 |
| 8' deep | " | 360.00 | 97.50 | 457.50 |
| **15440.20 LAVATORIES** | | | | |
| Lavatory, counter top, porcelain enamel on cast iron | | | | |
| Minimum | EA. | 150.00 | 78.00 | 228.00 |
| Average | " | 230.00 | 97.50 | 327.50 |
| Maximum | " | 420.00 | 130.00 | 550.00 |
| Wall hung, china | | | | |
| Minimum | EA. | 220.00 | 78.00 | 298.00 |
| Average | " | 260.00 | 97.50 | 357.50 |
| Maximum | " | 610.00 | 130.00 | 740.00 |
| Handicapped | | | | |
| Minimum | EA. | 370.00 | 97.50 | 467.50 |
| Average | " | 420.00 | 130.00 | 550.00 |
| Maximum | " | 670.00 | 200.00 | 870.00 |
| For trim and rough-in | | | | |
| Minimum | EA. | 150.00 | 97.50 | 247.50 |
| Average | " | 180.00 | 130.00 | 310.00 |
| Maximum | " | 310.00 | 200.00 | 510.00 |
| **15440.30 SHOWERS** | | | | |
| Shower, fiberglass, 36"x34"x84" | | | | |
| Minimum | EA. | 600.00 | 280.00 | 880.00 |
| Average | " | 830.00 | 390.00 | 1,220 |
| Maximum | " | 1,200 | 390.00 | 1,590 |
| Steel, 1 piece, 36"x36" | | | | |
| Minimum | EA. | 550.00 | 280.00 | 830.00 |
| Average | " | 830.00 | 390.00 | 1,220 |
| Maximum | " | 980.00 | 390.00 | 1,370 |
| Receptor, molded stone, 36"x36" | | | | |
| Minimum | EA. | 180.00 | 130.00 | 310.00 |
| Average | " | 310.00 | 200.00 | 510.00 |
| Maximum | " | 480.00 | 330.00 | 810.00 |
| For trim and rough-in | | | | |
| Minimum | EA. | 140.00 | 180.00 | 320.00 |
| Average | " | 180.00 | 220.00 | 400.00 |
| Maximum | " | 240.00 | 390.00 | 630.00 |
| **15440.40 SINKS** | | | | |
| Service sink, 24"x29" | | | | |
| Minimum | EA. | 550.00 | 97.50 | 647.50 |
| Average | " | 690.00 | 130.00 | 820.00 |
| Maximum | " | 920.00 | 200.00 | 1,120 |
| Kitchen sink, single, stainless steel, single bowl | | | | |
| Minimum | EA. | 210.00 | 78.00 | 288.00 |
| Average | " | 240.00 | 97.50 | 337.50 |
| Maximum | " | 420.00 | 130.00 | 550.00 |
| Double bowl | | | | |
| Minimum | EA. | 240.00 | 97.50 | 337.50 |
| Average | " | 260.00 | 130.00 | 390.00 |
| Maximum | " | 430.00 | 200.00 | 630.00 |

| PLUMBING FIXTURES | UNIT | MAT. | INST. | TOTAL |
|---|---|---|---|---|
| **15440.40 SINKS** | | | | |
| Porcelain enamel, cast iron, single bowl | | | | |
| Minimum | EA. | 170.00 | 78.00 | 248.00 |
| Average | " | 200.00 | 97.50 | 297.50 |
| Maximum | " | 290.00 | 130.00 | 420.00 |
| Double bowl | | | | |
| Minimum | EA. | 240.00 | 97.50 | 337.50 |
| Average | " | 260.00 | 130.00 | 390.00 |
| Maximum | " | 380.00 | 200.00 | 580.00 |
| Mop sink, 24"x36"x10" | | | | |
| Minimum | EA. | 400.00 | 78.00 | 478.00 |
| Average | " | 490.00 | 97.50 | 587.50 |
| Maximum | " | 650.00 | 130.00 | 780.00 |
| Washing machine box | | | | |
| Minimum | EA. | 150.00 | 97.50 | 247.50 |
| Average | " | 210.00 | 130.00 | 340.00 |
| Maximum | " | 260.00 | 200.00 | 460.00 |
| For trim and rough-in | | | | |
| Minimum | EA. | 220.00 | 130.00 | 350.00 |
| Average | " | 340.00 | 200.00 | 540.00 |
| Maximum | " | 420.00 | 260.00 | 680.00 |
| **15440.50 URINALS** | | | | |
| Urinal, flush valve, floor mounted | | | | |
| Minimum | EA. | 700.00 | 97.50 | 797.50 |
| Average | " | 830.00 | 130.00 | 960.00 |
| Maximum | " | 970.00 | 200.00 | 1,170 |
| Wall mounted | | | | |
| Minimum | EA. | 350.00 | 97.50 | 447.50 |
| Average | " | 490.00 | 130.00 | 620.00 |
| Maximum | " | 630.00 | 200.00 | 830.00 |
| For trim and rough-in | | | | |
| Minimum | EA. | 120.00 | 97.50 | 217.50 |
| Average | " | 180.00 | 200.00 | 380.00 |
| Maximum | " | 230.00 | 260.00 | 490.00 |
| **15440.60 WATER CLOSETS** | | | | |
| Water closet flush tank, floor mounted | | | | |
| Minimum | EA. | 280.00 | 97.50 | 377.50 |
| Average | " | 560.00 | 130.00 | 690.00 |
| Maximum | " | 1,100 | 200.00 | 1,300 |
| Handicapped | | | | |
| Minimum | EA. | 330.00 | 130.00 | 460.00 |
| Average | " | 570.00 | 200.00 | 770.00 |
| Maximum | " | 880.00 | 390.00 | 1,270 |
| Bowl, with flush valve, floor mounted | | | | |
| Minimum | EA. | 390.00 | 97.50 | 487.50 |
| Average | " | 440.00 | 130.00 | 570.00 |
| Maximum | " | 850.00 | 200.00 | 1,050 |
| Wall mounted | | | | |
| Minimum | EA. | 390.00 | 97.50 | 487.50 |
| Average | " | 470.00 | 130.00 | 600.00 |
| Maximum | " | 890.00 | 200.00 | 1,090 |

| PLUMBING FIXTURES | UNIT | MAT. | INST. | TOTAL |
|---|---|---|---|---|
| **15440.60 WATER CLOSETS** | | | | |
| For trim and rough-in | | | | |
| Minimum | EA. | 140.00 | 97.50 | 237.50 |
| Average | " | 170.00 | 130.00 | 300.00 |
| Maximum | " | 220.00 | 200.00 | 420.00 |
| **15440.70 WATER HEATERS** | | | | |
| Water heater, electric | | | | |
| 6 gal | EA. | 250.00 | 65.00 | 315.00 |
| 10 gal | " | 270.00 | 65.00 | 335.00 |
| 15 gal | " | 280.00 | 65.00 | 345.00 |
| 20 gal | " | 330.00 | 78.00 | 408.00 |
| 30 gal | " | 340.00 | 78.00 | 418.00 |
| 40 gal | " | 350.00 | 78.00 | 428.00 |
| 52 gal | " | 410.00 | 97.50 | 507.50 |
| 66 gal | " | 600.00 | 97.50 | 697.50 |
| 80 gal | " | 660.00 | 97.50 | 757.50 |
| 100 gal | " | 800.00 | 130.00 | 930.00 |
| 120 gal | " | 1,030 | 130.00 | 1,160 |
| Oil fired | | | | |
| 20 gal | EA. | 1,060 | 200.00 | 1,260 |
| 50 gal | " | 1,650 | 280.00 | 1,930 |
| **15440.90 MISCELLANEOUS FIXTURES** | | | | |
| Electric water cooler | | | | |
| Floor mounted | EA. | 600.00 | 130.00 | 730.00 |
| Wall mounted | " | 730.00 | 130.00 | 860.00 |
| Wash fountain | | | | |
| Wall mounted | EA. | 1,010 | 200.00 | 1,210 |
| Circular, floor supported | " | 1,900 | 390.00 | 2,290 |
| Deluge shower and eye wash | " | 720.00 | 200.00 | 920.00 |
| **15440.95 FIXTURE CARRIERS** | | | | |
| Water fountain, wall carrier | | | | |
| Minimum | EA. | 85.00 | 39.00 | 124.00 |
| Average | " | 110.00 | 48.80 | 158.80 |
| Maximum | " | 140.00 | 65.00 | 205.00 |
| Lavatory, wall carrier | | | | |
| Minimum | EA. | 120.00 | 39.00 | 159.00 |
| Average | " | 170.00 | 48.80 | 218.80 |
| Maximum | " | 210.00 | 65.00 | 275.00 |
| Sink, industrial, wall carrier | | | | |
| Minimum | EA. | 120.00 | 39.00 | 159.00 |
| Average | " | 150.00 | 48.80 | 198.80 |
| Maximum | " | 180.00 | 65.00 | 245.00 |
| Toilets, water closets, wall carrier | | | | |
| Minimum | EA. | 150.00 | 39.00 | 189.00 |
| Average | " | 170.00 | 48.80 | 218.80 |
| Maximum | " | 230.00 | 65.00 | 295.00 |
| Floor support | | | | |
| Minimum | EA. | 72.50 | 32.50 | 105.00 |

# 15 MECHANICAL

## PLUMBING FIXTURES

| 15440.95 FIXTURE CARRIERS | UNIT | MAT. | INST. | TOTAL |
|---|---|---|---|---|
| Average | EA. | 87.00 | 39.00 | 126.00 |
| Maximum | " | 100.00 | 48.80 | 148.80 |
| Urinals, wall carrier | | | | |
| Minimum | EA. | 94.00 | 39.00 | 133.00 |
| Average | " | 120.00 | 48.80 | 168.80 |
| Maximum | " | 150.00 | 65.00 | 215.00 |
| Floor support | | | | |
| Minimum | EA. | 79.50 | 32.50 | 112.00 |
| Average | " | 120.00 | 39.00 | 159.00 |
| Maximum | " | 140.00 | 48.80 | 188.80 |
| **15450.30 PUMPS** | | | | |
| In-line pump, bronze, centrifugal | | | | |
| 5 gpm, 20' head | EA. | 580.00 | 24.35 | 604.35 |
| 20 gpm, 40' head | " | 1,090 | 24.35 | 1,114 |
| 50 gpm | | | | |
| 50' head | EA. | 1,420 | 48.80 | 1,469 |
| 100' head | " | 2,170 | 48.80 | 2,219 |
| 70 gpm, 100' head | " | 2,170 | 65.00 | 2,235 |
| 100 gpm, 80' head | " | 2,170 | 65.00 | 2,235 |
| 250 gpm, 150' head | " | 6,800 | 97.50 | 6,898 |
| Cast iron, centrifugal | | | | |
| 50 gpm, 200' head | EA. | 2,990 | 48.80 | 3,039 |
| 100 gpm | | | | |
| 100' head | EA. | 2,760 | 65.00 | 2,825 |
| 200' head | " | 3,220 | 65.00 | 3,285 |
| 200 gpm | | | | |
| 100' head | EA. | 3,010 | 97.50 | 3,108 |
| 200' head | " | 4,060 | 97.50 | 4,158 |
| Centrifugal, close coupled, c.i., single stage | | | | |
| 50 gpm, 100' head | EA. | 1,210 | 48.80 | 1,259 |
| 100 gpm, 100' head | " | 1,470 | 65.00 | 1,535 |
| Base mounted | | | | |
| 50 gpm, 100' head | EA. | 1,980 | 48.80 | 2,029 |
| 100 gpm, 50' head | " | 2,240 | 65.00 | 2,305 |
| 200 gpm, 100' head | " | 2,870 | 97.50 | 2,968 |
| 300 gpm, 175' head | " | 3,730 | 97.50 | 3,828 |
| Sump pump, bronze, 1750 rpm, 25 gpm | | | | |
| 20' head | EA. | 3,970 | 490.00 | 4,460 |
| 150' head | " | 5,530 | 650.00 | 6,180 |
| 50 gpm | | | | |
| 100' head | EA. | 4,730 | 490.00 | 5,220 |
| 100 gpm | | | | |
| 50' head | EA. | 4,110 | 490.00 | 4,600 |
| Condensate pump, simplex | | | | |
| 1000 sf EDR, 2 gpm | EA. | 1,010 | 330.00 | 1,340 |
| 2000 sf EDR, 3 gpm | " | 1,030 | 330.00 | 1,360 |
| 4000 sf EDR, 6 gpm | " | 1,040 | 350.00 | 1,390 |
| 6000 sf EDR, 9 gpm | " | 1,060 | 350.00 | 1,410 |
| Duplex, bronze | | | | |
| 8000 sf EDR, 12 gpm | EA. | 1,450 | 350.00 | 1,800 |
| 10,000 sf EDR, 15 gpm | " | 1,500 | 490.00 | 1,990 |

| PLUMBING FIXTURES | UNIT | MAT. | INST. | TOTAL |
|---|---|---|---|---|
| **15450.30 PUMPS** | | | | |
| 15,000 sf EDR, 23 gpm | EA. | 1,810 | 560.00 | 2,370 |
| 20,000 sf EDR, 30 gpm | " | 2,100 | 780.00 | 2,880 |
| 25,000 sf EDR, 38 gpm | " | 2,170 | 780.00 | 2,950 |
| **15450.40 STORAGE TANKS** | | | | |
| Hot water storage tank, cement lined | | | | |
| 10 gallon | EA. | 3,510 | 130.00 | 3,640 |
| 70 gallon | " | 3,510 | 200.00 | 3,710 |
| 200 gallon | " | 3,650 | 280.00 | 3,930 |
| 900 gallon | " | 4,190 | 490.00 | 4,680 |
| 1100 gallon | " | 4,730 | 490.00 | 5,220 |
| 2000 gallon | " | 9,320 | 490.00 | 9,810 |
| **15480.10 SPECIAL SYSTEMS** | | | | |
| Air compressor, air cooled, two stage | | | | |
| 5.0 cfm, 175 psi | EA. | 3,050 | 780.00 | 3,830 |
| 10 cfm, 175 psi | " | 3,720 | 870.00 | 4,590 |
| 20 cfm, 175 psi | " | 5,140 | 930.00 | 6,070 |
| 50 cfm, 125 psi | " | 7,480 | 1,030 | 8,510 |
| 80 cfm, 125 psi | " | 10,720 | 1,110 | 11,830 |
| Single stage, 125 psi | | | | |
| 1.0 cfm | EA. | 2,810 | 560.00 | 3,370 |
| 1.5 cfm | " | 2,860 | 560.00 | 3,420 |
| 2.0 cfm | " | 2,920 | 560.00 | 3,480 |
| Automotive compressor, hose reel, air and water, 50' hose | " | 1,310 | 330.00 | 1,640 |
| Lube equipment, 3 reel, with pumps | " | 6,850 | 1,560 | 8,410 |
| Tire changer | | | | |
| Truck | EA. | 10,720 | 560.00 | 11,280 |
| Passenger car | " | 2,530 | 300.00 | 2,830 |
| Air hose reel, includes, 50' hose | " | 670.00 | 300.00 | 970.00 |
| Hose reel, 5 reel, motor oil, gear oil, lube, air & water | " | 6,260 | 1,560 | 7,820 |
| Water hose reel, 50' hose | " | 670.00 | 300.00 | 970.00 |
| Pump, air operated, for motor or gear oil, fits 55 gal drum | " | 880.00 | 39.00 | 919.00 |
| For chassis lube | " | 1,430 | 39.00 | 1,469 |
| Fuel dispensing pump, lighted dial, one product | | | | |
| One hose | EA. | 3,130 | 330.00 | 3,460 |
| Two hose | " | 5,510 | 330.00 | 5,840 |
| Two products, two hose | " | 5,810 | 330.00 | 6,140 |

| HEATING & VENTILATING | UNIT | MAT. | INST. | TOTAL |
|---|---|---|---|---|
| **15555.10 BOILERS** | | | | |
| Cast iron, gas fired, hot water | | | | |
| 115 mbh | EA. | 1,740 | 1,200 | 2,940 |
| 175 mbh | " | 2,030 | 1,310 | 3,340 |
| 235 mbh | " | 3,040 | 1,440 | 4,480 |
| 940 mbh | " | 8,970 | 2,880 | 11,850 |
| 1600 mbh | " | 12,300 | 3,600 | 15,900 |
| 3000 mbh | " | 26,040 | 4,800 | 30,840 |
| 6000 mbh | " | 60,760 | 7,200 | 67,960 |
| Steam | | | | |
| 115 mbh | EA. | 2,510 | 1,200 | 3,710 |
| 175 mbh | " | 3,190 | 1,310 | 4,500 |
| 235 mbh | " | 3,490 | 1,440 | 4,930 |
| 940 mbh | " | 8,820 | 2,880 | 11,700 |
| 1600 mbh | " | 15,910 | 3,600 | 19,510 |
| 3000 mbh | " | 30,380 | 4,800 | 35,180 |
| 6000 mbh | " | 65,100 | 7,200 | 72,300 |
| Electric, hot water | | | | |
| 115 mbh | EA. | 4,920 | 720.00 | 5,640 |
| 175 mbh | " | 5,900 | 720.00 | 6,620 |
| 235 mbh | " | 8,010 | 720.00 | 8,730 |
| 940 mbh | " | 14,050 | 1,440 | 15,490 |
| 1600 mbh | " | 21,070 | 2,880 | 23,950 |
| 3000 mbh | " | 36,520 | 3,600 | 40,120 |
| 6000 mbh | " | 54,780 | 4,800 | 59,580 |
| Steam | | | | |
| 115 mbh | EA. | 7,960 | 720.00 | 8,680 |
| 175 mbh | " | 9,110 | 720.00 | 9,830 |
| 235 mbh | " | 10,710 | 720.00 | 11,430 |
| 940 mbh | " | 21,700 | 1,440 | 23,140 |
| 1600 mbh | " | 36,170 | 2,880 | 39,050 |
| 3000 mbh | " | 41,950 | 3,600 | 45,550 |
| 6000 mbh | " | 65,100 | 4,800 | 69,900 |
| Oil fired, hot water | | | | |
| 115 mbh | EA. | 2,390 | 960.00 | 3,350 |
| 175 mbh | " | 2,950 | 1,110 | 4,060 |
| 235 mbh | " | 3,860 | 1,310 | 5,170 |
| 940 mbh | " | 11,380 | 2,400 | 13,780 |
| 1600 mbh | " | 19,730 | 2,880 | 22,610 |
| 3000 mbh | " | 33,710 | 3,600 | 37,310 |
| 6000 mbh | " | 63,200 | 7,200 | 70,400 |
| Steam | | | | |
| 115 mbh | EA. | 2,510 | 960.00 | 3,470 |
| 175 mbh | " | 3,020 | 1,110 | 4,130 |
| 235 mbh | " | 3,980 | 1,310 | 5,290 |
| 940 mbh | " | 12,830 | 2,400 | 15,230 |
| 1600 mbh | " | 20,940 | 2,880 | 23,820 |
| 3000 mbh | " | 36,280 | 3,600 | 39,880 |
| 6000 mbh | " | 64,610 | 7,200 | 71,810 |
| **15610.10 FURNACES** | | | | |
| Electric, hot air | | | | |
| 40 mbh | EA. | 660.00 | 200.00 | 860.00 |

# 15 MECHANICAL

## HEATING & VENTILATING

| 15610.10 FURNACES | UNIT | MAT. | INST. | TOTAL |
|---|---|---|---|---|
| 60 mbh | EA. | 770.00 | 210.00 | 980.00 |
| 80 mbh | " | 710.00 | 220.00 | 930.00 |
| 100 mbh | " | 1,150 | 230.00 | 1,380 |
| 125 mbh | " | 1,270 | 240.00 | 1,510 |
| 160 mbh | " | 1,840 | 240.00 | 2,080 |
| 200 mbh | " | 3,390 | 250.00 | 3,640 |
| 400 mbh | " | 6,050 | 260.00 | 6,310 |
| Gas fired hot air | | | | |
| 40 mbh | EA. | 660.00 | 200.00 | 860.00 |
| 60 mbh | " | 700.00 | 210.00 | 910.00 |
| 80 mbh | " | 810.00 | 220.00 | 1,030 |
| 100 mbh | " | 850.00 | 230.00 | 1,080 |
| 125 mbh | " | 920.00 | 240.00 | 1,160 |
| 160 mbh | " | 1,110 | 240.00 | 1,350 |
| 200 mbh | " | 1,970 | 250.00 | 2,220 |
| 400 mbh | " | 7,020 | 260.00 | 7,280 |
| Oil fired hot air | | | | |
| 40 mbh | EA. | 880.00 | 200.00 | 1,080 |
| 60 mbh | " | 1,060 | 210.00 | 1,270 |
| 80 mbh | " | 1,160 | 220.00 | 1,380 |
| 100 mbh | " | 1,350 | 230.00 | 1,580 |
| 125 mbh | " | 1,510 | 240.00 | 1,750 |
| 160 mbh | " | 1,630 | 240.00 | 1,870 |
| 200 mbh | " | 3,720 | 250.00 | 3,970 |
| 400 mbh | " | 7,370 | 260.00 | 7,630 |

## REFRIGERATION

| 15670.10 CONDENSING UNITS | UNIT | MAT. | INST. | TOTAL |
|---|---|---|---|---|
| Air cooled condenser, single circuit | | | | |
| 3 ton | EA. | 1,820 | 65.00 | 1,885 |
| 5 ton | " | 3,030 | 65.00 | 3,095 |
| 7.5 ton | " | 4,850 | 190.00 | 5,040 |
| 20 ton | " | 10,220 | 200.00 | 10,420 |
| 25 ton | " | 11,930 | 200.00 | 12,130 |
| 30 ton | " | 13,740 | 200.00 | 13,940 |
| With low ambient dampers | | | | |
| 3 ton | EA. | 2,120 | 97.50 | 2,218 |
| 5 ton | " | 3,330 | 97.50 | 3,428 |
| 7.5 ton | " | 5,150 | 200.00 | 5,350 |
| 20 ton | " | 10,600 | 260.00 | 10,860 |
| 25 ton | " | 12,270 | 260.00 | 12,530 |
| 30 ton | " | 14,090 | 260.00 | 14,350 |

| REFRIGERATION | UNIT | MAT. | INST. | TOTAL |
|---|---|---|---|---|
| **15670.10 CONDENSING UNITS** | | | | |
| Dual circuit | | | | |
| 10 ton | EA. | 4,540 | 200.00 | 4,740 |
| 15 ton | " | 6,660 | 280.00 | 6,940 |
| 20 ton | " | 11,360 | 280.00 | 11,640 |
| 25 ton | " | 13,030 | 280.00 | 13,310 |
| 30 ton | " | 15,150 | 280.00 | 15,430 |
| With low ambient dampers | | | | |
| 15 ton | EA. | 7,420 | 280.00 | 7,700 |
| 20 ton | " | 12,120 | 280.00 | 12,400 |
| 25 ton | " | 13,780 | 280.00 | 14,060 |
| 30 ton | " | 14,840 | 280.00 | 15,120 |
| **15680.10 CHILLERS** | | | | |
| Chiller, reciprocal | | | | |
| Air cooled, remote condenser, starter | | | | |
| 20 ton | EA. | 21,210 | 480.00 | 21,690 |
| 25 ton | " | 23,930 | 480.00 | 24,410 |
| 30 ton | " | 25,300 | 480.00 | 25,780 |
| 40 ton | " | 37,410 | 720.00 | 38,130 |
| Water cooled, with starter | | | | |
| 20 ton | EA. | 18,180 | 480.00 | 18,660 |
| 25 ton | " | 20,150 | 480.00 | 20,630 |
| 30 ton | " | 24,240 | 720.00 | 24,960 |
| 40 ton | " | 33,320 | 720.00 | 34,040 |
| Packaged, air cooled, with starter | | | | |
| 20 ton | EA. | 20,300 | 360.00 | 20,660 |
| 25 ton | " | 21,960 | 360.00 | 22,320 |
| 30 ton | " | 25,600 | 360.00 | 25,960 |
| 40 ton | " | 29,390 | 360.00 | 29,750 |
| **15710.10 COOLING TOWERS** | | | | |
| Cooling tower, propeller type | | | | |
| 100 ton | EA. | 9,270 | 480.00 | 9,750 |
| 200 ton | " | 15,450 | 720.00 | 16,170 |
| 300 ton | " | 23,170 | 1,200 | 24,370 |
| 400 ton | " | 30,900 | 1,440 | 32,340 |
| 600 ton | " | 46,350 | 2,060 | 48,410 |
| 800 ton | " | 61,800 | 2,880 | 64,680 |
| 1000 ton | " | 74,160 | 3,600 | 77,760 |
| Centrifugal | | | | |
| 100 ton | EA. | 9,270 | 480.00 | 9,750 |
| 200 ton | " | 18,540 | 720.00 | 19,260 |
| 300 ton | " | 27,810 | 1,200 | 29,010 |
| 400 ton | " | 37,080 | 1,440 | 38,520 |
| 600 ton | " | 55,620 | 2,060 | 57,680 |
| 800 ton | " | 74,160 | 2,880 | 77,040 |
| 1000 ton | " | 86,520 | 3,600 | 90,120 |

## HEAT TRANSFER

### 15780.10 COMPUTER ROOM A/C

| | UNIT | MAT. | INST. | TOTAL |
|---|---|---|---|---|
| Air cooled, alarm, high efficiency filter, elec. heat | | | | |
| 3 ton | EA. | 10,970 | 300.00 | 11,270 |
| 5 ton | " | 11,770 | 330.00 | 12,100 |
| 7.5 ton | " | 20,040 | 390.00 | 20,430 |
| 10 ton | " | 20,680 | 490.00 | 21,170 |
| 15 ton | " | 22,900 | 560.00 | 23,460 |
| Steam heat | | | | |
| 3 ton | EA. | 12,720 | 300.00 | 13,020 |
| 5 ton | " | 13,520 | 330.00 | 13,850 |
| 7.5 ton | " | 21,470 | 390.00 | 21,860 |
| 10 ton | " | 22,110 | 490.00 | 22,600 |
| 15 ton | " | 24,490 | 560.00 | 25,050 |
| Hot water heat | | | | |
| 3 ton | EA. | 12,720 | 300.00 | 13,020 |
| 5 ton | " | 13,520 | 330.00 | 13,850 |
| 7.5 ton | " | 21,470 | 390.00 | 21,860 |
| 10 ton | " | 22,110 | 490.00 | 22,600 |
| 15 ton | " | 24,490 | 560.00 | 25,050 |
| Air cooled condenser, low ambient damper | | | | |
| 3 ton | EA. | 2,860 | 78.00 | 2,938 |
| 5 ton | " | 3,180 | 97.50 | 3,278 |
| 7.5 ton | " | 4,610 | 200.00 | 4,810 |
| 10 ton | " | 4,610 | 280.00 | 4,890 |
| 15 ton | " | 5,090 | 230.00 | 5,320 |
| Water cooled, high efficiency filter, alarm, elec. heat | | | | |
| 3 ton | EA. | 12,560 | 280.00 | 12,840 |
| 5 ton | " | 13,520 | 330.00 | 13,850 |
| 7.5 ton | " | 21,470 | 490.00 | 21,960 |
| 10 ton | " | 22,270 | 560.00 | 22,830 |
| 15 ton | " | 26,080 | 650.00 | 26,730 |
| Steam heat | | | | |
| 3 ton | EA. | 14,310 | 280.00 | 14,590 |
| 5 ton | " | 16,360 | 330.00 | 16,690 |
| 7.5 ton | " | 23,060 | 490.00 | 23,550 |
| 10 ton | " | 23,860 | 560.00 | 24,420 |
| 15 ton | " | 27,670 | 650.00 | 28,320 |
| Hot water heat | | | | |
| 3 ton | EA. | 14,310 | 280.00 | 14,590 |
| 5 ton | " | 15,270 | 330.00 | 15,600 |
| 7.5 ton | " | 23,060 | 490.00 | 23,550 |
| 10 ton | " | 23,860 | 560.00 | 24,420 |
| 15 ton | " | 27,670 | 650.00 | 28,320 |
| Chilled water, alarm, high eff. filter, elec. heat | | | | |
| 7.5 ton | EA. | 10,020 | 350.00 | 10,370 |
| 10 ton | " | 10,500 | 430.00 | 10,930 |
| 15 ton | " | 11,930 | 490.00 | 12,420 |
| Steam heat | | | | |
| 7.5 ton | EA. | 11,450 | 350.00 | 11,800 |
| 10 ton | " | 11,930 | 430.00 | 12,360 |
| 15 ton | " | 13,360 | 490.00 | 13,850 |
| Hot water heat | | | | |
| 7.5 ton | EA. | 11,450 | 350.00 | 11,800 |

| HEAT TRANSFER | UNIT | MAT. | INST. | TOTAL |
|---|---|---|---|---|
| **15780.10 COMPUTER ROOM A/C** | | | | |
| 10 ton | EA. | 11,930 | 430.00 | 12,360 |
| 15 ton | " | 13,360 | 490.00 | 13,850 |
| **15780.20 ROOFTOP UNITS** | | | | |
| Packaged, single zone rooftop unit, with roof curb | | | | |
| 2 ton | EA. | 3,860 | 390.00 | 4,250 |
| 3 ton | " | 4,460 | 390.00 | 4,850 |
| 4 ton | " | 5,350 | 490.00 | 5,840 |
| 5 ton | " | 5,400 | 650.00 | 6,050 |
| 7.5 ton | " | 8,760 | 780.00 | 9,540 |
| **15820.10 DEHUMIDIFIERS** | | | | |
| Dessicant dehumidifier, 1125 cfm | EA. | | | 15,147 |
| **15830.10 RADIATION UNITS** | | | | |
| Baseboard radiation unit | | | | |
| 1.7 mbh/lf | L.F. | 55.00 | 15.60 | 70.60 |
| 2.1 mbh/lf | " | 70.00 | 19.50 | 89.50 |
| Enclosure only | | | | |
| Two tier | L.F. | 28.10 | 6.50 | 34.60 |
| Three tier | " | 36.50 | 6.50 | 43.00 |
| Copper element only, 3/4" dia. | | | | |
| Two tier | L.F. | 35.10 | 9.75 | 44.85 |
| Three tier | " | 50.50 | 13.00 | 63.50 |
| Fin-tube, 16 ga, sloping cover, 1-1/4" steel | | | | |
| One tier | L.F. | 39.30 | 13.00 | 52.30 |
| Two tier | " | 60.50 | 15.60 | 76.10 |
| 2" steel | | | | |
| Two tier | L.F. | 64.50 | 15.60 | 80.10 |
| Three tier | " | 90.00 | 19.50 | 109.50 |
| 1-1/4" copper | | | | |
| Two tier | L.F. | 67.50 | 13.00 | 80.50 |
| 18 ga flat cover, 1-1/4" steel | | | | |
| One tier | L.F. | 23.90 | 13.00 | 36.90 |
| Two tier | " | 42.10 | 15.60 | 57.70 |
| Three tier | " | 66.00 | 19.50 | 85.50 |
| 2" steel | | | | |
| One tier | L.F. | 29.50 | 13.00 | 42.50 |
| Two tier | " | 47.70 | 15.60 | 63.30 |
| Three tier | " | 67.50 | 19.50 | 87.00 |
| 1-1/4" copper | | | | |
| One tier | L.F. | 26.70 | 13.00 | 39.70 |
| Two tier | " | 46.30 | 15.60 | 61.90 |
| Three tier | " | 60.50 | 19.50 | 80.00 |
| **15830.20 FAN COIL UNITS** | | | | |
| Fan coil unit, 2 pipe, complete | | | | |
| 200 cfm ceiling hung | EA. | 710.00 | 130.00 | 840.00 |

| HEAT TRANSFER | UNIT | MAT. | INST. | TOTAL |
|---|---|---|---|---|
| **15830.20 FAN COIL UNITS** | | | | |
| Floor mounted | EA. | 670.00 | 97.50 | 767.50 |
| 300 cfm, ceiling hung | " | 760.00 | 160.00 | 920.00 |
| Floor mounted | " | 720.00 | 130.00 | 850.00 |
| 400 cfm, ceiling hung | " | 800.00 | 190.00 | 990.00 |
| Floor mounted | " | 770.00 | 130.00 | 900.00 |
| 500 cfm, ceiling hung | " | 930.00 | 200.00 | 1,130 |
| Floor mounted | " | 890.00 | 150.00 | 1,040 |
| 600 cfm, ceiling hung | " | 1,180 | 220.00 | 1,400 |
| Floor mounted | " | 1,090 | 180.00 | 1,270 |
| 800 cfm, ceiling hung | " | 1,370 | 240.00 | 1,610 |
| Floor mounted | " | 1,090 | 190.00 | 1,280 |
| 1000 cfm, ceiling hung | " | 1,560 | 280.00 | 1,840 |
| Floor mounted | " | 1,710 | 210.00 | 1,920 |
| 1200 cfm ceiling hung | " | 1,770 | 330.00 | 2,100 |
| Floor mounted | " | 1,860 | 240.00 | 2,100 |
| **15830.70 UNIT HEATERS** | | | | |
| Steam unit heater, horizontal | | | | |
| 12,500 btuh, 200 cfm | EA. | 470.00 | 65.00 | 535.00 |
| 17,000 btuh, 300 cfm | " | 540.00 | 65.00 | 605.00 |
| 40,000 btuh, 500 cfm | " | 680.00 | 65.00 | 745.00 |
| 60,000 btuh, 700 cfm | " | 760.00 | 65.00 | 825.00 |
| 70,000 btuh, 1000 cfm | " | 950.00 | 97.50 | 1,048 |
| Vertical | | | | |
| 12,500 btuh, 200 cfm | EA. | 410.00 | 65.00 | 475.00 |
| 17,000 btuh, 300 cfm | " | 680.00 | 65.00 | 745.00 |
| 40,000 btuh, 500 cfm | " | 810.00 | 65.00 | 875.00 |
| 60,000 btuh, 700 cfm | " | 950.00 | 65.00 | 1,015 |
| 70,000 btuh, 1000 cfm | " | 1,080 | 65.00 | 1,145 |
| Gas unit heater, horizontal | | | | |
| 27,400 btuh | EA. | 680.00 | 160.00 | 840.00 |
| 38,000 btuh | " | 740.00 | 160.00 | 900.00 |
| 56,000 btuh | " | 810.00 | 160.00 | 970.00 |
| 82,200 btuh | " | 890.00 | 160.00 | 1,050 |
| 103,900 btuh | " | 930.00 | 240.00 | 1,170 |
| 125,700 btuh | " | 960.00 | 240.00 | 1,200 |
| 133,200 btuh | " | 1,010 | 240.00 | 1,250 |
| 149,000 btuh | " | 1,180 | 240.00 | 1,420 |
| 172,000 btuh | " | 1,280 | 240.00 | 1,520 |
| 190,000 btuh | " | 1,350 | 240.00 | 1,590 |
| 225,000 btuh | " | 1,490 | 240.00 | 1,730 |
| Hot water unit heater, horizontal | | | | |
| 12,500 btuh, 200 cfm | EA. | 410.00 | 65.00 | 475.00 |
| 17,000 btuh, 300 cfm | " | 470.00 | 65.00 | 535.00 |
| 25,000 btuh, 500 cfm | " | 510.00 | 65.00 | 575.00 |
| 30,000 btuh, 700 cfm | " | 610.00 | 65.00 | 675.00 |
| 50,000 btuh, 1000 cfm | " | 680.00 | 97.50 | 777.50 |
| 60,000 btuh, 1300 cfm | " | 950.00 | 97.50 | 1,048 |
| Vertical | | | | |
| 12,500 btuh, 200 cfm | EA. | 500.00 | 65.00 | 565.00 |
| 17,000 btuh, 300 cfm | " | 540.00 | 65.00 | 605.00 |

# 15 MECHANICAL

| HEAT TRANSFER | UNIT | MAT. | INST. | TOTAL |
|---|---|---|---|---|
| **15830.70 UNIT HEATERS** | | | | |
| 25,000 btuh, 500 cfm | EA. | 570.00 | 65.00 | 635.00 |
| 30,000 btuh, 700 cfm | " | 620.00 | 65.00 | 685.00 |
| 50,000 btuh, 1000 cfm | " | 680.00 | 65.00 | 745.00 |
| 60,000 btuh, 1300 cfm | " | 740.00 | 65.00 | 805.00 |
| Cabinet unit heaters, ceiling, exposed, hot water | | | | |
| 200 cfm | EA. | 950.00 | 130.00 | 1,080 |
| 300 cfm | " | 1,010 | 160.00 | 1,170 |
| 400 cfm | " | 1,050 | 190.00 | 1,240 |
| 600 cfm | " | 1,080 | 210.00 | 1,290 |
| 800 cfm | " | 1,350 | 240.00 | 1,590 |
| 1000 cfm | " | 1,760 | 280.00 | 2,040 |
| 1200 cfm | " | 1,890 | 330.00 | 2,220 |
| 2000 cfm | " | 3,240 | 430.00 | 3,670 |

| AIR HANDLING | UNIT | MAT. | INST. | TOTAL |
|---|---|---|---|---|
| **15855.10 AIR HANDLING UNITS** | | | | |
| Air handling unit, medium pressure, single zone | | | | |
| 1500 cfm | EA. | 3,030 | 240.00 | 3,270 |
| 3000 cfm | " | 3,790 | 430.00 | 4,220 |
| 4000 cfm | " | 4,240 | 490.00 | 4,730 |
| 5000 cfm | " | 4,850 | 520.00 | 5,370 |
| 6000 cfm | " | 5,450 | 560.00 | 6,010 |
| 7000 cfm | " | 6,060 | 600.00 | 6,660 |
| 8500 cfm | " | 7,120 | 650.00 | 7,770 |
| 10,500 cfm | " | 8,030 | 780.00 | 8,810 |
| 12,500 cfm | " | 10,300 | 870.00 | 11,170 |
| 15,500 cfm | " | 18,120 | 1,110 | 19,230 |
| 17,500 cfm | " | 19,690 | 1,300 | 20,990 |
| 20,500 cfm | " | 22,720 | 1,560 | 24,280 |
| 25,000 cfm | " | 25,750 | 1,950 | 27,700 |
| 31,500 cfm | " | 31,810 | 2,600 | 34,410 |
| Rooftop air handling units | | | | |
| 4950 cfm | EA. | 9,110 | 430.00 | 9,540 |
| 7370 cfm | " | 10,160 | 560.00 | 10,720 |
| 9790 cfm | " | 12,290 | 650.00 | 12,940 |
| 14,300 cfm | " | 17,350 | 560.00 | 17,910 |
| 21,725 cfm | " | 24,580 | 560.00 | 25,140 |
| 33,000 cfm | " | 34,700 | 650.00 | 35,350 |
| **15870.20 EXHAUST FANS** | | | | |
| Belt drive roof exhaust fans | | | | |
| 640 cfm, 2618 fpm | EA. | 830.00 | 48.80 | 878.80 |

# 15 MECHANICAL

## AIR HANDLING

| 15870.20 EXHAUST FANS | UNIT | MAT. | INST. | TOTAL |
|---|---|---|---|---|
| 940 cfm, 2604 fpm | EA. | 1,070 | 48.80 | 1,119 |
| 1050 cfm, 3325 fpm | " | 950.00 | 48.80 | 998.80 |
| 1170 cfm, 2373 fpm | " | 1,380 | 48.80 | 1,429 |
| 2440 cfm, 4501 fpm | " | 1,090 | 48.80 | 1,139 |
| 2760 cfm, 4950 fpm | " | 1,200 | 48.80 | 1,249 |
| 3890 cfm, 6769 fpm | " | 1,360 | 48.80 | 1,409 |
| 2380 cfm, 3382 fpm | " | 1,510 | 48.80 | 1,559 |
| 2880 cfm, 3859 fpm | " | 1,580 | 48.80 | 1,629 |
| 3200 cfm, 4173 fpm | " | 1,600 | 65.00 | 1,665 |
| 3660 cfm, 3437 fpm | " | 1,620 | 65.00 | 1,685 |
| Direct drive fans | | | | |
| 60 to 390 cfm | EA. | 670.00 | 48.80 | 718.80 |
| 145 to 590 cfm | " | 810.00 | 48.80 | 858.80 |
| 295 to 860 cfm | " | 980.00 | 48.80 | 1,029 |
| 235 to 1300 cfm | " | 1,060 | 48.80 | 1,109 |
| 415 to 1630 cfm | " | 1,200 | 48.80 | 1,249 |
| 590 to 2045 cfm | " | 1,380 | 48.80 | 1,429 |

## AIR DISTRIBUTION

| 15890.10 METAL DUCTWORK | UNIT | MAT. | INST. | TOTAL |
|---|---|---|---|---|
| Rectangular duct | | | | |
| Galvanized steel | | | | |
| Minimum | Lb. | 1.45 | 3.55 | 5.00 |
| Average | " | 1.80 | 4.35 | 6.15 |
| Maximum | " | 2.80 | 6.50 | 9.30 |
| Aluminum | | | | |
| Minimum | Lb. | 3.65 | 7.80 | 11.45 |
| Average | " | 4.85 | 9.75 | 14.60 |
| Maximum | " | 6.10 | 13.00 | 19.10 |
| Fittings | | | | |
| Minimum | EA. | 12.15 | 13.00 | 25.15 |
| Average | " | 18.25 | 19.50 | 37.75 |
| Maximum | " | 26.70 | 39.00 | 65.70 |
| For work | | | | |
| 10-20' high, add per pound, $.30 | | | | |
| 30-50', add per pound, $.50 | | | | |

| 15890.30 FLEXIBLE DUCTWORK | UNIT | MAT. | INST. | TOTAL |
|---|---|---|---|---|
| Flexible duct, 1.25" fiberglass | | | | |
| 5" dia. | L.F. | 2.15 | 1.95 | 4.10 |
| 6" dia. | " | 2.55 | 2.15 | 4.70 |
| 7" dia. | " | 2.85 | 2.30 | 5.15 |

| AIR DISTRIBUTION | UNIT | MAT. | INST. | TOTAL |
|---|---|---|---|---|
| **15890.30 FLEXIBLE DUCTWORK** | | | | |
| 8" dia. | L.F. | 3.25 | 2.45 | 5.70 |
| 10" dia. | " | 3.70 | 2.80 | 6.50 |
| 12" dia. | " | 4.55 | 3.00 | 7.55 |
| 14" dia. | " | 5.55 | 3.25 | 8.80 |
| 16" dia. | " | 6.80 | 3.55 | 10.35 |
| Flexible duct connector, 3" wide fabric | " | 1.90 | 6.50 | 8.40 |
| **15895.10 ROOF CURBS** | | | | |
| 8" high, insulated, with liner and raised can | | | | |
| 15" x 15" | EA. | 75.00 | 19.50 | 94.50 |
| 17" x 17" | " | 81.00 | 19.50 | 100.50 |
| 19" x 19" | " | 85.50 | 19.50 | 105.00 |
| 21" x 21" | " | 94.00 | 19.50 | 113.50 |
| 25" x 25" | " | 100.00 | 24.35 | 124.35 |
| 28" x 28" | " | 110.00 | 26.00 | 136.00 |
| 32" x 32" | " | 120.00 | 27.90 | 147.90 |
| 36" x 36" | " | 140.00 | 27.90 | 167.90 |
| 40" x 40" | " | 160.00 | 27.90 | 187.90 |
| 44" x 44" | " | 170.00 | 30.00 | 200.00 |
| 48" x 48" | " | 380.00 | 30.00 | 410.00 |
| 52" x 52" | " | 470.00 | 32.50 | 502.50 |
| 56" x 56" | " | 590.00 | 32.50 | 622.50 |
| 60" x 60" | " | 710.00 | 39.00 | 749.00 |
| 64" x 64" | " | 850.00 | 39.00 | 889.00 |
| 68" x 68" | " | 980.00 | 43.30 | 1,023 |
| 72" x 72" | " | 1,130 | 48.80 | 1,179 |
| **15910.10 DAMPERS** | | | | |
| Horizontal parallel aluminum backdraft damper | | | | |
| 12" x 12" | EA. | 52.00 | 9.75 | 61.75 |
| 16" x 16" | " | 53.50 | 11.15 | 64.65 |
| 20" x 20" | " | 69.00 | 13.95 | 82.95 |
| 24" x 24" | " | 84.50 | 19.50 | 104.00 |
| 28" x 28" | " | 120.00 | 21.65 | 141.65 |
| 32" x 32" | " | 160.00 | 24.35 | 184.35 |
| 36" x 36" | " | 190.00 | 27.90 | 217.90 |
| 40" x 40" | " | 240.00 | 32.50 | 272.50 |
| 44" x 44" | " | 290.00 | 35.50 | 325.50 |
| 48" x 48" | " | 340.00 | 39.00 | 379.00 |
| "Up", parallel dampers | | | | |
| 12" x 12" | EA. | 67.50 | 9.75 | 77.25 |
| 16" x 16" | " | 95.00 | 11.15 | 106.15 |
| 20" x 20" | " | 100.00 | 13.95 | 113.95 |
| 24" x 24" | " | 120.00 | 19.50 | 139.50 |
| 28" x 28" | " | 170.00 | 21.65 | 191.65 |
| 32" x 32" | " | 190.00 | 24.35 | 214.35 |
| 36" x 36" | " | 200.00 | 27.90 | 227.90 |
| 40" x 40" | " | 260.00 | 32.50 | 292.50 |
| 44" x 44" | " | 320.00 | 35.50 | 355.50 |
| 48" x 48" | " | 390.00 | 39.00 | 429.00 |
| "Down", parallel dampers | | | | |
| 12" x 12" | EA. | 67.50 | 9.75 | 77.25 |

# 15 MECHANICAL

| AIR DISTRIBUTION | UNIT | MAT. | INST. | TOTAL |
|---|---|---|---|---|
| **15910.10 DAMPERS** | | | | |
| 16" x 16" | EA. | 95.00 | 11.15 | 106.15 |
| 20" x 20" | " | 100.00 | 13.95 | 113.95 |
| 24" x 24" | " | 120.00 | 19.50 | 139.50 |
| 28" x 28" | " | 170.00 | 21.65 | 191.65 |
| 32" x 32" | " | 190.00 | 24.35 | 214.35 |
| 36" x 36" | " | 200.00 | 27.90 | 227.90 |
| 40" x 40" | " | 270.00 | 32.50 | 302.50 |
| 44" x 44" | " | 320.00 | 35.50 | 355.50 |
| 48" x 48" | " | 390.00 | 39.00 | 429.00 |
| Fire damper, 1.5 hr rating | | | | |
| 12" x 12" | EA. | 23.00 | 19.50 | 42.50 |
| 16" x 16" | " | 36.80 | 19.50 | 56.30 |
| 20" x 20" | " | 39.90 | 19.50 | 59.40 |
| 24" x 24" | " | 46.00 | 19.50 | 65.50 |
| 28" x 28" | " | 55.00 | 27.90 | 82.90 |
| 32" x 32" | " | 64.50 | 32.50 | 97.00 |
| 36" x 36" | " | 76.50 | 39.00 | 115.50 |
| 40" x 40" | " | 92.00 | 43.30 | 135.30 |
| 44" x 44" | " | 110.00 | 48.80 | 158.80 |
| 48" x 48" | " | 140.00 | 55.50 | 195.50 |
| **15940.10 DIFFUSERS** | | | | |
| Ceiling diffusers, round, baked enamel finish | | | | |
| 6" dia. | EA. | 37.60 | 13.00 | 50.60 |
| 8" dia. | " | 45.10 | 16.25 | 61.35 |
| 10" dia. | " | 52.50 | 16.25 | 68.75 |
| 12" dia. | " | 67.50 | 16.25 | 83.75 |
| 14" dia. | " | 84.00 | 17.75 | 101.75 |
| 16" dia. | " | 97.50 | 17.75 | 115.25 |
| 18" dia. | " | 110.00 | 19.50 | 129.50 |
| 20" dia. | " | 140.00 | 19.50 | 159.50 |
| Rectangular | | | | |
| 6x6" | EA. | 41.50 | 13.00 | 54.50 |
| 9x9" | " | 50.00 | 19.50 | 69.50 |
| 12x12" | " | 73.00 | 19.50 | 92.50 |
| 15x15" | " | 89.00 | 19.50 | 108.50 |
| 18x18" | " | 110.00 | 19.50 | 129.50 |
| 21x21" | " | 140.00 | 24.35 | 164.35 |
| 24x24" | " | 210.00 | 24.35 | 234.35 |
| Lay in, flush mounted, perforated face, with grid | | | | |
| 6x6/24x24 | EA. | 68.50 | 15.60 | 84.10 |
| 8x8/24x24 | " | 68.50 | 15.60 | 84.10 |
| 9x9/24x24 | " | 68.50 | 15.60 | 84.10 |
| 10x10/24x24 | " | 74.50 | 15.60 | 90.10 |
| 12x12/24x24 | " | 77.50 | 15.60 | 93.10 |
| 15x15/24x24 | " | 77.50 | 15.60 | 93.10 |
| 18x6/24x24 | " | 77.50 | 15.60 | 93.10 |
| 18x18/24x24 | " | 80.00 | 15.60 | 95.60 |
| Two-way slot diffuser with balancing damper, 4' | " | 86.00 | 39.00 | 125.00 |

| AIR DISTRIBUTION | UNIT | MAT. | INST. | TOTAL |
|---|---|---|---|---|
| **15940.20 RELIEF VENTILATORS** | | | | |
| Intake ventilator, aluminum, with screen, no curbs | | | | |
| 12" x 12" | EA. | 140.00 | 32.50 | 172.50 |
| 16" x 16" | " | 200.00 | 39.00 | 239.00 |
| 20" x 20" | " | 330.00 | 39.00 | 369.00 |
| 30" x 30" | " | 510.00 | 55.50 | 565.50 |
| 36" x 36" | " | 780.00 | 65.00 | 845.00 |
| 42" x 42" | " | 1,060 | 65.00 | 1,125 |
| 48" x 48" | " | 1,280 | 78.00 | 1,358 |
| **15940.40 REGISTERS AND GRILLES** | | | | |
| Lay in flush mounted, perforated face, return | | | | |
| 6x6/24x24 | EA. | 34.70 | 15.60 | 50.30 |
| 8x8/24x24 | " | 34.70 | 15.60 | 50.30 |
| 9x9/24x24 | " | 37.60 | 15.60 | 53.20 |
| 10x10/24x24 | " | 40.50 | 15.60 | 56.10 |
| 12x12/24x24 | " | 40.50 | 15.60 | 56.10 |
| Rectangular, ceiling return, single deflection | | | | |
| 10x10 | EA. | 17.35 | 19.50 | 36.85 |
| 12x12 | " | 20.25 | 19.50 | 39.75 |
| 14x14 | " | 24.60 | 19.50 | 44.10 |
| 16x8 | " | 20.25 | 19.50 | 39.75 |
| 16x16 | " | 20.25 | 19.50 | 39.75 |
| 18x8 | " | 23.15 | 19.50 | 42.65 |
| 20x20 | " | 37.60 | 19.50 | 57.10 |
| 24x12 | " | 55.00 | 19.50 | 74.50 |
| 24x18 | " | 71.00 | 19.50 | 90.50 |
| 36x24 | " | 140.00 | 21.65 | 161.65 |
| 36x30 | " | 200.00 | 21.65 | 221.65 |
| Wall, return air register | | | | |
| 12x12 | EA. | 28.80 | 9.75 | 38.55 |
| 16x16 | " | 42.50 | 9.75 | 52.25 |
| 18x18 | " | 50.00 | 9.75 | 59.75 |
| 20x20 | " | 59.00 | 9.75 | 68.75 |
| 24x24 | " | 82.00 | 9.75 | 91.75 |
| Ceiling, return air grille | | | | |
| 6x6 | EA. | 10.65 | 13.00 | 23.65 |
| 8x8 | " | 12.15 | 15.60 | 27.75 |
| 10x10 | " | 13.65 | 15.60 | 29.25 |
| Ceiling, exhaust grille, aluminum egg crate | | | | |
| 6x6 | EA. | 13.65 | 13.00 | 26.65 |
| 8x8 | " | 13.65 | 15.60 | 29.25 |
| 10x10 | " | 15.20 | 15.60 | 30.80 |
| 12x12 | " | 19.75 | 19.50 | 39.25 |
| 14x14 | " | 25.80 | 19.50 | 45.30 |
| 16x16 | " | 30.40 | 19.50 | 49.90 |
| 18x18 | " | 36.40 | 19.50 | 55.90 |
| **15940.80 PENTHOUSE LOUVERS** | | | | |
| Penthouse louvers | | | | |
| 12" high, extruded aluminum, 4" louver | | | | |
| 6' perimeter | EA. | 340.00 | 97.50 | 437.50 |
| 8' perimeter | " | 460.00 | 97.50 | 557.50 |

## AIR DISTRIBUTION

### 15940.80 PENTHOUSE LOUVERS

| | UNIT | MAT. | INST. | TOTAL |
|---|---|---|---|---|
| 10' perimeter | EA. | 570.00 | 97.50 | 667.50 |
| 12' perimeter | " | 830.00 | 97.50 | 927.50 |
| 14' perimeter | " | 1,020 | 130.00 | 1,150 |
| 16' perimeter | " | 1,170 | 160.00 | 1,330 |
| 18' perimeter | " | 1,370 | 220.00 | 1,590 |
| 20' perimeter | " | 1,720 | 260.00 | 1,980 |
| 16" high x 4' perimeter | " | 290.00 | 97.50 | 387.50 |
| 6' perimeter | " | 400.00 | 97.50 | 497.50 |
| 8' perimeter | " | 520.00 | 97.50 | 617.50 |
| 10' perimeter | " | 630.00 | 97.50 | 727.50 |
| 12' perimeter | " | 930.00 | 97.50 | 1,028 |
| 14' perimeter | " | 1,150 | 130.00 | 1,280 |
| 16' perimeter | " | 1,290 | 160.00 | 1,450 |
| 18' perimeter | " | 1,580 | 220.00 | 1,800 |
| 20' perimeter | " | 2,000 | 260.00 | 2,260 |
| 22' perimeter | " | 2,150 | 330.00 | 2,480 |
| 24' perimeter | " | 2,290 | 430.00 | 2,720 |
| 20" high x 4' perimeter | " | 370.00 | 97.50 | 467.50 |
| 6' perimeter | " | 430.00 | 97.50 | 527.50 |
| 8' perimeter | " | 570.00 | 97.50 | 667.50 |
| 10' perimeter | " | 720.00 | 97.50 | 817.50 |
| 12' perimeter | " | 1,050 | 97.50 | 1,148 |
| 14' perimeter | " | 1,290 | 130.00 | 1,420 |
| 16' perimeter | " | 1,580 | 160.00 | 1,740 |
| 18' perimeter | " | 1,790 | 220.00 | 2,010 |
| 20' perimeter | " | 2,150 | 260.00 | 2,410 |
| 22' perimeter | " | 2,310 | 330.00 | 2,640 |
| 24' perimeter | " | 2,560 | 430.00 | 2,990 |
| 24" high x 4' perimeter | " | 400.00 | 97.50 | 497.50 |
| 6' perimeter | " | 480.00 | 97.50 | 577.50 |
| 8' perimeter | " | 640.00 | 97.50 | 737.50 |
| 10' perimeter | " | 790.00 | 97.50 | 887.50 |
| 12' perimeter | " | 1,170 | 97.50 | 1,268 |
| 16' perimeter | " | 1,720 | 160.00 | 1,880 |
| 18' perimeter | " | 2,000 | 220.00 | 2,220 |
| 20' perimeter | " | 2,420 | 260.00 | 2,680 |
| 22' perimeter | " | 2,680 | 330.00 | 3,010 |
| 24' perimeter | " | 2,860 | 430.00 | 3,290 |

## CONTROLS

### 15950.10 HVAC CONTROLS

| | UNIT | MAT. | INST. | TOTAL |
|---|---|---|---|---|
| Pressure gauge, direct reading gage cock and siphon | | | | |
| Control valve, 1", modulating | EA. | 65.00 | 24.35 | 89.35 |

| CONTROLS | UNIT | MAT. | INST. | TOTAL |
|---|---|---|---|---|
| **15950.10 HVAC CONTROLS** | | | | |
| 2-way | EA. | 650.00 | 32.50 | 682.50 |
| 3-way | " | 740.00 | 48.80 | 788.80 |
| Self contained control valve with sensing element, 3/4" | " | 120.00 | 24.35 | 144.35 |
| Instrument air system 2-2 hp compressors, rcvr refrigerant dryer | " | | | 5,494 |
| Thermostat primary control device | " | | | 118.56 |
| Humidistat primary control device | " | | | 89.64 |
| Timers primary control device, indoor/outdoor, 24 hour | " | | | 185.07 |
| Thermometer, dir. reading, 3 dial | " | | | 89.64 |
| Control dampers, round | | | | |
| 6" dia. | EA. | 66.50 | 15.60 | 82.10 |
| 8" dia | " | 97.00 | 15.60 | 112.60 |
| 10" dia | " | 130.00 | 15.60 | 145.60 |
| 12" dia | " | 170.00 | 15.60 | 185.60 |
| 12" dia | " | 250.00 | 19.50 | 269.50 |
| 18" dia | " | 260.00 | 19.50 | 279.50 |
| 20" dia | " | 350.00 | 19.50 | 369.50 |
| Rectangular, parallel blade standard leakage | | | | |
| 12" x 12" | EA. | 49.20 | 19.50 | 68.70 |
| 16" x 16" | " | 72.50 | 19.50 | 92.00 |
| 20" x 20" | " | 87.00 | 19.50 | 106.50 |
| 28" x 28" | " | 110.00 | 24.35 | 134.35 |
| 32" x 32" | " | 130.00 | 24.35 | 154.35 |
| 36" x 36" | " | 140.00 | 32.50 | 172.50 |
| 40" x 40" | " | 190.00 | 39.00 | 229.00 |
| 44" x 44" | " | 230.00 | 48.80 | 278.80 |
| 48" x 48" | " | 260.00 | 55.50 | 315.50 |
| 48" x 52" | " | 270.00 | 65.00 | 335.00 |
| 48" x 56" | " | 320.00 | 65.00 | 385.00 |
| 48" x 60" | " | 330.00 | 65.00 | 395.00 |
| 48" x 64" | " | 340.00 | 65.00 | 405.00 |
| 48" x 68" | " | 370.00 | 65.00 | 435.00 |
| 48" x 72" | " | 400.00 | 65.00 | 465.00 |
| Low leakage | | | | |
| 12" x 12" | EA. | 95.50 | 19.50 | 115.00 |
| 16" x 16" | " | 120.00 | 19.50 | 139.50 |
| 20" x 20" | " | 160.00 | 19.50 | 179.50 |
| 24" x 24" | " | 220.00 | 19.50 | 239.50 |
| 28" x 28" | " | 250.00 | 24.35 | 274.35 |
| 32" x 32" | " | 270.00 | 27.90 | 297.90 |
| 36" x 36" | " | 310.00 | 32.50 | 342.50 |
| 40" x 40" | " | 490.00 | 39.00 | 529.00 |
| 44" x 44" | " | 520.00 | 48.80 | 568.80 |
| 48" x 48" | " | 550.00 | 55.50 | 605.50 |
| 48" x 56" | " | 640.00 | 65.00 | 705.00 |
| 48" x 60" | " | 670.00 | 65.00 | 735.00 |
| 48" x 64" | " | 690.00 | 65.00 | 755.00 |
| 48" x 68" | " | 720.00 | 65.00 | 785.00 |
| 48" x 72" | " | 850.00 | 65.00 | 915.00 |
| Rectangular, opposed horizontal blade | | | | |
| 12" x 12" | EA. | 63.50 | 19.50 | 83.00 |
| 16" x 16" | " | 89.50 | 19.50 | 109.00 |
| 20" x 20" | " | 100.00 | 19.50 | 119.50 |

# 15 MECHANICAL

| CONTROLS | UNIT | MAT. | INST. | TOTAL |
|---|---|---|---|---|
| **15950.10** HVAC CONTROLS | | | | |
| 24" x 24" | EA. | 120.00 | 19.50 | 139.50 |
| 28" x 28" | " | 170.00 | 24.35 | 194.35 |
| 32" x 32" | " | 180.00 | 26.00 | 206.00 |
| 36" x 36" | " | 200.00 | 32.50 | 232.50 |
| 40" x 40" | " | 250.00 | 39.00 | 289.00 |
| 44" x 44" | " | 320.00 | 48.80 | 368.80 |
| 48" x 48" | " | 390.00 | 55.50 | 445.50 |
| 48" x 52" | " | 400.00 | 55.50 | 455.50 |
| 48" x 56" | " | 420.00 | 65.00 | 485.00 |
| 48" x 60" | " | 440.00 | 65.00 | 505.00 |
| 48" x 64" | " | 460.00 | 65.00 | 525.00 |
| 48" x 68" | " | 510.00 | 65.00 | 575.00 |
| 48" x 72" | " | 550.00 | 65.00 | 615.00 |

## BASIC MATERIALS

### 16110.20 CONDUIT SPECIALTIES

| BASIC MATERIALS | UNIT | MAT. | INST. | TOTAL |
|---|---|---|---|---|
| Rod beam clamp, 1/2" | EA. | 4.55 | 2.25 | 6.80 |
| Hanger rod | | | | |
| 3/8" | L.F. | 0.96 | 1.80 | 2.76 |
| 1/2" | " | 1.55 | 2.25 | 3.80 |
| All thread rod | | | | |
| 1/4" | L.F. | 0.58 | 1.35 | 1.93 |
| 3/8" | " | 0.95 | 1.80 | 2.75 |
| 1/2" | " | 1.55 | 2.25 | 3.80 |
| 5/8" | " | 2.50 | 3.65 | 6.15 |
| Hanger channel, 1-1/2" | | | | |
| No holes | EA. | 3.70 | 1.35 | 5.05 |
| Holes | " | 4.15 | 1.35 | 5.50 |
| Channel strap | | | | |
| 1/2" | EA. | 0.96 | 2.25 | 3.21 |
| 3/4" | " | 1.15 | 2.25 | 3.40 |
| 1" | " | 1.25 | 2.25 | 3.50 |
| 1-1/4" | " | 1.40 | 3.65 | 5.05 |
| 1-1/2" | " | 1.55 | 3.65 | 5.20 |
| 2" | " | 1.80 | 3.65 | 5.45 |
| 2-1/2" | " | 1.95 | 5.60 | 7.55 |
| 3" | " | 2.15 | 5.60 | 7.75 |
| 3-1/2" | " | 2.65 | 5.60 | 8.25 |
| 4" | " | 3.00 | 6.60 | 9.60 |
| 5" | " | 4.20 | 6.60 | 10.80 |
| 6" | " | 5.05 | 6.60 | 11.65 |
| Conduit penetrations, roof and wall, 8" thick | | | | |
| 1/2" | EA. | 0.00 | 27.90 | 27.90 |
| 3/4" | " | 0.00 | 27.90 | 27.90 |
| 1" | " | 0.00 | 36.30 | 36.30 |
| 1-1/4" | " | 0.00 | 36.30 | 36.30 |
| 1-1/2" | " | 0.00 | 36.30 | 36.30 |
| 2" | " | 0.00 | 72.50 | 72.50 |
| 2-1/2" | " | 0.00 | 72.50 | 72.50 |
| 3" | " | 0.00 | 72.50 | 72.50 |
| 3-1/2" | " | 0.00 | 91.00 | 91.00 |
| 4" | " | 0.00 | 91.00 | 91.00 |
| Fireproofing, for conduit penetrations | | | | |
| 1/2" | EA. | 2.15 | 22.70 | 24.85 |
| 3/4" | " | 2.20 | 22.70 | 24.90 |
| 1" | " | 2.25 | 22.70 | 24.95 |
| 1-1/4" | " | 3.10 | 35.50 | 38.60 |
| 1-1/2" | " | 3.15 | 33.00 | 36.15 |
| 2" | " | 3.25 | 33.00 | 36.25 |
| 2-1/2" | " | 6.15 | 40.80 | 46.95 |
| 3" | " | 6.35 | 43.80 | 50.15 |
| 3-1/2" | " | 7.45 | 56.50 | 63.95 |
| 4" | " | 9.00 | 68.50 | 77.50 |

### 16110.21 ALUMINUM CONDUIT

| BASIC MATERIALS | UNIT | MAT. | INST. | TOTAL |
|---|---|---|---|---|
| Aluminum conduit | | | | |
| 1/2" | L.F. | 1.10 | 1.35 | 2.45 |

| BASIC MATERIALS | UNIT | MAT. | INST. | TOTAL |
|---|---|---|---|---|
| **16110.21 ALUMINUM CONDUIT** | | | | |
| 3/4" | L.F. | 1.40 | 1.80 | 3.20 |
| 1" | " | 1.45 | 2.25 | 3.70 |
| 1-1/4" | " | 2.70 | 2.70 | 5.40 |
| 1-1/2" | " | 3.35 | 3.65 | 7.00 |
| 2" | " | 4.50 | 4.05 | 8.55 |
| 2-1/2" | " | 7.15 | 4.55 | 11.70 |
| 3" | " | 9.35 | 4.85 | 14.20 |
| 3-1/2" | " | 11.25 | 5.60 | 16.85 |
| 4" | " | 13.30 | 6.60 | 19.90 |
| 5" | " | 19.10 | 8.25 | 27.35 |
| 6" | " | 25.20 | 9.10 | 34.30 |
| 90 deg. elbow | | | | |
| 1/2" | EA. | 4.00 | 8.65 | 12.65 |
| 3/4" | " | 5.45 | 11.35 | 16.80 |
| 1" | " | 7.55 | 13.95 | 21.50 |
| 1-1/4" | " | 12.05 | 17.30 | 29.35 |
| 1-1/2" | " | 15.90 | 18.15 | 34.05 |
| 2" | " | 23.60 | 20.20 | 43.80 |
| 2-1/2" | " | 39.80 | 25.90 | 65.70 |
| 3" | " | 61.50 | 30.30 | 91.80 |
| 3-1/2" | " | 96.00 | 36.30 | 132.30 |
| 4" | " | 110.00 | 40.40 | 150.40 |
| 5" | " | 310.00 | 52.00 | 362.00 |
| 6" | " | 430.00 | 100.00 | 530.00 |
| Coupling | | | | |
| 1/2" | EA. | 1.30 | 2.25 | 3.55 |
| 3/4" | " | 1.95 | 2.70 | 4.65 |
| 1" | " | 2.55 | 3.65 | 6.20 |
| 1-1/4" | " | 3.15 | 4.05 | 7.20 |
| 1-1/2" | " | 3.60 | 4.55 | 8.15 |
| 2" | " | 5.15 | 4.85 | 10.00 |
| 2-1/2" | " | 11.70 | 5.60 | 17.30 |
| 3" | " | 15.25 | 5.60 | 20.85 |
| 3-1/2" | " | 20.95 | 6.60 | 27.55 |
| 4" | " | 25.20 | 7.25 | 32.45 |
| 5" | " | 64.00 | 7.25 | 71.25 |
| 6" | " | 100.00 | 8.65 | 108.65 |
| **16110.22 EMT CONDUIT** | | | | |
| EMT conduit | | | | |
| 1/2" | L.F. | 0.35 | 1.35 | 1.70 |
| 3/4" | " | 0.49 | 1.80 | 2.29 |
| 1" | " | 0.68 | 2.25 | 2.93 |
| 1-1/4" | " | 0.99 | 2.70 | 3.69 |
| 1-1/2" | " | 1.25 | 3.65 | 4.90 |
| 2" | " | 1.40 | 4.05 | 5.45 |
| 2-1/2" | " | 3.15 | 4.55 | 7.70 |
| 3" | " | 4.50 | 5.60 | 10.10 |
| 3-1/2" | " | 5.90 | 6.60 | 12.50 |
| 4" | " | 6.60 | 8.25 | 14.85 |
| 90 deg. elbow | | | | |

| BASIC MATERIALS | UNIT | MAT. | INST. | TOTAL |
|---|---|---|---|---|
| **16110.22 EMT CONDUIT** | | | | |
| 1/2" | EA. | 2.80 | 4.05 | 6.85 |
| 3/4" | " | 3.10 | 4.55 | 7.65 |
| 1" | " | 3.25 | 4.85 | 8.10 |
| 1-1/4" | " | 4.10 | 5.60 | 9.70 |
| 1-1/2" | " | 5.15 | 6.60 | 11.75 |
| 2" | " | 8.30 | 8.65 | 16.95 |
| 2-1/2" | " | 23.50 | 9.55 | 33.05 |
| 3" | " | 50.50 | 11.00 | 61.50 |
| 3-1/2" | " | 49.10 | 12.60 | 61.70 |
| 4" | " | 58.00 | 12.95 | 70.95 |
| Connector, die cast set screw | | | | |
| 1/2" | EA. | 0.24 | 2.70 | 2.94 |
| 3/4" | " | 0.39 | 2.70 | 3.09 |
| 1" | " | 0.63 | 2.70 | 3.33 |
| 1-1/4" | " | 1.10 | 4.05 | 5.15 |
| 1-1/2" | " | 1.70 | 4.85 | 6.55 |
| 2" | " | 2.20 | 6.60 | 8.80 |
| 2-1/2" | " | 7.10 | 9.10 | 16.20 |
| 3" | " | 8.65 | 10.10 | 18.75 |
| 3-1/2" | " | 12.20 | 11.35 | 23.55 |
| 4" | " | 14.05 | 12.95 | 27.00 |
| Coupling, steel set screw | | | | |
| 1/2" | EA. | 0.62 | 1.80 | 2.42 |
| 3/4" | " | 0.95 | 1.80 | 2.75 |
| 1" | " | 1.40 | 1.80 | 3.20 |
| 1-1/4" | " | 2.65 | 2.25 | 4.90 |
| 1-1/2" | " | 3.95 | 3.65 | 7.60 |
| 2" | " | 5.25 | 4.85 | 10.10 |
| 2-1/2" | " | 11.85 | 7.25 | 19.10 |
| 3" | " | 13.25 | 8.65 | 21.90 |
| 3-1/2" | " | 15.60 | 10.10 | 25.70 |
| 4" | " | 17.85 | 11.35 | 29.20 |
| **16110.23 FLEXIBLE CONDUIT** | | | | |
| Flexible conduit, steel | | | | |
| 3/8" | L.F. | 0.27 | 1.35 | 1.62 |
| 1/2 | " | 0.28 | 1.35 | 1.63 |
| 3/4" | " | 0.41 | 1.80 | 2.21 |
| 1" | " | 0.80 | 1.80 | 2.60 |
| 1-1/4" | " | 1.00 | 2.25 | 3.25 |
| 1-1/2" | " | 1.35 | 2.70 | 4.05 |
| 2" | " | 1.70 | 3.65 | 5.35 |
| 2-1/2" | " | 2.00 | 4.05 | 6.05 |
| 3" | " | 2.50 | 4.85 | 7.35 |
| Flexible conduit, liquid tight | | | | |
| 3/8" | L.F. | 1.15 | 1.35 | 2.50 |
| 1/2" | " | 1.20 | 1.35 | 2.55 |
| 3/4" | " | 1.35 | 1.80 | 3.15 |
| 1" | " | 2.60 | 1.80 | 4.40 |
| 1-1/4" | " | 3.55 | 2.25 | 5.80 |
| 1-1/2" | " | 4.90 | 2.70 | 7.60 |
| 2" | EA. | 6.20 | 3.65 | 9.85 |

| BASIC MATERIALS | UNIT | MAT. | INST. | TOTAL |
|---|---|---|---|---|
| **16110.23 FLEXIBLE CONDUIT** | | | | |
| 2-1/2" | EA. | 11.45 | 4.05 | 15.50 |
| 3" | " | 15.80 | 4.85 | 20.65 |
| 4" | " | 23.05 | 6.60 | 29.65 |
| Connector, straight | | | | |
| 3/8" | EA. | 0.69 | 3.65 | 4.34 |
| 1/2" | " | 1.75 | 3.65 | 5.40 |
| 3/4" | " | 1.40 | 4.05 | 5.45 |
| 1" | " | 2.80 | 4.55 | 7.35 |
| 1-1/4" | " | 4.35 | 4.85 | 9.20 |
| 1-1/2" | " | 7.55 | 5.60 | 13.15 |
| 2" | " | 10.50 | 6.60 | 17.10 |
| 2-1/2" | " | 19.65 | 8.25 | 27.90 |
| 3" | " | 27.50 | 8.65 | 36.15 |
| Flexible aluminum conduit | | | | |
| 3/8" | L.F. | 0.24 | 1.35 | 1.59 |
| 1/2" | " | 0.36 | 1.35 | 1.71 |
| 3/4" | " | 0.49 | 1.80 | 2.29 |
| 1" | " | 1.10 | 1.80 | 2.90 |
| 1-1/4" | " | 1.40 | 2.25 | 3.65 |
| 1-1/2" | " | 1.70 | 2.70 | 4.40 |
| 2" | " | 2.30 | 3.65 | 5.95 |
| 2-1/2" | " | 2.90 | 4.05 | 6.95 |
| 3" | " | 3.35 | 4.85 | 8.20 |
| 3-1/2" | " | 4.35 | 5.60 | 9.95 |
| 4" | " | 5.20 | 6.60 | 11.80 |
| **16110.24 GALVANIZED CONDUIT** | | | | |
| Galvanized rigid steel conduit | | | | |
| 1/2" | L.F. | 1.10 | 1.80 | 2.90 |
| 3/4" | " | 1.40 | 2.25 | 3.65 |
| 1" | " | 1.80 | 2.70 | 4.50 |
| 1-1/4" | " | 2.60 | 3.65 | 6.25 |
| 1-1/2" | " | 3.00 | 4.05 | 7.05 |
| 2" | " | 3.80 | 4.55 | 8.35 |
| 2-1/2" | " | 6.20 | 6.60 | 12.80 |
| 3" | " | 8.15 | 8.25 | 16.40 |
| 3-1/2" | " | 8.40 | 8.65 | 17.05 |
| 4" | " | 12.10 | 9.55 | 21.65 |
| 5" | " | 24.80 | 12.95 | 37.75 |
| 6" | " | 35.80 | 17.30 | 53.10 |
| 90 degree ell | | | | |
| 1/2" | EA. | 3.80 | 11.35 | 15.15 |
| 3/4" | " | 4.55 | 13.95 | 18.50 |
| 1" | " | 6.80 | 17.30 | 24.10 |
| 1-1/4" | " | 9.85 | 20.20 | 30.05 |
| 1-1/2" | " | 11.75 | 22.70 | 34.45 |
| 2" | " | 17.30 | 24.20 | 41.50 |
| 2-1/2" | " | 29.10 | 30.30 | 59.40 |
| 3" | " | 43.60 | 40.40 | 84.00 |
| 3-1/2" | " | 75.50 | 45.40 | 120.90 |
| 4" | " | 87.50 | 60.50 | 148.00 |

| BASIC MATERIALS | UNIT | MAT. | INST. | TOTAL |
|---|---|---|---|---|
| **16110.24 GALVANIZED CONDUIT** | | | | |
| 5" | EA. | 230.00 | 100.00 | 330.00 |
| 6" | " | 310.00 | 150.00 | 460.00 |
| Couplings, with set screws | | | | |
| 1/2" | EA. | 2.00 | 2.25 | 4.25 |
| 3/4" | " | 2.60 | 2.70 | 5.30 |
| 1" | " | 4.20 | 3.65 | 7.85 |
| 1-1/4" | " | 7.10 | 4.55 | 11.65 |
| 1-1/2" | " | 9.25 | 5.60 | 14.85 |
| 2" | " | 20.75 | 6.60 | 27.35 |
| 2-1/2" | " | 51.50 | 8.65 | 60.15 |
| 3" | " | 61.50 | 11.35 | 72.85 |
| 3-1/2" | " | 87.50 | 12.95 | 100.45 |
| 4" | " | 110.00 | 13.95 | 123.95 |
| 5" | " | 180.00 | 20.20 | 200.20 |
| 6" | " | 230.00 | 22.70 | 252.70 |
| Split couplings | | | | |
| 1/2" | EA. | 1.70 | 8.65 | 10.35 |
| 3/4" | " | 2.20 | 11.35 | 13.55 |
| 1" | " | 3.10 | 12.50 | 15.60 |
| 1-1/4" | " | 6.30 | 13.95 | 20.25 |
| 1-1/2" | " | 7.90 | 17.30 | 25.20 |
| 2" | " | 36.50 | 25.90 | 62.40 |
| 2-1/2" | " | 37.10 | 25.90 | 63.00 |
| 3" | " | 55.00 | 33.00 | 88.00 |
| 3-1/2" | " | 87.50 | 45.40 | 132.90 |
| 4" | " | 100.00 | 60.50 | 160.50 |
| 5" | " | 180.00 | 74.00 | 254.00 |
| 6" | " | 240.00 | 93.00 | 333.00 |
| Erickson couplings | | | | |
| 1/2" | EA. | 2.90 | 20.20 | 23.10 |
| 3/4" | " | 3.50 | 22.70 | 26.20 |
| 1" | " | 7.10 | 27.90 | 35.00 |
| 1-1/4" | " | 12.75 | 40.40 | 53.15 |
| 1-1/2" | " | 16.55 | 45.40 | 61.95 |
| 2" | " | 33.30 | 60.50 | 93.80 |
| 2-1/2" | " | 65.00 | 84.50 | 149.50 |
| 3" | " | 95.50 | 95.50 | 191.00 |
| 3-1/2" | " | 170.00 | 110.00 | 280.00 |
| 4" | " | 200.00 | 120.00 | 320.00 |
| 5" | " | 410.00 | 130.00 | 540.00 |
| 6" | " | 530.00 | 150.00 | 680.00 |
| Seal fittings | | | | |
| 1/2" | EA. | 12.15 | 30.30 | 42.45 |
| 3/4" | " | 14.05 | 36.30 | 50.35 |
| 1" | " | 17.20 | 45.40 | 62.60 |
| 1-1/4" | " | 20.85 | 52.00 | 72.85 |
| 1-1/2" | " | 30.90 | 60.50 | 91.40 |
| 2" | " | 40.30 | 72.50 | 112.80 |
| 2-1/2" | " | 62.00 | 86.50 | 148.50 |
| 3" | " | 74.50 | 95.50 | 170.00 |
| 3-1/2" | " | 200.00 | 110.00 | 310.00 |
| 4" | " | 310.00 | 130.00 | 440.00 |

## BASIC MATERIALS

### 16110.24 GALVANIZED CONDUIT

| BASIC MATERIALS | UNIT | MAT. | INST. | TOTAL |
|---|---|---|---|---|
| 5" | EA. | 470.00 | 200.00 | 670.00 |
| 6" | " | 640.00 | 230.00 | 870.00 |
| Entrance fitting, (weather head), threaded | | | | |
| 1/2" | EA. | 3.40 | 20.20 | 23.60 |
| 3/4" | " | 4.15 | 22.70 | 26.85 |
| 1" | " | 5.35 | 25.90 | 31.25 |
| 1-1/4" | " | 6.95 | 33.00 | 39.95 |
| 1-1/2" | " | 10.15 | 36.30 | 46.45 |
| 2" | " | 18.70 | 40.40 | 59.10 |
| 2-1/2" | " | 64.50 | 45.40 | 109.90 |
| 3" | " | 91.00 | 60.50 | 151.50 |
| 3-1/2" | " | 120.00 | 79.00 | 199.00 |
| 4" | " | 150.00 | 110.00 | 260.00 |
| 5" | " | 330.00 | 160.00 | 490.00 |
| 6" | " | 320.00 | 200.00 | 520.00 |
| Locknuts | | | | |
| 1/2" | EA. | 0.12 | 2.25 | 2.37 |
| 3/4" | " | 0.15 | 2.25 | 2.40 |
| 1" | " | 0.26 | 2.25 | 2.51 |
| 1-1/4" | " | 0.34 | 2.25 | 2.59 |
| 1-1/2" | " | 0.49 | 2.70 | 3.19 |
| 2" | " | 0.69 | 2.70 | 3.39 |
| 2-1/2" | " | 1.80 | 3.65 | 5.45 |
| 3" | " | 2.30 | 3.65 | 5.95 |
| 3-1/2" | " | 3.95 | 3.65 | 7.60 |
| 4" | " | 4.95 | 4.05 | 9.00 |
| 5" | " | 10.55 | 4.05 | 14.60 |
| 6" | " | 18.45 | 4.05 | 22.50 |
| Plastic conduit bushings | | | | |
| 1/2" | EA. | 0.12 | 5.60 | 5.72 |
| 3/4" | " | 0.19 | 6.60 | 6.79 |
| 1" | " | 0.26 | 8.65 | 8.91 |
| 1-1/4" | " | 0.35 | 10.10 | 10.45 |
| 1-1/2" | " | 0.47 | 11.35 | 11.82 |
| 2" | " | 0.84 | 13.95 | 14.79 |
| 2-1/2" | " | 2.15 | 22.70 | 24.85 |
| 3" | " | 2.60 | 30.30 | 32.90 |
| 3-1/2" | " | 2.75 | 36.30 | 39.05 |
| 4" | " | 3.15 | 40.40 | 43.55 |
| 5" | " | 7.25 | 52.00 | 59.25 |
| 6" | " | 13.90 | 72.50 | 86.40 |
| Conduit bushings, steel | | | | |
| 1/2" | EA. | 0.19 | 5.60 | 5.79 |
| 3/4" | " | 0.24 | 6.60 | 6.84 |
| 1" | " | 0.36 | 8.65 | 9.01 |
| 1-1/4" | " | 0.51 | 10.10 | 10.61 |
| 1-1/2" | " | 0.77 | 11.35 | 12.12 |
| 2" | " | 1.20 | 13.95 | 15.15 |
| 2-1/2" | " | 2.70 | 22.70 | 25.40 |
| 3" | " | 3.35 | 30.30 | 33.65 |
| 3-1/2" | " | 6.95 | 36.30 | 43.25 |
| 4" | " | 8.45 | 40.40 | 48.85 |

## BASIC MATERIALS

| BASIC MATERIALS | UNIT | MAT. | INST. | TOTAL |
|---|---|---|---|---|
| **16110.24 GALVANIZED CONDUIT** | | | | |
| 5" | EA. | 17.95 | 52.00 | 69.95 |
| 6" | " | 31.90 | 72.50 | 104.40 |
| Pipe cap | | | | |
| 1/2" | EA. | 0.22 | 2.25 | 2.47 |
| 3/4" | " | 0.23 | 2.25 | 2.48 |
| 1" | " | 0.38 | 2.25 | 2.63 |
| 1-1/4" | " | 0.65 | 3.65 | 4.30 |
| 1-1/2" | " | 1.00 | 3.65 | 4.65 |
| 2" | " | 1.15 | 3.65 | 4.80 |
| 2-1/2" | " | 1.95 | 4.05 | 6.00 |
| 3" | " | 2.45 | 4.05 | 6.50 |
| 3-1/2" | " | 3.30 | 4.05 | 7.35 |
| 4" | " | 4.25 | 4.85 | 9.10 |
| 5" | " | 5.65 | 6.60 | 12.25 |
| 6" | " | 7.10 | 9.10 | 16.20 |
| **16110.25 PLASTIC CONDUIT** | | | | |
| PVC conduit, schedule 40 | | | | |
| 1/2" | L.F. | 0.38 | 1.35 | 1.73 |
| 3/4" | " | 0.46 | 1.35 | 1.81 |
| 1" | " | 0.62 | 1.80 | 2.42 |
| 1-1/4" | " | 0.81 | 1.80 | 2.61 |
| 1-1/2" | " | 1.05 | 2.25 | 3.30 |
| 2" | " | 1.35 | 2.25 | 3.60 |
| 2-1/2" | " | 2.15 | 2.70 | 4.85 |
| 3" | " | 2.45 | 2.70 | 5.15 |
| 3-1/2" | " | 3.25 | 3.65 | 6.90 |
| 4" | " | 3.55 | 3.65 | 7.20 |
| 5" | " | 5.05 | 4.05 | 9.10 |
| 6" | " | 6.70 | 4.55 | 11.25 |
| Couplings | | | | |
| 1/2" | EA. | 0.27 | 2.25 | 2.52 |
| 3/4" | " | 0.28 | 2.25 | 2.53 |
| 1" | " | 0.45 | 2.25 | 2.70 |
| 1-1/4" | " | 0.57 | 2.70 | 3.27 |
| 1-1/2" | " | 0.76 | 2.70 | 3.46 |
| 2" | " | 1.05 | 2.70 | 3.75 |
| 2-1/2" | " | 1.90 | 2.70 | 4.60 |
| 3" | " | 3.00 | 3.65 | 6.65 |
| 3-1/2" | " | 3.45 | 3.65 | 7.10 |
| 4" | " | 4.70 | 4.55 | 9.25 |
| 5" | " | 11.60 | 4.55 | 16.15 |
| 6" | " | 14.95 | 4.55 | 19.50 |
| 90 degree elbows | | | | |
| 1/2" | EA. | 1.05 | 4.55 | 5.60 |
| 3/4" | " | 1.10 | 5.60 | 6.70 |
| 1" | " | 1.35 | 5.60 | 6.95 |
| 1-1/4" | " | 1.95 | 6.60 | 8.55 |
| 1-1/2" | " | 2.70 | 8.65 | 11.35 |
| 2" | " | 3.80 | 10.10 | 13.90 |
| 2-1/2" | " | 6.80 | 11.35 | 18.15 |

## BASIC MATERIALS

### 16110.25 PLASTIC CONDUIT

| | UNIT | MAT. | INST. | TOTAL |
|---|---|---|---|---|
| 3" | EA. | 11.90 | 13.95 | 25.85 |
| 3-1/2" | " | 16.60 | 17.30 | 33.90 |
| 4" | " | 24.70 | 22.70 | 47.40 |
| 5" | " | 36.20 | 27.90 | 64.10 |
| 6" | " | 65.50 | 33.00 | 98.50 |
| Terminal adapters | | | | |
| 1/2" | EA. | 0.35 | 4.55 | 4.90 |
| 3/4" | " | 0.58 | 4.55 | 5.13 |
| 1" | " | 0.70 | 4.55 | 5.25 |
| 1-1/4" | " | 0.90 | 7.25 | 8.15 |
| 1-1/2" | " | 1.15 | 7.25 | 8.40 |
| 2" | " | 1.60 | 7.25 | 8.85 |
| 2-1/2" | " | 2.70 | 10.10 | 12.80 |
| 3" | " | 3.80 | 10.10 | 13.90 |
| 3-1/2" | " | 4.95 | 10.10 | 15.05 |
| 4" | " | 6.35 | 17.30 | 23.65 |
| 5" | " | 12.75 | 17.30 | 30.05 |
| 6" | " | 15.30 | 17.30 | 32.60 |
| End bells | | | | |
| 1" | EA. | 1.90 | 4.55 | 6.45 |
| 1-1/4" | " | 2.10 | 7.25 | 9.35 |
| 1-1/2" | " | 2.15 | 7.25 | 9.40 |
| 2" | " | 3.20 | 7.25 | 10.45 |
| 2-1/2" | " | 3.45 | 10.10 | 13.55 |
| 3" | " | 3.90 | 10.10 | 14.00 |
| 3-1/2" | " | 4.00 | 10.10 | 14.10 |
| 4" | " | 4.60 | 17.30 | 21.90 |
| 5" | " | 7.00 | 17.30 | 24.30 |
| 6" | " | 8.30 | 17.30 | 25.60 |
| LB conduit body | | | | |
| 1/2" | EA. | 2.40 | 8.65 | 11.05 |
| 3/4" | " | 3.30 | 8.65 | 11.95 |
| 1 | " | 3.65 | 8.65 | 12.30 |
| 1-1/4" | " | 5.50 | 13.95 | 19.45 |
| 1-1/2" | " | 7.00 | 13.95 | 20.95 |
| 2" | " | 10.00 | 13.95 | 23.95 |
| 2" | L.F. | 0.58 | 2.25 | 2.83 |
| 3" | " | 1.05 | 2.70 | 3.75 |
| 4" | " | 2.10 | 3.65 | 5.75 |
| 5" | " | 3.00 | 4.05 | 7.05 |
| 6" | " | 4.30 | 4.55 | 8.85 |
| Encased burial conduit | | | | |
| 2" | L.F. | 0.59 | 2.25 | 2.84 |
| 3" | " | 0.99 | 2.70 | 3.69 |
| 4" | " | 1.50 | 3.65 | 5.15 |
| 5" | " | 2.15 | 4.05 | 6.20 |
| 6" | " | 2.80 | 4.55 | 7.35 |
| PVC cement | | | | |
| 1 pint | EA. | | | 7.41 |
| 1 quart | " | | | 10.77 |
| 1 gallon | " | | | 32.86 |

| BASIC MATERIALS | UNIT | MAT. | INST. | TOTAL |
|---|---|---|---|---|
| **16110.27 PLASTIC COATED CONDUIT** | | | | |
| Rigid steel conduit, plastic coated | | | | |
| 1/2" | L.F. | 2.95 | 2.25 | 5.20 |
| 3/4" | " | 3.45 | 2.70 | 6.15 |
| 1" | " | 4.50 | 3.65 | 8.15 |
| 1-1/4" | " | 5.60 | 4.55 | 10.15 |
| 1-1/2" | " | 6.85 | 5.60 | 12.45 |
| 2" | " | 9.00 | 6.60 | 15.60 |
| 2-1/2" | " | 13.70 | 8.65 | 22.35 |
| 3" | " | 17.20 | 10.10 | 27.30 |
| 3-1/2" | " | 20.90 | 11.35 | 32.25 |
| 4" | " | 25.60 | 13.95 | 39.55 |
| 5" | " | 43.80 | 17.30 | 61.10 |
| 90 degree elbows | | | | |
| 1/2" | EA. | 11.30 | 13.95 | 25.25 |
| 3/4" | " | 11.55 | 17.30 | 28.85 |
| 1" | " | 13.45 | 20.20 | 33.65 |
| 1-1/4" | " | 16.80 | 22.70 | 39.50 |
| 1-1/2" | " | 20.30 | 27.90 | 48.20 |
| 2" | " | 28.30 | 36.30 | 64.60 |
| 2-1/2" | " | 54.50 | 52.00 | 106.50 |
| 3" | " | 55.50 | 60.50 | 116.00 |
| 3-1/2" | " | 120.00 | 74.00 | 194.00 |
| 4" | " | 130.00 | 91.00 | 221.00 |
| 5" | " | 290.00 | 110.00 | 400.00 |
| Couplings | | | | |
| 1/2" | EA. | 3.30 | 2.70 | 6.00 |
| 3/4" | " | 3.50 | 3.65 | 7.15 |
| 1" | " | 4.70 | 4.05 | 8.75 |
| 1-1/4" | " | 5.45 | 4.85 | 10.30 |
| 1-1/2" | " | 7.60 | 5.60 | 13.20 |
| 2" | " | 9.55 | 6.60 | 16.15 |
| 2-1/2" | " | 23.85 | 8.25 | 32.10 |
| 3" | " | 27.80 | 8.65 | 36.45 |
| 3-1/2" | " | 38.30 | 9.10 | 47.40 |
| 4" | " | 46.90 | 10.10 | 57.00 |
| 5" | " | 130.00 | 11.35 | 141.35 |
| **16110.28 STEEL CONDUIT** | | | | |
| Intermediate metal conduit (IMC) | | | | |
| 1/2" | L.F. | 1.10 | 1.35 | 2.45 |
| 3/4" | " | 1.30 | 1.80 | 3.10 |
| 1" | " | 1.70 | 2.25 | 3.95 |
| 1-1/4" | " | 2.20 | 2.70 | 4.90 |
| 1-1/2" | " | 2.65 | 3.65 | 6.30 |
| 2" | " | 3.60 | 4.05 | 7.65 |
| 2-1/2" | " | 5.45 | 5.40 | 10.85 |
| 3" | " | 7.70 | 6.60 | 14.30 |
| 3-1/2" | " | 10.00 | 8.25 | 18.25 |
| 4" | " | 11.75 | 8.65 | 20.40 |
| 90 degree ell | | | | |
| 1/2" | EA. | 3.15 | 11.35 | 14.50 |
| 3/4" | " | 4.15 | 13.95 | 18.10 |

# 16 ELECTRICAL

| BASIC MATERIALS | UNIT | MAT. | INST. | TOTAL |
|---|---|---|---|---|
| **16110.28 STEEL CONDUIT** | | | | |
| 1" | EA. | 6.00 | 17.30 | 23.30 |
| 1-1/4" | " | 8.40 | 20.20 | 28.60 |
| 1-1/2" | " | 10.65 | 22.70 | 33.35 |
| 2" | " | 15.35 | 25.90 | 41.25 |
| 2-1/2" | " | 26.60 | 30.30 | 56.90 |
| 3" | " | 40.70 | 40.40 | 81.10 |
| 3-1/2" | " | 72.00 | 52.00 | 124.00 |
| 4" | " | 85.00 | 60.50 | 145.50 |
| **16110.35 WIREMOLD** | | | | |
| Wiremold raceway with fittings, surface mounted | | | | |
| #200 | L.F. | 0.61 | 1.35 | 1.96 |
| #500 | " | 0.73 | 1.35 | 2.08 |
| #700 | " | 0.80 | 1.80 | 2.60 |
| #800 | " | 1.30 | 1.80 | 3.10 |
| Fittings, #200, 90 degree flat elbow | EA. | 1.15 | 2.25 | 3.40 |
| Internal elbow | " | 2.30 | 2.25 | 4.55 |
| Extension adapter | " | 4.15 | 2.70 | 6.85 |
| #200, #500, #700 | | | | |
| Single pole switch and box | EA. | 7.80 | 18.15 | 25.95 |
| Duplex receptacle with box | " | 9.05 | 13.95 | 23.00 |
| #500, #700 | | | | |
| 90 deg. flat elbow | EA. | 1.00 | 3.65 | 4.65 |
| Internal elbow | " | 1.30 | 3.65 | 4.95 |
| Junction box | " | 6.65 | 6.05 | 12.70 |
| Fixture box | " | 6.70 | 6.05 | 12.75 |
| Shallow switch and receptacle box | " | 5.45 | 3.65 | 9.10 |
| #800 | | | | |
| 90 deg. flat elbow | EA. | 1.25 | 3.65 | 4.90 |
| Internal elbow | " | 1.25 | 3.65 | 4.90 |
| Junction box | " | 3.90 | 5.60 | 9.50 |
| **16110.80 WIREWAYS** | | | | |
| Wireway, hinge cover type | | | | |
| 2-1/2" x 2-1/2" | | | | |
| 1' section | EA. | 10.65 | 7.00 | 17.65 |
| 2' | " | 15.10 | 8.65 | 23.75 |
| 3' | " | 20.55 | 11.35 | 31.90 |
| 5' | " | 33.90 | 17.30 | 51.20 |
| 10' | " | 70.50 | 30.30 | 100.80 |
| 4" x 4" | | | | |
| 1' | EA. | 10.30 | 11.35 | 21.65 |
| 2' | " | 17.10 | 11.35 | 28.45 |
| 3' | " | 25.30 | 13.95 | 39.25 |
| 4' | " | 34.40 | 13.95 | 48.35 |
| 10' | " | 36.90 | 36.30 | 73.20 |
| 6" x 6" | | | | |
| 1' | EA. | 22.35 | 17.30 | 39.65 |
| 2' | " | 27.90 | 17.30 | 45.20 |
| 3' | " | 39.50 | 20.20 | 59.70 |
| 4' | " | 51.50 | 20.20 | 71.70 |

| BASIC MATERIALS | UNIT | MAT. | INST. | TOTAL |
|---|---|---|---|---|
| **16110.80 WIREWAYS** | | | | |
| 5' | EA. | 55.00 | 25.90 | 80.90 |
| 10' | " | 120.00 | 40.40 | 160.40 |
| 8" x 8" | | | | |
| 1' | EA. | 36.10 | 20.20 | 56.30 |
| 2' | " | 55.00 | 20.20 | 75.20 |
| 3' | " | 82.50 | 22.70 | 105.20 |
| 4' | " | 100.00 | 22.70 | 122.70 |
| 5' | " | 110.00 | 27.90 | 137.90 |
| 12" x 12" | | | | |
| 1' | EA. | 49.80 | 27.90 | 77.70 |
| 2' | " | 96.50 | 27.90 | 124.40 |
| 3' | " | 140.00 | 33.00 | 173.00 |
| 4' | " | 170.00 | 33.00 | 203.00 |
| 5' | " | 200.00 | 40.40 | 240.40 |
| **16120.43 COPPER CONDUCTORS** | | | | |
| Copper conductors, type THW, solid | | | | |
| #14 | L.F. | 0.08 | 0.18 | 0.26 |
| #12 | " | 0.09 | 0.23 | 0.32 |
| #10 | " | 0.11 | 0.27 | 0.38 |
| Stranded | | | | |
| #14 | L.F. | 0.11 | 0.18 | 0.29 |
| #12 | " | 0.12 | 0.23 | 0.35 |
| #10 | " | 0.15 | 0.27 | 0.42 |
| #8 | " | 0.23 | 0.36 | 0.59 |
| #6 | " | 0.31 | 0.41 | 0.72 |
| #4 | " | 0.45 | 0.45 | 0.90 |
| #3 | " | 0.62 | 0.45 | 1.07 |
| #2 | " | 0.77 | 0.55 | 1.32 |
| #1 | " | 1.00 | 0.64 | 1.64 |
| 1/0 | " | 1.20 | 0.73 | 1.93 |
| 2/0 | " | 1.40 | 0.91 | 2.31 |
| 3/0 | " | 1.75 | 1.15 | 2.90 |
| 4/0 | " | 2.15 | 1.25 | 3.40 |
| 250 MCM | " | 2.65 | 1.35 | 4.00 |
| 300 MCM | " | 3.20 | 1.50 | 4.70 |
| 350 MCM | " | 3.85 | 1.80 | 5.65 |
| 400 MCM | " | 4.35 | 2.00 | 6.35 |
| 500 MCM | " | 5.30 | 2.35 | 7.65 |
| 600 MCM | " | 7.50 | 2.70 | 10.20 |
| 750 MCM | " | 9.55 | 3.05 | 12.60 |
| 1000 MCM | " | 10.60 | 3.45 | 14.05 |
| THHN-THWN, solid | | | | |
| #14 | L.F. | 0.08 | 0.18 | 0.26 |
| #12 | " | 0.11 | 0.23 | 0.34 |
| #10 | " | 0.14 | 0.27 | 0.41 |
| Stranded | | | | |
| #14 | L.F. | 0.11 | 0.18 | 0.29 |
| #12 | " | 0.14 | 0.23 | 0.37 |
| #10 | " | 0.15 | 0.27 | 0.42 |
| #8 | " | 0.23 | 0.36 | 0.59 |
| #6 | " | 0.31 | 0.41 | 0.72 |

## BASIC MATERIALS

| | UNIT | MAT. | INST. | TOTAL |
|---|---|---|---|---|
| **16120.43 COPPER CONDUCTORS** | | | | |
| #4 | L.F. | 0.45 | 0.45 | 0.90 |
| #2 | " | 0.77 | 0.55 | 1.32 |
| #1 | " | 0.99 | 0.64 | 1.63 |
| 1/0 | " | 1.20 | 0.73 | 1.93 |
| 2/0 | " | 1.40 | 0.91 | 2.31 |
| 3/0 | " | 1.75 | 1.15 | 2.90 |
| 4/0 | " | 2.15 | 1.25 | 3.40 |
| 250 MCM | " | 2.65 | 1.35 | 4.00 |
| 350 MCM | " | 3.85 | 1.80 | 5.65 |
| XHHW | | | | |
| #14 | L.F. | 0.14 | 0.18 | 0.32 |
| #10 | " | 0.22 | 0.27 | 0.49 |
| #8 | " | 0.31 | 0.36 | 0.67 |
| #6 | " | 0.34 | 0.41 | 0.75 |
| #4 | " | 0.49 | 0.41 | 0.90 |
| #2 | " | 0.77 | 0.50 | 1.27 |
| #1 | " | 1.00 | 0.64 | 1.64 |
| 1/0 | " | 1.20 | 0.73 | 1.93 |
| 2/0 | " | 1.55 | 0.86 | 2.42 |
| 3/0 | " | 1.85 | 1.15 | 3.00 |
| XLP, 600v | | | | |
| #12 | L.F. | 0.15 | 0.23 | 0.38 |
| #10 | " | 0.22 | 0.27 | 0.49 |
| #8 | " | 0.31 | 0.36 | 0.67 |
| #6 | " | 0.36 | 0.41 | 0.77 |
| #4 | " | 0.51 | 0.45 | 0.96 |
| #3 | " | 0.63 | 0.50 | 1.13 |
| #2 | " | 0.78 | 0.55 | 1.33 |
| #1 | " | 1.05 | 0.64 | 1.69 |
| 1/0 | " | 1.25 | 0.73 | 1.98 |
| 2/0 | " | 1.55 | 0.91 | 2.46 |
| 3/0 | " | 1.90 | 1.15 | 3.05 |
| 4/0 | " | 2.25 | 1.25 | 3.50 |
| 250 MCM | " | 2.80 | 1.35 | 4.15 |
| 300 MCM | " | 3.30 | 1.50 | 4.80 |
| 350 MCM | " | 3.80 | 1.75 | 5.55 |
| 400 MCM | " | 4.50 | 2.00 | 6.50 |
| 500 MCM | " | 5.30 | 2.35 | 7.65 |
| 600 MCM | " | 7.65 | 2.70 | 10.35 |
| 750 MCM | " | 9.20 | 3.05 | 12.25 |
| 1000 MCM | " | 12.85 | 3.45 | 16.30 |
| **16120.47 SHEATHED CABLE** | | | | |
| Non-metallic sheathed cable | | | | |
| Type NM cable with ground | | | | |
| #14/2 | L.F. | 0.20 | 0.68 | 0.88 |
| #12/2 | " | 0.27 | 0.73 | 1.00 |
| #10/2 | " | 0.43 | 0.81 | 1.24 |
| #8/2 | " | 0.84 | 0.91 | 1.75 |
| #6/2 | " | 1.25 | 1.15 | 2.40 |
| #14/3 | " | 0.32 | 1.15 | 1.47 |

| BASIC MATERIALS | UNIT | MAT. | INST. | TOTAL |
|---|---|---|---|---|
| **16120.47 SHEATHED CABLE** | | | | |
| #12/3 | L.F. | 0.45 | 1.20 | 1.65 |
| #10/3 | " | 0.65 | 1.25 | 1.90 |
| #8/3 | " | 1.55 | 1.25 | 2.80 |
| #6/3 | " | 1.80 | 1.25 | 3.05 |
| #4/3 | " | 2.70 | 1.45 | 4.15 |
| #2/3 | " | 3.85 | 1.60 | 5.45 |
| Type U.F. cable with ground | | | | |
| #14/2 | L.F. | 0.26 | 0.73 | 0.99 |
| #12/2 | " | 0.32 | 0.86 | 1.19 |
| #10/2 | " | 0.41 | 0.91 | 1.32 |
| #8/2 | " | 1.30 | 1.05 | 2.35 |
| #6/2 | " | 1.55 | 1.25 | 2.80 |
| #14/3 | " | 0.34 | 0.91 | 1.25 |
| #12/3 | " | 0.46 | 1.00 | 1.46 |
| #10/3 | " | 0.65 | 1.15 | 1.80 |
| #8/3 | " | 1.65 | 1.25 | 2.90 |
| #6/3 | " | 2.45 | 1.45 | 3.90 |
| **16130.60 PULL AND JUNCTION BOXES** | | | | |
| 4" | | | | |
| Octagon box | EA. | 1.60 | 5.20 | 6.80 |
| Box extension | " | 2.30 | 2.70 | 5.00 |
| Plaster ring | " | 1.25 | 2.70 | 3.95 |
| Cover blank | " | 0.51 | 2.70 | 3.21 |
| Square box | " | 1.95 | 5.20 | 7.15 |
| Box extension | " | 2.10 | 2.70 | 4.80 |
| Plaster ring | " | 0.95 | 2.70 | 3.65 |
| Cover blank | " | 0.58 | 2.70 | 3.28 |
| 4-11/16" | | | | |
| Square box | EA. | 3.45 | 5.20 | 8.65 |
| Box extension | " | 4.60 | 2.70 | 7.30 |
| Plaster ring | " | 2.65 | 2.70 | 5.35 |
| Cover blank | " | 1.05 | 2.70 | 3.75 |
| Switch and device boxes | | | | |
| 2 gang | EA. | 10.00 | 5.20 | 15.20 |
| 3 gang | " | 12.35 | 5.20 | 17.55 |
| 4 gang | " | 17.45 | 7.25 | 24.70 |
| Device covers | | | | |
| 2 gang | EA. | 4.90 | 2.70 | 7.60 |
| 3 gang | " | 5.80 | 2.70 | 8.50 |
| 4 gang | " | 8.40 | 2.70 | 11.10 |
| Handy box | " | 1.50 | 5.20 | 6.70 |
| Extension | " | 2.05 | 2.70 | 4.75 |
| Switch cover | " | 0.58 | 2.70 | 3.28 |
| Switch box with knockout | " | 1.65 | 6.60 | 8.25 |
| Weatherproof cover, spring type | " | 5.80 | 3.65 | 9.45 |
| Cover plate, dryer receptacle 1 gang plastic | " | 2.65 | 4.55 | 7.20 |
| For 4" receptacle, 2 gang | " | 4.60 | 4.55 | 9.15 |
| Duplex receptacle cover plate, plastic | " | 0.51 | 2.70 | 3.21 |
| 4", vertical bracket box, 1-1/2" with RMX clamps | EA. | 2.45 | 6.60 | 9.05 |

# 16 ELECTRICAL

## BASIC MATERIALS

| BASIC MATERIALS | UNIT | MAT. | INST. | TOTAL |
|---|---|---|---|---|
| **16130.60 PULL AND JUNCTION BOXES** | | | | |
| BX clamps | EA. | 3.65 | 6.60 | 10.25 |
| 4", octagon device cover | | | | |
| 1 switch | EA. | 1.65 | 2.70 | 4.35 |
| 1 duplex recept | " | 1.55 | 2.70 | 4.25 |
| 4", square face bracket boxes, 1-1/2" | | | | |
| RMX | EA. | 4.00 | 6.60 | 10.60 |
| BX | " | 4.25 | 6.60 | 10.85 |
| 4" square to round plaster rings | " | 1.60 | 2.70 | 4.30 |
| 2 gang device plaster rings | " | 1.95 | 2.70 | 4.65 |
| Surface covers | | | | |
| 1 gang switch | EA. | 1.85 | 2.70 | 4.55 |
| 2 gang switch | " | 1.95 | 2.70 | 4.65 |
| 1 single recept | " | 1.85 | 2.70 | 4.55 |
| 1 20a twist lock recept | " | 2.30 | 2.70 | 5.00 |
| 1 30a twist lock recept | " | 2.40 | 2.70 | 5.10 |
| 1 duplex recept | " | 1.40 | 2.70 | 4.10 |
| 2 duplex recept | " | 1.65 | 2.70 | 4.35 |
| Switch and duplex recept | " | 2.05 | 2.70 | 4.75 |
| 4-11/16" square to round plaster rings | " | 4.25 | 2.70 | 6.95 |
| 2 gang device plaster rings | " | 3.35 | 2.70 | 6.05 |
| Surface covers | | | | |
| 1 gang switch | EA. | 5.15 | 2.70 | 7.85 |
| 2 gang switch | " | 5.55 | 2.70 | 8.25 |
| 1 single recept | " | 5.15 | 2.70 | 7.85 |
| 1 20a twist lock recept | " | 5.55 | 2.70 | 8.25 |
| 1 30a twist lock recept | " | 5.80 | 2.70 | 8.50 |
| 1 duplex recept | " | 5.15 | 2.70 | 7.85 |
| 2 duplex recept | " | 5.25 | 2.70 | 7.95 |
| Switch and duplex recept | " | 12.10 | 2.70 | 14.80 |
| 4" plastic round boxes, ground straps | | | | |
| Box only | EA. | 1.15 | 6.60 | 7.75 |
| Box w/clamps | " | 1.45 | 9.10 | 10.55 |
| Box w/16" bar | " | 2.05 | 10.40 | 12.45 |
| Box w/24" bar | " | 2.25 | 11.35 | 13.60 |
| 4" plastic round box covers | | | | |
| Blank cover | EA. | 0.68 | 2.70 | 3.38 |
| Plaster ring | " | 0.76 | 2.70 | 3.46 |
| 4" plastic square boxes | | | | |
| Box only | EA. | 0.86 | 6.60 | 7.46 |
| Box w/clamps | " | 1.10 | 9.10 | 10.20 |
| Box w/hanger | " | 1.35 | 11.35 | 12.70 |
| Box w/nails and clamp | " | 1.95 | 11.35 | 13.30 |
| 4" plastic square box covers | | | | |
| Blank cover | EA. | 0.76 | 2.70 | 3.46 |
| 1 gang ring | " | 0.72 | 2.70 | 3.42 |
| 2 gang ring | " | 0.76 | 2.70 | 3.46 |
| Round ring | " | 0.86 | 2.70 | 3.56 |
| **16130.65 PULL BOXES AND CABINETS** | | | | |
| Galvanized pull boxes, screw cover | | | | |
| 4x4x4 | EA. | 5.95 | 8.65 | 14.60 |

| BASIC MATERIALS | UNIT | MAT. | INST. | TOTAL |
|---|---|---|---|---|
| **16130.65 PULL BOXES AND CABINETS** | | | | |
| 4x6x4 | EA. | 7.45 | 8.65 | 16.10 |
| 6x6x4 | " | 9.65 | 8.65 | 18.30 |
| 6x8x4 | " | 11.60 | 8.65 | 20.25 |
| 8x8x4 | " | 14.40 | 11.35 | 25.75 |
| 8x10x4 | " | 16.65 | 11.00 | 27.65 |
| 8x12x4 | " | 18.30 | 11.35 | 29.65 |
| Screw cover | | | | |
| 10x10x4 | EA. | 18.25 | 13.95 | 32.20 |
| 12x12x6 | " | 27.90 | 20.20 | 48.10 |
| 12x15x6 | " | 31.60 | 20.20 | 51.80 |
| 12x18x6 | " | 36.90 | 22.70 | 59.60 |
| 15x18x6 | " | 41.80 | 25.90 | 67.70 |
| 18x24x6 | " | 75.50 | 27.90 | 103.40 |
| 18x30x6 | " | 83.50 | 33.00 | 116.50 |
| 24x36x6 | " | 130.00 | 33.00 | 163.00 |
| Cast iron junction box, unflanged | | | | |
| 6x6x4 | | | | |
| 3/4" tap | EA. | 96.50 | 22.70 | 119.20 |
| 1" tap | " | 98.00 | 22.70 | 120.70 |
| Two 1/2" taps | " | 100.00 | 22.70 | 122.70 |
| 3/4" taps | " | 100.00 | 22.70 | 122.70 |
| 6" adapter plate | " | 40.80 | 15.80 | 56.60 |
| 6" exterior collar | " | 66.50 | 15.80 | 82.30 |
| Screw cover cabinet | | | | |
| 12x12x4 | EA. | 78.00 | 27.90 | 105.90 |
| 12x16x4 | " | 100.00 | 27.90 | 127.90 |
| 12x16x6 | " | 120.00 | 27.90 | 147.90 |
| 12x18x4 | " | 110.00 | 30.30 | 140.30 |
| 12x18x6 | " | 130.00 | 30.30 | 160.30 |
| 18x18x4 | " | 130.00 | 45.40 | 175.40 |
| 18x18x6 | " | 150.00 | 45.40 | 195.40 |
| 18x24x6 | " | 220.00 | 52.00 | 272.00 |
| 24x24x6 | " | 250.00 | 60.50 | 310.50 |
| 24x36x6 | " | 360.00 | 75.50 | 435.50 |
| 36x48x6 | " | 720.00 | 110.00 | 830.00 |
| **16130.80 RECEPTACLES** | | | | |
| 125 volt, 20a, duplex, grounding type, standard grade | EA. | 6.75 | 9.10 | 15.85 |
| Ground fault interrupter type | " | 49.70 | 13.45 | 63.15 |
| 250 volt, 20a, 2 pole, single receptacle, ground type | " | 7.55 | 9.10 | 16.65 |
| 120/208v, 4 pole, single receptacle, twist lock | | | | |
| 20a | EA. | 16.30 | 15.80 | 32.10 |
| 50a | " | 31.10 | 15.80 | 46.90 |
| 125/250v, 3 pole, flush receptacle | | | | |
| 30a | EA. | 10.35 | 13.45 | 23.80 |
| 50a | " | 11.95 | 13.45 | 25.40 |
| 60a | " | 48.20 | 15.80 | 64.00 |
| 277 v, 20a, 2 pole, grounding type, twist lock | " | 11.95 | 9.10 | 21.05 |
| Dryer receptacle, 250v, 30a/50a, 3 wire | " | 20.00 | 13.45 | 33.45 |
| Clock receptacle, 2 pole, grounding type | " | 8.35 | 9.10 | 17.45 |
| 125v, 20a single recept. grounding type | | | | |

# BASIC MATERIALS

| BASIC MATERIALS | UNIT | MAT. | INST. | TOTAL |
|---|---|---|---|---|
| **16130.80 RECEPTACLES** | | | | |
| Standard grade | EA. | 6.10 | 9.10 | 15.20 |
| Specification | " | 9.40 | 9.10 | 18.50 |
| Hospital | " | 9.70 | 9.10 | 18.80 |
| Isolated ground orange | " | 16.95 | 11.35 | 28.30 |
| Duplex | | | | |
| Specification grade | EA. | 9.75 | 9.10 | 18.85 |
| Hospital | " | 12.65 | 9.10 | 21.75 |
| Isolated ground orange | " | 19.00 | 11.35 | 30.35 |
| 250v, 20a, duplex, 2 pole, grounding, spec. grade | " | 14.50 | 9.10 | 23.60 |
| Combination recepts, 20a, 125v and 250v, duplex | " | 12.95 | 9.10 | 22.05 |
| GFI hospital grade recepts, 20a, 125v, duplex | " | 57.50 | 13.45 | 70.95 |
| 125/250v, 3 pole, 3 wire surface recepts | | | | |
| 30a | EA. | 20.00 | 13.45 | 33.45 |
| 50a | " | 22.20 | 13.45 | 35.65 |
| 60a | " | 48.50 | 15.80 | 64.30 |
| Cord set, 3 wire, 6' cord | | | | |
| 30a | EA. | 21.65 | 13.45 | 35.10 |
| 50a | " | 30.20 | 13.45 | 43.65 |
| **16198.10 ELECTRIC MANHOLES** | | | | |
| Precast, handhole, 4' deep | | | | |
| 2'x2' | EA. | 410.00 | 160.00 | 570.00 |
| 3'x3' | " | 550.00 | 250.00 | 800.00 |
| 4'x4' | " | 1,190 | 470.00 | 1,660 |
| Power manhole, complete, precast, 8' deep | | | | |
| 4'x4' | EA. | 1,700 | 640.00 | 2,340 |
| 6'x6' | " | 2,270 | 910.00 | 3,180 |
| 8'x8' | " | 2,690 | 960.00 | 3,650 |
| 6' deep, 9' x 12' | " | 2,980 | 1,140 | 4,120 |
| Cast in place, power manhole, 8' deep | | | | |
| 4'x4' | EA. | 19,850 | 640.00 | 20,490 |
| 6'x6' | " | 2,550 | 910.00 | 3,460 |
| 8'x8' | " | 2,840 | 960.00 | 3,800 |
| **16199.10 UTILITY POLES & FITTINGS** | | | | |
| Wood pole, creosoted | | | | |
| 25' | EA. | 340.00 | 110.00 | 450.00 |
| 30' | " | 410.00 | 130.00 | 540.00 |
| 35' | " | 540.00 | 160.00 | 700.00 |
| 40' | " | 650.00 | 170.00 | 820.00 |
| 45' | " | 740.00 | 320.00 | 1,060 |
| 50' | " | 880.00 | 330.00 | 1,210 |
| 55' | " | 1,010 | 340.00 | 1,350 |
| Treated, wood preservative, 6"x6" | | | | |
| 8' | EA. | 33.80 | 22.70 | 56.50 |
| 10' | " | 50.00 | 36.30 | 86.30 |
| 12' | " | 52.50 | 40.40 | 92.90 |
| 14' | " | 66.00 | 60.50 | 126.50 |
| 16' | " | 79.50 | 72.50 | 152.00 |
| 18' | " | 94.50 | 91.00 | 185.50 |
| 20' | " | 110.00 | 91.00 | 201.00 |

## BASIC MATERIALS

### 16199.10 UTILITY POLES & FITTINGS

| BASIC MATERIALS | UNIT | MAT. | INST. | TOTAL |
|---|---|---|---|---|
| Aluminum, brushed, no base | | | | |
| 8' | EA. | 470.00 | 91.00 | 561.00 |
| 10' | " | 540.00 | 120.00 | 660.00 |
| 15' | " | 610.00 | 130.00 | 740.00 |
| 20' | " | 810.00 | 150.00 | 960.00 |
| 25' | " | 1,080 | 170.00 | 1,250 |
| 30' | " | 1,620 | 200.00 | 1,820 |
| 35' | " | 1,890 | 230.00 | 2,120 |
| 40' | " | 2,430 | 280.00 | 2,710 |
| Steel, no base | | | | |
| 10' | EA. | 550.00 | 110.00 | 660.00 |
| 15' | " | 610.00 | 130.00 | 740.00 |
| 20' | " | 680.00 | 170.00 | 850.00 |
| 25' | " | 880.00 | 210.00 | 1,090 |
| 30' | " | 1,170 | 230.00 | 1,400 |
| 35' | " | 1,620 | 280.00 | 1,900 |
| Concrete, no base | | | | |
| 13' | EA. | 670.00 | 250.00 | 920.00 |
| 16' | " | 990.00 | 330.00 | 1,320 |
| 18' | " | 1,200 | 400.00 | 1,600 |
| 25' | " | 1,460 | 450.00 | 1,910 |
| 30' | " | 1,950 | 550.00 | 2,500 |
| 35' | " | 2,500 | 640.00 | 3,140 |
| 40' | " | 2,920 | 730.00 | 3,650 |
| 45' | " | 3,480 | 770.00 | 4,250 |
| 50' | " | 4,310 | 830.00 | 5,140 |
| 55' | " | 4,800 | 860.00 | 5,660 |
| 60' | " | 5,490 | 910.00 | 6,400 |
| Pole line hardware | | | | |
| Wood crossarm | | | | |
| 4' | EA. | 34.80 | 60.50 | 95.30 |
| 8' | " | 69.50 | 75.50 | 145.00 |
| 10' | " | 140.00 | 93.00 | 233.00 |
| Angle steel brace | | | | |
| 1 piece | EA. | 8.30 | 11.35 | 19.65 |
| 2 piece | " | 17.70 | 15.80 | 33.50 |
| Eye nut, 5/8" | " | 4.10 | 2.25 | 6.35 |
| Bolt (14-16"), 5/8" | " | 10.85 | 9.10 | 19.95 |
| Transformer, ground connection | " | 4.80 | 11.35 | 16.15 |
| Stirrup | " | 14.80 | 13.95 | 28.75 |
| Secondary lead support | " | 16.50 | 18.15 | 34.65 |
| Spool insulator | " | 12.15 | 9.10 | 21.25 |
| Guy grip, preformed | | | | |
| 7/16" | EA. | 2.60 | 6.60 | 9.20 |
| 1/2" | " | 3.90 | 6.60 | 10.50 |
| Hook | " | 2.60 | 11.35 | 13.95 |
| Strain insulator | " | 21.20 | 16.50 | 37.70 |
| Wire | | | | |
| 5/16" | L.F. | 0.43 | 0.23 | 0.66 |
| 7/16" | " | 0.85 | 0.27 | 1.12 |
| 1/2" | " | 1.45 | 0.36 | 1.81 |
| Soft drawn ground, copper, #8 | " | 0.35 | 0.36 | 0.71 |

# 16 ELECTRICAL

## BASIC MATERIALS

### 16199.10 UTILITY POLES & FITTINGS

| BASIC MATERIALS | UNIT | MAT. | INST. | TOTAL |
|---|---|---|---|---|
| Ground clamp | EA. | 5.65 | 13.95 | 19.60 |
| Perforated strapping for conduit, 1-1/2" | L.F. | 2.15 | 6.60 | 8.75 |
| Hot line clamp | EA. | 16.50 | 36.30 | 52.80 |
| Lightning arrester | | | | |
| 3kv | EA. | 91.50 | 45.40 | 136.90 |
| 10kv | " | 150.00 | 72.50 | 222.50 |
| 30kv | " | 270.00 | 91.00 | 361.00 |
| 36kv | " | 550.00 | 110.00 | 660.00 |

## POWER GENERATION

### 16210.10 GENERATORS

| POWER GENERATION | UNIT | MAT. | INST. | TOTAL |
|---|---|---|---|---|
| Diesel generator, with auto transfer switch | | | | |
| 30kw | EA. | 20,950 | 1,400 | 22,350 |
| 50kw | " | 26,520 | 1,400 | 27,920 |
| 75kw | " | 33,680 | 1,910 | 35,590 |
| 100kw | " | 37,370 | 2,140 | 39,510 |
| 125kw | " | 39,930 | 2,270 | 42,200 |
| 150kw | " | 46,040 | 2,590 | 48,630 |
| 175kw | " | 47,980 | 3,030 | 51,010 |
| 200kw | " | 49,770 | 3,630 | 53,400 |
| 250kw | " | 54,390 | 4,040 | 58,430 |
| 300kw | " | 65,410 | 4,540 | 69,950 |
| 350kw | " | 70,030 | 5,190 | 75,220 |
| 400kw | " | 85,980 | 6,050 | 92,030 |
| 450kw | " | 92,090 | 6,600 | 98,690 |
| 500kw | " | 99,760 | 7,260 | |
| 600kw | " | 136,640 | 9,080 | |
| 750kw | " | 190,730 | 9,080 | |

### 16320.10 TRANSFORMERS

| TRANSFORMERS | UNIT | MAT. | INST. | TOTAL |
|---|---|---|---|---|
| Floor mounted, single phase, int. dry, 480v-120/240v | | | | |
| 3 kva | EA. | 290.00 | 82.50 | 372.50 |
| 5 kva | " | 400.00 | 140.00 | 540.00 |
| 7.5 kva | " | 600.00 | 160.00 | 760.00 |
| 10 kva | " | 690.00 | 170.00 | 860.00 |
| 15 kva | " | 930.00 | 200.00 | 1,130 |
| 25 kva | " | 1,190 | 340.00 | 1,530 |
| 37.5 kva | " | 1,620 | 430.00 | 2,050 |
| 50 kva | " | 2,030 | 470.00 | 2,500 |
| 75 kva | " | 2,700 | 480.00 | 3,180 |
| 100 kva | " | 3,240 | 530.00 | 3,770 |
| Three phase, 480v-120/208v | | | | |

## POWER GENERATION

| | UNIT | MAT. | INST. | TOTAL |
|---|---|---|---|---|
| **16320.10 TRANSFORMERS** | | | | |
| 15 kva | EA. | 1,110 | 270.00 | 1,380 |
| 30 kva | " | 1,430 | 430.00 | 1,860 |
| 45 kva | " | 1,910 | 490.00 | 2,400 |
| 75 kva | " | 2,710 | 500.00 | 3,210 |
| 112.5 kva | " | 3,670 | 580.00 | 4,250 |
| 150 kva | " | 4,620 | 620.00 | 5,240 |
| 225 kva | " | 6,300 | 700.00 | 7,000 |
| **16350.10 CIRCUIT BREAKERS** | | | | |
| Molded case, 240v, 15-60a, bolt-on | | | | |
| 1 pole | EA. | 12.40 | 11.35 | 23.75 |
| 2 pole | " | 24.80 | 15.80 | 40.60 |
| 70-100a, 2 pole | " | 69.00 | 24.20 | 93.20 |
| 15-60a, 3 pole | " | 82.50 | 18.15 | 100.65 |
| 70-100a, 3 pole | " | 110.00 | 27.90 | 137.90 |
| 480v, 2 pole | | | | |
| 15-60a | EA. | 180.00 | 13.45 | 193.45 |
| 70-100a | " | 220.00 | 18.15 | 238.15 |
| 3 pole | | | | |
| 15-60a | EA. | 220.00 | 18.15 | 238.15 |
| 70-100a | " | 260.00 | 20.20 | 280.20 |
| 70-225a | " | 550.00 | 27.90 | 577.90 |
| Draw out air circuit breakers | | | | |
| 600a | EA. | 9,530 | 730.00 | 10,260 |
| 800a | " | 12,320 | 830.00 | 13,150 |
| 1600a | " | 19,770 | 1,100 | 20,870 |
| 2000a | " | 26,500 | 1,250 | 27,750 |
| 3000a | " | 45,990 | 1,450 | 47,440 |
| 4000a | " | 70,630 | 1,730 | 72,360 |
| Load center circuit breakers, 240v | | | | |
| 1 pole, 10-60a | EA. | 12.40 | 11.35 | 23.75 |
| 2 pole | | | | |
| 10-60a | EA. | 24.80 | 18.15 | 42.95 |
| 70-100a | " | 41.30 | 30.30 | 71.60 |
| 110-150a | " | 150.00 | 33.00 | 183.00 |
| 3 pole | | | | |
| 10-60a | EA. | 82.50 | 22.70 | 105.20 |
| 70-100a | " | 120.00 | 33.00 | 153.00 |
| Load center, G.F.I. breakers, 240v | | | | |
| 1 pole, 15-30a | EA. | 82.50 | 13.45 | 95.95 |
| 2 pole, 15-30a | " | 140.00 | 18.15 | 158.15 |
| Key operated breakers, 240v, 1 pole, 10-30a | " | 13.75 | 13.45 | 27.20 |
| Tandem breakers, 240v | | | | |
| 1 pole, 15-30a | EA. | 23.40 | 18.15 | 41.55 |
| 2 pole, 15-30a | " | 42.70 | 24.20 | 66.90 |
| Bolt-on, G.F.I. breakers, 240v, 1 pole, 15-30a | " | 100.00 | 15.80 | 115.80 |
| **16360.10 SAFETY SWITCHES** | | | | |
| Fused, 3 phase, 30 amp, 600v, heavy duty | | | | |
| NEMA 1 | EA. | 150.00 | 52.00 | 202.00 |

# 16 ELECTRICAL

| POWER GENERATION | UNIT | MAT. | INST. | TOTAL |
|---|---|---|---|---|
| **16360.10 SAFETY SWITCHES** | | | | |
| NEMA 3r | EA. | 250.00 | 52.00 | 302.00 |
| NEMA 4 | " | 680.00 | 72.50 | 752.50 |
| NEMA 12 | " | 250.00 | 79.00 | 329.00 |
| 60a | | | | |
| NEMA 1 | EA. | 200.00 | 52.00 | 252.00 |
| NEMA 3r | " | 310.00 | 52.00 | 362.00 |
| NEMA 4 | " | 810.00 | 72.50 | 882.50 |
| NEMA 12 | " | 260.00 | 79.00 | 339.00 |
| 100a | | | | |
| NEMA 1 | EA. | 380.00 | 79.00 | 459.00 |
| NEMA 3r | " | 530.00 | 79.00 | 609.00 |
| NEMA 4 | " | 1,620 | 91.00 | 1,711 |
| NEMA 12 | " | 450.00 | 110.00 | 560.00 |
| 200a | | | | |
| NEMA 1 | EA. | 540.00 | 110.00 | 650.00 |
| NEMA 3r | " | 710.00 | 110.00 | 820.00 |
| NEMA 4 | " | 2,300 | 130.00 | 2,430 |
| NEMA 12 | " | 690.00 | 160.00 | 850.00 |
| 400a | | | | |
| NEMA 1 | EA. | 1,350 | 250.00 | 1,600 |
| NEMA 3r | " | 1,760 | 250.00 | 2,010 |
| NEMA 4 | " | 4,050 | 260.00 | 4,310 |
| NEMA 12 | " | 1,620 | 320.00 | 1,940 |
| NEMA 4 | " | 5,400 | 410.00 | 5,810 |
| NEMA 12 | " | 2,570 | 560.00 | 3,130 |
| Non-fused, 240-600v, heavy duty, 3 phase, 30 amp | | | | |
| NEMA 1 | EA. | 110.00 | 52.00 | 162.00 |
| NEMA 3r | " | 180.00 | 52.00 | 232.00 |
| NEMA 4 | " | 680.00 | 79.00 | 759.00 |
| NEMA 12 | " | 200.00 | 79.00 | 279.00 |
| 60a | | | | |
| NEMA1 | EA. | 180.00 | 52.00 | 232.00 |
| NEMA 3r | " | 280.00 | 52.00 | 332.00 |
| NEMA 4 | " | 810.00 | 79.00 | 889.00 |
| NEMA 12 | " | 260.00 | 79.00 | 339.00 |
| 100a | | | | |
| NEMA 1 | EA. | 270.00 | 79.00 | 349.00 |
| NEMA 3r | " | 410.00 | 79.00 | 489.00 |
| NEMA 4 | " | 1,620 | 110.00 | 1,730 |
| NEMA 12 | " | 350.00 | 110.00 | 460.00 |
| 200a, NEMA 1 | " | 420.00 | 110.00 | 530.00 |
| 600a, NEMA 12 | " | 1,820 | 560.00 | 2,380 |
| **16365.10 FUSES** | | | | |
| Fuse, one-time, 250v | | | | |
| 30a | EA. | 1.35 | 2.25 | 3.60 |
| 60a | " | 2.05 | 2.25 | 4.30 |
| 100a | " | 8.10 | 2.25 | 10.35 |
| 200a | " | 20.25 | 2.25 | 22.50 |
| 400a | " | 35.10 | 2.25 | 37.35 |
| 600a | " | 67.50 | 2.25 | 69.75 |
| 600v | | | | |

## POWER GENERATION

### 16365.10 FUSES

| | UNIT | MAT. | INST. | TOTAL |
|---|---|---|---|---|
| 30a | EA. | 6.10 | 2.25 | 8.35 |
| 60a | " | 6.75 | 2.25 | 9.00 |
| 100a | " | 18.90 | 2.25 | 21.15 |
| 200a | " | 40.50 | 2.25 | 42.75 |
| 400a | " | 81.00 | 2.25 | 83.25 |
| Fusetron, 600v | | | | |
| 200a | EA. | 29.70 | 2.25 | 31.95 |
| 400a | " | 54.00 | 2.25 | 56.25 |
| Fuse, amp-trap, K1, 250v | | | | |
| 30a | EA. | 9.45 | 2.25 | 11.70 |
| 60a | " | 19.90 | 2.25 | 22.15 |
| 100a | " | 45.90 | 2.25 | 48.15 |
| 200a | " | 85.00 | 2.25 | 87.25 |
| 400a | " | 170.00 | 2.25 | 172.25 |
| 600a | " | 220.00 | 2.25 | 222.25 |
| 600v | | | | |
| 30a | EA. | 19.60 | 2.25 | 21.85 |
| 60a | " | 35.10 | 2.25 | 37.35 |
| 100a | " | 70.00 | 2.25 | 72.25 |
| 200a | " | 110.00 | 2.25 | 112.25 |
| 400a | " | 220.00 | 2.25 | 222.25 |
| K5, 250v | | | | |
| 30a | EA. | 6.75 | 2.25 | 9.00 |
| 60a | " | 9.45 | 2.25 | 11.70 |
| 100a | " | 27.00 | 2.25 | 29.25 |
| 200a | " | 67.50 | 2.25 | 69.75 |
| 400a | " | 81.00 | 2.25 | 83.25 |
| 600a | " | 140.00 | 2.25 | 142.25 |
| 600v | | | | |
| 30a | EA. | 9.45 | 2.25 | 11.70 |
| 60a | " | 18.90 | 2.25 | 21.15 |
| 100a | " | 33.80 | 2.25 | 36.05 |
| 200a | " | 65.00 | 2.25 | 67.25 |
| 400a | " | 140.00 | 2.25 | 142.25 |
| 600a | " | 200.00 | 2.25 | 202.25 |
| J, 600v | | | | |
| 30a | EA. | 13.50 | 2.25 | 15.75 |
| 60a | " | 22.95 | 2.25 | 25.20 |
| 100a | " | 51.50 | 2.25 | 53.75 |
| 200a | " | 88.00 | 2.25 | 90.25 |
| 400a | " | 160.00 | 2.25 | 162.25 |
| L, 600v | | | | |
| 1200a | EA. | 330.00 | 18.15 | 348.15 |
| 1600a | " | 510.00 | 18.15 | 528.15 |
| 2000a | " | 680.00 | 18.15 | 698.15 |
| 2500a | " | 810.00 | 18.15 | 828.15 |
| 3000a | " | 950.00 | 18.15 | 968.15 |
| 4000a | " | 1,220 | 18.15 | 1,238 |
| 5000a | " | 2,030 | 18.15 | 2,048 |
| Fuse cl-ay 250v | | | | |
| 600a | EA. | 340.00 | 13.45 | 353.45 |
| 1200a | " | 340.00 | 13.45 | 353.45 |

# 16 ELECTRICAL

## POWER GENERATION

| POWER GENERATION | UNIT | MAT. | INST. | TOTAL |
|---|---|---|---|---|
| **16365.10 FUSES** | | | | |
| 1600a | EA. | 420.00 | 13.45 | 433.45 |
| 2000a | " | 540.00 | 13.45 | 553.45 |
| 600v | | | | |
| 1200a | EA. | 410.00 | 13.45 | 423.45 |
| 1600a | " | 470.00 | 13.45 | 483.45 |
| 2000a | " | 570.00 | 13.45 | 583.45 |
| **16395.10 GROUNDING** | | | | |
| Ground rods, copper clad, 1/2" x | | | | |
| 6' | EA. | 10.50 | 30.30 | 40.80 |
| 8' | " | 15.70 | 33.00 | 48.70 |
| 10' | " | 19.65 | 45.40 | 65.05 |
| 5/8" x | | | | |
| 5' | EA. | 14.40 | 27.90 | 42.30 |
| 6' | " | 15.70 | 33.00 | 48.70 |
| 8' | " | 19.65 | 45.40 | 65.05 |
| 10' | " | 22.25 | 56.50 | 78.75 |
| 3/4" x | | | | |
| 8' | EA. | 28.80 | 33.00 | 61.80 |
| 10' | " | 39.30 | 36.30 | 75.60 |
| Ground rod clamp | | | | |
| 5/8" | EA. | 4.60 | 5.60 | 10.20 |
| 3/4" | " | 5.90 | 5.60 | 11.50 |

## SERVICE AND DISTRIBUTION

| SERVICE AND DISTRIBUTION | UNIT | MAT. | INST. | TOTAL |
|---|---|---|---|---|
| **16425.10 SWITCHBOARDS** | | | | |
| Switchboard, 90" high, no main disconnect, 208/120v | | | | |
| 400a | EA. | 1,930 | 360.00 | 2,290 |
| 600a | " | 3,000 | 360.00 | 3,360 |
| 1000a | " | 3,780 | 360.00 | 4,140 |
| 1200a | " | 4,040 | 450.00 | 4,490 |
| 1600a | " | 4,430 | 540.00 | 4,970 |
| 2000a | " | 4,760 | 640.00 | 5,400 |
| 2500a | " | 3,900 | 730.00 | 4,630 |
| 277/480v | | | | |
| 600a | EA. | 3,450 | 370.00 | 3,820 |
| 800a | " | 3,780 | 370.00 | 4,150 |
| 1600a | " | 4,760 | 540.00 | 5,300 |
| 2000a | " | 5,090 | 640.00 | 5,730 |
| 2500a | " | 5,420 | 730.00 | 6,150 |
| 3000a | " | 6,240 | 1,250 | 7,490 |
| 4000a | " | 7,550 | 1,350 | 8,900 |

## SERVICE AND DISTRIBUTION

| SERVICE AND DISTRIBUTION | UNIT | MAT. | INST. | TOTAL |
|---|---|---|---|---|
| **16425.10 SWITCHBOARDS** | | | | |
| 1500 kva | EA. | 32,380 | 2,590 | 34,970 |
| **16470.10 PANELBOARDS** | | | | |
| 3 phase, 480/277v, main lugs only, 120a, 30 circuits | EA. | 810.00 | 160.00 | 970.00 |
| 277/480v, 4 wire, flush surface | | | | |
| 225a, 30 circuits | EA. | 860.00 | 180.00 | 1,040 |
| 400a, 30 circuits | " | 990.00 | 230.00 | 1,220 |
| 600a, 42 circuits | " | 1,230 | 270.00 | 1,500 |
| 208/120v, main circuit breaker, 3 phase, 4 wire | | | | |
| 100a | | | | |
| 12 circuits | EA. | 700.00 | 230.00 | 930.00 |
| 20 circuits | " | 870.00 | 290.00 | 1,160 |
| 30 circuits | " | 1,290 | 320.00 | 1,610 |
| 225a | | | | |
| 30 circuits | EA. | 1,100 | 350.00 | 1,450 |
| 42 circuits | " | 1,990 | 430.00 | 2,420 |
| 400a | | | | |
| 30 circuits | EA. | 2,710 | 670.00 | 3,380 |
| 42 circuits | " | 3,250 | 730.00 | 3,980 |
| 600a, 42 circuits | " | 6,310 | 830.00 | 7,140 |
| 120/208v, flush, 3 ph., 4 wire, main only | | | | |
| 100a | | | | |
| 12 circuits | EA. | 500.00 | 230.00 | 730.00 |
| 20 circuits | " | 680.00 | 290.00 | 970.00 |
| 30 circuits | " | 1,020 | 320.00 | 1,340 |
| 225a | | | | |
| 30 circuits | EA. | 1,040 | 350.00 | 1,390 |
| 42 circuits | " | 1,310 | 430.00 | 1,740 |
| 400a | | | | |
| 30 circuits | EA. | 1,990 | 670.00 | 2,660 |
| 42 circuits | " | 2,900 | 730.00 | 3,630 |
| 600a, 42 circuits | " | 4,510 | 830.00 | 5,340 |
| Panelboard accessories | | | | |
| Grounding bus | EA. | 19.90 | 15.80 | 35.70 |
| Handle lock device | " | 2.70 | 6.60 | 9.30 |
| Factory assembled panel | | | | |
| 1 pole space | EA. | 14.40 | 15.80 | 30.20 |
| 2 pole space | " | 30.70 | 6.60 | 37.30 |
| 3 pole space | " | 45.20 | 6.05 | 51.25 |
| Panelboards 1 phase, 240/120v main circuit breaker | | | | |
| Single phase, 3 wire, 120/240v flush | | | | |
| 100a, 20 circuits | EA. | 580.00 | 160.00 | 740.00 |
| 225a, 30 circuits | " | 1,170 | 180.00 | 1,350 |
| 240/120v, main lugs only | | | | |
| 100a | | | | |
| 8 circuits | EA. | 350.00 | 130.00 | 480.00 |
| 12 circuits | " | 380.00 | 130.00 | 510.00 |
| 20 circuits | " | 400.00 | 130.00 | 530.00 |
| 225a | | | | |
| 24 circuits | EA. | 420.00 | 160.00 | 580.00 |

## SERVICE AND DISTRIBUTION

| | UNIT | MAT. | INST. | TOTAL |
|---|---|---|---|---|
| **16470.10 PANELBOARDS** | | | | |
| 30 circuits | EA. | 480.00 | 170.00 | 650.00 |
| 42 circuits | " | 530.00 | 170.00 | 700.00 |
| **16480.10 MOTOR CONTROLS** | | | | |
| Motor generator set, 3 phase, 480/277v, w/controls | | | | |
| 10kw | EA. | 8,920 | 1,250 | 10,170 |
| 15kw | " | 11,630 | 1,400 | 13,030 |
| 20kw | " | 12,910 | 1,450 | 14,360 |
| 25kw | " | 14,890 | 1,580 | 16,470 |
| 30kw | " | 16,660 | 1,650 | 18,310 |
| 40kw | " | 18,200 | 1,730 | 19,930 |
| 50kw | " | 20,280 | 1,820 | 22,100 |
| 60kw | " | 22,980 | 2,020 | 25,000 |
| 75kw | " | 25,810 | 2,270 | 28,080 |
| 100kw | " | 29,780 | 2,790 | 32,570 |
| 125kw | " | 53,900 | 3,030 | 56,930 |
| 150kw | " | 59,570 | 3,030 | 62,600 |
| 200kw | " | 68,080 | 3,300 | 71,380 |
| 250kw | " | 73,750 | 3,300 | 77,050 |
| 300kw | " | 85,100 | 3,630 | 88,730 |
| 2 pole, 230 volt starter, w/NEMA-1 | | | | |
| 1 hp, 9 amp, size 00 | EA. | 130.00 | 45.40 | 175.40 |
| 2 hp, 18amp, size 0 | " | 170.00 | 45.40 | 215.40 |
| 3 hp, 27amp, size 1 | " | 180.00 | 45.40 | 225.40 |
| 5 hp, 45amp, size 1p | " | 240.00 | 45.40 | 285.40 |
| 7-1/2 hp, 45a, size 2 | " | 400.00 | 45.40 | 445.40 |
| 15 hp, 90a, size 3 | " | 610.00 | 45.40 | 655.40 |
| 2 pole, w/NEMA-4 enclosure | | | | |
| 2 hp, 18a, size 1 | EA. | 350.00 | 72.50 | 422.50 |
| 5 hp, 45amp, size 1p | " | 440.00 | 72.50 | 512.50 |
| 7-1/2 hp, 45a, size 2 | " | 690.00 | 72.50 | 762.50 |
| 3 pole, 2 hp, 9a, 200-575v starter | | | | |
| W/NEMA-1, size 00 | EA. | 170.00 | 60.50 | 230.50 |
| W/NEMA-4 enclosure, size 00 | " | 350.00 | 79.00 | 429.00 |
| 5hp, 18a | | | | |
| W/NEMA-1 enclosure, size 0 | EA. | 220.00 | 60.50 | 280.50 |
| W/NEMA-4 enclosure, size 0 | " | 380.00 | 79.00 | 459.00 |
| 7.5-10hp, 27a | | | | |
| 7.5-10hp 27a, w/NEMA-1 enclosure, size 1 | EA. | 230.00 | 60.50 | 290.50 |
| W/NEMA-4 enclosure size 1 | " | 380.00 | 79.00 | 459.00 |
| 10-25hp, 45a | | | | |
| W/NEMA-1 enclosure, size 2 | EA. | 400.00 | 60.50 | 460.50 |
| W/NEMA-4 enclosure, size 2 | " | 710.00 | 79.00 | 789.00 |
| 25-50hp, 90a | | | | |
| W/NEMA-1 enclosure, size 3 | EA. | 690.00 | 79.00 | 769.00 |
| W/NEMA-4 enclosure, size 3 | " | 680.00 | 110.00 | 790.00 |
| 40-100hp, 135a | | | | |
| W/NEMA-1 enclosure, size 4 | EA. | 1,500 | 110.00 | 1,610 |
| W/NEMA-4 enclosure, size 4 | " | 2,320 | 160.00 | 2,480 |
| 75-200hp, 270a | | | | |
| W/NEMA-1 enclosure, size 5 | EA. | 3,550 | 250.00 | 3,800 |
| W/NEMA-4 enclosure, size 5 | " | 4,490 | 320.00 | 4,810 |

# SERVICE AND DISTRIBUTION

## 16480.10 MOTOR CONTROLS

| | UNIT | MAT. | INST. | TOTAL |
|---|---|---|---|---|
| Magnetic starter accessories | | | | |
| On-off-auto selector switch kit | EA. | 45.80 | 14.55 | 60.35 |
| With pilot light | " | 83.50 | 15.80 | 99.30 |

## 16490.10 SWITCHES

| | UNIT | MAT. | INST. | TOTAL |
|---|---|---|---|---|
| Oil switches, medium voltage, bus components | | | | |
| Switches, 277/120v, toggle device only | EA. | 390.00 | 72.50 | 462.50 |
| With oil 35kv, g&w gram 44, 4 way switch | " | 12,780 | 360.00 | 13,140 |
| Weatherproof enclosure | | | | |
| 3 way switch | EA. | 15,790 | 450.00 | 16,240 |
| 4 way switch | " | 16,540 | 500.00 | 17,040 |
| Fused interrupter load, 35kv | | | | |
| 20A | | | | |
| 1 pole | EA. | 18,040 | 730.00 | 18,770 |
| 2 pole | " | 19,550 | 770.00 | 20,320 |
| 3 way | " | 21,050 | 770.00 | 21,820 |
| 4 way | " | 22,560 | 830.00 | 23,390 |
| 30a, 1 pole | " | 18,040 | 730.00 | 18,770 |
| 3 way | " | 21,050 | 770.00 | 21,820 |
| 4 way | " | 22,560 | 830.00 | 23,390 |
| Weatherproof switch, including box & cover, 20a | | | | |
| 1 pole | EA. | 19,550 | 730.00 | 20,280 |
| 2 pole | " | 21,050 | 770.00 | 21,820 |
| 3 way | " | 22,560 | 830.00 | 23,390 |
| 4 way | " | 24,060 | 830.00 | 24,890 |
| 3 way, oil switch, 15kv enclosure | " | 16,540 | 540.00 | 17,080 |
| Pedestal for 35kv double breaker switch | " | 650.00 | 230.00 | 880.00 |
| Bus terminal connector, 2 | " | 570.00 | 110.00 | 680.00 |
| 2 to 3 | " | 700.00 | 110.00 | 810.00 |
| Support connector, 3 | " | 370.00 | 72.50 | 442.50 |
| Tee connector, 2 to 3 | " | 530.00 | 91.00 | 621.00 |
| Flexible bus stud connector | " | 450.00 | 79.00 | 529.00 |
| End cap 3 | " | 490.00 | 60.50 | 550.50 |
| Weldment connection, 3 | " | 200.00 | 45.40 | 245.40 |
| Plate switch, 1 gang | " | 0.49 | 2.25 | 2.74 |
| Start stop stations, manual motor starters | " | 49.30 | 33.00 | 82.30 |
| Lockout switch | " | 8.20 | 11.35 | 19.55 |
| Forward-reverse switch | " | 57.50 | 33.00 | 90.50 |
| On-off switch | " | 57.50 | 33.00 | 90.50 |
| Open-close switch | " | 57.50 | 33.00 | 90.50 |
| Forward-reverse-stop switch | " | 65.50 | 45.40 | 110.90 |
| Standard 3 button switch any standard legend | " | 65.50 | 45.40 | 110.90 |
| Standard 3 button with lockout | " | 65.50 | 45.40 | 110.90 |
| Manual motor starters, tog, 115/230v | | | | |
| Size 1 gp | EA. | 98.50 | 45.40 | 143.90 |
| Size 2 | " | 130.00 | 45.40 | 175.40 |
| Button | | | | |
| Size 0 | EA. | 74.00 | 45.40 | 119.40 |
| Size 1 | " | 110.00 | 45.40 | 155.40 |
| Size 2 | " | 140.00 | 45.40 | 185.40 |
| 3-phase | | | | |

## SERVICE AND DISTRIBUTION

### 16490.10 SWITCHES

| Description | UNIT | MAT. | INST. | TOTAL |
|---|---|---|---|---|
| Size 0 | EA. | 120.00 | 60.50 | 180.50 |
| Size 1 | " | 160.00 | 60.50 | 220.50 |
| Time & float switches | " | 240.00 | 72.50 | 312.50 |
| Astronomical time switch, 40a, 240v | " | 380.00 | 45.40 | 425.40 |
| Timer switch 0-5 minute, with box | " | 24.60 | 22.70 | 47.30 |
| Single pole/single throw time, 277v, NEMA-1 | " | 82.50 | 33.00 | 115.50 |
| Single toggle switch, 20a, 120v, with pilot | " | 16.75 | 11.35 | 28.10 |
| 3-way toggle | " | 46.10 | 13.45 | 59.55 |
| Photo electric switches | | | | |
| 1000 watt | | | | |
| 105-135v | EA. | 25.60 | 33.00 | 58.60 |
| 208-277v | " | 34.50 | 33.00 | 67.50 |
| 3000 watt, 105-130v | | | | |
| Double throw | EA. | 100.00 | 45.40 | 145.40 |
| Single throw | " | 94.50 | 45.40 | 139.90 |
| Double pole/single throw, 210-250v | " | 120.00 | 60.50 | 180.50 |
| Dimmer switch and switch plate | | | | |
| 600 w | EA. | 14.20 | 13.95 | 28.15 |
| 1000 w | " | 71.00 | 15.80 | 86.80 |
| Dimmer switch incandescent | | | | |
| 1500w | EA. | 140.00 | 31.90 | 171.90 |
| 2000w | " | 200.00 | 33.90 | 233.90 |
| Fluorescent | | | | |
| 12 lamps | EA. | 60.50 | 22.70 | 83.20 |
| 20 lamps | " | 100.00 | 25.00 | 125.00 |
| 30 lamps | " | 170.00 | 27.30 | 197.30 |
| 40 lamps | " | 210.00 | 31.90 | 241.90 |
| Specification grade toggle switches, 20a, 120-277v | | | | |
| Single pole | EA. | 12.90 | 9.10 | 22.00 |
| Double pole | " | 14.80 | 13.45 | 28.25 |
| 3 way | " | 13.70 | 11.35 | 25.05 |
| 4 way | " | 34.80 | 13.45 | 48.25 |
| Switch plates, plastic ivory | | | | |
| 1 gang | EA. | 0.49 | 3.65 | 4.14 |
| 2 gang | " | 0.97 | 4.55 | 5.52 |
| 3 gang | " | 1.45 | 5.40 | 6.85 |
| 4 gang | " | 2.45 | 6.60 | 9.05 |
| 5 gang | " | 5.60 | 7.25 | 12.85 |
| 6 gang | " | 6.60 | 8.25 | 14.85 |
| Stainless steel | | | | |
| 1 gang | EA. | 1.65 | 3.65 | 5.30 |
| 2 gang | " | 3.20 | 4.55 | 7.75 |
| 3 gang | " | 5.60 | 5.60 | 11.20 |
| 4 gang | " | 7.75 | 6.60 | 14.35 |
| 5 gang | " | 10.70 | 7.25 | 17.95 |
| 6 gang | " | 13.25 | 8.25 | 21.50 |
| Brass | | | | |
| 1 gang | EA. | 3.55 | 3.65 | 7.20 |
| 2 gang | " | 7.10 | 4.55 | 11.65 |
| 3 gang | " | 12.90 | 5.60 | 18.50 |
| 4 gang | " | 17.40 | 6.60 | 24.00 |
| 5 gang | " | 21.55 | 7.25 | 28.80 |

# 16 ELECTRICAL

## SERVICE AND DISTRIBUTION

| 16490.10 SWITCHES | UNIT | MAT. | INST. | TOTAL |
|---|---|---|---|---|
| 6 gang | EA. | 25.70 | 8.25 | 33.95 |

| 16490.20 TRANSFER SWITCHES | UNIT | MAT. | INST. | TOTAL |
|---|---|---|---|---|
| Automatic transfer switch 600v, 3 pole | | | | |
| 30a | EA. | 1,810 | 160.00 | 1,970 |
| 60a | " | 2,560 | 160.00 | 2,720 |
| 100a | " | 3,760 | 220.00 | 3,980 |
| 150a | " | 4,740 | 270.00 | 5,010 |
| 225a | " | 5,720 | 360.00 | 6,080 |
| 260a | " | 6,170 | 360.00 | 6,530 |
| 400a | " | 8,270 | 450.00 | 8,720 |
| 600a | " | 11,290 | 690.00 | 11,980 |
| 800a | " | 13,840 | 830.00 | 14,670 |
| 1000a | " | 19,860 | 960.00 | 20,820 |
| 1200a | " | 22,570 | 1,040 | 23,610 |
| 1600a | " | 27,840 | 1,140 | 28,980 |
| 2000a | " | 29,790 | 1,350 | 31,140 |
| 2600a | " | 52,660 | 1,910 | 54,570 |
| 3000a | " | 82,760 | 2,270 | 85,030 |

| 16490.80 SAFETY SWITCHES | UNIT | MAT. | INST. | TOTAL |
|---|---|---|---|---|
| Safety switch, 600v, 3 pole, heavy duty, NEMA-1 | | | | |
| 30a | EA. | 150.00 | 45.40 | 195.40 |
| 60a | " | 200.00 | 52.00 | 252.00 |
| 100a | " | 380.00 | 72.50 | 452.50 |
| 200a | " | 540.00 | 110.00 | 650.00 |
| 400a | " | 1,350 | 250.00 | 1,600 |
| 600a | " | 2,430 | 360.00 | 2,790 |
| 800a | " | 3,740 | 480.00 | 4,220 |
| 1200a | " | 4,630 | 650.00 | 5,280 |

## LIGHTING

| 16510.05 INTERIOR LIGHTING | UNIT | MAT. | INST. | TOTAL |
|---|---|---|---|---|
| Recessed fluorescent fixtures, 2'x2' | | | | |
| 2 lamp | EA. | 63.50 | 33.00 | 96.50 |
| 4 lamp | " | 85.50 | 33.00 | 118.50 |
| 2 lamp w/flange | " | 80.00 | 45.40 | 125.40 |
| 4 lamp w/flange | " | 100.00 | 45.40 | 145.40 |
| 1'x4' | | | | |
| 2 lamp | EA. | 65.00 | 30.30 | 95.30 |

# 16 ELECTRICAL

| LIGHTING | UNIT | MAT. | INST. | TOTAL |
|---|---|---|---|---|
| **16510.05 INTERIOR LIGHTING** | | | | |
| 3 lamp | EA. | 89.50 | 30.30 | 119.80 |
| 2 lamp w/flange | " | 80.00 | 33.00 | 113.00 |
| 3 lamp w/flange | " | 100.00 | 33.00 | 133.00 |
| 2'x4' | | | | |
| 2 lamp | EA. | 80.00 | 33.00 | 113.00 |
| 3 lamp | " | 93.50 | 33.00 | 126.50 |
| 4 lamp | " | 91.50 | 33.00 | 124.50 |
| 2 lamp w/flange | " | 95.50 | 45.40 | 140.90 |
| 3 lamp w/flange | " | 110.00 | 45.40 | 155.40 |
| 4 lamp w/flange | " | 110.00 | 45.40 | 155.40 |
| 4'x4' | | | | |
| 4 lamp | EA. | 190.00 | 45.40 | 235.40 |
| 6 lamp | " | 220.00 | 45.40 | 265.40 |
| 8 lamp | " | 240.00 | 45.40 | 285.40 |
| 4 lamp w/flange | " | 230.00 | 68.50 | 298.50 |
| 6 lamp w/flange | " | 290.00 | 68.50 | 358.50 |
| 8 lamp, w/flange | " | 320.00 | 68.50 | 388.50 |
| Surface mounted incandescent fixtures | | | | |
| 40w | EA. | 58.00 | 30.30 | 88.30 |
| 75w | " | 63.00 | 30.30 | 93.30 |
| 100w | " | 74.50 | 30.30 | 104.80 |
| 150w | " | 88.00 | 30.30 | 118.30 |
| Pendant | | | | |
| 40w | EA. | 63.00 | 36.30 | 99.30 |
| 75w | " | 69.50 | 36.30 | 105.80 |
| 100w | " | 79.50 | 36.30 | 115.80 |
| 150w | " | 91.00 | 36.30 | 127.30 |
| Recessed incandescent fixtures | | | | |
| 40w | EA. | 110.00 | 68.50 | 178.50 |
| 75w | " | 110.00 | 68.50 | 178.50 |
| 100w | " | 110.00 | 68.50 | 178.50 |
| 150w | " | 130.00 | 68.50 | 198.50 |
| Exit lights, 120v | | | | |
| Recessed | EA. | 83.00 | 56.50 | 139.50 |
| Back mount | " | 58.00 | 33.00 | 91.00 |
| Universal mount | " | 66.50 | 33.00 | 99.50 |
| Emergency battery units, 6v-120v, 50 unit | " | 120.00 | 68.50 | 188.50 |
| With 1 head | " | 140.00 | 68.50 | 208.50 |
| With 2 heads | " | 170.00 | 68.50 | 238.50 |
| Light track single circuit | | | | |
| 2' | EA. | 29.40 | 22.70 | 52.10 |
| 4' | " | 55.00 | 22.70 | 77.70 |
| 8' | " | 98.00 | 45.40 | 143.40 |
| 12' | " | 150.00 | 68.50 | 218.50 |
| Fittings and accessories | | | | |
| Dead end | EA. | 5.25 | 6.60 | 11.85 |
| Starter kit | " | 27.50 | 11.35 | 38.85 |
| Conduit feed | " | 33.30 | 6.60 | 39.90 |
| Straight connector | " | 18.50 | 6.60 | 25.10 |
| Center feed | " | 37.70 | 6.60 | 44.30 |
| L-connector | " | 29.40 | 6.60 | 36.00 |
| T-connector | " | 45.10 | 6.60 | 51.70 |

## LIGHTING

### 16510.05 INTERIOR LIGHTING

| LIGHTING | UNIT | MAT. | INST. | TOTAL |
|---|---|---|---|---|
| X-connector | EA. | 59.00 | 9.10 | 68.10 |
| Cord and plug | " | 43.20 | 4.55 | 47.75 |
| Rigid corner | " | 57.00 | 6.60 | 63.60 |
| Flex connector | " | 88.50 | 6.60 | 95.10 |
| 2 way connector | " | 130.00 | 9.10 | 139.10 |
| Spacer clip | " | 1.95 | 2.25 | 4.20 |
| Grid box | " | 10.75 | 6.60 | 17.35 |
| T-bar clip | " | 3.65 | 2.25 | 5.90 |
| Utility hook | " | 11.85 | 6.60 | 18.45 |
| Fixtures, square | | | | |
| R-20 | EA. | 78.50 | 6.60 | 85.10 |
| R-30 | " | 82.00 | 6.60 | 88.60 |
| 40w flood | " | 130.00 | 6.60 | 136.60 |
| 40w spot | " | 140.00 | 6.60 | 146.60 |
| 100w flood | " | 150.00 | 6.60 | 156.60 |
| 100w spot | " | 150.00 | 6.60 | 156.60 |
| Mini spot | " | 110.00 | 6.60 | 116.60 |
| Mini flood | " | 110.00 | 6.60 | 116.60 |
| Quartz, 500w | " | 300.00 | 6.60 | 306.60 |
| R-20 sphere | " | 110.00 | 6.60 | 116.60 |
| R-30 sphere | " | 130.00 | 6.60 | 136.60 |
| R-20 cylinder | " | 61.00 | 6.60 | 67.60 |
| R-30 cylinder | " | 68.50 | 6.60 | 75.10 |
| R-40 cylinder | " | 98.00 | 6.60 | 104.60 |
| R-30 wall wash | " | 120.00 | 6.60 | 126.60 |
| R-40 wall wash | " | 150.00 | 6.60 | 156.60 |
| Energy saving rapid start fluor. lamps | | | | |
| F30cw | EA. | 8.30 | 4.55 | 12.85 |
| F40 cw | " | 4.30 | 4.55 | 8.85 |
| F40 cwx | " | 7.40 | 4.55 | 11.95 |
| F30 ww | " | 10.00 | 4.55 | 14.55 |
| F40 ww | " | 5.40 | 4.55 | 9.95 |
| F40 wwx | " | 7.40 | 4.55 | 11.95 |
| Incandescent lamps | | | | |
| 200w | EA. | 4.30 | 4.55 | 8.85 |
| 300w | " | 4.70 | 4.55 | 9.25 |
| 500w | " | 10.50 | 4.55 | 15.05 |
| 750w | " | 25.70 | 6.60 | 32.30 |
| 1000w | " | 28.00 | 9.10 | 37.10 |
| 1500w | " | 41.00 | 9.10 | 50.10 |
| Energy saving reflector floodlight lamps | | | | |
| 25w | EA. | 9.90 | 4.55 | 14.45 |
| 30w | " | 8.00 | 4.55 | 12.55 |
| 50w | " | 8.85 | 4.55 | 13.40 |
| 75w | " | 7.75 | 4.55 | 12.30 |
| 120w | " | 9.30 | 4.55 | 13.85 |
| 150w | " | 6.45 | 4.55 | 11.00 |
| 200w | " | 13.45 | 4.55 | 18.00 |
| 300w | " | 13.70 | 4.55 | 18.25 |
| 500w | " | 31.00 | 6.60 | 37.60 |
| 750w | " | 39.20 | 6.60 | 45.80 |
| Reflector spotlight | | | | |

| LIGHTING | UNIT | MAT. | INST. | TOTAL |
|---|---|---|---|---|
| **16510.05 INTERIOR LIGHTING** | | | | |
| 75w | EA. | 7.05 | 4.55 | 11.60 |
| 100w | " | 14.65 | 4.55 | 19.20 |
| 125w | " | 19.30 | 4.55 | 23.85 |
| 150w | " | 25.30 | 4.55 | 29.85 |
| 250w | " | 25.30 | 4.55 | 29.85 |
| 300w | " | 25.80 | 4.55 | 30.35 |
| 400w | " | 26.00 | 6.60 | 32.60 |
| 500w | " | 32.00 | 6.60 | 38.60 |
| 1000w | " | 71.50 | 9.10 | 80.60 |
| Medium par flood lamps | | | | |
| 75w | EA. | 6.95 | 4.55 | 11.50 |
| 100w | " | 7.75 | 4.55 | 12.30 |
| 150w | " | 8.35 | 4.55 | 12.90 |
| 200w | " | 32.00 | 4.55 | 36.55 |
| 300w | " | 35.60 | 4.55 | 40.15 |
| 500w | " | 69.50 | 6.60 | 76.10 |
| Medium par spot lamps | | | | |
| 75w | EA. | 8.65 | 4.55 | 13.20 |
| 120w | " | 8.90 | 4.55 | 13.45 |
| 150w | " | 9.30 | 4.55 | 13.85 |
| Ballast replacements rapid start fluor | | | | |
| 1f-40-120v | EA. | 18.80 | 33.00 | 51.80 |
| 1f-40-277v | " | 31.60 | 33.00 | 64.60 |
| 1f-96-120v | " | 97.50 | 33.00 | 130.50 |
| 1f-96-277v | " | 120.00 | 33.00 | 153.00 |
| 2f-40-120v | " | 26.80 | 33.00 | 59.80 |
| 2f-40-277v | " | 31.60 | 33.00 | 64.60 |
| 2f-96-120v | " | 83.50 | 33.00 | 116.50 |
| 2f-96-277v | " | 87.00 | 33.00 | 120.00 |
| **16510.10 LIGHTING INDUSTRIAL** | | | | |
| Surface mounted fluorescent, wrap around lens | | | | |
| 1 lamp | EA. | 59.00 | 36.30 | 95.30 |
| 2 lamps | " | 64.00 | 40.40 | 104.40 |
| 4 lamps | " | 100.00 | 45.40 | 145.40 |
| Wall mounted fluorescent | | | | |
| 2-20w lamps | EA. | 47.30 | 22.70 | 70.00 |
| 2-30w lamps | " | 50.50 | 22.70 | 73.20 |
| 2-40w lamps | " | 59.00 | 30.30 | 89.30 |
| Indirect, with wood shielding, 2049w lamps | | | | |
| 4' | EA. | 100.00 | 45.40 | 145.40 |
| 8' | " | 140.00 | 72.50 | 212.50 |
| Industrial fluorescent, 2 lamp | | | | |
| 4' | EA. | 100.00 | 33.00 | 133.00 |
| 8' | " | 160.00 | 60.50 | 220.50 |
| Strip fluorescent | | | | |
| 4' | | | | |
| 1 lamp | EA. | 33.80 | 30.30 | 64.10 |
| 2 lamps | " | 42.20 | 30.30 | 72.50 |
| 8' | | | | |
| 1 lamp | EA. | 67.50 | 33.00 | 100.50 |

## LIGHTING

### 16510.10 LIGHTING INDUSTRIAL

| LIGHTING | UNIT | MAT. | INST. | TOTAL |
|---|---|---|---|---|
| 2 lamps | EA. | 76.00 | 40.40 | 116.40 |
| Wire guard for strip fixture, 4' long | " | 16.90 | 15.80 | 32.70 |
| Strip fluorescent, 8' long, two 4' lamps | " | 100.00 | 60.50 | 160.50 |
| With four 4' lamps | " | 140.00 | 72.50 | 212.50 |
| Wet location fluorescent, plastic housing | | | | |
| 4' long | | | | |
| 1 lamp | EA. | 120.00 | 45.40 | 165.40 |
| 2 lamps | " | 130.00 | 60.50 | 190.50 |
| 8' long | | | | |
| 2 lamps | EA. | 210.00 | 72.50 | 282.50 |
| 4 lamps | " | 290.00 | 79.00 | 369.00 |
| Parabolic troffer, 2'x2' | | | | |
| With 2 "U" lamps | EA. | 140.00 | 45.40 | 185.40 |
| With 3 "U" lamps | " | 170.00 | 52.00 | 222.00 |
| 2'x4' | | | | |
| With 2 40w lamps | EA. | 190.00 | 52.00 | 242.00 |
| With 3 40w lamps | " | 220.00 | 60.50 | 280.50 |
| With 4 40w lamps | " | 230.00 | 60.50 | 290.50 |
| 1'x4' | | | | |
| With 1 T-12 lamp, 9 cell | EA. | 140.00 | 33.00 | 173.00 |
| With 2 T-12 lamps | " | 160.00 | 40.40 | 200.40 |
| With 1 T-12 lamp, 20 cell | " | 160.00 | 33.00 | 193.00 |
| With 2 T-12 lamps | " | 170.00 | 40.40 | 210.40 |
| Steel sided surface fluorescent, 2'x4' | | | | |
| 3 lamps | EA. | 130.00 | 60.50 | 190.50 |
| 4 lamps | " | 140.00 | 60.50 | 200.50 |
| Outdoor sign fluor., 1 lamp, remote ballast | | | | |
| 4' long | EA. | 2,420 | 270.00 | 2,690 |
| 6' long | " | 2,900 | 360.00 | 3,260 |
| Recess mounted, commercial, 2'x2', 13" high | | | | |
| 100w | EA. | 800.00 | 180.00 | 980.00 |
| 250w | " | 880.00 | 200.00 | 1,080 |
| High pressure sodium, hi-bay open | | | | |
| 400w | EA. | 430.00 | 79.00 | 509.00 |
| 1000w | " | 740.00 | 110.00 | 850.00 |
| Enclosed | | | | |
| 400w | EA. | 670.00 | 110.00 | 780.00 |
| 1000w | " | 980.00 | 130.00 | 1,110 |
| Metal halide hi-bay, open | | | | |
| 400w | EA. | 390.00 | 79.00 | 469.00 |
| 1000w | " | 660.00 | 110.00 | 770.00 |
| Enclosed | | | | |
| 400w | EA. | 620.00 | 110.00 | 730.00 |
| 1000w | " | 940.00 | 130.00 | 1,070 |
| High pressure sodium, low bay, surface mounted | | | | |
| 100w | EA. | 270.00 | 45.40 | 315.40 |
| 150w | " | 300.00 | 52.00 | 352.00 |
| 250w | " | 330.00 | 60.50 | 390.50 |
| 400w | " | 430.00 | 72.50 | 502.50 |
| Metal halide, low bay, pendant mounted | | | | |
| 175w | EA. | 300.00 | 60.50 | 360.50 |
| 250w | " | 390.00 | 72.50 | 462.50 |

# 16 ELECTRICAL

## LIGHTING

### 16510.10 LIGHTING INDUSTRIAL

| | UNIT | MAT. | INST. | TOTAL |
|---|---|---|---|---|
| 400w | EA. | 480.00 | 100.00 | 580.00 |
| Indirect luminare, square, metal halide, freestanding | | | | |
| 175w | EA. | 780.00 | 45.40 | 825.40 |
| 250w | " | 800.00 | 45.40 | 845.40 |
| 400w | " | 860.00 | 45.40 | 905.40 |
| High pressure sodium | | | | |
| 150w | EA. | 800.00 | 45.40 | 845.40 |
| 250w | " | 840.00 | 45.40 | 885.40 |
| 400w | " | 900.00 | 45.40 | 945.40 |
| Round, metal halide | | | | |
| 175w | EA. | 740.00 | 45.40 | 785.40 |
| 250w | " | 780.00 | 45.40 | 825.40 |
| 400w | " | 820.00 | 45.40 | 865.40 |
| High pressure sodium | | | | |
| 150w | EA. | 750.00 | 45.40 | 795.40 |
| 250w | " | 810.00 | 45.40 | 855.40 |
| 400w | " | 840.00 | 45.40 | 885.40 |
| Wall mounted, metal halide | | | | |
| 175w | EA. | 610.00 | 110.00 | 720.00 |
| 250w | " | 650.00 | 110.00 | 760.00 |
| 400w | " | 1,080 | 150.00 | 1,230 |
| High pressure sodium | | | | |
| 150w | EA. | 700.00 | 110.00 | 810.00 |
| 250w | " | 860.00 | 110.00 | 970.00 |
| 400w | " | 1,120 | 150.00 | 1,270 |
| Wall pack lithonia, high pressure sodium | | | | |
| 35w | EA. | 190.00 | 40.40 | 230.40 |
| 55w | " | 210.00 | 45.40 | 255.40 |
| 150w | " | 230.00 | 72.50 | 302.50 |
| 250w | " | 300.00 | 79.00 | 379.00 |
| Low pressure sodium | | | | |
| 35w | EA. | 260.00 | 79.00 | 339.00 |
| 55w | " | 400.00 | 91.00 | 491.00 |
| Wall pack hubbell, high pressure sodium | | | | |
| 35w | EA. | 200.00 | 40.40 | 240.40 |
| 150w | " | 250.00 | 72.50 | 322.50 |
| 250w | " | 330.00 | 79.00 | 409.00 |
| Compact fluorescent | | | | |
| 2-7w | EA. | 120.00 | 45.40 | 165.40 |
| 2-13w | " | 140.00 | 60.50 | 200.50 |
| 1-18w | " | 170.00 | 60.50 | 230.50 |
| Handball & racquet ball court, 2'x2', metal halide | | | | |
| 250w | EA. | 440.00 | 110.00 | 550.00 |
| 400w | " | 550.00 | 130.00 | 680.00 |
| High pressure sodium | | | | |
| 250w | EA. | 480.00 | 110.00 | 590.00 |
| 400w | " | 530.00 | 130.00 | 660.00 |
| Bollard light, 42" w/found., high pressure sodium | | | | |
| 70w | EA. | 740.00 | 120.00 | 860.00 |
| 100w | " | 760.00 | 120.00 | 880.00 |
| 150w | " | 760.00 | 120.00 | 880.00 |
| Light fixture lamps | | | | |

## LIGHTING

### 16510.10 LIGHTING INDUSTRIAL

| | UNIT | MAT. | INST. | TOTAL |
|---|---|---|---|---|
| Lamp | | | | |
| 20w med. bipin base, cool white, 24" | EA. | 5.45 | 6.60 | 12.05 |
| 30w cool white, rapid start, 36" | " | 7.10 | 6.60 | 13.70 |
| 40w cool white "U", 3" | " | 13.60 | 6.60 | 20.20 |
| 40w cool white, rapid start, 48" | " | 3.25 | 6.60 | 9.85 |
| 70w high pressure sodium, mogul base | " | 70.50 | 9.10 | 79.60 |
| 75w slimline, 96" | " | 9.05 | 9.10 | 18.15 |
| 100w | | | | |
| Incandescent, 100a, inside frost | EA. | 2.80 | 4.55 | 7.35 |
| Mercury vapor, clear, mogul base | " | 34.40 | 9.10 | 43.50 |
| High pressure sodium, mogul base | " | 75.00 | 9.10 | 84.10 |
| 150w | | | | |
| Par 38 flood or spot, incandescent | EA. | 8.40 | 4.55 | 12.95 |
| High pressure sodium, 1/2 mogul base | " | 78.00 | 9.10 | 87.10 |
| 175w | | | | |
| Mercury vapor, clear, mogul base | EA. | 59.50 | 9.10 | 68.60 |
| Metal halide, clear, mogul base | " | 26.60 | 9.10 | 35.70 |
| 250w | | | | |
| High pressure sodium, mogul base | EA. | 83.00 | 9.10 | 92.10 |
| Mercury vapor, clear, mogul base | " | 46.90 | 9.10 | 56.00 |
| Metal halide, clear, mogul base | " | 75.00 | 9.10 | 84.10 |
| High pressure sodium, mogul base | " | 83.00 | 9.10 | 92.10 |
| 400w | | | | |
| Mercury vapor, clear, mogul base | EA. | 37.50 | 9.10 | 46.60 |
| Metal halide, clear, mogul base | " | 70.50 | 9.10 | 79.60 |
| High pressure sodium, mogul base | " | 89.00 | 9.10 | 98.10 |
| 1000w | | | | |
| Mercury vapor, clear, mogul base | EA. | 81.50 | 11.35 | 92.85 |
| High pressure sodium, mogul base | " | 210.00 | 11.35 | 221.35 |

### 16510.30 EXTERIOR LIGHTING

| | UNIT | MAT. | INST. | TOTAL |
|---|---|---|---|---|
| Exterior light fixtures | | | | |
| Rectangle, high pressure sodium | | | | |
| 70w | EA. | 480.00 | 110.00 | 590.00 |
| 100w | " | 490.00 | 120.00 | 610.00 |
| 150w | " | 530.00 | 120.00 | 650.00 |
| 250w | " | 720.00 | 130.00 | 850.00 |
| 400w | " | 800.00 | 160.00 | 960.00 |
| Flood, rectangular, high pressure sodium | | | | |
| 70w | EA. | 490.00 | 110.00 | 600.00 |
| 100w | " | 520.00 | 120.00 | 640.00 |
| 150w | " | 570.00 | 120.00 | 690.00 |
| 400w | " | 850.00 | 160.00 | 1,010 |
| 1000w | " | 1,290 | 200.00 | 1,490 |
| Round | | | | |
| 400w | EA. | 880.00 | 160.00 | 1,040 |
| 1000w | " | 1,380 | 200.00 | 1,580 |
| Round, metal halide | | | | |
| 400w | EA. | 970.00 | 160.00 | 1,130 |
| 1000w | " | 1,440 | 200.00 | 1,640 |
| Light fixture arms, cobra head, 6', high press. sodium | | | | |

# 16 ELECTRICAL

| LIGHTING | UNIT | MAT. | INST. | TOTAL |
|---|---|---|---|---|
| **16510.30 EXTERIOR LIGHTING** | | | | |
| 100w | EA. | 530.00 | 91.00 | 621.00 |
| 150w | " | 830.00 | 110.00 | 940.00 |
| 250w | " | 870.00 | 110.00 | 980.00 |
| 400w | " | 900.00 | 130.00 | 1,030 |
| Flood, metal halide | | | | |
| 400w | EA. | 930.00 | 160.00 | 1,090 |
| 1000w | " | 1,400 | 200.00 | 1,600 |
| 1500w | " | 1,900 | 270.00 | 2,170 |
| Mercury vapor | | | | |
| 250w | EA. | 550.00 | 130.00 | 680.00 |
| 400w | " | 940.00 | 160.00 | 1,100 |
| Incandescent | | | | |
| 300w | EA. | 98.50 | 79.00 | 177.50 |
| 500w | " | 170.00 | 91.00 | 261.00 |
| 1000w | " | 180.00 | 150.00 | 330.00 |
| **16510.90 POWER LINE FILTERS** | | | | |
| Heavy duty power line filter, 240v | | | | |
| 100a | EA. | 3,440 | 450.00 | 3,890 |
| 300a | " | 11,430 | 730.00 | 12,160 |
| 600a | " | 15,840 | 1,100 | 16,940 |
| **16610.30 UNINTERRUPTIBLE POWER** | | | | |
| Uninterruptible power systems, (U.P.S.), 3kva | EA. | 12,150 | 360.00 | 12,510 |
| 5 kva | " | 13,500 | 500.00 | 14,000 |
| 7.5 kva | " | 16,200 | 730.00 | 16,930 |
| 10 kva | " | 20,250 | 1,000 | 21,250 |
| 15 kva | " | 24,300 | 1,040 | 25,340 |
| 20 kva | " | 33,750 | 1,090 | 34,840 |
| 25 kva | " | 43,200 | 1,140 | 44,340 |
| 30 kva | " | 44,550 | 1,180 | 45,730 |
| 35 kva | " | 47,250 | 1,230 | 48,480 |
| 40 kva | " | 51,300 | 1,270 | 52,570 |
| 45 kva | " | 54,000 | 1,320 | 55,320 |
| 50 kva | " | 58,050 | 1,360 | 59,410 |
| 62.5 kva | " | 68,850 | 1,450 | 70,300 |
| 75 kva | " | 79,650 | 1,590 | 81,240 |
| 100 kva | " | 106,650 | 1,640 | |
| 150 kva | " | 162,000 | 2,270 | |
| 200 kva | " | 216,000 | 2,500 | |
| 300 kva | " | 324,000 | 3,390 | |
| 400 kva | " | 484,700 | 4,080 | |
| 500 kva | " | 605,880 | 4,980 | |
| **16670.10 LIGHTNING PROTECTION** | | | | |
| Lightning protection | | | | |
| Copper point, nickel plated, 12' | | | | |
| 1/2" dia. | EA. | 40.50 | 45.40 | 85.90 |
| 5/8" dia. | " | 48.60 | 45.40 | 94.00 |

## COMMUNICATIONS

### 16720.10 FIRE ALARM SYSTEMS

| COMMUNICATIONS | UNIT | MAT. | INST. | TOTAL |
|---|---|---|---|---|
| Master fire alarm box, pedestal mounted | " | 5,480 | 730.00 | 6,210 |
| Master fire alarm box | " | 2,820 | 270.00 | 3,090 |
| Box light | " | 160.00 | 22.70 | 182.70 |
| Ground assembly for box | " | 78.00 | 30.30 | 108.30 |
| Bracket for pole type box | " | 100.00 | 33.00 | 133.00 |
| Pull station | | | | |
| Waterproof | EA. | 78.00 | 22.70 | 100.70 |
| Manual | " | 46.90 | 18.15 | 65.05 |
| Horn, waterproof | " | 70.50 | 45.40 | 115.90 |
| Interior alarm | " | 46.90 | 33.00 | 79.90 |
| Coded transmitter, automatic | " | 830.00 | 91.00 | 921.00 |
| Control panel, 8 zone | " | 1,650 | 360.00 | 2,010 |
| Battery charger and cabinet | " | 550.00 | 91.00 | 641.00 |
| Batteries, nickel cadmium or lead calcium | " | 410.00 | 230.00 | 640.00 |
| CO2 pressure switch connection | " | 78.00 | 33.00 | 111.00 |
| Annunciator panels | | | | |
| Fire detection annunciator, remote type, 8 zone | EA. | 280.00 | 82.50 | 362.50 |
| 12 zone | " | 360.00 | 91.00 | 451.00 |
| 16 zone | " | 450.00 | 110.00 | 560.00 |
| Fire alarm systems | | | | |
| Bell | EA. | 86.00 | 27.90 | 113.90 |
| Weatherproof bell | " | 110.00 | 30.30 | 140.30 |
| Horn | " | 50.00 | 33.00 | 83.00 |
| Siren | " | 510.00 | 91.00 | 601.00 |
| Chime | " | 62.50 | 27.90 | 90.40 |
| Audio/visual | " | 90.50 | 33.00 | 123.50 |
| Strobe light | " | 83.50 | 33.00 | 116.50 |
| Smoke detector | " | 130.00 | 30.30 | 160.30 |
| Heat detection | " | 23.45 | 22.70 | 46.15 |
| Thermal detector | " | 21.90 | 22.70 | 44.60 |
| Ionization detector | " | 110.00 | 24.20 | 134.20 |
| Duct detector | " | 370.00 | 130.00 | 500.00 |
| Test switch | " | 62.50 | 22.70 | 85.20 |
| Remote indicator | " | 31.60 | 25.90 | 57.50 |
| Door holder | " | 140.00 | 33.00 | 173.00 |
| Telephone jack | " | 23.45 | 13.45 | 36.90 |
| Fireman phone | " | 340.00 | 45.40 | 385.40 |
| Speaker | " | 67.00 | 36.30 | 103.30 |
| Remote fire alarm annunciator panel | | | | |
| 24 zone | EA. | 1,720 | 300.00 | 2,020 |
| 48 zone | " | 3,440 | 590.00 | 4,030 |
| Control panel | | | | |
| 12 zone | EA. | 1,170 | 130.00 | 1,300 |
| 16 zone | " | 1,530 | 200.00 | 1,730 |
| 24 zone | " | 2,350 | 300.00 | 2,650 |
| 48 zone | " | 4,380 | 730.00 | 5,110 |
| Power supply | " | 270.00 | 68.50 | 338.50 |
| Status command | " | 7,430 | 230.00 | 7,660 |
| Printer | " | 2,890 | 68.50 | 2,959 |
| Transponder | " | 240.00 | 40.80 | 280.80 |
| Transformer | " | 170.00 | 30.30 | 200.30 |
| Transceiver | " | 230.00 | 33.00 | 263.00 |

# 16 ELECTRICAL

## COMMUNICATIONS

### 16720.10 FIRE ALARM SYSTEMS

| | UNIT | MAT. | INST. | TOTAL |
|---|---|---|---|---|
| Relays | EA. | 94.00 | 22.70 | 116.70 |
| Flow switch | " | 300.00 | 91.00 | 391.00 |
| Tamper switch | " | 180.00 | 130.00 | 310.00 |
| End of line resistor | " | 13.25 | 15.80 | 29.05 |
| Printed ckt. card | " | 120.00 | 22.70 | 142.70 |
| Central processing unit | " | 15,320 | 280.00 | 15,600 |
| UPS backup to c.p.u. | " | 14,860 | 410.00 | 15,270 |
| Smoke detector, fixed temp. & rate of rise comb. | " | 230.00 | 72.50 | 302.50 |

### 16720.50 SECURITY SYSTEMS

| | UNIT | MAT. | INST. | TOTAL |
|---|---|---|---|---|
| Sensors | EA. | | | |
| Balanced magnetic door switch, surface mounted | " | 110.00 | 22.70 | 132.70 |
| With remote test | " | 150.00 | 45.40 | 195.40 |
| Flush mounted | " | 110.00 | 84.50 | 194.50 |
| Mounted bracket | " | 8.25 | 15.80 | 24.05 |
| Mounted bracket spacer | " | 7.45 | 15.80 | 23.25 |
| Photoelectric sensor, for fence | " | | | |
| 6 beam | " | 12,140 | 130.00 | 12,270 |
| 9 beam | " | 14,840 | 190.00 | 15,030 |
| Photoelectric sensor, 12 volt dc | " | | | |
| 500' range | " | 360.00 | 72.50 | 432.50 |
| 800' range | " | 400.00 | 91.00 | 491.00 |
| Capacitance wire grid kit | " | | | |
| Surface | " | 100.00 | 45.40 | 145.40 |
| Duct | " | 74.50 | 72.50 | 147.00 |
| Tube grid kit | " | 120.00 | 22.70 | 142.70 |
| Vibration sensor, 30 max per zone | " | 150.00 | 22.70 | 172.70 |
| Audio sensor, 30 max per zone | " | 160.00 | 22.70 | 182.70 |
| Inertia sensor | " | | | |
| Outdoor | " | 120.00 | 33.00 | 153.00 |
| Indoor | " | 74.50 | 22.70 | 97.20 |
| Ultrasonic transmitter, 20 max per zone | | | | |
| Omni-directional | EA. | 83.00 | 72.50 | 155.50 |
| Directional | " | 91.00 | 60.50 | 151.50 |
| Transceiver | | | | |
| Omni-directional | EA. | 91.00 | 45.40 | 136.40 |
| Directional | " | 100.00 | 45.40 | 145.40 |
| Passive infra-red sensor, 20 max per zone | " | 650.00 | 72.50 | 722.50 |
| Access/secure control unit, balanced magnetic switch | " | 400.00 | 72.50 | 472.50 |
| Photoelectric sensor | " | 660.00 | 72.50 | 732.50 |
| Photoelectric fence sensor | " | 680.00 | 72.50 | 752.50 |
| Capacitance sensor | " | 790.00 | 79.00 | 869.00 |
| Audio and vibration sensor | " | 690.00 | 72.50 | 762.50 |
| Inertia sensor | " | 930.00 | 72.50 | 1,003 |
| Ultrasonic sensor | " | 1,060 | 79.00 | 1,139 |
| Infra-red sensor | " | 660.00 | 91.00 | 751.00 |
| Monitor panel, with access/secure tone, standard | " | 450.00 | 79.00 | 529.00 |
| High security | " | 660.00 | 91.00 | 751.00 |
| Emergency power indicator | " | 270.00 | 22.70 | 292.70 |
| Monitor rack with 115v power supply | | | | |
| 1 zone | EA. | 370.00 | 45.40 | 415.40 |
| 10 zone | " | 1,910 | 110.00 | 2,020 |

# 16 ELECTRICAL

## COMMUNICATIONS

| COMMUNICATIONS | UNIT | MAT. | INST. | TOTAL |
|---|---|---|---|---|
| **16720.50 SECURITY SYSTEMS** | | | | |
| Monitor cabinet, wall mounted | | | | |
| 1 zone | EA. | 580.00 | 45.40 | 625.40 |
| 5 zone | " | 2,070 | 72.50 | 2,143 |
| 10 zone | " | 910.00 | 79.00 | 989.00 |
| 20 zone | " | 2,880 | 91.00 | 2,971 |
| Floor mounted, 50 zone | " | 3,000 | 180.00 | 3,180 |
| Security system accessories | | | | |
| Tamper assembly for monitor cabinet | EA. | 74.50 | 20.20 | 94.70 |
| Monitor panel blank | " | 9.90 | 15.80 | 25.70 |
| Audible alarm | " | 84.50 | 22.70 | 107.20 |
| Audible alarm control | " | 350.00 | 15.80 | 365.80 |
| Termination screw, terminal cabinet | | | | |
| 25 pair | EA. | 250.00 | 72.50 | 322.50 |
| 50 pair | " | 400.00 | 110.00 | 510.00 |
| 150 pair | " | 650.00 | 230.00 | 880.00 |
| Universal termination, cabinets & panel | | | | |
| Remote test | EA. | 58.00 | 79.00 | 137.00 |
| No remote test | " | 41.40 | 33.00 | 74.40 |
| High security line supervision termination | " | 300.00 | 45.40 | 345.40 |
| Door cord for capacitance sensor, 12" | " | 9.95 | 22.70 | 32.65 |
| Insulation block kit for capacitance sensor | " | 46.40 | 15.80 | 62.20 |
| Termination block for capacitance sensor | " | 10.10 | 15.80 | 25.90 |
| Guard alert display | " | 1,060 | 27.90 | 1,088 |
| Uninterrupted power supply | | | | 1,080 |
| Plug-in 40kva transformer | | | | |
| 12 volt | EA. | 42.30 | 15.80 | 58.10 |
| 18 volt | " | 28.20 | 15.80 | 44.00 |
| 24 volt | " | 19.90 | 15.80 | 35.70 |
| Test relay | " | 64.00 | 15.80 | 79.80 |
| Coaxial cable, 50 ohm | L.F. | 0.26 | 0.27 | 0.53 |
| Door openers | EA. | 74.50 | 22.70 | 97.20 |
| Push buttons | | | | |
| Standard | EA. | 16.55 | 15.80 | 32.35 |
| Weatherproof | " | 24.85 | 20.20 | 45.05 |
| Bells | " | 63.00 | 33.00 | 96.00 |
| Horns | | | | |
| Standard | EA. | 66.50 | 45.40 | 111.90 |
| Weatherproof | " | 120.00 | 56.50 | 176.50 |
| Chimes | " | 96.00 | 30.30 | 126.30 |
| Flasher | " | 69.50 | 27.90 | 97.40 |
| Motion detectors | " | 290.00 | 68.50 | 358.50 |
| Intercom units | " | 66.50 | 33.00 | 99.50 |
| Remote annunciator | " | 2,980 | 230.00 | 3,210 |
| **16740.10 TELEPHONE SYSTEMS** | | | | |
| Communication cable | | | | |
| 25 pair | L.F. | 0.54 | 1.15 | 1.69 |
| 100 pair | " | 2.55 | 1.30 | 3.85 |
| 150 pair | " | 3.90 | 1.50 | 5.40 |
| 200 pair | " | 5.25 | 1.80 | 7.05 |
| 300 pair | " | 6.70 | 1.90 | 8.60 |

# 16 ELECTRICAL

## COMMUNICATIONS

| 16740.10 TELEPHONE SYSTEMS | UNIT | MAT. | INST. | TOTAL |
|---|---|---|---|---|
| 400 pair | L.F. | 9.10 | 2.00 | 11.10 |
| Cable tap in manhole or junction box | | | | |
| 25 pair cable | EA. | 4.20 | 170.00 | 174.20 |
| 50 pair cable | " | 8.45 | 340.00 | 348.45 |
| 75 pair cable | " | 12.65 | 510.00 | 522.65 |
| 100 pair cable | " | 16.85 | 690.00 | 706.85 |
| 150 pair cable | " | 25.30 | 1,010 | 1,035 |
| 200 pair cable | " | 33.80 | 1,350 | 1,384 |
| 300 pair cable | " | 50.50 | 2,020 | 2,071 |
| 400 pair cable | " | 67.50 | 2,790 | 2,858 |
| Cable terminations, manhole or junction box | | | | |
| 25 pair cable | EA. | 4.20 | 170.00 | 174.20 |
| 50 pair cable | " | 8.45 | 340.00 | 348.45 |
| 100 pair cable | " | 16.85 | 690.00 | 706.85 |
| 150 pair cable | " | 25.30 | 1,010 | 1,035 |
| 200 pair cable | " | 33.80 | 1,350 | 1,384 |
| 300 pair cable | " | 50.50 | 2,020 | 2,071 |
| 400 pair cable | " | 52.50 | 2,790 | 2,843 |

| 16780.10 ANTENNAS AND TOWERS | UNIT | MAT. | INST. | TOTAL |
|---|---|---|---|---|
| Guy cable, alumaweld | | | | |
| 1x3, 7/32" | L.F. | 0.36 | 2.25 | 2.61 |
| 1x3, 1/4" | " | 0.43 | 2.25 | 2.68 |
| 1x3, 25/64" | " | 0.61 | 2.70 | 3.31 |
| 1x19, 1/2" | " | 1.55 | 3.15 | 4.70 |
| 1x7, 35/64" | " | 1.75 | 3.65 | 5.40 |
| 1x19, 13/16" | " | 1.90 | 4.55 | 6.45 |
| Preformed alumaweld end grip | | | | |
| 1/4" cable | EA. | 2.70 | 4.55 | 7.25 |
| 3/8" cable | " | 3.60 | 4.55 | 8.15 |
| 1/2" cable | " | 4.55 | 6.60 | 11.15 |
| 9/16" cable | " | 5.65 | 9.10 | 14.75 |
| 5/8" cable | " | 6.90 | 11.35 | 18.25 |
| Fiberglass guy rod, white epoxy coated | | | | |
| 1/4" dia. | L.F. | 1.80 | 6.60 | 8.40 |
| 3/8" dia | " | 2.70 | 6.60 | 9.30 |
| 1/2" dia | " | 3.60 | 9.10 | 12.70 |
| 5/8" dia | " | 4.55 | 11.35 | 15.90 |
| Preformed glass grip end grip, guy rod | | | | |
| 1/4" dia. | EA. | 10.05 | 6.60 | 16.65 |
| 3/8" dia. | " | 11.60 | 9.10 | 20.70 |
| 1/2" dia. | " | 14.20 | 11.35 | 25.55 |
| 5/8" dia. | " | 15.95 | 11.35 | 27.30 |
| Spelter socket end grip, 1/4" dia. guy rod | | | | |
| Standard strength | EA. | 29.60 | 22.70 | 52.30 |
| High performance | " | 36.50 | 22.70 | 59.20 |
| 3/8" dia. guy rod | | | | |
| Standard strength | EA. | 28.80 | 15.80 | 44.60 |
| High performance | " | 36.50 | 22.70 | 59.20 |
| Timber pole, Douglas Fir | | | | |
| 80-85 ft | EA. | 2,250 | 890.00 | 3,140 |
| 90-95 ft | " | 2,810 | 1,010 | 3,820 |

# COMMUNICATIONS

| | UNIT | MAT. | INST. | TOTAL |
|---|---|---|---|---|
| **16780.10 ANTENNAS AND TOWERS** | | | | |
| Southern yellow pine | | | | |
| 35-45 ft | EA. | 1,400 | 500.00 | 1,900 |
| 50-55 ft | " | 1,970 | 640.00 | 2,610 |
| **16780.50 TELEVISION SYSTEMS** | | | | |
| TV outlet, self terminating, w/cover plate | EA. | 9.00 | 13.95 | 22.95 |
| Thru splitter | " | 19.65 | 72.50 | 92.15 |
| End of line | " | 16.40 | 60.50 | 76.90 |
| In line splitter multitap | | | | |
| 4 way | EA. | 32.80 | 82.50 | 115.30 |
| 2 way | " | 24.60 | 77.50 | 102.10 |
| Equipment cabinet | " | 82.00 | 72.50 | 154.50 |
| Antenna | | | | |
| Broad band uhf | EA. | 160.00 | 160.00 | 320.00 |
| Lightning arrester | " | 44.30 | 33.00 | 77.30 |
| TV cable | L.F. | 0.38 | 0.23 | 0.61 |
| Coaxial cable rg | " | 0.24 | 0.23 | 0.47 |
| Cable drill, with replacement tip | EA. | 8.20 | 22.70 | 30.90 |
| Cable blocks for in-line taps | " | 16.40 | 33.00 | 49.40 |
| In-line taps ptu-series 36 tv system | " | 19.65 | 52.00 | 71.65 |
| **16850.10 ELECTRIC HEATING** | | | | |
| Baseboard heater | | | | |
| 2', 375w | EA. | 49.70 | 45.40 | 95.10 |
| 3', 500w | " | 58.00 | 45.40 | 103.40 |
| 4', 750w | " | 63.50 | 52.00 | 115.50 |
| 5', 935w | " | 88.00 | 60.50 | 148.50 |
| 6', 1125w | " | 93.50 | 72.50 | 166.00 |
| 7', 1310w | " | 110.00 | 82.50 | 192.50 |
| 8', 1500w | " | 120.00 | 91.00 | 211.00 |
| 9', 1680w | " | 170.00 | 100.00 | 270.00 |
| 10', 1875w | " | 180.00 | 100.00 | 280.00 |
| Unit heater, wall mounted | | | | |
| 1500w | EA. | 230.00 | 75.50 | 305.50 |
| 2500w | " | 250.00 | 82.50 | 332.50 |
| 4000w | " | 330.00 | 100.00 | 430.00 |
| Thermostat | | | | |
| Integral | EA. | 38.10 | 22.70 | 60.80 |
| Line voltage | " | 38.60 | 22.70 | 61.30 |
| Electric heater connection | " | 1.65 | 11.35 | 13.00 |

BNi Building News

# Man-Hour Tables

The man-hour productivities used to develop the labor costs are listed in the following section of this book. These productivities represent typical installation labor for thousands of construction items. The data takes into account all activities involved in normal construction under commonly experienced working conditions. As with the Costbook pages, these items are listed according to the CSI MASTERFORMAT. In order to best use the information in this book, please review this sample page and read the "Features in this Book" section.

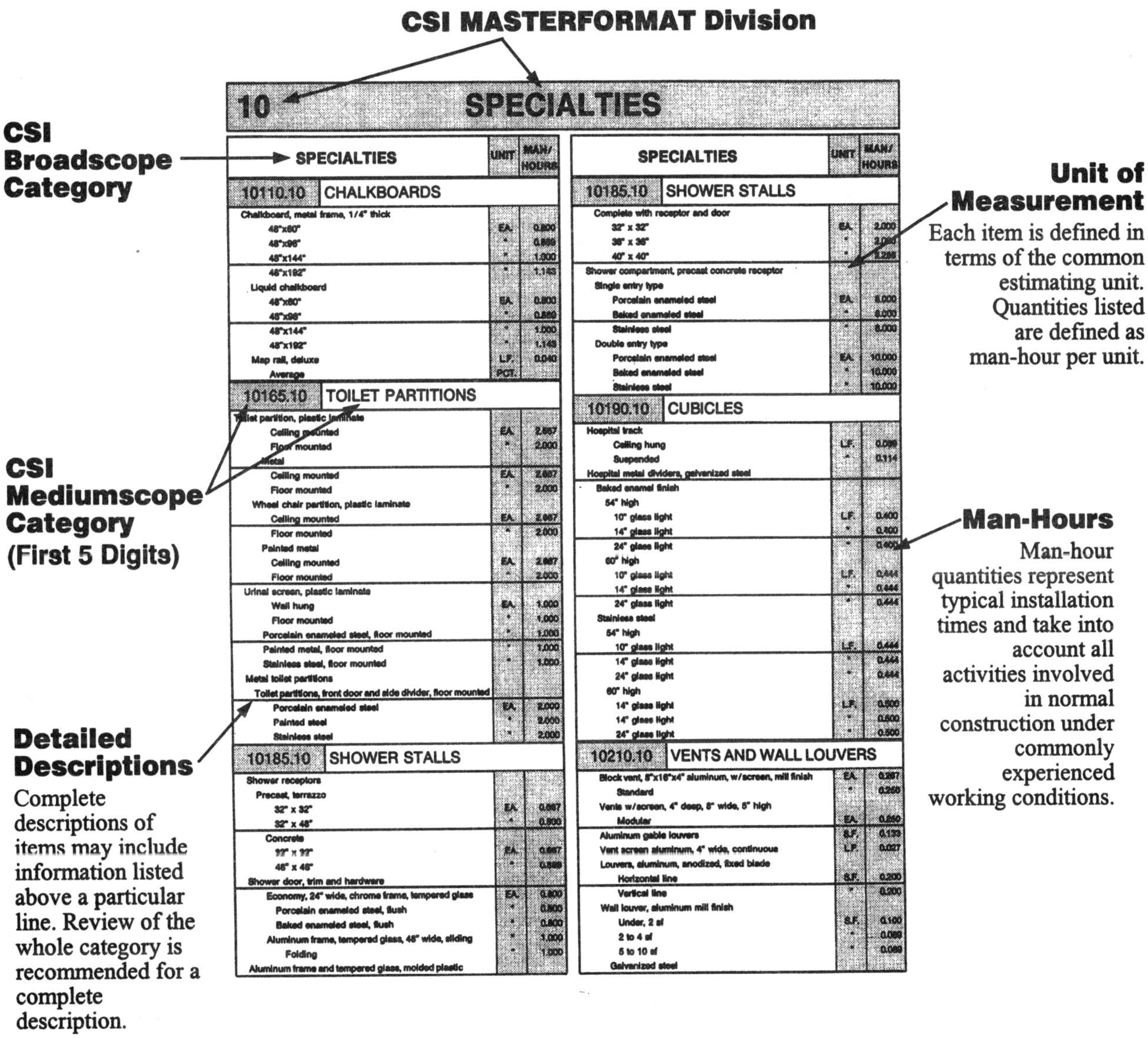

BNi Building News

# 02 SITEWORK

| DEMOLITION | UNIT | MAN/ HOURS |
|---|---|---|
| **02060.10 BUILDING DEMOLITION** | | |
| Building, complete with disposal | | |
| Wood frame | C.F. | 0.003 |
| Concrete | " | 0.004 |
| Steel frame | " | 0.005 |
| Partition removal | | |
| Concrete block partitions | | |
| 4" thick | S.F. | 0.040 |
| 8" thick | " | 0.053 |
| 12" thick | " | 0.073 |
| Brick masonry partitions | | |
| 4" thick | S.F. | 0.040 |
| 8" thick | " | 0.050 |
| 12" thick | " | 0.067 |
| 16" thick | " | 0.100 |
| Cast in place concrete partitions | | |
| Unreinforced | | |
| 6" thick | S.F. | 0.160 |
| 8" thick | " | 0.171 |
| 10" thick | " | 0.200 |
| 12" thick | " | 0.240 |
| Reinforced | | |
| 6" thick | S.F. | 0.185 |
| 8" thick | " | 0.240 |
| 10" thick | " | 0.267 |
| 12" thick | " | 0.320 |
| Terra cotta | | |
| To 6" thick | S.F. | 0.040 |
| Stud partitions | | |
| Metal or wood, with drywall both sides | S.F. | 0.040 |
| Metal studs, both sides, lath and plaster | " | 0.053 |
| Door and frame removal | | |
| Hollow metal in masonry wall | | |
| Single | | |
| 2'6"x6'8" | EA. | 1.000 |
| 3'x7' | " | 1.333 |
| Double | | |
| 3'x7' | EA. | 1.600 |
| 4'x8' | " | 1.600 |
| Wood in framed wall | | |
| Single | | |
| 2'6"x6'8" | EA. | 0.571 |
| 3'x6'8" | " | 0.667 |
| Double | | |
| 2'6"x6'8" | EA. | 0.800 |
| 3'x6'8" | " | 0.889 |
| Remove for re-use | | |
| Hollow metal | EA. | 2.000 |
| Wood | " | 1.333 |
| Floor removal | | |
| Brick flooring | S.F. | 0.032 |
| Ceramic or quarry tile | " | 0.018 |
| Terrazzo | " | 0.036 |
| Heavy wood | " | 0.021 |
| Residential wood | " | 0.023 |

| DEMOLITION | UNIT | MAN/ HOURS |
|---|---|---|
| **02060.10 BUILDING DEMOLITION** | | |
| Resilient tile or linoleum | S.F. | 0.008 |
| Ceiling removal | | |
| Acoustical tile ceiling | | |
| Adhesive fastened | S.F. | 0.008 |
| Furred and glued | " | 0.007 |
| Suspended grid | " | 0.005 |
| Drywall ceiling | | |
| Furred and nailed | S.F. | 0.009 |
| Nailed to framing | " | 0.008 |
| Plastered ceiling | | |
| Furred on framing | S.F. | 0.020 |
| Suspended system | " | 0.027 |
| Roofing removal | | |
| Steel frame | | |
| Corrugated metal roofing | S.F. | 0.016 |
| Built-up roof on metal deck | " | 0.027 |
| Wood frame | | |
| Built up roof on wood deck | S.F. | 0.025 |
| Roof shingles | " | 0.013 |
| Roof tiles | " | 0.027 |
| Concrete frame | C.F. | 0.053 |
| Concrete plank | S.F. | 0.040 |
| Built-up roof on concrete | " | 0.023 |
| Cut-outs | | |
| Concrete, elevated slabs, mesh reinforcing | | |
| Under 5 cf | C.F. | 0.800 |
| Over 5 cf | " | 0.667 |
| Bar reinforcing | | |
| Under 5 cf | C.F. | 1.333 |
| Over 5 cf | " | 1.000 |
| Window removal | | |
| Metal windows, trim included | | |
| 2'x3' | EA. | 0.800 |
| 2'x4' | " | 0.889 |
| 2'x6' | " | 1.000 |
| 3'x4' | " | 1.000 |
| 3'x6' | " | 1.143 |
| 3'x8' | " | 1.333 |
| 4'x4' | " | 1.333 |
| 4'x6' | " | 1.600 |
| 4'x8' | " | 2.000 |
| Wood windows, trim included | | |
| 2'x3' | EA. | 0.444 |
| 2'x4' | " | 0.471 |
| 2'x6' | " | 0.500 |
| 3'x4' | " | 0.533 |
| 3'x6' | " | 0.571 |
| 3'x8' | " | 0.615 |
| 6'x4' | " | 0.667 |
| 6'x6' | " | 0.727 |
| 6'x8' | " | 0.800 |
| Walls, concrete, bar reinforcing | | |
| Small jobs | C.F. | 0.533 |
| Large jobs | " | 0.444 |

# 02 SITEWORK

## DEMOLITION

### 02060.10 BUILDING DEMOLITION

| | UNIT | MAN/HOURS |
|---|---|---|
| Brick walls, not including toothing | | |
| 4" thick | S.F. | 0.040 |
| 8" thick | " | 0.050 |
| 12" thick | " | 0.067 |
| 16" thick | " | 0.100 |
| Concrete block walls, not including toothing | | |
| 4" thick | S.F. | 0.044 |
| 6" thick | " | 0.047 |
| 8" thick | " | 0.050 |
| 10" thick | " | 0.057 |
| 12" thick | " | 0.067 |
| Rubbish handling | | |
| Load in dumpster or truck | | |
| Minimum | C.F. | 0.018 |
| Maximum | " | 0.027 |
| For use of elevators, add | | |
| Minimum | C.F. | 0.004 |
| Maximum | " | 0.008 |
| Rubbish hauling | | |
| Hand loaded on trucks, 2 mile trip | C.Y. | 0.320 |
| Machine loaded on trucks, 2 mile trip | " | 0.240 |

## HIGHWAY DEMOLITION

### 02065.10 PAVEMENT DEMOLITION

| | UNIT | MAN/HOURS |
|---|---|---|
| Bituminous pavement, up to 3" thick | | |
| On streets | S.Y. | 0.096 |
| On pipe trench | " | 0.120 |
| Concrete pavement, 6" thick | | |
| No reinforcement | S.Y. | 0.160 |
| With wire mesh | " | 0.240 |
| With rebars | " | 0.300 |
| 9" thick | | |
| No reinforcement | S.Y. | 0.200 |
| With wire mesh | " | 0.300 |
| With rebars | " | 0.400 |
| 12" thick | | |
| No reinforcement | S.Y. | 0.240 |
| With wire mesh | " | 0.343 |
| With rebars | " | 0.480 |
| Sidewalk, 4" thick, with disposal | " | 0.080 |
| Removal of pavement markings by waterblasting | S.F. | 0.004 |

## HIGHWAY DEMOLITION

### 02065.15 SAW CUTTING PAVEMENT

| | UNIT | MAN/HOURS |
|---|---|---|
| Pavement, bituminous | | |
| 2" thick | L.F. | 0.016 |
| 3" thick | " | 0.020 |
| Concrete pavement, with wire mesh | | |
| 4" thick | L.F. | 0.031 |
| 5" thick | " | 0.033 |
| Plain concrete, unreinforced | | |
| 4" thick | L.F. | 0.027 |
| 5" thick | " | 0.031 |

### 02065.80 CURB & GUTTER

| | UNIT | MAN/HOURS |
|---|---|---|
| Removal, plain concrete curb | L.F. | 0.060 |
| Plain concrete curb and 2' gutter | " | 0.083 |

### 02065.85 GUARDRAILS

| | UNIT | MAN/HOURS |
|---|---|---|
| Remove standard guardrail | | |
| Steel | L.F. | 0.080 |
| Wood | " | 0.062 |

### 02075.80 CORE DRILLING

| | UNIT | MAN/HOURS |
|---|---|---|
| Concrete | | |
| 6" thick | | |
| 3" dia. | EA. | 0.571 |
| 4" dia. | " | 0.667 |
| 6" dia. | " | 0.800 |
| 8" dia. | " | 1.333 |
| 8" thick | | |
| 3" dia. | EA. | 0.800 |
| 4" dia. | " | 1.000 |
| 6" dia. | " | 1.143 |
| 8" dia. | " | 1.600 |
| 10" thick | | |
| 3" dia. | EA. | 1.000 |
| 4" dia. | " | 1.143 |
| 6" dia. | " | 1.333 |
| 8" dia. | " | 2.000 |
| 12" thick | | |
| 3" dia. | EA. | 1.333 |
| 4" dia. | " | 1.600 |
| 6" dia. | " | 2.000 |
| 8" dia. | " | 2.667 |

# 02 SITEWORK

## HAZARDOUS WASTE

| HAZARDOUS WASTE | UNIT | MAN/ HOURS |
|---|---|---|
| **02080.10 ASBESTOS REMOVAL** | | |
| Enclosure using wood studs & poly, install & remove | S.F. | 0.020 |
| **02080.12 DUCT INSULATION REMOVAL** | | |
| Remove duct insulation, duct size | | |
| 6" x 12" | L.F. | 0.044 |
| x 18" | " | 0.062 |
| x 24" | " | 0.089 |
| 8" x 12" | " | 0.067 |
| x 18" | " | 0.073 |
| x 24" | " | 0.100 |
| 12" x 12" | " | 0.067 |
| x 18" | " | 0.089 |
| x 24" | " | 0.114 |
| **02080.15 PIPE INSULATION REMOVAL** | | |
| Removal, asbestos insulation | | |
| 2" thick, pipe | | |
| 1" to 3" dia. | L.F. | 0.067 |
| 4" to 6" dia. | " | 0.076 |
| 3" thick | | |
| 7" to 8" dia. | L.F. | 0.080 |
| 9" to 10" dia. | " | 0.084 |
| 11" to 12" dia. | " | 0.089 |
| 13" to 14" dia. | " | 0.094 |
| 15" to 18" dia. | " | 0.100 |

## SITE DEMOLITION

| SITE DEMOLITION | UNIT | MAN/ HOURS |
|---|---|---|
| **02105.10 CATCH BASINS/MANHOLES** | | |
| Abandon catch basin or manhole (fill with sand) | EA. | 4.800 |
| Remove and reset frame and cover | " | 2.400 |
| Remove catch basin, to 10' deep | | |
| Masonry | EA. | 6.000 |
| Concrete | " | 8.000 |
| **02105.20 FENCES** | | |
| Remove fencing | | |
| Chain link, 8' high | | |
| For disposal | L.F. | 0.040 |
| For reuse | " | 0.100 |
| Wood | | |
| 4' high | S.F. | 0.027 |
| 6' high | " | 0.032 |
| 8' high | " | 0.040 |

## SITE DEMOLITION

| SITE DEMOLITION | UNIT | MAN/ HOURS |
|---|---|---|
| **02105.20 FENCES** | | |
| Masonry | | |
| 8" thick | | |
| 4' high | S.F. | 0.080 |
| 6' high | " | 0.100 |
| 8' high | " | 0.114 |
| 12" thick | | |
| 4' high | S.F. | 0.133 |
| 6' high | " | 0.160 |
| 8' high | " | 0.200 |
| 12' high | " | 0.267 |
| **02105.42 DRAINAGE PIPING** | | |
| Remove drainage pipe, not including excavation | | |
| 12" dia. | L.F. | 0.100 |
| 18" dia. | " | 0.126 |
| 24" dia. | " | 0.160 |
| 36" dia. | " | 0.200 |
| **02105.43 GAS PIPING** | | |
| Remove welded steel pipe, not including excavation | | |
| 4" dia. | L.F. | 0.150 |
| 5" dia. | " | 0.240 |
| 6" dia. | " | 0.300 |
| 8" dia. | " | 0.480 |
| 10" dia. | " | 0.600 |
| **02105.45 SANITARY PIPING** | | |
| Remove sewer pipe, not including excavation | | |
| 4" dia. | L.F. | 0.096 |
| 6" dia. | " | 0.109 |
| 8" dia. | " | 0.120 |
| 10" dia. | " | 0.126 |
| 12" dia. | " | 0.133 |
| 15" dia. | " | 0.141 |
| 18" dia. | " | 0.160 |
| 24" dia. | " | 0.200 |
| 30" dia. | " | 0.240 |
| 36" dia. | " | 0.300 |
| **02105.48 WATER PIPING** | | |
| Remove water pipe, not including excavation | | |
| 4" dia. | L.F. | 0.109 |
| 6" dia. | " | 0.114 |
| 8" dia. | " | 0.126 |
| 10" dia. | " | 0.133 |
| 12" dia. | " | 0.141 |
| 14" dia. | " | 0.150 |
| 16" dia. | " | 0.160 |
| 18" dia. | " | 0.171 |
| 20" dia. | " | 0.185 |
| Remove valves | | |
| 6" | EA. | 1.200 |

# 02 SITEWORK

## SITE DEMOLITION

| 02105.48 WATER PIPING | UNIT | MAN/HOURS |
|---|---|---|
| 10" | EA. | 1.333 |
| 14" | " | 1.500 |
| 18" | " | 2.000 |

| 02105.60 UNDERGROUND TANKS | UNIT | MAN/HOURS |
|---|---|---|
| Remove underground storage tank, and backfill | | |
| 50 to 250 gals | EA. | 8.000 |
| 600 gals | " | 8.000 |
| 1000 gals | " | 12.000 |
| 4000 gals | " | 19.200 |
| 5000 gals | " | 19.200 |
| 10,000 gals | " | 32.000 |
| 12,000 gals | " | 40.000 |
| 15,000 gals | " | 48.000 |
| 20,000 gals | " | 60.000 |

| 02105.66 SEPTIC TANKS | UNIT | MAN/HOURS |
|---|---|---|
| Remove septic tank | | |
| 1000 gals | EA. | 2.000 |
| 2000 gals | " | 2.400 |
| 5000 gals | " | 3.000 |
| 15,000 gals | " | 24.000 |
| 25,000 gals | " | 32.000 |
| 40,000 gals | " | 48.000 |

| 02105.80 WALLS, EXTERIOR | UNIT | MAN/HOURS |
|---|---|---|
| Concrete wall | | |
| Light reinforcing | | |
| 6" thick | S.F. | 0.120 |
| 8" thick | " | 0.126 |
| 10" thick | " | 0.133 |
| 12" thick | " | 0.150 |
| Medium reinforcing | | |
| 6" thick | S.F. | 0.126 |
| 8" thick | " | 0.133 |
| 10" thick | " | 0.150 |
| 12" thick | " | 0.171 |
| Heavy reinforcing | | |
| 6" thick | S.F. | 0.141 |
| 8" thick | " | 0.150 |
| 10" thick | " | 0.171 |
| 12" thick | " | 0.200 |
| Masonry | | |
| No reinforcing | | |
| 8" thick | S.F. | 0.053 |
| 12" thick | " | 0.060 |
| 16" thick | " | 0.069 |
| Horizontal reinforcing | | |
| 8" thick | S.F. | 0.060 |
| 12" thick | " | 0.065 |
| 16" thick | " | 0.077 |
| Vertical reinforcing | | |

## SITE DEMOLITION

| 02105.80 WALLS, EXTERIOR | UNIT | MAN/HOURS |
|---|---|---|
| 8" thick | S.F. | 0.077 |
| 12" thick | " | 0.089 |
| 16" thick | " | 0.109 |
| Remove concrete headwall | | |
| 15" pipe | EA. | 1.714 |
| 24" pipe | " | 2.182 |

| 02162.10 TRENCH SHEETING | UNIT | MAN/HOURS |
|---|---|---|
| Closed timber, including pull and salvage, excavation | | |
| 8' deep | S.F. | 0.064 |
| 20' deep | " | 0.098 |

## EARTHWORK

| 02210.10 HAULING MATERIAL | UNIT | MAN/HOURS |
|---|---|---|
| Haul material by 10 cy dump truck, round trip distance | | |
| 1 mile | C.Y. | 0.044 |
| 2 mile | " | 0.053 |
| 5 mile | " | 0.073 |
| 10 mile | " | 0.080 |
| 20 mile | " | 0.089 |
| 30 mile | " | 0.107 |
| Site grading, cut & fill, sandy clay, 200' haul, 75 hp dozer | " | 0.032 |
| Spread topsoil by equipment on site | " | 0.036 |
| Site grading (cut and fill to 6") less than 1 acre | | |
| 75 hp dozer | C.Y. | 0.053 |
| 1.5 cy backhoe/loader | " | 0.080 |

| 02210.30 BULK EXCAVATION | UNIT | MAN/HOURS |
|---|---|---|
| Hydraulic excavator | | |
| 1 cy capacity | | |
| Light material | C.Y. | 0.040 |
| Medium material | " | 0.048 |
| Wet material | " | 0.060 |
| Blasted rock | " | 0.069 |
| Wheel mounted front-end loader | | |
| 7/8 cy capacity | | |
| Light material | C.Y. | 0.020 |
| Medium material | " | 0.023 |
| Wet material | " | 0.027 |
| Blasted rock | " | 0.032 |
| Track mounted front-end loader | | |
| 1-1/2 cy capacity | | |
| Light material | C.Y. | 0.013 |
| Medium material | " | 0.015 |
| Wet material | " | 0.016 |

| EARTHWORK | UNIT | MAN/ HOURS |
|---|---|---|
| **02210.30 BULK EXCAVATION** | | |
| Blasted rock | C.Y. | 0.018 |
| **02220.40 BUILDING EXCAVATION** | | |
| Structural excavation, unclassified earth | | |
| 3/8 cy backhoe | C.Y. | 0.107 |
| 3/4 cy backhoe | " | 0.080 |
| 1 cy backhoe | " | 0.067 |
| Foundation backfill and compaction by machine | " | 0.160 |
| **02220.50 UTILITY EXCAVATION** | | |
| Trencher, sandy clay, 8" wide trench | | |
| 18" deep | L.F. | 0.018 |
| 24" deep | " | 0.020 |
| 36" deep | " | 0.023 |
| Trench backfill, 95% compaction | | |
| Tamp by hand | C.Y. | 0.500 |
| Vibratory compaction | " | 0.400 |
| Trench backfilling, with borrow sand, place & compact | " | 0.400 |
| **02220.60 TRENCHING** | | |
| Trenching and continuous footing excavation | | |
| By gradall | | |
| 1 cy capacity | | |
| Light soil | C.Y. | 0.023 |
| Medium soil | " | 0.025 |
| Heavy/wet soil | " | 0.027 |
| Loose rock | " | 0.029 |
| Blasted rock | " | 0.031 |
| By hydraulic excavator | | |
| 1/2 cy capacity | | |
| Light soil | C.Y. | 0.027 |
| Medium soil | " | 0.029 |
| Heavy/wet soil | " | 0.032 |
| Loose rock | " | 0.036 |
| Blasted rock | " | 0.040 |
| Hand excavation | | |
| Bulk, wheeled 100' | | |
| Normal soil | C.Y. | 0.889 |
| Sand or gravel | " | 0.800 |
| Medium clay | " | 1.143 |
| Heavy clay | " | 1.600 |
| Loose rock | " | 2.000 |
| Trenches, up to 2' deep | | |
| Normal soil | C.Y. | 1.000 |
| Sand or gravel | " | 0.889 |
| Medium clay | " | 1.333 |
| Heavy clay | " | 2.000 |
| Loose rock | " | 2.667 |
| Trenches, to 6' deep | | |
| Normal soil | C.Y. | 1.143 |
| Sand or gravel | " | 1.000 |

| EARTHWORK | UNIT | MAN/ HOURS |
|---|---|---|
| **02220.60 TRENCHING** | | |
| Medium clay | C.Y. | 1.600 |
| Heavy clay | " | 2.667 |
| Loose rock | " | 4.000 |
| Backfill trenches | | |
| With compaction | | |
| By hand | C.Y. | 0.667 |
| By 60 hp tracked dozer | " | 0.020 |
| By 200 hp tracked dozer | " | 0.009 |
| By small front-end loader | " | 0.023 |
| Backfill trenches, sand bedding, no compaction | | |
| By hand | C.Y. | 0.667 |
| By small front-end loader | " | 0.023 |
| **02220.90 HAND EXCAVATION** | | |
| Excavation | | |
| To 2' deep | | |
| Normal soil | C.Y. | 0.889 |
| Sand and gravel | " | 0.800 |
| Medium clay | " | 1.000 |
| Heavy clay | " | 1.143 |
| Loose rock | " | 1.333 |
| To 6' deep | | |
| Normal soil | C.Y. | 1.143 |
| Sand and gravel | " | 1.000 |
| Medium clay | " | 1.333 |
| Heavy clay | " | 1.600 |
| Loose rock | " | 2.000 |
| Backfilling foundation without compaction, 6" lifts | " | 0.500 |
| Compaction of backfill around structures or in trench | | |
| By hand with air tamper | C.Y. | 0.571 |
| By hand with vibrating plate tamper | " | 0.533 |
| 1 ton roller | " | 0.400 |
| Miscellaneous hand labor | | |
| Trim slopes, sides of excavation | S.F. | 0.001 |
| Trim bottom of excavation | " | 0.002 |
| Excavation around obstructions and services | C.Y. | 2.667 |
| **02240.05 SOIL STABILIZATION** | | |
| Straw bale secured with rebar | L.F. | 0.027 |
| Filter barrier, 18" high filter fabric | " | 0.080 |
| Sediment fence, 36" fabric with 6" mesh | " | 0.100 |
| **02280.20 SOIL TREATMENT** | | |
| Soil treatment, termite control pretreatment | | |
| Under slabs | S.F. | 0.004 |
| By walls | " | 0.005 |
| **02290.30 WEED CONTROL** | | |
| Weed control, bromicil, 15 lb./acre, wettable powder | ACRE | 4.000 |
| Vegetation control, by application of plant killer | S.Y. | 0.003 |
| Weed killer, lawns and fields | " | 0.002 |

# 02 SITEWORK

## PILES AND CAISSONS

| 02360.50 PRESTRESSED PILING | UNIT | MAN/ HOURS |
|---|---|---|
| Prestressed concrete piling, less than 60' long | | |
| 10" sq. | L.F. | 0.040 |
| 12" sq. | " | 0.042 |
| 14" sq. | " | 0.043 |
| 16" sq. | " | 0.044 |
| 18" sq. | " | 0.047 |
| 20" sq. | " | 0.048 |
| 24" sq. | " | 0.049 |

| 02360.60 STEEL PILES | UNIT | MAN/ HOURS |
|---|---|---|
| H-section piles | | |
| 8x8 | | |
| 36 lb/ft | | |
| 30' long | L.F. | 0.080 |
| 40' long | " | 0.064 |
| 50' long | " | 0.053 |
| 10x10 | | |
| 42 lb/ft | | |
| 30' long | L.F. | 0.080 |
| 40' long | " | 0.064 |
| 50' long | " | 0.053 |
| 57 lb/ft | | |
| 30' long | L.F. | 0.080 |
| 40' long | " | 0.064 |
| 50' long | " | 0.053 |
| Splice | | |
| 8" | EA. | 1.333 |
| 10" | " | 1.600 |

| 02360.65 STEEL PIPE PILES | UNIT | MAN/ HOURS |
|---|---|---|
| Concrete filled, 3000# concrete, up to 40' | | |
| 8" dia. | L.F. | 0.069 |
| 10" dia. | " | 0.071 |
| 12" dia. | " | 0.074 |
| Pipe piles, non-filled | | |
| 8" dia. | L.F. | 0.053 |
| 10" dia. | " | 0.055 |
| 12" dia. | " | 0.056 |
| Splice | | |
| 8" dia. | EA. | 1.600 |
| 10" dia. | " | 1.600 |
| 12" dia. | " | 2.000 |
| Standard point | | |
| 8" dia. | EA. | 1.600 |
| 10" dia. | " | 1.600 |
| 12" dia. | " | 2.000 |
| Heavy duty point | | |
| 8" dia. | EA. | 2.000 |
| 10" dia. | " | 2.000 |
| 12" dia. | " | 2.667 |

| 02360.70 STEEL SHEET PILING | UNIT | MAN/ HOURS |
|---|---|---|
| Steel sheet piling,12" wide | | |
| 20' long | S.F. | 0.096 |
| 35' long | S.F. | 0.069 |

| 02360.80 WOOD AND TIMBER PILES | UNIT | MAN/ HOURS |
|---|---|---|
| Treated wood piles, 12" butt, 8" tip | | |
| 25' long | L.F. | 0.096 |
| 30' long | " | 0.080 |
| 35' long | " | 0.069 |
| 40' long | " | 0.060 |

| 02380.10 CAISSONS | UNIT | MAN/ HOURS |
|---|---|---|
| Caisson, including 3000# concrete, in stable ground | | |
| 18" dia. | L.F. | 0.192 |
| 24" dia. | " | 0.200 |
| Wet ground, casing required but pulled | | |
| 18" dia. | L.F. | 0.240 |
| 24" dia. | " | 0.267 |
| Soft rock | | |
| 18" dia. | L.F. | 0.686 |
| 24" dia. | " | 1.200 |

## PAVING AND SURFACING

| 02510.20 ASPHALT SURFACES | UNIT | MAN/ HOURS |
|---|---|---|
| Asphalt wearing surface, for flexible pavement | | |
| 1" thick | S.Y. | 0.016 |
| 1-1/2" thick | " | 0.019 |
| 2" thick | " | 0.024 |
| 3" thick | " | 0.032 |
| Binder course | | |
| 1-1/2" thick | S.Y. | 0.018 |
| 2" thick | " | 0.022 |
| 3" thick | " | 0.029 |
| 4" thick | " | 0.032 |
| 5" thick | " | 0.036 |
| 6" thick | " | 0.040 |
| Bituminous sidewalk, no base | | |
| 2" thick | S.Y. | 0.028 |
| 3" thick | " | 0.030 |

## PAVING AND SURFACING

### 02520.10 CONCRETE PAVING

| | UNIT | MAN/HOURS |
|---|---|---|
| Concrete paving, reinforced, 5000 psi concrete | | |
| 6" thick | S.Y. | 0.150 |
| 7" thick | " | 0.160 |
| 8" thick | " | 0.171 |
| Concrete paving, for pipe trench, reinforced | | |
| 7" thick | S.Y. | 0.240 |
| 8" thick | " | 0.267 |

### 02545.10 ASPHALT REPAIR

| | UNIT | MAN/HOURS |
|---|---|---|
| Coal tar emulsion seal coat, rubber additive, fuel resistant | S.Y. | 0.011 |
| Bituminous surface treatment, single | " | 0.008 |
| Double | " | 0.001 |
| Bituminous prime coat | " | 0.001 |
| Tack coat | " | 0.001 |
| Crack sealing, concrete paving | L.F. | 0.005 |
| Bituminous paving for pipe trench, 4" thick | S.Y. | 0.160 |
| Polypropylene, nonwoven paving fabric | " | 0.004 |
| Rubberized asphalt | " | 0.073 |
| Asphalt slurry seal | " | 0.047 |

### 02580.10 PAVEMENT MARKINGS

| | UNIT | MAN/HOURS |
|---|---|---|
| Pavement line marking, paint | | |
| 4" wide | L.F. | 0.002 |
| 6" wide | " | 0.004 |
| 8" wide | " | 0.007 |
| Reflective paint, 4" wide | " | 0.007 |
| Preformed tape, 4" wide | | |
| Inlaid reflective | L.F. | 0.001 |
| Reflective paint | " | 0.002 |
| Thermoplastic | | |
| White | L.F. | 0.004 |
| Yellow | " | 0.004 |
| 12" wide, thermoplastic, white | " | 0.011 |
| Directional arrows, reflective preformed tape | EA. | 0.800 |
| Messages, reflective preformed tape (per letter) | " | 0.400 |
| Handicap symbol, preformed tape | " | 0.800 |
| Parking stall painting | " | 0.160 |

## SITE IMPROVEMENTS

### 02830.10 CHAIN LINK FENCE

| | UNIT | MAN/HOURS |
|---|---|---|
| Chain link fence, 9 ga., galvanized, with posts 10' o.c. | | |
| 4' high | L.F. | 0.057 |
| 5' high | " | 0.073 |
| 6' high | " | 0.100 |
| 7' high | " | 0.123 |
| 8' high | L.F. | 0.160 |
| For barbed wire with hangers, add | | |
| 3 strand | L.F. | 0.040 |
| 6 strand | " | 0.067 |
| Corner or gate post, 3" post | | |
| 4' high | EA. | 0.267 |
| 5' high | " | 0.296 |
| 6' high | " | 0.348 |
| 7' high | " | 0.400 |
| 8' high | " | 0.444 |
| 4" post | | |
| 4' high | EA. | 0.296 |
| 5' high | " | 0.348 |
| 6' high | " | 0.400 |
| 7' high | " | 0.444 |
| 8' high | " | 0.500 |
| Gate with gate posts, galvanized, 3' wide | | |
| 4' high | EA. | 2.000 |
| 5' high | " | 2.667 |
| 6' high | " | 2.667 |
| 7' high | " | 4.000 |
| 8' high | " | 4.000 |
| Fabric, galvanized chain link, 2" mesh, 9 ga. | | |
| 4' high | L.F. | 0.027 |
| 5' high | " | 0.032 |
| 6' high | " | 0.040 |
| 8' high | " | 0.053 |
| Line post, no rail fitting, galvanized, 2-1/2" dia. | | |
| 4' high | EA. | 0.229 |
| 5' high | " | 0.250 |
| 6' high | " | 0.267 |
| 7' high | " | 0.320 |
| 8' high | " | 0.400 |
| 1-7/8" H beam | | |
| 4' high | EA. | 0.229 |
| 5' high | " | 0.250 |
| 6' high | " | 0.267 |
| 7' high | " | 0.320 |
| 8' high | " | 0.400 |
| 2-1/4" H beam | | |
| 4' high | EA. | 0.229 |
| 5' high | " | 0.250 |
| 6' high | " | 0.267 |
| 7' high | " | 0.320 |
| 8' high | " | 0.400 |
| Vinyl coated, 9 ga., with posts 10' o.c. | | |
| 4' high | L.F. | 0.057 |
| 5' high | " | 0.073 |
| 6' high | " | 0.100 |
| 7' high | " | 0.123 |
| 8' high | " | 0.160 |
| For barbed wire w/hangers, add | | |
| 3 strand | L.F. | 0.040 |
| 6 Strand | " | 0.067 |

## SITE IMPROVEMENTS

### 02830.10 CHAIN LINK FENCE

| SITE IMPROVEMENTS | UNIT | MAN/ HOURS |
|---|---|---|
| Corner, or gate post, 4' high | | |
| 3" dia. | EA. | 0.267 |
| 4" dia. | " | 0.267 |
| 6" dia. | " | 0.320 |
| Gate, with posts, 3' wide | | |
| 4' high | EA. | 2.000 |
| 5' high | " | 2.667 |
| 6' high | " | 2.667 |
| 7' high | " | 4.000 |
| 8' high | " | 4.000 |
| Line post, no rail fitting, 2-1/2" dia. | | |
| 4' high | EA. | 0.229 |
| 5' high | " | 0.250 |
| 6' high | " | 0.267 |
| 7' high | " | 0.320 |
| 8' high | " | 0.400 |
| Corner post, no top rail fitting, 4" dia. | | |
| 4' high | EA. | 0.267 |
| 5' high | " | 0.296 |
| 6' high | " | 0.348 |
| 7' high | " | 0.400 |
| 8' high | " | 0.444 |
| Fabric, vinyl, chain link, 2" mesh, 9 ga. | | |
| 4' high | L.F. | 0.027 |
| 5' high | " | 0.032 |
| 6' high | " | 0.040 |
| 8' high | " | 0.053 |
| Swing gates, galvanized, 4' high | | |
| Single gate | | |
| 3' wide | EA. | 2.000 |
| 4' wide | " | 2.000 |
| Double gate | | |
| 10' wide | EA. | 3.200 |
| 12' wide | " | 3.200 |
| 14' wide | " | 3.200 |
| 16' wide | " | 3.200 |
| 18' wide | " | 4.571 |
| 20' wide | " | 4.571 |
| 22' wide | " | 4.571 |
| 24' wide | " | 5.333 |
| 26' wide | " | 5.333 |
| 28' wide | " | 6.400 |
| 30' wide | " | 6.400 |
| 5' high | | |
| Single gate | | |
| 3' wide | EA. | 2.667 |
| 4' wide | " | 2.667 |
| Double gate | | |
| 10' wide | EA. | 4.000 |
| 12' wide | " | 4.000 |
| 14' wide | " | 4.000 |
| 16' wide | " | 4.000 |
| 18' wide | " | 4.571 |
| 20' wide | " | 4.571 |

## SITE IMPROVEMENTS

### 02830.10 CHAIN LINK FENCE

| SITE IMPROVEMENTS | UNIT | MAN/ HOURS |
|---|---|---|
| 22' wide | EA. | 4.571 |
| 24' wide | " | 5.333 |
| 26' wide | " | 5.333 |
| 28' wide | " | 6.400 |
| 30' wide | " | 6.400 |
| 6' high | | |
| Single gate | | |
| 3' wide | EA. | 2.667 |
| 4' wide | " | 2.667 |
| Double gate | | |
| 10' wide | EA. | 4.000 |
| 12' wide | " | 4.000 |
| 14' wide | " | 4.000 |
| 16' wide | " | 4.000 |
| 18' wide | " | 4.571 |
| 20' wide | " | 4.571 |
| 22' wide | " | 4.571 |
| 24' wide | " | 5.333 |
| 26' wide | " | 5.333 |
| 28' wide | " | 6.400 |
| 30' wide | " | 6.400 |
| 7' high | | |
| Single gate | | |
| 3' wide | EA. | 4.000 |
| 4' wide | " | 4.000 |
| Double gate | | |
| 10' wide | EA. | 5.333 |
| 12' wide | " | 5.333 |
| 14' wide | " | 5.333 |
| 16' wide | " | 5.333 |
| 18' wide | " | 6.400 |
| 20' wide | " | 6.400 |
| 22' wide | " | 6.400 |
| 24' wide | " | 8.000 |
| 26' wide | " | 8.000 |
| 28' wide | " | 10.000 |
| 30' wide | " | 10.000 |
| 8' high | | |
| Single gate | | |
| 3' wide | EA. | 4.000 |
| 4' wide | " | 4.000 |
| Double gate | | |
| 10' wide | EA. | 5.333 |
| 12' wide | " | 5.333 |
| 14' wide | " | 5.333 |
| 16' wide | " | 5.333 |
| 18' wide | " | 6.400 |
| 20' wide | " | 6.400 |
| 22' wide | " | 6.400 |
| 24' wide | " | 8.000 |
| 26' wide | " | 8.000 |
| 28' wide | " | 10.000 |
| 30' wide | " | 10.000 |
| Vinyl coated swing gates, 4' high | | |

| SITE IMPROVEMENTS | UNIT | MAN/ HOURS |
|---|---|---|
| **02830.10 CHAIN LINK FENCE** | | |
| Single gate | | |
| 3' wide | EA. | 2.000 |
| 4' wide | " | 2.000 |
| Double gate | | |
| 10' wide | EA. | 3.200 |
| 12' wide | " | 3.200 |
| 14' wide | " | 3.200 |
| 16' wide | " | 3.200 |
| 18' wide | " | 4.571 |
| 20' wide | " | 4.571 |
| 22' wide | " | 4.571 |
| 24' wide | " | 5.333 |
| 26' wide | " | 5.333 |
| 28' wide | " | 6.400 |
| 30' wide | " | 6.400 |
| 5' high | | |
| Single gate | | |
| 3' wide | EA. | 2.667 |
| 4' wide | " | 2.667 |
| Double gate | | |
| 10' wide | EA. | 4.000 |
| 12' wide | " | 4.000 |
| 14' wide | " | 4.000 |
| 16' wide | " | 4.000 |
| 18' wide | " | 4.571 |
| 20' wide | " | 4.571 |
| 22' wide | " | 4.571 |
| 24' wide | " | 5.333 |
| 26' wide | " | 5.333 |
| 28' wide | " | 6.400 |
| 30' wide | " | 6.400 |
| 6' high | | |
| Single gate | | |
| 3' wide | EA. | 2.667 |
| 4' wide | " | 2.667 |
| Double gate | | |
| 10' wide | EA. | 4.000 |
| 12' wide | " | 4.000 |
| 14' wide | " | 4.000 |
| 16' wide | " | 4.000 |
| 18' wide | " | 4.571 |
| 20' wide | " | 4.571 |
| 22' wide | " | 4.571 |
| 24' wide | " | 5.333 |
| 26' wide | " | 5.333 |
| 28' wide | " | 6.400 |
| 30' wide | " | 6.400 |
| 7' high | | |
| Single gate | | |
| 3' wide | EA. | 4.000 |
| 4' wide | " | 4.000 |
| Double gate | | |
| 10' wide | EA. | 5.333 |
| 12' wide | " | 5.333 |

| SITE IMPROVEMENTS | UNIT | MAN/ HOURS |
|---|---|---|
| **02830.10 CHAIN LINK FENCE** | | |
| 14' wide | EA. | 5.333 |
| 16' wide | " | 5.333 |
| 18' wide | " | 6.400 |
| 20' wide | " | 6.400 |
| 22' wide | " | 6.400 |
| 24' wide | " | 8.000 |
| 26' wide | " | 8.000 |
| 28' wide | " | 10.000 |
| 30' wide | " | 10.000 |
| 8' high | | |
| Single gate | | |
| 3' wide | EA. | 4.000 |
| 4' wide | " | 4.000 |
| Double gate | | |
| 10' wide | EA. | 5.333 |
| 12' wide | " | 5.333 |
| 14' wide | " | 5.333 |
| 16' wide | " | 5.333 |
| 18' wide | " | 6.400 |
| 20' wide | " | 6.400 |
| 22' wide | " | 6.400 |
| 24' wide | " | 8.000 |
| 28' wide | " | 8.000 |
| 30' wide | " | 10.000 |
| Motor operator for gates, no wiring | " | |
| Drilling fence post holes | | |
| In soil | | |
| By hand | EA. | 0.400 |
| By machine auger | " | 0.200 |
| In rock | | |
| By jackhammer | EA. | 2.007 |
| By rock drill | " | 0.800 |
| Aluminum privacy slats, installed vertically | S.F. | 0.020 |
| Post hole, dig by hand | EA. | 0.533 |
| Set fence post in concrete | " | 0.400 |
| **02840.30 GUARDRAILS** | | |
| Pipe bollard, steel pipe, concrete filled, painted | | |
| 6" dia. | EA. | 0.667 |
| 8" dia. | " | 1.000 |
| 12" dia. | " | 2.667 |
| Corrugated steel, guardrail, galvanized | L.F. | 0.040 |
| End section, wrap around or flared | EA. | 0.800 |
| Timber guardrail, 4" x 8" | L.F. | 0.030 |
| Guard rail, 3 cables, 3/4" dia. | | |
| Steel posts | L.F. | 0.120 |
| Wood posts | " | 0.096 |
| Steel box beam | | |
| 6" x 6" | L.F. | 0.133 |
| 6" x 8" | " | |
| Concrete posts | EA. | 0.400 |
| Barrel type impact barrier | " | 0.800 |
| Light shield, 6' high | L.F. | 0.160 |

| SITE IMPROVEMENTS | UNIT | MAN/ HOURS |
|---|---|---|
| **02840.40 PARKING BARRIERS** | | |
| Timber, treated, 4' long | | |
| 4" x 4" | EA. | 0.667 |
| 6" x 6" | " | 0.800 |
| Precast concrete, 6' long, with dowels | | |
| 12" x 6" | EA. | 0.400 |
| 12" x 8" | " | 0.444 |
| **02870.10 PREFABRICATED PLANTERS** | | |
| Concrete precast, circular | | |
| 24" dia., 18" high | EA. | 0.800 |
| 42" dia., 30" high | " | 1.000 |
| Fiberglass, circular | | |
| 36" dia., 27" high | EA. | 0.400 |
| 60" dia., 39" high | " | 0.444 |
| Tapered, circular | | |
| 24" dia., 36" high | EA. | 0.364 |
| 40" dia., 36" high | " | 0.400 |
| Square | | |
| 2' by 2', 17" high | EA. | 0.364 |
| 4' by 4', 39" high | " | 0.444 |
| Rectangular | | |
| 4' by 1', 18" high | EA. | 0.400 |

# 03 CONCRETE

## FORMWORK

| FORMWORK | UNIT | MAN/HOURS |
|---|---|---|
| **03110.05 BEAM FORMWORK** | | |
| Beam forms, job built | | |
| Beam bottoms | | |
| 1 use | S.F. | 0.133 |
| 5 uses | " | 0.114 |
| Beam sides | | |
| 1 use | S.F. | 0.089 |
| 5 uses | " | 0.073 |
| **03110.15 COLUMN FORMWORK** | | |
| Column, square forms, job built | | |
| 8" x 8" columns | | |
| 1 use | S.F. | 0.160 |
| 5 uses | " | 0.138 |
| 12" x 12" columns | | |
| 1 use | S.F. | 0.145 |
| 5 uses | " | 0.127 |
| 16" x 16" columns | | |
| 1 use | S.F. | 0.133 |
| 5 uses | " | 0.118 |
| Round fiber forms, 1 use | | |
| 10" dia. | L.F. | 0.160 |
| 12" dia. | " | 0.163 |
| 18" dia. | " | 0.190 |
| 24" dia. | " | 0.205 |
| 36" dia. | " | 0.242 |
| **03110.18 CURB FORMWORK** | | |
| Curb forms | | |
| Straight, 6" high | | |
| 1 use | L.F. | 0.080 |
| 5 uses | " | 0.067 |
| Curved, 6" high | | |
| 1 use | L.F. | 0.100 |
| 5 uses | " | 0.082 |
| **03110.20 ELEVATED SLAB FORMWORK** | | |
| Elevated slab formwork | | |
| Slab, with drop panels | | |
| 1 use | S.F. | 0.064 |
| 5 uses | " | 0.055 |
| Floor slab, hung from steel beams | | |
| 1 use | S.F. | 0.062 |
| 5 uses | " | 0.053 |
| Floor slab, with pans or domes | | |
| 1 use | S.F. | 0.073 |
| 5 uses | " | 0.062 |
| Equipment curbs, 12" high | | |
| 1 use | L.F. | 0.080 |
| 5 uses | " | 0.067 |

| FORMWORK | UNIT | MAN/HOURS |
|---|---|---|
| **03110.25 EQUIPMENT PAD FORMWORK** | | |
| Equipment pad, job built | | |
| 1 use | S.F. | 0.100 |
| 2 uses | " | 0.094 |
| 3 uses | " | 0.089 |
| 4 uses | " | 0.084 |
| 5 uses | " | 0.080 |
| **03110.35 FOOTING FORMWORK** | | |
| Wall footings, job built, continuous | | |
| 1 use | S.F. | 0.080 |
| 5 uses | " | 0.067 |
| Column footings, spread | | |
| 1 use | S.F. | 0.100 |
| 5 uses | " | 0.080 |
| **03110.50 GRADE BEAM FORMWORK** | | |
| Grade beams, job built | | |
| 1 use | S.F. | 0.080 |
| 5 uses | " | 0.067 |
| **03110.53 PILE CAP FORMWORK** | | |
| Pile cap forms, job built | | |
| Square | | |
| 1 use | S.F. | 0.100 |
| 5 uses | " | 0.080 |
| Triangular | | |
| 1 use | S.F. | 0.114 |
| 5 uses | " | 0.089 |
| **03110.55 SLAB/MAT FORMWORK** | | |
| Mat foundations, job built | | |
| 1 use | S.F. | 0.100 |
| 5 uses | " | 0.080 |
| Edge forms | | |
| 6" high | | |
| 1 use | L.F. | 0.073 |
| 5 uses | " | 0.062 |
| 12" high | | |
| 1 use | L.F. | 0.080 |
| 5 uses | " | 0.067 |
| Formwork for openings | | |
| 1 use | S.F. | 0.160 |
| 5 uses | " | 0.114 |
| **03110.60 STAIR FORMWORK** | | |
| Stairway forms, job built | | |
| 1 use | S.F. | 0.160 |
| 5 uses | " | 0.114 |
| Stairs, elevated | | |
| 1 use | S.F. | 0.160 |
| 5 uses | " | 0.100 |

## FORMWORK

### 03110.65 WALL FORMWORK

| FORMWORK | UNIT | MAN/HOURS |
|---|---|---|
| Wall forms, exterior, job built | | |
| Up to 8' high wall | | |
| 1 use | S.F. | 0.080 |
| 5 uses | " | 0.067 |
| Over 8' high wall | | |
| 1 use | S.F. | 0.100 |
| 5 uses | " | 0.080 |
| Column pier and pilaster | | |
| 1 use | S.F. | 0.160 |
| 5 uses | " | 0.114 |
| Interior wall forms | | |
| Up to 8' high | | |
| 1 use | S.F. | 0.073 |
| 5 uses | " | 0.062 |
| Over 8' high | | |
| 1 use | S.F. | 0.089 |
| 5 uses | " | 0.073 |
| PVC form liner, per side, smooth finish | | |
| 1 use | S.F. | 0.067 |
| 5 uses | " | 0.053 |

### 03110.90 MISCELLANEOUS FORMWORK

| FORMWORK | UNIT | MAN/HOURS |
|---|---|---|
| Keyway forms (5 uses) | | |
| 2 x 4 | L.F. | 0.040 |
| 2 x 6 | " | 0.044 |
| Bulkheads | | |
| Walls, with keyways | | |
| 2 piece | L.F. | 0.073 |
| 3 piece | " | 0.080 |
| Elevated slab, with keyway | | |
| 2 piece | L.F. | 0.067 |
| 3 piece | " | 0.073 |
| Ground slab, with keyway | | |
| 2 piece | L.F. | 0.057 |
| 3 piece | " | 0.062 |
| Chamfer strips | | |
| Wood | | |
| 1/2" wide | L.F. | 0.018 |
| 3/4" wide | " | 0.018 |
| 1" wide | " | 0.018 |
| PVC | | |
| 1/2" wide | L.F. | 0.018 |
| 3/4" wide | " | 0.018 |
| 1" wide | " | 0.018 |
| Radius | | |
| 1" | L.F. | 0.019 |
| 1-1/2" | " | 0.019 |
| Reglets | | |
| Galvanized steel, 24 ga. | L.F. | 0.032 |
| Metal formwork | | |
| Straight edge forms | | |
| 4" high | L.F. | 0.050 |
| 6" high | " | 0.053 |
| 8" high | " | 0.057 |
| 12" high | L.F. | 0.062 |
| 16" high | " | 0.067 |
| Curb form, S-shape | | |
| 12" x | | |
| 1'-6" | L.F. | 0.114 |
| 2' | " | 0.107 |
| 2'-6" | " | 0.100 |
| 3' | " | 0.089 |

## REINFORCEMENT

### 03210.05 BEAM REINFORCING

| REINFORCEMENT | UNIT | MAN/HOURS |
|---|---|---|
| Beam-girders | | |
| #3 - #4 | TON | 20.000 |
| #5 - #6 | " | 16.000 |
| #7 - #8 | " | 13.333 |
| Galvanized | | |
| #3 - #4 | TON | 20.000 |
| #5 - #6 | " | 16.000 |
| #7 - #8 | " | 13.333 |
| Bond Beams | | |
| #3 - #4 | TON | 26.667 |
| #5 - #6 | " | 20.000 |
| #7 - #8 | " | 17.778 |
| Galvanized | | |
| #3 - #4 | TON | 26.667 |
| #5 - #6 | " | 20.000 |
| #7 - #8 | " | 17.778 |

### 03210.15 COLUMN REINFORCING

| REINFORCEMENT | UNIT | MAN/HOURS |
|---|---|---|
| Columns | | |
| #3 - #4 | TON | 22.857 |
| #5 - #6 | " | 17.778 |
| #7 - #8 | " | 16.000 |
| Galvanized | | |
| #3 - #4 | TON | 22.857 |
| #5 - #6 | " | 17.778 |
| #7 - #8 | " | 16.000 |

### 03210.20 ELEVATED SLAB REINFORCING

| REINFORCEMENT | UNIT | MAN/HOURS |
|---|---|---|
| Elevated slab | | |
| #3 - #4 | TON | 10.000 |
| #5 - #6 | " | 8.889 |
| #7 - #8 | " | 8.000 |
| Galvanized | | |
| #3 - #4 | TON | 10.000 |
| #5 - #6 | " | 8.889 |

# 03 CONCRETE

| REINFORCEMENT | UNIT | MAN/ HOURS |
|---|---|---|
| **03210.25 EQUIP. PAD REINFORCING** | | |
| Equipment pad | | |
| #3 - #4 | TON | 16.000 |
| #5 - #6 | " | 14.545 |
| #7 - #8 | " | 13.333 |
| #9 - #10 | " | 12.308 |
| #11 - #12 | " | 11.429 |
| **03210.35 FOOTING REINFORCING** | | |
| Footings | | |
| Grade 50 | | |
| #3 - #4 | TON | 13.333 |
| #5 - #6 | " | 11.429 |
| #7 - #8 | " | 10.000 |
| #9 - #10 | " | 8.889 |
| Grade 60 | | |
| #3 - #4 | TON | 13.333 |
| #5 - #6 | " | 11.429 |
| #7 - #8 | " | 10.000 |
| Straight dowels, 24" long | | |
| 1" dia. (#8) | EA. | 0.080 |
| 3/4" dia. (#6) | " | 0.080 |
| 5/8" dia. (#5) | " | 0.067 |
| 1/2" dia. (#4) | " | 0.057 |
| **03210.45 FOUNDATION REINFORCING** | | |
| Foundations | | |
| #3 - #4 | TON | 13.333 |
| #5 - #6 | " | 11.429 |
| #7 - #8 | " | 10.000 |
| Galvanized | | |
| #3 - #4 | TON | 13.333 |
| #5 - #6 | " | 11.429 |
| #7 - #8 | " | 10.000 |
| **03210.50 GRADE BEAM REINFORCING** | | |
| Grade beams | | |
| #3 - #4 | TON | 12.308 |
| #5 - #6 | " | 10.667 |
| #7 - #8 | " | 9.412 |
| Galvanized | | |
| #3 - #4 | TON | 12.308 |
| #5 - #6 | " | 10.667 |
| #7 - #8 | " | 9.412 |
| **03210.53 PILE CAP REINFORCING** | | |
| Pile caps | | |
| #3 - #4 | TON | 20.000 |
| #5 - #6 | " | 17.778 |
| #7 - #8 | " | 16.000 |
| Galvanized | | |
| #3 - #4 | TON | 20.000 |
| #5 - #6 | " | 17.778 |
| #7 - #8 | " | 16.000 |

| REINFORCEMENT | UNIT | MAN/ HOURS |
|---|---|---|
| **03210.55 SLAB/MAT REINFORCING** | | |
| Bars, slabs | | |
| #3 - #4 | TON | 13.333 |
| #5 - #6 | " | 11.429 |
| #7 - #8 | " | 10.000 |
| Galvanized | | |
| #3 - #4 | TON | 13.333 |
| #5 - #6 | " | 11.429 |
| #7 - #8 | " | 10.000 |
| Wire mesh, slabs | | |
| Galvanized | | |
| 4x4 | | |
| W1.4xW1.4 | S.F. | 0.005 |
| W2.0xW2.0 | " | 0.006 |
| W2.9xW2.9 | " | 0.006 |
| W4.0xW4.0 | " | 0.007 |
| 6x6 | | |
| W1.4xW1.4 | S.F. | 0.004 |
| W2.0xW2.0 | " | 0.004 |
| W2.9xW2.9 | " | 0.005 |
| W4.0xW4.0 | " | 0.005 |
| Standard | | |
| 2x2 | | |
| W.9xW.9 | S.F. | 0.005 |
| 4x4 | | |
| W1.4xW1.4 | S.F. | 0.005 |
| W4.0xW4.0 | " | 0.007 |
| 6x6 | | |
| W1.4xW1.4 | S.F. | 0.004 |
| W4.0xW4.0 | " | 0.005 |
| **03210.60 STAIR REINFORCING** | | |
| Stairs | | |
| #3 - #4 | TON | 16.000 |
| #5 - #6 | " | 13.333 |
| Galvanized | | |
| #3 - #4 | TON | 16.000 |
| #5 - #6 | " | 13.333 |
| **03210.65 WALL REINFORCING** | | |
| Walls | | |
| #3 - #4 | TON | 11.429 |
| #5 - #6 | " | 10.000 |
| #7 - #8 | " | 8.889 |
| Galvanized | | |
| #3 - #4 | TON | 11.429 |
| #5 - #6 | " | 10.000 |
| #7 - #8 | " | 8.889 |
| Masonry wall (horizontal) | | |
| #3 - #4 | TON | 32.000 |
| #5 - #6 | " | 26.667 |
| Galvanized | | |
| #3 - #4 | TON | 32.000 |
| #5 - #6 | " | 26.667 |

## REINFORCEMENT

### 03210.65 WALL REINFORCING

| | UNIT | MAN/HOURS |
|---|---|---|
| Masonry wall (vertical) | | |
| #3 - #4 | TON | 40.000 |
| #5 - #6 | " | 32.000 |
| Galvanized | | |
| #3 - #4 | TON | 40.000 |
| #5 - #6 | " | 32.000 |

## ACCESSORIES

### 03250.40 CONCRETE ACCESSORIES

| | UNIT | MAN/HOURS |
|---|---|---|
| Expansion joint, poured | | |
| Asphalt | | |
| 1/2" x 1" | L.F. | 0.016 |
| 1" x 2" | " | 0.017 |
| Liquid neoprene, cold applied | | |
| 1/2" x 1" | L.F. | 0.016 |
| 1" x 2" | " | 0.018 |
| Polyurethane, 2 parts | | |
| 1/2" x 1" | L.F. | 0.027 |
| 1" x 2" | " | 0.029 |
| Rubberized asphalt, cold | | |
| 1/2" x 1" | L.F. | 0.016 |
| 1" x 2" | " | 0.017 |
| Hot, fuel resistant | | |
| 1/2" x 1" | L.F. | 0.016 |
| 1" x 2" | " | 0.017 |
| Expansion joint, premolded, in slabs | | |
| Asphalt | | |
| 1/2" x 6" | L.F. | 0.020 |
| 1" x 12" | " | 0.027 |
| Cork | | |
| 1/2" x 6" | L.F. | 0.020 |
| 1" x 12" | " | 0.027 |
| Neoprene sponge | | |
| 1/2" x 6" | L.F. | 0.020 |
| 1" x 12" | " | 0.027 |
| Polyethylene foam | | |
| 1/2" x 6" | L.F. | 0.020 |
| 1" x 12" | " | 0.027 |
| Polyurethane foam | | |
| 1/2" x 6" | L.F. | 0.020 |
| 1" x 12" | " | 0.027 |
| Polyvinyl chloride foam | | |
| 1/2" x 6" | L.F. | 0.020 |
| 1" x 12" | " | 0.027 |
| Rubber, gray sponge | | |
| 1/2" x 6" | L.F. | 0.020 |

## ACCESSORIES

### 03250.40 CONCRETE ACCESSORIES

| | UNIT | MAN/HOURS |
|---|---|---|
| 1" x 12" | L.F. | 0.027 |
| Asphalt felt control joints or bond breaker, screed joints | | |
| 4" slab | L.F. | 0.016 |
| 6" slab | " | 0.018 |
| 8" slab | " | 0.020 |
| 10" slab | " | 0.023 |
| Keyed cold expansion and control joints, 24 ga. | | |
| 4" slab | L.F. | 0.050 |
| 5" slab | " | 0.050 |
| 6" slab | " | 0.053 |
| 8" slab | " | 0.057 |
| 10" slab | " | 0.062 |
| Waterstops | | |
| Polyvinyl chloride | | |
| Ribbed | | |
| 3/16" thick x | | |
| 4" wide | L.F. | 0.040 |
| 6" wide | " | 0.044 |
| 1/2" thick x | | |
| 9" wide | L.F. | 0.050 |
| Ribbed with center bulb | | |
| 3/16" thick x 9" wide | L.F. | 0.050 |
| 3/8" thick x 9" wide | " | 0.050 |
| Dumbbell type, 3/8" thick x 6" wide | " | 0.044 |
| Plain, 3/8" thick x 9" wide | " | 0.050 |
| Center bulb, 3/8" thick x 9" wide | " | 0.050 |
| Rubber | | |
| Vapor barrier | | |
| 4 mil polyethylene | S.F. | 0.003 |
| 6 mil polyethylene | " | 0.003 |
| Gravel porous fill, under floor slabs, 3/4" stone | C.Y. | 1.333 |

## CAST-IN-PLACE CONCRETE

### 03350.10 CONCRETE FINISHES

| | UNIT | MAN/HOURS |
|---|---|---|
| Floor finishes | | |
| Broom | S.F. | 0.011 |
| Screed | " | 0.010 |
| Darby | " | 0.010 |
| Steel float | " | 0.013 |
| Wall finishes | | |
| Burlap rub, with cement paste | S.F. | 0.013 |
| Break ties and patch holes | " | 0.016 |
| Carborundum | | |
| Dry rub | S.F. | 0.027 |
| Wet rub | " | 0.040 |

# 03 CONCRETE

## CAST-IN-PLACE CONCRETE

| 03350.10 CONCRETE FINISHES | UNIT | MAN/HOURS |
|---|---|---|
| Floor hardeners | | |
| Metallic | | |
| Light service | S.F. | 0.010 |
| Heavy service | " | 0.013 |
| Non-metallic | | |
| Light service | S.F. | 0.010 |
| Heavy service | " | 0.013 |

| 03360.10 PNEUMATIC CONCRETE | UNIT | MAN/HOURS |
|---|---|---|
| Pneumatic applied concrete (gunite) | | |
| 2" thick | S.F. | 0.030 |
| 3" thick | " | 0.040 |
| 4" thick | " | 0.048 |
| Finish surface | | |
| Minimum | S.F. | 0.040 |
| Maximum | " | 0.080 |

| 03370.10 CURING CONCRETE | UNIT | MAN/HOURS |
|---|---|---|
| Sprayed membrane | | |
| Slabs | S.F. | 0.002 |
| Walls | " | 0.002 |
| Curing paper | | |
| Slabs | S.F. | 0.002 |
| Walls | " | 0.002 |
| Burlap | | |
| 7.5 oz. | S.F. | 0.003 |
| 12 oz. | " | 0.003 |

## PLACING CONCRETE

| 03380.05 BEAM CONCRETE | UNIT | MAN/HOURS |
|---|---|---|
| Beams and girders | | |
| 2500# or 3000# concrete | | |
| By pump | C.Y. | 0.873 |
| By hand buggy | " | 0.800 |
| 5000# concrete | | |
| By pump | C.Y. | 0.873 |
| By hand buggy | " | 0.800 |
| Bond beam, 3000# concrete | | |
| By pump | | |
| 8" high | | |
| 4" wide | L.F. | 0.019 |
| 6" wide | " | 0.022 |
| 8" wide | " | 0.024 |
| 10" wide | L.F. | 0.027 |
| 12" wide | " | 0.030 |

| 03380.15 COLUMN CONCRETE | UNIT | MAN/HOURS |
|---|---|---|
| Columns | | |
| 2500# or 3000# concrete | | |
| By pump | C.Y. | 0.800 |
| 5000# concrete | | |
| By pump | C.Y. | 0.800 |
| 5000# concrete | | |

| 03380.25 EQUIPMENT PAD CONCRETE | UNIT | MAN/HOURS |
|---|---|---|
| Equipment pad | | |
| 2500# or 3000# concrete | | |
| By chute | C.Y. | 0.267 |
| By pump | " | 0.686 |
| 3500# or 4000# concrete | | |
| By chute | C.Y. | 0.267 |
| By pump | " | 0.686 |
| 5000# concrete | | |
| By chute | C.Y. | 0.267 |
| By pump | " | 0.686 |

| 03380.35 FOOTING CONCRETE | UNIT | MAN/HOURS |
|---|---|---|
| Continuous footing | | |
| 2500# or 3000# concrete | | |
| By chute | C.Y. | 0.267 |
| By pump | " | 0.600 |
| 5000# concrete | | |
| By chute | C.Y. | 0.267 |
| By pump | " | 0.600 |
| Spread footing | | |
| 2500# or 3000# concrete | | |
| Under 5 cy | | |
| By chute | C.Y. | 0.267 |
| By pump | " | 0.640 |
| 5000# concrete | | |
| Under 5 c.y. | | |
| By chute | C.Y. | 0.267 |
| By pump | " | 0.640 |

| 03380.50 GRADE BEAM CONCRETE | UNIT | MAN/HOURS |
|---|---|---|
| Grade beam | | |
| 2500# or 3000# concrete | | |
| By chute | C.Y. | 0.267 |
| By pump | " | 0.600 |
| By hand buggy | " | 0.800 |
| 5000# concrete | | |
| By chute | C.Y. | 0.267 |
| By pump | " | 0.600 |
| By hand buggy | " | 0.800 |

## PLACING CONCRETE

| PLACING CONCRETE | UNIT | MAN/ HOURS |
|---|---|---|
| **03380.53 PILE CAP CONCRETE** | | |
| Pile cap | | |
| 2500# or 3000 concrete | | |
| By chute | C.Y. | 0.267 |
| By pump | " | 0.686 |
| By hand buggy | " | 0.800 |
| 5000# concrete | | |
| By chute | C.Y. | 0.267 |
| By pump | " | 0.686 |
| By hand buggy | " | 0.800 |
| **03380.55 SLAB/MAT CONCRETE** | | |
| Slab on grade | | |
| 2500# or 3000# concrete | | |
| By chute | C.Y. | 0.200 |
| By pump | " | 0.343 |
| By hand buggy | " | 0.533 |
| 5000# concrete | | |
| By chute | C.Y. | 0.200 |
| By pump | " | 0.343 |
| By hand buggy | " | 0.533 |
| **03380.58 SIDEWALKS** | | |
| Walks, cast in place with wire mesh, base not incl. | | |
| 4" thick | S.F. | 0.027 |
| 5" thick | " | 0.032 |
| 6" thick | " | 0.040 |
| **03380.60 STAIR CONCRETE** | | |
| Stairs | | |
| 2500# or 3000# concrete | | |
| By chute | C.Y. | 0.267 |
| By pump | " | 0.686 |
| By hand buggy | " | 0.800 |
| 3500# or 4000# concrete | | |
| By chute | C.Y. | 0.267 |
| By pump | " | 0.686 |
| By hand buggy | " | 0.800 |
| 5000# concrete | | |
| By chute | C.Y. | 0.267 |
| By pump | " | 0.686 |
| By hand buggy | " | 0.800 |
| **03380.65 WALL CONCRETE** | | |
| Walls | | |
| 2500# or 3000# concrete | | |
| To 4' | | |
| By chute | C.Y. | 0.229 |
| By pump | " | 0.738 |
| To 8' | | |
| By pump | C.Y. | 0.800 |
| 3500# or 4000# concrete | | |

| PLACING CONCRETE | UNIT | MAN/ HOURS |
|---|---|---|
| **03380.65 WALL CONCRETE** | | |
| To 4' | | |
| By chute | C.Y. | 0.229 |
| By pump | " | 0.738 |
| To 8' | | |
| By pump | C.Y. | 0.800 |
| Filled block (CMU) | | |
| 3000# concrete, by pump | | |
| 4" wide | S.F. | 0.034 |
| 6" wide | " | 0.040 |
| 8" wide | " | 0.048 |
| 10" wide | " | 0.056 |
| 12" wide | " | 0.069 |
| Pilasters, 3000# concrete | C.F. | 0.960 |
| Wall cavity, 2" thick, 3000# concrete | S.F. | 0.032 |

## CEMENTITOUS TOPPINGS

| CEMENTITOUS TOPPINGS | UNIT | MAN/ HOURS |
|---|---|---|
| **03550.10 CONCRETE TOPPINGS** | | |
| Gypsum fill | | |
| 2" thick | S.F. | 0.005 |
| 2-1/2" thick | " | 0.005 |
| 3" thick | " | 0.005 |
| 3-1/2" thick | " | 0.005 |
| 4" thick | " | 0.006 |
| Formboard | | |
| Mineral fiber board | | |
| 1" thick | S.F. | 0.020 |
| 1-1/2" thick | " | 0.023 |
| Cement fiber board | | |
| 1" thick | S.F. | 0.027 |
| 1-1/2" thick | " | 0.031 |
| Glass fiber board | | |
| 1" thick | S.F. | 0.020 |
| 1-1/2" thick | " | 0.023 |
| Poured deck | | |
| Vermiculite or perlite | | |
| 1 to 4 mix | C.Y. | 0.800 |
| 1 to 6 mix | " | 0.738 |
| Vermiculite or perlite | | |
| 2" thick | | |
| 1 to 4 mix | S.F. | 0.005 |
| 1 to 6 mix | " | 0.005 |
| 3" thick | | |
| 1 to 4 mix | S.F. | 0.007 |
| 1 to 6 mix | " | 0.007 |
| Concrete plank, lightweight | | |
| 2" thick | S.F. | 0.024 |

# 03 CONCRETE

## CEMENTITOUS TOPPINGS

### 03550.10 CONCRETE TOPPINGS

| CEMENTITOUS TOPPINGS | UNIT | MAN/HOURS |
|---|---|---|
| 2-1/2" thick | S.F. | 0.024 |
| 3-1/2" thick | " | 0.027 |
| 4" thick | " | 0.027 |
| Channel slab, lightweight, straight | | |
| 2-3/4" thick | S.F. | 0.024 |
| 3-1/2" thick | " | 0.024 |
| 3-3/4" thick | " | 0.024 |
| 4-3/4" thick | " | 0.027 |
| Gypsum plank | | |
| 2" thick | S.F. | 0.024 |
| 3" thick | " | 0.024 |
| Cement fiber, T and G planks | | |
| 1" thick | S.F. | 0.022 |
| 1-1/2" thick | " | 0.022 |
| 2" thick | " | 0.024 |
| 2-1/2" thick | " | 0.024 |
| 3" thick | " | 0.024 |
| 3-1/2" thick | " | 0.027 |
| 4" thick | " | 0.027 |

## CONCRETE RESTORATION

### 03730.10 CONCRETE REPAIR

| CONCRETE RESTORATION | UNIT | MAN/HOURS |
|---|---|---|
| Epoxy grout floor patch, 1/4" thick | S.F. | 0.080 |
| Epoxy gel grout | " | 0.800 |
| Injection valve, 1 way, threaded plastic | EA. | 0.160 |
| Grout crack seal, 2 component | C.F. | 0.800 |
| Grout, non shrink | " | 0.800 |
| Concrete, epoxy modified | | |
| Sand mix | C.F. | 0.320 |
| Gravel mix | " | 0.296 |
| Concrete repair | | |
| Soffit repair | | |
| 16" wide | L.F. | 0.160 |
| 18" wide | " | 0.167 |
| 24" wide | " | 0.178 |
| 30" wide | " | 0.190 |
| 32" wide | " | 0.200 |
| Edge repair | | |
| 2" spall | L.F. | 0.200 |
| 3" spall | " | 0.211 |
| 4" spall | " | 0.216 |
| 6" spall | " | 0.222 |
| 8" spall | " | 0.235 |
| 9" spall | " | 0.267 |
| Crack repair, 1/8" crack | " | 0.080 |

## CONCRETE RESTORATION

### 03730.10 CONCRETE REPAIR

| CONCRETE RESTORATION | UNIT | MAN/HOURS |
|---|---|---|
| Reinforcing steel repair | | |
| 1 bar, 4 ft | | |
| #4 bar | L.F. | 0.100 |
| #5 bar | " | 0.100 |
| #6 bar | " | 0.107 |
| #8 bar | " | 0.107 |
| #9 bar | " | 0.114 |
| #11 bar | " | 0.114 |
| Pile repairs | | |
| Polyethylene wrap | | |
| 30 mil thick | | |
| 60" wide | S.F. | 0.267 |
| 72" wide | " | 0.320 |
| 60 mil thick | | |
| 60" wide | S.F. | 0.267 |
| 80" wide | " | 0.364 |
| Pile spall, average repair 3' | | |
| 18" x 18" | EA. | 0.667 |
| 20" x 20" | " | 0.800 |

# 04 MASONRY

| MORTAR AND GROUT | UNIT | MAN/ HOURS |
|---|---|---|
| **04100.10 MASONRY GROUT** | | |
| Grout, non shrink, non-metallic, trowelable | C.F. | 0.016 |
| Grout door frame, hollow metal | | |
| Single | EA. | 0.600 |
| Double | " | 0.632 |
| Grout-filled concrete block (CMU) | | |
| 4" wide | S.F. | 0.020 |
| 6" wide | " | 0.022 |
| 8" wide | " | 0.024 |
| 12" wide | " | 0.025 |
| Grout-filled individual CMU cells | | |
| 4" wide | L.F. | 0.012 |
| 6" wide | " | 0.012 |
| 8" wide | " | 0.012 |
| 10" wide | " | 0.014 |
| 12" wide | " | 0.014 |
| Bond beams or lintels, 8" deep | | |
| 6" thick | L.F. | 0.022 |
| 8" thick | " | 0.024 |
| 10" thick | " | 0.027 |
| 12" thick | " | 0.030 |
| Cavity walls | | |
| 2" thick | S.F. | 0.032 |
| 3" thick | " | 0.032 |
| 4" thick | " | 0.034 |
| 6" thick | " | 0.040 |
| **04150.10 MASONRY ACCESSORIES** | | |
| Foundation vents | EA. | 0.320 |
| Bar reinforcing | | |
| Horizontal | | |
| #3 - #4 | Lb. | 0.032 |
| #5 - #6 | " | 0.027 |
| Vertical | | |
| #3 - #4 | Lb. | 0.040 |
| #5 - #6 | " | 0.032 |
| Horizontal joint reinforcing | | |
| Truss type | | |
| 4" wide, 6" wall | L.F. | 0.003 |
| 6" wide, 8" wall | " | 0.003 |
| 8" wide, 10" wall | " | 0.003 |
| 10" wide, 12" wall | " | 0.004 |
| 12" wide, 14" wall | " | 0.004 |
| Ladder type | | |
| 4" wide, 6" wall | L.F. | 0.003 |
| 6" wide, 8" wall | " | 0.003 |
| 8" wide, 10" wall | " | 0.003 |
| 10" wide, 12" wall | " | 0.003 |
| Rectangular wall ties | | |
| 3/16" dia., galvanized | | |
| 2" x 6" | EA. | 0.013 |
| 2" x 8" | " | 0.013 |
| 2" x 10" | " | 0.013 |
| 2" x 12" | " | 0.013 |

| MORTAR AND GROUT | UNIT | MAN/ HOURS |
|---|---|---|
| **04150.10 MASONRY ACCESSORIES** | | |
| 4" x 6" | EA. | 0.016 |
| 4" x 8" | " | 0.016 |
| 4" x 10" | " | 0.016 |
| 4" x 12" | " | 0.016 |
| 1/4" dia., galvanized | | |
| 2" x 6" | EA. | 0.013 |
| 2" x 8" | " | 0.013 |
| 2" x 10" | " | 0.013 |
| 2" x 12" | " | 0.013 |
| 4" x 6" | " | 0.016 |
| 4" x 8" | " | 0.016 |
| 4" x 10" | " | 0.016 |
| 4" x 12" | " | 0.016 |
| "Z" type wall ties, galvanized | | |
| 6" long | | |
| 1/8" dia. | EA. | 0.013 |
| 3/16" dia. | " | 0.013 |
| 1/4" dia. | " | 0.013 |
| 8" long | | |
| 1/8" dia. | EA. | 0.013 |
| 3/16" dia. | " | 0.013 |
| 1/4" dia. | " | 0.013 |
| 10" long | | |
| 1/8" dia. | EA. | 0.013 |
| 3/16" dia. | " | 0.013 |
| 1/4" dia. | " | 0.013 |
| Dovetail anchor slots | | |
| Galvanized steel, filled | | |
| 24 ga. | L.F. | 0.020 |
| 20 ga. | " | 0.020 |
| 16 oz. copper, foam filled | " | 0.020 |
| Dovetail anchors | | |
| 16 ga. | | |
| 3-1/2" long | EA. | 0.013 |
| 5-1/2" long | " | 0.013 |
| 12 ga. | | |
| 3-1/2" long | EA. | 0.013 |
| 5-1/2" long | " | 0.013 |
| Dovetail, triangular galvanized ties, 12 ga. | | |
| 3" x 3" | EA. | 0.013 |
| 5" x 5" | " | 0.013 |
| 7" x 7" | " | 0.013 |
| 7" x 9" | " | 0.013 |
| Brick anchors | | |
| Corrugated, 3-1/2" long | | |
| 16 ga. | EA. | 0.013 |
| 12 ga. | " | 0.013 |
| Non-corrugated, 3-1/2" long | | |
| 16 ga. | EA. | 0.013 |
| 12 ga. | " | 0.013 |
| Cavity wall anchors, corrugated, galvanized | | |
| 5" long | | |
| 16 ga. | EA. | 0.013 |
| 12 ga. | " | 0.013 |

# 04 MASONRY

## MORTAR AND GROUT

### 04150.10 MASONRY ACCESSORIES

| | UNIT | MAN/HOURS |
|---|---|---|
| 7" long | | |
| 28 ga. | EA. | 0.013 |
| 24 ga. | " | 0.013 |
| 22 ga. | " | 0.013 |
| 16 ga. | " | 0.013 |
| Mesh ties, 16 ga., 3" wide | | |
| 8" long | EA. | 0.013 |
| 12" long | " | 0.013 |
| 20" long | " | 0.013 |
| 24" long | " | 0.013 |

### 04150.20 MASONRY CONTROL JOINTS

| | UNIT | MAN/HOURS |
|---|---|---|
| Control joint, cross shaped PVC | L.F. | 0.020 |
| Closed cell joint filler | | |
| 1/2" | L.F. | 0.020 |
| 3/4" | " | 0.020 |
| Rubber, for | | |
| 4" wall | L.F. | 0.020 |
| 6" wall | " | 0.021 |
| 8" wall | " | 0.022 |
| PVC, for | | |
| 4" wall | L.F. | 0.020 |
| 6" wall | " | 0.021 |
| 8" wall | " | 0.022 |

### 04150.50 MASONRY FLASHING

| | UNIT | MAN/HOURS |
|---|---|---|
| Through-wall flashing | | |
| 5 oz. coated copper | S.F. | 0.067 |
| 0.030" elastomeric | " | 0.053 |

## UNIT MASONRY

### 04210.10 BRICK MASONRY

| | UNIT | MAN/HOURS |
|---|---|---|
| Standard size brick, running bond | | |
| Face brick, red (6.4/sf) | | |
| Veneer | S.F. | 0.133 |
| Cavity wall | " | 0.114 |
| 9" solid wall | " | 0.229 |
| Common brick (6.4/sf) | | |
| Select common for veneers | S.F. | 0.133 |
| Back-up | | |
| 4" thick | S.F. | 0.100 |
| 8" thick | " | 0.160 |
| Firewall | | |
| 12" thick | S.F. | 0.267 |
| 16" thick | " | 0.364 |

## UNIT MASONRY

### 04210.10 BRICK MASONRY

| | UNIT | MAN/HOURS |
|---|---|---|
| Glazed brick (7.4/sf) | | |
| Veneer | S.F. | 0.145 |
| Buff or gray face brick (6.4/sf) | | |
| Veneer | S.F. | 0.133 |
| Cavity wall | " | 0.114 |
| Jumbo or oversize brick (3/sf) | | |
| 4" veneer | S.F. | 0.080 |
| 4" back-up | " | 0.067 |
| 8" back-up | " | 0.114 |
| 12" firewall | " | 0.200 |
| 16" firewall | " | 0.267 |
| Norman brick, red face, (4.5/sf) | | |
| 4" veneer | S.F. | 0.100 |
| Cavity wall | " | 0.089 |
| Chimney, standard brick, including flue | | |
| 16" x 16" | L.F. | 0.800 |
| 16" x 20" | " | 0.800 |
| 16" x 24" | " | 0.800 |
| 20" x 20" | " | 1.000 |
| 20" x 24" | " | 1.000 |
| 20" x 32" | " | 1.143 |
| Window sill, face brick on edge | " | 0.200 |

### 04210.20 STRUCTURAL TILE

| | UNIT | MAN/HOURS |
|---|---|---|
| Structural glazed tile | | |
| 6T series, 5-1/2" x 12" | | |
| Glazed on one side | | |
| 2" thick | S.F. | 0.080 |
| 4" thick | " | 0.080 |
| 6" thick | " | 0.089 |
| 8" thick | " | 0.100 |
| Glazed on two sides | | |
| 4" thick | S.F. | 0.100 |
| 6" thick | " | 0.114 |
| Special shapes | | |
| Group 1 | S.F. | 0.160 |
| Group 2 | " | 0.160 |
| Group 3 | " | 0.160 |
| Group 4 | " | 0.160 |
| Group 5 | " | 0.160 |
| Fire rated | | |
| 4" thick, 1 hr rating | S.F. | 0.080 |
| 6" thick, 2 hr rating | " | 0.089 |
| 8W series, 8" x 16" | | |
| Glazed on one side | | |
| 2" thick | S.F. | 0.053 |
| 4" thick | " | 0.053 |
| 6" thick | " | 0.062 |
| 8" thick | " | 0.062 |
| Glazed on two sides | | |
| 4" thick | S.F. | 0.067 |
| 6" thick | " | 0.080 |
| 8" thick | " | 0.080 |
| Special shapes | | |

# 04 MASONRY

## UNIT MASONRY

### 04210.20 STRUCTURAL TILE

| UNIT MASONRY | UNIT | MAN/ HOURS |
|---|---|---|
| Group 1 | S.F. | 0.114 |
| Group 2 | " | 0.114 |
| Group 3 | " | 0.114 |
| Group 4 | " | 0.114 |
| Group 5 | " | 0.114 |
| Fire rated | | |
| 4" thick, 1 hr rating | S.F. | 0.114 |
| 6" thick, 2 hr rating | " | 0.114 |

### 04210.60 PAVERS, MASONRY

| UNIT MASONRY | UNIT | MAN/ HOURS |
|---|---|---|
| Brick walk laid on sand, sand joints | | |
| Laid flat, (4.5 per sf) | S.F. | 0.089 |
| Laid on edge, (7.2 per sf) | " | 0.133 |
| Precast concrete patio blocks | | |
| 2" thick | | |
| Natural | S.F. | 0.027 |
| Colors | " | 0.027 |
| Exposed aggregates, local aggregate | | |
| Natural | S.F. | 0.027 |
| Colors | " | 0.027 |
| Granite or limestone aggregate | " | 0.027 |
| White tumblestone aggregate | " | 0.027 |
| Stone pavers, set in mortar | | |
| Bluestone | | |
| 1" thick | | |
| Irregular | S.F. | 0.200 |
| Snapped rectangular | " | 0.160 |
| 1-1/2" thick, random rectangular | " | 0.200 |
| 2" thick, random rectangular | " | 0.229 |
| Slate | | |
| Natural cleft | | |
| Irregular, 3/4" thick | S.F. | 0.229 |
| Random rectangular | | |
| 1-1/4" thick | S.F. | 0.200 |
| 1-1/2" thick | " | 0.222 |
| Granite blocks | | |
| 3" thick, 3" to 6" wide | | |
| 4" to 12" long | S.F. | 0.267 |
| 6" to 15" long | " | 0.229 |
| Crushed stone, white marble, 3" thick | " | 0.016 |

### 04220.10 CONCRETE MASONRY UNITS

| UNIT MASONRY | UNIT | MAN/ HOURS |
|---|---|---|
| Hollow, load bearing | | |
| 4" | S.F. | 0.059 |
| 6" | " | 0.062 |
| 8" | " | 0.067 |
| 10" | " | 0.073 |
| 12" | " | 0.080 |
| Solid, load bearing | | |
| 4" | S.F. | 0.059 |
| 6" | " | 0.062 |
| 8" | " | 0.067 |
| 10" | " | 0.073 |
| 12" | " | 0.080 |

## UNIT MASONRY

### 04220.10 CONCRETE MASONRY UNITS

| UNIT MASONRY | UNIT | MAN/ HOURS |
|---|---|---|
| Back-up block, 8" x 16" | | |
| 2" | S.F. | 0.046 |
| 4" | " | 0.047 |
| 6" | " | 0.050 |
| 8" | " | 0.053 |
| 10" | " | 0.057 |
| 12" | " | 0.062 |
| Foundation wall, 8" x 16" | | |
| 6" | S.F. | 0.057 |
| 8" | " | 0.062 |
| 10" | " | 0.067 |
| 12" | " | 0.073 |
| Solid | | |
| 6" | S.F. | 0.062 |
| 8" | " | 0.067 |
| 10" | " | 0.073 |
| 12" | " | 0.080 |
| Exterior, styrofoam inserts, standard weight, 8" x 16" | | |
| 6" | S.F. | 0.062 |
| 8" | " | 0.067 |
| 10" | " | 0.073 |
| 12" | " | 0.080 |
| Lightweight | | |
| 6" | S.F. | 0.062 |
| 8" | " | 0.067 |
| 10" | " | 0.073 |
| 12" | " | 0.080 |
| Acoustical slotted block | | |
| 4" | S.F. | 0.073 |
| 6" | " | 0.073 |
| 8" | " | 0.080 |
| Filled cavities | | |
| 4" | S.F. | 0.089 |
| 6" | " | 0.094 |
| 8" | " | 0.100 |
| Hollow, split face | | |
| 4" | S.F. | 0.059 |
| 6" | " | 0.062 |
| 8" | " | 0.067 |
| 10" | " | 0.073 |
| 12" | " | 0.080 |
| Split rib profile | | |
| 4" | S.F. | 0.073 |
| 6" | " | 0.073 |
| 8" | " | 0.080 |
| 10" | " | 0.080 |
| 12" | " | 0.080 |
| High strength block, 3500 psi | | |
| 2" | S.F. | 0.059 |
| 4" | " | 0.062 |
| 6" | " | 0.062 |
| 8" | " | 0.067 |
| 10" | " | 0.073 |
| 12" | " | 0.080 |

# 04 MASONRY

## UNIT MASONRY

### 04220.10 CONCRETE MASONRY UNITS

| UNIT MASONRY | UNIT | MAN/HOURS |
|---|---|---|
| Solar screen concrete block | | |
| 4" thick | | |
| 6" x 6" | S.F. | 0.178 |
| 8" x 8" | " | 0.160 |
| 12" x 12" | " | 0.123 |
| 8" thick | | |
| 8" x 16" | S.F. | 0.114 |
| Glazed block | | |
| Cove base, glazed 1 side, 2" | L.F. | 0.089 |
| 4" | " | 0.089 |
| 6" | " | 0.100 |
| 8" | " | 0.100 |
| Single face | | |
| 2" | S.F. | 0.067 |
| 4" | " | 0.067 |
| 6" | " | 0.073 |
| 8" | " | 0.080 |
| 10" | " | 0.089 |
| 12" | " | 0.094 |
| Double face | | |
| 4" | S.F. | 0.084 |
| 6" | " | 0.089 |
| 8" | " | 0.100 |
| Corner or bullnose | | |
| 2" | EA. | 0.100 |
| 4" | " | 0.114 |
| 6" | " | 0.114 |
| 8" | " | 0.133 |
| 10" | " | 0.145 |
| 12" | " | 0.160 |
| Gypsum unit masonry | | |
| Partition blocks (12"x30") | | |
| Solid | | |
| 2" | S.F. | 0.032 |
| Hollow | | |
| 3" | S.F. | 0.032 |
| 4" | " | 0.033 |
| 6" | " | 0.036 |
| Vertical reinforcing | | |
| 4' o.c., add 5% to labor | | |
| 2'8" o.c., add 15% to labor | | |
| Interior partitions, add 10% to labor | | |

### 04220.90 BOND BEAMS & LINTELS

| UNIT MASONRY | UNIT | MAN/HOURS |
|---|---|---|
| Bond beam, no grout or reinforcement | | |
| 8" x 16" x | | |
| 4" thick | L.F. | 0.062 |
| 6" thick | " | 0.064 |
| 8" thick | " | 0.067 |
| 10" thick | " | 0.070 |
| 12" thick | " | 0.073 |
| Beam lintel, no grout or reinforcement | | |
| 8" x 16" x | | |
| 10" thick | L.F. | 0.080 |
| 12" thick | L.F. | 0.089 |
| Precast masonry lintel | | |
| 6 lf, 8" high x | | |
| 4" thick | L.F. | 0.133 |
| 6" thick | " | 0.133 |
| 8" thick | " | 0.145 |
| 10" thick | " | 0.145 |
| 10 lf, 8" high x | | |
| 4" thick | L.F. | 0.080 |
| 6" thick | " | 0.080 |
| 8" thick | " | 0.089 |
| 10" thick | " | 0.089 |
| Steel angles and plates | | |
| Minimum | Lb. | 0.011 |
| Maximum | " | 0.020 |
| Various size angle lintels | | |
| 1/4" stock | | |
| 3" x 3" | L.F. | 0.050 |
| 3" x 3-1/2" | " | 0.050 |
| 3/8" stock | | |
| 3" x 4" | L.F. | 0.050 |
| 3-1/2" x 4" | " | 0.050 |
| 4" x 4" | " | 0.050 |
| 5" x 3-1/2" | " | 0.050 |
| 6" x 3-1/2" | " | 0.050 |
| 1/2" stock | | |
| 6" x 4" | L.F. | 0.050 |

### 04240.10 CLAY TILE

| UNIT MASONRY | UNIT | MAN/HOURS |
|---|---|---|
| Hollow clay tile, for back-up, 12" x 12" | | |
| Scored face | | |
| Load bearing | | |
| 4" thick | S.F. | 0.057 |
| 6" thick | " | 0.059 |
| 8" thick | " | 0.062 |
| 10" thick | " | 0.064 |
| 12" thick | " | 0.067 |
| Non-load bearing | | |
| 3" thick | S.F. | 0.055 |
| 4" thick | " | 0.057 |
| 6" thick | " | 0.059 |
| 8" thick | " | 0.062 |
| 12" thick | " | 0.067 |
| Partition, 12" x 12" | | |
| In walls | | |
| 3" thick | S.F. | 0.067 |
| 4" thick | " | 0.067 |
| 6" thick | " | 0.070 |
| 8" thick | " | 0.073 |
| 10" thick | " | 0.076 |
| 12" thick | " | 0.080 |
| Clay tile floors | | |
| 4" thick | S.F. | 0.044 |

## UNIT MASONRY

### 04240.10 CLAY TILE

| Description | UNIT | MAN/HOURS |
|---|---|---|
| 6" thick | S.F. | 0.047 |
| 8" thick | " | 0.050 |
| 10" thick | " | 0.053 |
| 12" thick | " | 0.057 |
| Terra cotta | | |
| Coping, 10" or 12" wide, 3" thick | L.F. | 0.160 |

### 04270.10 GLASS BLOCK

| Description | UNIT | MAN/HOURS |
|---|---|---|
| Glass block, 4" thick | | |
| 6" x 6" | S.F. | 0.267 |
| 8" x 8" | " | 0.200 |
| 12" x 12" | " | 0.160 |
| Replacement glass blocks, 4" x 8" x 8" | | |
| Minimum | S.F. | 0.800 |
| Maximum | " | 1.600 |

### 04295.10 PARGING/MASONRY PLASTER

| Description | UNIT | MAN/HOURS |
|---|---|---|
| Parging | | |
| 1/2" thick | S.F. | 0.053 |
| 3/4" thick | " | 0.067 |
| 1" thick | " | 0.080 |

## STONE

### 04400.10 STONE

| Description | UNIT | MAN/HOURS |
|---|---|---|
| Rubble stone | | |
| Walls set in mortar | | |
| 8" thick | S.F. | 0.200 |
| 12" thick | " | 0.320 |
| 18" thick | " | 0.400 |
| 24" thick | " | 0.533 |
| Dry set wall | | |
| 8" thick | S.F. | 0.133 |
| 12" thick | " | 0.200 |
| 18" thick | " | 0.267 |
| 24" thick | " | 0.320 |
| Cut stone | | |
| Imported marble | | |
| Facing panels | | |
| 3/4" thick | S.F. | 0.320 |
| 1-1/2" thick | " | 0.364 |
| 2-1/4" thick | " | 0.444 |
| Base | | |
| 1" thick | | |
| 4" high | L.F. | 0.400 |
| 6" high | " | 0.400 |

## STONE

### 04400.10 STONE

| Description | UNIT | MAN/HOURS |
|---|---|---|
| Columns, solid | | |
| Plain faced | C.F. | 5.333 |
| Fluted | " | 5.333 |
| Flooring, travertine, minimum | S.F. | 0.123 |
| Average | " | 0.160 |
| Maximum | " | 0.178 |
| Domestic marble | | |
| Facing panels | | |
| 7/8" thick | S.F. | 0.320 |
| 1-1/2" thick | " | 0.364 |
| 2-1/4" thick | " | 0.444 |
| Stairs | | |
| 12" treads | L.F. | 0.400 |
| 6" risers | " | 0.267 |
| Thresholds, 7/8" thick, 3' long, 4" to 6" wide | | |
| Plain | EA. | 0.667 |
| Beveled | " | 0.667 |
| Window sill | | |
| 6" wide, 2" thick | L.F. | 0.320 |
| Stools | | |
| 5" wide, 7/8" thick | L.F. | 0.320 |
| Limestone panels up to 12' x 5', smooth finish | | |
| 2" thick | S.F. | 0.096 |
| 3" thick | " | 0.096 |
| 4" thick | " | 0.096 |
| Miscellaneous limestone items | | |
| Steps, 14" wide, 6" deep | L.F. | 0.533 |
| Coping, smooth finish | C.F. | 0.267 |
| Sills, lintels, jambs, smooth finish | " | 0.320 |
| Granite veneer facing panels, polished | | |
| 7/8" thick | | |
| Black | S.F. | 0.320 |
| Gray | " | 0.320 |
| Base | | |
| 4" high | L.F. | 0.160 |
| 6" high | " | 0.178 |
| Curbing, straight, 6" x 16" | " | 0.400 |
| Radius curbs, radius over 5' | " | 0.533 |
| Ashlar veneer | | |
| 4" thick, random | S.F. | 0.320 |
| Pavers, 4" x 4" split | | |
| Gray | S.F. | 0.160 |
| Pink | " | 0.160 |
| Black | " | 0.160 |
| Slate, panels | | |
| 1" thick | S.F. | 0.320 |
| 2" thick | " | 0.364 |
| Sills or stools | | |
| 1" thick | | |
| 6" wide | L.F. | 0.320 |
| 10" wide | " | 0.348 |
| 2" thick | | |
| 6" wide | L.F. | 0.364 |
| 10" wide | " | 0.400 |

| MASONRY RESTORATION | UNIT | MAN/ HOURS |
|---|---|---|
| **04520.10 RESTORATION AND CLEANING** | | |
| Masonry cleaning | | |
| Washing brick | | |
| Smooth surface | S.F. | 0.013 |
| Rough surface | " | 0.018 |
| Steam clean masonry | | |
| Smooth face | | |
| Minimum | S.F. | 0.010 |
| Maximum | " | 0.015 |
| Rough face | | |
| Minimum | S.F. | 0.013 |
| Maximum | " | 0.020 |
| Sandblast masonry | | |
| Minimum | S.F. | 0.016 |
| Maximum | " | 0.027 |
| Pointing masonry | | |
| Brick | S.F. | 0.032 |
| Concrete block | " | 0.023 |
| Cut and repoint | | |
| Brick | | |
| Minimum | S.F. | 0.040 |
| Maximum | " | 0.080 |
| Stone work | L.F. | 0.062 |
| Cut and recaulk | | |
| Oil base caulks | L.F. | 0.053 |
| Butyl caulks | " | 0.053 |
| Polysulfides and acrylics | " | 0.053 |
| Silicones | " | 0.053 |
| Cement and sand grout on walls, to 1/8" thick | | |
| Minimum | S.F. | 0.032 |
| Maximum | " | 0.040 |
| Brick removal and replacement | | |
| Minimum | EA. | 0.100 |
| Average | " | 0.133 |
| Maximum | " | 0.400 |
| **04550.10 REFRACTORIES** | | |
| Flue liners | | |
| Rectangular | | |
| 8" x 12" | L.F. | 0.133 |
| 12" x 12" | " | 0.145 |
| 12" x 18" | " | 0.160 |
| 16" x 16" | " | 0.178 |
| 18" x 18" | " | 0.190 |
| 20" x 20" | " | 0.200 |
| 24" x 24" | " | 0.229 |
| Round | | |
| 18" dia. | L.F. | 0.190 |
| 24" dia. | " | 0.229 |

# 05 METALS

## METAL FASTENING

| 05050.10 STRUCTURAL WELDING | UNIT | MAN/ HOURS |
|---|---|---|
| Welding | | |
| Single pass | | |
| 1/8" | L.F. | 0.040 |
| 3/16" | " | 0.053 |
| 1/4" | " | 0.067 |
| Miscellaneous steel shapes | | |
| Plain | Lb. | 0.002 |
| Galvanized | " | 0.003 |
| Plates | | |
| Plain | Lb. | 0.002 |
| Galvanized | " | 0.003 |
| **05050.95 METAL LINTELS** | | |
| Lintels, steel | | |
| Plain | Lb. | 0.020 |
| Galvanized | " | 0.020 |
| **05300.10 METAL DECKING** | | |
| Roof, 1-1/2" deep, non-composite | | |
| 16 ga. | | |
| Primed | S.F. | 0.008 |
| Galvanized | " | 0.008 |
| Open type decking, galvanized | | |
| 1-1/2" deep | | |
| 18 ga. | S.F. | 0.008 |
| 20 ga. | " | 0.008 |
| 3" deep | | |
| 16 ga. | S.F. | 0.009 |
| 18 ga. | " | 0.009 |
| 20 ga. | " | 0.009 |
| Cellular type | | |
| 1-1/2" deep, galvanized | | |
| 18-18 ga. | S.F. | 0.010 |
| 22-18 ga. | " | 0.010 |
| 3" deep, galvanized | | |
| 16-16 ga. | S.F. | 0.011 |
| 18-16 ga. | " | 0.011 |
| 18-18 ga. | " | 0.011 |
| 20-18 ga. | " | 0.011 |
| Composite deck, non-cellular, galvanized | | |
| 1-1/2" deep | | |
| 18 ga. | S.F. | 0.009 |
| 20 ga. | " | 0.009 |
| 3" deep | | |
| 18 ga. | S.F. | 0.009 |
| 20 ga. | " | 0.009 |

## COLD FORMED FRAMING

| 05410.10 METAL FRAMING | UNIT | MAN/ HOURS |
|---|---|---|
| Furring channel, galvanized | | |
| Beams and columns, 3/4" | | |
| 12" o.c. | S.F. | 0.080 |
| 16" o.c. | " | 0.073 |
| Walls, 3/4" | | |
| 12" o.c. | S.F. | 0.040 |
| 16" o.c. | " | 0.033 |
| 24" o.c. | " | 0.027 |
| 1-1/2" | | |
| 12" o.c. | S.F. | 0.040 |
| 16" o.c. | " | 0.033 |
| 24" o.c. | " | 0.027 |
| Stud, load bearing | | |
| 16" o.c. | | |
| 16 ga. | | |
| 2-1/2" | S.F. | 0.036 |
| 3-5/8" | " | 0.036 |
| 4" | " | 0.036 |
| 6" | " | 0.040 |
| 18 ga. | | |
| 2-1/2" | S.F. | 0.036 |
| 3-5/8" | " | 0.036 |
| 4" | " | 0.036 |
| 6" | " | 0.040 |
| 8" | " | 0.040 |
| 20 ga. | | |
| 2-1/2" | S.F. | 0.036 |
| 3-5/8" | " | 0.036 |
| 4" | " | 0.036 |
| 6" | " | 0.040 |
| 8" | " | 0.040 |
| 24" o.c. | | |
| 16 ga. | | |
| 2-1/2" | S.F. | 0.031 |
| 3-5/8" | " | 0.031 |
| 4" | " | 0.031 |
| 6" | " | 0.033 |
| 8" | " | 0.033 |
| 18 ga. | | |
| 2-1/2" | S.F. | 0.031 |
| 3-5/8" | " | 0.031 |
| 4" | " | 0.031 |
| 6" | " | 0.033 |
| 8" | " | 0.033 |
| 20 ga. | | |
| 2-1/2" | S.F. | 0.031 |
| 3-5/8" | " | 0.031 |
| 4" | " | 0.031 |
| 6" | " | 0.033 |
| 8" | " | 0.033 |

# 05 METALS

| METAL FABRICATIONS | UNIT | MAN/ HOURS |
|---|---|---|
| **05510.10 STAIRS** | | |
| Stock unit, steel, complete, per riser | | |
| Tread | | |
| 3'-6" wide | EA. | 1.000 |
| 4' wide | " | 1.143 |
| 5' wide | " | 1.333 |
| Metal pan stair, cement filled, per riser | | |
| 3'-6" wide | EA. | 0.800 |
| 4' wide | " | 0.889 |
| 5' wide | " | 1.000 |
| Landing, steel pan | S.F. | 0.200 |
| Cast iron tread, steel stringers, stock units, per riser | | |
| Tread | | |
| 3'-6" wide | EA. | 1.000 |
| 4' wide | " | 1.143 |
| 5' wide | " | 1.333 |
| Stair treads, abrasive, 12" x 3'-6" | | |
| Cast iron | | |
| 3/8" | EA. | 0.400 |
| 1/2" | " | 0.400 |
| Cast aluminum | | |
| 5/16" | EA. | 0.400 |
| 3/8" | " | 0.400 |
| 1/2" | " | 0.400 |
| **05515.10 LADDERS** | | |
| Ladder, 18" wide | | |
| With cage | L.F. | 0.533 |
| Without cage | " | 0.400 |
| **05520.10 RAILINGS** | | |
| Railing, pipe | | |
| 1-1/4" diameter, welded steel | | |
| 2-rail | | |
| Primed | L.F. | 0.160 |
| Galvanized | " | 0.160 |
| 3-rail | | |
| Primed | L.F. | 0.200 |
| Galvanized | " | 0.200 |
| Wall mounted, single rail, welded steel | | |
| Primed | L.F. | 0.123 |
| Galvanized | " | 0.123 |
| 1-1/2" diameter, welded steel | | |
| 2-rail | | |
| Primed | L.F. | 0.160 |
| Galvanized | " | 0.160 |
| 3-rail | | |
| Primed | L.F. | 0.200 |
| Galvanized | " | 0.200 |
| Wall mounted, single rail, welded steel | | |
| Primed | L.F. | 0.123 |
| Galvanized | " | 0.123 |
| 2" diameter, welded steel | | |

| METAL FABRICATIONS | UNIT | MAN/ HOURS |
|---|---|---|
| **05520.10 RAILINGS** | | |
| 2-rail | | |
| Primed | L.F. | 0.178 |
| Galvanized | " | 0.178 |
| 3-rail | | |
| Primed | L.F. | 0.229 |
| Galvanized | " | 0.229 |
| Wall mounted, single rail, welded steel | | |
| Primed | L.F. | 0.133 |
| Galvanized | " | 0.133 |
| **05530.10 METAL GRATING** | | |
| Floor plate, checkered, steel | | |
| 1/4" | | |
| Primed | S.F. | 0.011 |
| Galvanized | " | 0.011 |
| 3/8" | | |
| Primed | S.F. | 0.012 |
| Galvanized | " | 0.012 |
| Aluminum grating, pressure-locked bearing bars | | |
| 3/4" x 1/8" | S.F. | 0.020 |
| 1" x 1/8" | " | 0.020 |
| 1-1/4" x 1/8" | " | 0.020 |
| 1-1/4" x 3/16" | " | 0.020 |
| 1-1/2" x 1/8" | " | 0.020 |
| 1-3/4" x 3/16" | " | 0.020 |
| Miscellaneous expenses | | |
| Cutting | | |
| Minimum | L.F. | 0.053 |
| Maximum | " | 0.080 |
| Banding | | |
| Minimum | L.F. | 0.100 |
| Maximum | " | 0.160 |
| Toe plates | | |
| Minimum | L.F. | 0.160 |
| Maximum | " | 0.200 |
| Steel grating, primed | | |
| 3/4" x 1/8" | S.F. | 0.027 |
| 1" x 1/8" | " | 0.027 |
| 1-1/4" x 1/8" | " | 0.027 |
| 1-1/4" x 3/16" | " | 0.027 |
| 1-1/2" x 1/8" | " | 0.027 |
| 1-3/4" x 3/16" | " | 0.027 |
| Galvanized | | |
| 3/4" x 1/8" | S.F. | 0.027 |
| 1" x 1/8" | " | 0.027 |
| 1-1/4" x 1/8" | " | 0.027 |
| 1-1/4" x 3/16" | " | 0.027 |
| 1-1/2" x 1/8" | " | 0.027 |
| 1-3/4" x 3/16" | " | 0.027 |
| Miscellaneous expenses | | |
| Cutting | | |
| Minimum | L.F. | 0.057 |
| Maximum | " | 0.089 |

# 05 METALS

## METAL FABRICATIONS

| 05530.10 METAL GRATING | UNIT | MAN/ HOURS |
|---|---|---|
| Banding | | |
| Minimum | L.F. | 0.145 |
| Maximum | " | 0.178 |
| Toe plates | | |
| Minimum | L.F. | 0.178 |
| Maximum | " | 0.229 |

| 05540.10 CASTINGS | UNIT | MAN/ HOURS |
|---|---|---|
| Miscellaneous castings | | |
| Light sections | Lb. | 0.016 |
| Heavy sections | " | 0.011 |

## MISC. FABRICATIONS

| 05580.10 METAL SPECIALTIES | UNIT | MAN/ HOURS |
|---|---|---|
| Kick plate | | |
| 4" high x 1/4" thick | | |
| Primed | L.F. | 0.160 |
| Galvanized | " | 0.160 |
| 6" high x 1/4" thick | | |
| Primed | L.F. | 0.178 |
| Galvanized | " | 0.178 |

| 05700.10 ORNAMENTAL METAL | UNIT | MAN/ HOURS |
|---|---|---|
| Railings, vertical square bars, 6" o.c., with shaped top rails | | |
| Steel | L.F. | 0.400 |
| Aluminum | " | 0.400 |
| Bronze | " | 0.533 |
| Stainless steel | " | 0.533 |
| Laminated metal or wood handrails with metal supports | | |
| 2-1/2" round or oval shape | L.F. | 0.400 |
| Grilles and louvers | | |
| Fixed type louvers | | |
| 4 through 10 sf | S.F. | 0.133 |
| Over 10 sf | " | 0.100 |
| Movable type louvers | | |
| 4 through 10 sf | S.F. | 0.133 |
| Over 10 sf | " | 0.100 |
| Aluminum louvers | | |
| Residential use, fixed type, with screen | | |
| 8" x 8" | EA. | 0.400 |
| 12" x 12" | " | 0.400 |
| 12" x 18" | " | 0.400 |
| 14" x 24" | " | 0.400 |
| 18" x 24" | " | 0.400 |

## MISC. FABRICATIONS

| 05700.10 ORNAMENTAL METAL | UNIT | MAN/ HOURS |
|---|---|---|
| 30" x 24" | EA. | 0.444 |

| 05800.10 EXPANSION CONTROL | UNIT | MAN/ HOURS |
|---|---|---|
| Expansion joints with covers, floor assembly type | | |
| With 1" space | | |
| Aluminum | L.F. | 0.133 |
| Bronze | " | 0.133 |
| Stainless steel | " | 0.133 |
| Aluminum | " | 0.133 |
| Ceiling and wall assembly type | | |
| With 1" space | | |
| Aluminum | L.F. | 0.160 |
| Bronze | " | 0.160 |
| Stainless steel | " | 0.160 |
| Exterior roof and wall, aluminum | | |
| Roof to roof | | |
| With 1" space | L.F. | 0.133 |
| With 2" space | " | 0.133 |
| Roof to wall | | |
| With 1" space | L.F. | 0.145 |
| With 2" space | " | 0.145 |
| Flat wall to wall | | |
| With 1" space | L.F. | 0.133 |
| With 2" space | " | 0.133 |
| Corner to flat wall | | |
| With 1" space | L.F. | 0.160 |
| With 2" in space | " | 0.160 |

## FASTENERS AND ADHESIVES

### 06050.10 ACCESSORIES

| | UNIT | MAN/HOURS |
|---|---|---|
| Column/post base, cast aluminum | | |
| 4" x 4" | EA. | 0.200 |
| 6" x 6" | " | 0.200 |
| Bridging, metal, per pair | | |
| 12" o.c. | EA. | 0.080 |
| 16" o.c. | " | 0.073 |
| Anchors | | |
| Bolts, threaded two ends, with nuts and washers | | |
| 1/2" dia. | | |
| 4" long | EA. | 0.050 |
| 7-1/2" long | " | 0.050 |
| 3/4" dia. | | |
| 7-1/2" long | EA. | 0.050 |
| 15" long | " | 0.050 |
| Framing anchors | | |
| 10 gauge | EA. | 0.067 |
| Bolts, carriage | | |
| 1/4 x 4 | EA. | 0.080 |
| 5/16 x 6 | " | 0.084 |
| 3/8 x 6 | " | 0.084 |
| 1/2 x 6 | " | 0.084 |
| Joist and beam hangers | | |
| 18 ga. | | |
| 2 x 4 | EA. | 0.080 |
| 2 x 6 | " | 0.080 |
| 2 x 8 | " | 0.080 |
| 2 x 10 | " | 0.089 |
| 2 x 12 | " | 0.100 |
| 16 ga. | | |
| 3 x 6 | EA. | 0.089 |
| 3 x 8 | " | 0.089 |
| 3 x 10 | " | 0.094 |
| 3 x 12 | " | 0.107 |
| 3 x 14 | " | 0.114 |
| 4 x 6 | " | 0.089 |
| 4 x 8 | " | 0.089 |
| 4 x 10 | " | 0.094 |
| 4 x 12 | " | 0.107 |
| 4 x 14 | " | 0.114 |
| Rafter anchors, 18 ga., 1-1/2" wide | | |
| 5-1/4" long | EA. | 0.067 |
| 10-3/4" long | " | 0.067 |
| Shear plates | | |
| 2-5/8" dia. | EA. | 0.062 |
| 4" dia. | " | 0.067 |
| Sill anchors | | |
| Embedded in concrete | EA. | 0.080 |
| Split rings | | |
| 2-1/2" dia. | EA. | 0.089 |
| 4" dia. | " | 0.100 |
| Strap ties, 14 ga., 1-3/8" wide | | |
| 12" long | EA. | 0.067 |
| 18" long | " | 0.073 |
| 24" long | " | 0.080 |
| 36" long | EA. | 0.089 |
| Toothed rings | | |
| 2-5/8" dia. | EA. | 0.133 |
| 4" dia. | " | 0.160 |

## ROUGH CARPENTRY

### 06110.10 BLOCKING

| | UNIT | MAN/HOURS |
|---|---|---|
| Steel construction | | |
| Walls | | |
| 2x4 | L.F. | 0.053 |
| 2x6 | " | 0.062 |
| 2x8 | " | 0.067 |
| 2x10 | " | 0.073 |
| 2x12 | " | 0.080 |
| Ceilings | | |
| 2x4 | L.F. | 0.062 |
| 2x6 | " | 0.073 |
| 2x8 | " | 0.080 |
| 2x10 | " | 0.089 |
| 2x12 | " | 0.100 |
| Wood construction | | |
| Walls | | |
| 2x4 | L.F. | 0.044 |
| 2x6 | " | 0.050 |
| 2x8 | " | 0.053 |
| 2x10 | " | 0.057 |
| 2x12 | " | 0.062 |
| Ceilings | | |
| 2x4 | L.F. | 0.050 |
| 2x6 | " | 0.057 |
| 2x8 | " | 0.062 |
| 2x10 | " | 0.067 |
| 2x12 | " | 0.073 |

### 06110.20 CEILING FRAMING

| | UNIT | MAN/HOURS |
|---|---|---|
| Ceiling joists | | |
| 12" o.c. | | |
| 2x4 | S.F. | 0.019 |
| 2x6 | " | 0.020 |
| 2x8 | " | 0.021 |
| 2x10 | " | 0.022 |
| 2x12 | " | 0.024 |
| 16" o.c. | | |
| 2x4 | S.F. | 0.015 |
| 2x6 | " | 0.016 |
| 2x8 | " | 0.017 |

# 06 WOOD AND PLASTICS

| ROUGH CARPENTRY | UNIT | MAN/ HOURS |
|---|---|---|
| **06110.20 CEILING FRAMING** | | |
| 2x10 | S.F. | 0.017 |
| 2x12 | " | 0.018 |
| 24" o.c. | | |
| 2x4 | S.F. | 0.013 |
| 2x6 | " | 0.013 |
| 2x8 | " | 0.014 |
| 2x10 | " | 0.015 |
| 2x12 | " | 0.016 |
| Headers and nailers | | |
| 2x4 | L.F. | 0.026 |
| 2x6 | " | 0.027 |
| 2x8 | " | 0.029 |
| 2x10 | " | 0.031 |
| 2x12 | " | 0.033 |
| Sister joists for ceilings | | |
| 2x4 | L.F. | 0.057 |
| 2x6 | " | 0.067 |
| 2x8 | " | 0.080 |
| 2x10 | " | 0.100 |
| 2x12 | " | 0.133 |
| **06110.30 FLOOR FRAMING** | | |
| Floor joists | | |
| 12" o.c. | | |
| 2x6 | S.F. | 0.016 |
| 2x8 | " | 0.016 |
| 2x10 | " | 0.017 |
| 2x12 | " | 0.017 |
| 2x14 | " | 0.017 |
| 3x6 | " | 0.017 |
| 3x8 | " | 0.017 |
| 3x10 | " | 0.018 |
| 3x12 | " | 0.019 |
| 3x14 | " | 0.020 |
| 4x6 | " | 0.017 |
| 4x8 | " | 0.017 |
| 4x10 | " | 0.018 |
| 4x12 | " | 0.019 |
| 4x14 | " | 0.020 |
| 16" o.c. | | |
| 2x6 | S.F. | 0.013 |
| 2x8 | " | 0.014 |
| 2x10 | " | 0.014 |
| 2x12 | " | 0.014 |
| 2x14 | " | 0.015 |
| 3x6 | " | 0.014 |
| 3x8 | " | 0.014 |
| 3x10 | " | 0.015 |
| 3x12 | " | 0.015 |
| 3x14 | " | 0.016 |
| 4x6 | " | 0.014 |
| 4x8 | " | 0.014 |
| 4x10 | " | 0.015 |
| 4x12 | " | 0.015 |

| ROUGH CARPENTRY | UNIT | MAN/ HOURS |
|---|---|---|
| **06110.30 FLOOR FRAMING** | | |
| 4x14 | S.F. | 0.016 |
| Sister joists for floors | | |
| 2x4 | L.F. | 0.050 |
| 2x6 | " | 0.057 |
| 2x8 | " | 0.067 |
| 2x10 | " | 0.080 |
| 2x12 | " | 0.100 |
| 3x6 | " | 0.080 |
| 3x8 | " | 0.089 |
| 3x10 | " | 0.100 |
| 3x12 | " | 0.114 |
| 4x6 | " | 0.080 |
| 4x8 | " | 0.089 |
| 4x10 | " | 0.100 |
| 4x12 | " | 0.114 |
| **06110.40 FURRING** | | |
| Furring, wood strips | | |
| Walls | | |
| On masonry or concrete walls | | |
| 1x2 furring | | |
| 12" o.c. | S.F. | 0.025 |
| 16" o.c. | " | 0.023 |
| 24" o.c. | " | 0.021 |
| 1x3 furring | | |
| 12" o.c. | S.F. | 0.025 |
| 16" o.c. | " | 0.023 |
| 24" o.c. | " | 0.021 |
| On wood walls | | |
| 1x2 furring | | |
| 12" o.c. | S.F. | 0.018 |
| 16" o.c. | " | 0.016 |
| 24" o.c. | " | 0.015 |
| 1x3 furring | | |
| 12" o.c. | S.F. | 0.018 |
| 16" o.c. | " | 0.016 |
| 24" o.c. | " | 0.015 |
| Ceilings | | |
| On masonry or concrete ceilings | | |
| 1x2 furring | | |
| 12" o.c. | S.F. | 0.044 |
| 16" o.c. | " | 0.040 |
| 24" o.c. | " | 0.036 |
| 1x3 furring | | |
| 12" o.c. | S.F. | 0.044 |
| 16" o.c. | " | 0.040 |
| 24" o.c. | " | 0.036 |
| On wood ceilings | | |
| 1x2 furring | | |
| 12" o.c. | S.F. | 0.030 |
| 16" o.c. | " | 0.027 |
| 24" o.c. | " | 0.024 |
| 1x3 | | |

| ROUGH CARPENTRY | UNIT | MAN/ HOURS |
|---|---|---|
| **06110.40 FURRING** | | |
| 12" o.c. | S.F. | 0.030 |
| 16" o.c. | " | 0.027 |
| 24" o.c. | " | 0.024 |
| **06110.50 ROOF FRAMING** | | |
| Roof framing | | |
| Rafters, gable end | | |
| 0-2 pitch (flat to 2-in-12) | | |
| 12" o.c. | | |
| 2x4 | S.F. | 0.017 |
| 2x6 | " | 0.017 |
| 2x8 | " | 0.018 |
| 2x10 | " | 0.019 |
| 2x12 | " | 0.020 |
| 16" o.c. | | |
| 2x6 | S.F. | 0.014 |
| 2x8 | " | 0.015 |
| 2x10 | " | 0.015 |
| 2x12 | " | 0.016 |
| 24" o.c. | | |
| 2x6 | S.F. | 0.012 |
| 2x8 | " | 0.013 |
| 2x10 | " | 0.013 |
| 2x12 | " | 0.013 |
| 4-6 pitch (4-in-12 to 6-in-12) | | |
| 12" o.c. | | |
| 2x4 | S.F. | 0.017 |
| 2x6 | " | 0.018 |
| 2x8 | " | 0.019 |
| 2x10 | " | 0.020 |
| 2x12 | " | 0.021 |
| 16" o.c. | | |
| 2x6 | S.F. | 0.015 |
| 2x8 | " | 0.015 |
| 2x10 | " | 0.016 |
| 2x12 | " | 0.017 |
| 24" o.c. | | |
| 2x6 | S.F. | 0.013 |
| 2x8 | " | 0.013 |
| 2x10 | " | 0.014 |
| 2x12 | " | 0.015 |
| 8-12 pitch (8-in-12 to 12-in-12) | | |
| 12" o.c. | | |
| 2x4 | S.F. | 0.018 |
| 2x6 | " | 0.019 |
| 2x8 | " | 0.020 |
| 2x10 | " | 0.021 |
| 2x12 | " | 0.022 |
| 16" o.c. | | |
| 2x6 | S.F. | 0.015 |
| 2x8 | " | 0.016 |
| 2x10 | " | 0.017 |
| 2x12 | " | 0.017 |

| ROUGH CARPENTRY | UNIT | MAN/ HOURS |
|---|---|---|
| **06110.50 ROOF FRAMING** | | |
| 24" o.c. | | |
| 2x6 | S.F. | 0.013 |
| 2x8 | " | 0.013 |
| 2x10 | " | 0.014 |
| 2x12 | " | 0.014 |
| Ridge boards | | |
| 2x6 | L.F. | 0.040 |
| 2x8 | " | 0.044 |
| 2x10 | " | 0.050 |
| 2x12 | " | 0.057 |
| Hip rafters | | |
| 2x6 | L.F. | 0.029 |
| 2x8 | " | 0.030 |
| 2x10 | " | 0.031 |
| 2x12 | " | 0.032 |
| Jack rafters | | |
| 4-6 pitch (4-in-12 to 6-in-12) | | |
| 16" o.c. | | |
| 2x6 | S.F. | 0.024 |
| 2x8 | " | 0.024 |
| 2x10 | " | 0.026 |
| 2x12 | " | 0.027 |
| 24" o.c. | | |
| 2x6 | S.F. | 0.018 |
| 2x8 | " | 0.019 |
| 2x10 | " | 0.020 |
| 2x12 | " | 0.020 |
| 8-12 pitch (8-in-12 to 12-in-12) | | |
| 16" o.c. | | |
| 2x6 | S.F. | 0.025 |
| 2x8 | " | 0.026 |
| 2x10 | " | 0.027 |
| 2x12 | " | 0.028 |
| 24" o.c. | | |
| 2x6 | S.F. | 0.019 |
| 2x8 | " | 0.020 |
| 2x10 | " | 0.020 |
| 2x12 | " | 0.021 |
| Sister rafters | | |
| 2x4 | L.F. | 0.057 |
| 2x6 | " | 0.067 |
| 2x8 | " | 0.080 |
| 2x10 | " | 0.100 |
| 2x12 | " | 0.133 |
| Fascia boards | | |
| 2x4 | L.F. | 0.040 |
| 2x6 | " | 0.040 |
| 2x8 | " | 0.044 |
| 2x10 | " | 0.044 |
| 2x12 | " | 0.050 |
| Cant strips | | |
| Fiber | | |
| 3x3 | L.F. | 0.023 |
| 4x4 | " | 0.024 |

# 06 WOOD AND PLASTICS

| ROUGH CARPENTRY | UNIT | MAN/ HOURS |
|---|---|---|
| **06110.50 ROOF FRAMING** | | |
| Wood | | |
| 3x3 | L.F. | 0.024 |
| **06110.60 SLEEPERS** | | |
| Sleepers, over concrete | | |
| 12" o.c. | | |
| 1x2 | S.F. | 0.018 |
| 1x3 | " | 0.019 |
| 2x4 | " | 0.022 |
| 2x6 | " | 0.024 |
| 16" o.c. | | |
| 1x2 | S.F. | 0.016 |
| 1x3 | " | 0.016 |
| 2x4 | " | 0.019 |
| 2x6 | " | 0.020 |
| **06110.65 SOFFITS** | | |
| Soffit framing | | |
| 2x3 | L.F. | 0.057 |
| 2x4 | " | 0.062 |
| 2x6 | " | 0.067 |
| 2x8 | " | 0.073 |
| **06110.70 WALL FRAMING** | | |
| Framing wall, studs | | |
| 12" o.c. | | |
| 2x3 | S.F. | 0.015 |
| 2x4 | " | 0.015 |
| 2x6 | " | 0.016 |
| 2x8 | " | 0.017 |
| 16" o.c. | | |
| 2x3 | S.F. | 0.013 |
| 2x4 | " | 0.013 |
| 2x6 | " | 0.013 |
| 2x8 | " | 0.014 |
| 24" o.c. | | |
| 2x3 | S.F. | 0.011 |
| 2x4 | " | 0.011 |
| 2x6 | " | 0.011 |
| 2x8 | " | 0.012 |
| Plates, top or bottom | | |
| 2x3 | L.F. | 0.024 |
| 2x4 | " | 0.025 |
| 2x6 | " | 0.027 |
| 2x8 | " | 0.029 |
| Headers, door or window | | |
| 2x6 | | |
| Single | | |
| 3' long | EA. | 0.400 |
| 6' long | " | 0.500 |
| Double | | |
| 3' long | EA. | 0.444 |

| ROUGH CARPENTRY | UNIT | MAN/ HOURS |
|---|---|---|
| **06110.70 WALL FRAMING** | | |
| 6' long | EA. | 0.571 |
| 2x8 | | |
| Single | | |
| 4' long | EA. | 0.500 |
| 8' long | " | 0.615 |
| Double | | |
| 4' long | EA. | 0.571 |
| 8' long | " | 0.727 |
| 2x10 | | |
| Single | | |
| 5' long | EA. | 0.615 |
| 10' long | " | 0.800 |
| Double | | |
| 5' long | EA. | 0.667 |
| 10' long | " | 0.800 |
| 2x12 | | |
| Single | | |
| 6' long | EA. | 0.615 |
| 12' long | " | 0.800 |
| Double | | |
| 6' long | EA. | 0.727 |
| 12' long | " | 0.889 |
| **06115.10 FLOOR SHEATHING** | | |
| Sub-flooring, plywood, CDX | | |
| 1/2" thick | S.F. | 0.010 |
| 5/8" thick | " | 0.011 |
| 3/4" thick | " | 0.013 |
| Structural plywood | | |
| 1/2" thick | S.F. | 0.010 |
| 5/8" thick | " | 0.011 |
| 3/4" thick | " | 0.012 |
| Board type subflooring | | |
| 1x6 | | |
| Minimum | S.F. | 0.018 |
| Maximum | " | 0.020 |
| 1x8 | | |
| Minimum | S.F. | 0.017 |
| Maximum | " | 0.019 |
| 1x10 | | |
| Minimum | S.F. | 0.016 |
| Maximum | " | 0.018 |
| Underlayment | | |
| Hardboard, 1/4" tempered | S.F. | 0.010 |
| Plywood, CDX | | |
| 3/8" thick | S.F. | 0.010 |
| 1/2" thick | " | 0.011 |
| 5/8" thick | " | 0.011 |
| 3/4" thick | " | 0.012 |
| **06115.20 ROOF SHEATHING** | | |
| Sheathing | | |
| Plywood, CDX | | |

# 06 WOOD AND PLASTICS

## ROUGH CARPENTRY

### 06115.20 ROOF SHEATHING

| ROUGH CARPENTRY | UNIT | MAN/HOURS |
|---|---|---|
| 3/8" thick | S.F. | 0.010 |
| 1/2" thick | " | 0.011 |
| 5/8" thick | " | 0.011 |
| 3/4" thick | " | 0.012 |
| Structural plywood | | |
| 3/8" thick | S.F. | 0.010 |
| 1/2" thick | " | 0.011 |
| 5/8" thick | " | 0.011 |
| 3/4" thick | " | 0.012 |

### 06115.30 WALL SHEATHING

| ROUGH CARPENTRY | UNIT | MAN/HOURS |
|---|---|---|
| Sheathing | | |
| Plywood, CDX | | |
| 3/8" thick | S.F. | 0.012 |
| 1/2" thick | " | 0.012 |
| 5/8" thick | " | 0.013 |
| 3/4" thick | " | 0.015 |
| Waferboard | | |
| 3/8" thick | S.F. | 0.012 |
| 1/2" thick | " | 0.012 |
| 5/8" thick | " | 0.013 |
| 3/4" thick | " | 0.015 |
| Structural plywood | | |
| 3/8" thick | S.F. | 0.012 |
| 1/2" thick | " | 0.012 |
| 5/8" thick | " | 0.013 |
| 3/4" thick | " | 0.015 |
| Gypsum, 1/2" thick | " | 0.012 |
| Asphalt impregnated fiberboard, 1/2" thick | " | 0.012 |

### 06125.10 WOOD DECKING

| ROUGH CARPENTRY | UNIT | MAN/HOURS |
|---|---|---|
| Decking, T&G solid | | |
| Cedar | | |
| 3" thick | S.F. | 0.020 |
| 4" thick | " | 0.021 |
| Fir | | |
| 3" thick | S.F. | 0.020 |
| 4" thick | " | 0.021 |
| Southern yellow pine | | |
| 3" thick | S.F. | 0.023 |
| 4" thick | " | 0.025 |
| White pine | | |
| 3" thick | S.F. | 0.020 |
| 4" thick | " | 0.021 |

### 06130.10 HEAVY TIMBER

| ROUGH CARPENTRY | UNIT | MAN/HOURS |
|---|---|---|
| Mill framed structures | | |
| Beams to 20' long | | |
| Douglas fir | | |
| 6x8 | L.F. | 0.080 |
| 6x10 | " | 0.083 |
| 6x12 | " | 0.089 |
| 6x14 | L.F. | 0.092 |
| 6x16 | " | 0.096 |
| 8x10 | " | 0.083 |
| 8x12 | " | 0.089 |
| 8x14 | " | 0.092 |
| 8x16 | " | 0.096 |
| Southern yellow pine | | |
| 6x8 | L.F. | 0.080 |
| 6x10 | " | 0.083 |
| 6x12 | " | 0.089 |
| 6x14 | " | 0.092 |
| 6x16 | " | 0.096 |
| 8x10 | " | 0.083 |
| 8x12 | " | 0.089 |
| 8x14 | " | 0.092 |
| 8x16 | " | 0.096 |
| Columns to 12' high | | |
| Douglas fir | | |
| 6x6 | L.F. | 0.120 |
| 8x8 | " | 0.120 |
| 10x10 | " | 0.133 |
| 12x12 | " | 0.133 |
| Southern yellow pine | | |
| 6x6 | L.F. | 0.120 |
| 8x8 | " | 0.120 |
| 10x10 | " | 0.133 |
| 12x12 | " | 0.133 |
| Posts, treated | | |
| 4x4 | L.F. | 0.032 |
| 6x6 | " | 0.040 |

### 06190.20 WOOD TRUSSES

| ROUGH CARPENTRY | UNIT | MAN/HOURS |
|---|---|---|
| Truss, fink, 2x4 members | | |
| 3-in-12 slope | | |
| 24' span | EA. | 0.686 |
| 26' span | " | 0.686 |
| 28' span | " | 0.727 |
| 30' span | " | 0.727 |
| 34' span | " | 0.774 |
| 38' span | " | 0.774 |
| 5-in-12 slope | | |
| 24' span | EA. | 0.706 |
| 28' span | " | 0.727 |
| 30' span | " | 0.750 |
| 32' span | " | 0.750 |
| 40' span | " | 0.800 |
| Gable, 2x4 members | | |
| 5-in-12 slope | | |
| 24' span | EA. | 0.706 |
| 26' span | " | 0.706 |
| 28' span | " | 0.727 |
| 30' span | " | 0.750 |
| 32' span | " | 0.750 |
| 36' span | " | 0.774 |

# 06 WOOD AND PLASTICS

## ROUGH CARPENTRY

| 06190.20 WOOD TRUSSES | UNIT | MAN/ HOURS |
|---|---|---|
| 40' span | EA. | 0.800 |
| King post type, 2x4 members | | |
| 4-in-12 slope | | |
| 16' span | EA. | 0.649 |
| 18' span | " | 0.667 |
| 24' span | " | 0.706 |
| 26' span | " | 0.706 |
| 30' span | " | 0.750 |
| 34' span | " | 0.750 |
| 38' span | " | 0.774 |
| 42' span | " | 0.828 |

## FINISH CARPENTRY

| 06200.10 FINISH CARPENTRY | UNIT | MAN/ HOURS |
|---|---|---|
| Mouldings and trim | | |
| Apron, flat | | |
| 9/16 x 2 | L.F. | 0.040 |
| 9/16 x 3-1/2 | " | 0.042 |
| Base | | |
| Colonial | | |
| 7/16 x 2-1/4 | L.F. | 0.040 |
| 7/16 x 3 | " | 0.040 |
| 7/16 x 3-1/4 | " | 0.040 |
| 9/16 x 3 | " | 0.042 |
| 9/16 x 3-1/4 | " | 0.042 |
| 11/16 x 2-1/4 | " | 0.044 |
| Ranch | | |
| 7/16 x 2-1/4 | L.F. | 0.040 |
| 7/16 x 3-1/4 | " | 0.040 |
| 9/16 x 2-1/4 | " | 0.042 |
| 9/16 x 3 | " | 0.042 |
| 9/16 x 3-1/4 | " | 0.042 |
| Casing | | |
| 11/16 x 2-1/2 | L.F. | 0.036 |
| 11/16 x 3-1/2 | " | 0.038 |
| Chair rail | | |
| 9/16 x 2-1/2 | L.F. | 0.040 |
| 9/16 x 3-1/2 | " | 0.040 |
| Closet pole | | |
| 1-1/8" dia. | L.F. | 0.053 |
| 1-5/8" dia. | " | 0.053 |
| Cove | | |
| 9/16 x 1-3/4 | L.F. | 0.040 |
| 11/16 x 2-3/4 | " | 0.040 |
| Crown | | |

## FINISH CARPENTRY

| 06200.10 FINISH CARPENTRY | UNIT | MAN/ HOURS |
|---|---|---|
| 9/16 x 1-5/8 | L.F. | 0.053 |
| 9/16 x 2-5/8 | " | 0.062 |
| 11/16 x 3-5/8 | " | 0.067 |
| 11/16 x 4-1/4 | " | 0.073 |
| 11/16 x 5-1/4 | " | 0.080 |
| Drip cap | | |
| 1-1/16 x 1-5/8 | L.F. | 0.040 |
| Glass bead | | |
| 3/8 x 3/8 | L.F. | 0.050 |
| 1/2 x 9/16 | " | 0.050 |
| 5/8 x 5/8 | " | 0.050 |
| 3/4 x 3/4 | " | 0.050 |
| Half round | | |
| 1/2 | L.F. | 0.032 |
| 5/8 | " | 0.032 |
| 3/4 | " | 0.032 |
| Lattice | | |
| 1/4 x 7/8 | L.F. | 0.032 |
| 1/4 x 1-1/8 | " | 0.032 |
| 1/4 x 1-3/8 | " | 0.032 |
| 1/4 x 1-3/4 | " | 0.032 |
| 1/4 x 2 | " | 0.032 |
| Ogee molding | | |
| 5/8 x 3/4 | L.F. | 0.040 |
| 11/16 x 1-1/8 | " | 0.040 |
| 11/16 x 1-3/8 | " | 0.040 |
| Parting bead | | |
| 3/8 x 7/8 | L.F. | 0.050 |
| Quarter round | | |
| 1/4 x 1/4 | L.F. | 0.032 |
| 3/8 x 3/8 | " | 0.032 |
| 1/2 x 1/2 | " | 0.032 |
| 11/16 x 11/16 | " | 0.035 |
| 3/4 x 3/4 | " | 0.035 |
| 1-1/16 x 1-1/16 | " | 0.036 |
| Railings, balusters | | |
| 1-1/8 x 1-1/8 | L.F. | 0.080 |
| 1-1/2 x 1-1/2 | " | 0.073 |
| Screen moldings | | |
| 1/4 x 3/4 | L.F. | 0.067 |
| 5/8 x 5/16 | " | 0.067 |
| Shoe | | |
| 7/16 x 11/16 | L.F. | 0.032 |
| Sash beads | | |
| 1/2 x 3/4 | L.F. | 0.067 |
| 1/2 x 7/8 | " | 0.067 |
| 1/2 x 1-1/8 | " | 0.073 |
| 5/8 x 7/8 | " | 0.073 |
| Stop | | |
| 5/8 x 1-5/8 | | |
| Colonial | L.F. | 0.050 |
| Ranch | " | 0.050 |
| Stools | | |
| 11/16 x 2-1/4 | L.F. | 0.089 |

# 06 WOOD AND PLASTICS

## FINISH CARPENTRY

| 06200.10 FINISH CARPENTRY | UNIT | MAN/HOURS |
|---|---|---|
| 11/16 x 2-1/2 | L.F. | 0.089 |
| 11/16 x 5-1/4 | " | 0.100 |
| Exterior trim, casing, select pine, 1x3 | " | 0.040 |
| Douglas fir | | |
| 1x3 | L.F. | 0.040 |
| 1x4 | " | 0.040 |
| 1x6 | " | 0.044 |
| 1x8 | " | 0.050 |
| Cornices, white pine, #2 or better | | |
| 1x2 | L.F. | 0.040 |
| 1x4 | " | 0.040 |
| 1x6 | " | 0.044 |
| 1x8 | " | 0.047 |
| 1x10 | " | 0.050 |
| 1x12 | " | 0.053 |
| Shelving, pine | | |
| 1x8 | L.F. | 0.062 |
| 1x10 | " | 0.064 |
| 1x12 | " | 0.067 |
| Plywood shelf, 3/4", with edge band, 12" wide | " | 0.080 |
| Adjustable shelf, and rod, 12" wide | | |
| 3' to 4' long | EA. | 0.200 |
| 5' to 8' long | " | 0.267 |
| Prefinished wood shelves with brackets and supports | | |
| 8" wide | | |
| 3' long | EA. | 0.200 |
| 4' long | " | 0.200 |
| 6' long | " | 0.200 |
| 10" wide | | |
| 3' long | EA. | 0.200 |
| 4' long | " | 0.200 |
| 6' long | " | 0.200 |

| 06220.10 MILLWORK | UNIT | MAN/HOURS |
|---|---|---|
| Countertop, laminated plastic | | |
| 25" x 7/8" thick | | |
| Minimum | L.F. | 0.200 |
| Average | " | 0.267 |
| Maximum | " | 0.320 |
| 25" x 1-1/4" thick | | |
| Minimum | L.F. | 0.267 |
| Average | " | 0.320 |
| Maximum | " | 0.400 |
| Add for cutouts | EA. | 0.500 |
| Backsplash, 4" high, 7/8" thick | L.F. | 0.160 |
| Plywood, sanded, A-C | | |
| 1/4" thick | S.F. | 0.027 |
| 3/8" thick | " | 0.029 |
| 1/2" thick | " | 0.031 |
| A-D | | |
| 1/4" thick | S.F. | 0.027 |
| 3/8" thick | " | 0.029 |
| 1/2" thick | " | 0.031 |
| Base cabinets, 34-1/2" high, 24" deep, hardwood, no tops | | |

## FINISH CARPENTRY

| 06220.10 MILLWORK | UNIT | MAN/HOURS |
|---|---|---|
| Minimum | L.F. | 0.320 |
| Average | " | 0.400 |
| Maximum | " | 0.533 |
| Wall cabinets | | |
| Minimum | L.F. | 0.267 |
| Average | " | 0.320 |
| Maximum | " | 0.400 |
| Oil borne | | |
| Water borne | | |

## ARCHITECTURAL WOODWORK

| 06420.10 PANEL WORK | UNIT | MAN/HOURS |
|---|---|---|
| Hardboard, tempered, 1/4" thick | | |
| Natural faced | S.F. | 0.020 |
| Plastic faced | " | 0.023 |
| Pegboard, natural | " | 0.020 |
| Plastic faced | " | 0.023 |
| Untempered, 1/4" thick | | |
| Natural faced | S.F. | 0.020 |
| Plastic faced | " | 0.023 |
| Pegboard, natural | " | 0.020 |
| Plastic faced | " | 0.023 |
| Plywood unfinished, 1/4" thick | | |
| Birch | | |
| Natural | S.F. | 0.027 |
| Select | " | 0.027 |
| Knotty pine | " | 0.027 |
| Cedar (closet lining) | | |
| Standard boards T&G | S.F. | 0.027 |
| Particle board | " | 0.027 |
| Plywood, prefinished, 1/4" thick, premium grade | | |
| Birch veneer | S.F. | 0.032 |
| Cherry veneer | " | 0.032 |
| Chestnut veneer | " | 0.032 |
| Lauan veneer | " | 0.032 |
| Mahogany veneer | " | 0.032 |
| Oak veneer (red) | " | 0.032 |
| Pecan veneer | " | 0.032 |
| Rosewood veneer | " | 0.032 |
| Teak veneer | " | 0.032 |
| Walnut veneer | " | 0.032 |

# 06 WOOD AND PLASTICS

| ARCHITECTURAL WOODWORK | UNIT | MAN/ HOURS |
|---|---|---|
| **06430.10 STAIRWORK** | | |
| Risers, 1x8, 42" wide | | |
| White oak | EA. | 0.400 |
| Pine | " | 0.400 |
| Treads, 1-1/16" x 9-1/2" x 42" | | |
| White oak | EA. | 0.500 |
| **06440.10 COLUMNS** | | |
| Column, hollow, round wood | | |
| 12" diameter | | |
| 10' high | EA. | 0.800 |
| 12' high | " | 0.857 |
| 14' high | " | 0.960 |
| 16' high | " | 1.200 |
| 24" diameter | | |
| 16' high | EA. | 1.200 |
| 18' high | " | 1.263 |
| 20' high | " | 1.263 |
| 22' high | " | 1.333 |
| 24' high | " | 1.333 |

# 07 THERMAL AND MOISTURE

| MOISTURE PROTECTION | UNIT | MAN/ HOURS |
|---|---|---|
| **07100.10 WATERPROOFING** | | |
| Membrane waterproofing, elastomeric | | |
| Butyl | | |
| 1/32" thick | S.F. | 0.032 |
| 1/16" thick | " | 0.033 |
| Butyl with nylon | | |
| 1/32" thick | S.F. | 0.032 |
| 1/16" thick | " | 0.033 |
| Neoprene | | |
| 1/32" thick | S.F. | 0.032 |
| 1/16" thick | " | 0.033 |
| Neoprene with nylon | | |
| 1/32" thick | S.F. | 0.032 |
| 1/16" thick | " | 0.033 |
| Bituminous membrane waterproofing, asphalt felt, 15 lb. | | |
| One ply | S.F. | 0.020 |
| Two ply | " | 0.024 |
| Three ply | " | 0.029 |
| Four ply | " | 0.033 |
| Five ply | " | 0.042 |
| Modified asphalt membrane waterproofing, fibrous asphalt | | |
| One ply | S.F. | 0.033 |
| Two ply | " | 0.040 |
| Three ply | " | 0.044 |
| Four ply | " | 0.053 |
| Five ply | " | 0.064 |
| Asphalt coated protective board | | |
| 1/8" thick | S.F. | 0.020 |
| 1/4" thick | " | 0.020 |
| 3/8" thick | " | 0.020 |
| 1/2" thick | " | 0.021 |
| Cement protective board | | |
| 3/8" thick | S.F. | 0.027 |
| 1/2" thick | " | 0.027 |
| Fluid applied, neoprene | | |
| 50 mil | S.F. | 0.027 |
| 90 mil | " | 0.027 |
| Tab extended polyurethane | | |
| .050" thick | S.F. | 0.020 |
| Fluid applied rubber based polyurethane | | |
| 6 mil | S.F. | 0.025 |
| 15 mil | " | 0.020 |
| Bentonite waterproofing, panels | | |
| 3/16" thick | S.F. | 0.020 |
| 1/4" thick | " | 0.020 |
| 5/8" thick | " | 0.021 |
| Granular admixtures, trowel on, 3/8" thick | " | 0.020 |
| Metallic oxide waterproofing, iron compound, troweled | | |
| 5/8" thick | S.F. | 0.020 |
| 3/4" thick | " | 0.023 |
| **07150.10 DAMPROOFING** | | |
| Silicone dampproofing, sprayed on | | |
| Concrete surface | | |
| 1 coat | S.F. | 0.004 |
| 2 coats | " | 0.006 |
| Concrete block | | |
| 1 coat | S.F. | 0.005 |
| 2 coats | " | 0.007 |
| Brick | | |
| 1 coat | S.F. | 0.006 |
| 2 coats | " | 0.008 |
| **07160.10 BITUMINOUS DAMPPROOFING** | | |
| Building paper, asphalt felt | | |
| 15 lb | S.F. | 0.032 |
| 30 lb | " | 0.033 |
| Asphalt dampproofing, troweled, cold, primer plus | | |
| 1 coat | S.F. | 0.027 |
| 2 coats | " | 0.040 |
| 3 coats | " | 0.050 |
| Fibrous asphalt dampproofing, hot troweled, primer plus | | |
| 1 coat | S.F. | 0.032 |
| 2 coats | " | 0.044 |
| 3 coats | " | 0.057 |
| Asphaltic paint dampproofing, per coat | | |
| Brush on | S.F. | 0.011 |
| Spray on | " | 0.009 |
| **07190.10 VAPOR BARRIERS** | | |
| Vapor barrier, polyethylene | | |
| 2 mil | S.F. | 0.004 |
| 6 mil | " | 0.004 |
| 8 mil | " | 0.004 |
| 10 mil | " | 0.004 |

| INSULATION | UNIT | MAN/ HOURS |
|---|---|---|
| **07210.10 BATT INSULATION** | | |
| Ceiling, fiberglass, unfaced | | |
| 3-1/2" thick, R11 | S.F. | 0.009 |
| 6" thick, R19 | " | 0.011 |
| 9" thick, R30 | " | 0.012 |
| Suspended ceiling, unfaced | | |
| 3-1/2" thick, R11 | S.F. | 0.009 |
| 6" thick, R19 | " | 0.010 |
| 9" thick, R30 | " | 0.011 |
| Crawl space, unfaced | | |
| 3-1/2" thick, R11 | S.F. | 0.012 |
| 6" thick, R19 | " | 0.013 |

# 07 THERMAL AND MOISTURE

| INSULATION | UNIT | MAN/ HOURS |
|---|---|---|
| **07210.10 BATT INSULATION** | | |
| 9" thick, R30 | S.F. | 0.015 |
| Wall, fiberglass | | |
| Paper backed | | |
| 2" thick, R7 | S.F. | 0.008 |
| 3" thick, R8 | " | 0.009 |
| 4" thick, R11 | " | 0.009 |
| 6" thick, R19 | " | 0.010 |
| Foil backed, 1 side | | |
| 2" thick, R7 | S.F. | 0.008 |
| 3" thick, R11 | " | 0.009 |
| 4" thick, R14 | " | 0.009 |
| 6" thick, R21 | " | 0.010 |
| Foil backed, 2 sides | | |
| 2" thick, R7 | S.F. | 0.009 |
| 3" thick, R11 | " | 0.010 |
| 4" thick, R14 | " | 0.011 |
| 6" thick, R21 | " | 0.011 |
| Unfaced | | |
| 2" thick, R7 | S.F. | 0.008 |
| 3" thick, R9 | " | 0.009 |
| 4" thick, R11 | " | 0.009 |
| 6" thick, R19 | " | 0.010 |
| Mineral wool batts | | |
| Paper backed | | |
| 2" thick, R6 | S.F. | 0.008 |
| 4" thick, R12 | " | 0.009 |
| 6" thick, R19 | " | 0.010 |
| Fasteners, self adhering, attached to ceiling deck | | |
| 2-1/2" long | EA. | 0.013 |
| 4-1/2" long | " | 0.015 |
| Capped, self-locking washers for fastening insulation | " | 0.008 |
| **07210.20 BOARD INSULATION** | | |
| Insulation, rigid | | |
| Fiberglass, roof | | |
| 0.75" thick, R2.78 | S.F. | 0.007 |
| 1.06" thick, R4.17 | " | 0.008 |
| 1.31" thick, R5.26 | " | 0.008 |
| 1.63" thick, R6.67 | " | 0.008 |
| 2.25" thick, R8.33 | " | 0.009 |
| Composite board, roof | | |
| 1-1/2" thick, R6.67 | S.F. | 0.008 |
| 1-5/8" thick, R7.69 | " | 0.008 |
| 2" thick, R10.0 | " | 0.009 |
| 2-1/4" thick, R12.50 | " | 0.009 |
| 2-1/2" thick, R14.29 | " | 0.010 |
| 2-3/4" thick, R16.67 | " | 0.011 |
| 3-1/4" thick, R20.00 | " | 0.011 |
| Perlite board, roof | | |
| 1.00" thick, R2.78 | S.F. | 0.007 |
| 1.50" thick, R4.17 | " | 0.007 |
| 2.00" thick, R5.92 | " | 0.007 |
| 2.50" thick, R6.67 | " | 0.008 |

| INSULATION | UNIT | MAN/ HOURS |
|---|---|---|
| **07210.20 BOARD INSULATION** | | |
| 3.00" thick, R8.33 | S.F. | 0.008 |
| 4.00" thick, R10.00 | " | 0.008 |
| 5.25" thick, R14.29 | " | 0.009 |
| Rigid urethane | | |
| Roof | | |
| 1" thick, R6.67 | S.F. | 0.007 |
| 1.20" thick, R8.33 | " | 0.007 |
| 1.50" thick, R11.11 | " | 0.007 |
| 2" thick, R14.29 | " | 0.007 |
| 2.25" thick, R16.67 | " | 0.008 |
| Wall | | |
| 1" thick, R6.67 | S.F. | 0.008 |
| 1.5" thick, R11.11 | " | 0.009 |
| 2" thick, R14.29 | " | 0.009 |
| Polystyrene | | |
| Roof | | |
| 1.0" thick, R4.17 | S.F. | 0.007 |
| 1.5" thick, R6.26 | " | 0.007 |
| 2.0" thick, R8.33 | " | 0.007 |
| Wall | | |
| 1.0" thick, R4.17 | S.F. | 0.008 |
| 1.5" thick, R6.26 | " | 0.009 |
| 2.0" thick, R8.33 | " | 0.009 |
| Rigid board insulation, deck | | |
| Mineral fiberboard | | |
| 1" thick, R3.0 | S.F. | 0.007 |
| 2" thick, R5.26 | " | 0.007 |
| Fiberglass | | |
| 1" thick, R4.3 | S.F. | 0.007 |
| 2" thick, R8.5 | " | 0.007 |
| Polystyrene | | |
| 1" thick, R5.4 | S.F. | 0.007 |
| 2" thick, R10.8 | " | 0.007 |
| Urethane | | |
| .75" thick, R5.4 | S.F. | 0.007 |
| 1" thick, R6.4 | " | 0.007 |
| 1.5" thick, R10.7 | " | 0.007 |
| 2" thick, R14.3 | " | 0.007 |
| Foamglass | | |
| 1" thick, R1.8 | S.F. | 0.007 |
| 2" thick, R5.26 | " | 0.007 |
| Wood fiber | | |
| 1" thick, R3.85 | S.F. | 0.007 |
| 2" thick, R7.7 | " | 0.007 |
| Particle board | | |
| 3/4" thick, R2.08 | S.F. | 0.007 |
| 1" thick, R2.77 | " | 0.007 |
| 2" thick, R5.50 | " | 0.007 |
| **07210.60 LOOSE FILL INSULATION** | | |
| Blown-in type | | |
| Fiberglass | | |
| 5" thick, R11 | S.F. | 0.007 |
| 6" thick, R13 | " | 0.008 |

# 07 THERMAL AND MOISTURE

| INSULATION | UNIT | MAN/ HOURS |
|---|---|---|
| **07210.60 LOOSE FILL INSULATION** | | |
| 9" thick, R19 | S.F. | 0.011 |
| Rockwool, attic application | | |
| 6" thick, R13 | S.F. | 0.008 |
| 8" thick, R19 | " | 0.010 |
| 10" thick, R22 | " | 0.012 |
| 12" thick, R26 | " | 0.013 |
| 15" thick, R30 | " | 0.016 |
| Poured type | | |
| Fiberglass | | |
| 1" thick, R4 | S.F. | 0.005 |
| 2" thick, R8 | " | 0.006 |
| 3" thick, R12 | " | 0.007 |
| 4" thick, R16 | " | 0.008 |
| Mineral wool | | |
| 1" thick, R3 | S.F. | 0.005 |
| 2" thick, R6 | " | 0.006 |
| 3" thick, R9 | " | 0.007 |
| 4" thick, R12 | " | 0.008 |
| Vermiculite or perlite | | |
| 2" thick, R4.8 | S.F. | 0.006 |
| 3" thick, R7.2 | " | 0.007 |
| 4" thick, R9.6 | " | 0.008 |
| Masonry, poured vermiculite or perlite | | |
| 4" block | S.F. | 0.004 |
| 6" block | " | 0.005 |
| 8" block | " | 0.006 |
| 10" block | " | 0.006 |
| 12" block | " | 0.007 |
| **07210.70 SPRAYED INSULATION** | | |
| Foam, sprayed on | | |
| Polystyrene | | |
| 1" thick, R4 | S.F. | 0.008 |
| 2" thick, R8 | " | 0.011 |
| Urethane | | |
| 1" thick, R4 | S.F. | 0.008 |
| 2" thick, R8 | " | 0.011 |
| **07250.10 FIREPROOFING** | | |
| Sprayed on | | |
| 1" thick | | |
| On beams | S.F. | 0.018 |
| On columns | " | 0.016 |
| On decks | | |
| Flat surface | S.F. | 0.008 |
| Fluted surface | " | 0.010 |
| 1-1/2" thick | | |
| On beams | S.F. | 0.023 |
| On columns | " | 0.020 |
| On decks | | |
| Flat surface | S.F. | 0.010 |
| Fluted surface | " | 0.013 |

| SHINGLES AND TILES | UNIT | MAN/ HOURS |
|---|---|---|
| **07310.10 ASPHALT SHINGLES** | | |
| Standard asphalt shingles, strip shingles | | |
| 210 lb/square | SQ. | 0.800 |
| 235 lb/square | " | 0.889 |
| 240 lb/square | " | 1.000 |
| 260 lb/square | " | 1.143 |
| 300 lb/square | " | 1.333 |
| 385 lb/square | " | 1.600 |
| Roll roofing, mineral surface | | |
| 90 lb | SQ. | 0.571 |
| 110 lb | " | 0.667 |
| 140 lb | " | 0.800 |
| **07310.30 METAL SHINGLES** | | |
| Aluminum, .020" thick | | |
| Plain | SQ. | 1.600 |
| Colors | " | 1.600 |
| Steel, galvanized | | |
| 26 ga. | | |
| Plain | SQ. | 1.600 |
| Colors | " | 1.600 |
| 24 ga. | | |
| Plain | SQ. | 1.600 |
| Colors | " | 1.600 |
| Porcelain enamel, 22 ga. | | |
| Minimum | SQ. | 2.000 |
| Average | " | 2.000 |
| Maximum | " | 2.000 |
| **07310.60 SLATE SHINGLES** | | |
| Slate shingles | | |
| Pennsylvania | | |
| Ribbon | SQ. | 4.000 |
| Clear | " | 4.000 |
| Vermont | | |
| Black | SQ. | 4.000 |
| Gray | " | 4.000 |
| Green | " | 4.000 |
| Red | " | 4.000 |
| Replacement shingles | | |
| Small jobs | EA. | 0.267 |
| Large jobs | S.F. | 0.133 |
| **07310.70 WOOD SHINGLES** | | |
| Wood shingles, on roofs | | |
| White cedar, #1 shingles | | |
| 4" exposure | SQ. | 2.667 |
| 5" exposure | " | 2.000 |
| #2 shingles | | |
| 4" exposure | SQ. | 2.667 |
| 5" exposure | " | 2.000 |
| Resquared and rebutted | | |
| 4" exposure | SQ. | 2.667 |
| 5" exposure | " | 2.000 |

# 07 THERMAL AND MOISTURE

| SHINGLES AND TILES | UNIT | MAN/ HOURS |
|---|---|---|
| **07310.70 WOOD SHINGLES** | | |
| On walls | | |
| White cedar, #1 shingles | | |
| 4" exposure | SQ. | 4.000 |
| 5" exposure | " | 3.200 |
| 6" exposure | " | 2.667 |
| #2 shingles | | |
| 4" exposure | SQ. | 4.000 |
| 5" exposure | " | 3.200 |
| 6" exposure | " | 2.667 |
| **07310.80 WOOD SHAKES** | | |
| Shakes, hand split, 24" red cedar, on roofs | | |
| 5" exposure | SQ. | 4.000 |
| 7" exposure | " | 3.200 |
| 9" exposure | " | 2.667 |
| On walls | | |
| 6" exposure | SQ. | 4.000 |
| 8" exposure | " | 3.200 |
| 10" exposure | " | 2.667 |

| ROOFING AND SIDING | UNIT | MAN/ HOURS |
|---|---|---|
| **07410.10 MANUFACTURED ROOFS** | | |
| Aluminum roof panels, for structural steel framing | | |
| Corrugated | | |
| Unpainted finish | | |
| .024" | S.F. | 0.020 |
| .030" | " | 0.020 |
| Painted finish | | |
| .024" | S.F. | 0.020 |
| .030" | " | 0.020 |
| V-beam | | |
| Unpainted finish | | |
| .032" | S.F. | 0.020 |
| .040" | " | 0.020 |
| .050" | " | 0.020 |
| Painted finish | | |
| .032" | S.F. | 0.020 |
| .040" | " | 0.020 |
| .050" | " | 0.020 |
| Steel roof panels, for structural steel framing | | |
| Corrugated, painted | | |
| 18 ga. | S.F. | 0.020 |
| 20 ga. | " | 0.020 |
| 22 ga. | " | 0.020 |
| Box rib, painted | | |

| ROOFING AND SIDING | UNIT | MAN/ HOURS |
|---|---|---|
| **07410.10 MANUFACTURED ROOFS** | | |
| 18 ga. | S.F. | 0.021 |
| 20 ga. | " | 0.021 |
| 22 ga. | " | 0.021 |
| 4" rib, painted | | |
| 18 ga. | S.F. | 0.022 |
| 20 ga. | " | 0.022 |
| 22 ga. | " | 0.022 |
| Standing seam roof | | |
| 2" high seam, painted | | |
| 22 ga. | S.F. | 0.032 |
| 24 ga. | " | 0.032 |
| 26 ga. | " | 0.032 |
| **07410.30 MANUFACTURED WALLS** | | |
| Sandwich panels with 1-1/2" fiberglass insulation | | |
| Galvanized 18 ga. steel interior panels | | |
| Exterior panels | | |
| 16 ga. aluminum | S.F. | 0.107 |
| 18 ga. galvanized steel | " | 0.107 |
| 20 ga. painted steel | " | 0.107 |
| 20 ga. stainless steel | " | 0.107 |
| Metal liner panels, 1-3/8" thick, 24" wide | | |
| Galvanized | | |
| 22 ga. | S.F. | 0.027 |
| 20 ga. | " | 0.027 |
| 18 ga. | " | 0.027 |
| Primed | | |
| 22 ga. | S.F. | 0.027 |
| 20 ga. | " | 0.027 |
| 18 ga. | " | 0.027 |
| **07440.10 AGGREGATE COATED PANELS** | | |
| Dryvit type system | | |
| 1" thick | S.F. | 0.027 |
| 1-1/2" thick | " | 0.029 |
| 2" thick | " | 0.033 |
| **07460.10 METAL SIDING PANELS** | | |
| Aluminum siding panels | | |
| Corrugated | | |
| Plain finish | | |
| .024" | S.F. | 0.032 |
| .032" | " | 0.032 |
| Painted finish | | |
| .024" | S.F. | 0.032 |
| .032" | " | 0.032 |
| V. beam | | |
| Plain finish | | |
| .032" | S.F. | 0.032 |
| .040" | " | 0.032 |
| .050" | " | 0.032 |
| Painted finish | | |
| .032" | S.F. | 0.032 |
| .040" | " | 0.032 |

## ROOFING AND SIDING

### 07460.10 METAL SIDING PANELS

| ROOFING AND SIDING | UNIT | MAN/HOURS |
|---|---|---|
| .050" | S.F. | 0.032 |
| 4" rib | | |
| Plain finish | | |
| .032" | S.F. | 0.036 |
| .040" | " | 0.036 |
| .050" | " | 0.036 |
| Painted finish | | |
| .032" | S.F. | 0.036 |
| .040" | " | 0.036 |
| .050" | " | 0.036 |
| Steel siding panels | | |
| Corrugated | | |
| 22 ga. | S.F. | 0.053 |
| 24 ga. | " | 0.053 |
| 26 ga. | " | 0.053 |
| Box rib | | |
| 20 ga. | S.F. | 0.053 |
| 22 ga. | " | 0.053 |
| 24 ga. | " | 0.053 |
| 26 ga. | " | 0.053 |

### 07460.50 PLASTIC SIDING

| ROOFING AND SIDING | UNIT | MAN/HOURS |
|---|---|---|
| Horizontal vinyl siding, solid | | |
| 8" wide | | |
| Standard | S.F. | 0.031 |
| Insulated | " | 0.031 |
| 10" wide | | |
| Standard | S.F. | 0.029 |
| Insulated | " | 0.029 |
| Vinyl moldings for doors and windows | L.F. | 0.032 |

### 07460.60 PLYWOOD SIDING

| ROOFING AND SIDING | UNIT | MAN/HOURS |
|---|---|---|
| Rough sawn cedar, 3/8" thick | S.F. | 0.027 |
| Fir, 3/8" thick | " | 0.027 |
| Texture 1-11, 5/8" thick | | |
| Cedar | S.F. | 0.029 |
| Fir | " | 0.029 |
| Redwood | " | 0.029 |
| Southern Yellow Pine | " | 0.029 |

### 07460.70 STEEL SIDING

| ROOFING AND SIDING | UNIT | MAN/HOURS |
|---|---|---|
| Ribbed, sheets, galvanized | | |
| 22 ga. | S.F. | 0.032 |
| 24 ga. | " | 0.032 |
| 26 ga. | " | 0.032 |
| 28 ga. | " | 0.032 |
| Primed | | |
| 24 ga. | S.F. | 0.032 |
| 26 ga. | " | 0.032 |
| 28 ga. | " | 0.032 |

## ROOFING AND SIDING

### 07460.80 WOOD SIDING

| ROOFING AND SIDING | UNIT | MAN/HOURS |
|---|---|---|
| Beveled siding, cedar | | |
| A grade | | |
| 1/2 x 6 | S.F. | 0.040 |
| 1/2 x 8 | " | 0.032 |
| 3/4 x 10 | " | 0.027 |
| Clear | | |
| 1/2 x 6 | S.F. | 0.040 |
| 1/2 x 8 | " | 0.032 |
| 3/4 x 10 | " | 0.027 |
| B grade | | |
| 1/2 x 6 | S.F. | 0.040 |
| 1/2 x 8 | " | 0.320 |
| 3/4 x 10 | " | 0.027 |
| Board and batten | | |
| Cedar | | |
| 1x6 | S.F. | 0.040 |
| 1x8 | " | 0.032 |
| 1x10 | " | 0.029 |
| 1x12 | " | 0.026 |
| Pine | | |
| 1x6 | S.F. | 0.040 |
| 1x8 | " | 0.032 |
| 1x10 | " | 0.029 |
| 1x12 | " | 0.026 |
| Redwood | | |
| 1x6 | S.F. | 0.040 |
| 1x8 | " | 0.032 |
| 1x10 | " | 0.029 |
| 1x12 | " | 0.026 |
| Tongue and groove | | |
| Cedar | | |
| 1x4 | S.F. | 0.044 |
| 1x6 | " | 0.042 |
| 1x8 | " | 0.040 |
| 1x10 | " | 0.038 |
| Pine | | |
| 1x4 | S.F. | 0.044 |
| 1x6 | " | 0.042 |
| 1x8 | " | 0.040 |
| 1x10 | " | 0.038 |
| Redwood | | |
| 1x4 | S.F. | 0.044 |
| 1x6 | " | 0.042 |
| 1x8 | " | 0.040 |
| 1x10 | " | 0.038 |

# 07 THERMAL AND MOISTURE

## MEMBRANE ROOFING

### 07510.10 BUILT-UP ASPHALT ROOFING

| | UNIT | MAN/ HOURS |
|---|---|---|
| Built-up roofing, asphalt felt, including gravel | | |
| 2 ply | SQ. | 2.000 |
| 3 ply | " | 2.667 |
| 4 ply | " | 3.200 |
| Walkway, for built-up roofs | | |
| 3' x 3' x | | |
| 1/2" thick | S.F. | 0.027 |
| 3/4" thick | " | 0.027 |
| 1" thick | " | 0.027 |
| Cant strip, 4" x 4" | | |
| Treated wood | L.F. | 0.023 |
| Foamglass | " | 0.020 |
| Mineral fiber | " | 0.020 |
| New gravel for built-up roofing, 400 lb/sq | SQ. | 1.600 |
| Roof gravel (ballast) | C.Y. | 4.000 |
| Aluminum coating, top surfacing, for built-up roofing | SQ. | 1.333 |
| Remove 4-ply built-up roof (includes gravel) | " | 4.000 |
| Remove & replace gravel, includes flood coat | " | 2.667 |

### 07530.10 SINGLE-PLY ROOFING

| | UNIT | MAN/ HOURS |
|---|---|---|
| Elastic sheet roofing | | |
| Neoprene, 1/16" thick | S.F. | 0.010 |
| EPDM rubber | | |
| 45 mil | S.F. | 0.010 |
| 60 mil | " | 0.010 |
| PVC | | |
| 45 mil | S.F. | 0.010 |
| 60 mil | " | 0.010 |
| Flashing | | |
| Pipe flashing, 90 mil thick | | |
| 1" pipe | EA. | 0.200 |
| 2" pipe | " | 0.200 |
| 3" pipe | " | 0.211 |
| 4" pipe | " | 0.211 |
| 5" pipe | " | 0.222 |
| 6" pipe | " | 0.222 |
| 8" pipe | " | 0.235 |
| 10" pipe | " | 0.267 |
| 12" pipe | " | 0.267 |
| Neoprene flashing, 60 mil thick strip | | |
| 6" wide | L.F. | 0.067 |
| 12" wide | " | 0.100 |
| 18" wide | " | 0.133 |
| 24" wide | " | 0.200 |
| Adhesives | | |
| Mastic sealer, applied at joints only | | |
| 1/4" bead | L.F. | 0.004 |
| Fluid applied roofing | | |
| Urethane, 2 components, elastomeric top membrane | | |
| 1" thick | S.F. | 0.013 |
| Vinyl liquid roofing, 2 coats, 2 mils per coat | " | 0.011 |
| Silicone roofing, 2 coats sprayed, 16 mil per coat | " | 0.013 |
| Inverted roof system | | |
| Insulated membrane with coarse gravel ballast | | |
| 3 ply with 2" polystyrene | S.F. | 0.013 |
| Ballast, 3/4" through 1-1/2" dia. river gravel, 100lb/sf | " | 0.800 |
| Walkway for membrane roofs, 1/2" thick | " | 0.027 |

## FLASHING AND SHEET METAL

### 07610.10 METAL ROOFING

| | UNIT | MAN/ HOURS |
|---|---|---|
| Sheet metal roofing, copper, 16 oz, batten seam | SQ. | 5.333 |
| Standing seam | " | 5.000 |
| Aluminum roofing, natural finish | | |
| Corrugated, on steel frame | | |
| .0175" thick | SQ. | 2.286 |
| .0215" thick | " | 2.286 |
| .024" thick | " | 2.286 |
| .032" thick | " | 2.286 |
| V-beam, on steel frame | | |
| .032" thick | SQ. | 2.286 |
| .040" thick | " | 2.286 |
| .050" thick | " | 2.286 |
| Ridge cap | | |
| .019" thick | L.F. | 0.027 |
| Corrugated galvanized steel roofing, on steel frame | | |
| 28 ga. | SQ. | 2.286 |
| 26 ga. | " | 2.286 |
| 24 ga. | " | 2.286 |
| 22 ga. | " | 2.286 |
| 26 ga., factory insulated with 1" polystyrene | " | 3.200 |
| Ridge roll | | |
| 10" wide | L.F. | 0.027 |
| 20" wide | " | 0.032 |

### 07620.10 FLASHING AND TRIM

| | UNIT | MAN/ HOURS |
|---|---|---|
| Counter flashing | | |
| Aluminum, .032" | S.F. | 0.080 |
| Stainless steel, .015" | " | 0.080 |
| Copper | | |
| 16 oz. | S.F. | 0.080 |
| 20 oz. | " | 0.080 |
| 24 oz. | " | 0.080 |
| 32 oz. | " | 0.080 |
| Valley flashing | | |
| Aluminum, .032" | S.F. | 0.050 |
| Stainless steel, .015 | " | 0.050 |
| Copper | | |
| 16 oz. | S.F. | 0.050 |

# 07 THERMAL AND MOISTURE

| FLASHING AND SHEET METAL | UNIT | MAN/ HOURS |
|---|---|---|
| **07620.10 FLASHING AND TRIM** | | |
| 20 oz. | S.F. | 0.067 |
| 24 oz. | " | 0.050 |
| 32 oz. | " | 0.050 |
| Base flashing | | |
| Aluminum, .040" | S.F. | 0.067 |
| Stainless steel, .018" | " | 0.067 |
| Copper | | |
| 16 oz. | S.F. | 0.067 |
| 20 oz. | " | 0.050 |
| 24 oz. | " | 0.067 |
| 32 oz. | " | 0.067 |
| Waterstop, "T" section, 22 ga. | | |
| 1-1/2" x 3" | L.F. | 0.040 |
| 2" x 2" | " | 0.040 |
| 4" x 3" | " | 0.040 |
| 6" x 4" | " | 0.040 |
| 8" x 4" | " | 0.040 |
| Scupper outlets | | |
| 10" x 10" x 4" | EA. | 0.200 |
| 22" x 4" x 4" | " | 0.200 |
| 8" x 8" x 5" | " | 0.200 |
| Flashing and trim, aluminum | | |
| .019" thick | S.F. | 0.057 |
| .032" thick | " | 0.057 |
| .040" thick | " | 0.062 |
| Neoprene sheet flashing, .060" thick | " | 0.050 |
| Copper, paper backed | | |
| 2 oz. | S.F. | 0.080 |
| Drainage boots, roof, cast iron | | |
| 2 x 3 | L.F. | 0.100 |
| 3 x 4 | " | 0.100 |
| 4 x 5 | " | 0.107 |
| 4 x 6 | " | 0.107 |
| 5 x 7 | " | 0.114 |
| Pitch pocket, copper, 16 oz. | | |
| 4 x 4 | EA. | 0.200 |
| 6 x 6 | " | 0.200 |
| 8 x 8 | " | 0.200 |
| 8 x 10 | " | 0.200 |
| 8 x 12 | " | 0.200 |
| Reglets, copper 10 oz. | L.F. | 0.053 |
| Stainless steel, .020" | " | 0.053 |
| Gravel stop | | |
| Aluminum, .032" | | |
| 4" | L.F. | 0.027 |
| 6" | " | 0.027 |
| 8" | " | 0.031 |
| 10" | " | 0.031 |
| Copper, 16 oz. | | |
| 4" | L.F. | 0.027 |
| 6" | " | 0.027 |
| 8" | " | 0.031 |
| 10" | " | 0.031 |

| FLASHING AND SHEET METAL | UNIT | MAN/ HOURS |
|---|---|---|
| **07620.20 GUTTERS AND DOWNSPOUTS** | | |
| Copper gutter and downspout | | |
| Downspouts, 16 oz. copper | | |
| Round | | |
| 3" dia. | L.F. | 0.053 |
| 4" dia. | " | 0.053 |
| Rectangular, corrugated | | |
| 2" x 3" | L.F. | 0.050 |
| 3" x 4" | " | 0.050 |
| Rectangular, flat surface | | |
| 2" x 3" | L.F. | 0.053 |
| 3" x 4" | " | 0.053 |
| Lead-coated copper downspouts | | |
| Round | | |
| 3" dia. | L.F. | 0.050 |
| 4" dia. | " | 0.057 |
| Rectangular, corrugated | | |
| 2" x 3" | L.F. | 0.053 |
| 3" x 4" | " | 0.053 |
| Rectangular, plain | | |
| 2" x 3" | L.F. | 0.053 |
| 3" x 4" | " | 0.053 |
| Gutters, 16 oz. copper | | |
| Half round | | |
| 4" wide | L.F. | 0.080 |
| 5" wide | " | 0.089 |
| Type K | | |
| 4" wide | L.F. | 0.080 |
| 5" wide | " | 0.089 |
| Lead-coated copper gutters | | |
| Half round | | |
| 4" wide | L.F. | 0.080 |
| 6" wide | " | 0.089 |
| Type K | | |
| 4" wide | L.F. | 0.080 |
| 5" wide | " | 0.089 |
| Aluminum gutter and downspout | | |
| Downspouts | | |
| 2" x 3" | L.F. | 0.053 |
| 3" x 4" | " | 0.057 |
| 4" x 5" | " | 0.062 |
| Round | | |
| 3" dia. | L.F. | 0.053 |
| 4" dia. | " | 0.057 |
| Gutters, stock units | | |
| 4" wide | L.F. | 0.084 |
| 5" wide | " | 0.089 |
| Galvanized steel gutter and downspout | | |
| Downspouts, round corrugated | | |
| 3" dia. | L.F. | 0.053 |
| 4" dia. | " | 0.053 |
| 5" dia. | " | 0.057 |
| 6" dia. | " | 0.057 |
| Rectangular | | |
| 2" x 3" | L.F. | 0.053 |

## FLASHING AND SHEET METAL

### 07620.20 GUTTERS AND DOWNSPOUTS

| | UNIT | MAN/HOURS |
|---|---|---|
| 3" x 4" | L.F. | 0.050 |
| 4" x 4" | " | 0.050 |
| Gutters, stock units | | |
| 5" wide | | |
| Plain | L.F. | 0.089 |
| Painted | " | 0.089 |
| 6" wide | | |
| Plain | L.F. | 0.094 |
| Painted | " | 0.094 |

## ROOFING SPECIALTIES

### 07700.10 MANUFACTURED SPECIALTIES

| | UNIT | MAN/HOURS |
|---|---|---|
| Moisture relief vent | | |
| Aluminum | EA. | 0.114 |
| Copper | " | 0.114 |
| Expansion joint | | |
| Aluminum | | |
| Opening to 2.5" | L.F. | 0.057 |
| Opening to 3.5" | " | 0.062 |
| Copper, 16 oz. | | |
| Opening to 2.5" | L.F. | 0.057 |
| Opening to 3.5" | " | 0.062 |
| Butyl or neoprene | | |
| 4" wide | | |
| 16 oz. copper bellows | L.F. | 0.067 |
| 28 ga. stainless steel bellows | " | 0.067 |
| 6" wide | | |
| Copper bellows | L.F. | 0.073 |
| Stainless steel | | |
| Opening to 2.5" | L.F. | 0.057 |
| Opening to 3.5" | " | 0.062 |
| Smoke vent, 48" x 48" | | |
| Aluminum | EA. | 2.000 |
| Galvanized steel | " | 2.000 |
| Heat/smoke vent, 48" x 96" | | |
| Aluminum | EA. | 2.667 |
| Galvanized steel | " | 2.667 |
| Ridge vent strips | | |
| Mill finish | L.F. | 0.053 |
| Connectors | EA. | 0.200 |
| End cap | " | 0.229 |
| Soffit vents | | |
| Mill finish | | |
| 2-1/2" wide | L.F. | 0.032 |
| 3" wide | " | 0.032 |

## ROOFING SPECIALTIES

### 07700.10 MANUFACTURED SPECIALTIES

| | UNIT | MAN/HOURS |
|---|---|---|
| 6" wide | L.F. | 0.032 |
| Roof hatches | | |
| Steel, plain, primed | | |
| 2'6" x 3'0" | EA. | 2.000 |
| 2'6" x 4'6" | " | 2.667 |
| 2'6" x 8'0" | " | 4.000 |
| Galvanized steel | | |
| 2'6" x 3'0" | EA. | 2.000 |
| 2'6" x 4'6" | " | 2.667 |
| 2'6" x 8'0" | " | 4.000 |
| Aluminum | | |
| 2'6" x 3'0" | EA. | 2.000 |
| 2'6" x 4'6" | " | 2.667 |
| 2'6" x 8'0" | " | 4.000 |
| Ceiling access doors | | |
| Swing up model, metal frame | | |
| Steel door | | |
| 2'6" x 2'6" | EA. | 0.800 |
| 2'6" x 3'0" | " | 0.800 |
| Aluminum door | | |
| 2'6" x 2'6" | EA. | 0.800 |
| 2'6" x 3'0" | " | 0.800 |
| Swing down model, metal frame | | |
| Steel door | | |
| 2'6" x 2'6" | EA. | 0.800 |
| 2'6" x 3'0" | " | 0.800 |
| Aluminum door | | |
| 2'6" x 2'6" | EA. | 0.800 |
| 2'6" x 3'0" | " | 0.800 |
| Gravity ventilators, with curb, base, damper and screen | | |
| Stationary siphon | | |
| 6" dia. | EA. | 0.533 |
| 12" dia. | " | 0.533 |
| 24" dia. | " | 0.800 |
| 36" dia. | " | 0.800 |
| Wind driven spinner | | |
| 6" dia. | EA. | 0.533 |
| 12" dia. | " | 0.533 |
| 24" dia. | " | 0.800 |
| 36" dia. | " | 0.800 |
| Stationary mushroom | | |
| 16" dia. | EA. | 0.800 |
| 30" dia. | " | 1.000 |
| 36" dia. | " | 1.333 |
| 42" dia. | " | 1.600 |

# 07 THERMAL AND MOISTURE

## SKYLIGHTS

| 07810.10 PLASTIC SKYLIGHTS | UNIT | MAN/ HOURS |
|---|---|---|
| Single thickness, not including mounting curb | | |
| 2' x 4' | EA. | 1.000 |
| 4' x 4' | " | 1.333 |
| 5' x 5' | " | 2.000 |
| 6' x 8' | " | 2.667 |
| Double thickness, not including mounting curb | | |
| 2' x 4' | EA. | 1.000 |
| 4' x 4' | " | 1.333 |
| 5' x 5' | " | 2.000 |
| 6' x 8' | " | 2.667 |
| Metal framed skylights | | |
| Translucent panels, 2-1/2" thick | S.F. | 0.080 |
| Continuous vaults, 8' wide | | |
| Single glazed | S.F. | 0.100 |
| Double glazed | " | 0.114 |

## JOINT SEALERS

| 07920.10 CAULKING | UNIT | MAN/ HOURS |
|---|---|---|
| Caulk exterior, two component | | |
| 1/4 x 1/2 | L.F. | 0.040 |
| 3/8 x 1/2 | " | 0.044 |
| 1/2 x 1/2 | " | 0.050 |
| Caulk interior, single component | | |
| 1/4 x 1/2 | L.F. | 0.038 |
| 3/8 x 1/2 | " | 0.042 |
| 1/2 x 1/2 | " | 0.047 |
| Butyl rubber fillers | | |
| 1/4" x 1/4" | L.F. | 0.016 |
| 1/2" x 1/2" | " | 0.027 |
| 1/2" x 3/4" | " | 0.032 |
| 3/4" x 3/4" | " | 0.032 |
| 1" x 1" | " | 0.036 |
| Seals, "O" ring type cord | | |
| 1/4" dia. | L.F. | 0.020 |
| 1/2" dia. | " | 0.021 |
| 1" dia. | " | 0.022 |
| 1-1/4" dia. | " | 0.024 |
| 1-1/2" dia. | " | 0.025 |
| 1-3/4" dia. | " | 0.026 |
| 2" dia. | " | 0.027 |
| Polyvinyl chloride, closed cell | | |
| 1/4" x 2" | L.F. | 0.029 |
| 1/4" x 6" | " | 0.036 |
| Silicon foam penetration seal | | |
| 1/4" x 1/2" | L.F. | 0.010 |

## JOINT SEALERS

| 07920.10 CAULKING | UNIT | MAN/ HOURS |
|---|---|---|
| 1/2" x 1/2" | L.F. | 0.013 |
| 1/2" x 3/4" | " | 0.016 |
| 3/4" x 3/4" | " | 0.020 |
| 1/8" x 1" | " | 0.010 |
| 1/8" x 3" | " | 0.016 |
| 1/4" x 3" | " | 0.020 |
| 1/4" x 6" | " | 0.027 |
| 1/2" x 6" | " | 0.062 |
| 1/2" x 9" | " | 0.100 |
| 1/2" x 12" | " | 0.145 |
| Oil base sealants and caulking | | |
| 1/4" x 1/4" | L.F. | 0.020 |
| 1/4" x 3/8" | " | 0.021 |
| 1/4" x 1/2" | " | 0.022 |
| 3/8" x 3/8" | " | 0.023 |
| 3/8" x 1/2" | " | 0.024 |
| 3/8" x 5/8" | " | 0.026 |
| 3/8" x 3/4" | " | 0.028 |
| 1/2" x 1/2" | " | 0.031 |
| 1/2" x 5/8" | " | 0.035 |
| 1/2" x 3/4" | " | 0.040 |
| 1/2" x 7/8" | " | 0.041 |
| 1/2" x 1" | " | 0.042 |
| 3/4" x 3/4" | " | 0.043 |
| 1" x 1" | " | 0.044 |
| Polyurethane compounds | | |
| 1/4" x 1/4" | L.F. | 0.020 |
| 1/4" x 3/8" | " | 0.021 |
| 1/4" x 1/2" | " | 0.022 |
| 3/8" x 3/8" | " | 0.023 |
| 3/8" x 1/2" | " | 0.024 |
| 3/8" x 5/8" | " | 0.026 |
| 3/8" x 3/4" | " | 0.028 |
| 1/2" x 1/2" | " | 0.031 |
| 1/2" x 5/8" | " | 0.035 |
| 1/2" x 3/4" | " | 0.040 |
| 1/2" x 7/8" | " | 0.041 |
| 1/2" x 1" | " | 0.044 |
| 3/4" x 3/4" | " | 0.043 |
| 3/4" x 1" | " | 0.044 |
| Backer rod, polyethylene | | |
| 1/4" | L.F. | 0.020 |
| 1/2" | " | 0.021 |
| 3/4" | " | 0.022 |
| 1" | " | 0.024 |

# 08 DOORS AND WINDOWS

| METAL | UNIT | MAN/ HOURS |
|---|---|---|
| **08110.10 METAL DOORS** | | |
| Flush hollow metal, standard duty, 20 ga., 1-3/8" thick | | |
| 2-6 x 6-8 | EA. | 0.889 |
| 2-8 x 6-8 | " | 0.889 |
| 3-0 x 6-8 | " | 0.889 |
| 1-3/4" thick | | |
| 2-6 x 6-8 | EA. | 0.889 |
| 2-8 x 6-8 | " | 0.889 |
| 3-0 x 6-8 | " | 0.889 |
| 2-6 x 7-0 | " | 0.889 |
| 2-8 x 7-0 | " | 0.889 |
| 3-0 x 7-0 | " | 0.889 |
| Heavy duty, 20 ga., unrated, 1-3/4" | | |
| 2-8 x 6-8 | EA. | 0.889 |
| 3-0 x 6-8 | " | 0.889 |
| 2-8 x 7-0 | " | 0.889 |
| 3-0 x 7-0 | " | 0.889 |
| 3-4 x 7-0 | " | 0.889 |
| 18 ga., 1-3/4", unrated door | | |
| 2-0 x 7-0 | EA. | 0.889 |
| 2-4 x 7-0 | " | 0.889 |
| 2-6 x 7-0 | " | 0.889 |
| 2-8 x 7-0 | " | 0.889 |
| 3-0 x 7-0 | " | 0.889 |
| 3-4 x 7-0 | " | 0.889 |
| 2", unrated door | | |
| 2-0 x 7-0 | EA. | 1.000 |
| 2-4 x 7-0 | " | 1.000 |
| 2-6 x 7-0 | " | 1.000 |
| 2-8 x 7-0 | " | 1.000 |
| 3-0 x 7-0 | " | 1.000 |
| 3-4 x 7-0 | " | 1.000 |
| Galvanized metal door | | |
| 3-0 x 7-0 | EA. | 1.000 |
| **08110.40 METAL DOOR FRAMES** | | |
| Hollow metal, stock, 18 ga., 4-3/4" x 1-3/4" | | |
| 2-0 x 7-0 | EA. | 1.000 |
| 2-4 x 7-0 | " | 1.000 |
| 2-6 x 7-0 | " | 1.000 |
| 2-8 x 7-0 | " | 1.000 |
| 3-0 x 7-0 | " | 1.000 |
| 4-0 x 7-0 | " | 1.333 |
| 5-0 x 7-0 | " | 1.333 |
| 6-0 x 7-0 | " | 1.333 |
| 16 ga., 6-3/4" x 1-3/4" | | |
| 2-0 x 7-0 | EA. | 1.000 |
| 2-4 x 7-0 | " | 1.000 |
| 2-6 x 7-0 | " | 1.000 |
| 2-8 x 7-0 | " | 1.000 |
| 3-0 x 7-0 | " | 1.000 |
| 4-0 x 7-0 | " | 1.333 |
| 6-0 x 7-0 | " | 1.333 |
| Transom frame | | |

| METAL | UNIT | MAN/ HOURS |
|---|---|---|
| **08110.40 METAL DOOR FRAMES** | | |
| 3-4 x 1-6 | EA. | 1.000 |
| 3-8 x 1-6 | " | 1.000 |
| 6-4 x 1-6 | " | 1.000 |
| Transom sash | | |
| 3-0 x 1-4 | EA. | 1.000 |
| 3-4 x 1-4 | " | 1.000 |
| 6-0 x 1-4 | " | 1.000 |
| Sidelights, complete | | |
| 1-0 x 7-2 | EA. | 1.000 |
| 1-4 x 7-2 | " | 1.000 |
| 1-0 x 8-8 | " | 1.000 |
| 1-6 x 8-8 | " | 1.000 |
| 16 ga., 4-3/4" x 1-3/4" | | |
| 2-0 x 7-0 | EA. | 1.000 |
| 2-4 x 7-0 | " | 1.000 |
| 2-6 x 7-0 | " | 1.000 |
| 2-8 x 7-0 | " | 1.000 |
| 3-0 x 7-0 | " | 1.000 |
| 4-0 x 7-0 | " | 1.333 |
| 6-0 x 7-0 | " | 1.333 |
| 3-4 x 1-6 | " | 1.000 |
| 3-8 x 1-6 | " | 1.000 |
| 6-4 x 1-6 | " | 1.000 |
| Transom sash | | |
| 3-0 x 1-4 | EA. | 1.000 |
| 3-4 x 1-4 | " | 1.000 |
| 6-0 x 1-4 | " | 1.000 |
| Sidelights, complete | | |
| 1-0 x 7-2 | EA. | 1.000 |
| 1-4 x 7-2 | " | 1.000 |
| 1-0 x 8-8 | " | 1.000 |
| 1-4 x 8-8 | " | 1.000 |
| 16 ga., 5-3/4" x 1-3/4" | | |
| 2-0 x 7-0 | EA. | 1.000 |
| 2-4 x 7-0 | " | 1.000 |
| 2-6 x 7-0 | " | 1.000 |
| 2-8 x 7-0 | " | 1.000 |
| 3-0 x 7-0 | " | 1.000 |
| 4-0 x 7-0 | " | 1.333 |
| 5-0 x 7-0 | " | 1.333 |
| 6-0 x 7-0 | " | 1.333 |
| Mullions, vertical | | |
| 5-1/4" x 1-3/4" | L.F. | 0.100 |
| 5-1/4" x 2" | " | 0.100 |
| Horizontal | | |
| 5-1/4" x 1-3/4" | L.F. | 0.100 |
| 5-1/4" x 2" | " | 0.100 |
| **08120.10 ALUMINUM DOORS** | | |
| Aluminum doors, commercial | | |
| Narrow stile | | |
| 2-6 x 7-0 | EA. | 4.000 |
| 3-0 x 7-0 | " | 4.000 |

# 08 DOORS AND WINDOWS

| METAL | UNIT | MAN/ HOURS |
|---|---|---|
| **08120.10 ALUMINUM DOORS** | | |
| 3-6 x 7-0 | EA. | 4.000 |
| Pair | | |
| 5-0 x 7-0 | EA. | 8.000 |
| 6-0 x 7-0 | " | 8.000 |
| 7-0 x 7-0 | " | 8.000 |
| Wide stile | | |
| 2-6 x 7-0 | EA. | 4.000 |
| 3-0 x 7-0 | " | 4.000 |
| 3-6 x 7-0 | " | 4.000 |
| Pair | | |
| 5-0 x 7-0 | EA. | 8.000 |
| 6-0 x 7-0 | " | 8.000 |
| 7-0 x 7-0 | " | 8.000 |

| WOOD AND PLASTIC | UNIT | MAN/ HOURS |
|---|---|---|
| **08210.10 WOOD DOORS** | | |
| Solid core, 1-3/8" thick | | |
| Birch faced | | |
| 2-4 x 7-0 | EA. | 1.000 |
| 2-8 x 7-0 | " | 1.000 |
| 3-0 x 7-0 | " | 1.000 |
| 3-4 x 7-0 | " | 1.000 |
| 2-4 x 6-8 | " | 1.000 |
| 2-6 x 6-8 | " | 1.000 |
| 2-8 x 6-8 | " | 1.000 |
| 3-0 x 6-8 | " | 1.000 |
| Lauan faced | | |
| 2-4 x 6-8 | EA. | 1.000 |
| 2-8 x 6-8 | " | 1.000 |
| 3-0 x 6-8 | " | 1.000 |
| 3-4 x 6-8 | " | 1.000 |
| Tempered hardboard faced | | |
| 2-4 x 7-0 | EA. | 1.000 |
| 2-8 x 7-0 | " | 1.000 |
| 3-0 x 7-0 | " | 1.000 |
| 3-4 x 7-0 | " | 1.000 |
| Hollow core, 1-3/8" thick | | |
| Birch faced | | |
| 2-4 x 7-0 | EA. | 1.000 |
| 2-8 x 7-0 | " | 1.000 |
| 3-0 x 7-0 | " | 1.000 |
| 3-4 x 7-0 | " | 1.000 |
| Lauan faced | | |
| 2-4 x 6-8 | EA. | 1.000 |
| 2-6 x 6-8 | " | 1.000 |

| WOOD AND PLASTIC | UNIT | MAN/ HOURS |
|---|---|---|
| **08210.10 WOOD DOORS** | | |
| 2-8 x 6-8 | EA. | 1.000 |
| 3-0 x 6-8 | " | 1.000 |
| 3-4 x 6-8 | " | 1.000 |
| Tempered hardboard faced | | |
| 2-4 x 7-0 | EA. | 1.000 |
| 2-6 x 7-0 | " | 1.000 |
| 2-8 x 7-0 | " | 1.000 |
| 3-0 x 7-0 | " | 1.000 |
| 3-4 x 7-0 | " | 1.000 |
| Solid core, 1-3/4" thick | | |
| Birch faced | | |
| 2-4 x 7-0 | EA. | 1.000 |
| 2-6 x 7-0 | " | 1.000 |
| 2-8 x 7-0 | " | 1.000 |
| 3-0 x 7-0 | " | 1.000 |
| 3-4 x 7-0 | " | 1.000 |
| Lauan faced | | |
| 2-4 x 7-0 | EA. | 1.000 |
| 2-6 x 7-0 | " | 1.000 |
| 2-8 x 7-0 | " | 1.000 |
| 3-0 x 7-0 | " | 1.000 |
| 3-0 x 7-0 | " | 1.000 |
| Tempered hardboard faced | | |
| 2-4 x 7-0 | EA. | 1.000 |
| 2-6 x 7-0 | " | 1.000 |
| 2-8 x 7-0 | " | 1.000 |
| 3-0 x 7-0 | " | 1.000 |
| 3-4 x 7-0 | " | 1.000 |
| Hollow core, 1-3/4" thick | | |
| Birch faood | | |
| 2-4 x 7-0 | EA. | 1.000 |
| 2-6 x 7-0 | " | 1.000 |
| 2-8 x 7-0 | " | 1.000 |
| 3-0 x 7-0 | " | 1.000 |
| 3-4 x 7-0 | " | 1.000 |
| Lauan faced | | |
| 2-4 x 6-8 | EA. | 1.000 |
| 2-6 x 6-8 | " | 1.000 |
| 2-8 x 6-8 | " | 1.000 |
| 3-0 x 6-8 | " | 1.000 |
| 3-4 x 6-8 | " | 1.000 |
| Tempered hardboard | | |
| 2-4 x 7-0 | EA. | 1.000 |
| 2-6 x 7-0 | " | 1.000 |
| 2-8 x 7-0 | " | 1.000 |
| 3-0 x 7-0 | " | 1.000 |
| 3-4 x 7-0 | " | 1.000 |
| Add-on, louver | " | 0.800 |
| Glass | " | 0.800 |
| Exterior doors, 3-0 x 7-0 x 2-1/2", solid core | | |
| Carved | | |
| One face | EA. | 2.000 |
| Two faces | " | 2.000 |
| Closet doors, 1-3/4" thick | | |

# 08 DOORS AND WINDOWS

| WOOD AND PLASTIC | UNIT | MAN/HOURS |
|---|---|---|
| **08210.10 WOOD DOORS** | | |
| Bi-fold or bi-passing, includes frame and trim | | |
| Paneled | | |
| 4-0 x 6-8 | EA. | 1.333 |
| 6-0 x 6-8 | " | 1.333 |
| Louvered | | |
| 4-0 x 6-8 | EA. | 1.333 |
| 6-0 x 6-8 | " | 1.333 |
| Flush | | |
| 4-0 x 6-8 | EA. | 1.333 |
| 6-0 x 6-8 | " | 1.333 |
| Primed | | |
| 4-0 x 6-8 | EA. | 1.333 |
| 6-0 x 6-8 | " | 1.333 |
| **08210.90 WOOD FRAMES** | | |
| Frame, interior, pine | | |
| 2-6 x 6-8 | EA. | 1.143 |
| 2-8 x 6-8 | " | 1.143 |
| 3-0 x 6-8 | " | 1.143 |
| 5-0 x 6-8 | " | 1.143 |
| 6-0 x 6-8 | " | 1.143 |
| 2-6 x 7-0 | " | 1.143 |
| 2-8 x 7-0 | " | 1.143 |
| 3-0 x 7-0 | " | 1.143 |
| 5-0 x 7-0 | " | 1.600 |
| 6-0 x 7-0 | " | 1.600 |
| Exterior, custom, with threshold, including trim | | |
| Walnut | | |
| 3-0 x 7-0 | EA. | 2.000 |
| 6-0 x 7-0 | " | 2.000 |
| Oak | | |
| 3-0 x 7-0 | EA. | 2.000 |
| 6-0 x 7-0 | " | 2.000 |
| Pine | | |
| 2-4 x 7-0 | EA. | 1.600 |
| 2-6 x 7-0 | " | 1.600 |
| 2-8 x 7-0 | " | 1.600 |
| 3-0 x 7-0 | " | 1.600 |
| 3-4 x 7-0 | " | 1.600 |
| 6-0 x 7-0 | " | 2.667 |
| **08300.10 SPECIAL DOORS** | | |
| Vault door and frame, class 5, steel | EA. | 8.000 |
| Overhead door, coiling insulated | | |
| Chain gear, no frame, 12' x 12' | EA. | 10.000 |
| Aluminum, bronze glass panels, 12-9 x 13-0 | " | 8.000 |
| Garage door, flush insulated metal, primed, 9-0 x 7-0 | " | 2.667 |
| Sliding metal fire doors, motorized, fusible link, 3 hr. | | |
| 3-0 x 6-8 | EA. | 16.000 |
| 3-8 x 6-8 | " | 16.000 |
| 4-0 x 8-0 | " | 16.000 |
| 5-0 x 8-0 | " | 16.000 |

| WOOD AND PLASTIC | UNIT | MAN/HOURS |
|---|---|---|
| **08300.10 SPECIAL DOORS** | | |
| Metal clad doors, including electric motor | | |
| Light duty | | |
| Minimum | S.F. | 0.133 |
| Maximum | " | 0.320 |
| Heavy duty | | |
| Minimum | S.F. | 0.400 |
| Maximum | " | 0.500 |
| Hangar doors, based on 150' openings | | |
| To 20' high | S.F. | 0.096 |
| 20' to 40' high | " | 0.060 |
| 40' to 60' high | " | 0.040 |
| 60' to 80' high | " | 0.024 |
| Over 80' high | " | 0.019 |
| Counter doors, (roll-up shutters), standard, manual | | |
| Opening, 4' high | | |
| 4' wide | EA. | 6.667 |
| 6' wide | " | 6.667 |
| 8' wide | " | 7.273 |
| 10' wide | " | 10.000 |
| 14' wide | " | 10.000 |
| 6' high | | |
| 4' wide | EA. | 6.667 |
| 6' wide | " | 7.273 |
| 8' wide | " | 8.000 |
| 10' wide | " | 10.000 |
| 14' wide | " | 11.429 |
| Service doors, (roll up shutters), standard, manual | | |
| Opening | | |
| 8' high x 8' wide | EA. | 4.444 |
| 10' high x 10' wide | " | 6.667 |
| 12' high x 12' wide | " | 10.000 |
| 14' high x 14' wide | " | 13.333 |
| 16' high x 14' wide | " | 13.333 |
| 20' high x 14' wide | " | 20.000 |
| 24' high x 16' wide | " | 17.778 |
| Roll-up doors | | |
| 13-0 high x 14-0 wide | EA. | 11.429 |
| 12-0 high x 14-0 wide | " | 11.429 |
| Top coiling grilles, manually operated, steel or aluminum | | |
| Opening, 4' high x | | |
| 4' wide | EA. | 3.200 |
| 6' wide | " | 3.200 |
| 8' wide | " | 4.444 |
| 12' wide | " | 4.444 |
| 16' wide | " | 6.667 |
| 6' high x | | |
| 4' wide | EA. | 6.667 |
| 6' wide | " | 7.273 |
| 8' wide | " | 8.000 |
| 12' wide | " | 8.889 |
| 16' wide | " | 11.429 |
| Side coiling grilles, manually operated, aluminum | | |
| Opening, 8' high x | | |
| 18' wide | EA. | 60.000 |

## WOOD AND PLASTIC

| 08300.10 SPECIAL DOORS | UNIT | MAN/ HOURS |
|---|---|---|
| 24' wide | EA. | 68.571 |
| 12' high x | | |
| 12' wide | EA. | 60.000 |
| 18' wide | " | 68.571 |
| 24' wide | " | 80.000 |
| Accordion folding doors, tracks and fittings included | | |
| Vinyl covered, 2 layers | S.F. | 0.320 |
| Woven mahogany and vinyl | " | 0.320 |
| Economy vinyl | " | 0.320 |
| Rigid polyvinyl chloride | " | 0.320 |
| Sectional wood overhead doors, frames not included | | |
| Commercial grade, heavy duty, 1-3/4" thick, manual | | |
| 8' x 8' | EA. | 6.667 |
| 10' x 10' | " | 7.273 |
| 12' x 12' | " | 8.000 |
| Chain hoist | | |
| 12' x 16' high | EA. | 13.333 |
| 14' x 14' high | " | 10.000 |
| 20' x 8' high | " | 16.000 |
| 16' high | " | 20.000 |
| Sectional metal overhead doors, complete | | |
| Residential grade, manual | | |
| 9' x 7' | EA. | 3.200 |
| 16' x 7' | " | 4.000 |
| Commercial grade | | |
| 8' x 8' | EA. | 6.667 |
| 10' x 10' | " | 7.273 |
| 12' x 12' | " | 8.000 |
| 20' x 14', with chain hoist | " | 16.000 |
| Sliding glass doors | | |
| Tempered plate glass, 1/4" thick | | |
| 6' wide | | |
| Economy grade | EA. | 2.667 |
| Premium grade | " | 2.667 |
| 12' wide | | |
| Economy grade | EA. | 4.000 |
| Premium grade | " | 4.000 |
| Insulating glass, 5/8" thick | | |
| 6' wide | | |
| Economy grade | EA. | 2.667 |
| Premium grade | " | 2.667 |
| 12' wide | | |
| Economy grade | EA. | 4.000 |
| Premium grade | " | 4.000 |
| 1" thick | | |
| 6' wide | | |
| Economy grade | EA. | 2.667 |
| Premium grade | " | 2.667 |
| 12' wide | | |
| Economy grade | EA. | 4.000 |
| Premium grade | " | 4.000 |
| Vertical lift doors, channel frame construction | | |
| 20' high x | | |
| 10' wide | EA. | 9.600 |

## WOOD AND PLASTIC

| 08300.10 SPECIAL DOORS | UNIT | MAN/ HOURS |
|---|---|---|
| 15' wide | EA. | 9.600 |
| 20' wide | " | 17.143 |
| 25' wide | " | 17.143 |
| 25' high | | |
| 20' wide | EA. | 17.143 |
| 25' wide | " | 20.000 |
| 30' high | | |
| 25' wide | EA. | 20.000 |
| 30' wide | " | 20.000 |
| 35' wide | " | 20.000 |
| Residential storm door | | |
| Minimum | EA. | 1.333 |
| Average | " | 1.333 |
| Maximum | " | 2.000 |

## STOREFRONTS

| 08410.10 STOREFRONTS | UNIT | MAN/ HOURS |
|---|---|---|
| Storefront, aluminum and glass | | |
| Minimum | S.F. | 0.100 |
| Average | " | 0.114 |
| Maximum | " | 0.133 |
| Entrance doors, premium, incl. glass, closers, panic dev.,etc. | | |
| 1/2" thick glass | | |
| 3' x 7' | EA. | 6.667 |
| 6' x 7' | " | 10.000 |
| 3/4" thick glass | | |
| 3' x 7' | EA. | 6.667 |
| 6' x 7' | " | 10.000 |
| 1" thick glass | | |
| 3' x 7' | EA. | 6.667 |
| 6' x 7' | " | 10.000 |
| Revolving doors | | |
| 7'diameter, 7' high | | |
| Minimum | EA. | 60.000 |
| Average | " | 96.000 |
| Maximum | " | 120 |

# 08 DOORS AND WINDOWS

## METAL WINDOWS

### 08510.10 STEEL WINDOWS

| METAL WINDOWS | UNIT | MAN/ HOURS |
|---|---|---|
| Steel windows, primed | | |
| Casements | | |
| Operable | | |
| Minimum | S.F. | 0.047 |
| Maximum | " | 0.053 |
| Fixed sash | " | 0.040 |
| Double hung | " | 0.044 |
| Industrial windows | | |
| Horizontally pivoted sash | S.F. | 0.053 |
| Fixed sash | " | 0.044 |
| Security sash | | |
| Operable | S.F. | 0.053 |
| Fixed | " | 0.044 |
| Picture window | " | 0.044 |
| Projecting sash | | |
| Minimum | S.F. | 0.050 |
| Maximum | " | 0.050 |
| Mullions | L.F. | 0.040 |

### 08520.10 ALUMINUM WINDOWS

| METAL WINDOWS | UNIT | MAN/ HOURS |
|---|---|---|
| Jalousie | | |
| 3-0 x 4-0 | EA. | 1.000 |
| 3-0 x 5-0 | " | 1.000 |
| Fixed window | | |
| 6 sf to 8 sf | S.F. | 0.114 |
| 12 sf to 16 sf | " | 0.089 |
| Projecting window | | |
| 6 sf to 8 sf | S.F. | 0.200 |
| 12 sf to 16 sf | " | 0.133 |
| Horizontal sliding | | |
| 6 sf to 8 sf | S.F. | 0.100 |
| 12 sf to 16 sf | " | 0.080 |
| Double hung | | |
| 6 sf to 8 sf | S.F. | 0.160 |
| 10 sf to 12 sf | " | 0.133 |
| Storm window, 0.5 cfm, up to | | |
| 60 u.i. (united inches) | EA. | 0.400 |
| 70 u.i. | " | 0.400 |
| 80 u.i. | " | 0.400 |
| 90 u.i. | " | 0.444 |
| 100 u.i. | " | 0.444 |
| 2.0 cfm, up to | | |
| 60 u.i. | EA. | 0.400 |
| 70 u.i. | " | 0.400 |
| 80 u.i. | " | 0.400 |
| 90 u.i. | " | 0.444 |
| 100 u.i. | " | 0.444 |

## WOOD AND PLASTIC

### 08600.10 WOOD WINDOWS

| WOOD AND PLASTIC | UNIT | MAN/ HOURS |
|---|---|---|
| Double hung | | |
| 24" x 36" | | |
| Minimum | EA. | 0.800 |
| Average | " | 1.000 |
| Maximum | " | 1.333 |
| 24" x 48" | | |
| Minimum | EA. | 0.800 |
| Average | " | 1.000 |
| Maximum | " | 1.333 |
| 30" x 48" | | |
| Minimum | EA. | 0.889 |
| Average | " | 1.143 |
| Maximum | " | 1.600 |
| 30" x 60" | | |
| Minimum | EA. | 0.889 |
| Average | " | 1.143 |
| Maximum | " | 1.600 |
| Casement | | |
| 1 leaf, 22" x 38" high | | |
| Minimum | EA. | 0.800 |
| Average | " | 1.000 |
| Maximum | " | 1.333 |
| 2 leaf, 50" x 50" high | | |
| Minimum | EA. | 1.000 |
| Average | " | 1.333 |
| Maximum | " | 2.000 |
| 3 leaf, 71" x 62" high | | |
| Minimum | EA. | 1.000 |
| Average | " | 1.333 |
| Maximum | " | 2.000 |
| 4 leaf, 95" x 75" high | | |
| Minimum | EA. | 1.143 |
| Average | " | 1.600 |
| Maximum | " | 2.667 |
| 5 leaf, 119" x 75" high | | |
| Minimum | EA. | 1.143 |
| Average | " | 1.600 |
| Maximum | " | 2.667 |
| Picture window, fixed glass, 54" x 54" high | | |
| Minimum | EA. | 1.000 |
| Average | " | 1.143 |
| Maximum | " | 1.333 |
| 68" x 55" high | | |
| Minimum | EA. | 1.000 |
| Average | " | 1.143 |
| Maximum | " | 1.333 |
| Sliding, 40" x 31" high | | |
| Minimum | EA. | 0.800 |
| Average | " | 1.000 |
| Maximum | " | 1.333 |
| 52" x 39" high | | |
| Minimum | EA. | 1.000 |
| Average | " | 1.143 |
| Maximum | " | 1.333 |

# 08 DOORS AND WINDOWS

## WOOD AND PLASTIC

| 08600.10 WOOD WINDOWS | UNIT | MAN/ HOURS |
|---|---|---|
| 64" x 72" high | | |
| Minimum | EA. | 1.000 |
| Average | " | 1.333 |
| Maximum | " | 1.600 |
| Awning windows | | |
| 34" x 21" high | | |
| Minimum | EA. | 0.800 |
| Average | " | 1.000 |
| Maximum | " | 1.333 |
| 40" x 21" high | | |
| Minimum | EA. | 0.889 |
| Average | " | 1.143 |
| Maximum | " | 1.600 |
| 48" x 27" high | | |
| Minimum | EA. | 0.889 |
| Average | " | 1.143 |
| Maximum | " | 1.600 |
| 60" x 36" high | | |
| Minimum | EA. | 1.000 |
| Average | " | 1.333 |
| Maximum | " | 1.600 |
| Window frame, milled | | |
| Minimum | L.F. | 0.160 |
| Average | " | 0.200 |
| Maximum | " | 0.267 |

## HARDWARE

| 08710.20 LOCKSETS | UNIT | MAN/ HOURS |
|---|---|---|
| Latchset, heavy duty | | |
| Cylindrical | EA. | 0.500 |
| Mortise | " | 0.800 |
| Lockset, heavy duty | | |
| Cylindrical | EA. | 0.500 |
| Mortise | " | 0.800 |
| Mortise locks and latchsets, chrome | | |
| Latchset passage or closet latch | EA. | 0.667 |
| Privacy (bath or bedroom) | " | 0.667 |
| Entry lockset | " | 0.667 |
| Classroom lockset (outside key operated) | " | 0.667 |
| Storeroom lock | " | 0.667 |
| Front door lock | " | 0.667 |
| Dormitory or exit lock | " | 0.667 |
| Preassembled locks and latches, brass | | |
| Latchset, passage or closet latch | EA. | 0.667 |
| Lockset | | |

## HARDWARE

| 08710.20 LOCKSETS | UNIT | MAN/ HOURS |
|---|---|---|
| Privacy (bath or bathroom) | EA. | 0.667 |
| Entry lock | " | 0.667 |
| Classroom lock (outside key, operated) | " | 0.667 |
| Storeroom lock | " | 0.667 |
| Bored locks and latches, satin chrome plated | | |
| Latchset passage or closet latch | EA. | 0.667 |
| Lockset | | |
| Privacy (bath or bedroom) | EA. | 0.667 |
| Entry lock | " | 0.667 |
| Classroom lock | " | 0.667 |
| Corridor lock | " | 0.667 |
| Miscellaneous locks | | |
| Exit lock with alarm, single door | EA. | 3.200 |
| Electric strike | | |
| Rim mounted wrought steel | EA. | 2.000 |
| Mortised, wrought steel with bronze plating | " | 3.200 |
| Dead bolt | | |
| Bored, wrought brass, keyed both sides | EA. | 1.333 |
| Mortised, cast brass | " | 1.333 |
| Lockset, cipher, mechanical | " | 0.800 |

| 08710.30 CLOSERS | UNIT | MAN/ HOURS |
|---|---|---|
| Door closers | | |
| Surface mounted, traditional type, parallel arm | | |
| Standard | EA. | 1.000 |
| Heavy duty | " | 1.000 |
| Modern type, parallel arm, standard duty | " | 1.000 |
| Overhead, concealed, pivot hung, single acting | | |
| Interior | EA. | 1.000 |
| Exterior | " | 1.000 |
| Floor concealed, single acting, offset, pivoted | | |
| Interior | EA. | 2.667 |
| Exterior | " | 2.667 |

| 08710.40 DOOR TRIM | UNIT | MAN/ HOURS |
|---|---|---|
| Door bumper, bronze, wall type | EA. | 0.160 |
| Wall type, 4" dia. with convex rubber pad, aluminum | " | 0.160 |
| Floor type | | |
| Aluminum | EA. | 0.160 |
| Brass | " | 0.160 |
| Door holders | | |
| Wall type, bronze | EA. | 0.160 |
| Overhead | " | 0.400 |
| Floor type | " | 0.400 |
| Plunger type | " | 0.400 |
| Wall type, aluminum | " | 0.400 |
| Surface bolt | " | 0.160 |
| Panic device | | |
| Rim type with thumb piece | EA. | 2.000 |
| Mortise | " | 2.000 |
| Vertical rod | " | 2.000 |
| Labelled, rim type | " | 2.000 |
| Mortise | " | 2.000 |

# 08 DOORS AND WINDOWS

## HARDWARE

| 08710.40 DOOR TRIM | UNIT | MAN/HOURS |
|---|---|---|
| Vertical rod | EA. | 2.000 |
| Silencers, rubber type | " | 0.016 |
| Dust proof strike with plate, brass | " | 0.267 |
| Flush bolt, lever extension, brass, rated | " | 0.160 |
| Surface bolt with strike, brass, 6" long | " | 0.160 |
| Door coordinator, labeled, brass, satin chrome | " | 0.571 |
| Door plates | | |
| Kick plate, aluminum, 3 beveled edges | | |
| 10" x 28" | EA. | 0.400 |
| 10" x 30" | " | 0.400 |
| 10" x 34" | " | 0.400 |
| 10" x 38" | " | 0.400 |
| Push plate, 4" x 16" | | |
| Aluminum | EA. | 0.160 |
| Bronze | " | 0.160 |
| Stainless steel | " | 0.160 |
| Armor plate, 40" x 34" | " | 0.320 |
| Pull handle, 4" x 16" | | |
| Aluminum | EA. | 0.160 |
| Bronze | " | 0.160 |
| Stainless steel | " | 0.160 |
| Hasp assembly | | |
| 3" | EA. | 0.133 |
| 4-1/2" | " | 0.178 |
| 6" | " | 0.229 |
| Electro-magnetic door holder | | |
| Wall mounted | EA. | 2.667 |
| Floor mounted | " | 2.667 |
| Smoke detector door holder | | |
| Photo electric type | EA. | 2.667 |
| Ionization type | " | 2.667 |
| Pneumatic operators, activated by rubber mats | | |
| Swing | | |
| Single | EA. | 6.667 |
| Double | " | 10.000 |
| Sliding | | |
| Single | EA. | 6.667 |
| Double | " | 10.000 |

| 08710.60 WEATHERSTRIPPING | UNIT | MAN/HOURS |
|---|---|---|
| Weatherstrip, head and jamb, metal strip, neoprene bulb | | |
| Standard duty | L.F. | 0.044 |
| Heavy duty | " | 0.050 |
| Spring type | | |
| Metal doors | EA. | 2.000 |
| Wood doors | " | 2.667 |
| Sponge type with adhesive backing | " | 0.800 |
| Astragal | | |
| 1-3/4" x 13 ga., aluminum | L.F. | 0.067 |
| 1-3/8" x 5/8", oak | " | 0.053 |
| Thresholds | | |
| Bronze | L.F. | 0.200 |
| Aluminum | | |
| Plain | L.F. | 0.200 |

## HARDWARE

| 08710.60 WEATHERSTRIPPING | UNIT | MAN/HOURS |
|---|---|---|
| Vinyl insert | L.F. | 0.200 |
| Aluminum with grit | " | 0.200 |
| Steel | | |
| Plain | L.F. | 0.200 |
| Interlocking | " | 0.667 |

## GLAZING

| 08810.10 GLAZING | UNIT | MAN/HOURS |
|---|---|---|
| Sheet glass, 1/8" thick | S.F. | 0.044 |
| Plate glass, bronze or grey, 1/4" thick | " | 0.073 |
| Clear | " | 0.073 |
| Polished | " | 0.073 |
| Plexiglass | | |
| 1/8" thick | S.F. | 0.073 |
| 1/4" thick | " | 0.044 |
| Float glass, clear | | |
| 3/16" thick | S.F. | 0.067 |
| 1/4" thick | " | 0.073 |
| 5/16" thick | " | 0.080 |
| 3/8" thick | " | 0.100 |
| 1/2" thick | " | 0.133 |
| 5/8" thick | " | 0.160 |
| 3/4" thick | " | 0.200 |
| 1" thick | " | 0.267 |
| Tinted glass, polished plate, twin ground | | |
| 3/16" thick | S.F. | 0.067 |
| 1/4" thick | " | 0.073 |
| 3/8" thick | " | 0.100 |
| 1/2" thick | " | 0.133 |
| Total, full vision, all glass window system | | |
| To 10' high | | |
| Minimum | S.F. | 0.200 |
| Average | " | 0.200 |
| Maximum | " | 0.200 |
| 10' to 20' high | | |
| Minimum | S.F. | 0.200 |
| Average | " | 0.200 |
| Maximum | " | 0.200 |
| Insulated glass, bronze or gray | | |
| 1/2" thick | S.F. | 0.133 |
| 1" thick | " | 0.200 |
| Spandrel glass, polished bronze/grey, 1 side, 1/4" thick | " | 0.073 |
| Tempered glass (safety) | | |
| Clear sheet glass | | |
| 1/8" thick | S.F. | 0.044 |

| GLAZING | UNIT | MAN/ HOURS |
|---|---|---|
| **08810.10 GLAZING** | | |
| 3/16" thick | S.F. | 0.062 |
| Clear float glass | | |
| 1/4" thick | S.F. | 0.067 |
| 5/16" thick | " | 0.080 |
| 3/8" thick | " | 0.100 |
| 1/2" thick | " | 0.133 |
| 5/8" thick | " | 0.160 |
| 3/4" thick | " | 0.267 |
| Tinted float glass | | |
| 3/16" thick | S.F. | 0.062 |
| 1/4" thick | " | 0.067 |
| 3/8" thick | " | 0.100 |
| 1/2" thick | " | 0.133 |
| Laminated glass | | |
| Float safety glass with polyvinyl plastic interlayer | | |
| 1/4", sheet or float | | |
| Two lites, 1/8" thick, clear glass | S.F. | 0.067 |
| 1/2" thick, float glass | | |
| Two lites, 1/4" thick, clear glass | S.F. | 0.133 |
| Tinted glass | " | 0.133 |
| Insulating glass, two lites, clear float glass | | |
| 1/2" thick | S.F. | 0.133 |
| 5/8" thick | " | 0.160 |
| 3/4" thick | " | 0.200 |
| 7/8" thick | " | 0.229 |
| 1" thick | " | 0.267 |
| Glass seal edge | | |
| 3/8" thick | S.F. | 0.133 |
| Tinted glass | | |
| 1/2" thick | S.F. | 0.133 |
| 1" thick | " | 0.267 |
| Tempered, clear | | |
| 1" thick | S.F. | 0.267 |
| Wire reinforced | " | 0.267 |
| Plate mirror glass | | |
| 1/4" thick | | |
| 15 sf | S.F. | 0.080 |
| Over 15 sf | " | 0.073 |
| Transparent, one way vision, 1/4" thick | " | 0.080 |
| Sheet mirror glass | | |
| 3/16" thick | S.F. | 0.080 |
| 1/4" thick | " | 0.067 |
| Wall tiles, 12" x 12" | | |
| Clear glass | S.F. | 0.044 |
| Veined glass | " | 0.044 |
| Wire glass, 1/4" thick | | |
| Clear | S.F. | 0.267 |
| Hammered | " | 0.267 |
| Obscure | " | 0.267 |
| Bullet resistant glass, plate glass with inter-leaved vinyl | | |
| 1-3/16" thick | | |
| To 15 sf | S.F. | 0.400 |
| Over 15 sf | " | 0.400 |
| 2" thick | | |

| GLAZING | UNIT | MAN/ HOURS |
|---|---|---|
| **08810.10 GLAZING** | | |
| To 15 sf | S.F. | 0.667 |
| Over 15 sf | " | 0.667 |
| Glazing accessories | | |
| Neoprene glazing gaskets | | |
| 1/4" glass | L.F. | 0.032 |
| 3/8" glass | " | 0.033 |
| 1/2" glass | " | 0.035 |
| 3/4" glass | " | 0.036 |
| 1" glass | " | 0.040 |
| Mullion section | | |
| 1/4" glass | L.F. | 0.016 |
| 3/8" glass | " | 0.020 |
| 1/2" glass | " | 0.023 |
| 3/4" glass | " | 0.027 |
| 1" glass | " | 0.032 |
| Molded corners | EA. | 0.533 |

| GLAZED CURTAIN WALLS | UNIT | MAN/ HOURS |
|---|---|---|
| **08910.10 GLAZED CURTAIN WALLS** | | |
| Curtain wall, aluminum system, framing sections | | |
| 2" x 3" | | |
| Jamb | L.F. | 0.067 |
| Horizontal | " | 0.067 |
| Mullion | " | 0.067 |
| 2" x 4" | | |
| Jamb | L.F. | 0.100 |
| Horizontal | " | 0.100 |
| Mullion | " | 0.100 |
| 3" x 5-1/2" | | |
| Jamb | L.F. | 0.100 |
| Horizontal | " | 0.100 |
| Mullion | " | 0.100 |
| 4" corner mullion | " | 0.133 |
| Coping sections | | |
| 1/8" x 8" | L.F. | 0.133 |
| 1/8" x 9" | " | 0.133 |
| 1/8" x 12-1/2" | " | 0.160 |
| Sill section | | |
| 1/8" x 6" | L.F. | 0.080 |
| 1/8" x 7" | " | 0.080 |
| 1/8" x 8-1/2" | " | 0.080 |
| Column covers, aluminum | | |
| 1/8" x 26" | L.F. | 0.200 |
| 1/8" x 34" | " | 0.211 |
| 1/8" x 38" | " | 0.211 |
| Doors | | |

| GLAZED CURTAIN WALLS | UNIT | MAN/ HOURS |
|---|---|---|
| **08910.10 GLAZED CURTAIN WALLS** | | |
| Aluminum framed, standard hardware | | |
| Narrow stile | | |
| 2-6 x 7-0 | EA. | 4.000 |
| 3-0 x 7-0 | " | 4.000 |
| 3-6 x 7-0 | " | 4.000 |
| Wide stile | | |
| 2-6 x 7-0 | EA. | 4.000 |
| 3-0 x 7-0 | " | 4.000 |
| 3-6 x 7-0 | " | 4.000 |
| Flush panel doors, to match adjacent wall panels | | |
| 2-6 x 7-0 | EA. | 5.000 |
| 3-0 x 7-0 | " | 5.000 |
| 3-6 x 7-0 | " | 5.000 |
| Wall panel, insulated | | |
| "U"=.08 | S.F. | 0.067 |
| "U"=.10 | " | 0.067 |
| "U"=.15 | " | 0.067 |
| Window wall system, complete | | |
| Minimum | S.F. | 0.080 |
| Average | " | 0.089 |
| Maximum | " | 0.114 |

# 09 FINISHES

## SUPPORT SYSTEMS

### 09110.10 METAL STUDS

| | UNIT | MAN/HOURS |
|---|---|---|
| Studs, non load bearing, galvanized | | |
| 2-1/2", 20 ga. | | |
| 12" o.c. | S.F. | 0.017 |
| 16" o.c. | " | 0.013 |
| 25 ga. | | |
| 12" o.c. | S.F. | 0.017 |
| 16" o.c. | " | 0.013 |
| 24" o.c. | " | 0.011 |
| 3-5/8", 20 ga. | | |
| 12" o.c. | S.F. | 0.020 |
| 16" o.c. | " | 0.016 |
| 24" o.c. | " | 0.013 |
| 25 ga. | | |
| 12" o.c. | S.F. | 0.020 |
| 16" o.c. | " | 0.016 |
| 24" o.c. | " | 0.013 |
| 4", 20 ga. | | |
| 12" o.c. | S.F. | 0.020 |
| 16" o.c. | " | 0.016 |
| 24" o.c. | " | 0.013 |
| 25 ga. | | |
| 12" o.c. | S.F. | 0.020 |
| 16" o.c. | " | 0.016 |
| 24" o.c. | " | 0.013 |
| 6", 20 ga. | | |
| 12" o.c. | S.F. | 0.025 |
| 16" o.c. | " | 0.020 |
| 24" o.c. | " | 0.017 |
| 25 ga. | | |
| 12" o.c. | S.F. | 0.025 |
| 16" o.c. | " | 0.020 |
| 24" o.c. | " | 0.017 |
| Load bearing studs, galvanized | | |
| 3-5/8", 16 ga. | | |
| 12" o.c. | S.F. | 0.020 |
| 16" o.c. | " | 0.016 |
| 18 ga. | | |
| 12" o.c. | S.F. | 0.013 |
| 16" o.c. | " | 0.016 |
| 4", 16 ga. | | |
| 12" o.c. | S.F. | 0.020 |
| 16" o.c. | " | 0.016 |
| 6", 16 ga. | | |
| 12" o.c. | S.F. | 0.025 |
| 16" o.c. | " | 0.020 |
| Furring | | |
| On beams and columns | | |
| 7/8" channel | L.F. | 0.053 |
| 1-1/2" channel | " | 0.062 |
| On ceilings | | |
| 3/4" furring channels | | |
| 12" o.c. | S.F. | 0.033 |
| 16" o.c. | " | 0.032 |
| 24" o.c. | " | 0.029 |
| 1-1/2" furring channels | | |
| 12" o.c. | S.F. | 0.036 |
| 16" o.c. | " | 0.033 |
| 24" o.c. | " | 0.031 |
| On walls | | |
| 3/4" furring channels | | |
| 12" o.c. | S.F. | 0.027 |
| 16" o.c. | " | 0.025 |
| 24" o.c. | " | 0.024 |
| 1-1/2" furring channels | | |
| 12" o.c. | S.F. | 0.029 |
| 16" o.c. | " | 0.027 |
| 24" o.c. | " | 0.025 |

## LATH AND PLASTER

### 09205.10 GYPSUM LATH

| | UNIT | MAN/HOURS |
|---|---|---|
| Gypsum lath, 1/2" thick | | |
| Clipped | S.Y. | 0.044 |
| Nailed | " | 0.050 |

### 09205.20 METAL LATH

| | UNIT | MAN/HOURS |
|---|---|---|
| Diamond expanded, galvanized | | |
| 2.5 lb., on walls | | |
| Nailed | S.Y. | 0.100 |
| Wired | " | 0.114 |
| On ceilings | | |
| Nailed | S.Y. | 0.114 |
| Wired | " | 0.133 |
| 3.4 lb., on walls | | |
| Nailed | S.Y. | 0.100 |
| Wired | " | 0.114 |
| On ceilings | | |
| Nailed | S.Y. | 0.114 |
| Wired | " | 0.133 |
| Flat rib | | |
| 2.75 lb., on walls | | |
| Nailed | S.Y. | 0.100 |
| Wired | " | 0.114 |
| On ceilings | | |
| Nailed | S.Y. | 0.114 |
| Wired | " | 0.133 |
| 3.4 lb., on walls | | |
| Nailed | S.Y. | 0.100 |
| Wired | " | 0.114 |

# 09 FINISHES

## LATH AND PLASTER

| LATH AND PLASTER | UNIT | MAN/ HOURS |
|---|---|---|
| **09205.20 METAL LATH** | | |
| On ceilings | | |
| Nailed | S.Y. | 0.114 |
| Wired | " | 0.133 |
| Stucco lath | | |
| 1.8 lb. | S.Y. | 0.100 |
| 3.6 lb. | " | 0.100 |
| Paper backed | | |
| Minimum | S.Y. | 0.080 |
| Maximum | " | 0.114 |
| **09205.60 PLASTER ACCESSORIES** | | |
| Expansion joint, 3/4", 26 ga., galvanized, one piece | L.F. | 0.020 |
| Plaster corner beads, 3/4", galvanized | " | 0.023 |
| Casing bead, expanded flange, galvanized | " | 0.020 |
| Expanded wing, 1-1/4" wide, galvanized | " | 0.020 |
| Joint clips for lath | EA. | 0.004 |
| Metal base, galvanized, 2-1/2" high | L.F. | 0.027 |
| Stud clips for gypsum lath | EA. | 0.004 |
| Sound deadening board, 1/4", nailed or clipped | S.F. | 0.013 |
| **09210.10 PLASTER** | | |
| Gypsum plaster, trowel finish, 2 coats | | |
| Ceilings | S.Y. | 0.250 |
| Walls | " | 0.235 |
| 3 coats | | |
| Ceilings | S.Y. | 0.348 |
| Walls | " | 0.308 |
| Vermiculite plaster | | |
| 2 coats | | |
| Ceilings | S.Y. | 0.381 |
| Walls | " | 0.348 |
| 3 coats | | |
| Ceilings | S.Y. | 0.471 |
| Walls | " | 0.421 |
| Keenes cement plaster | | |
| 2 coats | | |
| Ceilings | S.Y. | 0.308 |
| Walls | " | 0.267 |
| 3 coats | | |
| Ceilings | S.Y. | 0.348 |
| Walls | " | 0.308 |
| On columns, add to installation, 50% | " | |
| Chases, fascia, and soffits, add to installation, 50% | " | |
| Beams, add to installation, 50% | " | |
| Patch holes, average size holes | | |
| 1 sf to 5 sf | | |
| Minimum | S.F. | 0.133 |
| Average | " | 0.160 |
| Maximum | " | 0.200 |
| Over 5 sf | | |
| Minimum | S.F. | 0.080 |
| Average | " | 0.114 |
| Maximum | " | 0.133 |

| LATH AND PLASTER | UNIT | MAN/ HOURS |
|---|---|---|
| **09210.10 PLASTER** | | |
| Patch cracks | | |
| Minimum | S.F. | 0.027 |
| average | " | 0.040 |
| Maximum | " | 0.080 |
| **09220.10 PORTLAND CEMENT PLASTER** | | |
| Stucco, portland, gray, 3 coat, 1" thick | | |
| Sand finish | S.Y. | 0.348 |
| Trowel finish | " | 0.364 |
| White cement | | |
| Sand finish | S.Y. | 0.364 |
| Trowel finish | " | 0.400 |
| Scratch coat | | |
| For ceramic tile | S.Y. | 0.080 |
| For quarry tile | " | 0.080 |
| Portland cement plaster | | |
| 2 coats, 1/2" | S.Y. | 0.160 |
| 3 coats, 7/8" | " | 0.200 |
| **09250.10 GYPSUM BOARD** | | |
| Drywall, plasterboard, 3/8" clipped to | | |
| Metal furred ceiling | S.F. | 0.009 |
| Columns and beams | " | 0.020 |
| Walls | " | 0.008 |
| Nailed or screwed to | | |
| Wood framed ceiling | S.F. | 0.008 |
| Columns and beams | " | 0.018 |
| Walls | " | 0.007 |
| 1/2", clipped to | | |
| Metal furred ceiling | S.F. | 0.009 |
| Columns and beams | " | 0.020 |
| Walls | " | 0.008 |
| Nailed or screwed to | | |
| Wood framed ceiling | S.F. | 0.008 |
| Columns and beams | " | 0.018 |
| Walls | " | 0.007 |
| 5/8", clipped to | | |
| Metal furred ceiling | S.F. | 0.010 |
| Columns and beams | " | 0.022 |
| Walls | " | 0.009 |
| Nailed or screwed to | | |
| Wood framed ceiling | S.F. | 0.010 |
| Columns and beams | " | 0.022 |
| Walls | " | 0.009 |
| Vinyl faced, clipped to metal studs | | |
| 1/2" | S.F. | 0.010 |
| 5/8" | " | 0.010 |
| Taping and finishing joints | | |
| Minimum | S.F. | 0.005 |
| Average | " | 0.007 |
| Maximum | " | 0.008 |
| Casing bead | | |
| Minimum | L.F. | 0.023 |
| Average | " | 0.027 |

## LATH AND PLASTER

| 09250.10 GYPSUM BOARD | UNIT | MAN/HOURS |
|---|---|---|
| Maximum | L.F. | 0.040 |
| Corner bead | | |
| Minimum | L.F. | 0.023 |
| Average | " | 0.027 |
| Maximum | " | 0.040 |

## TILE

| 09310.10 CERAMIC TILE | UNIT | MAN/HOURS |
|---|---|---|
| Glazed wall tile, 4-1/4" x 4-1/4" | | |
| Minimum | S.F. | 0.057 |
| Average | " | 0.067 |
| Maximum | " | 0.080 |
| Base, 4-1/4" high | | |
| Minimum | L.F. | 0.100 |
| Average | " | 0.100 |
| Maximum | " | 0.100 |
| Unglazed floor tile | | |
| Portland cement bed, cushion edge, face mounted | | |
| 1" x 1" | S.F. | 0.073 |
| 1" x 2" | " | 0.070 |
| 2" x 2" | " | 0.067 |
| Adhesive bed, with white grout | | |
| 1" x 1" | S.F. | 0.073 |
| 1" x 2" | " | 0.070 |
| 2" x 2" | " | 0.067 |
| Organic adhesive bed, thin set, back mounted | | |
| 1" x 1" | S.F. | 0.073 |
| 1" x 2" | " | 0.070 |
| 2" x 2" | " | 0.067 |
| Unglazed wall tile | | |
| Organic adhesive, face mounted cushion edge | | |
| 1" x 1" | | |
| Minimum | S.F. | 0.067 |
| Average | " | 0.073 |
| Maximum | " | 0.080 |
| 1" x 2" | | |
| Minimum | S.F. | 0.064 |
| Average | " | 0.070 |
| Maximum | " | 0.076 |
| 2" x 2" | | |
| Minimum | S.F. | 0.062 |
| Average | " | 0.067 |
| Maximum | " | 0.073 |
| Back mounted | | |
| 1" x 1" | | |

## TILE

| 09310.10 CERAMIC TILE | UNIT | MAN/HOURS |
|---|---|---|
| Minimum | S.F. | 0.067 |
| Average | " | 0.073 |
| Maximum | " | 0.080 |
| 1" x 2" | | |
| Minimum | S.F. | 0.064 |
| Average | " | 0.070 |
| Maximum | " | 0.076 |
| 2" x 2" | | |
| Minimum | S.F. | 0.062 |
| Average | " | 0.067 |
| Maximum | " | 0.073 |
| Conductive floor tile, unglazed square edged | | |
| Portland cement bed | | |
| 1 x 1 | S.F. | 0.100 |
| 1-9/16 x 1-9/16 | " | 0.100 |
| Dry set | | |
| 1 x 1 | S.F. | 0.100 |
| 1-9/16 x 1-9/16 | " | 0.100 |
| Epoxy bed with epoxy joints | | |
| 1 x 1 | S.F. | 0.100 |
| 1-9/16 x 1-9/16 | " | 0.100 |
| Ceramic accessories | | |
| Towel bar, 24" long | | |
| Minimum | EA. | 0.320 |
| Average | " | 0.400 |
| Maximum | " | 0.533 |
| Soap dish | | |
| Minimum | EA. | 0.533 |
| Average | " | 0.667 |
| Maximum | " | 0.800 |

| 09330.10 QUARRY TILE | UNIT | MAN/HOURS |
|---|---|---|
| Floor | | |
| 4 x 4 x 1/2" | S.F. | 0.107 |
| 6 x 6 x 1/2" | " | 0.100 |
| 6 x 6 x 3/4" | " | 0.100 |
| Wall, applied to 3/4" portland cement bed | | |
| 4 x 4 x 1/2" | S.F. | 0.160 |
| 6 x 6 x 3/4" | " | 0.133 |
| Cove base | | |
| 5 x 6 x 1/2" straight top | L.F. | 0.133 |
| 6 x 6 x 3/4" round top | " | 0.133 |
| Stair treads 6 x 6 x 3/4" | " | 0.200 |
| Window sill 6 x 8 x 3/4" | " | 0.160 |
| For abrasive surface, add to material, 25% | | |

| 09410.10 TERRAZZO | UNIT | MAN/HOURS |
|---|---|---|
| Floors on concrete, 1-3/4" thick, 5/8" topping | | |
| Gray cement | S.F. | 0.114 |
| White cement | " | 0.114 |
| Sand cushion, 3" thick, 5/8" top, 1/4" | | |
| Gray cement | S.F. | 0.133 |
| White cement | " | 0.133 |

# 09 FINISHES

## TILE

### 09410.10 TERRAZZO

| TILE | UNIT | MAN/HOURS |
|---|---|---|
| Monolithic terrazzo, 3-1/2" base slab, 5/8" topping | S.F. | 0.100 |
| Terrazzo wainscot, cast-in-place, 1/2" thick | " | 0.200 |
| Base, cast in place, terrazzo cove type, 6" high | L.F. | 0.114 |
| Curb, cast in place, 6" wide x 6" high, polished top | " | 0.400 |
| Stairs, cast-in-place, topping on concrete or metal | | |
| 1-1/2" thick treads, 12" wide | L.F. | 0.400 |
| Combined tread and riser | " | 1.000 |
| Precast terrazzo, thin set | | |
| Terrazzo tiles, non-slip surface | | |
| 9" x 9" x 1" thick | S.F. | 0.114 |
| 12" x 12" | | |
| 1" thick | S.F. | 0.107 |
| 1-1/2" thick | " | 0.114 |
| 18" x 18" x 1-1/2" thick | " | 0.114 |
| 24" x 24" x 1-1/2" thick | " | 0.094 |
| Terrazzo wainscot | | |
| 12" x 12" x 1" thick | S.F. | 0.200 |
| 18" x 18" x 1-1/2" thick | " | 0.229 |
| Base | | |
| 6" high | | |
| Straight | L.F. | 0.062 |
| Coved | " | 0.062 |
| 8" high | | |
| Straight | L.F. | 0.067 |
| Coved | " | 0.067 |
| Terrazzo curbs | | |
| 8" wide x 8" high | L.F. | 0.320 |
| 6" wide x 6" high | " | 0.267 |
| Precast terrazzo stair treads, 12" wide | | |
| 1-1/2" thick | | |
| Diamond pattern | L.F. | 0.145 |
| Non-slip surface | " | 0.145 |
| 2" thick | | |
| Diamond pattern | L.F. | 0.145 |
| Non-slip surface | " | 0.160 |
| Stair risers, 1" thick to 6" high | | |
| Straight sections | L.F. | 0.080 |
| Cove sections | " | 0.080 |
| Combined tread and riser | | |
| Straight sections | | |
| 1-1/2" tread, 3/4" riser | L.F. | 0.229 |
| 3" tread, 1" riser | " | 0.229 |
| Curved sections | | |
| 2" tread, 1" riser | L.F. | 0.267 |
| 3" tread, 1" riser | " | 0.267 |
| Stair stringers, notched for treads and risers | | |
| 1" thick | L.F. | 0.200 |
| 2" thick | " | 0.267 |
| Landings, structural, nonslip | | |
| 1-1/2" thick | S.F. | 0.133 |
| 3" thick | " | 0.160 |
| Conductive terrazzo, spark proof industrial floor | | |
| Epoxy terrazzo | | |
| Floor | S.F. | 0.050 |
| Base | S.F. | 0.067 |
| Polyacrylate | | |
| Floor | S.F. | 0.050 |
| Base | " | 0.067 |
| Polyester | | |
| Floor | S.F. | 0.032 |
| Base | " | 0.040 |
| Synthetic latex mastic | | |
| Floor | S.F. | 0.050 |
| Base | " | 0.067 |

## ACOUSTICAL TREATMENT

### 09510.10 CEILINGS AND WALLS

| ACOUSTICAL TREATMENT | UNIT | MAN/HOURS |
|---|---|---|
| Acoustical panels, suspension system not included | | |
| Fiberglass panels | | |
| 5/8" thick | | |
| 2' x 2' | S.F. | 0.011 |
| 2' x 4' | " | 0.009 |
| 3/4" thick | | |
| 2' x 2' | S.F. | 0.011 |
| 2' x 4' | " | 0.009 |
| Glass cloth faced fiberglass panels | | |
| 3/4" thick | S.F. | 0.013 |
| 1" thick | " | 0.013 |
| Mineral fiber panels | | |
| 5/8" thick | | |
| 2' x 2' | S.F. | 0.011 |
| 2' x 4' | " | 0.009 |
| 3/4" thick | | |
| 2' x 2' | S.F. | 0.011 |
| 2' x 4' | " | 0.009 |
| Wood fiber panels | | |
| 2" thick | | |
| 2' x 2' | S.F. | 0.013 |
| 2' x 4' | " | 0.010 |
| Air distributing panels | | |
| 3/4" thick | S.F. | 0.020 |
| 5/8" thick | " | 0.016 |
| Acoustical tiles, suspension system not included | | |
| Fiberglass tile, 12" x 12" | | |
| 5/8" thick | S.F. | 0.015 |
| 3/4" thick | " | 0.018 |
| Glass cloth faced fiberglass tile | | |
| 3/4" thick | S.F. | 0.018 |
| 3" thick | " | 0.020 |
| Mineral fiber tile, 12" x 12" | | |

# 09 FINISHES

| ACOUSTICAL TREATMENT | UNIT | MAN/ HOURS |
|---|---|---|
| **09510.10 CEILINGS AND WALLS** | | |
| 5/8" thick | | |
| Standard | S.F. | 0.016 |
| Vinyl faced | " | 0.016 |
| 3/4" thick | | |
| Standard | S.F. | 0.016 |
| Vinyl faced | " | 0.016 |
| Fire rated | " | 0.016 |
| Aluminum or mylar faced | " | 0.016 |
| Wood fiber tile, 12" x 12" | | |
| 1/2" thick | S.F. | 0.016 |
| 3/4" thick | " | 0.016 |
| Metal pan units, 24 ga. steel | | |
| 12" x 12" | S.F. | 0.032 |
| 12" x 24" | " | 0.027 |
| Aluminum, .025" thick | | |
| 12" x 12" | S.F. | 0.032 |
| 12" x 24" | " | 0.027 |
| Anodized aluminum, 0.25" thick | | |
| 12" x 12" | S.F. | 0.032 |
| 12" x 24" | " | 0.027 |
| Stainless steel, 24 ga. | | |
| 12" x 12" | S.F. | 0.032 |
| 12" x 24" | " | 0.027 |
| Metal ceiling systems | | |
| .020" thick panels | | |
| 10', 12', and 16' lengths | S.F. | 0.023 |
| Custom lengths, 3' to 20' | " | 0.023 |
| .025" thick panels | | |
| 32 sf, 38 sf, and 52 sf pieces | S.F. | 0.027 |
| Custom lengths, 10 sf to 65 sf | " | 0.027 |
| Sound absorption walls, with fabric cover | | |
| 2-6" x 9' x 3/4" | S.F. | 0.027 |
| 2' x 9' x 1" | " | 0.027 |
| Starter spline | L.F. | 0.020 |
| Internal spline | " | 0.020 |
| Acoustical treatment | | |
| Barriers for plenums | | |
| Leaded vinyl | | |
| 0.48 lb per sf | S.F. | 0.038 |
| 0.87 lb per sf | " | 0.040 |
| Aluminum foil, fiberglass reinforcement | | |
| Minimum | S.F. | 0.027 |
| Maximum | " | 0.040 |
| Aluminum mesh, paper backed | " | 0.027 |
| Fibered cement sheet, 3/16" thick | " | 0.029 |
| Sheet lead, 1/64" thick | " | 0.020 |
| Sound attenuation blanket | | |
| 1" thick | S.F. | 0.080 |
| 1-1/2" thick | " | 0.080 |
| 2" thick | " | 0.080 |
| 3" thick | " | 0.089 |
| Ceiling suspension systems | | |
| T bar system | | |
| 2' x 4' | S.F. | 0.008 |
| 2' x 2' | S.F. | 0.009 |
| Concealed Z bar suspension system, 12" module | " | 0.013 |

| FLOORING | UNIT | MAN/ HOURS |
|---|---|---|
| **09550.10 WOOD FLOORING** | | |
| Wood strip flooring, unfinished | | |
| Fir floor | | |
| C and better | | |
| Vertical grain | S.F. | 0.027 |
| Flat grain | " | 0.027 |
| Oak floor | | |
| Minimum | S.F. | 0.038 |
| Average | " | 0.038 |
| Maximum | " | 0.038 |
| Maple floor | | |
| 25/32" x 2-1/4" | | |
| Minimum | S.F. | 0.038 |
| Maximum | " | 0.038 |
| 33/32" x 3-1/4" | | |
| Minimum | S.F. | 0.038 |
| Maximum | " | 0.038 |
| Wood block industrial flooring | | |
| Creosoted | | |
| 2" thick | S.F. | 0.021 |
| 2-1/2" thick | " | 0.025 |
| 3" thick | " | 0.027 |
| Parquet, 5/16", white oak | | |
| Finished | S.F. | 0.040 |
| Unfinished | " | 0.040 |
| Gym floor, 2 ply felt, 25/32" maple, finished, in mastic | " | 0.044 |
| Over wood sleepers | " | 0.050 |
| Finishing, sand, fill, finish, and wax | " | 0.020 |
| Refinish sand, seal, and 2 coats of polyurethane | " | 0.027 |
| Clean and wax floors | " | 0.004 |
| **09630.10 UNIT MASONRY FLOORING** | | |
| Clay brick | | |
| 9 x 4-1/2 x 3" thick | | |
| Glazed | S.F. | 0.067 |
| Unglazed | " | 0.067 |
| 8 x 4 x 3/4" thick | | |
| Glazed | S.F. | 0.070 |
| Unglazed | " | 0.070 |

# 09 FINISHES

| FLOORING | UNIT | MAN/HOURS |
|---|---|---|
| **09660.10 RESILIENT TILE FLOORING** | | |
| Solid vinyl tile, 1/8" thick, 12" x 12" | | |
| Marble patterns | S.F. | 0.020 |
| Solid colors | " | 0.020 |
| Travertine patterns | " | 0.020 |
| Conductive resilient flooring, vinyl tile | | |
| 1/8" thick, 12" x 12" | S.F. | 0.023 |
| **09665.10 RESILIENT SHEET FLOORING** | | |
| Vinyl sheet flooring | | |
| Minimum | S.F. | 0.008 |
| Average | " | 0.010 |
| Maximum | " | 0.013 |
| Cove, to 6" | L.F. | 0.016 |
| Fluid applied resilient flooring | | |
| Polyurethane, poured in place, 3/8" thick | S.F. | 0.067 |
| Vinyl sheet goods, backed | | |
| 0.070" thick | S.F. | 0.010 |
| 0.093" thick | " | 0.010 |
| 0.125" thick | " | 0.010 |
| 0.250" thick | " | 0.010 |
| Wall base, vinyl | | |
| Group 1 | | |
| 4" high | L.F. | 0.027 |
| 6" high | " | 0.027 |
| Group 2 | | |
| 4" high | L.F. | 0.027 |
| 6" high | " | 0.027 |
| Group 3 | | |
| 4" high | L.F. | 0.027 |
| 6" high | " | 0.027 |
| Stair accessories | | |
| Treads, 1/4" x 12", rubber diamond surface | | |
| Marbled | L.F. | 0.067 |
| Plain | " | 0.067 |
| Grit strip safety tread, 12" wide, colors | | |
| 3/16" thick | L.F. | 0.067 |
| 5/16" thick | " | 0.067 |
| Risers, 7" high, 1/8" thick, colors | | |
| Flat | L.F. | 0.040 |
| Coved | " | 0.040 |
| Nosing, rubber | | |
| 3/16" thick, 3" wide | | |
| Black | L.F. | 0.040 |
| Colors | " | 0.040 |
| 6" wide | | |
| Black | L.F. | 0.067 |
| Colors | " | 0.067 |

| CARPET | UNIT | MAN/HOURS |
|---|---|---|
| **09680.10 FLOOR LEVELING** | | |
| Repair and level floors to receive new flooring | | |
| Minimum | S.Y. | 0.027 |
| Average | " | 0.067 |
| Maximum | " | 0.080 |
| **09682.10 CARPET PADDING** | | |
| Carpet padding | | |
| Foam rubber, waffle type, 0.3" thick | S.Y. | 0.040 |
| Jute padding | | |
| Minimum | S.Y. | 0.036 |
| Average | " | 0.040 |
| Maximum | " | 0.044 |
| Sponge rubber cushion | | |
| Minimum | S.Y. | 0.036 |
| Average | " | 0.040 |
| Maximum | " | 0.044 |
| Urethane cushion, 3/8" thick | | |
| Minimum | S.Y. | 0.036 |
| Average | " | 0.040 |
| Maximum | " | 0.044 |
| **09685.10 CARPET** | | |
| Carpet, acrylic | | |
| 24 oz., light traffic | S.Y. | 0.044 |
| 28 oz., medium traffic | " | 0.044 |
| Residential | | |
| Nylon | | |
| 15 oz., light traffic | S.Y. | 0.044 |
| 28 oz., medium traffic | " | 0.044 |
| Commercial | | |
| Nylon | | |
| 28 oz., medium traffic | S.Y. | 0.044 |
| 35 oz., heavy traffic | " | 0.044 |
| Wool | | |
| 30 oz., medium traffic | S.Y. | 0.044 |
| 36 oz., medium traffic | " | 0.044 |
| 42 oz., heavy traffic | " | 0.044 |
| Carpet tile | | |
| Foam backed | | |
| Minimum | S.F. | 0.008 |
| Average | " | 0.009 |
| Maximum | " | 0.010 |
| Tufted loop or shag | | |
| Minimum | S.F. | 0.008 |
| Average | " | 0.009 |
| Maximum | " | 0.010 |
| Clean and vacuum carpet | | |
| Minimum | S.Y. | 0.004 |
| Average | " | 0.005 |
| Maximum | " | 0.008 |
| **09700.10 SPECIAL FLOORING** | | |
| Epoxy flooring, marble chips | | |
| Epoxy with colored quartz chips in 1/4" base | S.F. | 0.044 |

# 09 FINISHES

## CARPET

| 09700.10 SPECIAL FLOORING | UNIT | MAN/HOURS |
|---|---|---|
| Heavy duty epoxy topping, 3/16" thick | S.F. | 0.044 |
| Epoxy terrazzo | | |
| 1/4" thick chemical resistant | S.F. | 0.050 |

## PAINTING

| 09910.10 EXTERIOR PAINTING | UNIT | MAN/HOURS |
|---|---|---|
| Exterior painting | | |
| Wood surfaces, 1 coat primer, two coats paint | | |
| Door and frame | EA. | 1.333 |
| Windows | S.F. | 0.020 |
| Wood trim | " | 0.020 |
| Wood siding | " | 0.010 |
| Hardboard surfaces | | |
| One coat primer, two coats paint | S.F. | 0.010 |
| Asbestos cement surfaces | | |
| One coat primer, two coats paint | S.F. | 0.010 |
| Galvanized surfaces, galvanized primer | | |
| One coat primer, two coats paint | S.F. | 0.009 |
| Stucco surfaces, acrylic primer, acrylic latex paint | | |
| One coat primer, two coats paint | S.F. | 0.013 |
| Concrete masonry unit surfaces, brush work | | |
| One coat filler, one coat paint | S.F. | 0.010 |
| Two coats epoxy | " | 0.013 |
| Texture coating | " | 0.008 |
| Concrete surfaces | | |
| One coat filler, one coat paint | S.F. | 0.010 |
| Two coats paint | " | 0.013 |
| Structural steel | | |
| One field coat paint, brush work | | |
| Light framing | S.F. | 0.007 |
| Heavy framing | " | 0.004 |
| One field coat paint, spray work | | |
| Light framing | S.F. | 0.002 |
| Heavy framing | " | 0.002 |
| Miscellaneous steel items, spray work, one coat | | |
| Exposed decking | S.F. | 0.007 |
| Joist | " | 0.010 |
| Columns | " | 0.010 |
| Pipes, one coat primer, one coat paint | | |
| 4" dia. | L.F. | 0.010 |
| 8" dia. | " | 0.013 |
| 12" dia. | " | 0.020 |
| Paint letters on pipe with brush | " | 0.020 |
| Paint pipe insulation cloth cover | S.F. | 0.010 |
| Sprinkler system piping | L.F. | 0.007 |

## PAINTING

| 09910.10 EXTERIOR PAINTING | UNIT | MAN/HOURS |
|---|---|---|
| Miscellaneous surfaces | | |
| Stair pipe rails | | |
| Two rails | L.F. | 0.027 |
| One rail | " | 0.016 |
| Stair to 4' wide, including rails, per riser | EA. | 0.114 |
| Gratings and frames | S.F. | 0.027 |
| Ladders | L.F. | 0.023 |
| Miscellaneous exposed metal | S.F. | 0.008 |

| 09920.10 INTERIOR PAINTING | UNIT | MAN/HOURS |
|---|---|---|
| Walls, concrete and masonry, brush, primer, acrylic | | |
| One coat primer, one coat paint | S.F. | 0.010 |
| Two coats paint | " | 0.013 |
| Plywood, paint | " | 0.004 |
| Natural finish | " | 0.005 |
| Wood, paint | " | 0.005 |
| Natural finish | " | 0.006 |
| Metal | | |
| One coat filler | S.F. | 0.005 |
| One coat primer, one coat paint | " | 0.010 |
| Two coats paint | " | 0.013 |
| Plaster or gypsum board, paint | " | 0.004 |
| Epoxy | " | 0.005 |
| Ceilings, one coat paint, wood | " | 0.006 |
| Concrete | " | 0.005 |
| Plaster | " | 0.004 |
| Miscellaneous metal, brushwork | " | 0.010 |

| 09920.30 DOORS AND MILLWORK | UNIT | MAN/HOURS |
|---|---|---|
| Painting, doors | | |
| Minimum | S.F. | 0.027 |
| Average | " | 0.040 |
| Maximum | " | 0.053 |
| Cabinets, shelves, and millwork | | |
| Minimum | S.F. | 0.013 |
| Average | " | 0.023 |
| Maximum | " | 0.040 |

| 09920.60 WINDOWS | UNIT | MAN/HOURS |
|---|---|---|
| Painting, windows | | |
| Minimum | S.F. | 0.016 |
| Average | " | 0.020 |
| Maximum | " | 0.032 |

| 09955.10 WALL COVERING | UNIT | MAN/HOURS |
|---|---|---|
| Vinyl wall covering | | |
| Medium duty | S.F. | 0.011 |
| Heavy duty | " | 0.013 |
| Over pipes and irregular shapes | | |
| Lightweight, 13 oz. | S.F. | 0.016 |
| Medium weight, 25 oz. | " | 0.018 |
| Heavy weight, 34 oz. | " | 0.020 |
| Cork wall covering | | |

# 09 FINISHES

| PAINTING | UNIT | MAN/ HOURS |
|---|---|---|
| **09955.10 WALL COVERING** | | |
| 1' x 1' squares | | |
| 1/4" thick | S.F. | 0.020 |
| 1/2" thick | " | 0.020 |
| 3/4" thick | " | 0.020 |
| Wall fabrics | | |
| Natural fabrics, grass cloths | | |
| Minimum | S.F. | 0.012 |
| Average | " | 0.013 |
| Maximum | " | 0.016 |
| Flexible gypsum coated wall fabric, fire resistant | " | 0.008 |
| Vinyl corner guards | | |
| 3/4" x 3/4" x 8' | EA. | 0.100 |
| 2-3/4" x 2-3/4" x 4' | " | 0.100 |
| **09980.10 PAINTING PREPARATION** | | |
| Cleaning, light | | |
| Wood | S.F. | 0.002 |
| Plaster or gypsum wallboard | " | 0.002 |
| Prepare sprinkler piping for painting | L.F. | 0.002 |
| Prepare insulated pipe for painting | S.F. | 0.002 |
| Normal painting prep, masonry and concrete | | |
| Unpainted | S.F. | 0.001 |
| Painted | " | 0.002 |
| Plaster or gypsum | | |
| Unpainted | S.F. | 0.001 |
| Painted | " | 0.002 |
| Wood | | |
| Unpainted | S.F. | 0.001 |
| Painted | " | 0.002 |
| Painted steel, light rusting | " | 0.003 |
| Sandblasting | | |
| Brush off blast | S.F. | 0.005 |
| Commercial blast | " | 0.013 |
| Near white metal blast | " | 0.023 |
| White metal blast | " | 0.027 |

| SPECIALTIES | UNIT | MAN/HOURS |
|---|---|---|
| **10110.10 CHALKBOARDS** | | |
| Chalkboard, metal frame, 1/4" thick | | |
| 48"x60" | EA. | 0.800 |
| 48"x96" | " | 0.889 |
| 48"x144" | " | 1.000 |
| 48"x192" | " | 1.143 |
| Liquid chalkboard | | |
| 48"x60" | EA. | 0.800 |
| 48"x96" | " | 0.889 |
| 48"x144" | " | 1.000 |
| 48"x192" | " | 1.143 |
| Map rail, deluxe | L.F. | 0.040 |
| **10165.10 TOILET PARTITIONS** | | |
| Toilet partition, plastic laminate | | |
| Ceiling mounted | EA. | 2.667 |
| Floor mounted | " | 2.000 |
| Metal | | |
| Ceiling mounted | EA. | 2.667 |
| Floor mounted | " | 2.000 |
| Wheel chair partition, plastic laminate | | |
| Ceiling mounted | EA. | 2.667 |
| Floor mounted | " | 2.000 |
| Painted metal | | |
| Ceiling mounted | EA. | 2.667 |
| Floor mounted | " | 2.000 |
| Urinal screen, plastic laminate | | |
| Wall hung | EA. | 1.000 |
| Floor mounted | " | 1.000 |
| Porcelain enameled steel, floor mounted | " | 1.000 |
| Painted metal, floor mounted | " | 1.000 |
| Stainless steel, floor mounted | " | 1.000 |
| Metal toilet partitions | | |
| Toilet partitions, front door and side divider, floor mounted | | |
| Porcelain enameled steel | EA. | 2.000 |
| Painted steel | " | 2.000 |
| Stainless steel | " | 2.000 |
| **10185.10 SHOWER STALLS** | | |
| Shower receptors | | |
| Precast, terrazzo | | |
| 32" x 32" | EA. | 0.667 |
| 32" x 48" | " | 0.800 |
| Concrete | | |
| 32" x 32" | EA. | 0.667 |
| 48" x 48" | " | 0.889 |
| Shower door, trim and hardware | | |
| Economy, 24" wide, chrome frame, tempered glass | EA. | 0.800 |
| Porcelain enameled steel, flush | " | 0.800 |
| Baked enameled steel, flush | " | 0.800 |
| Aluminum frame, tempered glass, 48" wide, sliding | " | 1.000 |
| Folding | " | 1.000 |
| Aluminum frame and tempered glass, molded plastic | | |

| SPECIALTIES | UNIT | MAN/HOURS |
|---|---|---|
| **10185.10 SHOWER STALLS** | | |
| Complete with receptor and door | | |
| 32" x 32" | EA. | 2.000 |
| 36" x 36" | " | 2.000 |
| 40" x 40" | " | 2.286 |
| Shower compartment, precast concrete receptor | | |
| Single entry type | | |
| Porcelain enameled steel | EA. | 8.000 |
| Baked enameled steel | " | 8.000 |
| Stainless steel | " | 8.000 |
| Double entry type | | |
| Porcelain enameled steel | EA. | 10.000 |
| Baked enameled steel | " | 10.000 |
| Stainless steel | " | 10.000 |
| **10190.10 CUBICLES** | | |
| Hospital track | | |
| Ceiling hung | L.F. | 0.089 |
| Suspended | " | 0.114 |
| Hospital metal dividers, galvanized steel | | |
| Baked enamel finish | | |
| 54" high | | |
| 10" glass light | L.F. | 0.400 |
| 14" glass light | " | 0.400 |
| 24" glass light | " | 0.400 |
| 60" high | | |
| 10" glass light | L.F. | 0.444 |
| 14" glass light | " | 0.444 |
| 24" glass light | " | 0.444 |
| Stainless steel | | |
| 54" high | | |
| 10" glass light | L.F. | 0.444 |
| 14" glass light | " | 0.444 |
| 24" glass light | " | 0.444 |
| 60" high | | |
| 14" glass light | L.F. | 0.500 |
| 14" glass light | " | 0.500 |
| 24" glass light | " | 0.500 |
| **10210.10 VENTS AND WALL LOUVERS** | | |
| Block vent, 8"x16"x4" aluminum, w/screen, mill finish | EA. | 0.267 |
| Standard | " | 0.250 |
| Vents w/screen, 4" deep, 8" wide, 5" high | | |
| Modular | EA. | 0.250 |
| Aluminum gable louvers | S.F. | 0.133 |
| Vent screen aluminum, 4" wide, continuous | L.F. | 0.027 |
| Louvers, aluminum, anodized, fixed blade | | |
| Horizontal line | S.F. | 0.200 |
| Vertical line | " | 0.200 |
| Wall louver, aluminum mill finish | | |
| Under, 2 sf | S.F. | 0.100 |
| 2 to 4 sf | " | 0.089 |
| 5 to 10 sf | " | 0.089 |
| Galvanized steel | | |

| SPECIALTIES | UNIT | MAN/ HOURS |
|---|---|---|
| **10210.10 VENTS AND WALL LOUVERS** | | |
| Under 2 sf | S.F. | 0.100 |
| 2 to 4 sf | " | 0.089 |
| 5 to 10 sf | " | 0.089 |
| **10225.10 DOOR LOUVERS** | | |
| Fixed, 1" thick, enameled steel | | |
| 8"x8" | EA. | 0.100 |
| 12"x8" | " | 0.100 |
| 12"x12" | " | 0.114 |
| 16"x12" | " | 0.123 |
| 18"x12" | " | 0.200 |
| 20"x8" | " | 0.114 |
| 20"x12" | " | 0.229 |
| 20"x16" | " | 0.267 |
| 20"x20" | " | 0.320 |
| 24"x12" | " | 0.267 |
| 24"x16" | " | 0.286 |
| 24"x18" | " | 0.308 |
| 24"x20" | " | 0.333 |
| 24"x24" | " | 0.364 |
| 26"x26" | " | 0.500 |
| **10270.40 ACCESS & PEDESTAL FLOOR** | | |
| Panels, no covering, 2'x2' | | |
| Plain | S.F. | 0.010 |
| Perforated | " | 0.400 |
| Pedestals | | |
| For 6" to 12" clearance | EA. | 0.080 |
| Stringers | | |
| 2' | L.F. | 0.038 |
| 6' | " | 0.027 |
| Accessories | | |
| Ramp assembly | S.F. | 0.032 |
| Elevated floor assembly | " | 0.030 |
| Handrail | L.F. | 0.400 |
| Fascia plate | " | 0.200 |
| RF shielding components, floor liner | | |
| Hot rolled steel sheet | | |
| 14 ga. | S.F. | 0.020 |
| 11 ga. | " | 0.062 |
| **10290.10 PEST CONTROL** | | |
| Termite control | | |
| Under slab spraying | | |
| Minimum | S.F. | 0.002 |
| Average | " | 0.004 |
| Maximum | " | 0.008 |
| **10350.10 FLAGPOLES** | | |
| Installed in concrete base | | |
| Fiberglass | | |

| SPECIALTIES | UNIT | MAN/ HOURS |
|---|---|---|
| **10350.10 FLAGPOLES** | | |
| 25' high | EA. | 5.333 |
| 50' high | " | 13.333 |
| Aluminum | | |
| 25' high | EA. | 5.333 |
| 50' high | " | 13.333 |
| Bonderized steel | | |
| 25' high | EA. | 6.154 |
| 50' high | " | 16.000 |
| Freestanding tapered, fiberglass | | |
| 30' high | EA. | 5.714 |
| 40' high | " | 7.273 |
| 50' high | " | 8.000 |
| 60' high | " | 9.412 |
| Wall mounted, with collar, brushed aluminum finish | | |
| 15' long | EA. | 4.000 |
| 18' long | " | 4.000 |
| 20' long | " | 4.211 |
| 24' long | " | 4.706 |
| Outrigger, wall, including base | | |
| 10' long | EA. | 5.333 |
| 20' long | " | 6.667 |
| **10400.10 IDENTIFYING DEVICES** | | |
| Directory and bulletin boards | | |
| Open face boards | | |
| Chrome plated steel frame | S.F. | 0.400 |
| Aluminum framed | " | 0.400 |
| Bronze framed | " | 0.400 |
| Stainless steel framed | " | 0.400 |
| Tack board, aluminum framed | " | 0.400 |
| Visual aid board, aluminum framed | " | 0.400 |
| Glass encased boards, hinged and keyed | | |
| Aluminum framed | S.F. | 1.000 |
| Bronze framed | " | 1.000 |
| Stainless steel framed | " | 1.000 |
| Chrome plated steel framed | " | 1.000 |
| Metal plaque | | |
| Cast bronze | S.F. | 0.667 |
| Aluminum | " | 0.667 |
| Metal engraved plaque | | |
| Porcelain steel | S.F. | 0.667 |
| Stainless steel | " | 0.667 |
| Brass | " | 0.667 |
| Aluminum | " | 0.667 |
| Metal built-up plaque | | |
| Bronze | S.F. | 0.800 |
| Copper and bronze | " | 0.800 |
| Copper and aluminum | " | 0.800 |
| Metal nameplate plaques | | |
| Cast bronze | S.F. | 0.500 |
| Cast aluminum | " | 0.500 |
| Engraved, 1-1/2" x 6" | | |
| Bronze | EA. | 0.500 |

# 10 SPECIALTIES

| SPECIALTIES | UNIT | MAN/ HOURS |
|---|---|---|
| **10400.10 IDENTIFYING DEVICES** | | |
| Aluminum | EA. | 0.500 |
| Letters, on masonry or concrete, aluminum, satin finish | | |
| 1/2" thick | | |
| 2" high | EA. | 0.320 |
| 4" high | " | 0.400 |
| 6" high | " | 0.444 |
| 3/4" thick | | |
| 8" high | EA. | 0.500 |
| 10" high | " | 0.571 |
| 1" thick | | |
| 12" high | EA. | 0.667 |
| 14" high | " | 0.800 |
| 16" high | " | 1.000 |
| 3/8" thick | | |
| 2" high | EA. | 0.320 |
| 4" high | " | 0.400 |
| 1/2" thick, 6" high | " | 0.444 |
| 5/8" thick, 8" high | " | 0.500 |
| 1" thick | | |
| 10" high | EA. | 0.571 |
| 12" high | " | 0.667 |
| 14" high | " | 0.800 |
| 16" high | " | 1.000 |
| Interior door signs, adhesive, flexible | | |
| 2" x 8" | EA. | 0.200 |
| 4" x 4" | " | 0.200 |
| 6" x 7" | " | 0.200 |
| 6" x 9" | " | 0.200 |
| 10" x 9" | " | 0.200 |
| 10" x 12" | " | 0.200 |
| Hard plastic type, no frame | | |
| 3" x 8" | EA. | 0.200 |
| 4" x 4" | " | 0.200 |
| 4" x 12" | " | 0.200 |
| Hard plastic type, with frame | | |
| 3" x 8" | EA. | 0.200 |
| 4" x 4" | " | 0.200 |
| 4" x 12" | " | 0.200 |
| **10450.10 CONTROL** | | |
| Access control, 7' high, indoor or outdoor impenetrability | | |
| Remote or card control, type B | EA. | 10.667 |
| Free passage, type B | " | 10.667 |
| Remote or card control, type AA | " | 10.667 |
| Free passage, type AA | " | 10.667 |
| **10500.10 LOCKERS** | | |
| Locker bench, floor mounted, laminated maple | | |
| 4' | EA. | 0.667 |
| 6' | " | 0.667 |
| Wardrobe locker, single tier type | | |
| 12" x 15" x 72" | EA. | 0.800 |
| 18" x 15" x 72" | " | 0.842 |

| SPECIALTIES | UNIT | MAN/ HOURS |
|---|---|---|
| **10500.10 LOCKERS** | | |
| 12" x 18" x 72" | EA. | 0.889 |
| 18" x 18" x 72" | " | 0.941 |
| Double tier type | | |
| 12" x 15" x 36" | EA. | 0.400 |
| 18" x 15" x 36" | " | 0.400 |
| 12" x 18" x 36" | " | 0.400 |
| 18" x 18" x 36" | " | 0.400 |
| Two person unit | | |
| 18" x 15" x 72" | EA. | 1.333 |
| 18" x 18" x 72" | " | 1.600 |
| Duplex unit | | |
| 15" x 15" x 72" | EA. | 0.800 |
| 15" x 21" x 72" | " | 0.800 |
| Basket lockers, basket sets with baskets | | |
| 24 basket set | SET | 4.000 |
| 30 basket set | " | 5.000 |
| 36 basket set | " | 6.667 |
| 42 basket set | " | 8.000 |
| **10520.10 FIRE PROTECTION** | | |
| Portable fire extinguishers | | |
| Water pump tank type | | |
| 2.5 gal. | | |
| Red enameled galvanized | EA. | 0.533 |
| Red enameled copper | " | 0.533 |
| Polished copper | " | 0.533 |
| Carbon dioxide type, red enamel steel | | |
| Squeeze grip with hose and horn | | |
| 2.5 lb | EA. | 0.533 |
| 5 lb | " | 0.615 |
| 10 lb | " | 0.800 |
| 15 lb | " | 1.000 |
| 20 lb | " | 1.000 |
| Wheeled type | | |
| 125 lb | EA. | 1.600 |
| 250 lb | " | 1.600 |
| 500 lb | " | 1.600 |
| Dry chemical, pressurized type | | |
| Red enameled steel | | |
| 2.5 lb | EA. | 0.533 |
| 5 lb | " | 0.615 |
| 10 lb | " | 0.800 |
| 20 lb | " | 1.000 |
| 30 lb | " | 1.000 |
| Chrome plated steel, 2.5 lb | " | 0.533 |
| Other type extinguishers | | |
| 2.5 gal, stainless steel, pressurized water tanks | EA. | 0.533 |
| Soda and acid type | " | 0.533 |
| Cartridge operated, water type | " | 0.533 |
| Loaded stream, water type | " | 0.533 |
| Foam type | " | 0.533 |
| 40 gal, wheeled foam type | " | 1.600 |
| Fire extinguisher cabinets | | |
| Enameled steel | | |

# 10 SPECIALTIES

| SPECIALTIES | UNIT | MAN/HOURS |
|---|---|---|
| **10520.10 FIRE PROTECTION** | | |
| 8" x 12" x 27" | EA. | 1.600 |
| 8" x 16" x 38" | " | 1.600 |
| Aluminum | | |
| 8" x 12" x 27" | EA. | 1.600 |
| 8" x 16" x 38" | " | 1.600 |
| 8" x 12" x 27" | " | 1.600 |
| Stainless steel | | |
| 8" x 16" x 38" | EA. | 1.600 |
| **10550.10 POSTAL SPECIALTIES** | | |
| Mail chutes | | |
| Single mail chute | | |
| Finished aluminum | L.F. | 2.000 |
| Bronze | " | 2.000 |
| Single mail chute receiving box | | |
| Finished aluminum | EA. | 4.000 |
| Bronze | " | 4.000 |
| Twin mail chute, double parallel | | |
| Finished aluminum | FLR | 4.000 |
| Bronze | " | 4.000 |
| Receiving box, 36" x 20" x 12" | | |
| Finished aluminum | EA. | 6.667 |
| Bronze | " | 6.667 |
| Locked receiving mail box | | |
| Finished aluminum | EA. | 4.000 |
| Bronze | " | 4.000 |
| Commercial postal accessories for mail chutes | | |
| Letter slot, brass | EA. | 1.333 |
| Bulk mail slot, brass | " | 1.333 |
| Mail boxes | | |
| Residential postal accessories | | |
| Letter slot | EA. | 0.400 |
| Rural letter box | " | 1.000 |
| Apartment house, keyed, 3.5" x 4.5" x 16" | " | 0.267 |
| Ranch style | " | 0.400 |
| Commercial postal accessories | | |
| Letter box, with combination lock | EA. | 0.286 |
| Key lock | " | 0.286 |
| Mail box, aluminum w/glass front, 4x5 | | |
| Horizontal rear load | EA. | 0.229 |
| Vertical front load | " | 0.229 |
| **10600.10 MOVABLE PARTITIONS** | | |
| Partition, movable office type, 2-1/2" thick, vinyl-gypsum | S.F. | 0.040 |
| Enameled steel frame, with 1/4" thick clear glass | " | 0.040 |
| Door frame and hardware for movable partitions | EA. | 2.667 |
| Add for acoustic movable partition | S.F. | |
| Accordion partition, 12' high | | |
| Vinyl | S.F. | 0.133 |
| Acoustical | " | 0.133 |
| Standard office cubicles, 8' high, steel framed | | |
| Baked enamel finish | | |
| 100% flush | L.F. | 0.200 |

| SPECIALTIES | UNIT | MAN/HOURS |
|---|---|---|
| **10600.10 MOVABLE PARTITIONS** | | |
| 75% flush and 25% glass | L.F. | 0.222 |
| 50% flush and 50% glass | " | 0.222 |
| 100% glass | " | 0.267 |
| Natural hardwood panels | | |
| 100% flush | L.F. | 0.267 |
| 50% flush and 50% glass | " | 0.286 |
| Plastic laminated panels | | |
| 100% flush | L.F. | 0.267 |
| 75% flush and 25% glass | " | 0.286 |
| 50% flush and 50% glass | " | 0.286 |
| Vinyl covered panels | | |
| 100% flush | L.F. | 0.276 |
| 75% flush and 25% glass | " | 0.296 |
| 50% and 50% glass | " | 0.296 |
| Aluminum framed | | |
| Enameled or anodized aluminum panels | | |
| 100% flush | L.F. | 0.200 |
| 75% flush and 25% glass | " | 0.222 |
| 50% flush and 50% glass | " | 0.222 |
| Vinyl covered panels | | |
| 100% flush | L.F. | 0.276 |
| 75% flush and 25% glass | " | 0.296 |
| 50% flush and 50% glass | " | 0.296 |
| 60" high partitions, steel framed | | |
| Enameled panels | L.F. | 0.178 |
| Natural hardwood panels, two sides | " | 0.186 |
| Plastic laminated panels | " | 0.186 |
| Vinyl covered panels | " | 0.178 |
| Aluminum framed | | |
| Anodized or baked enamel panels | L.F. | 0.178 |
| Natural hardwood panels | " | 0.186 |
| Plastic laminated panels | " | 0.186 |
| Vinyl covered panels | " | 0.178 |
| Wire mesh partitions | | |
| Wall panels | | |
| 4' x 7' | EA. | 0.500 |
| 4' x 8' | " | 0.533 |
| 4' x 10' | " | 0.615 |
| Wall filler panels | | |
| 1' x 7' | EA. | 0.500 |
| 1' x 8' | " | 0.533 |
| 1' x 10' | " | 0.571 |
| 2' x 7' | " | 0.500 |
| 2' x 8' | " | 0.533 |
| 2' x 10' | " | 0.571 |
| 3' x 7' | " | 0.500 |
| 3' x 8' | " | 0.533 |
| 3' x 10' | " | 0.571 |
| Ceiling panels | | |
| 10' x 2' | EA. | 1.143 |
| 10' x 4' | " | 1.600 |
| Wall panel with service window | | |
| 5' wide | | |
| 7' high | EA. | 0.500 |

| SPECIALTIES | UNIT | MAN/ HOURS |
|---|---|---|
| **10600.10 MOVABLE PARTITIONS** | | |
| 8' high | EA. | 0.533 |
| 10' high | " | 0.571 |
| Doors | | |
| Sliding | | |
| 3' x 7' | EA. | 2.000 |
| 3' x 8' | " | 2.286 |
| 3' x 10' | " | 3.200 |
| 4' x 7' | " | 2.286 |
| 4' x 8' | " | 3.200 |
| 4' x 10' | " | 4.000 |
| 5' x 7' | " | 3.200 |
| 5' x 8' | " | 4.000 |
| 5' x 10' | " | 4.000 |
| Swing door | | |
| 3' x 7' | EA. | 2.000 |
| 4' x 7' | " | 2.286 |
| Swing door, with 1' transom | | |
| 3' x 7' | EA. | 2.286 |
| 4' x 7' | " | 3.200 |
| Swing door, with 3' transom | | |
| 3' x 7' | EA. | 3.200 |
| 4' x 7' | " | 4.000 |
| **10670.10 SHELVING** | | |
| Shelving, enamel, closed side and back, 12" x 36" | | |
| 5 shelves | EA. | 1.333 |
| 8 shelves | " | 1.778 |
| Open | | |
| 5 shelves | EA. | 1.333 |
| 8 shelves | " | 1.778 |
| Metal storage shelving, baked enamel | | |
| 7 shelf unit, 72" or 84" high | | |
| 10" shelf | L.F. | 0.800 |
| 12" shelf | " | 0.842 |
| 15" shelf | " | 0.889 |
| 18" shelf | " | 0.941 |
| 24" shelf | " | 1.000 |
| 30" shelf | " | 1.067 |
| 36" shelf | " | 1.143 |
| 4 shelf unit, 40" high | | |
| 10" shelf | L.F. | 0.667 |
| 12" shelf | " | 0.727 |
| 15" shelf | " | 0.800 |
| 18" shelf | " | 0.842 |
| 24" shelf | " | 0.889 |
| 3 shelf unit, 32" high | | |
| 10" shelf | L.F. | 0.400 |
| 12" shelf | " | 0.421 |
| 15" shelf | " | 0.444 |
| 18" shelf | " | 0.471 |
| 24" shelf | " | 0.500 |
| Single shelf unit, attached to masonry | | |
| 10" shelf | L.F. | 0.133 |
| 12" shelf | " | 0.145 |

| SPECIALTIES | UNIT | MAN/ HOURS |
|---|---|---|
| **10670.10 SHELVING** | | |
| 15" shelf | L.F. | 0.154 |
| 18" shelf | " | 0.163 |
| 24" shelf | " | 0.174 |
| Built-in wood shelves | | |
| Posts and trimmed plywood | L.F. | 0.114 |
| Solid clear pine | " | 0.123 |
| Closet shelf, pine with rod | " | 0.123 |
| **10750.10 TELEPHONE ENCLOSURES** | | |
| Telephone enclosure, wall mounted, shelf, 28" x 30" x 15" | EA. | 2.000 |
| Directory shelf, stainless steel, 3 binders | " | 1.333 |
| **10800.10 BATH ACCESSORIES** | | |
| Ash receiver, wall mounted, aluminum | EA. | 0.400 |
| Grab bar, 1-1/2" dia., stainless steel, wall mounted | | |
| 24" long | EA. | 0.400 |
| 36" long | " | 0.421 |
| 42" long | " | 0.444 |
| 48" long | " | 0.471 |
| 52" long | " | 0.500 |
| 1" dia., stainless steel | | |
| 12" long | EA. | 0.348 |
| 18" long | " | 0.364 |
| 24" long | " | 0.400 |
| 30" long | " | 0.421 |
| 36" long | " | 0.444 |
| 48" long | " | 0.471 |
| Hand dryer, surface mounted, 110 volt | " | 1.000 |
| Medicine cabinet, 16 x 22, baked enamel, steel, lighted | " | 0.320 |
| With mirror, lighted | " | 0.533 |
| Mirror, 1/4" plate glass, up to 10 sf | S.F. | 0.080 |
| Mirror, stainless steel frame | | |
| 18"x24" | EA. | 0.267 |
| 18"x32" | " | 0.320 |
| 18"x36" | " | 0.400 |
| 24"x30" | " | 0.400 |
| 24"x36" | " | 0.444 |
| 24"x48" | " | 0.667 |
| 24"x60" | " | 0.800 |
| 30"x30" | " | 0.800 |
| 30"x72" | " | 1.000 |
| 48"x72" | " | 1.333 |
| With shelf, 18"x24" | " | 0.320 |
| Sanitary napkin dispenser, stainless steel, wall mounted | " | 0.533 |
| Shower rod, 1" diameter | | |
| Chrome finish over brass | EA. | 0.400 |
| Stainless steel | " | 0.400 |
| Soap dish, stainless steel, wall mounted | " | 0.533 |
| Toilet tissue dispenser, stainless, wall mounted | | |
| Single roll | EA. | 0.200 |
| Double roll | " | 0.229 |
| Towel dispenser, stainless steel | | |
| Flush mounted | EA. | 0.444 |

# 10 SPECIALTIES

| SPECIALTIES | UNIT | MAN/ HOURS |
|---|---|---|
| **10800.10 BATH ACCESSORIES** | | |
| Surface mounted | EA. | 0.400 |
| Combination towel dispenser and waste receptacle | " | 0.533 |
| Towel bar, stainless steel | | |
| 18" long | EA. | 0.320 |
| 24" long | " | 0.364 |
| 30" long | " | 0.400 |
| 36" long | " | 0.444 |
| Toothbrush and tumbler holder | " | 0.267 |
| Waste receptacle, stainless steel, wall mounted | " | 0.667 |
| **10900.10 WARDROBE SPECIALTIES** | | |
| Hospital wardrobe units, 24" x 24" x 76", with door | | |
| Baked enameled steel | EA. | 4.444 |
| Hardwood | " | 4.444 |
| Stainless steel | " | 4.444 |
| Plastic laminated | " | 4.444 |
| Dormitory wardrobe units, 24" x 76", with door | | |
| Hardwood | EA. | 4.444 |
| Plastic laminated | " | 4.444 |
| Hat and coat rack | | |
| Single tier | | |
| Baked enameled steel | L.F. | 0.200 |
| Stainless steel | " | 0.200 |
| Aluminum | " | 0.200 |
| Double tier | | |
| Baked enameled steel | L.F. | 0.229 |
| Stainless steel | " | 0.229 |
| Aluminum | " | 0.229 |

# 11 EQUIPMENT

| ARCHITECTURAL EQUIPMENT | UNIT | MAN/ HOURS |
|---|---|---|
| **11010.10 MAINTENANCE EQUIPMENT** | | |
| Vacuum cleaning system | | |
| 3 valves | | |
| 1.5 hp | EA. | 8.889 |
| 2.5 hp | " | 11.429 |
| 5 valves | " | 16.000 |
| 7 valves | " | 20.000 |
| **11020.10 SECURITY EQUIPMENT** | | |
| Bulletproof teller window | | |
| 4' x 4' | EA. | 13.333 |
| 5' x 4' | " | 16.000 |
| Bulletproof partitions | | |
| Up to 12' high, 2.5" thick | S.F. | 0.053 |
| Counter for banks | | |
| Minimum | L.F. | 1.600 |
| Maximum | " | 2.667 |
| Drive-up window | | |
| Minimum | EA. | 11.429 |
| Maximum | " | 26.667 |
| Night depository | | |
| Minimum | EA. | 11.429 |
| Maximum | " | 26.667 |
| Office safes, 30" x 20" x 20", 1 hr rating | " | 2.000 |
| 30" x 16" x 15", 2 hr rating | " | 1.600 |
| 30" x 28" x 20", H&G rating | " | 1.000 |
| Service windows, pass through painted steel | | |
| 24" x 36" | EA. | 8.000 |
| 48" x 40" | " | 10.000 |
| 72" x 40" | " | 16.000 |
| Special doors and windows | | |
| 3' x 7' bulletproof door with frame | EA. | 11.429 |
| 12" x 12" vision panel | " | 5.714 |
| Surveillance system | | |
| Minimum | EA. | 16.000 |
| Maximum | " | 80.000 |
| Vault door, 3' wide, 6'6" high | | |
| 3-1/2" thick | EA. | 100 |
| 7" thick | " | 133 |
| 10" thick | " | 160 |
| Insulated vault door | | |
| 2 hr rating | | |
| 32" wide | EA. | 8.000 |
| 40" wide | " | 8.421 |
| 4 hr rating | | |
| 32" wide | EA. | 8.889 |
| 40" wide | " | 10.000 |
| 6 hr rating | | |
| 32" wide | EA. | 8.889 |
| 40" wide | " | 10.000 |
| Insulated file room door | | |
| 1 hr rating | | |
| 32" wide | EA. | 8.000 |
| 40" wide | " | 8.889 |

| ARCHITECTURAL EQUIPMENT | UNIT | MAN/ HOURS |
|---|---|---|
| **11060.10 THEATER EQUIPMENT** | | |
| Roll out stage, steel frame, wood floor | | |
| Manual | S.F. | 0.050 |
| Electric | " | 0.080 |
| Portable stages | | |
| 8" high | S.F. | 0.040 |
| 18" high | " | 0.044 |
| 36" high | " | 0.047 |
| 48" high | " | 0.050 |
| Band risers | | |
| Minimum | S.F. | 0.040 |
| Maximum | " | 0.040 |
| Chairs for risers | | |
| Minimum | EA. | 0.036 |
| Maximum | " | 0.036 |
| **11080.10 POLICE EQUIPMENT** | | |
| Firing range equipment, rifle | | |
| 3 position | EA. | 26.667 |
| 4 position | " | 40.000 |
| 5 position | " | 44.444 |
| 6 position | " | 47.059 |
| **11090.10 CHECKROOM EQUIPMENT** | | |
| Motorized checkroom equipment | | |
| No shelf system, 6'4" height | | |
| 7'6" length | EA. | 8.000 |
| 14'6" length | " | 8.000 |
| 28' length | " | 8.000 |
| One shelf, 6'8" height | | |
| 7'6" length | EA. | 8.000 |
| 14'6" length | " | 8.000 |
| 28' length | " | 8.000 |
| Two shelves, 7'5" height | | |
| 7'6" length | EA. | 8.000 |
| 14'6" length | " | 8.000 |
| 28' length | " | 8.000 |
| Three shelves, 8' height | | |
| 7'6" length | EA. | 16.000 |
| 14'6" length | " | 16.000 |
| 28' length | " | 16.000 |
| Four shelves, 8'7" height | | |
| 7'6" length | EA. | 16.000 |
| 14'6" length | " | 16.000 |
| 28' length | " | 16.000 |
| 200 lb | | |
| **11110.10 LAUNDRY EQUIPMENT** | | |
| High capacity, heavy duty | | |
| Washer extractors | | |
| 135 lb | | |
| Standard | EA. | 6.667 |
| Pass through | " | 6.667 |
| 200 lb | | |

# 11 EQUIPMENT

| ARCHITECTURAL EQUIPMENT | UNIT | MAN/ HOURS |
|---|---|---|
| **11110.10 LAUNDRY EQUIPMENT** | | |
| Standard | EA. | 6.667 |
| Pass through | " | 6.667 |
| 110 lb dryer | " | 6.667 |
| Hand operated presser | " | 8.889 |
| Mushroom press | " | 8.889 |
| Spreader feeders | | |
| 2 station | EA. | 8.889 |
| 4 station | " | 16.000 |
| Delivery carts | | |
| 12 bushel | EA. | 0.100 |
| 16 bushel | " | 0.107 |
| 18 bushel | " | 0.114 |
| 30 bushel | " | 0.133 |
| 40 bushel | " | 0.160 |
| Low capacity | | |
| Pressers | | |
| Air operated | EA. | 3.200 |
| Hand operated | " | 3.200 |
| Extractor, low capacity | " | 3.200 |
| Ironer, 48" | " | 1.600 |
| Coin washers | | |
| 10 lb capacity | EA. | 1.600 |
| 20 lb capacity | " | 1.600 |
| Coin dryer | " | 1.000 |
| Coin dry cleaner, 20 lb | " | 3.200 |
| **11161.10 LOADING DOCK EQUIPMENT** | | |
| Dock leveler, 10 ton capacity | | |
| 6' x 8' | EA. | 8.000 |
| 7' x 8' | " | 8.000 |
| Bumpers, laminated rubber | | |
| 4-1/2" thick | | |
| 6" x 14" | EA. | 0.160 |
| 6" x 36" | " | 0.178 |
| 10" x 14" | " | 0.200 |
| 10" x 24" | " | 0.229 |
| 10" x 36" | " | 0.267 |
| 12" x 14" | " | 0.211 |
| 12" x 24" | " | 0.250 |
| 12" x 36" | " | 0.296 |
| 6" thick | | |
| 10" x 14" | EA. | 0.229 |
| 10" x 24" | " | 0.276 |
| 10" x 36" | " | 0.400 |
| Extruded rubber bumpers | | |
| T-section, 22" x 22" x 3" | EA. | 0.160 |
| Molded rubber bumpers | | |
| 24" x 12" x 3" thick | EA. | 0.400 |
| Door seal, 12" x 12", vinyl covered | L.F. | 0.200 |
| Dock boards, heavy duty, 5' x 5' | | |
| 5000 lb | | |
| Minimum | EA. | 6.667 |
| Maximum | " | 6.667 |

| ARCHITECTURAL EQUIPMENT | UNIT | MAN/ HOURS |
|---|---|---|
| **11161.10 LOADING DOCK EQUIPMENT** | | |
| 9000 lb | | |
| Minimum | EA. | 6.667 |
| Maximum | " | 7.273 |
| 15,000 lb | " | 7.273 |
| Truck shelters | | |
| Minimum | EA. | 6.154 |
| Maximum | " | 11.429 |
| **11170.10 WASTE HANDLING** | | |
| Incinerator, electric | | |
| 100 lb/hr | | |
| Minimum | EA. | 8.000 |
| Maximum | " | 8.000 |
| 400 lb/hr | | |
| Minimum | EA. | 16.000 |
| Maximum | " | 16.000 |
| 1000 lb/hr | | |
| Minimum | EA. | 24.242 |
| Maximum | " | 24.242 |
| Incinerator, medical-waste | | |
| 25 lb/hr, 2-7 x 4-0 | EA. | 16.000 |
| 50 lb/hr, 2-11 x 4-11 | " | 16.000 |
| 75 lb/hr, 3-8 x 5-0 | " | 32.000 |
| 100 lb/hr, 3-8 x 6-0 | " | 32.000 |
| Industrial compactor | | |
| 1 cy | EA. | 8.889 |
| 3 cy | " | 11.429 |
| 5 cy | " | 16.000 |
| Trash chutes steel, including sprinklers | | |
| 18" dia. | L.F. | 4.000 |
| 24" dia. | " | 4.211 |
| 30" dia. | " | 4.444 |
| 36" dia. | " | 4.706 |
| Refuse bottom hopper | EA. | 4.444 |
| **11400.10 FOOD SERVICE EQUIPMENT** | | |
| Unit kitchens | | |
| 30" compact kitchen | | |
| Refrigerator, with range, sink | EA. | 4.000 |
| Sink only | " | 2.667 |
| Range only | " | 2.000 |
| Cabinet for upper wall section | " | 1.143 |
| Stainless shield, for rear wall | " | 0.320 |
| Side wall | " | 0.320 |
| 42" compact kitchen | | |
| Refrigerator with range, sink | EA. | 4.444 |
| Sink only | " | 4.000 |
| Cabinet for upper wall section | " | 1.333 |
| Stainless shield, for rear wall | " | 0.333 |
| Side wall | " | 0.333 |
| 54" compact kitchen | | |
| Refrigerator, oven, range, sink | EA. | 5.714 |

| ARCHITECTURAL EQUIPMENT | UNIT | MAN/HOURS |
|---|---|---|
| **11400.10 FOOD SERVICE EQUIPMENT** | | |
| Cabinet for upper wall section | EA. | 1.600 |
| Stainless shield, for | | |
| Rear wall | EA. | 0.364 |
| Side wall | " | 0.364 |
| 60" compact kitchen | | |
| Refrigerator, oven, range, sink | EA. | 5.714 |
| Cabinet for upper wall section | " | 1.600 |
| Stainless shield, for | | |
| Rear wall | EA. | 0.364 |
| Side wall | " | 0.364 |
| 72" compact kitchen | | |
| Refrigerator, oven, range, sink | EA. | 6.667 |
| Cabinet for upper wall section | " | 1.600 |
| Stainless shield for | | |
| Rear wall | EA. | 0.400 |
| Side wall | " | 0.400 |
| Bake oven | | |
| Single deck | | |
| Minimum | EA. | 1.000 |
| Maximum | " | 2.000 |
| Double deck | | |
| Minimum | EA. | 1.333 |
| Maximum | " | 2.000 |
| Triple deck | | |
| Minimum | EA. | 1.333 |
| Maximum | " | 2.667 |
| Convection type oven, electric, 40" x 45" x 57" | | |
| Minimum | EA. | 1.000 |
| Maximum | " | 2.000 |
| Broiler, without oven, 60" x 26" x 30" | | |
| Minimum | EA. | 1.000 |
| Maximum | " | 1.333 |
| Coffee urns, 10 gallons | | |
| Minimum | EA. | 2.667 |
| Maximum | " | 4.000 |
| Fryer, with submerger | | |
| Single | | |
| Minimum | EA. | 1.600 |
| Maximum | " | 2.667 |
| Double | | |
| Minimum | EA. | 2.000 |
| Maximum | " | 2.667 |
| Griddle, counter | | |
| 3' long | | |
| Minimum | EA. | 1.333 |
| Maximum | " | 1.600 |
| 5' long | | |
| Minimum | EA. | 2.000 |
| Maximum | " | 2.667 |
| Kettles, steam, jacketed | | |
| 20 gallons | | |
| Minimum | EA. | 2.000 |
| Maximum | " | 4.000 |
| 40 gallons | | |

| ARCHITECTURAL EQUIPMENT | UNIT | MAN/HOURS |
|---|---|---|
| **11400.10 FOOD SERVICE EQUIPMENT** | | |
| Minimum | EA. | 2.000 |
| Maximum | " | 4.000 |
| 60 gallons | | |
| Minimum | EA. | 2.000 |
| Maximum | " | 4.000 |
| Range | | |
| Heavy duty, single oven, open top | | |
| Minimum | EA. | 1.000 |
| Maximum | " | 2.667 |
| Fry top | | |
| Minimum | EA. | 1.000 |
| Maximum | " | 2.667 |
| Steamers, electric | | |
| 27 kw | | |
| Minimum | EA. | 2.000 |
| Maximum | " | 2.667 |
| 18 kw | | |
| Minimum | EA. | 2.000 |
| Maximum | " | 2.667 |
| Dishwasher, rack type | | |
| Single tank, 190 racks/hr | EA. | 4.000 |
| Double tank | | |
| 234 racks/hr | EA. | 4.444 |
| 265 racks/hr | " | 5.333 |
| Dishwasher, automatic 100 meals/hr | " | 2.667 |
| Disposals | | |
| 100 gal/hr | EA. | 2.667 |
| 120 gal/hr | " | 2.759 |
| 250 gal/hr | " | 2.857 |
| Exhaust hood for dishwasher, gutter 4 sides, s-steel | | |
| 4'x4'x2' | EA. | 2.963 |
| 4'x7'x2' | " | 3.200 |
| Food preparation machines | | |
| Vertical cutter mixers | | |
| 25 quart | EA. | 2.667 |
| 40 quart | " | 2.667 |
| 80 quart | " | 4.000 |
| 130 quart | " | 6.667 |
| Choppers | | |
| 5 lb | EA. | 2.000 |
| 16 lb | " | 2.667 |
| 40 lb | " | 4.000 |
| Mixers, floor models | | |
| 20 quart | EA. | 1.000 |
| 60 quart | " | 1.000 |
| 80 quart | " | 1.143 |
| 140 quart | " | 1.600 |
| Ice cube maker | | |
| 50 lb per day | | |
| Minimum | EA. | 8.000 |
| Maximum | " | 8.000 |
| 500 lb per day | | |
| Minimum | EA. | 13.333 |
| Maximum | " | 13.333 |

# 11 EQUIPMENT

| ARCHITECTURAL EQUIPMENT | UNIT | MAN/ HOURS |
|---|---|---|
| **11400.10 FOOD SERVICE EQUIPMENT** | | |
| Ice flakers | | |
| 300 lb per day | EA. | 8.000 |
| 600 lb per day | " | 13.333 |
| 1000 lb per day | " | 17.778 |
| 2000 lb per day | " | 20.000 |
| Refrigerated cases | | |
| Dairy products | | |
| Multi deck type | L.F. | 0.533 |
| Delicatessen case, service deli | | |
| Single deck | L.F. | 4.000 |
| Multi deck | " | 5.000 |
| Meat case | | |
| Single deck | L.F. | 4.706 |
| Multi deck | " | 5.000 |
| Produce case | | |
| Single deck | L.F. | 4.706 |
| Multi deck | " | 5.000 |
| Bottle coolers | | |
| 6' long | | |
| Minimum | EA. | 16.000 |
| Maximum | " | 16.000 |
| 10' long | | |
| Minimum | EA. | 26.667 |
| Maximum | " | 26.667 |
| Frozen food cases | | |
| Chest type | L.F. | 4.706 |
| Reach-in, glass door | " | 5.000 |
| Island case, single | " | 4.706 |
| Multi deck | " | 5.000 |
| Ice storage bins | | |
| 500 lb capacity | EA. | 11.429 |
| 1000 lb capacity | " | 22.857 |
| **11450.10 RESIDENTIAL EQUIPMENT** | | |
| Compactor, 4 to 1 compaction | EA. | 2.000 |
| Dishwasher, built-in | | |
| 2 cycles | EA. | 4.000 |
| 4 or more cycles | " | 4.000 |
| Disposal | | |
| Garbage disposer | EA. | 2.667 |
| Heaters, electric, built-in | | |
| Ceiling type | EA. | 2.667 |
| Wall type | | |
| Minimum | EA. | 2.000 |
| Maximum | " | 2.667 |
| Hood for range, 2-speed, vented | | |
| 30" wide | EA. | 2.667 |
| 42" wide | " | 2.667 |
| Ice maker, automatic | | |
| 30 lb per day | EA. | 1.143 |
| 50 lb per day | " | 4.000 |
| Folding access stairs, disappearing metal stair | | |
| 8' long | EA. | 1.143 |
| 11' long | " | 1.143 |

| ARCHITECTURAL EQUIPMENT | UNIT | MAN/ HOURS |
|---|---|---|
| **11450.10 RESIDENTIAL EQUIPMENT** | | |
| 12' long | EA. | 1.143 |
| Wood frame, wood stair | | |
| 22" x 54" x 8'9" long | EA. | 0.800 |
| 25" x 54" x 10' long | " | 0.800 |
| Ranges electric | | |
| Built-in, 30", 1 oven | EA. | 2.667 |
| 2 oven | " | 2.667 |
| Counter top, 4 burner, standard | " | 2.000 |
| With grill | " | 2.000 |
| Free standing, 21", 1 oven | " | 2.667 |
| 30", 1 oven | " | 1.600 |
| 2 oven | " | 1.600 |
| Water softener | | |
| 30 grains per gallon | EA. | 2.667 |
| 70 grains per gallon | " | 4.000 |
| **11470.10 DARKROOM EQUIPMENT** | | |
| Dryers | | |
| 36" x 25" x 68" | EA. | 4.000 |
| 48" x 25" x 68" | " | 4.000 |
| Processors, film | | |
| Black and white | EA. | 4.000 |
| Color negatives | " | 4.000 |
| Prints | " | 4.000 |
| Transparencies | " | 4.000 |
| Sinks with cabinet and/or stand | | |
| 5" sink with stand | | |
| 24" x 48" | EA. | 2.000 |
| 32" x 64" | " | 2.667 |
| 38" x 52" | " | 2.667 |
| 42" x 132" | " | 4.000 |
| 48" x 52" | " | 4.000 |
| 5" sink with cabinet | | |
| 24" x 48" | EA. | 2.000 |
| 32" x 64" | " | 2.667 |
| 38" x 52" | " | 2.667 |
| 42" x 132" | " | 4.000 |
| 48" x 52" | " | 4.000 |
| 10" sink with stand | | |
| 24" x 48" | EA. | 2.000 |
| 32" x 64" | " | 2.667 |
| 38" x 52" | " | 2.667 |
| 10" sink with cabinet | | |
| 24" x 48" | EA. | 2.000 |
| 38" x 52" | " | 2.667 |
| **11480.10 ATHLETIC EQUIPMENT** | | |
| Basketball backboard | | |
| Fixed | EA. | 10.000 |
| Swing-up | " | 16.000 |
| Portable, hydraulic | " | 4.000 |
| Suspended type, standard | " | 16.000 |
| Bleacher, telescoping, manual | | |

# 11 EQUIPMENT

| ARCHITECTURAL EQUIPMENT | UNIT | MAN/ HOURS |
|---|---|---|
| **11480.10 ATHLETIC EQUIPMENT** | | |
| 15 tier, minimum | SEAT | 0.160 |
| Maximum | " | 0.160 |
| 20 tier, minimum | " | 0.178 |
| Maximum | " | 0.178 |
| 30 tier, minimum | " | 0.267 |
| Maximum | " | 0.267 |
| Boxing ring elevated, complete, 22' x 22' | EA. | 80.000 |
| Gym divider curtain | | |
| Minimum | S.F. | 0.011 |
| Maximum | " | 0.011 |
| Scoreboards, single face | | |
| Minimum | EA. | 8.000 |
| Maximum | " | 40.000 |
| Parallel bars | | |
| Minimum | EA. | 8.000 |
| Maximum | " | 13.333 |
| **11500.10 INDUSTRIAL EQUIPMENT** | | |
| Vehicular paint spray booth, solid back, 14'4" x 9'6" | | |
| 24' deep | EA. | 8.000 |
| 26'6" deep | " | 8.000 |
| 28'6" deep | " | 8.000 |
| Drive through, 14'9" x 9'6" | | |
| 24' deep | EA. | 8.000 |
| 26'6" deep | " | 8.000 |
| 28'6" deep | " | 8.000 |
| Water wash, paint spray booth | | |
| 5' x 11'2" x 10'8" | EA. | 8.000 |
| 6' x 11'2" x 10'8" | " | 8.000 |
| 8' x 11'2" x 10'8" | " | 8.000 |
| 10' x 11'2" x 11'2" | " | 8.000 |
| 12' x 12'2" x 11'2" | " | 8.000 |
| 14' x 12'2" x 11'2" | " | 8.000 |
| 16' x 12'2" x 11'2" | " | 8.000 |
| 20' x 12'2" x 11'2" | " | 8.000 |
| Dry type spray booth, with paint arrestors | | |
| 5'4" x 7'2" x 6'8" | EA. | 8.000 |
| 6'4" x 7'2" x 6'8" | " | 8.000 |
| 8'4" x 7'2" x 9'2" | " | 8.000 |
| 10'4" x 7'2" x 9'2" | " | 8.000 |
| 12'4" x 7'6" x 9'2" | " | 8.000 |
| 14'4" x 7'6" x 9'8" | " | 8.000 |
| 16'4" x 7'7" x 9'8" | " | 8.000 |
| 20'4" x 7'7" x 10'8" | " | 8.000 |
| Air compressor, electric | | |
| 1 hp | | |
| 115 volt | EA. | 5.333 |
| 5 hp | | |
| 115 volt | EA. | 8.000 |
| 230 volt | " | 8.000 |
| Hydraulic lifts | | |
| 8,000 lb capacity | EA. | 20.000 |
| 11,000 lb capacity | " | 32.000 |
| 24,000 lb capacity | " | 53.333 |

| ARCHITECTURAL EQUIPMENT | UNIT | MAN/ HOURS |
|---|---|---|
| **11500.10 INDUSTRIAL EQUIPMENT** | | |
| Power tools | | |
| Band saws | | |
| 10" | EA. | 0.667 |
| 14" | " | 0.800 |
| Motorized shaper | " | 0.615 |
| Motorized lathe | " | 0.667 |
| Bench saws | | |
| 9" saw | EA. | 0.533 |
| 10" saw | " | 0.571 |
| 12" saw | " | 0.667 |
| Electric grinders | | |
| 1/3 hp | EA. | 0.320 |
| 1/2 hp | " | 0.348 |
| 3/4 hp | " | 0.348 |
| **11600.10 LABORATORY EQUIPMENT** | | |
| Cabinets, base | | |
| Minimum | L.F. | 0.667 |
| Maximum | " | 0.667 |
| Full storage, 7' high | | |
| Minimum | L.F. | 0.667 |
| Maximum | " | 0.667 |
| Wall | | |
| Minimum | L.F. | 0.800 |
| Maximum | " | 0.800 |
| Counter tops | | |
| Minimum | S.F. | 0.100 |
| Average | " | 0.114 |
| Maximum | " | 0.133 |
| Tables | | |
| Open underneath | S.F. | 0.400 |
| Doors underneath | " | 0.500 |
| Medical laboratory equipment | | |
| Analyzer | | |
| Chloride | EA. | 0.400 |
| Blood | " | 0.667 |
| Bath, water, utility, countertop unit | " | 0.800 |
| Hot plate, lab, countertop | " | 0.727 |
| Stirrer | " | 0.727 |
| Incubator, anaerobic, 23x23x36" | " | 4.000 |
| Dry heat bath | " | 1.333 |
| Incinerator, for sterilizing | " | 0.080 |
| Meter, serum protein | " | 0.100 |
| Ph analog, general purpose | " | 0.114 |
| Refrigerator, blood bank, undercounter type 153 litres | " | 1.333 |
| 5.4 cf, undercounter type | " | 1.333 |
| Refrigerator/freezer, 4.4 cf, undercounter type | " | 1.333 |
| Sealer, impulse, free standing, 20x12x4" | " | 0.267 |
| Timer, electric, 1-60 minutes, bench or wall mounted | " | 0.444 |
| Glassware washer - dryer, undercounter | " | 10.000 |
| Balance, torsion suspension, tabletop, 4.5 lb capacity | " | 0.444 |
| Binocular microscope, with in base illuminator | " | 0.308 |
| Centrifuge, table model, 19x16x13" | " | 0.320 |
| Clinical model, with four place head | " | 0.178 |

# 11 EQUIPMENT

## ARCHITECTURAL EQUIPMENT

### 11700.10 MEDICAL EQUIPMENT

| ARCHITECTURAL EQUIPMENT | UNIT | MAN/ HOURS |
|---|---|---|
| Hospital equipment, lights | | |
| Examination, portable | EA. | 0.667 |
| Meters | | |
| Air flow meter | EA. | 0.444 |
| Oxygen flow meters | " | 0.333 |
| Racks | | |
| 40 chart, revolving open frame; mobile caddy | EA. | 0.667 |
| Scales. | | |
| Clinical, metric with measure rod, 350 lb | EA. | 0.727 |
| Physical therapy | | |
| Chair, hydrotherapy | EA. | 0.133 |
| Diathermy, shortwave, portable, on casters | " | 0.320 |
| Exercise bicycle, floor standing, 35" x 15" | " | 0.267 |
| Hydrocollator, 4 pack, portable, 129 x 90 x 160" | " | 0.114 |
| Lamp, infrared, mobile with variable heat control | " | 0.615 |
| Ultra violet, base mounted | " | 0.615 |
| Mirror, posture training, 27" wide and 72" high | " | 0.200 |
| Parallel bars, adjustable | " | 1.000 |
| Platform mat 10'x6', 1" thick | " | 0.200 |
| Pulley, duplex, wall mounted | " | 2.667 |
| Rack, crutch, wall mounted, 66 x 16 x 13" | " | 0.800 |
| Stimulator, galvanic-faradic, hand held | " | 0.053 |
| Ultrasound, muscle stimulator, portable, 13x13x8" | " | 0.067 |
| Sandbag set, velcro straps, saddle bag type | " | 0.114 |
| Whirlpool, 85 gallon | " | 4.000 |
| 65 gallon capacity | " | 4.000 |
| Radiology | | |
| Radiographic table, motor driven tilting table | EA. | 80.000 |
| Fluoroscope image/tv system | " | 160 |
| Processor for washing and drying radiographs | | |
| Water filter unit, 30" x 48-1/2" x 37-1/2" | EA. | 13.333 |
| Cassette transfer cabinet | " | 0.667 |
| Base storage cabinets, sectional design | | |
| With back splash, 24" deep and 35" high | L.F. | 0.667 |
| Wall storage cabinets | " | 1.000 |
| Steam sterilizers | | |
| For heat and moisture stable materials | EA. | 0.800 |
| For fast drying after sterilization | " | 1.000 |
| Compact unit | " | 1.000 |
| Semi-automatic | " | 4.000 |
| Floor loading | | |
| Single door | EA. | 6.667 |
| Double door | " | 8.000 |
| Utensil washer, sanitizer | " | 6.154 |
| Automatic washer/sterilizer | " | 16.000 |
| 16 x 16 x 26", including generator & accessories | " | 26.667 |
| Steam generator, elec., 10 kw to 180 kw | " | 16.000 |
| Surgical scrub | | |
| Minimum | EA. | 2.667 |
| Maximum | " | 2.667 |
| Gas sterilizers | | |
| Automatic, free standing, 21x19x29" | EA. | 8.000 |
| Surgical tables | | |
| Minimum | EA. | 11.429 |

## ARCHITECTURAL EQUIPMENT

### 11700.10 MEDICAL EQUIPMENT

| ARCHITECTURAL EQUIPMENT | UNIT | MAN/ HOURS |
|---|---|---|
| Maximum | EA. | 16.000 |
| Surgical lights, ceiling mounted | | |
| Minimum | EA. | 13.333 |
| Maximum | " | 16.000 |
| Water stills | | |
| 4 liters/hr | EA. | 2.667 |
| 8 liters/hr | " | 2.667 |
| 19 liters/hr | " | 6.667 |
| X-ray equipment | | |
| Mobile unit | | |
| Minimum | EA. | 4.000 |
| Maximum | " | 8.000 |
| Film viewers | | |
| Minimum | EA. | 1.333 |
| Maximum | " | 2.667 |
| Autopsy table | | |
| Minimum | EA. | 8.000 |
| Maximum | " | 8.000 |
| Incubators | | |
| 15 cf | EA. | 4.000 |
| 29 cf | " | 6.667 |
| Infant transport, portable | " | 4.211 |
| Beds | | |
| Stretcher, with pad, 30" x 78" | EA. | 2.000 |
| Transfer, for patient transport | " | 2.000 |
| Headwall | | |
| Aluminum, with back frame and console | EA. | 4.000 |
| Hospital ground detection system | | |
| Power ground module | EA. | 2.286 |
| Ground slave module | " | 1.739 |
| Master ground module | " | 1.509 |
| Remote indicator | " | 1.600 |
| X-ray indicator | " | 1.739 |
| Micro ammeter | " | 2.000 |
| Supervisory module | " | 1.739 |
| Ground cords | " | 0.296 |
| Hospital isolation monitors, 5 ma | | |
| 120v | EA. | 3.478 |
| 208v | " | 3.478 |
| 240v | " | 3.478 |
| Digital clock-timers separate display | " | 1.600 |
| One display | " | 1.600 |
| Remote control | " | 1.250 |
| Battery pack | " | 1.250 |
| Surgical chronometer clock and 3 timers | " | 2.500 |
| Auxilary control | " | 1.159 |

### 11700.20 DENTAL EQUIPMENT

| ARCHITECTURAL EQUIPMENT | UNIT | MAN/ HOURS |
|---|---|---|
| Dental care equipment | | |
| Drill console with accessories | EA. | 13.333 |
| Amalgamator | " | 0.400 |
| Lathe | " | 0.267 |
| Finish polisher | " | 0.533 |
| Model trimmer | " | 0.364 |

| ARCHITECTURAL EQUIPMENT | UNIT | MAN/ HOURS |
|---|---|---|
| **11700.20 DENTAL EQUIPMENT** | | |
| Motor, wall mounted | EA. | 0.364 |
| Cleaner, ultrasonic | " | 0.800 |
| Curing unit, bench mounted | " | 1.333 |
| Oral evacuation system, dual pump | " | 1.000 |
| Sterilizer, table top, self contained | " | 0.444 |
| Dental lights | | |
| Light, floor or ceiling mounted | EA. | 4.000 |
| X-ray unit | | |
| Portable | EA. | 2.000 |
| Wall mounted with remote control | " | 6.667 |
| Illuminator, single panel | " | 11.429 |
| X-ray film processor | " | 6.667 |
| Shield, portable x-ray, lead lined | " | 0.533 |

| INTERIOR | UNIT | MAN/HOURS |
|---|---|---|
| **12302.10 CASEWORK** | | |
| Kitchen base cabinet, prefinished, 24" deep, 35" high | | |
| 12"wide | EA. | 0.800 |
| 18" wide | " | 0.800 |
| 24" wide | " | 0.889 |
| 27" wide | " | 0.889 |
| 36" wide | " | 1.000 |
| 48" wide | " | 1.000 |
| Corner cabinet, 36" wide | " | 1.000 |
| Wall cabinet, 12" deep, 12" high | | |
| 30" wide | EA. | 0.800 |
| 36" wide | " | 0.800 |
| 15" high | | |
| 30" wide | EA. | 0.889 |
| 36" wide | " | 0.889 |
| 24" high | | |
| 30" wide | EA. | 0.889 |
| 36" wide | " | 0.889 |
| 30" high | | |
| 12" wide | EA. | 1.000 |
| 18" wide | " | 1.000 |
| 24" wide | " | 1.000 |
| 27" wide | " | 1.000 |
| 30" wide | " | 1.143 |
| 36" wide | " | 1.143 |
| Corner cabinet, 30" high | | |
| 24" wide | EA. | 1.333 |
| 30" wide | " | 1.333 |
| 36" wide | " | 1.333 |
| Wardrobe | " | 2.000 |
| Vanity with top, laminated plastic | | |
| 24" wide | EA. | 2.000 |
| 30" wide | " | 2.000 |
| 36" wide | " | 2.667 |
| 48" wide | " | 3.200 |
| **12390.10 COUNTER TOPS** | | |
| Stainless steel, counter top, with backsplash | S.F. | 0.200 |
| Acid-proof, kemrock surface | " | 0.133 |
| **12500.10 WINDOW TREATMENT** | | |
| Drapery tracks, wall or ceiling mounted | | |
| Basic traverse rod | | |
| 50 to 90" | EA. | 0.400 |
| 84 to 156" | " | 0.444 |
| 136 to 250" | " | 0.444 |
| 165 to 312" | " | 0.500 |
| Traverse rod with stationary curtain rod | | |
| 30 to 50" | EA. | 0.400 |
| 50 to 90" | " | 0.400 |
| 84 to 156" | " | 0.444 |
| 136 to 250" | " | 0.500 |
| Double traverse rod | | |
| 30 to 50" | EA. | 0.400 |
| 50 to 84" | " | 0.400 |

| INTERIOR | UNIT | MAN/HOURS |
|---|---|---|
| **12500.10 WINDOW TREATMENT** | | |
| 84 to 156" | EA. | 0.444 |
| 136 to 250" | " | 0.500 |
| **12510.10 BLINDS** | | |
| Venetian blinds | | |
| 2" slats | S.F. | 0.020 |
| 1" slats | " | 0.020 |
| **12690.40 FLOOR MATS** | | |
| Recessed entrance mat, 3/8" thick, aluminum link | S.F. | 0.400 |
| Steel, flexible | " | 0.400 |

| CONSTRUCTION | UNIT | MAN/ HOURS |
|---|---|---|
| **13056.10 VAULTS** | | |
| Floor safes | | |
| Class C | | |
| 1.0 cf | EA. | 0.667 |
| 1.3 cf | " | 1.000 |
| 1.9 cf | " | 1.333 |
| 5.2 cf | " | 1.333 |
| **13121.10 PRE-ENGINEERED BUILDINGS** | | |
| Pre-engineered metal building, 40'x100' | | |
| 14' eave height | S.F. | 0.032 |
| 16' eave height | " | 0.037 |
| 20' eave height | " | 0.048 |
| 60'x100' | | |
| 14' eave height | S.F. | 0.032 |
| 16' eave height | " | 0.037 |
| 20' eave height | " | 0.048 |
| 80'x100' | | |
| 14' eave height | S.F. | 0.032 |
| 16' eave height | " | 0.037 |
| 20' eave height | " | 0.048 |
| 100'x100' | | |
| 14' eave height | S.F. | 0.032 |
| 16' eave height | " | 0.037 |
| 20' eave height | " | 0.048 |
| 100'x150' | | |
| 14' eave height | S.F. | 0.032 |
| 16' eave height | " | 0.037 |
| 20' eave height | " | 0.048 |
| 120'x150' | | |
| 14' eave height | S.F. | 0.032 |
| 16' eave height | " | 0.037 |
| 20' eave height | " | 0.048 |
| 140'x150' | | |
| 14' eave height | S.F. | 0.032 |
| 16' eave height | " | 0.037 |
| 20' eave height | " | 0.048 |
| 160'x200' | | |
| 14' eave height | S.F. | 0.032 |
| 16' eave height | " | 0.037 |
| 20' eave height | " | 0.048 |
| 200'x200' | | |
| 14' eave height | S.F. | 0.032 |
| 16' eave height | " | 0.037 |
| 20' eave height | " | 0.048 |
| Liner panel, 26 ga, painted steel | " | 0.020 |
| Wall panel insulated, 26 ga. steel, foam core | " | 0.020 |
| Roof panel, 26 ga. painted steel | " | 0.011 |
| Plastic (sky light) | " | 0.011 |
| Insulation, 3-1/2" thick blanket, R11 | " | 0.005 |
| **13152.10 SWIMMING POOL EQUIPMENT** | | |
| Diving boards | | |
| 14' long | | |

| CONSTRUCTION | UNIT | MAN/ HOURS |
|---|---|---|
| **13152.10 SWIMMING POOL EQUIPMENT** | | |
| Aluminum | EA. | 4.444 |
| Fiberglass | " | 4.444 |
| Ladders, heavy duty | | |
| 2 steps | | |
| Minimum | EA. | 1.600 |
| Maximum | " | 1.600 |
| 4 steps | | |
| Minimum | EA. | 2.000 |
| Maximum | " | 2.000 |
| Lifeguard chair | | |
| Minimum | EA. | 8.000 |
| Maximum | " | 8.000 |
| Lights, underwater | | |
| 12 volt, with transformer | EA. | 2.000 |
| 110 volt | | |
| Minimum | EA. | 2.000 |
| Maximum | " | 2.000 |
| Ground fault interrupter for 110 volt, each light | " | 0.667 |
| Pool covers | | |
| Reinforced polyethylene | S.F. | 0.062 |
| Vinyl water tube | | |
| Minimum | S.F. | 0.062 |
| Maximum | " | 0.062 |
| Slides with water tube | | |
| Minimum | EA. | 6.667 |
| Maximum | " | 6.667 |

# 14 CONVEYING

| ELEVATORS | UNIT | MAN/ HOURS |
|---|---|---|
| **14210.10 ELEVATORS** | | |
| Passenger elevators, electric, geared | | |
| Based on a shaft of 6 stops and 6 openings | | |
| 50 fpm, 2000 lb | EA. | 24.000 |
| 100 fpm, 2000 lb | " | 26.667 |
| 150 fpm | | |
| 2000 lb | EA. | 30.000 |
| 3000 lb | " | 34.286 |
| 4000 lb | " | 40.000 |
| 200 fpm | | |
| 2500 lb | EA. | 34.286 |
| 3000 lb | " | 36.923 |
| 4000 lb | " | 40.000 |
| 250 fpm | | |
| 2500 lb | EA. | 34.286 |
| 3000 lb | " | 36.923 |
| 4000 lb | " | 40.000 |
| 300 fpm | | |
| 2500 lb | EA. | 34.286 |
| 3000 lb | " | 36.923 |
| 4000 lb | " | 24.000 |
| Based on a shaft of 8 stops and 8 openings | | |
| 300 fpm | | |
| 3000 lb | EA. | 48.000 |
| 3500 lb | " | 48.000 |
| 4000 lb | " | 53.333 |
| 5000 lb | " | 57.143 |
| 400 fpm | | |
| 3000 lb | EA. | 48.000 |
| 3500 lb | " | 48.000 |
| 4000 lb | " | 53.333 |
| 5000 lb | " | 57.143 |
| 600 fpm | | |
| 3000 lb | EA. | 53.333 |
| 3500 lb | " | 57.143 |
| 4000 lb | " | 58.537 |
| 5000 lb | " | 60.000 |
| 800 fpm | | |
| 3000 lb | EA. | 53.333 |
| 3500 lb | " | 57.143 |
| 4000 lb | " | 58.537 |
| 5000 lb | " | 60.000 |
| Hydraulic, based on a shaft of 3 stops, 3 openings | | |
| 50 fpm | | |
| 2000 lb | EA. | 20.000 |
| 2500 lb | " | 20.000 |
| 3000 lb | " | 20.870 |
| 100 fpm | | |
| 2000 lb | EA. | 20.000 |
| 2500 lb | " | 20.870 |
| 3000 lb | " | 21.818 |
| 150 fpm | | |
| 2000 lb | EA. | 20.000 |
| 2500 lb | " | 20.870 |
| 3000 lb | " | 22.857 |

| ELEVATORS | UNIT | MAN/ HOURS |
|---|---|---|
| **14210.10 ELEVATORS** | | |
| Small elevators, 4 to 6 passenger capacity | | |
| Electric, push | | |
| 2 stops | EA. | 20.000 |
| 3 stops | " | 21.818 |
| 4 stops | " | 24.000 |
| Freight elevators, electric | | |
| Based on a shaft of 6 stops and 6 openings | | |
| 50 fpm | | |
| 3500 lb | EA. | 26.667 |
| 4000 lb | " | 26.667 |
| 5000 lb | " | 30.000 |
| 100 fpm | | |
| 3500 lb | EA. | 30.000 |
| 4000 lb | " | 30.000 |
| 5000 lb | " | 34.286 |
| 200 fpm | | |
| 3500 lb | EA. | 34.286 |
| 4000 lb | " | 34.286 |
| 5000 lb | " | 40.000 |
| Based on shaft of 8 stops and 8 openings | | |
| 100 fpm | | |
| 4000 lb | EA. | 30.000 |
| 6000 lb | " | 30.769 |
| 8000 lb | " | 32.432 |
| 150 fpm | | |
| 4000 lb | EA. | 34.286 |
| 6000 lb | " | 35.294 |
| 8000 lb | " | 37.500 |
| 200 fpm | | |
| 4000 lb | EA. | 40.000 |
| 6000 lb | " | 41.379 |
| 8000 lb | " | 43.636 |
| Hydraulic, based on 3 stops and 3 openings | | |
| 50 fpm | | |
| 3000 lb | EA. | 17.143 |
| 4000 lb | " | 17.778 |
| 6000 lb | " | 18.462 |
| 100 fpm | | |
| 3000 lb | EA. | 17.143 |
| 4000 lb | " | 17.778 |
| 6000 lb | " | 18.462 |
| 150 fpm | | |
| 3000 lb | EA. | 17.143 |
| 4000 lb | " | 17.778 |
| 6000 lb | " | 18.462 |
| **14300.10 ESCALATORS** | | |
| Escalators | | |
| 32" wide, floor to floor | | |
| 12' high | EA. | 40.000 |
| 15' high | " | 48.000 |
| 18' high | " | 60.000 |
| 22' high | " | 80.000 |

# 14 CONVEYING

## ELEVATORS

| | UNIT | MAN/ HOURS |
|---|---|---|
| **14300.10 ESCALATORS** | | |
| 25' high | EA. | 96.000 |
| 48" wide | | |
| 12' high | EA. | 41.379 |
| 15' high | " | 50.000 |
| 18' high | " | 63.158 |
| 22' high | " | 85.714 |
| 25' high | " | 96.000 |

## LIFTS

| | UNIT | MAN/ HOURS |
|---|---|---|
| **14410.10 PERSONNEL LIFTS** | | |
| Residential stair climber, per story | EA. | 6.667 |
| **14450.10 VEHICLE LIFTS** | | |
| Automotive hoist, one post, semi-hydraulic, 8,000 lb | EA. | 24.000 |
| Full hydraulic, 8,000 lb | " | 24.000 |
| 2 post, semi-hydraulic, 10,000 lb | " | 34.286 |
| Full hydraulic | | |
| 10,000 lb | EA. | 34.286 |
| 13,000 lb | " | 60.000 |
| 18,500 lb | " | 60.000 |
| 24,000 lb | " | 60.000 |
| 26,000 lb | " | 60.000 |
| Pneumatic hoist, fully hydraulic | | |
| 11,000 lb | EA. | 80.000 |
| 24,000 lb | " | 80.000 |

## MATERIAL HANDLING

| | UNIT | MAN/ HOURS |
|---|---|---|
| **14560.10 CHUTES** | | |
| Linen chutes, stainless steel, with supports | | |
| 18" dia. | L.F. | 0.057 |
| 24" dia. | " | 0.062 |
| 30" dia. | " | 0.067 |
| Hopper | EA. | 0.533 |
| Skylight | " | 0.800 |

## MATERIAL HANDLING

| | UNIT | MAN/ HOURS |
|---|---|---|
| **14560.10 CHUTES** | | |
| Sprinkler unit at top | EA. | 0.889 |
| **14580.10 PNEUMATIC SYSTEMS** | | |
| Pneumatic message tube system | | |
| Average, 20 station job | | |
| 3" round system | EA. | 72.727 |
| 4" round system | " | 80.000 |
| 6" round system | " | 88.889 |
| 4" x 7" oval system | " | 160 |
| Trash and linen tube system | | |
| 10 stations | EA. | 120 |
| 15 stations | " | 160 |
| 20 stations | " | 185 |
| 30 stations | " | 218 |

## HOISTS AND CRANES

| | UNIT | MAN/ HOURS |
|---|---|---|
| **14600.10 INDUSTRIAL HOISTS** | | |
| Industrial hoists, electric, light to medium duty | | |
| 500 lb | EA. | 4.000 |
| 1000 lb | " | 4.211 |
| 2000 lb | " | 4.444 |
| 3000 lb | " | 4.706 |
| 4000 lb | " | 5.000 |
| 5000 lb | " | 5.333 |
| 6000 lb | " | 5.517 |
| 7500 lb | " | 5.714 |
| 10,000 lb | " | 5.926 |
| 15,000 lb | " | 6.154 |
| 20,000 lb | " | 6.667 |
| 25,000 lb | " | 7.273 |
| 30,000 lb | " | 8.000 |
| Heavy duty | | |
| 500 lb | EA. | 4.000 |
| 1000 lb | " | 4.211 |
| 2000 lb | " | 4.444 |
| 3000 lb | " | 4.706 |
| 4000 lb | " | 5.000 |
| 5000 lb | " | 5.333 |
| 6000 lb | " | 5.517 |
| 7500 lb | " | 5.714 |
| 10,000 lb | " | 5.926 |
| 15,000 lb | " | 6.154 |
| 20,000 lb | " | 6.667 |

# 14 CONVEYING

## HOISTS AND CRANES

| HOISTS AND CRANES | UNIT | MAN/ HOURS |
|---|---|---|
| **14600.10 INDUSTRIAL HOISTS** | | |
| 25,000 lb | EA. | 7.273 |
| 30,000 lb | " | 8.000 |
| Air powered hoists | | |
| 500 lb | EA. | 4.000 |
| 1000 lb | " | 4.000 |
| 2000 lb | " | 4.211 |
| 4000 lb | " | 4.706 |
| 6000 lb | " | 6.154 |
| Overhead traveling bridge crane | | |
| Single girder, 20' span | | |
| 3 ton | EA. | 12.000 |
| 5 ton | " | 12.000 |
| 7.5 ton | " | 12.000 |
| 10 ton | " | 15.000 |
| 15 ton | " | 15.000 |
| 30' span | | |
| 3 ton | EA. | 12.000 |
| 5 ton | " | 12.000 |
| 10 ton | " | 15.000 |
| 15 ton | " | 15.000 |
| Double girder, 40' span | | |
| 3 ton | EA. | 26.667 |
| 5 ton | " | 26.667 |
| 7.5 ton | " | 26.667 |
| 10 ton | " | 34.286 |
| 15 ton | " | 34.286 |
| 25 ton | " | 34.286 |
| 50' span | | |
| 3 ton | EA. | 26.667 |
| 5 ton | " | 26.667 |
| 7.5 ton | " | 26.667 |
| 10 ton | " | 34.286 |
| 15 ton | " | 34.286 |
| 25 ton | " | 34.286 |
| **14650.10 JIB CRANES** | | |
| Self supporting, swinging 8' boom, 200 deg rotation | | |
| 1000 lb | EA. | 6.667 |
| 2000 lb | " | 6.667 |
| 3000 lb | " | 13.333 |
| 4000 lb | " | 13.333 |
| 6000 lb | " | 13.333 |
| 10,000 lb | " | 13.333 |
| Wall mounted, 180 deg rotation | | |
| 2000 lb | EA. | 6.667 |
| 3000 lb | " | 6.667 |
| 4000 lb | " | 13.333 |
| 6000 lb | " | 13.333 |
| 10,000 lb | " | 13.333 |

| BASIC MATERIALS | UNIT | MAN/ HOURS |
|---|---|---|
| **15100.10 SPECIALTIES** | | |
| Wall penetration | | |
| Concrete wall, 6" thick | | |
| 2" dia. | EA. | 0.267 |
| 4" dia. | " | 0.400 |
| 8" dia. | " | 0.571 |
| 12" thick | | |
| 2" dia. | EA. | 0.364 |
| 4" dia. | " | 0.571 |
| 8" dia. | " | 0.889 |
| Non-destructive testing, piping systems | | |
| X-ray of welds | | |
| 3" dia. pipe | EA. | 0.800 |
| 4" dia. pipe | " | 0.800 |
| 6" dia. pipe | " | 0.800 |
| 8" dia. pipe | " | 1.000 |
| 10" dia. pipe | " | 1.000 |
| Liquid penetration of welds | | |
| 2" dia. pipe | EA. | 0.500 |
| 3" dia. pipe | " | 0.500 |
| 4" dia. pipe | " | 0.500 |
| 6" dia. pipe | " | 0.500 |
| 8" dia. pipe | " | 0.500 |
| 10" dia. pipe | " | 0.500 |
| **15120.10 BACKFLOW PREVENTERS** | | |
| Backflow preventer, flanged, cast iron, with valves | | |
| 3" pipe | EA. | 4.000 |
| 4" pipe | " | 4.444 |
| 6" pipe | " | 6.667 |
| 8" pipe | " | 8.000 |
| Threaded | | |
| 3/4" pipe | EA. | 0.500 |
| 2" pipe | " | 0.800 |
| Reduced pressure assembly, bronze, threaded | | |
| 3/4" | EA. | 0.500 |
| 1" | " | 0.571 |
| 1-1/4" | " | 0.667 |
| 1-1/2" | " | 0.800 |
| **15140.10 PIPE HANGERS, HEAVY** | | |
| Hangers | | |
| 1/2" pipe, clevis pipe hanger | | |
| Black steel | EA. | 0.267 |
| Galvanized | " | 0.267 |
| U bolt | " | 0.080 |
| 3/4" pipe, clevis pipe hanger | | |
| Black steel | EA. | 0.267 |
| Galvanized | " | 0.267 |
| U bolt | " | 0.080 |
| 1" pipe, clevis pipe hanger | | |
| Black steel | EA. | 0.267 |
| Galvanized | " | 0.267 |

| BASIC MATERIALS | UNIT | MAN/ HOURS |
|---|---|---|
| **15140.10 PIPE HANGERS, HEAVY** | | |
| U bolt | EA. | 0.080 |
| 1-1/4" pipe, clevis pipe hanger | | |
| Black steel | EA. | 0.267 |
| Galvanized | " | 0.267 |
| U bolt | " | 0.080 |
| 1-1/2" pipe, clevis pipe hanger | | |
| Black steel | EA. | 0.267 |
| Galvanized | " | 0.267 |
| U bolt | " | 0.080 |
| 2" pipe, clevis pipe hanger | | |
| Black steel | EA. | 0.267 |
| Galvanized | " | 0.267 |
| U bolt | " | 0.080 |
| 2-1/2" pipe, clevis pipe hanger | | |
| Black steel | EA. | 0.267 |
| Galvanized | " | 0.267 |
| 3" pipe, clevis pipe hanger | | |
| Black steel | EA. | 0.267 |
| Galvanized | " | 0.267 |
| U bolt | " | 0.080 |
| 3-1/2" pipe, clevis pipe hanger | | |
| Black steel | EA. | 0.267 |
| Galvanized | " | 0.267 |
| U bolt | " | 0.080 |
| 4" pipe, clevis pipe hanger | | |
| Black steel | EA. | 0.267 |
| Galvanized | " | 0.267 |
| U bolt | " | 0.080 |
| 5" pipe, clevis pipe hanger | | |
| Black steel | EA. | 0.320 |
| Galvanized | " | 0.320 |
| U bolt | " | 0.080 |
| 6" pipe, clevis pipe hanger | | |
| Black steel | EA. | 0.320 |
| Galvanized | " | 0.320 |
| U bolt | " | 0.100 |
| 8" pipe, clevis pipe hanger | | |
| Black steel | EA. | 0.320 |
| Galvanized | " | 0.320 |
| U bolt | " | 0.100 |
| 10" clevis pipe hanger | | |
| Black steel | EA. | 0.320 |
| Galvanized | " | 0.320 |
| 12" pipe, clevis pipe hanger | | |
| Black steel | EA. | 0.320 |
| Galvanized | " | 0.320 |
| Threaded rod, galvanized | | |
| C-clamp, steel, with lock nut | | |
| 3/8" | EA. | 0.100 |
| 1/2" | " | 0.100 |
| 5/8" | " | 0.100 |
| 3/4" | " | 0.100 |
| 7/8" | " | 0.100 |
| Angle support, medium, welded steel | | |

| BASIC MATERIALS | UNIT | MAN/ HOURS |
|---|---|---|
| **15140.10 PIPE HANGERS, HEAVY** | | |
| 12"x18" | EA. | 0.800 |
| 18"x24" | " | 0.800 |
| 24"x30" | " | 0.800 |
| Heavy, welded steel | | |
| 12"x18" | EA. | 0.800 |
| 18"x24" | " | 0.800 |
| 24"x30" | " | 0.800 |
| **15140.11 PIPE HANGERS, LIGHT** | | |
| A band, black iron | | |
| 1/2" | EA. | 0.057 |
| 1" | " | 0.059 |
| 1-1/4" | " | 0.062 |
| 1-1/2" | " | 0.067 |
| 2" | " | 0.073 |
| 2-1/2" | " | 0.080 |
| 3" | " | 0.089 |
| 4" | " | 0.100 |
| 5" | " | 0.107 |
| 6" | " | 0.114 |
| 8" | " | 0.133 |
| Copper | | |
| 1/2" | EA. | 0.057 |
| 3/4" | " | 0.059 |
| 1" | " | 0.059 |
| 1-1/4" | " | 0.062 |
| 1-1/2" | " | 0.067 |
| 2" | " | 0.073 |
| 2-1/2" | " | 0.080 |
| 3" | " | 0.089 |
| 4" | " | 0.100 |
| Black riser friction hangers | | |
| 3/4" | EA. | 0.067 |
| 1" | " | 0.070 |
| 1-1/4" | " | 0.073 |
| 1-1/2" | " | 0.076 |
| 2" | " | 0.080 |
| 2-1/2" | " | 0.089 |
| 3" | " | 0.100 |
| 4" | " | 0.114 |
| 5" | " | 0.123 |
| 6" | " | 0.133 |
| 8" | " | 0.145 |
| 10" | " | 0.160 |
| 2 hole clips, galvanized | | |
| 3/4" | EA. | 0.053 |
| 1" | " | 0.055 |
| 1-1/4" | " | 0.057 |
| 1-1/2" | " | 0.059 |
| 2" | " | 0.062 |
| 2-1/2" | " | 0.064 |
| 3" | " | 0.067 |
| 4" | " | 0.073 |

| BASIC MATERIALS | UNIT | MAN/ HOURS |
|---|---|---|
| **15140.11 PIPE HANGERS, LIGHT** | | |
| Perforated strap | | |
| 3/4" | | |
| Galvanized, 20 ga. | L.F. | 0.040 |
| Copper, 22 ga. | " | 0.040 |
| J-Hooks | | |
| 1/2" | EA. | 0.036 |
| 3/4" | " | 0.036 |
| 1" | " | 0.038 |
| 1-1/4" | " | 0.039 |
| 1-1/2" | " | 0.040 |
| 2" | " | 0.040 |
| 3" | " | 0.042 |
| 4" | " | 0.042 |
| PVC coated hangers, galvanized, 28 ga. | | |
| 1-1/2" x 12" | EA. | 0.053 |
| 2" x 12" | " | 0.057 |
| 3" x 12" | " | 0.062 |
| 4" x 12" | " | 0.067 |
| Copper, 30 ga. | | |
| 1-1/2" x 12" | EA. | 0.053 |
| 2" x 12" | " | 0.057 |
| 3" x 12" | " | 0.062 |
| 4" x 12" | " | 0.067 |
| 2" x 24" | " | 0.062 |
| 3" x 24" | " | 0.067 |
| 4" x 24" | " | 0.073 |
| Wire hook hangers | | |
| Black wire, 1/2" x | | |
| 4" | EA. | 0.040 |
| 6" | " | 0.042 |
| 8" | " | 0.044 |
| 10" | " | 0.044 |
| 12" | " | 0.047 |
| 3/4" x | | |
| 4" | EA. | 0.042 |
| 6" | " | 0.044 |
| 8" | " | 0.047 |
| 10" | " | 0.050 |
| 12" | " | 0.053 |
| 1" x | | |
| 4" | EA. | 0.044 |
| 6" | " | 0.047 |
| 8" | " | 0.050 |
| 10" | " | 0.053 |
| 12" | " | 0.057 |
| 1-1/4" x | | |
| 4" | EA. | 0.047 |
| 6" | " | 0.050 |
| 8" | " | 0.053 |
| 10" | " | 0.057 |
| 12" | " | 0.062 |
| 1-1/2" x | | |
| 6" | EA. | 0.053 |
| 8" | " | 0.057 |

# 15 MECHANICAL

## BASIC MATERIALS

### 15140.11 PIPE HANGERS, LIGHT

| BASIC MATERIALS | UNIT | MAN/HOURS |
|---|---|---|
| 10" | EA. | 0.062 |
| 12" | " | 0.067 |
| 2" x | | |
| 6" | EA. | 0.057 |
| 8" | " | 0.062 |
| 10" | " | 0.067 |
| 12" | " | 0.073 |
| Copper wire hooks | | |
| 1/2" x | | |
| 4" | EA. | 0.040 |
| 6" | " | 0.042 |
| 8" | " | 0.044 |
| 10" | " | 0.047 |
| 12" | " | 0.050 |
| 3/4" x | | |
| 4" | EA. | 0.042 |
| 6" | " | 0.044 |
| 8" | " | 0.047 |
| 10" | " | 0.050 |
| 12" | " | 0.053 |
| 1" x | | |
| 4" | EA. | 0.044 |
| 6" | " | 0.047 |
| 8" | " | 0.050 |
| 10" | " | 0.053 |
| 12" | " | 0.057 |
| 1-1/4" x | | |
| 6" | EA. | 0.047 |
| 8" | " | 0.050 |
| 10" | " | 0.053 |
| 12" | " | 0.057 |
| 1-1/2" x | | |
| 6" | EA. | 0.053 |
| 8" | " | 0.057 |
| 10" | " | 0.062 |
| 12" | " | 0.067 |
| 2" x | | |
| 6" | EA. | 0.057 |
| 8" | " | 0.062 |
| 10" | " | 0.067 |
| 12" | " | 0.073 |

### 15175.60 EXPANSION TANKS

| BASIC MATERIALS | UNIT | MAN/HOURS |
|---|---|---|
| Expansion tank, 125 psi, steel | | |
| 20 gallon | EA. | 1.000 |
| 30 gallon | " | 1.333 |
| 65 gallon | " | 2.000 |
| 80 gallon | " | 2.286 |

### 15240.10 VIBRATION CONTROL

| BASIC MATERIALS | UNIT | MAN/HOURS |
|---|---|---|
| Vibration isolator, in-line, stainless connector, screwed | | |
| 1/2" | EA. | 0.444 |
| 3/4" | EA. | 0.471 |
| 1" | " | 0.500 |
| 1-1/4" | " | 0.533 |
| 1-1/2" | " | 0.571 |
| 2" | " | 0.615 |
| 2-1/2" | " | 0.667 |
| 3" | " | 0.727 |
| 4" | " | 0.800 |
| 6" | " | 0.889 |
| Flanged | | |
| 8" | EA. | 1.000 |
| 10" | " | 1.143 |
| 12" | " | 1.333 |

## INSULATION

### 15260.10 FIBERGLASS PIPE INSULATION

| INSULATION | UNIT | MAN/HOURS |
|---|---|---|
| Fiberglass insulation on 1/2" pipe | | |
| 1" thick | L.F. | 0.027 |
| 1-1/2" thick | " | 0.033 |
| 3/4" pipe | | |
| 1" thick | L.F. | 0.027 |
| 1-1/2" thick | " | 0.033 |
| 1" pipe | | |
| 1" thick | L.F. | 0.027 |
| 1-1/2" thick | " | 0.033 |
| 2" pipe | | |
| 1" thick | L.F. | 0.033 |
| 1-1/2" thick | " | 0.036 |
| 3" pipe | | |
| 1" thick | L.F. | 0.038 |
| 1-1/2" thick | " | 0.040 |
| 6" pipe | | |
| 1" thick | L.F. | 0.042 |
| 2" thick | " | 0.044 |
| 10" pipe | | |
| 2" thick | L.F. | 0.042 |
| 3" thick | " | 0.044 |

### 15260.20 CALCIUM SILICATE

| INSULATION | UNIT | MAN/HOURS |
|---|---|---|
| Calcium silicate insulation, 6" pipe | | |
| 2" thick | L.F. | 0.057 |
| 2-1/2" thick | " | 0.062 |
| 3" thick | " | 0.067 |
| 4" thick | " | 0.073 |

# 15 MECHANICAL

| INSULATION | UNIT | MAN/ HOURS |
|---|---|---|
| **15260.20 CALCIUM SILICATE** | | |
| 6" thick | L.F. | 0.080 |
| 12" pipe | | |
| 2" thick | L.F. | 0.062 |
| 2-1/2" thick | " | 0.067 |
| 3" thick | " | 0.073 |
| 4" thick | " | 0.080 |
| 6" thick | " | 0.089 |
| **15260.60 EXTERIOR PIPE INSULATION** | | |
| Fiberglass insulation, aluminum jacket | | |
| 1/2" pipe | | |
| 1" thick | L.F. | 0.062 |
| 1-1/2" thick | " | 0.067 |
| 3/4" pipe | | |
| 1" thick | L.F. | 0.062 |
| 1-1/2" thick | " | 0.067 |
| 1" pipe | | |
| 1" thick | L.F. | 0.062 |
| 1-1/2" thick | " | 0.067 |
| 2" pipe | | |
| 1" thick | L.F. | 0.073 |
| 1-1/2" thick | " | 0.076 |
| 3" pipe | | |
| 1" thick | L.F. | 0.080 |
| 1-1/2" thick | " | 0.084 |
| 6" pipe | | |
| 1" thick | L.F. | 0.089 |
| 2" thick | " | 0.094 |
| 10" pipe | | |
| 2" thick | L.F. | 0.089 |
| 3" thick | " | 0.094 |
| **15260.90 PIPE INSULATION FITTINGS** | | |
| Insulation protection saddle | | |
| 1" thick covering | | |
| 1/2" pipe | EA. | 0.320 |
| 3/4" pipe | " | 0.320 |
| 1" pipe | " | 0.320 |
| 2" pipe | " | 0.320 |
| 3" pipe | " | 0.364 |
| 6" pipe | " | 0.500 |
| 1-1/2" thick covering | | |
| 3/4" pipe | EA. | 0.320 |
| 1" pipe | " | 0.320 |
| 2" pipe | " | 0.320 |
| 3" pipe | " | 0.320 |
| 6" pipe | " | 0.500 |
| 10" pipe | " | 0.667 |
| **15280.10 EQUIPMENT INSULATION** | | |
| Equipment insulation, 2" thick, cellular glass | S.F. | 0.050 |
| Urethane, rigid, field applied jacket, plastered finish | " | 0.100 |

| INSULATION | UNIT | MAN/ HOURS |
|---|---|---|
| **15280.10 EQUIPMENT INSULATION** | | |
| Fiberglass, rigid, with vapor barrier | S.F. | 0.044 |
| **15290.10 DUCTWORK INSULATION** | | |
| Fiberglass duct insulation, plain blanket | | |
| 1-1/2" thick | S.F. | 0.010 |
| 2" thick | " | 0.013 |
| With vapor barrier | | |
| 1-1/2" thick | S.F. | 0.010 |
| 2" thick | " | 0.013 |
| Rigid with vapor barrier | | |
| 2" thick | S.F. | 0.027 |
| 3" thick | " | 0.032 |
| 4" thick | " | 0.040 |
| 6" thick | " | 0.053 |
| Weatherproof, polystyrene, 3" thick, w/vapor barrier | " | 0.080 |
| Urethane board with vapor barrier | " | 0.100 |

| FIRE PROTECTION | UNIT | MAN/ HOURS |
|---|---|---|
| **15330.10 WET SPRINKLER SYSTEM** | | |
| Sprinkler head, 212 deg, brass, exposed piping | EA. | 0.320 |
| Chrome, concealed piping | " | 0.444 |
| Water motor alarm | " | 1.333 |
| Fire department inlet connection | " | 1.600 |
| Wall plate for fire dept connection | " | 0.667 |
| Swing check valve flanged iron body, 4" | " | 2.667 |
| Check valve, 6" | " | 4.000 |
| Wet pipe valve, flange to groove, 4" | " | 0.889 |
| Flange to flange | | |
| 6" | EA. | 1.333 |
| 8" | " | 2.667 |
| Alarm valve, flange to flange, (wet valve) | | |
| 4" | EA. | 0.889 |
| 8" | " | 6.667 |
| Inspector's test connection | " | 0.667 |
| Wall hydrant, polished brass, 2-1/2" x 2-1/2", single | " | 0.571 |
| 2-way | " | 0.571 |
| 3-way | " | 0.571 |
| Wet valve trim, includes retard chamber & gauges, 4"-6" | " | 0.667 |
| Retard pressure switch for wet systems | " | 1.600 |
| Air maintenance device | " | 0.667 |
| Wall hydrant non-freeze, 8" thick wall, vacuum breaker | " | 0.400 |
| 12" thick wall | " | 0.400 |

## FIRE PROTECTION

| 15330.50 DRY SPRINKLER SYSTEM | UNIT | MAN/HOURS |
|---|---|---|
| Dry pipe valve, flange to flange | | |
| 4" | EA. | 1.600 |
| 6" | " | 2.000 |
| Trim, 4" and 6", includes gauges | " | 0.667 |
| Field testing and flushing | " | 6.667 |
| Disinfection | " | 6.667 |
| Pressure switch double circuit, open/close contacts | " | 2.000 |
| Low air | | |
| Supervisory unit | EA. | 1.333 |
| Pressure switch | " | 0.667 |

| 15330.70 CO2 SYSTEM | UNIT | MAN/HOURS |
|---|---|---|
| CO2 system, high pressure, 75# cylinder with | | |
| Valve assemblies | EA. | 1.600 |
| Storage rack | " | 1.143 |
| Manifold | " | 5.714 |
| Flexible loops | " | 0.100 |
| Beam scale for cylinders | " | 1.333 |
| Mechanically control head | " | 0.533 |
| Electrically control head | " | 0.533 |
| Stop valves | " | 0.800 |
| Check valves | " | 1.000 |
| Activation station | " | 0.800 |
| Nozzles | " | 0.667 |
| Hose reel with 75' of 3/4" hose | " | 4.000 |
| Main/reserve transfer switch | " | 1.333 |
| Pressure switch | " | 0.800 |
| Heat responsive device | " | 1.333 |
| Battery and charger | " | 4.000 |
| Low pressure | | |
| Battery and charger | EA. | 4.000 |
| Pressure switch | " | 0.889 |
| Nozzles | " | 0.727 |
| Master selector valve | " | 1.333 |
| Selector valve | " | 1.333 |
| Low pressure hose reel with 75' of 3/4" hose | " | 4.000 |
| Tank fill lines | " | 1.000 |
| Activation stations | " | 0.667 |
| Electro manual pilot panels | " | 1.000 |
| Halon type fire protection system, per computer room | | |

## PLUMBING

| 15410.05 C.I. PIPE, ABOVE GROUND | UNIT | MAN/HOURS |
|---|---|---|
| No hub pipe | | |
| 1-1/2" pipe | L.F. | 0.057 |
| 2" pipe | " | 0.067 |

## PLUMBING

| 15410.05 C.I. PIPE, ABOVE GROUND | UNIT | MAN/HOURS |
|---|---|---|
| 3" pipe | L.F. | 0.080 |
| 4" pipe | " | 0.133 |
| 6" pipe | " | 0.160 |
| 8" pipe | " | 0.267 |
| 10" pipe | " | 0.320 |
| No hub fittings, 1-1/2" pipe | | |
| 1/4 bend | EA. | 0.267 |
| 1/8 bend | " | 0.267 |
| Sanitary tee | " | 0.400 |
| Sanitary cross | " | 0.400 |
| Wye | " | 0.400 |
| 2" pipe | | |
| 1/4 bend | EA. | 0.320 |
| 1/8 bend | " | 0.320 |
| Sanitary tee | " | 0.533 |
| Sanitary cross | " | 0.533 |
| Wye | " | 0.667 |
| 3" pipe | | |
| 1/4 bend | EA. | 0.400 |
| 1/8 bend | " | 0.400 |
| Sanitary tee | " | 0.500 |
| 3"x2" sanitary tee | " | 0.500 |
| 3"x1-1/2" sanitary tee | " | 0.500 |
| Sanitary cross | " | 0.667 |
| 3x2" sanitary cross | " | 0.667 |
| Wye | " | 0.667 |
| 4" pipe | | |
| 1/4 bend | EA. | 0.400 |
| 1/8 bend | " | 0.400 |
| Sanitary tee | " | 0.667 |
| 4x3" sanitary tee | " | 0.667 |
| 4x2" sanitary tee | " | 0.667 |
| Sanitary cross | " | 0.800 |
| 4x3" sanitary cross | " | 0.800 |
| 4x2" sanitary cross | " | 0.800 |
| Wye | " | 0.667 |
| 8" deep | " | 0.400 |
| 6" pipe | | |
| 1/4 bend | EA. | 0.667 |
| 1/8 bend | " | 0.667 |
| Sanitary tee | " | 0.800 |
| 6x4" sanitary tee | " | 0.800 |
| Wye | " | 0.800 |
| 8" pipe | | |
| 1/4 bend | EA. | 0.667 |
| 1/8 bend | " | 0.667 |
| Sanitary tee | " | 1.000 |
| 8x6" sanitary tee | " | 1.000 |
| Wye | " | 0.800 |
| 10" pipe | | |
| 1/4 bend | EA. | 0.667 |
| 1/8 bend | " | 0.667 |
| Wye | " | 1.333 |

# 15 MECHANICAL

| PLUMBING | UNIT | MAN/ HOURS |
|---|---|---|
| **15410.06 C.I. PIPE, BELOW GROUND** | | |
| No hub pipe | | |
| 1-1/2" pipe | L.F. | 0.040 |
| 2" pipe | " | 0.044 |
| 3" pipe | " | 0.050 |
| 4" pipe | " | 0.067 |
| 6" pipe | " | 0.073 |
| 8" pipe | " | 0.089 |
| 10" pipe | " | 0.100 |
| Fittings, 1-1/2" | | |
| 1/4 bend | EA. | 0.229 |
| 1/8 bend | " | 0.229 |
| Wye | " | 0.320 |
| Wye & 1/8 bend | " | 0.229 |
| P-trap | " | 0.229 |
| 2" | | |
| 1/4 bend | EA. | 0.267 |
| 1/8 bend | " | 0.267 |
| Double wye | " | 0.500 |
| Wye & 1/8 bend | " | 0.400 |
| Double wye & 1/8 bend | " | 0.500 |
| P-trap | " | 0.267 |
| 3" | | |
| 1/4 bend | EA. | 0.320 |
| 1/8 bend | " | 0.320 |
| Wye | " | 0.500 |
| 3x2" wye | " | 0.500 |
| Wye & 1/8 bend | " | 0.500 |
| Double wye & 1/8 bend | " | 0.500 |
| 3x2" double wye & 1/8 bend | " | 0.500 |
| 3x2" reducer | " | 0.320 |
| P-trap | " | 0.320 |
| 4" | | |
| 1/4 bend | EA. | 0.320 |
| 1/8 bend | " | 0.320 |
| Wye | " | 0.500 |
| 4x3" wye | " | 0.500 |
| 4x2" wye | " | 0.500 |
| Double wye | " | 0.667 |
| 4x3" double wye | " | 0.667 |
| 4x2" double wye | " | 0.667 |
| Wye & 1/8 bend | " | 0.500 |
| Double wye & 1/8 bend | " | 0.667 |
| 6" | | |
| 1/4 bend | EA. | 0.500 |
| 1/8 bend | " | 0.500 |
| Wye & 1/8 bend | " | 0.667 |
| P-trap | " | 0.400 |
| 8" | | |
| 1/4 bend | EA. | 0.500 |
| 1/8 bend | " | 0.500 |
| Wye | " | 0.667 |
| 10" | | |
| 1/4 bend | EA. | 0.500 |
| 1/8 bend | " | 0.500 |

| PLUMBING | UNIT | MAN/ HOURS |
|---|---|---|
| **15410.06 C.I. PIPE, BELOW GROUND** | | |
| Wye | EA. | 1.000 |
| **15410.09 SERVICE WEIGHT PIPE** | | |
| Service weight pipe, single hub | | |
| 3" x 5' | EA. | 0.170 |
| 4" x 5' | " | 0.178 |
| 6" x 5' | " | 0.200 |
| 1/8 bend | | |
| 3" | EA. | 0.320 |
| 4" | " | 0.364 |
| 6" | " | 0.400 |
| 1/4 bend | | |
| 3" | EA. | 0.320 |
| 4" | " | 0.364 |
| 6" | " | 0.400 |
| Sweep | | |
| 3" | EA. | 0.320 |
| 4" | " | 0.364 |
| 6" | " | 0.400 |
| Sanitary T | | |
| 3" | EA. | 0.571 |
| 4" | " | 0.667 |
| 6" | " | 0.727 |
| Wye | | |
| 3" | EA. | 0.444 |
| 4" | " | 0.471 |
| 6" | " | 0.571 |
| **15410.10 COPPER PIPE** | | |
| Type "K" copper | | |
| 1/2" | L.F. | 0.025 |
| 3/4" | " | 0.027 |
| 1" | " | 0.029 |
| 1-1/4" | " | 0.031 |
| 1-1/2" | " | 0.033 |
| 2" | " | 0.036 |
| 2-1/2" | " | 0.040 |
| 3" | " | 0.042 |
| 4" | " | 0.044 |
| DWV, copper | | |
| 1-1/4" | L.F. | 0.033 |
| 1-1/2" | " | 0.036 |
| 2" | " | 0.040 |
| 3" | " | 0.044 |
| 4" | " | 0.050 |
| 6" | " | 0.057 |
| Refrigeration tubing, copper, sealed | | |
| 1/8" | L.F. | 0.032 |
| 3/16" | " | 0.033 |
| 1/4" | " | 0.035 |
| 5/16" | " | 0.036 |
| 3/8" | " | 0.038 |

| PLUMBING | UNIT | MAN/ HOURS |
|---|---|---|
| **15410.10 COPPER PIPE** | | |
| 1/2" | L.F. | 0.040 |
| 7/8" | " | 0.046 |
| 1-1/8" | " | 0.053 |
| 1-3/8" | " | 0.062 |
| Type "L" copper | | |
| 1/4" | L.F. | 0.024 |
| 3/8" | " | 0.024 |
| 1/2" | " | 0.025 |
| 3/4" | " | 0.027 |
| 1" | " | 0.029 |
| 1-1/4" | " | 0.031 |
| 1-1/2" | " | 0.033 |
| 2" | " | 0.036 |
| Type "M" copper | | |
| 1/2" | L.F. | 0.025 |
| 3/4" | " | 0.027 |
| 1" | " | 0.029 |
| 1-1/4" | " | 0.031 |
| 2" | " | 0.036 |
| **15410.11 COPPER FITTINGS** | | |
| Coupling, with stop | | |
| 1/4" | EA. | 0.267 |
| 3/8" | " | 0.320 |
| 1/2" | " | 0.348 |
| 5/8" | " | 0.400 |
| 3/4" | " | 0.444 |
| 1" | " | 0.471 |
| 3" | " | 0.800 |
| 4" | " | 1.000 |
| Reducing coupling | | |
| 1/4" x 1/8" | EA. | 0.320 |
| 3/8" x 1/4" | " | 0.348 |
| 1/2" x | | |
| 3/8" | EA. | 0.400 |
| 1/4" | " | 0.400 |
| 1/8" | " | 0.400 |
| 3/4" x | | |
| 3/8" | EA. | 0.444 |
| 1/2" | " | 0.444 |
| 1" x | | |
| 3/8" | EA. | 0.500 |
| 1" x 1/2" | " | 0.500 |
| 1" x 3/4" | " | 0.500 |
| 1-1/4" x | | |
| 1/2" | EA. | 0.533 |
| 3/4" | " | 0.533 |
| 1" | " | 0.533 |
| 1-1/2" x | | |
| 1/2" | EA. | 0.571 |
| 3/4" | " | 0.571 |
| 1" | " | 0.571 |
| 1-1/4" | " | 0.571 |

| PLUMBING | UNIT | MAN/ HOURS |
|---|---|---|
| **15410.11 COPPER FITTINGS** | | |
| 2" x | | |
| 1/2" | EA. | 0.667 |
| 3/4" | " | 0.667 |
| 1" | " | 0.667 |
| 1-1/4" | " | 0.667 |
| 1-1/2" | " | 0.667 |
| 2-1/2" x | | |
| 1" | EA. | 0.800 |
| 1-1/4" | " | 0.800 |
| 1-1/2" | " | 0.800 |
| 2" | " | 0.800 |
| 3" x | | |
| 1-1/2" | EA. | 1.000 |
| 2" | " | 1.000 |
| 2-1/2" | " | 1.000 |
| 4" x | | |
| 2" | EA. | 1.143 |
| 2-1/2" | " | 1.143 |
| 3" | " | 1.143 |
| Slip coupling | | |
| 1/4" | EA. | 0.267 |
| 1/2" | " | 0.320 |
| 3/4" | " | 0.400 |
| 1" | " | 0.444 |
| 1-1/4" | " | 0.500 |
| 1-1/2" | " | 0.533 |
| 2" | " | 0.667 |
| 2-1/2" | " | 0.667 |
| 3" | " | 0.800 |
| 4" | " | 1.000 |
| Coupling with drain | | |
| 1/2" | EA. | 0.400 |
| 3/4" | " | 0.444 |
| 1" | " | 0.500 |
| Reducer | | |
| 3/8" x 1/4" | EA. | 0.320 |
| 1/2" x 3/8" | " | 0.320 |
| 3/4" x | | |
| 1/4" | EA. | 0.364 |
| 3/8" | " | 0.364 |
| 1/2" | " | 0.364 |
| 1" x | | |
| 1/2" | EA. | 0.400 |
| 3/4" | " | 0.400 |
| 1-1/4" x | | |
| 1/2" | EA. | 0.444 |
| 3/4" | " | 0.444 |
| 1" | " | 0.444 |
| 1-1/2" x | | |
| 1/2" | EA. | 0.500 |
| 3/4" | " | 0.500 |
| 1" | " | 0.500 |
| 1-1/4" | " | 0.500 |
| 2" x | | |

| PLUMBING | UNIT | MAN/HOURS |
|---|---|---|
| **15410.11 COPPER FITTINGS** | | |
| 1/2" | EA. | 0.571 |
| 3/4" | " | 0.571 |
| 1" | " | 0.571 |
| 1-1/4" | " | 0.571 |
| 1-1/2" | " | 0.571 |
| 2-1/2" x | | |
| 1" | EA. | 0.667 |
| 1-1/4" | " | 0.667 |
| 1-1/2" | " | 0.667 |
| 2" | " | 0.667 |
| 3" x | | |
| 1-1/4" | EA. | 0.800 |
| 1-1/2" | " | 0.800 |
| 2" | " | 0.800 |
| 2-1/2" | " | 0.800 |
| 4" x | | |
| 2" | EA. | 1.000 |
| 3" | " | 1.000 |
| Female adapters | | |
| 1/4" | EA. | 0.320 |
| 3/8" | " | 0.364 |
| 1/2" | " | 0.400 |
| 3/4" | " | 0.444 |
| 1" | " | 0.444 |
| 1-1/4" | " | 0.500 |
| 1-1/2" | " | 0.500 |
| 2" | " | 0.533 |
| 2-1/2" | " | 0.571 |
| 3" | " | 0.667 |
| 4" | " | 0.800 |
| Increasing female adapters | | |
| 1/8" x | | |
| 3/8" | EA. | 0.320 |
| 1/2" | " | 0.320 |
| 1/4" x 1/2" | " | 0.348 |
| 3/8" x 1/2" | " | 0.364 |
| 1/2" X | | |
| 3/4" | EA. | 0.400 |
| 1" | " | 0.400 |
| 3/4" X | | |
| 1" | EA. | 0.444 |
| 1-1/4" | " | 0.444 |
| 1" x | | |
| 1-1/4" | EA. | 0.444 |
| 1-1/2" | " | 0.444 |
| 1-1/4" x | | |
| 1-1/2" | EA. | 0.500 |
| 2" | " | 0.500 |
| 1-1/2" x 2" | " | 0.533 |
| Reducing female adapters | | |
| 3/8" x 1/4" | EA. | 0.364 |
| 1/2" x | | |
| 1/4" | EA. | 0.400 |
| 3/8" | " | 0.400 |

| PLUMBING | UNIT | MAN/HOURS |
|---|---|---|
| **15410.11 COPPER FITTINGS** | | |
| 3/4" x 1/2" | EA. | 0.444 |
| 1" x | | |
| 1/2" | EA. | 0.444 |
| 3/4" | " | 0.444 |
| 1-1/4" x | | |
| 1/2" | EA. | 0.500 |
| 3/4" | " | 0.500 |
| 1" | " | 0.500 |
| 1-1/2" x | | |
| 1" | EA. | 0.533 |
| 1-1/4" | " | 0.533 |
| 2" x | | |
| 1" | EA. | 0.571 |
| 1-1/4" | " | 0.571 |
| 1-1/2" | " | 0.571 |
| Female fitting adapters | | |
| 1/2" | EA. | 0.400 |
| 3/4" | " | 0.400 |
| 3/4" x 1/2" | " | 0.421 |
| 1" | " | 0.444 |
| 1-1/4" | " | 0.471 |
| 1-1/2" | " | 0.500 |
| 2" | " | 0.533 |
| Male adapters | | |
| 1/4" | EA. | 0.364 |
| 3/8" | " | 0.364 |
| 3" | " | 0.667 |
| 4" | " | 0.800 |
| Increasing male adapters | | |
| 3/8" x 1/2" | EA. | 0.364 |
| 1/2" x | | |
| 3/4" | EA. | 0.400 |
| 1" | " | 0.400 |
| 3/4" x | | |
| 1" | EA. | 0.421 |
| 1-1/4" | " | 0.421 |
| 1" x 1-1/4" | " | 0.444 |
| 1-1/2" x | | |
| 3/4" | EA. | 0.471 |
| 1" | " | 0.471 |
| 1-1/4" | " | 0.471 |
| 2" x | | |
| 1" | EA. | 0.500 |
| 1-1/4" | " | 0.500 |
| 1-1/2" | " | 0.500 |
| 2" x 2-1/2" | " | 0.533 |
| Copper pipe fittings | | |
| 1/2" | | |
| 90 deg ell | EA. | 0.178 |
| 45 deg ell | " | 0.178 |
| Tee | " | 0.229 |
| Cap | " | 0.089 |
| Coupling | " | 0.178 |
| Union | " | 0.200 |

## PLUMBING

### 15410.11 COPPER FITTINGS

| PLUMBING | UNIT | MAN/HOURS |
|---|---|---|
| 3/4" | | |
| 90 deg ell | EA. | 0.200 |
| 45 deg ell | " | 0.200 |
| Tee | " | 0.267 |
| Cap | " | 0.094 |
| Coupling | " | 0.200 |
| Union | " | 0.229 |
| 1" | | |
| 90 deg ell | EA. | 0.267 |
| 45 deg ell | " | 0.267 |
| Tee | " | 0.320 |
| Cap | " | 0.133 |
| Coupling | " | 0.267 |
| Union | " | 0.267 |
| 1-1/4" | | |
| 90 deg ell | EA. | 0.229 |
| 45 deg ell | " | 0.229 |
| Tee | " | 0.400 |
| Cap | " | 0.133 |
| Union | " | 0.286 |
| 1-1/2" | | |
| 90 deg ell | EA. | 0.286 |
| 45 deg ell | " | 0.286 |
| Tee | " | 0.444 |
| Cap | " | 0.133 |
| Coupling | " | 0.267 |
| Union | " | 0.364 |
| 2" | | |
| 90 deg ell | EA. | 0.320 |
| 45 deg ell | " | 0.500 |
| Tee | " | 0.500 |
| Cap | " | 0.160 |
| Coupling | " | 0.320 |
| Union | " | 0.400 |
| 2-1/2" | | |
| 90 deg ell | EA. | 0.400 |
| 45 deg ell | " | 0.400 |
| Tee | " | 0.571 |
| Cap | " | 0.200 |
| Coupling | " | 0.400 |
| Union | " | 0.444 |

### 15410.15 BRASS FITTINGS

| PLUMBING | UNIT | MAN/HOURS |
|---|---|---|
| Compression fittings, union | | |
| 3/8" | EA. | 0.133 |
| 1/2" | " | 0.133 |
| 5/8" | " | 0.133 |
| Union elbow | | |
| 3/8" | EA. | 0.133 |
| 1/2" | " | 0.133 |
| 5/8" | " | 0.133 |
| Union tee | | |
| 3/8" | EA. | 0.133 |

## PLUMBING

### 15410.15 BRASS FITTINGS

| PLUMBING | UNIT | MAN/HOURS |
|---|---|---|
| 1/2" | EA. | 0.133 |
| 5/8" | " | 0.133 |
| Male connector | | |
| 3/8" | EA. | 0.133 |
| 1/2" | " | 0.133 |
| 5/8" | " | 0.133 |
| Female connector | | |
| 3/8" | EA. | 0.133 |
| 1/2" | " | 0.133 |
| 5/8" | " | 0.133 |
| Brass flare fittings, union | | |
| 3/8" | EA. | 0.129 |
| 1/2" | " | 0.129 |
| 5/8" | " | 0.129 |
| 90 deg elbow union | | |
| 3/8" | EA. | 0.129 |
| 1/2" | " | 0.129 |
| 5/8" | " | 0.129 |
| Three way tee | | |
| 3/8" | EA. | 0.216 |
| 1/2" | " | 0.216 |
| 5/8" | " | 0.216 |
| Cross | | |
| 3/8" | EA. | 0.286 |
| 1/2" | " | 0.286 |
| 5/8" | " | 0.286 |
| Male connector, half union | | |
| 3/8" | EA. | 0.129 |
| 1/2" | " | 0.129 |
| 5/8" | " | 0.120 |
| Female connector, half union | | |
| 3/8" | EA. | 0.129 |
| 1/2" | " | 0.129 |
| 5/8" | " | 0.129 |
| Long forged nut | | |
| 3/8" | EA. | 0.129 |
| 1/2" | " | 0.129 |
| 5/8" | " | 0.129 |
| Short forged nut | | |
| 3/8" | EA. | 0.129 |
| 1/2" | " | 0.129 |
| 5/8" | " | 0.129 |

### 15410.18 GLASS PIPE

| PLUMBING | UNIT | MAN/HOURS |
|---|---|---|
| Glass pipe | | |
| 1-1/2" dia. | L.F. | 0.160 |
| 2" dia. | " | 0.178 |
| 3" dia. | " | 0.200 |
| 4" dia. | " | 0.229 |
| 6" dia. | " | 0.267 |

### 15410.30 PVC/CPVC PIPE

| PLUMBING | UNIT | MAN/HOURS |
|---|---|---|
| PVC schedule 40 | | |
| 1/2" pipe | L.F. | 0.033 |

| PLUMBING | UNIT | MAN/ HOURS |
|---|---|---|
| **15410.30 PVC/CPVC PIPE** | | |
| 3/4" pipe | L.F. | 0.036 |
| 1" pipe | " | 0.040 |
| 1-1/4" pipe | " | 0.044 |
| 1-1/2" pipe | " | 0.050 |
| 2" pipe | " | 0.057 |
| 2-1/2" pipe | " | 0.067 |
| 3" pipe | " | 0.080 |
| 4" pipe | " | 0.100 |
| 6" pipe | " | 0.200 |
| 8" pipe | " | 0.267 |
| Fittings, 1/2" | | |
| 90 deg ell | EA. | 0.100 |
| 45 deg ell | " | 0.100 |
| Tee | " | 0.114 |
| 3/4" | | |
| 90 deg elbow | EA. | 0.133 |
| 45 deg elbow | " | 0.133 |
| Tee | " | 0.160 |
| 1" | | |
| 90 deg elbow | EA. | 0.160 |
| 45 deg elbow | " | 0.160 |
| Tee | " | 0.178 |
| 1-1/4" | | |
| 90 deg elbow | EA. | 0.229 |
| 45 deg elbow | " | 0.229 |
| Tee | " | 0.267 |
| 1-1/2" | | |
| 90 deg elbow | EA. | 0.229 |
| 45 deg elbow | " | 0.229 |
| Tee | " | 0.267 |
| 2" | | |
| 90 deg elbow | EA. | 0.267 |
| 45 deg elbow | " | 0.267 |
| Tee | " | 0.320 |
| 2-1/2" | | |
| 90 deg elbow | EA. | 0.500 |
| 45 deg elbow | " | 0.500 |
| Tee | " | 0.533 |
| 3" | | |
| 90 deg elbow | EA. | 0.667 |
| 45 deg elbow | " | 0.667 |
| Tee | " | 0.727 |
| 4" | | |
| 90 deg elbow | EA. | 0.800 |
| 45 deg elbow | " | 0.800 |
| Tee | " | 0.889 |
| PVC schedule 80 pipe | | |
| 1-1/2" pipe | L.F. | 0.050 |
| 2" pipe | " | 0.057 |
| 3" pipe | " | 0.080 |
| 4" pipe | " | 0.100 |
| Fittings, 1-1/2" | | |
| 90 deg elbow | EA. | 0.267 |
| 45 deg elbow | " | 0.267 |

| PLUMBING | UNIT | MAN/ HOURS |
|---|---|---|
| **15410.30 PVC/CPVC PIPE** | | |
| Tee | EA. | 0.400 |
| 2" | | |
| 90 deg elbow | EA. | 0.320 |
| 45 deg elbow | " | 0.320 |
| Tee | " | 0.500 |
| 2-1/2" | | |
| 90 deg elbow | EA. | 0.500 |
| 45 deg elbow | " | 0.500 |
| Tee | " | 0.667 |
| 3" | | |
| 90 deg elbow | EA. | 0.667 |
| 45 deg elbow | " | 0.667 |
| Tee | " | 0.800 |
| 4" | | |
| 90 deg elbow | EA. | 0.800 |
| 45 deg elbow | " | 0.800 |
| Tee | " | 1.000 |
| CPVC schedule 40 | | |
| 1/2" pipe | L.F. | 0.033 |
| 3/4" pipe | " | 0.036 |
| 1" pipe | " | 0.040 |
| 1-1/4" pipe | " | 0.044 |
| 1-1/2" pipe | " | 0.050 |
| 2" pipe | " | 0.057 |
| Fittings, CPVC, schedule 80 | | |
| 1/2", 90 deg ell | EA. | 0.080 |
| Tee | " | 0.133 |
| 3/4", 90 deg ell | " | 0.080 |
| Tee | " | 0.133 |
| 1", 90 deg ell | " | 0.089 |
| Tee | " | 0.145 |
| 1-1/4", 90 deg ell | " | 0.089 |
| Tee | " | 0.145 |
| 1-1/2", 90 deg ell | " | 0.160 |
| Tee | " | 0.200 |
| 2", 90 deg ell | " | 0.160 |
| Tee | " | 0.200 |
| Polypropylene, acid resistant, DWV pipe | | |
| Schedule 40 | | |
| 1-1/2" pipe | L.F. | 0.057 |
| 2" pipe | " | 0.067 |
| 3" pipe | " | 0.080 |
| 4" pipe | " | 0.100 |
| 6" pipe | " | 0.200 |
| Fittings | | |
| 1-1/2" | | |
| 1/4 bend | EA. | 0.200 |
| 1/8 bend | " | 0.200 |
| Sanitary tee | " | 0.400 |
| Cleanout with plug | " | 0.400 |
| Wye | " | 0.400 |
| 2" | | |
| 1/4 bend | EA. | 0.229 |
| 1/8 bend | " | 0.229 |

| PLUMBING | UNIT | MAN/ HOURS |
|---|---|---|
| **15410.30 PVC/CPVC PIPE** | | |
| Sanitary tee | EA. | 0.471 |
| Cleanout with plug | " | 0.471 |
| Wye | " | 0.471 |
| 3" | | |
| 1/4 bend | EA. | 0.267 |
| 1/8 bend | " | 0.267 |
| Sanitary tee | " | 0.533 |
| Cleanout with plug | " | 0.533 |
| Wye | " | 0.533 |
| 4" | | |
| 1/4 bend | EA. | 0.400 |
| 1/8 bend | " | 0.400 |
| Sanitary tee | " | 0.800 |
| Cleanout with plug | " | 0.800 |
| Wye | " | 0.800 |
| 6" | | |
| 1/4 bend | EA. | 0.667 |
| 1/8 bend | " | 0.667 |
| Sanitary tee | " | 1.333 |
| Cleanout with plug | " | 1.333 |
| Wye | " | 1.333 |
| Polyethylene pipe and fittings | | |
| SDR-21 | | |
| 3" pipe | L.F. | 0.100 |
| 4" pipe | " | 0.133 |
| 6" pipe | " | 0.200 |
| 8" pipe | " | 0.229 |
| 10" pipe | " | 0.267 |
| 12" pipe | " | 0.320 |
| 14" pipe | " | 0.400 |
| 16" pipe | " | 0.500 |
| 18" pipe | " | 0.615 |
| 20" pipe | " | 0.800 |
| 22" pipe | " | 0.889 |
| 24" pipe | " | 1.000 |
| Fittings, 3" | | |
| 90 deg elbow | EA. | 0.400 |
| 45 deg elbow | " | 0.400 |
| Tee | " | 0.667 |
| 4" | | |
| 90 deg elbow | EA. | 0.500 |
| 45 deg elbow | " | 0.500 |
| Tee | " | 0.800 |
| 8" | | |
| 90 deg elbow | EA. | 1.000 |
| 45 deg elbow | " | 1.000 |
| Tee | " | 1.600 |
| 10" | | |
| 90 deg elbow | EA. | 1.333 |
| 45 deg elbow | " | 1.333 |
| Tee | " | 2.000 |
| 12" | | |
| 90 deg elbow | EA. | 1.600 |
| 45 deg elbow | " | 1.600 |

| PLUMBING | UNIT | MAN/ HOURS |
|---|---|---|
| **15410.30 PVC/CPVC PIPE** | | |
| Tee | EA. | 2.667 |
| 14" | | |
| 90 deg elbow | EA. | 2.000 |
| 45 deg elbow | " | 2.000 |
| Tee | " | 3.200 |
| 16" | | |
| 90 deg elbow | EA. | 2.000 |
| 45 deg elbow | " | 2.000 |
| Tee | " | 3.200 |
| 18" | | |
| 90 deg elbow | EA. | 2.667 |
| 45 deg elbow | " | 2.667 |
| Tee | " | 4.000 |
| 20" | | |
| 90 deg elbow | EA. | 2.667 |
| 45 deg elbow | " | 2.667 |
| **15410.33 ABS DWV PIPE** | | |
| Schedule 40 ABS | | |
| 1-1/2" pipe | L.F. | 0.040 |
| 2" pipe | " | 0.044 |
| 3" pipe | " | 0.057 |
| 4" pipe | " | 0.080 |
| 6" pipe | " | 0.100 |
| Fittings | | |
| 1/8 bend | | |
| 1-1/2" | EA. | 0.160 |
| 2" | " | 0.200 |
| 3" | " | 0.267 |
| 4" | " | 0.320 |
| 6" | " | 0.400 |
| Tee, sanitary | | |
| 1-1/2" | EA. | 0.267 |
| 2" | " | 0.320 |
| 3" | " | 0.400 |
| 4" | " | 0.500 |
| 6" | " | 0.667 |
| Tee, sanitary reducing | | |
| 2 x 1-1/2 x 1-1/2 | EA. | 0.320 |
| 2 x 1-1/2 x 2 | " | 0.333 |
| 2 x 2 x 1-1/2 | " | 0.364 |
| 3 x 3 x 1-1/2 | " | 0.400 |
| 3 x 3 x 2 | " | 0.444 |
| 4 x 4 x 1-1/2 | " | 0.500 |
| 4 x 4 x 2 | " | 0.571 |
| 4 x 4 x 3 | " | 0.615 |
| 6 x 6 x 4 | " | 0.667 |
| Wye | | |
| 1-1/2" | EA. | 0.229 |
| 2" | " | 0.320 |
| 3" | " | 0.400 |
| 4" | " | 0.500 |
| 6" | " | 0.667 |
| Reducer | | |

| PLUMBING | UNIT | MAN/HOURS |
|---|---|---|
| **15410.33 ABS DWV PIPE** | | |
| 2 x 1-1/2 | EA. | 0.200 |
| 3 x 1-1/2 | " | 0.267 |
| 3 x 2 | " | 0.267 |
| 4 x 2 | " | 0.320 |
| 4 x 3 | " | 0.320 |
| 6 x 4 | " | 0.400 |
| P-trap | | |
| 1-1/2" | EA. | 0.267 |
| 2" | " | 0.296 |
| 3" | " | 0.348 |
| 4" | " | 0.400 |
| 6" | " | 0.500 |
| Double sanitary, tee | | |
| 1-1/2" | EA. | 0.320 |
| 2" | " | 0.400 |
| 3" | " | 0.500 |
| 4" | " | 0.667 |
| Long sweep, 1/4 bend | | |
| 1-1/2" | EA. | 0.160 |
| 2" | " | 0.200 |
| 3" | " | 0.267 |
| 4" | " | 0.400 |
| Wye, standard | | |
| 1-1/2" | EA. | 0.267 |
| 2" | " | 0.320 |
| 3" | " | 0.400 |
| 4" | " | 0.500 |
| **15410.35 PLASTIC PIPE** | | |
| Fiberglass reinforced pipe | | |
| 2" pipe | L.F. | 0.062 |
| 3" pipe | " | 0.067 |
| 4" pipe | " | 0.073 |
| 6" pipe | " | 0.080 |
| 8" pipe | " | 0.133 |
| 10" pipe | " | 0.160 |
| 12" pipe | " | 0.200 |
| Fittings | | |
| 90 deg elbow, flanged | | |
| 2" | EA. | 0.800 |
| 3" | " | 0.889 |
| 4" | " | 1.000 |
| 6" | " | 1.333 |
| 8" | " | 1.600 |
| 10" | " | 2.000 |
| 12" | " | 2.667 |
| 45 deg elbow, flanged | | |
| 2" | EA. | 0.667 |
| 3" | " | 0.800 |
| 4" | " | 1.000 |
| 6" | " | 1.333 |
| 8" | " | 1.600 |
| 10" | " | 2.000 |

| PLUMBING | UNIT | MAN/HOURS |
|---|---|---|
| **15410.35 PLASTIC PIPE** | | |
| 12" | EA. | 2.667 |
| Tee, flanged | | |
| 2" | EA. | 1.000 |
| 3" | " | 1.143 |
| 4" | " | 1.333 |
| 6" | " | 1.600 |
| 8" | " | 2.000 |
| 10" | " | 2.667 |
| 12" | " | 4.000 |
| Wye, flanged | | |
| 2" | EA. | 1.000 |
| 3" | " | 1.143 |
| 4" | " | 1.333 |
| 6" | " | 1.600 |
| 8" | " | 2.000 |
| 10" | " | 2.667 |
| 12" | " | 4.000 |
| **15410.70 STAINLESS STEEL PIPE** | | |
| Stainless steel, schedule 40, threaded | | |
| 1/2" pipe | L.F. | 0.114 |
| 1" pipe | " | 0.123 |
| 1-1/2" pipe | " | 0.133 |
| 2" pipe | " | 0.145 |
| 2-1/2" pipe | " | 0.160 |
| 3" pipe | " | 0.178 |
| 4" pipe | " | 0.200 |
| Fittings, 1/2" | | |
| 90 deg ell | EA. | 1.000 |
| 45 deg ell | " | 1.000 |
| Tee | " | 1.333 |
| 3/4" | | |
| 90 deg ell | EA. | 1.000 |
| 45 deg ell | " | 1.000 |
| Tee | " | 1.333 |
| 1" | | |
| 90 deg ell | EA. | 1.000 |
| 45 deg ell | " | 1.000 |
| Tee | " | 1.333 |
| 1-1/4" | | |
| 90 deg ell | EA. | 1.000 |
| 45 deg ell | " | 1.000 |
| Tee | " | 1.333 |
| 1-1/2" | | |
| 90 deg ell | EA. | 1.333 |
| 45 deg ell | " | 1.333 |
| Tee | " | 1.600 |
| 2" | | |
| 90 deg ell | EA. | 1.600 |
| 45 deg ell | " | 1.600 |
| Tee | " | 2.667 |
| Type 304, sch 10 pipe | | |
| 1" pipe | L.F. | 0.100 |

| PLUMBING | UNIT | MAN/ HOURS |
|---|---|---|
| **15410.70 STAINLESS STEEL PIPE** | | |
| 1-1/4" pipe | L.F. | 0.133 |
| 1-1/2" pipe | " | 0.145 |
| 2" pipe | " | 0.178 |
| 2-1/2" pipe | " | 0.200 |
| 3" pipe | " | 0.229 |
| 4" pipe | " | 0.267 |
| 6" pipe | " | 0.320 |
| Fittings, 1" | | |
| 90 deg elbow | EA. | 1.600 |
| 45 deg elbow | " | 1.600 |
| Tee | " | 2.667 |
| 1-1/4" | | |
| 90 deg elbow | EA. | 1.600 |
| 45 deg elbow | " | 1.600 |
| Tee | " | 2.667 |
| 1-1/2" | | |
| 90 deg elbow | EA. | 1.333 |
| 45 deg elbow | " | 1.333 |
| Tee | " | 2.667 |
| 2" | | |
| 90 deg elbow | EA. | 1.600 |
| 45 deg elbow | " | 1.600 |
| Tee | " | 4.000 |
| 2-1/2" | | |
| 90 deg elbow | EA. | 1.600 |
| 45 deg elbow | " | 1.600 |
| Tee | " | 4.000 |
| 3" | | |
| 90 deg elbow | EA. | 2.000 |
| 45 deg elbow | " | 2.000 |
| Tee | " | 5.333 |
| 4" | | |
| 90 deg elbow | EA. | 2.000 |
| 45 deg elbow | " | 2.000 |
| 6" | | |
| 90 deg elbow | EA. | 2.667 |
| 45 deg elbow | " | 2.667 |
| Type 304 tubing | | |
| .035 wall | | |
| 1/4" | L.F. | 0.044 |
| 3/8" | " | 0.050 |
| 1/2" | " | 0.057 |
| 5/8" | " | 0.067 |
| 3/4" | " | 0.080 |
| 7/8" | " | 0.089 |
| 1" | " | 0.100 |
| .049 wall | | |
| 1/4" | L.F. | 0.047 |
| 3/8" | " | 0.053 |
| 1/2" | " | 0.062 |
| 5/8" | " | 0.073 |
| 3/4" | " | 0.089 |
| 7/8" | " | 0.100 |
| 1" | " | 0.114 |

| PLUMBING | UNIT | MAN/ HOURS |
|---|---|---|
| **15410.70 STAINLESS STEEL PIPE** | | |
| .065 wall | | |
| 1/4" | L.F. | 0.053 |
| 3/8" | " | 0.067 |
| 1/2" | " | 0.073 |
| 5/8" | " | 0.089 |
| 3/4" | " | 0.114 |
| 7/8" | " | 0.133 |
| 1" | " | 0.160 |
| Type 316 tubing | | |
| .035 wall | | |
| 1/4" | L.F. | 0.044 |
| 3/8" | " | 0.050 |
| 1/2" | " | 0.057 |
| 5/8" | " | 0.067 |
| 3/4" | " | 0.080 |
| 7/8" | " | 0.089 |
| 1" | " | 0.100 |
| .049 wall | | |
| 1/4" | L.F. | 0.053 |
| 3/8" | " | 0.067 |
| 1/2" | " | 0.073 |
| 5/8" | " | 0.089 |
| 3/4" | " | 0.114 |
| 7/8" | " | 0.133 |
| 1" | " | 0.160 |
| .065 wall | | |
| 1/4" | L.F. | 0.053 |
| 3/8" | " | 0.067 |
| 1/2" | " | 0.073 |
| 5/8" | " | 0.089 |
| 3/4" | " | 0.114 |
| 7/8" | " | 0.133 |
| 1" | " | 0.160 |
| Fittings, 1/4" | | |
| 90 deg elbow | EA. | 0.160 |
| Union tee | " | 0.267 |
| Union | " | 0.267 |
| Male connector | " | 0.200 |
| 3/8" | | |
| 90 deg elbow | EA. | 0.200 |
| Union tee | " | 0.308 |
| Union | " | 0.308 |
| Male connector | " | 0.200 |
| 1/2" | | |
| 90 deg elbow | EA. | 0.211 |
| Union tee | " | 0.333 |
| Union | " | 0.333 |
| Male connector | " | 0.200 |
| 5/8" | | |
| 90 deg elbow | EA. | 0.267 |
| Union tee | " | 0.400 |
| Union | " | 0.400 |
| Male connector | " | 0.267 |
| 3/4" | | |

| PLUMBING | UNIT | MAN/ HOURS |
|---|---|---|
| **15410.70 STAINLESS STEEL PIPE** | | |
| 90 deg elbow | EA. | 0.267 |
| Union tee | " | 0.400 |
| Union | " | 0.400 |
| Male connector | " | 0.267 |
| 7/8" | | |
| 90 deg elbow | EA. | 0.286 |
| Union tee | " | 0.444 |
| Union | " | 0.444 |
| Male connector | " | 0.286 |
| 1" | | |
| 90 deg elbow | EA. | 0.364 |
| Union tee | " | 0.500 |
| Union | " | 0.500 |
| Male connector | " | 0.400 |
| Type 316 valves | | |
| Gate valves | | |
| 1/4" | EA. | 0.267 |
| 3/8" | " | 0.320 |
| 1/2" | " | 0.348 |
| 3/4" | " | 0.400 |
| 1" | " | 0.533 |
| Globe valves | | |
| 1/4" | EA. | 0.267 |
| 3/8" | " | 0.320 |
| 1/2" | " | 0.348 |
| 3/4" | " | 0.400 |
| 1" | " | 0.533 |
| Check valves | | |
| 1/4" | EA. | 0.267 |
| 3/8" | " | 0.320 |
| 1/2" | " | 0.348 |
| 3/4" | " | 0.400 |
| 1" | " | 0.533 |
| **15410.80 STEEL PIPE** | | |
| Black steel, extra heavy pipe, threaded | | |
| 1/2" pipe | L.F. | 0.032 |
| 3/4" pipe | " | 0.032 |
| 1" pipe | " | 0.040 |
| 1-1/2" pipe | " | 0.044 |
| 2-1/2" pipe | " | 0.100 |
| 3" pipe | " | 0.133 |
| 4" pipe | " | 0.160 |
| 5" pipe | " | 0.200 |
| 6" pipe | " | 0.200 |
| 8" pipe | " | 0.267 |
| 10" pipe | " | 0.320 |
| 12" pipe | " | 0.400 |
| Fittings, malleable iron, threaded, 1/2" pipe | | |
| 90 deg ell | EA. | 0.267 |
| 45 deg ell | " | 0.267 |
| Tee | " | 0.400 |
| Reducing tee | " | 0.400 |

| PLUMBING | UNIT | MAN/ HOURS |
|---|---|---|
| **15410.80 STEEL PIPE** | | |
| Cap | EA. | 0.160 |
| Coupling | " | 0.320 |
| Union | " | 0.267 |
| Nipple, 4" long | " | 0.267 |
| 3/4" pipe | | |
| 90 deg ell | EA. | 0.267 |
| 45 deg ell | " | 0.400 |
| Tee | " | 0.400 |
| Reducing tee | " | 0.267 |
| Cap | " | 0.160 |
| Coupling | " | 0.267 |
| Union | " | 0.267 |
| Nipple, 4" long | " | 0.267 |
| 1" pipe | | |
| 90 deg ell | EA. | 0.320 |
| 45 deg ell | " | 0.320 |
| Tee | " | 0.444 |
| Reducing tee | " | 0.444 |
| Cap | " | 0.160 |
| Coupling | " | 0.320 |
| Union | " | 0.320 |
| Nipple, 4" long | " | 0.320 |
| 1-1/2" pipe | | |
| 90 deg ell | EA. | 0.400 |
| 45 deg ell | " | 0.400 |
| Tee | " | 0.571 |
| Reducing tee | " | 0.571 |
| Cap | " | 0.200 |
| Coupling | " | 0.400 |
| Union | " | 0.400 |
| Nipple, 4" long | " | 0.400 |
| 2-1/2" pipe | | |
| 90 deg ell | EA. | 1.000 |
| 45 deg ell | " | 1.000 |
| Tee | " | 1.333 |
| Reducing tee | " | 1.333 |
| Cap | " | 0.500 |
| Coupling | " | 1.333 |
| Union | " | 1.333 |
| Nipple, 4" long | " | 1.333 |
| 3" pipe | | |
| 90 deg ell | EA. | 1.333 |
| 45 deg ell | " | 1.333 |
| Tee | " | 2.000 |
| Reducing tee | " | 2.000 |
| Cap | " | 0.667 |
| Coupling | " | 1.333 |
| Union | " | 1.333 |
| Nipple, 4" long | " | 1.333 |
| 4" pipe | | |
| 90 deg ell | EA. | 1.600 |
| 45 deg ell | " | 1.600 |
| Tee | " | 2.667 |
| Reducing tee | " | 2.667 |

| PLUMBING | UNIT | MAN/ HOURS |
|---|---|---|
| **15410.80 STEEL PIPE** | | |
| Cap | EA. | 2.667 |
| Coupling | " | 0.800 |
| Union | " | 2.667 |
| Nipple, 4" long | " | 2.667 |
| 6" pipe | | |
| 90 deg ell | EA. | 1.600 |
| 45 deg ell | " | 1.600 |
| Tee | " | 2.667 |
| Reducing tee | " | 2.667 |
| Cap | " | 0.800 |
| 8" pipe | | |
| 90 deg ell | EA. | 3.200 |
| 45 deg ell | " | 3.200 |
| Tee | " | 5.000 |
| Reducing tee | " | 4.211 |
| Cap | " | 1.600 |
| 10" pipe | | |
| 90 deg ell | EA. | 4.000 |
| 45 deg ell | " | 4.000 |
| Tee | " | 5.000 |
| Reducing tee | " | 2.000 |
| Cap | " | 2.000 |
| 12" pipe | | |
| 90 deg ell | EA. | 5.000 |
| 45 deg ell | " | 5.000 |
| Tee | " | 6.667 |
| Reducing tee | " | 6.667 |
| Cap | " | 2.667 |
| **15410.82 GALVANIZED STEEL PIPE** | | |
| Galvanized pipe | | |
| 1/2" pipe | L.F. | 0.080 |
| 3/4" pipe | " | 0.100 |
| 1" pipe | " | 0.114 |
| 1-1/4" pipe | " | 0.133 |
| 1-1/2" pipe | " | 0.160 |
| 2" pipe | " | 0.200 |
| 2-1/2" pipe | " | 0.267 |
| 3" pipe | " | 0.286 |
| 4" pipe | " | 0.333 |
| 6" pipe | " | 0.667 |
| 90 degree ell, 150 lb malleable iron, galvanized | | |
| 1/2" | EA. | 0.160 |
| 3/4" | " | 0.200 |
| 1" | " | 0.211 |
| 1-1/4" | " | 0.235 |
| 1-1/2" | " | 0.267 |
| 2" | " | 0.320 |
| 2-1/2" | " | 0.500 |
| 3" | " | 0.615 |
| 4" | " | 0.667 |
| 5" | " | 0.800 |
| 6" | " | 0.800 |
| 45 degree ell, 150 lb m.i., galv. | | |

| PLUMBING | UNIT | MAN/ HOURS |
|---|---|---|
| **15410.82 GALVANIZED STEEL PIPE** | | |
| 1/2" | EA. | 0.160 |
| 3/4" | " | 0.200 |
| 1" | " | 0.211 |
| 1-1/4" | " | 0.235 |
| 1-1/2" | " | 0.267 |
| 2" | " | 0.320 |
| 2-1/2" | " | 0.500 |
| 3" | " | 0.615 |
| 4" | " | 0.800 |
| 5" | " | 0.800 |
| 6" | " | 1.000 |
| Tees, straight, 150 lb m.i., galv. | | |
| 1/2" | EA. | 0.200 |
| 3/4" | " | 0.229 |
| 1" | " | 0.267 |
| 1-1/4" | " | 0.320 |
| 1-1/2" | " | 0.400 |
| 2" | " | 0.500 |
| 2-1/2" | " | 0.667 |
| 3" | " | 0.800 |
| 4" | " | 1.000 |
| 5" | " | 1.143 |
| 6" | " | 1.333 |
| Couplings, straight, 150 lb m.i., galv. | | |
| 1/2" | EA. | 0.160 |
| 3/4" | " | 0.178 |
| 1" | " | 0.200 |
| 1-1/4" | " | 0.229 |
| 1-1/2" | " | 0.267 |
| 2" | " | 0.320 |
| 2-1/2" | " | 0.500 |
| 3" | " | 0.667 |
| 4" | " | 0.727 |
| 5" | " | 0.800 |
| 6" | " | 0.800 |
| Caps, 150 lb m.i., galv. | | |
| 1/2" | EA. | 0.080 |
| 3/4" | " | 0.084 |
| 1" | " | 0.089 |
| 1-1/4" | " | 0.094 |
| 1-1/2" | " | 0.100 |
| 2" | " | 0.114 |
| 2-1/2" | " | 0.145 |
| 3" | " | 0.200 |
| 4" | " | 0.250 |
| 5" | " | 0.308 |
| 6" | " | 0.400 |
| Unions, 150 lb m.i., galv. | | |
| 1/2" | EA. | 0.200 |
| 3/4" | " | 0.229 |
| 1" | " | 0.267 |
| 1-1/4" | " | 0.320 |
| 1-1/2" | " | 0.400 |
| 2" | " | 0.444 |

| PLUMBING | UNIT | MAN/ HOURS |
|---|---|---|
| **15410.82 GALVANIZED STEEL PIPE** | | |
| 2-1/2" | EA. | 0.533 |
| 3" | " | 0.667 |
| Nipples, galvanized steel, 4" long | | |
| 1/2" | EA. | 0.100 |
| 3/4" | " | 0.107 |
| 1" | " | 0.114 |
| 1-1/4" | " | 0.123 |
| 1-1/2" | " | 0.133 |
| 2" | " | 0.145 |
| 2-1/2" | " | 0.160 |
| 3" | " | 0.200 |
| 4" | " | 0.267 |
| Square head plug (C.I.) | | |
| 1/2" | EA. | 0.089 |
| 3/4" | " | 0.100 |
| 1" | " | 0.107 |
| 1-1/4" | " | 0.114 |
| 1-1/2" | " | 0.123 |
| 2" | " | 0.133 |
| 2-1/2" | " | 0.178 |
| 3" | " | 0.200 |
| 4" | " | 0.267 |
| 5" | " | 0.320 |
| 6" | " | 0.400 |
| Screwed flanges, galv. | | |
| 1" | EA. | 0.400 |
| 1-1/4" | " | 0.444 |
| 1-1/2" | " | 0.500 |
| 2" | " | 0.500 |
| 2-1/2" | " | 0.533 |
| 3" | " | 0.727 |
| 4" | " | 1.000 |
| 5" | " | 1.333 |
| 6" | " | 1.333 |
| **15430.23 CLEANOUTS** | | |
| Cleanout, wall | | |
| 2" | EA. | 0.533 |
| 3" | " | 0.533 |
| 4" | " | 0.667 |
| 6" | " | 0.800 |
| 8" | " | 1.000 |
| Floor | | |
| 2" | EA. | 0.667 |
| 3" | " | 0.667 |
| 4" | " | 0.800 |
| 6" | " | 1.000 |
| 8" | " | 1.143 |
| **15430.24 GREASE TRAPS** | | |
| Grease traps, cast iron, 3" pipe | | |
| 35 gpm, 70 lb capacity | EA. | 8.000 |
| 50 gpm, 100 lb capacity | " | 10.000 |

| PLUMBING | UNIT | MAN/ HOURS |
|---|---|---|
| **15430.25 HOSE BIBBS** | | |
| Hose bibb | | |
| 1/2" | EA. | 0.267 |
| 3/4" | " | 0.267 |
| **15430.60 VALVES** | | |
| Gate valve, 125 lb, bronze, soldered | | |
| 1/2" | EA. | 0.200 |
| 3/4" | " | 0.200 |
| 1" | " | 0.267 |
| 1-1/2" | " | 0.320 |
| 2" | " | 0.400 |
| 2-1/2" | " | 0.500 |
| Threaded | | |
| 1/4", 125 lb | EA. | 0.320 |
| 1/2" | | |
| 125 lb | EA. | 0.320 |
| 150 lb | " | 0.320 |
| 300 lb | " | 0.320 |
| 3/4" | | |
| 125 lb | EA. | 0.320 |
| 150 lb | " | 0.320 |
| 300 lb | " | 0.320 |
| 1" | | |
| 125 lb | EA. | 0.320 |
| 150 lb | " | 0.320 |
| 300 lb | " | 0.400 |
| 1-1/2" | | |
| 125 lb | EA. | 0.400 |
| 150 lb | " | 0.400 |
| 300 lb | " | 0.444 |
| 2" | | |
| 125 lb | EA. | 0.571 |
| 150 lb | " | 0.571 |
| 300 lb | " | 0.667 |
| Cast iron, flanged | | |
| 2", 150 lb | EA. | 0.667 |
| 2-1/2" | | |
| 125 lb | EA. | 0.667 |
| 150 lb | " | 0.667 |
| 250 lb | " | 0.667 |
| 3" | | |
| 125 lb | EA. | 0.800 |
| 150 lb | " | 0.800 |
| 250 lb | " | 0.800 |
| 4" | | |
| 125 lb | EA. | 1.143 |
| 150 lb | " | 1.143 |
| 250 lb | " | 1.143 |
| 6" | | |
| 125 lb | EA. | 1.600 |
| 250 lb | " | 1.600 |
| 8" | | |
| 125 lb | EA. | 2.000 |

| PLUMBING | UNIT | MAN/ HOURS |
|---|---|---|
| **15430.60 VALVES** | | |
| 250 lb | EA. | 2.000 |
| OS&Y, flanged | | |
| 2" | | |
| 125 lb | EA. | 0.667 |
| 250 lb | " | 0.667 |
| 2-1/2" | | |
| 125 lb | EA. | 0.667 |
| 250 lb | " | 0.800 |
| 3" | | |
| 125 lb | EA. | 0.800 |
| 250 lb | " | 0.800 |
| 4" | | |
| 125 lb | EA. | 1.333 |
| 250 lb | " | 1.333 |
| 6" | | |
| 125 lb | EA. | 1.600 |
| 250 lb | " | 1.600 |
| Check valve, bronze, soldered, 125 lb | | |
| 1/2" | EA. | 0.200 |
| 3/4" | " | 0.200 |
| 1" | " | 0.267 |
| 1-1/4" | " | 0.320 |
| 1-1/2" | " | 0.320 |
| 2" | " | 0.400 |
| Threaded | | |
| 1/2" | | |
| 125 lb | EA. | 0.267 |
| 150 lb | " | 0.267 |
| 200 lb | " | 0.267 |
| 3/4" | | |
| 125 lb | EA. | 0.320 |
| 150 lb | " | 0.320 |
| 200 lb | " | 0.320 |
| 1" | | |
| 125 lb | EA. | 0.400 |
| 150 lb | " | 0.400 |
| 200 lb | " | 0.400 |
| Flow check valve, cast iron, threaded | | |
| 1" | EA. | 0.320 |
| 1-1/4" | " | 0.400 |
| 1-1/2" | | |
| 125 lb | EA. | 0.400 |
| 150 lb | " | 0.400 |
| 200 lb | " | 0.444 |
| 2" | | |
| 125 lb | EA. | 0.444 |
| 150 lb | " | 0.444 |
| 200 lb | " | 0.500 |
| 2-1/2" | | |
| 125 lb | EA. | 0.667 |
| 250 lb | " | 0.800 |
| 3" | | |
| 125 lb | EA. | 0.800 |
| 250 lb | " | 1.000 |

| PLUMBING | UNIT | MAN/ HOURS |
|---|---|---|
| **15430.60 VALVES** | | |
| 4" | | |
| 125 lb | EA. | 1.143 |
| 250 lb | " | 1.333 |
| 6" | | |
| 125 lb | EA. | 1.600 |
| 250 lb | " | 1.600 |
| Vertical check valve, bronze, 125 lb, threaded | | |
| 1/2" | EA. | 0.320 |
| 3/4" | " | 0.364 |
| 1" | " | 0.400 |
| 1-1/4" | " | 0.444 |
| 1-1/2" | " | 0.500 |
| 2" | " | 0.571 |
| Cast iron, flanged | | |
| 2-1/2" | EA. | 0.800 |
| 3" | " | 1.000 |
| 4" | " | 1.333 |
| 6 | " | 1.600 |
| 8" | " | 2.000 |
| 10" | " | 2.667 |
| 12" | " | 3.200 |
| Globe valve, bronze, soldered, 125 lb | | |
| 1/2" | EA. | 0.229 |
| 3/4" | " | 0.250 |
| 1" | " | 0.267 |
| 1-1/4" | " | 0.286 |
| 1-1/2" | " | 0.333 |
| 2" | " | 0.400 |
| Threaded | | |
| 1/2" | | |
| 125 lb | EA. | 0.267 |
| 150 lb | " | 0.267 |
| 300 lb | " | 0.267 |
| 3/4" | | |
| 125 lb | EA. | 0.320 |
| 150 lb | " | 0.320 |
| 300 lb | " | 0.320 |
| 1" | | |
| 125 lb | EA. | 0.400 |
| 150 lb | " | 0.400 |
| 300 lb | " | 0.400 |
| 1-1/4" | | |
| 125 lb | EA. | 0.400 |
| 150 lb | " | 0.400 |
| 300 lb | " | 0.400 |
| 1-1/2" | | |
| 125 lb | EA. | 0.444 |
| 150 lb | " | 0.444 |
| 300 lb | " | 0.444 |
| 2" | | |
| 125 lb | EA. | 0.533 |
| 150 lb | " | 0.533 |
| 300 lb | " | 0.533 |
| Cast iron flanged | | |

| PLUMBING | UNIT | MAN/ HOURS |
|---|---|---|
| **15430.60 VALVES** | | |
| 2-1/2" | | |
| 125 lb | EA. | 0.800 |
| 250 lb | " | 0.800 |
| 3" | | |
| 125 lb | EA. | 1.000 |
| 250 lb | " | 1.000 |
| 4" | | |
| 125 lb | EA. | 1.333 |
| 250 lb | " | 1.333 |
| 6" | | |
| 125 lb | EA. | 1.600 |
| 250 lb | " | 1.600 |
| 8" | | |
| 125 lb | EA. | 2.000 |
| 250 lb | " | 2.000 |
| Butterfly valve, cast iron, wafer type | | |
| 2" | | |
| 150 lb | EA. | 0.571 |
| 200 lb | " | 0.667 |
| 2-1/2" | | |
| 150 lb | EA. | 0.667 |
| 200 lb | " | 0.727 |
| 3" | | |
| 150 lb | EA. | 0.800 |
| 200 lb | " | 0.889 |
| 4" | | |
| 150 lb | EA. | 1.143 |
| 200 lb | " | 1.333 |
| 6" | | |
| 150 lb | EA. | 1.600 |
| 200 lb | " | 1.600 |
| 8" | | |
| 150 lb | EA. | 1.778 |
| 200 lb | " | 2.000 |
| 10" | | |
| 150 lb | EA. | 2.000 |
| 200 lb | " | 2.667 |
| Ball valve, bronze, 250 lb, threaded | | |
| 1/2" | EA. | 0.320 |
| 3/4" | " | 0.320 |
| 1" | " | 0.400 |
| 1-1/4" | " | 0.444 |
| 1-1/2" | " | 0.500 |
| 2" | " | 0.571 |
| Angle valve, bronze, 150 lb, threaded | | |
| 1/2" | EA. | 0.286 |
| 3/4" | " | 0.320 |
| 1" | " | 0.320 |
| 1-1/4" | " | 0.400 |
| 1-1/2" | " | 0.444 |
| Balancing valve, with meter connections, circuit setter | | |
| 1/2" | EA. | 0.320 |
| 3/4" | " | 0.364 |
| 1" | " | 0.400 |

| PLUMBING | UNIT | MAN/ HOURS |
|---|---|---|
| **15430.60 VALVES** | | |
| 1-1/4" | EA. | 0.444 |
| 1-1/2" | " | 0.533 |
| 2" | " | 0.667 |
| 2-1/2" | " | 0.800 |
| 3" | " | 1.000 |
| 4" | " | 1.333 |
| Balancing valve, straight type | | |
| 1/2" | EA. | 0.320 |
| 3/4" | " | 0.320 |
| Angle type | | |
| 1/2" | EA. | 0.320 |
| 3/4" | " | 0.320 |
| Square head cock, 125 lb, bronze body | | |
| 1/2" | EA. | 0.267 |
| 3/4" | " | 0.320 |
| 1" | " | 0.364 |
| 1-1/4" | " | 0.400 |
| Radiator temp control valve, with control and sensor | | |
| 1/2" valve | EA. | 0.500 |
| 1" valve | " | 0.500 |
| Pressure relief valve, 1/2", bronze | | |
| Low pressure | EA. | 0.320 |
| High pressure | " | 0.320 |
| Press and temperature relief valve | | |
| Bronze, 3/4" | EA. | 0.320 |
| Cast iron, 3/4" | | |
| High pressure | EA. | 0.320 |
| Temperature relief | " | 0.320 |
| Pressure & temp relief valve | " | 0.320 |
| Pressure reducing valve, bronze, threaded, 250 lb | | |
| 1/2" | EA. | 0.500 |
| 3/4" | " | 0.500 |
| 1" | " | 0.500 |
| 1-1/4" | " | 0.571 |
| 1-1/2" | " | 0.667 |
| Pressure regulating valve, bronze, class 300 | | |
| 1" | EA. | 0.500 |
| 1-1/2" | " | 0.615 |
| 2" | " | 0.800 |
| 3" | " | 1.143 |
| 4" | " | 1.600 |
| 5" | " | 2.000 |
| 6" | " | 2.667 |
| Solar water temperature regulating valve | | |
| 3/4" | EA. | 0.667 |
| 1" | " | 0.800 |
| 1-1/4" | " | 0.889 |
| 1-1/2" | " | 1.000 |
| 2" | " | 1.143 |
| 2-1/2" | " | 2.000 |
| Tempering valve, threaded | | |
| 3/4" | EA. | 0.267 |
| 1" | " | 0.320 |
| 1-1/4" | " | 0.400 |

| PLUMBING | UNIT | MAN/ HOURS |
|---|---|---|
| **15430.60 VALVES** | | |
| 1-1/2" | EA. | 0.400 |
| 2" | " | 0.500 |
| 2-1/2" | " | 0.667 |
| 3" | " | 0.800 |
| 4" | " | 1.143 |
| Thermostatic mixing valve, threaded | | |
| 1/2" | EA. | 0.286 |
| 3/4" | " | 0.320 |
| 1" | " | 0.348 |
| 1-1/2" | " | 0.400 |
| 2" | " | 0.500 |
| Sweat connection | | |
| 1/2" | EA. | 0.286 |
| 3/4" | " | 0.320 |
| Mixing valve, sweat connection | | |
| 1/2" | EA. | 0.286 |
| 3/4" | " | 0.320 |
| Liquid level gauge, aluminum body | | |
| 3/4" | EA. | 0.320 |
| 4125 psi, pvc body | | |
| 3/4" | EA. | 0.320 |
| 150 psi, crs body | | |
| 3/4" | EA. | 0.320 |
| 1" | " | 0.320 |
| 175 psi, bronze body, 1/2" | " | 0.286 |
| **15430.65 VACUUM BREAKERS** | | |
| Vacuum breaker, atmospheric, threaded connection | | |
| 3/4" | EA. | 0.320 |
| 1" | " | 0.320 |
| Anti-siphon, brass | | |
| 3/4" | EA. | 0.320 |
| 1" | " | 0.320 |
| 1-1/4" | " | 0.400 |
| 1-1/2" | " | 0.444 |
| 2" | " | 0.500 |
| **15430.68 STRAINERS** | | |
| Strainer, Y pattern, 125 psi, cast iron body, threaded | | |
| 3/4" | EA. | 0.286 |
| 1" | " | 0.320 |
| 1-1/4" | " | 0.400 |
| 1-1/2" | " | 0.400 |
| 2" | " | 0.500 |
| 250 psi, brass body, threaded | | |
| 3/4" | EA. | 0.320 |
| 1" | " | 0.320 |
| 1-1/4" | " | 0.400 |
| 1-1/2" | " | 0.400 |
| 2" | " | 0.500 |
| Cast iron body, threaded | | |
| 3/4" | EA. | 0.320 |

| PLUMBING | UNIT | MAN/ HOURS |
|---|---|---|
| **15430.68 STRAINERS** | | |
| 1" | EA. | 0.320 |
| 1-1/4" | " | 0.400 |
| 1-1/2" | " | 0.400 |
| 2" | " | 0.500 |
| **15430.70 DRAINS, ROOF & FLOOR** | | |
| Floor drain, cast iron, with cast iron top | | |
| 2" | EA. | 0.667 |
| 3" | " | 0.667 |
| 4" | " | 0.667 |
| 6" | " | 0.800 |
| Roof drain, cast iron | | |
| 2" | EA. | 0.667 |
| 3" | " | 0.667 |
| 4" | " | 0.667 |
| 5" | " | 0.800 |
| 6" | " | 0.800 |
| **15430.80 TRAPS** | | |
| Bucket trap, threaded | | |
| 3/4" | EA. | 0.500 |
| 1" | " | 0.533 |
| 1-1/4" | " | 0.615 |
| 1-1/2" | " | 0.727 |
| Inverted bucket steam trap, threaded | | |
| 3/4" | EA. | 0.500 |
| 1" | " | 0.500 |
| 1-1/4" | " | 0.444 |
| 1-1/2" | " | 0.667 |
| Float trap, 15 psi | | |
| 3/4" | EA. | 0.500 |
| 1" | " | 0.533 |
| 1-1/4" | " | 0.571 |
| 1-1/2" | " | 0.667 |
| 2" | " | 0.800 |
| Float and thermostatic trap, 15 psi | | |
| 3/4" | EA. | 0.500 |
| 1" | " | 0.533 |
| 1-1/4" | " | 0.571 |
| 1-1/2" | " | 0.667 |
| 2" | " | 0.800 |
| Steam trap, cast iron body, threaded, 125 psi | | |
| 3/4" | EA. | 0.500 |
| 1" | " | 0.533 |
| 1-1/4" | " | 0.571 |
| 1-1/2" | " | 0.667 |
| Thermostatic trap, low pressure, angle type, 25 psi | | |
| 1/2" | EA. | 0.500 |
| 3/4" | " | 0.500 |
| 1" | " | 0.533 |
| 50 psi | | |
| 1/2" | EA. | 0.500 |
| 3/4" | " | 0.500 |

# 15 MECHANICAL

## PLUMBING

| 15430.80 TRAPS | UNIT | MAN/HOURS |
|---|---|---|
| 1" | EA. | 0.533 |
| Cast iron body, threaded, 125 psi | | |
| 3/4" | EA. | 0.500 |
| 1" | " | 0.571 |
| 1-1/4" | " | 0.615 |
| 1-1/2" | " | 0.667 |

## PLUMBING FIXTURES

| 15440.10 BATHS | UNIT | MAN/HOURS |
|---|---|---|
| Bath tub, 5' long | | |
| Minimum | EA. | 2.667 |
| Average | " | 4.000 |
| Maximum | " | 8.000 |
| 6' long | | |
| Minimum | EA. | 2.667 |
| Average | " | 4.000 |
| Maximum | " | 8.000 |
| Square tub, whirlpool, 4'x4' | | |
| Minimum | EA. | 4.000 |
| Average | " | 8.000 |
| Maximum | " | 10.000 |
| 5'x5' | | |
| Minimum | EA. | 4.000 |
| Average | " | 8.000 |
| Maximum | " | 10.000 |
| 6'x6' | | |
| Minimum | EA. | 4.000 |
| Average | " | 8.000 |
| Maximum | " | 10.000 |
| For trim and rough-in | | |
| Minimum | EA. | 2.667 |
| Average | " | 4.000 |
| Maximum | " | 8.000 |

| 15440.15 FAUCETS | UNIT | MAN/HOURS |
|---|---|---|
| Kitchen | | |
| Minimum | EA. | 1.333 |
| Average | " | 1.600 |
| Maximum | " | 2.000 |
| Bath | | |
| Minimum | EA. | 1.333 |
| Average | " | 1.600 |
| Maximum | " | 2.000 |
| Lavatory, domestic | | |
| Minimum | EA. | 1.333 |
| Average | " | 1.600 |

## PLUMBING FIXTURES

| 15440.15 FAUCETS | UNIT | MAN/HOURS |
|---|---|---|
| Maximum | EA. | 2.000 |
| Hospital, patient rooms | | |
| Minimum | EA. | 2.000 |
| Average | " | 2.667 |
| Maximum | " | 4.000 |
| Operating room | | |
| Minimum | EA. | 2.000 |
| Average | " | 2.667 |
| Maximum | " | 4.000 |
| Washroom | | |
| Minimum | EA. | 1.333 |
| Average | " | 1.600 |
| Maximum | " | 2.000 |
| Handicapped | | |
| Minimum | EA. | 1.600 |
| Average | " | 2.000 |
| Maximum | " | 2.667 |
| Shower | | |
| Minimum | EA. | 1.333 |
| Average | " | 1.600 |
| Maximum | " | 2.000 |
| For trim and rough-in | | |
| Minimum | EA. | 1.600 |
| Average | " | 2.000 |
| Maximum | " | 4.000 |

| 15440.18 HYDRANTS | UNIT | MAN/HOURS |
|---|---|---|
| Wall hydrant | | |
| 8" thick | EA. | 1.333 |
| 12" thick | " | 1.600 |
| 18" thick | " | 1.778 |
| 24" thick | " | 2.000 |
| Ground hydrant | | |
| 2' deep | EA. | 1.000 |
| 4' deep | " | 1.143 |
| 6' deep | " | 1.333 |
| 8' deep | " | 2.000 |

| 15440.20 LAVATORIES | UNIT | MAN/HOURS |
|---|---|---|
| Lavatory, counter top, porcelain enamel on cast iron | | |
| Minimum | EA. | 1.600 |
| Average | " | 2.000 |
| Maximum | " | 2.667 |
| Wall hung, china | | |
| Minimum | EA. | 1.600 |
| Average | " | 2.000 |
| Maximum | " | 2.667 |
| Handicapped | | |
| Minimum | EA. | 2.000 |
| Average | " | 2.667 |
| Maximum | " | 4.000 |
| For trim and rough-in | | |
| Minimum | EA. | 2.000 |

# 15 MECHANICAL

| PLUMBING FIXTURES | UNIT | MAN/ HOURS |
|---|---|---|
| **15440.20 LAVATORIES** | | |
| Average | EA. | 2.667 |
| Maximum | " | 4.000 |
| **15440.30 SHOWERS** | | |
| Shower, fiberglass, 36"x34"x84" | | |
| Minimum | EA. | 5.714 |
| Average | " | 8.000 |
| Maximum | " | 8.000 |
| Steel, 1 piece, 36"x36" | | |
| Minimum | EA. | 5.714 |
| Average | " | 8.000 |
| Maximum | " | 8.000 |
| Receptor, molded stone, 36"x36" | | |
| Minimum | EA. | 2.667 |
| Average | " | 4.000 |
| Maximum | " | 6.667 |
| For trim and rough-in | | |
| Minimum | EA. | 3.636 |
| Average | " | 4.444 |
| Maximum | " | 8.000 |
| **15440.40 SINKS** | | |
| Service sink, 24"x29" | | |
| Minimum | EA. | 2.000 |
| Average | " | 2.667 |
| Maximum | " | 4.000 |
| Kitchen sink, single, stainless steel, single bowl | | |
| Minimum | EA. | 1.600 |
| Average | " | 2.000 |
| Maximum | " | 2.667 |
| Double bowl | | |
| Minimum | EA. | 2.000 |
| Average | " | 2.667 |
| Maximum | " | 4.000 |
| Porcelain enamel, cast iron, single bowl | | |
| Minimum | EA. | 1.600 |
| Average | " | 2.000 |
| Maximum | " | 2.667 |
| Double bowl | | |
| Minimum | EA. | 2.000 |
| Average | " | 2.667 |
| Maximum | " | 4.000 |
| Mop sink, 24"x36"x10" | | |
| Minimum | EA. | 1.600 |
| Average | " | 2.000 |
| Maximum | " | 2.667 |
| Washing machine box | | |
| Minimum | EA. | 2.000 |
| Average | " | 2.667 |
| Maximum | " | 4.000 |
| For trim and rough-in | | |
| Minimum | EA. | 2.667 |
| Average | " | 4.000 |
| Maximum | " | 5.333 |

| PLUMBING FIXTURES | UNIT | MAN/ HOURS |
|---|---|---|
| **15440.50 URINALS** | | |
| Urinal, flush valve, floor mounted | | |
| Minimum | EA. | 2.000 |
| Average | " | 2.667 |
| Maximum | " | 4.000 |
| Wall mounted | | |
| Minimum | EA. | 2.000 |
| Average | " | 2.667 |
| Maximum | " | 4.000 |
| For trim and rough-in | | |
| Minimum | EA. | 2.000 |
| Average | " | 4.000 |
| Maximum | " | 5.333 |
| **15440.60 WATER CLOSETS** | | |
| Water closet flush tank, floor mounted | | |
| Minimum | EA. | 2.000 |
| Average | " | 2.667 |
| Maximum | " | 4.000 |
| Handicapped | | |
| Minimum | EA. | 2.667 |
| Average | " | 4.000 |
| Maximum | " | 8.000 |
| Bowl, with flush valve, floor mounted | | |
| Minimum | EA. | 2.000 |
| Average | " | 2.667 |
| Maximum | " | 4.000 |
| Wall mounted | | |
| Minimum | EA. | 2.000 |
| Average | " | 2.667 |
| Maximum | " | 4.000 |
| For trim and rough-in | | |
| Minimum | EA. | 2.000 |
| Average | " | 2.667 |
| Maximum | " | 4.000 |
| **15440.70 WATER HEATERS** | | |
| Water heater, electric | | |
| 6 gal | EA. | 1.333 |
| 10 gal | " | 1.333 |
| 15 gal | " | 1.333 |
| 20 gal | " | 1.600 |
| 30 gal | " | 1.600 |
| 40 gal | " | 1.600 |
| 52 gal | " | 2.000 |
| 66 gal | " | 2.000 |
| 80 gal | " | 2.000 |
| 100 gal | " | 2.667 |
| 120 gal | " | 2.667 |
| Oil fired | | |
| 20 gal | EA. | 4.000 |
| 50 gal | " | 5.714 |

# 15 MECHANICAL

## PLUMBING FIXTURES

| PLUMBING FIXTURES | UNIT | MAN/ HOURS |
|---|---|---|
| **15440.90 MISCELLANEOUS FIXTURES** | | |
| Electric water cooler | | |
| Floor mounted | EA. | 2.667 |
| Wall mounted | " | 2.667 |
| Wash fountain | | |
| Wall mounted | EA. | 4.000 |
| Circular, floor supported | " | 8.000 |
| Deluge shower and eye wash | " | 4.000 |
| **15440.95 FIXTURE CARRIERS** | | |
| Water fountain, wall carrier | | |
| Minimum | EA. | 0.800 |
| Average | " | 1.000 |
| Maximum | " | 1.333 |
| Lavatory, wall carrier | | |
| Minimum | EA. | 0.800 |
| Average | " | 1.000 |
| Maximum | " | 1.333 |
| Sink, industrial, wall carrier | | |
| Minimum | EA. | 0.800 |
| Average | " | 1.000 |
| Maximum | " | 1.333 |
| Toilets, water closets, wall carrier | | |
| Minimum | EA. | 0.800 |
| Average | " | 1.000 |
| Maximum | " | 1.333 |
| Floor support | | |
| Minimum | EA. | 0.667 |
| Average | " | 0.800 |
| Maximum | " | 1.000 |
| Urinals, wall carrier | | |
| Minimum | EA. | 0.800 |
| Average | " | 1.000 |
| Maximum | " | 1.333 |
| Floor support | | |
| Minimum | EA. | 0.667 |
| Average | " | 0.800 |
| Maximum | " | 1.000 |
| **15450.30 PUMPS** | | |
| In-line pump, bronze, centrifugal | | |
| 5 gpm, 20' head | EA. | 0.500 |
| 20 gpm, 40' head | " | 0.500 |
| 50 gpm | | |
| 50' head | EA. | 1.000 |
| 100' head | " | 1.000 |
| 70 gpm, 100' head | " | 1.333 |
| 100 gpm, 80' head | " | 1.333 |
| 250 gpm, 150' head | " | 2.000 |
| Cast iron, centrifugal | | |
| 50 gpm, 200' head | EA. | 1.000 |
| 100 gpm | | |
| 100' head | EA. | 1.333 |
| 200' head | " | 1.333 |

| PLUMBING FIXTURES | UNIT | MAN/ HOURS |
|---|---|---|
| **15450.30 PUMPS** | | |
| 200 gpm | | |
| 100' head | EA. | 2.000 |
| 200' head | " | 2.000 |
| Centrifugal, close coupled, c.i., single stage | | |
| 50 gpm, 100' head | EA. | 1.000 |
| 100 gpm, 100' head | " | 1.333 |
| Base mounted | | |
| 50 gpm, 100' head | EA. | 1.000 |
| 100 gpm, 50' head | " | 1.333 |
| 200 gpm, 100' head | " | 2.000 |
| 300 gpm, 175' head | " | 2.000 |
| Sump pump, bronze, 1750 rpm, 25 gpm | | |
| 20' head | EA. | 10.000 |
| 150' head | " | 13.333 |
| 50 gpm | | |
| 100' head | EA. | 10.000 |
| 100 gpm | | |
| 50' head | EA. | 10.000 |
| Condensate pump, simplex | | |
| 1000 sf EDR, 2 gpm | EA. | 6.667 |
| 2000 sf EDR, 3 gpm | " | 6.667 |
| 4000 sf EDR, 6 gpm | " | 7.273 |
| 6000 sf EDR, 9 gpm | " | 7.273 |
| Duplex, bronze | | |
| 8000 sf EDR, 12 gpm | EA. | 7.273 |
| 10,000 sf EDR, 15 gpm | " | 10.000 |
| 15,000 sf EDR, 23 gpm | " | 11.429 |
| 20,000 sf EDR, 30 gpm | " | 16.000 |
| 25,000 sf EDR, 38 gpm | " | 16.000 |
| **15450.40 STORAGE TANKS** | | |
| Hot water storage tank, cement lined | | |
| 10 gallon | EA. | 2.667 |
| 70 gallon | " | 4.000 |
| 200 gallon | " | 5.714 |
| 900 gallon | " | 10.000 |
| 1100 gallon | " | 10.000 |
| 2000 gallon | " | 10.000 |
| **15480.10 SPECIAL SYSTEMS** | | |
| Air compressor, air cooled, two stage | | |
| 5.0 cfm, 175 psi | EA. | 16.000 |
| 10 cfm, 175 psi | " | 17.778 |
| 20 cfm, 175 psi | " | 19.048 |
| 50 cfm, 125 psi | " | 21.053 |
| 80 cfm, 125 psi | " | 22.857 |
| Single stage, 125 psi | | |
| 1.0 cfm | EA. | 11.429 |
| 1.5 cfm | " | 11.429 |
| 2.0 cfm | " | 11.429 |
| Automotive compressor, hose reel, air and water, 50' hose | " | 6.667 |
| Lube equipment, 3 reel, with pumps | " | 32.000 |

## PLUMBING FIXTURES

### 15480.10 SPECIAL SYSTEMS

| | UNIT | MAN/ HOURS |
|---|---|---|
| Tire changer | | |
| Truck | EA. | 11.429 |
| Passenger car | " | 6.154 |
| Air hose reel, includes, 50' hose | " | 6.154 |
| Hose reel, 5 reel, motor oil, gear oil, lube, air & water | " | 32.000 |
| Water hose reel, 50' hose | " | 6.154 |
| Pump, air operated, for motor or gear oil, fits 55 gal drum | " | 0.800 |
| For chassis lube | " | 0.800 |
| Fuel dispensing pump, lighted dial, one product | | |
| One hose | EA. | 6.667 |
| Two hose | " | 6.667 |
| Two products, two hose | " | 6.667 |

## HEATING & VENTILATING

### 15555.10 BOILERS

| | UNIT | MAN/ HOURS |
|---|---|---|
| Cast iron, gas fired, hot water | | |
| 115 mbh | EA. | 20.000 |
| 175 mbh | " | 21.818 |
| 235 mbh | " | 24.000 |
| 940 mbh | " | 48.000 |
| 1600 mbh | " | 60.000 |
| 3000 mbh | " | 80.000 |
| 6000 mbh | " | 120 |
| Steam | | |
| 115 mbh | EA. | 20.000 |
| 175 mbh | " | 21.818 |
| 235 mbh | " | 24.000 |
| 940 mbh | " | 48.000 |
| 1600 mbh | " | 60.000 |
| 3000 mbh | " | 80.000 |
| 6000 mbh | " | 120 |
| Electric, hot water | | |
| 115 mbh | EA. | 12.000 |
| 175 mbh | " | 12.000 |
| 235 mbh | " | 12.000 |
| 940 mbh | " | 24.000 |
| 1600 mbh | " | 48.000 |
| 3000 mbh | " | 60.000 |
| 6000 mbh | " | 80.000 |
| Steam | | |
| 115 mbh | EA. | 12.000 |
| 175 mbh | " | 12.000 |
| 235 mbh | " | 12.000 |
| 940 mbh | " | 24.000 |
| 1600 mbh | " | 48.000 |
| 3000 mbh | " | 60.000 |

## HEATING & VENTILATING

### 15555.10 BOILERS

| | UNIT | MAN/ HOURS |
|---|---|---|
| 6000 mbh | EA. | 80.000 |
| Oil fired, hot water | | |
| 115 mbh | EA. | 16.000 |
| 175 mbh | " | 18.462 |
| 235 mbh | " | 21.818 |
| 940 mbh | " | 40.000 |
| 1600 mbh | " | 48.000 |
| 3000 mbh | " | 60.000 |
| 6000 mbh | " | 120 |
| Steam | | |
| 115 mbh | EA. | 16.000 |
| 175 mbh | " | 18.462 |
| 235 mbh | " | 21.818 |
| 940 mbh | " | 40.000 |
| 1600 mbh | " | 48.000 |
| 3000 mbh | " | 60.000 |
| 6000 mbh | " | 120 |

### 15610.10 FURNACES

| | UNIT | MAN/ HOURS |
|---|---|---|
| Electric, hot air | | |
| 40 mbh | EA. | 4.000 |
| 60 mbh | " | 4.211 |
| 80 mbh | " | 4.444 |
| 100 mbh | " | 4.706 |
| 125 mbh | " | 4.848 |
| 160 mbh | " | 5.000 |
| 200 mbh | " | 5.161 |
| 400 mbh | " | 5.333 |
| Gas fired hot air | | |
| 40 mbh | EA. | 4.000 |
| 60 mbh | " | 4.211 |
| 80 mbh | " | 4.444 |
| 100 mbh | " | 4.706 |
| 125 mbh | " | 4.848 |
| 160 mbh | " | 5.000 |
| 200 mbh | " | 5.161 |
| 400 mbh | " | 5.333 |
| Oil fired hot air | | |
| 40 mbh | EA. | 4.000 |
| 60 mbh | " | 4.211 |
| 80 mbh | " | 4.444 |
| 100 mbh | " | 4.706 |
| 125 mbh | " | 4.848 |
| 160 mbh | " | 5.000 |
| 200 mbh | " | 5.161 |
| 400 mbh | " | 5.333 |

# 15 MECHANICAL

## REFRIGERATION

| 15670.10 CONDENSING UNITS | UNIT | MAN/ HOURS |
|---|---|---|
| Air cooled condenser, single circuit | | |
| 3 ton | EA. | 1.333 |
| 5 ton | " | 1.333 |
| 7.5 ton | " | 3.810 |
| 20 ton | " | 4.000 |
| 25 ton | " | 4.000 |
| 30 ton | " | 4.000 |
| With low ambient dampers | | |
| 3 ton | EA. | 2.000 |
| 5 ton | " | 2.000 |
| 7.5 ton | " | 4.000 |
| 20 ton | " | 5.333 |
| 25 ton | " | 5.333 |
| 30 ton | " | 5.333 |
| Dual circuit | | |
| 10 ton | EA. | 4.000 |
| 15 ton | " | 5.714 |
| 20 ton | " | 5.714 |
| 25 ton | " | 5.714 |
| 30 ton | " | 5.714 |
| With low ambient dampers | | |
| 15 ton | EA. | 5.714 |
| 20 ton | " | 5.714 |
| 25 ton | " | 5.714 |
| 30 ton | " | 5.714 |

| 15680.10 CHILLERS | UNIT | MAN/ HOURS |
|---|---|---|
| Chiller, reciprocal | | |
| Air cooled, remote condenser, starter | | |
| 20 ton | EA. | 8.000 |
| 25 ton | " | 8.000 |
| 30 ton | " | 8.000 |
| 40 ton | " | 12.000 |
| Water cooled, with starter | | |
| 20 ton | EA. | 8.000 |
| 25 ton | " | 8.000 |
| 30 ton | " | 12.000 |
| 40 ton | " | 12.000 |
| Packaged, air cooled, with starter | | |
| 20 ton | EA. | 6.000 |
| 25 ton | " | 6.000 |
| 30 ton | " | 6.000 |
| 40 ton | " | 6.000 |

| 15710.10 COOLING TOWERS | UNIT | MAN/ HOURS |
|---|---|---|
| Cooling tower, propeller type | | |
| 100 ton | EA. | 8.000 |
| 200 ton | " | 12.000 |
| 300 ton | " | 20.000 |
| 400 ton | " | 24.000 |
| 600 ton | " | 34.286 |
| 800 ton | " | 48.000 |
| 1000 ton | " | 60.000 |

## REFRIGERATION

| 15710.10 COOLING TOWERS | UNIT | MAN/ HOURS |
|---|---|---|
| Centrifugal | | |
| 100 ton | EA. | 8.000 |
| 200 ton | " | 12.000 |
| 300 ton | " | 20.000 |
| 400 ton | " | 24.000 |
| 600 ton | " | 34.286 |
| 800 ton | " | 48.000 |
| 1000 ton | " | 60.000 |

## HEAT TRANSFER

| 15780.10 COMPUTER ROOM A/C | UNIT | MAN/ HOURS |
|---|---|---|
| Air cooled, alarm, high efficiency filter, elec. heat | | |
| 3 ton | EA. | 6.154 |
| 5 ton | " | 6.667 |
| 7.5 ton | " | 8.000 |
| 10 ton | " | 10.000 |
| 15 ton | " | 11.429 |
| Steam heat | | |
| 3 ton | EA. | 6.154 |
| 5 ton | " | 6.667 |
| 7.5 ton | " | 8.000 |
| 10 ton | " | 10.000 |
| 15 ton | " | 11.429 |
| Hot water heat | | |
| 3 ton | EA. | 6.154 |
| 5 ton | " | 6.667 |
| 7.5 ton | " | 8.000 |
| 10 ton | " | 10.000 |
| 15 ton | " | 11.429 |
| Air cooled condenser, low ambient damper | | |
| 3 ton | EA. | 1.600 |
| 5 ton | " | 2.000 |
| 7.5 ton | " | 4.000 |
| 10 ton | " | 5.714 |
| 15 ton | " | 4.706 |
| Water cooled, high efficiency filter, alarm, elec. heat | | |
| 3 ton | EA. | 5.714 |
| 5 ton | " | 6.667 |
| 7.5 ton | " | 10.000 |
| 10 ton | " | 11.429 |
| 15 ton | " | 13.333 |
| Steam heat | | |
| 3 ton | EA. | 5.714 |
| 5 ton | " | 6.667 |
| 7.5 ton | " | 10.000 |
| 10 ton | " | 11.429 |

# 15 MECHANICAL

| HEAT TRANSFER | UNIT | MAN/ HOURS |
|---|---|---|
| **15780.10 COMPUTER ROOM A/C** | | |
| 15 ton | EA. | 13.333 |
| Hot water heat | | |
| 3 ton | EA. | 5.714 |
| 5 ton | " | 6.667 |
| 7.5 ton | " | 10.000 |
| 10 ton | " | 11.429 |
| 15 ton | " | 13.333 |
| Chilled water, alarm, high eff. filter, elec. heat | | |
| 7.5 ton | EA. | 7.273 |
| 10 ton | " | 8.889 |
| 15 ton | " | 10.000 |
| Steam heat | | |
| 7.5 ton | EA. | 7.273 |
| 10 ton | " | 8.889 |
| 15 ton | " | 10.000 |
| Hot water heat | | |
| 7.5 ton | EA. | 7.273 |
| 10 ton | " | 8.889 |
| 15 ton | " | 10.000 |
| **15780.20 ROOFTOP UNITS** | | |
| Packaged, single zone rooftop unit, with roof curb | | |
| 2 ton | EA. | 8.000 |
| 3 ton | " | 8.000 |
| 4 ton | " | 10.000 |
| 5 ton | " | 13.333 |
| 7.5 ton | " | 16.000 |
| **15830.10 RADIATION UNITS** | | |
| Baseboard radiation unit | | |
| 1.7 mbh/lf | L.F. | 0.320 |
| 2.1 mbh/lf | " | 0.400 |
| Enclosure only | | |
| Two tier | L.F. | 0.133 |
| Three tier | " | 0.133 |
| Copper element only, 3/4" dia. | | |
| Two tier | L.F. | 0.200 |
| Three tier | " | 0.267 |
| Fin-tube, 16 ga, sloping cover, 1-1/4" steel | | |
| One tier | L.F. | 0.267 |
| Two tier | " | 0.320 |
| 2" steel | | |
| Two tier | L.F. | 0.320 |
| Three tier | " | 0.400 |
| 1-1/4" copper | | |
| Two tier | L.F. | 0.267 |
| 18 ga flat cover, 1-1/4" steel | | |
| One tier | L.F. | 0.267 |
| Two tier | " | 0.320 |
| Three tier | " | 0.400 |
| 2" steel | | |
| One tier | L.F. | 0.267 |
| Two tier | " | 0.320 |

| HEAT TRANSFER | UNIT | MAN/ HOURS |
|---|---|---|
| **15830.10 RADIATION UNITS** | | |
| Three tier | L.F. | 0.400 |
| 1-1/4" copper | | |
| One tier | L.F. | 0.267 |
| Two tier | " | 0.320 |
| Three tier | " | 0.400 |
| **15830.20 FAN COIL UNITS** | | |
| Fan coil unit, 2 pipe, complete | | |
| 200 cfm ceiling hung | EA. | 2.667 |
| Floor mounted | " | 2.000 |
| 300 cfm, ceiling hung | " | 3.200 |
| Floor mounted | " | 2.667 |
| 400 cfm, ceiling hung | " | 3.810 |
| Floor mounted | " | 2.667 |
| 500 cfm, ceiling hung | " | 4.000 |
| Floor mounted | " | 3.077 |
| 600 cfm, ceiling hung | " | 4.420 |
| Floor mounted | " | 3.636 |
| 800 cfm, ceiling hung | " | 5.000 |
| Floor mounted | " | 3.810 |
| 1000 cfm, ceiling hung | " | 5.714 |
| Floor mounted | " | 4.211 |
| 1200 cfm ceiling hung | " | 6.667 |
| Floor mounted | " | 5.000 |
| **15830.70 UNIT HEATERS** | | |
| Steam unit heater, horizontal | | |
| 12,500 btuh, 200 cfm | EA. | 1.333 |
| 17,000 btuh, 300 cfm | " | 1.333 |
| 40,000 btuh, 500 cfm | " | 1.333 |
| 60,000 btuh, 700 cfm | " | 1.333 |
| 70,000 btuh, 1000 cfm | " | 2.000 |
| Vertical | | |
| 12,500 btuh, 200 cfm | EA. | 1.333 |
| 17,000 btuh, 300 cfm | " | 1.333 |
| 40,000 btuh, 500 cfm | " | 1.333 |
| 60,000 btuh, 700 cfm | " | 1.333 |
| 70,000 btuh, 1000 cfm | " | 1.333 |
| Gas unit heater, horizontal | | |
| 27,400 btuh | EA. | 3.200 |
| 38,000 btuh | " | 3.200 |
| 56,000 btuh | " | 3.200 |
| 82,200 btuh | " | 3.200 |
| 103,900 btuh | " | 5.000 |
| 125,700 btuh | " | 5.000 |
| 133,200 btuh | " | 5.000 |
| 149,000 btuh | " | 5.000 |
| 172,000 btuh | " | 5.000 |
| 190,000 btuh | " | 5.000 |
| 225,000 btuh | " | 5.000 |
| Hot water unit heater, horizontal | | |
| 12,500 btuh, 200 cfm | EA. | 1.333 |

# 15 MECHANICAL

| HEAT TRANSFER | UNIT | MAN/ HOURS |
|---|---|---|
| **15830.70 UNIT HEATERS** | | |
| 17,000 btuh, 300 cfm | EA. | 1.333 |
| 25,000 btuh, 500 cfm | " | 1.333 |
| 30,000 btuh, 700 cfm | " | 1.333 |
| 50,000 btuh, 1000 cfm | " | 2.000 |
| 60,000 btuh, 1300 cfm | " | 2.000 |
| Vertical | | |
| 12,500 btuh, 200 cfm | EA. | 1.333 |
| 17,000 btuh, 300 cfm | " | 1.333 |
| 25,000 btuh, 500 cfm | " | 1.333 |
| 30,000 btuh, 700 cfm | " | 1.333 |
| 50,000 btuh, 1000 cfm | " | 1.333 |
| 60,000 btuh, 1300 cfm | " | 1.333 |
| Cabinet unit heaters, ceiling, exposed, hot water | | |
| 200 cfm | EA. | 2.667 |
| 300 cfm | " | 3.200 |
| 400 cfm | " | 3.810 |
| 600 cfm | " | 4.211 |
| 800 cfm | " | 5.000 |
| 1000 cfm | " | 5.714 |
| 1200 cfm | " | 6.667 |
| 2000 cfm | " | 8.889 |

| AIR HANDLING | UNIT | MAN/ HOURS |
|---|---|---|
| **15855.10 AIR HANDLING UNITS** | | |
| Air handling unit, medium pressure, single zone | | |
| 1500 cfm | EA. | 5.000 |
| 3000 cfm | " | 8.889 |
| 4000 cfm | " | 10.000 |
| 5000 cfm | " | 10.667 |
| 6000 cfm | " | 11.429 |
| 7000 cfm | " | 12.308 |
| 8500 cfm | " | 13.333 |
| 10,500 cfm | " | 16.000 |
| 12,500 cfm | " | 17.778 |
| 15,500 cfm | " | 22.857 |
| 17,500 cfm | " | 26.667 |
| 20,500 cfm | " | 32.000 |
| 25,000 cfm | " | 40.000 |
| 31,500 cfm | " | 53.333 |
| Rooftop air handling units | | |
| 4950 cfm | EA. | 8.889 |
| 7370 cfm | " | 11.429 |
| 9790 cfm | " | 13.333 |
| 14,300 cfm | " | 11.429 |
| 21,725 cfm | " | 11.429 |

| AIR HANDLING | UNIT | MAN/ HOURS |
|---|---|---|
| **15855.10 AIR HANDLING UNITS** | | |
| 33,000 cfm | EA. | 13.333 |
| **15870.20 EXHAUST FANS** | | |
| Belt drive roof exhaust fans | | |
| 640 cfm, 2618 fpm | EA. | 1.000 |
| 940 cfm, 2604 fpm | " | 1.000 |
| 1050 cfm, 3325 fpm | " | 1.000 |
| 1170 cfm, 2373 fpm | " | 1.000 |
| 2440 cfm, 4501 fpm | " | 1.000 |
| 2760 cfm, 4950 fpm | " | 1.000 |
| 3890 cfm, 6769 fpm | " | 1.000 |
| 2380 cfm, 3382 fpm | " | 1.000 |
| 2880 cfm, 3859 fpm | " | 1.000 |
| 3200 cfm, 4173 fpm | " | 1.333 |
| 3660 cfm, 3437 fpm | " | 1.333 |
| Direct drive fans | | |
| 60 to 390 cfm | EA. | 1.000 |
| 145 to 590 cfm | " | 1.000 |
| 295 to 860 cfm | " | 1.000 |
| 235 to 1300 cfm | " | 1.000 |
| 415 to 1630 cfm | " | 1.000 |
| 590 to 2045 cfm | " | 1.000 |

| AIR DISTRIBUTION | UNIT | MAN/ HOURS |
|---|---|---|
| **15890.10 METAL DUCTWORK** | | |
| Rectangular duct | | |
| Galvanized steel | | |
| Minimum | Lb. | 0.073 |
| Average | " | 0.089 |
| Maximum | " | 0.133 |
| Aluminum | | |
| Minimum | Lb. | 0.160 |
| Average | " | 0.200 |
| Maximum | " | 0.267 |
| Fittings | | |
| Minimum | EA. | 0.267 |
| Average | " | 0.400 |
| Maximum | " | 0.800 |

## AIR DISTRIBUTION

### 15890.30 FLEXIBLE DUCTWORK

| AIR DISTRIBUTION | UNIT | MAN/HOURS |
|---|---|---|
| Flexible duct, 1.25" fiberglass | | |
| 5" dia. | L.F. | 0.040 |
| 6" dia. | " | 0.044 |
| 7" dia. | " | 0.047 |
| 8" dia. | " | 0.050 |
| 10" dia. | " | 0.057 |
| 12" dia. | " | 0.062 |
| 14" dia. | " | 0.067 |
| 16" dia. | " | 0.073 |
| Flexible duct connector, 3" wide fabric | " | 0.133 |

### 15895.10 ROOF CURBS

| AIR DISTRIBUTION | UNIT | MAN/HOURS |
|---|---|---|
| 8" high, insulated, with liner and raised can | | |
| 15" x 15" | EA. | 0.400 |
| 17" x 17" | " | 0.400 |
| 19" x 19" | " | 0.400 |
| 21" x 21" | " | 0.400 |
| 25" x 25" | " | 0.500 |
| 28" x 28" | " | 0.533 |
| 32" x 32" | " | 0.571 |
| 36" x 36" | " | 0.571 |
| 40" x 40" | " | 0.571 |
| 44" x 44" | " | 0.615 |
| 48" x 48" | " | 0.615 |
| 52" x 52" | " | 0.667 |
| 56" x 56" | " | 0.667 |
| 60" x 60" | " | 0.800 |
| 64" x 64" | " | 0.800 |
| 68" x 68" | " | 0.889 |
| 72" x 72" | " | 1.000 |

### 15910.10 DAMPERS

| AIR DISTRIBUTION | UNIT | MAN/HOURS |
|---|---|---|
| Horizontal parallel aluminum backdraft damper | | |
| 12" x 12" | EA. | 0.200 |
| 16" x 16" | " | 0.229 |
| 20" x 20" | " | 0.286 |
| 24" x 24" | " | 0.400 |
| 28" x 28" | " | 0.444 |
| 32" x 32" | " | 0.500 |
| 36" x 36" | " | 0.571 |
| 40" x 40" | " | 0.667 |
| 44" x 44" | " | 0.727 |
| 48" x 48" | " | 0.800 |
| "Up", parallel dampers | | |
| 12" x 12" | EA. | 0.200 |
| 16" x 16" | " | 0.229 |
| 20" x 20" | " | 0.286 |
| 24" x 24" | " | 0.400 |
| 28" x 28" | " | 0.444 |
| 32" x 32" | " | 0.500 |
| 36" x 36" | " | 0.571 |
| 40" x 40" | " | 0.667 |
| 44" x 44" | " | 0.727 |
| 48" x 48" | " | 0.800 |

## AIR DISTRIBUTION

### 15910.10 DAMPERS

| AIR DISTRIBUTION | UNIT | MAN/HOURS |
|---|---|---|
| "Down", parallel dampers | | |
| 12" x 12" | EA. | 0.200 |
| 16" x 16" | " | 0.229 |
| 20" x 20" | " | 0.286 |
| 24" x 24" | " | 0.400 |
| 28" x 28" | " | 0.444 |
| 32" x 32" | " | 0.500 |
| 36" x 36" | " | 0.571 |
| 40" x 40" | " | 0.667 |
| 44" x 44" | " | 0.727 |
| 48" x 48" | " | 0.800 |
| Fire damper, 1.5 hr rating | | |
| 12" x 12" | EA. | 0.400 |
| 16" x 16" | " | 0.400 |
| 20" x 20" | " | 0.400 |
| 24" x 24" | " | 0.400 |
| 28" x 28" | " | 0.571 |
| 32" x 32" | " | 0.667 |
| 36" x 36" | " | 0.800 |
| 40" x 40" | " | 0.889 |
| 44" x 44" | " | 1.000 |
| 48" x 48" | " | 1.143 |

### 15940.10 DIFFUSERS

| AIR DISTRIBUTION | UNIT | MAN/HOURS |
|---|---|---|
| Ceiling diffusers, round, baked enamel finish | | |
| 6" dia. | EA. | 0.267 |
| 8" dia. | " | 0.333 |
| 10" dia. | " | 0.333 |
| 12" dia. | " | 0.333 |
| 14" dia. | " | 0.364 |
| 16" dia. | " | 0.364 |
| 18" dia. | " | 0.400 |
| 20" dia. | " | 0.400 |
| Rectangular | | |
| 6x6" | EA. | 0.267 |
| 9x9" | " | 0.400 |
| 12x12" | " | 0.400 |
| 15x15" | " | 0.400 |
| 18x18" | " | 0.400 |
| 21x21" | " | 0.500 |
| 24x24" | " | 0.500 |
| Lay in, flush mounted, perforated face, with grid | | |
| 6x6/24x24 | EA. | 0.320 |
| 8x8/24x24 | " | 0.320 |
| 9x9/24x24 | " | 0.320 |
| 10x10/24x24 | " | 0.320 |
| 12x12/24x24 | " | 0.320 |
| 15x15/24x24 | " | 0.320 |
| 18x6/24x24 | " | 0.320 |
| 18x18/24x24 | " | 0.320 |
| Two-way slot diffuser with balancing damper, 4' | " | 0.800 |

## AIR DISTRIBUTION

| AIR DISTRIBUTION | UNIT | MAN/ HOURS |
|---|---|---|
| **15940.20 RELIEF VENTILATORS** | | |
| Intake ventilator, aluminum, with screen, no curbs | | |
| 12" x 12" | EA. | 0.667 |
| 16" x 16" | " | 0.800 |
| 20" x 20" | " | 0.800 |
| 30" x 30" | " | 1.143 |
| 36" x 36" | " | 1.333 |
| 42" x 42" | " | 1.333 |
| 48" x 48" | " | 1.600 |
| **15940.40 REGISTERS AND GRILLES** | | |
| Lay in flush mounted, perforated face, return | | |
| 6x6/24x24 | EA. | 0.320 |
| 8x8/24x24 | " | 0.320 |
| 9x9/24x24 | " | 0.320 |
| 10x10/24x24 | " | 0.320 |
| 12x12/24x24 | " | 0.320 |
| Rectangular, ceiling return, single deflection | | |
| 10x10 | EA. | 0.400 |
| 12x12 | " | 0.400 |
| 14x14 | " | 0.400 |
| 16x8 | " | 0.400 |
| 16x16 | " | 0.400 |
| 18x8 | " | 0.400 |
| 20x20 | " | 0.400 |
| 24x12 | " | 0.400 |
| 24x18 | " | 0.400 |
| 36x24 | " | 0.444 |
| 36x30 | " | 0.444 |
| Wall, return air register | | |
| 12x12 | EA. | 0.200 |
| 16x16 | " | 0.200 |
| 18x18 | " | 0.200 |
| 20x20 | " | 0.200 |
| 24x24 | " | 0.200 |
| Ceiling, return air grille | | |
| 6x6 | EA. | 0.267 |
| 8x8 | " | 0.320 |
| 10x10 | " | 0.320 |
| Ceiling, exhaust grille, aluminum egg crate | | |
| 6x6 | EA. | 0.267 |
| 8x8 | " | 0.320 |
| 10x10 | " | 0.320 |
| 12x12 | " | 0.400 |
| 14x14 | " | 0.400 |
| 16x16 | " | 0.400 |
| 18x18 | " | 0.400 |
| **15940.80 PENTHOUSE LOUVERS** | | |
| Penthouse louvers | | |
| 12" high, extruded aluminum, 4" louver | | |
| 6' perimeter | EA. | 2.000 |
| 8' perimeter | " | 2.000 |
| 10' perimeter | " | 2.000 |
| 12' perimeter | " | 2.000 |
| 14' perimeter | EA. | 2.667 |
| 16' perimeter | " | 3.200 |
| 18' perimeter | " | 4.444 |
| 20' perimeter | " | 5.333 |
| 16" high x 4' perimeter | " | 2.000 |
| 6' perimeter | " | 2.000 |
| 8' perimeter | " | 2.000 |
| 10' perimeter | " | 2.000 |
| 12' perimeter | " | 2.000 |
| 14' perimeter | " | 2.667 |
| 16' perimeter | " | 3.200 |
| 18' perimeter | " | 4.444 |
| 20' perimeter | " | 5.333 |
| 22' perimeter | " | 6.667 |
| 24' perimeter | " | 8.889 |
| 20" high x 4' perimeter | " | 2.000 |
| 6' perimeter | " | 2.000 |
| 8' perimeter | " | 2.000 |
| 10' perimeter | " | 2.000 |
| 12' perimeter | " | 2.000 |
| 14' perimeter | " | 2.667 |
| 16' perimeter | " | 3.200 |
| 18' perimeter | " | 4.444 |
| 20' perimeter | " | 5.333 |
| 22' perimeter | " | 6.667 |
| 24' perimeter | " | 8.889 |
| 24" high x 4' perimeter | " | 2.000 |
| 6' perimeter | " | 2.000 |
| 8' perimeter | " | 2.000 |
| 10' perimeter | " | 2.000 |
| 12' perimeter | " | 2.000 |
| 16' perimeter | " | 3.200 |
| 18' perimeter | " | 4.444 |
| 20' perimeter | " | 5.333 |
| 22' perimeter | " | 6.667 |
| 24' perimeter | " | 8.889 |

## CONTROLS

| CONTROLS | UNIT | MAN/ HOURS |
|---|---|---|
| **15950.10 HVAC CONTROLS** | | |
| Pressure gauge, direct reading gage cock and siphon | EA. | 0.500 |
| Control valve, 1", modulating | | |
| 2-way | EA. | 0.667 |
| 3-way | " | 1.000 |
| Self contained control valve with sensing element, 3/4" | " | 0.500 |
| Control dampers, round | | |
| 6" dia. | EA. | 0.320 |

| CONTROLS | UNIT | MAN/ HOURS |
|---|---|---|
| **15950.10 HVAC CONTROLS** | | |
| 8" dia | EA. | 0.320 |
| 10" dia | " | 0.320 |
| 12" dia | " | 0.320 |
| 12" dia | " | 0.400 |
| 18" dia | " | 0.400 |
| 20" dia | " | 0.400 |
| Rectangular, parallel blade standard leakage | | |
| 12" x 12" | EA. | 0.400 |
| 16" x 16" | " | 0.400 |
| 20" x 20" | " | 0.400 |
| 28" x 28" | " | 0.500 |
| 32" x 32" | " | 0.500 |
| 36" x 36" | " | 0.667 |
| 40" x 40" | " | 0.800 |
| 44" x 44" | " | 1.000 |
| 48" x 48" | " | 1.143 |
| 48" x 52" | " | 1.333 |
| 48" x 56" | " | 1.333 |
| 48" x 60" | " | 1.333 |
| 48" x 64" | " | 1.333 |
| 48" x 68" | " | 1.333 |
| 48" x 72" | " | 1.333 |
| Low leakage | | |
| 12" x 12" | EA. | 0.400 |
| 16" x 16" | " | 0.400 |
| 20" x 20" | " | 0.400 |
| 24" x 24" | " | 0.400 |
| 28" x 28" | " | 0.500 |
| 32" x 32" | " | 0.571 |
| 36" x 36" | " | 0.667 |
| 40" x 40" | " | 0.800 |
| 44" x 44" | " | 1.000 |
| 48" x 48" | " | 1.143 |
| 48" x 56" | " | 1.333 |
| 48" x 60" | " | 1.333 |
| 48" x 64" | " | 1.333 |
| 48" x 68" | " | 1.333 |
| 48" x 72" | " | 1.333 |
| Rectangular, opposed horizontal blade | | |
| 12" x 12" | EA. | 0.400 |
| 16" x 16" | " | 0.400 |
| 20" x 20" | " | 0.400 |
| 24" x 24" | " | 0.400 |
| 28" x 28" | " | 0.500 |
| 32" x 32" | " | 0.533 |
| 36" x 36" | " | 0.667 |
| 40" x 40" | " | 0.800 |
| 44" x 44" | " | 1.000 |
| 48" x 48" | " | 1.143 |
| 48" x 52" | " | 1.143 |
| 48" x 56" | " | 1.333 |
| 48" x 60" | " | 1.333 |
| 48" x 64" | " | 1.333 |
| 48" x 68" | " | 1.333 |

| CONTROLS | UNIT | MAN/ HOURS |
|---|---|---|
| **15950.10 HVAC CONTROLS** | | |
| 48" x 72" | EA. | 1.333 |

| BASIC MATERIALS | UNIT | MAN/ HOURS |
|---|---|---|
| **16110.20 CONDUIT SPECIALTIES** | | |
| Rod beam clamp, 1/2" | EA. | 0.050 |
| Hanger rod | | |
| 3/8" | L.F. | 0.040 |
| 1/2" | " | 0.050 |
| All thread rod | | |
| 1/4" | L.F. | 0.030 |
| 3/8" | " | 0.040 |
| 1/2" | " | 0.050 |
| 5/8" | " | 0.080 |
| Hanger channel, 1-1/2" | | |
| No holes | EA. | 0.030 |
| Holes | " | 0.030 |
| Channel strap | | |
| 1/2" | EA. | 0.050 |
| 3/4" | " | 0.050 |
| 1" | " | 0.050 |
| 1-1/4" | " | 0.080 |
| 1-1/2" | " | 0.080 |
| 2" | " | 0.080 |
| 2-1/2" | " | 0.123 |
| 3" | " | 0.123 |
| 3-1/2" | " | 0.123 |
| 4" | " | 0.145 |
| 5" | " | 0.145 |
| 6" | " | 0.145 |
| Conduit penetrations, roof and wall, 8" thick | | |
| 1/2" | EA. | 0.615 |
| 3/4" | " | 0.615 |
| 1" | " | 0.800 |
| 1-1/4" | " | 0.800 |
| 1-1/2" | " | 0.800 |
| 2" | " | 1.600 |
| 2-1/2" | " | 1.600 |
| 3" | " | 1.600 |
| 3-1/2" | " | 2.000 |
| 4" | " | 2.000 |
| Fireproofing, for conduit penetrations | | |
| 1/2" | EA. | 0.500 |
| 3/4" | " | 0.500 |
| 1" | " | 0.500 |
| 1-1/4" | " | 0.727 |
| 1-1/2" | " | 0.727 |
| 2" | " | 0.727 |
| 2-1/2" | " | 0.899 |
| 3" | " | 0.899 |
| 3-1/2" | " | 1.250 |
| 4" | " | 1.509 |
| **16110.21 ALUMINUM CONDUIT** | | |
| Aluminum conduit | | |
| 1/2" | L.F. | 0.030 |
| 3/4" | " | 0.040 |
| 1" | " | 0.050 |

| BASIC MATERIALS | UNIT | MAN/ HOURS |
|---|---|---|
| **16110.21 ALUMINUM CONDUIT** | | |
| 1-1/4" | L.F. | 0.059 |
| 1-1/2" | " | 0.080 |
| 2" | " | 0.089 |
| 2-1/2" | " | 0.100 |
| 3" | " | 0.107 |
| 3-1/2" | " | 0.123 |
| 4" | " | 0.145 |
| 5" | " | 0.182 |
| 6" | " | 0.200 |
| 90 deg. elbow | | |
| 1/2" | EA. | 0.190 |
| 3/4" | " | 0.250 |
| 1" | " | 0.308 |
| 1-1/4" | " | 0.381 |
| 1-1/2" | " | 0.400 |
| 2" | " | 0.444 |
| 2-1/2" | " | 0.571 |
| 3" | " | 0.667 |
| 3-1/2" | " | 0.800 |
| 4" | " | 0.889 |
| 5" | " | 1.143 |
| 6" | " | 2.222 |
| Coupling | | |
| 1/2" | EA. | 0.050 |
| 3/4" | " | 0.059 |
| 1" | " | 0.080 |
| 1-1/4" | " | 0.089 |
| 1-1/2" | " | 0.100 |
| 2" | " | 0.107 |
| 2-1/2" | " | 0.123 |
| 3" | " | 0.123 |
| 3-1/2" | " | 0.145 |
| 4" | " | 0.160 |
| 5" | " | 0.160 |
| 6" | " | 0.190 |
| **16110.22 EMT CONDUIT** | | |
| EMT conduit | | |
| 1/2" | L.F. | 0.030 |
| 3/4" | " | 0.040 |
| 1" | " | 0.050 |
| 1-1/4" | " | 0.059 |
| 1-1/2" | " | 0.080 |
| 2" | " | 0.089 |
| 2-1/2" | " | 0.100 |
| 3" | " | 0.123 |
| 3-1/2" | " | 0.145 |
| 4" | " | 0.182 |
| 90 deg. elbow | | |
| 1/2" | EA. | 0.089 |
| 3/4" | " | 0.100 |
| 1" | " | 0.107 |
| 1-1/4" | " | 0.123 |

# 16 ELECTRICAL

| BASIC MATERIALS | UNIT | MAN/ HOURS |
|---|---|---|
| **16110.22 EMT CONDUIT** | | |
| 1-1/2" | EA. | 0.145 |
| 2" | " | 0.190 |
| 2-1/2" | " | 0.211 |
| 3" | " | 0.242 |
| 3-1/2" | " | 0.258 |
| 4" | " | 0.286 |
| Connector, die cast set screw | | |
| 1/2" | EA. | 0.059 |
| 3/4" | " | 0.059 |
| 1" | " | 0.059 |
| 1-1/4" | " | 0.089 |
| 1-1/2" | " | 0.107 |
| 2" | " | 0.145 |
| 2-1/2" | " | 0.200 |
| 3" | " | 0.222 |
| 3-1/2" | " | 0.250 |
| 4" | " | 0.286 |
| Coupling, steel set screw | | |
| 1/2" | EA. | 0.040 |
| 3/4" | " | 0.040 |
| 1" | " | 0.040 |
| 1-1/4" | " | 0.050 |
| 1-1/2" | " | 0.080 |
| 2" | " | 0.107 |
| 2-1/2" | " | 0.160 |
| 3" | " | 0.190 |
| 3-1/2" | " | 0.222 |
| 4" | " | 0.250 |
| **16110.23 FLEXIBLE CONDUIT** | | |
| Flexible conduit, steel | | |
| 3/8" | L.F. | 0.030 |
| 1/2 | " | 0.030 |
| 3/4" | " | 0.040 |
| 1" | " | 0.040 |
| 1-1/4" | " | 0.050 |
| 1-1/2" | " | 0.059 |
| 2" | " | 0.080 |
| 2-1/2" | " | 0.089 |
| 3" | " | 0.107 |
| Flexible conduit, liquid tight | | |
| 3/8" | L.F. | 0.030 |
| 1/2" | " | 0.030 |
| 3/4" | " | 0.040 |
| 1" | " | 0.040 |
| 1-1/4" | " | 0.050 |
| 1-1/2" | " | 0.059 |
| 2" | EA. | 0.080 |
| 2-1/2" | " | 0.089 |
| 3" | " | 0.107 |
| 4" | " | 0.145 |
| Connector, straight | | |
| 3/8" | EA. | 0.080 |
| 1/2" | " | 0.080 |

| BASIC MATERIALS | UNIT | MAN/ HOURS |
|---|---|---|
| **16110.23 FLEXIBLE CONDUIT** | | |
| 3/4" | EA. | 0.089 |
| 1" | " | 0.100 |
| 1-1/4" | " | 0.107 |
| 1-1/2" | " | 0.123 |
| 2" | " | 0.145 |
| 2-1/2" | " | 0.182 |
| 3" | " | 0.190 |
| Flexible aluminum conduit | | |
| 3/8" | L.F. | 0.030 |
| 1/2" | " | 0.030 |
| 3/4" | " | 0.040 |
| 1" | " | 0.040 |
| 1-1/4" | " | 0.050 |
| 1-1/2" | " | 0.059 |
| 2" | " | 0.080 |
| 2-1/2" | " | 0.089 |
| 3" | " | 0.107 |
| 3-1/2" | " | 0.123 |
| 4" | " | 0.145 |
| **16110.24 GALVANIZED CONDUIT** | | |
| Galvanized rigid steel conduit | | |
| 1/2" | L.F. | 0.040 |
| 3/4" | " | 0.050 |
| 1" | " | 0.059 |
| 1-1/4" | " | 0.080 |
| 1-1/2" | " | 0.089 |
| 2" | " | 0.100 |
| 2-1/2" | " | 0.145 |
| 3" | " | 0.182 |
| 3-1/2" | " | 0.190 |
| 4" | " | 0.211 |
| 5" | " | 0.286 |
| 6" | " | 0.381 |
| 90 degree ell | | |
| 1/2" | EA. | 0.250 |
| 3/4" | " | 0.308 |
| 1" | " | 0.381 |
| 1-1/4" | " | 0.444 |
| 1-1/2" | " | 0.500 |
| 2" | " | 0.533 |
| 2-1/2" | " | 0.667 |
| 3" | " | 0.889 |
| 3-1/2" | " | 1.000 |
| 4" | " | 1.333 |
| 5" | " | 2.222 |
| 6" | " | 3.333 |
| Couplings, with set screws | | |
| 1/2" | EA. | 0.050 |
| 3/4" | " | 0.059 |
| 1" | " | 0.080 |
| 1-1/4" | " | 0.100 |
| 1-1/2" | " | 0.123 |

| BASIC MATERIALS | UNIT | MAN/ HOURS |
|---|---|---|
| **16110.24 GALVANIZED CONDUIT** | | |
| 2" | EA. | 0.145 |
| 2-1/2" | " | 0.190 |
| 3" | " | 0.250 |
| 3-1/2" | " | 0.286 |
| 4" | " | 0.308 |
| 5" | " | 0.444 |
| 6" | " | 0.500 |
| Split couplings | | |
| 1/2" | EA. | 0.190 |
| 3/4" | " | 0.250 |
| 1" | " | 0.276 |
| 1-1/4" | " | 0.308 |
| 1-1/2" | " | 0.381 |
| 2" | " | 0.571 |
| 2-1/2" | " | 0.571 |
| 3" | " | 0.727 |
| 3-1/2" | " | 1.000 |
| 4" | " | 1.333 |
| 5" | " | 1.633 |
| 6" | " | 2.051 |
| Erickson couplings | | |
| 1/2" | EA. | 0.444 |
| 3/4" | " | 0.500 |
| 1" | " | 0.615 |
| 1-1/4" | " | 0.889 |
| 1-1/2" | " | 1.000 |
| 2" | " | 1.333 |
| 2-1/2" | " | 1.860 |
| 3" | " | 2.105 |
| 3-1/2" | " | 2.500 |
| 4" | " | 2.667 |
| 5" | " | 2.963 |
| 6" | " | 3.200 |
| Seal fittings | | |
| 1/2" | EA. | 0.667 |
| 3/4" | " | 0.800 |
| 1" | " | 1.000 |
| 1-1/4" | " | 1.143 |
| 1-1/2" | " | 1.333 |
| 2" | " | 1.600 |
| 2-1/2" | " | 1.905 |
| 3" | " | 2.105 |
| 3-1/2" | " | 2.500 |
| 4" | " | 2.963 |
| 5" | " | 4.444 |
| 6" | " | 5.000 |
| Entrance fitting, (weather head), threaded | | |
| 1/2" | EA. | 0.444 |
| 3/4" | " | 0.500 |
| 1" | " | 0.571 |
| 1-1/4" | " | 0.727 |
| 1-1/2" | " | 0.800 |
| 2" | " | 0.889 |
| 2-1/2" | " | 1.000 |

| BASIC MATERIALS | UNIT | MAN/ HOURS |
|---|---|---|
| **16110.24 GALVANIZED CONDUIT** | | |
| 3" | EA. | 1.333 |
| 3-1/2" | " | 1.739 |
| 4" | " | 2.500 |
| 5" | " | 3.478 |
| 6" | " | 4.444 |
| Locknuts | | |
| 1/2" | EA. | 0.050 |
| 3/4" | " | 0.050 |
| 1" | " | 0.050 |
| 1-1/4" | " | 0.050 |
| 1-1/2" | " | 0.059 |
| 2" | " | 0.059 |
| 2-1/2" | " | 0.080 |
| 3" | " | 0.080 |
| 3-1/2" | " | 0.080 |
| 4" | " | 0.089 |
| 5" | " | 0.089 |
| 6" | " | 0.089 |
| Plastic conduit bushings | | |
| 1/2" | EA. | 0.123 |
| 3/4" | " | 0.145 |
| 1" | " | 0.190 |
| 1-1/4" | " | 0.222 |
| 1-1/2" | " | 0.250 |
| 2" | " | 0.308 |
| 2-1/2" | " | 0.500 |
| 3" | " | 0.667 |
| 3-1/2" | " | 0.800 |
| 4" | " | 0.889 |
| 5" | " | 1.143 |
| 6" | " | 1.600 |
| Conduit bushings, steel | | |
| 1/2" | EA. | 0.123 |
| 3/4" | " | 0.145 |
| 1" | " | 0.190 |
| 1-1/4" | " | 0.222 |
| 1-1/2" | " | 0.250 |
| 2" | " | 0.308 |
| 2-1/2" | " | 0.500 |
| 3" | " | 0.667 |
| 3-1/2" | " | 0.800 |
| 4" | " | 0.889 |
| 5" | " | 1.143 |
| 6" | " | 1.600 |
| Pipe cap | | |
| 1/2" | EA. | 0.050 |
| 3/4" | " | 0.050 |
| 1" | " | 0.050 |
| 1-1/4" | " | 0.080 |
| 1-1/2" | " | 0.080 |
| 2" | " | 0.080 |
| 2-1/2" | " | 0.089 |
| 3" | " | 0.089 |
| 3-1/2" | " | 0.089 |

| BASIC MATERIALS | UNIT | MAN/ HOURS |
|---|---|---|
| **16110.24 GALVANIZED CONDUIT** | | |
| 4" | EA. | 0.107 |
| 5" | " | 0.145 |
| 6" | " | 0.200 |
| **16110.25 PLASTIC CONDUIT** | | |
| PVC conduit, schedule 40 | | |
| 1/2" | L.F. | 0.030 |
| 3/4" | " | 0.030 |
| 1" | " | 0.040 |
| 1-1/4" | " | 0.040 |
| 1-1/2" | " | 0.050 |
| 2" | " | 0.050 |
| 2-1/2" | " | 0.059 |
| 3" | " | 0.059 |
| 3-1/2" | " | 0.080 |
| 4" | " | 0.080 |
| 5" | " | 0.089 |
| 6" | " | 0.100 |
| Couplings | | |
| 1/2" | EA. | 0.050 |
| 3/4" | " | 0.050 |
| 1" | " | 0.050 |
| 1-1/4" | " | 0.059 |
| 1-1/2" | " | 0.059 |
| 2" | " | 0.059 |
| 2-1/2" | " | 0.059 |
| 3" | " | 0.080 |
| 3-1/2" | " | 0.080 |
| 4" | " | 0.100 |
| 5" | " | 0.100 |
| 6" | " | 0.100 |
| 90 degree elbows | | |
| 1/2" | EA. | 0.100 |
| 3/4" | " | 0.123 |
| 1" | " | 0.123 |
| 1-1/4" | " | 0.145 |
| 1-1/2" | " | 0.190 |
| 2" | " | 0.222 |
| 2-1/2" | " | 0.250 |
| 3" | " | 0.308 |
| 3-1/2" | " | 0.381 |
| 4" | " | 0.500 |
| 5" | " | 0.615 |
| 6" | " | 0.727 |
| Terminal adapters | | |
| 1/2" | EA. | 0.100 |
| 3/4" | " | 0.100 |
| 1" | " | 0.100 |
| 1-1/4" | " | 0.160 |
| 1-1/2" | " | 0.160 |
| 2" | " | 0.160 |
| 2-1/2" | " | 0.222 |
| 3" | " | 0.222 |

| BASIC MATERIALS | UNIT | MAN/ HOURS |
|---|---|---|
| **16110.25 PLASTIC CONDUIT** | | |
| 3-1/2" | EA. | 0.222 |
| 4" | " | 0.381 |
| 5" | " | 0.381 |
| 6" | " | 0.381 |
| End bells | | |
| 1" | EA. | 0.100 |
| 1-1/4" | " | 0.160 |
| 1-1/2" | " | 0.160 |
| 2" | " | 0.160 |
| 2-1/2" | " | 0.222 |
| 3" | " | 0.222 |
| 3-1/2" | " | 0.222 |
| 4" | " | 0.381 |
| 5" | " | 0.381 |
| 6" | " | 0.381 |
| LB conduit body | | |
| 1/2" | EA. | 0.190 |
| 3/4" | " | 0.190 |
| 1 | " | 0.190 |
| 1-1/4" | " | 0.308 |
| 1-1/2" | " | 0.308 |
| 2" | " | 0.308 |
| 2" | L.F. | 0.050 |
| 3" | " | 0.059 |
| 4" | " | 0.080 |
| 5" | " | 0.089 |
| 6" | " | 0.100 |
| Encased burial conduit | | |
| 2" | L.F. | 0.050 |
| 3" | " | 0.059 |
| 4" | " | 0.080 |
| 5" | " | 0.089 |
| 6" | " | 0.100 |
| **16110.27 PLASTIC COATED CONDUIT** | | |
| Rigid steel conduit, plastic coated | | |
| 1/2" | L.F. | 0.050 |
| 3/4" | " | 0.059 |
| 1" | " | 0.080 |
| 1-1/4" | " | 0.100 |
| 1-1/2" | " | 0.123 |
| 2" | " | 0.145 |
| 2-1/2" | " | 0.190 |
| 3" | " | 0.222 |
| 3-1/2" | " | 0.250 |
| 4" | " | 0.308 |
| 5" | " | 0.381 |
| 90 degree elbows | | |
| 1/2" | EA. | 0.308 |
| 3/4" | " | 0.381 |
| 1" | " | 0.444 |
| 1-1/4" | " | 0.500 |
| 1-1/2" | " | 0.615 |

# 16 ELECTRICAL

## BASIC MATERIALS

### 16110.27 PLASTIC COATED CONDUIT

| | UNIT | MAN/HOURS |
|---|---|---|
| 2" | EA. | 0.800 |
| 2-1/2" | " | 1.143 |
| 3" | " | 1.333 |
| 3-1/2" | " | 1.633 |
| 4" | " | 2.000 |
| 5" | " | 2.500 |
| Couplings | | |
| 1/2" | EA. | 0.059 |
| 3/4" | " | 0.080 |
| 1" | " | 0.089 |
| 1-1/4" | " | 0.107 |
| 1-1/2" | " | 0.123 |
| 2" | " | 0.145 |
| 2-1/2" | " | 0.182 |
| 3" | " | 0.190 |
| 3-1/2" | " | 0.200 |
| 4" | " | 0.222 |
| 5" | " | 0.250 |

### 16110.28 STEEL CONDUIT

| | UNIT | MAN/HOURS |
|---|---|---|
| Intermediate metal conduit (IMC) | | |
| 1/2" | L.F. | 0.030 |
| 3/4" | " | 0.040 |
| 1" | " | 0.050 |
| 1-1/4" | " | 0.059 |
| 1-1/2" | " | 0.080 |
| 2" | " | 0.089 |
| 2-1/2" | " | 0.119 |
| 3" | " | 0.145 |
| 3-1/2" | " | 0.182 |
| 4" | " | 0.190 |
| 90 degree ell | | |
| 1/2" | EA. | 0.250 |
| 3/4" | " | 0.308 |
| 1" | " | 0.381 |
| 1-1/4" | " | 0.444 |
| 1-1/2" | " | 0.500 |
| 2" | " | 0.571 |
| 2-1/2" | " | 0.667 |
| 3" | " | 0.889 |
| 3-1/2" | " | 1.143 |
| 4" | " | 1.333 |

### 16110.35 WIREMOLD

| | UNIT | MAN/HOURS |
|---|---|---|
| Wiremold raceway with fittings, surface mounted | | |
| #200 | L.F. | 0.030 |
| #500 | " | 0.030 |
| #700 | " | 0.040 |
| #800 | " | 0.040 |
| Fittings, #200, 90 degree flat elbow | EA. | 0.050 |
| Internal elbow | " | 0.050 |
| Extension adapter | " | 0.059 |
| #200, #500, #700 | | |
| Single pole switch and box | EA. | 0.400 |

## BASIC MATERIALS

### 16110.35 WIREMOLD

| | UNIT | MAN/HOURS |
|---|---|---|
| Duplex receptacle with box | EA. | 0.308 |
| #500, #700 | | |
| 90 deg. flat elbow | EA. | 0.080 |
| Internal elbow | " | 0.080 |
| Junction box | " | 0.133 |
| Fixture box | " | 0.133 |
| Shallow switch and receptacle box | " | 0.080 |
| #800 | | |
| 90 deg. flat elbow | EA. | 0.080 |
| Internal elbow | " | 0.080 |
| Junction box | " | 0.123 |

### 16110.80 WIREWAYS

| | UNIT | MAN/HOURS |
|---|---|---|
| Wireway, hinge cover type | | |
| 2-1/2" x 2-1/2" | | |
| 1' section | EA. | 0.154 |
| 2' | " | 0.190 |
| 3' | " | 0.250 |
| 5' | " | 0.381 |
| 10' | " | 0.667 |
| 4" x 4" | | |
| 1' | EA. | 0.250 |
| 2' | " | 0.250 |
| 3' | " | 0.308 |
| 4' | " | 0.308 |
| 10' | " | 0.800 |
| 6" x 6" | | |
| 1' | EA. | 0.381 |
| 2' | " | 0.381 |
| 3' | " | 0.444 |
| 4' | " | 0.444 |
| 5' | " | 0.571 |
| 10' | " | 0.889 |
| 8" x 8" | | |
| 1' | EA. | 0.444 |
| 2' | " | 0.444 |
| 3' | " | 0.500 |
| 4' | " | 0.500 |
| 5' | " | 0.615 |
| 12" x 12" | | |
| 1' | EA. | 0.615 |
| 2' | " | 0.615 |
| 3' | " | 0.727 |
| 4' | " | 0.727 |
| 5' | " | 0.889 |

### 16120.43 COPPER CONDUCTORS

| | UNIT | MAN/HOURS |
|---|---|---|
| Copper conductors, type THW, solid | | |
| #14 | L.F. | 0.004 |
| #12 | " | 0.005 |
| #10 | " | 0.006 |
| Stranded | | |
| #14 | L.F. | 0.004 |

# 16 ELECTRICAL

## BASIC MATERIALS

### 16120.43 COPPER CONDUCTORS

| BASIC MATERIALS | UNIT | MAN/ HOURS |
|---|---|---|
| #12 | L.F. | 0.005 |
| #10 | " | 0.006 |
| #8 | " | 0.008 |
| #6 | " | 0.009 |
| #4 | " | 0.010 |
| #3 | " | 0.010 |
| #2 | " | 0.012 |
| #1 | " | 0.014 |
| 1/0 | " | 0.016 |
| 2/0 | " | 0.020 |
| 3/0 | " | 0.025 |
| 4/0 | " | 0.028 |
| 250 MCM | " | 0.030 |
| 300 MCM | " | 0.033 |
| 350 MCM | " | 0.040 |
| 400 MCM | " | 0.044 |
| 500 MCM | " | 0.052 |
| 600 MCM | " | 0.059 |
| 750 MCM | " | 0.067 |
| 1000 MCM | " | 0.076 |
| THHN-THWN, solid | | |
| #14 | L.F. | 0.004 |
| #12 | " | 0.005 |
| #10 | " | 0.006 |
| Stranded | | |
| #14 | L.F. | 0.004 |
| #12 | " | 0.005 |
| #10 | " | 0.006 |
| #8 | " | 0.008 |
| #6 | " | 0.009 |
| #4 | " | 0.010 |
| #2 | " | 0.012 |
| #1 | " | 0.014 |
| 1/0 | " | 0.016 |
| 2/0 | " | 0.020 |
| 3/0 | " | 0.025 |
| 4/0 | " | 0.028 |
| 250 MCM | " | 0.030 |
| 350 MCM | " | 0.040 |
| XHHW | | |
| #14 | L.F. | 0.004 |
| #10 | " | 0.006 |
| #8 | " | 0.008 |
| #6 | " | 0.009 |
| #4 | " | 0.009 |
| #2 | " | 0.011 |
| #1 | " | 0.014 |
| 1/0 | " | 0.016 |
| 2/0 | " | 0.019 |
| 3/0 | " | 0.025 |
| XLP, 600v | | |
| #12 | L.F. | 0.005 |
| #10 | " | 0.006 |
| #8 | " | 0.008 |

## BASIC MATERIALS

### 16120.43 COPPER CONDUCTORS

| BASIC MATERIALS | UNIT | MAN/ HOURS |
|---|---|---|
| #6 | L.F. | 0.009 |
| #4 | " | 0.010 |
| #3 | " | 0.011 |
| #2 | " | 0.012 |
| #1 | " | 0.014 |
| 1/0 | " | 0.016 |
| 2/0 | " | 0.020 |
| 3/0 | " | 0.026 |
| 4/0 | " | 0.028 |
| 250 MCM | " | 0.030 |
| 300 MCM | " | 0.033 |
| 350 MCM | " | 0.039 |
| 400 MCM | " | 0.044 |
| 500 MCM | " | 0.052 |
| 600 MCM | " | 0.059 |
| 750 MCM | " | 0.067 |
| 1000 MCM | " | 0.076 |

### 16120.47 SHEATHED CABLE

| BASIC MATERIALS | UNIT | MAN/ HOURS |
|---|---|---|
| Non-metallic sheathed cable | | |
| Type NM cable with ground | | |
| #14/2 | L.F. | 0.015 |
| #12/2 | " | 0.016 |
| #10/2 | " | 0.018 |
| #8/2 | " | 0.020 |
| #6/2 | " | 0.025 |
| #14/3 | " | 0.026 |
| #12/3 | " | 0.027 |
| #10/3 | " | 0.027 |
| #8/3 | " | 0.020 |
| #6/3 | " | 0.028 |
| #4/3 | " | 0.032 |
| #2/3 | " | 0.035 |
| Type U.F. cable with ground | | |
| #14/2 | L.F. | 0.016 |
| #12/2 | " | 0.019 |
| #10/2 | " | 0.020 |
| #8/2 | " | 0.023 |
| #6/2 | " | 0.027 |
| #14/3 | " | 0.020 |
| #12/3 | " | 0.022 |
| #10/3 | " | 0.025 |
| #8/3 | " | 0.028 |
| #6/3 | " | 0.032 |

### 16130.60 PULL AND JUNCTION BOXES

| BASIC MATERIALS | UNIT | MAN/ HOURS |
|---|---|---|
| 4" | | |
| Octagon box | EA. | 0.114 |
| Box extension | " | 0.059 |
| Plaster ring | " | 0.059 |
| Cover blank | " | 0.059 |
| Square box | " | 0.114 |

# 16 ELECTRICAL

| BASIC MATERIALS | UNIT | MAN/ HOURS |
|---|---|---|
| **16130.60 PULL AND JUNCTION BOXES** | | |
| Box extension | EA. | 0.059 |
| Plaster ring | " | 0.059 |
| Cover blank | " | 0.059 |
| 4-11/16" | | |
| Square box | EA. | 0.114 |
| Box extension | " | 0.059 |
| Plaster ring | " | 0.059 |
| Cover blank | " | 0.059 |
| Switch and device boxes | | |
| 2 gang | EA. | 0.114 |
| 3 gang | " | 0.114 |
| 4 gang | " | 0.160 |
| Device covers | | |
| 2 gang | EA. | 0.059 |
| 3 gang | " | 0.059 |
| 4 gang | " | 0.059 |
| Handy box | " | 0.114 |
| Extension | " | 0.059 |
| Switch cover | " | 0.059 |
| Switch box with knockout | " | 0.145 |
| Weatherproof cover, spring type | " | 0.080 |
| Cover plate, dryer receptacle 1 gang plastic | " | 0.100 |
| For 4" receptacle, 2 gang | " | 0.100 |
| Duplex receptacle cover plate, plastic | " | 0.059 |
| 4", vertical bracket box, 1-1/2" with | | |
| RMX clamps | EA. | 0.145 |
| BX clamps | " | 0.145 |
| 4", octagon device cover | | |
| 1 switch | EA. | 0.059 |
| 1 duplex recept | " | 0.059 |
| 4", square face bracket boxes, 1-1/2" | | |
| RMX | EA. | 0.145 |
| BX | " | 0.145 |
| 4" square to round plaster rings | " | 0.059 |
| 2 gang device plaster rings | " | 0.059 |
| Surface covers | | |
| 1 gang switch | EA. | 0.059 |
| 2 gang switch | " | 0.059 |
| 1 single recept | " | 0.059 |
| 1 20a twist lock recept | " | 0.059 |
| 1 30a twist lock recept | " | 0.059 |
| 1 duplex recept | " | 0.059 |
| 2 duplex recept | " | 0.059 |
| Switch and duplex recept | " | 0.059 |
| 4-11/16" square to round plaster rings | " | 0.059 |
| 2 gang device plaster rings | " | 0.059 |
| Surface covers | | |
| 1 gang switch | EA. | 0.059 |
| 2 gang switch | " | 0.059 |
| 1 single recept | " | 0.059 |
| 1 20a twist lock recept | " | 0.059 |
| 1 30a twist lock recept | " | 0.059 |
| 1 duplex recept | " | 0.059 |
| 2 duplex recept | " | 0.059 |

| BASIC MATERIALS | UNIT | MAN/ HOURS |
|---|---|---|
| **16130.60 PULL AND JUNCTION BOXES** | | |
| Switch and duplex recept | EA. | 0.059 |
| 4" plastic round boxes, ground straps | | |
| Box only | EA. | 0.145 |
| Box w/clamps | " | 0.200 |
| Box w/16" bar | " | 0.229 |
| Box w/24" bar | " | 0.250 |
| 4" plastic round box covers | | |
| Blank cover | EA. | 0.059 |
| Plaster ring | " | 0.059 |
| 4" plastic square boxes | | |
| Box only | EA. | 0.145 |
| Box w/clamps | " | 0.200 |
| Box w/hanger | " | 0.250 |
| Box w/nails and clamp | " | 0.250 |
| 4" plastic square box covers | | |
| Blank cover | EA. | 0.059 |
| 1 gang ring | " | 0.059 |
| 2 gang ring | " | 0.059 |
| Round ring | " | 0.059 |
| **16130.65 PULL BOXES AND CABINETS** | | |
| Galvanized pull boxes, screw cover | | |
| 4x4x4 | EA. | 0.190 |
| 4x6x4 | " | 0.190 |
| 6x6x4 | " | 0.190 |
| 6x8x4 | " | 0.190 |
| 8x8x4 | " | 0.250 |
| 8x10x4 | " | 0.242 |
| 8x12x4 | " | 0.250 |
| Screw cover | | |
| 10x10x4 | EA. | 0.308 |
| 12x12x6 | " | 0.444 |
| 12x15x6 | " | 0.444 |
| 12x18x6 | " | 0.500 |
| 15x18x6 | " | 0.571 |
| 18x24x6 | " | 0.615 |
| 18x30x6 | " | 0.727 |
| 24x36x6 | " | 0.727 |
| Cast iron junction box, unflanged | | |
| 6x6x4 | | |
| 3/4" tap | EA. | 0.500 |
| 1" tap | " | 0.500 |
| Two 1/2" taps | " | 0.500 |
| 3/4" taps | " | 0.500 |
| 6" adapter plate | " | 0.348 |
| 6" exterior collar | " | 0.348 |
| Screw cover cabinet | | |
| 12x12x4 | EA. | 0.615 |
| 12x16x4 | " | 0.615 |
| 12x16x6 | " | 0.615 |
| 12x18x4 | " | 0.667 |
| 12x18x6 | " | 0.667 |
| 18x18x4 | " | 1.000 |

| BASIC MATERIALS | UNIT | MAN/ HOURS |
|---|---|---|
| **16130.65 PULL BOXES AND CABINETS** | | |
| 18x18x6 | EA. | 1.000 |
| 18x24x6 | " | 1.143 |
| 24x24x6 | " | 1.333 |
| 24x36x6 | " | 1.667 |
| 36x48x6 | " | 2.500 |
| **16130.80 RECEPTACLES** | | |
| 125 volt, 20a, duplex, grounding type, standard grade | EA. | 0.200 |
| Ground fault interrupter type | " | 0.296 |
| 250 volt, 20a, 2 pole, single receptacle, ground type | " | 0.200 |
| 120/208v, 4 pole, single receptacle, twist lock | | |
| 20a | EA. | 0.348 |
| 50a | " | 0.348 |
| 125/250v, 3 pole, flush receptacle | | |
| 30a | EA. | 0.296 |
| 50a | " | 0.296 |
| 60a | " | 0.348 |
| 277 v, 20a, 2 pole, grounding type, twist lock | " | 0.200 |
| Dryer receptacle, 250v, 30a/50a, 3 wire | " | 0.296 |
| Clock receptacle, 2 pole, grounding type | " | 0.200 |
| 125v, 20a single recept. grounding type | | |
| Standard grade | EA. | 0.200 |
| Specification | " | 0.200 |
| Hospital | " | 0.200 |
| Isolated ground orange | " | 0.250 |
| Duplex | | |
| Specification grade | EA. | 0.200 |
| Hospital | " | 0.200 |
| Isolated ground orange | " | 0.250 |
| 250v, 20a, duplex, 2 pole, grounding, spec. grade | " | 0.200 |
| Combination recepts, 20a, 125v and 250v, duplex | " | 0.200 |
| GFI hospital grade recepts, 20a, 125v, duplex | " | 0.296 |
| 125/250v, 3 pole, 3 wire surface recepts | | |
| 30a | EA. | 0.296 |
| 50a | " | 0.296 |
| 60a | " | 0.348 |
| Cord set, 3 wire, 6' cord | | |
| 30a | EA. | 0.296 |
| 50a | " | 0.296 |
| **16198.10 ELECTRIC MANHOLES** | | |
| Precast, handhole, 4' deep | | |
| 2'x2' | EA. | 3.478 |
| 3'x3' | " | 5.556 |
| 4'x4' | " | 10.256 |
| Power manhole, complete, precast, 8' deep | | |
| 4'x4' | EA. | 14.035 |
| 6'x6' | " | 20.000 |
| 8'x8' | " | 21.053 |
| 6' deep, 9' x 12' | " | 25.000 |
| Cast in place, power manhole, 8' deep | | |
| 4'x4' | EA. | 14.035 |
| 6'x6' | " | 20.000 |

| BASIC MATERIALS | UNIT | MAN/ HOURS |
|---|---|---|
| **16198.10 ELECTRIC MANHOLES** | | |
| 8'x8' | EA. | 21.053 |
| **16199.10 UTILITY POLES & FITTINGS** | | |
| Wood pole, creosoted | | |
| 25' | EA. | 2.353 |
| 30' | " | 2.963 |
| 35' | " | 3.478 |
| 40' | " | 3.791 |
| 45' | " | 6.957 |
| 50' | " | 7.207 |
| 55' | " | 7.547 |
| Treated, wood preservative, 6"x6" | | |
| 8' | EA. | 0.500 |
| 10' | " | 0.800 |
| 12' | " | 0.889 |
| 14' | " | 1.333 |
| 16' | " | 1.600 |
| 18' | " | 2.000 |
| 20' | " | 2.000 |
| Aluminum, brushed, no base | | |
| 8' | EA. | 2.000 |
| 10' | " | 2.667 |
| 15' | " | 2.759 |
| 20' | " | 3.200 |
| 25' | " | 3.810 |
| 30' | " | 4.396 |
| 35' | " | 5.000 |
| 40' | " | 6.250 |
| Steel, no base | | |
| 10' | EA. | 2.500 |
| 15' | " | 2.963 |
| 20' | " | 3.810 |
| 25' | " | 4.520 |
| 30' | " | 5.096 |
| 35' | " | 6.250 |
| Concrete, no base | | |
| 13' | EA. | 5.517 |
| 16' | " | 7.273 |
| 18' | " | 8.791 |
| 25' | " | 10.000 |
| 30' | " | 12.121 |
| 35' | " | 14.035 |
| 40' | " | 16.000 |
| 45' | " | 17.021 |
| 50' | " | 18.182 |
| 55' | " | 19.048 |
| 60' | " | 20.000 |
| Pole line hardware | | |
| Wood crossarm | | |
| 4' | EA. | 1.333 |
| 8' | " | 1.667 |
| 10' | " | 2.051 |

# 16 ELECTRICAL

## BASIC MATERIALS

| 16199.10 UTILITY POLES & FITTINGS | UNIT | MAN/HOURS |
|---|---|---|
| Angle steel brace | | |
| 1 piece | EA. | 0.250 |
| 2 piece | " | 0.348 |
| Eye nut, 5/8" | " | 0.050 |
| Bolt (14-16"), 5/8" | " | 0.200 |
| Transformer, ground connection | " | 0.250 |
| Stirrup | " | 0.308 |
| Secondary lead support | " | 0.400 |
| Spool insulator | " | 0.200 |
| Guy grip, preformed | | |
| 7/16" | EA. | 0.145 |
| 1/2" | " | 0.145 |
| Hook | " | 0.250 |
| Strain insulator | " | 0.364 |
| Wire | | |
| 5/16" | L.F. | 0.005 |
| 7/16" | " | 0.006 |
| 1/2" | " | 0.008 |
| Soft drawn ground, copper, #8 | " | 0.008 |
| Ground clamp | EA. | 0.308 |
| Perforated strapping for conduit, 1-1/2" | L.F. | 0.145 |
| Hot line clamp | EA. | 0.800 |
| Lightning arrester | | |
| 3kv | EA. | 1.000 |
| 10kv | " | 1.600 |
| 30kv | " | 2.000 |
| 36kv | " | 2.500 |

## POWER GENERATION

| 16210.10 GENERATORS | UNIT | MAN/HOURS |
|---|---|---|
| Diesel generator, with auto transfer switch | | |
| 30kw | EA. | 30.769 |
| 50kw | " | 30.769 |
| 75kw | " | 42.105 |
| 100kw | " | 47.059 |
| 125kw | " | 50.000 |
| 150kw | " | 57.143 |
| 175kw | " | 66.667 |
| 200kw | " | 80.000 |
| 250kw | " | 88.889 |
| 300kw | " | 100 |
| 350kw | " | 114 |
| 400kw | " | 133 |
| 450kw | " | 145 |
| 500kw | " | 160 |

## POWER GENERATION

| 16210.10 GENERATORS | UNIT | MAN/HOURS |
|---|---|---|
| 600kw | EA. | 200 |
| 750kw | " | 200 |

| 16320.10 TRANSFORMERS | UNIT | MAN/HOURS |
|---|---|---|
| Floor mounted, single phase, int. dry, 480v-120/240v | | |
| 3 kva | EA. | 1.818 |
| 5 kva | " | 3.077 |
| 7.5 kva | " | 3.478 |
| 10 kva | " | 3.810 |
| 15 kva | " | 4.301 |
| 25 kva | " | 7.547 |
| 37.5 kva | " | 9.412 |
| 50 kva | " | 10.256 |
| 75 kva | " | 10.667 |
| 100 kva | " | 11.594 |
| Three phase, 480v-120/208v | | |
| 15 kva | EA. | 6.015 |
| 30 kva | " | 9.412 |
| 45 kva | " | 10.811 |
| 75 kva | " | 10.959 |
| 112.5 kva | " | 12.698 |
| 150 kva | " | 13.559 |
| 225 kva | " | 15.385 |

| 16350.10 CIRCUIT BREAKERS | UNIT | MAN/HOURS |
|---|---|---|
| Molded case, 240v, 15-60a, bolt-on | | |
| 1 pole | EA. | 0.250 |
| 2 pole | " | 0.348 |
| 70-100a, 2 pole | " | 0.533 |
| 15-60a, 3 pole | " | 0.400 |
| 70-100a, 3 pole | " | 0.615 |
| 480v, 2 pole | | |
| 15-60a | EA. | 0.296 |
| 70-100a | " | 0.400 |
| 3 pole | | |
| 15-60a | EA. | 0.400 |
| 70-100a | " | 0.444 |
| 70-225a | " | 0.615 |
| Draw out air circuit breakers | | |
| 600a | EA. | 16.000 |
| 800a | " | 18.182 |
| 1600a | " | 24.242 |
| 2000a | " | 27.586 |
| 3000a | " | 32.000 |
| 4000a | " | 38.095 |
| Load center circuit breakers, 240v | | |
| 1 pole, 10-60a | EA. | 0.250 |
| 2 pole | | |
| 10-60a | EA. | 0.400 |
| 70-100a | " | 0.667 |
| 110-150a | " | 0.727 |
| 3 pole | | |
| 10-60a | EA. | 0.500 |

## POWER GENERATION

### 16350.10 CIRCUIT BREAKERS

| POWER GENERATION | UNIT | MAN/ HOURS |
|---|---|---|
| 70-100a | EA. | 0.727 |
| Load center, G.F.I. breakers, 240v | | |
| 1 pole, 15-30a | EA. | 0.296 |
| 2 pole, 15-30a | " | 0.400 |
| Key operated breakers, 240v, 1 pole, 10-30a | " | 0.296 |
| Tandem breakers, 240v | | |
| 1 pole, 15-30a | EA. | 0.400 |
| 2 pole, 15-30a | " | 0.533 |
| Bolt-on, G.F.I. breakers, 240v, 1 pole, 15-30a | " | 0.348 |

### 16360.10 SAFETY SWITCHES

| POWER GENERATION | UNIT | MAN/ HOURS |
|---|---|---|
| Fused, 3 phase, 30 amp, 600v, heavy duty | | |
| NEMA 1 | EA. | 1.143 |
| NEMA 3r | " | 1.143 |
| NEMA 4 | " | 1.600 |
| NEMA 12 | " | 1.739 |
| 60a | | |
| NEMA 1 | EA. | 1.143 |
| NEMA 3r | " | 1.143 |
| NEMA 4 | " | 1.600 |
| NEMA 12 | " | 1.739 |
| 100a | | |
| NEMA 1 | EA. | 1.739 |
| NEMA 3r | " | 1.739 |
| NEMA 4 | " | 2.000 |
| NEMA 12 | " | 2.500 |
| 200a | | |
| NEMA 1 | EA. | 2.500 |
| NEMA 3r | " | 2.500 |
| NEMA 4 | " | 2.759 |
| NEMA 12 | " | 3.478 |
| 400a | | |
| NEMA 1 | EA. | 5.517 |
| NEMA 3r | " | 5.517 |
| NEMA 4 | " | 5.755 |
| NEMA 12 | " | 7.018 |
| NEMA 4 | " | 8.989 |
| NEMA 12 | " | 12.308 |
| Non-fused, 240-600v, heavy duty, 3 phase, 30 amp | | |
| NEMA 1 | EA. | 1.143 |
| NEMA 3r | " | 1.143 |
| NEMA 4 | " | 1.739 |
| NEMA 12 | " | 1.739 |
| 60a | | |
| NEMA1 | EA. | 1.143 |
| NEMA 3r | " | 1.143 |
| NEMA 4 | " | 1.739 |
| NEMA 12 | " | 1.739 |
| 100a | | |
| NEMA 1 | EA. | 1.739 |
| NEMA 3r | " | 1.739 |
| NEMA 4 | " | 2.500 |
| NEMA 12 | " | 2.500 |

## POWER GENERATION

### 16360.10 SAFETY SWITCHES

| POWER GENERATION | UNIT | MAN/ HOURS |
|---|---|---|
| 200a, NEMA 1 | EA. | 2.500 |
| 600a, NEMA 12 | " | 12.308 |

### 16365.10 FUSES

| POWER GENERATION | UNIT | MAN/ HOURS |
|---|---|---|
| Fuse, one-time, 250v | | |
| 30a | EA. | 0.050 |
| 60a | " | 0.050 |
| 100a | " | 0.050 |
| 200a | " | 0.050 |
| 400a | " | 0.050 |
| 600a | " | 0.050 |
| 600v | | |
| 30a | EA. | 0.050 |
| 60a | " | 0.050 |
| 100a | " | 0.050 |
| 200a | " | 0.050 |
| 400a | " | 0.050 |
| Fusetron, 600v | | |
| 200a | EA. | 0.050 |
| 400a | " | 0.050 |
| Fuse, amp-trap, K1, 250v | | |
| 30a | EA. | 0.050 |
| 60a | " | 0.050 |
| 100a | " | 0.050 |
| 200a | " | 0.050 |
| 400a | " | 0.050 |
| 600a | " | 0.050 |
| 600v | | |
| 30a | EA. | 0.050 |
| 60a | " | 0.050 |
| 100a | " | 0.050 |
| 200a | " | 0.050 |
| 400a | " | 0.050 |
| K5, 250v | | |
| 30a | EA. | 0.050 |
| 60a | " | 0.050 |
| 100a | " | 0.050 |
| 200a | " | 0.050 |
| 400a | " | 0.050 |
| 600a | " | 0.050 |
| 600v | | |
| 30a | EA. | 0.050 |
| 60a | " | 0.050 |
| 100a | " | 0.050 |
| 200a | " | 0.050 |
| 400a | " | 0.050 |
| 600a | " | 0.050 |
| J, 600v | | |
| 30a | EA. | 0.050 |
| 60a | " | 0.050 |
| 100a | " | 0.050 |
| 200a | " | 0.050 |
| 400a | " | 0.050 |
| L, 600v | | |

# 16 ELECTRICAL

## POWER GENERATION

| POWER GENERATION | UNIT | MAN/HOURS |
|---|---|---|
| **16365.10 FUSES** | | |
| 1200a | EA. | 0.400 |
| 1600a | " | 0.400 |
| 2000a | " | 0.400 |
| 2500a | " | 0.400 |
| 3000a | " | 0.400 |
| 4000a | " | 0.400 |
| 5000a | " | 0.400 |
| Fuse cl-ay 250v | | |
| 600a | EA. | 0.296 |
| 1200a | " | 0.296 |
| 1600a | " | 0.296 |
| 2000a | " | 0.296 |
| 600v | | |
| 1200a | EA. | 0.296 |
| 1600a | " | 0.296 |
| 2000a | " | 0.296 |
| **16395.10 GROUNDING** | | |
| Ground rods, copper clad, 1/2" x | | |
| 6' | EA. | 0.667 |
| 8' | " | 0.727 |
| 10' | " | 1.000 |
| 5/8" x | | |
| 5' | EA. | 0.615 |
| 6' | " | 0.727 |
| 8' | " | 1.000 |
| 10' | " | 1.250 |
| 3/4" x | | |
| 8' | EA. | 0.727 |
| 10' | " | 0.800 |
| Ground rod clamp | | |
| 5/8" | EA. | 0.123 |
| 3/4" | " | 0.123 |

## SERVICE AND DISTRIBUTION

| SERVICE AND DISTRIBUTION | UNIT | MAN/HOURS |
|---|---|---|
| **16425.10 SWITCHBOARDS** | | |
| Switchboard, 90" high, no main disconnect, 208/120v | | |
| 400a | EA. | 7.921 |
| 600a | " | 8.000 |
| 1000a | " | 8.000 |
| 1200a | " | 10.000 |
| 1600a | " | 11.940 |
| 2000a | " | 14.035 |
| 2500a | " | 16.000 |
| 277/480v | | |
| 600a | EA. | 8.163 |
| 800a | " | 8.163 |
| 1600a | " | 11.940 |
| 2000a | " | 14.035 |
| 2500a | " | 16.000 |
| 3000a | " | 27.586 |
| 4000a | " | 29.630 |
| 1500 kva | " | 57.143 |
| **16470.10 PANELBOARDS** | | |
| 3 phase, 480/277v, main lugs only, 120a, 30 circuits | EA. | 3.478 |
| 277/480v, 4 wire, flush surface | | |
| 225a, 30 circuits | EA. | 4.000 |
| 400a, 30 circuits | " | 5.000 |
| 600a, 42 circuits | " | 6.015 |
| 208/120v, main circuit breaker, 3 phase, 4 wire | | |
| 100a | | |
| 12 circuits | EA. | 5.096 |
| 20 circuits | " | 6.299 |
| 30 circuits | " | 7.018 |
| 225a | | |
| 30 circuits | EA. | 7.767 |
| 42 circuits | " | 9.524 |
| 400a | | |
| 30 circuits | EA. | 14.815 |
| 42 circuits | " | 16.000 |
| 600a, 42 circuits | " | 18.182 |
| 120/208v, flush, 3 ph., 4 wire, main only | | |
| 100a | | |
| 12 circuits | EA. | 5.096 |
| 20 circuits | " | 6.299 |
| 30 circuits | " | 7.018 |
| 225a | | |
| 30 circuits | EA. | 7.767 |
| 42 circuits | " | 9.524 |
| 400a | | |
| 30 circuits | EA. | 14.815 |
| 42 circuits | " | 16.000 |
| 600a, 42 circuits | " | 18.182 |
| Panelboard accessories | | |
| Grounding bus | EA. | 0.348 |
| Handle lock device | " | 0.145 |
| Factory assembled panel | | |
| 1 pole space | EA. | 0.348 |
| 2 pole space | " | 0.145 |
| 3 pole space | " | 0.133 |
| Panelboards 1 phase, 240/120v main circuit breaker | | |
| Single phase, 3 wire, 120/240v flush | | |
| 100a, 20 circuits | EA. | 3.478 |
| 225a, 30 circuits | " | 4.000 |
| 240/120v, main lugs only | | |
| 100a | | |
| 8 circuits | EA. | 2.963 |
| 12 circuits | " | 2.963 |

# 16 ELECTRICAL

| SERVICE AND DISTRIBUTION | UNIT | MAN/HOURS |
|---|---|---|
| **16470.10 PANELBOARDS** | | |
| 20 circuits | EA. | 2.963 |
| 225a | | |
| 24 circuits | EA. | 3.478 |
| 30 circuits | " | 3.810 |
| 42 circuits | " | 3.810 |
| **16480.10 MOTOR CONTROLS** | | |
| Motor generator set, 3 phase, 480/277v, w/controls | | |
| 10kw | EA. | 27.586 |
| 15kw | " | 30.769 |
| 20kw | " | 32.000 |
| 25kw | " | 34.783 |
| 30kw | " | 36.364 |
| 40kw | " | 38.095 |
| 50kw | " | 40.000 |
| 60kw | " | 44.444 |
| 75kw | " | 50.000 |
| 100kw | " | 61.538 |
| 125kw | " | 66.667 |
| 150kw | " | 66.667 |
| 200kw | " | 72.727 |
| 250kw | " | 72.727 |
| 300kw | " | 80.000 |
| 2 pole, 230 volt starter, w/NEMA-1 | | |
| 1 hp, 9 amp, size 00 | EA. | 1.000 |
| 2 hp, 18amp, size 0 | " | 1.000 |
| 3 hp, 27amp, size 1 | " | 1.000 |
| 5 hp, 45amp, size 1p | " | 1.000 |
| 7-1/2 hp, 45a, size 2 | " | 1.000 |
| 15 hp, 90a, size 3 | " | 1.000 |
| 2 pole, w/NEMA-4 enclosure | | |
| 2 hp, 18a, size 1 | EA. | 1.600 |
| 5 hp, 45amp, size 1p | " | 1.600 |
| 7-1/2 hp, 45a, size 2 | " | 1.600 |
| 3 pole, 2 hp, 9a, 200-575v starter | | |
| W/NEMA-1, size 00 | EA. | 1.333 |
| W/NEMA-4 enclosure, size 00 | " | 1.739 |
| 5hp, 18a | | |
| W/NEMA-1 enclosure, size 0 | EA. | 1.333 |
| W/NEMA-4 enclosure, size 0 | " | 1.739 |
| 7.5-10hp, 27a | | |
| 7.5-10hp 27a, w/NEMA-1 enclosure, size 1 | EA. | 1.333 |
| W/NEMA-4 enclosure size 1 | " | 1.739 |
| 10-25hp, 45a | | |
| W/NEMA-1 enclosure, size 2 | EA. | 1.333 |
| W/NEMA-4 enclosure, size 2 | " | 1.739 |
| 25-50hp, 90a | | |
| W/NEMA-1 enclosure, size 3 | EA. | 1.739 |
| W/NEMA-4 enclosure, size 3 | " | 2.500 |
| 40-100hp, 135a | | |
| W/NEMA-1 enclosure, size 4 | EA. | 2.500 |
| W/NEMA-4 enclosure, size 4 | " | 3.478 |
| 75-200hp, 270a | | |

| SERVICE AND DISTRIBUTION | UNIT | MAN/HOURS |
|---|---|---|
| **16480.10 MOTOR CONTROLS** | | |
| W/NEMA-1 enclosure, size 5 | EA. | 5.517 |
| W/NEMA-4 enclosure, size 5 | " | 7.018 |
| Magnetic starter accessories | | |
| On-off-auto selector switch kit | EA. | 0.320 |
| With pilot light | " | 0.348 |
| **16490.10 SWITCHES** | | |
| Oil switches, medium voltage, bus components | | |
| Switches, 277/120v, toggle device only | EA. | 1.600 |
| With oil 35kv, g&w gram 44, 4 way switch | " | 8.000 |
| Weatherproof enclosure | | |
| 3 way switch | EA. | 10.000 |
| 4 way switch | " | 10.959 |
| Fused interrupter load, 35kv | | |
| 20A | | |
| 1 pole | EA. | 16.000 |
| 2 pole | " | 17.021 |
| 3 way | " | 17.021 |
| 4 way | " | 18.182 |
| 30a, 1 pole | " | 16.000 |
| 3 way | " | 17.021 |
| 4 way | " | 18.182 |
| Weatherproof switch, including box & cover, 20a | | |
| 1 pole | EA. | 16.000 |
| 2 pole | " | 17.021 |
| 3 way | " | 18.182 |
| 4 way | " | 18.182 |
| 3 way, oil switch, 15kv enclosure | " | 11.940 |
| Pedestal for 35kv double breaker switch | " | 5.000 |
| Bus terminal connector, 2 | " | 2.500 |
| 2 to 3 | " | 2.500 |
| Support connector, 3 | " | 1.600 |
| Tee connector, 2 to 3 | " | 2.000 |
| Flexible bus stud connector | " | 1.739 |
| End cap 3 | " | 1.333 |
| Weldment connection, 3 | " | 1.000 |
| Plate switch, 1 gang | " | 0.050 |
| Start stop stations, manual motor starters | " | 0.727 |
| Lockout switch | " | 0.250 |
| Forward-reverse switch | " | 0.727 |
| On-off switch | " | 0.727 |
| Open-close switch | " | 0.727 |
| Forward-reverse-stop switch | " | 1.000 |
| Standard 3 button switch any standard legend | " | 1.000 |
| Standard 3 button with lockout | " | 1.000 |
| Manual motor starters, tog, 115/230v | | |
| Size 1 gp | EA. | 1.000 |
| Size 2 | " | 1.000 |
| Button | | |
| Size 0 | EA. | 1.000 |
| Size 1 | " | 1.000 |
| Size 2 | " | 1.000 |
| 3-phase | | |

# 16 ELECTRICAL

## SERVICE AND DISTRIBUTION

### 16490.10 SWITCHES

| | UNIT | MAN/ HOURS |
|---|---|---|
| Size 0 | EA. | 1.333 |
| Size 1 | " | 1.333 |
| Time & float switches | " | 1.600 |
| Astronomical time switch, 40a, 240v | " | 1.000 |
| Timer switch 0-5 minute, with box | " | 0.500 |
| Single pole/single throw time, 277v, NEMA-1 | " | 0.727 |
| Single toggle switch, 20a, 120v, with pilot | " | 0.250 |
| 3-way toggle | " | 0.296 |
| Photo electric switches | | |
| 1000 watt | | |
| 105-135v | EA. | 0.727 |
| 208-277v | " | 0.727 |
| 3000 watt, 105-130v | | |
| Double throw | EA. | 1.000 |
| Single throw | " | 1.000 |
| Double pole/single throw, 210-250v | " | 1.333 |
| Dimmer switch and switch plate | | |
| 600 w | EA. | 0.308 |
| 1000 w | " | 0.348 |
| Dimmer switch incandescent | | |
| 1500w | EA. | 0.702 |
| 2000w | " | 0.748 |
| Fluorescent | | |
| 12 lamps | EA. | 0.500 |
| 20 lamps | " | 0.552 |
| 30 lamps | " | 0.602 |
| 40 lamps | " | 0.702 |
| Specification grade toggle switches, 20a, 120-277v | | |
| Single pole | EA. | 0.200 |
| Double pole | " | 0.296 |
| 3 way | " | 0.250 |
| 4 way | " | 0.296 |
| Switch plates, plastic ivory | | |
| 1 gang | EA. | 0.080 |
| 2 gang | " | 0.100 |
| 3 gang | " | 0.119 |
| 4 gang | " | 0.145 |
| 5 gang | " | 0.160 |
| 6 gang | " | 0.182 |
| Stainless steel | | |
| 1 gang | EA. | 0.080 |
| 2 gang | " | 0.100 |
| 3 gang | " | 0.123 |
| 4 gang | " | 0.145 |
| 5 gang | " | 0.160 |
| 6 gang | " | 0.182 |
| Brass | | |
| 1 gang | EA. | 0.080 |
| 2 gang | " | 0.100 |
| 3 gang | " | 0.123 |
| 4 gang | " | 0.145 |
| 5 gang | " | 0.160 |
| 6 gang | " | 0.182 |

## SERVICE AND DISTRIBUTION

### 16490.20 TRANSFER SWITCHES

| | UNIT | MAN/ HOURS |
|---|---|---|
| Automatic transfer switch 600v, 3 pole | | |
| 30a | EA. | 3.478 |
| 60a | " | 3.478 |
| 100a | " | 4.762 |
| 150a | " | 6.015 |
| 225a | " | 8.000 |
| 260a | " | 8.000 |
| 400a | " | 10.000 |
| 600a | " | 15.094 |
| 800a | " | 18.182 |
| 1000a | " | 21.053 |
| 1200a | " | 22.857 |
| 1600a | " | 25.000 |
| 2000a | " | 29.630 |
| 2600a | " | 42.105 |
| 3000a | " | 50.000 |

### 16490.80 SAFETY SWITCHES

| | UNIT | MAN/ HOURS |
|---|---|---|
| Safety switch, 600v, 3 pole, heavy duty, NEMA-1 | | |
| 30a | EA. | 1.000 |
| 60a | " | 1.143 |
| 100a | " | 1.600 |
| 200a | " | 2.500 |
| 400a | " | 5.517 |
| 600a | " | 8.000 |
| 800a | " | 10.526 |
| 1200a | " | 14.286 |

## LIGHTING

### 16510.05 INTERIOR LIGHTING

| | UNIT | MAN/ HOURS |
|---|---|---|
| Recessed fluorescent fixtures, 2'x2' | | |
| 2 lamp | EA. | 0.727 |
| 4 lamp | " | 0.727 |
| 2 lamp w/flange | " | 1.000 |
| 4 lamp w/flange | " | 1.000 |
| 1'x4' | | |
| 2 lamp | EA. | 0.667 |
| 3 lamp | " | 0.667 |
| 2 lamp w/flange | " | 0.727 |
| 3 lamp w/flange | " | 0.727 |
| 2'x4' | | |
| 2 lamp | EA. | 0.727 |
| 3 lamp | " | 0.727 |
| 4 lamp | " | 0.727 |
| 2 lamp w/flange | " | 1.000 |

| LIGHTING | UNIT | MAN/ HOURS |
|---|---|---|
| **16510.05 INTERIOR LIGHTING** | | |
| 3 lamp w/flange | EA. | 1.000 |
| 4 lamp w/flange | " | 1.000 |
| 4'x4' | | |
| 4 lamp | EA. | 1.000 |
| 6 lamp | " | 1.000 |
| 8 lamp | " | 1.000 |
| 4 lamp w/flange | " | 1.509 |
| 6 lamp w/flange | " | 1.509 |
| 8 lamp, w/flange | " | 1.509 |
| Surface mounted incandescent fixtures | | |
| 40w | EA. | 0.667 |
| 75w | " | 0.667 |
| 100w | " | 0.667 |
| 150w | " | 0.667 |
| Pendant | | |
| 40w | EA. | 0.800 |
| 75w | " | 0.800 |
| 100w | " | 0.800 |
| 150w | " | 0.800 |
| Recessed incandescent fixtures | | |
| 40w | EA. | 1.509 |
| 75w | " | 1.509 |
| 100w | " | 1.509 |
| 150w | " | 1.509 |
| Exit lights, 120v | | |
| Recessed | EA. | 1.250 |
| Back mount | " | 0.727 |
| Universal mount | " | 0.727 |
| Emergency battery units, 6v-120v, 50 unit | " | 1.509 |
| With 1 head | " | 1.509 |
| With 2 heads | " | 1.509 |
| Light track single circuit | | |
| 2' | EA. | 0.500 |
| 4' | " | 0.500 |
| 8' | " | 1.000 |
| 12' | " | 1.509 |
| Fittings and accessories | | |
| Dead end | EA. | 0.145 |
| Starter kit | " | 0.250 |
| Conduit feed | " | 0.145 |
| Straight connector | " | 0.145 |
| Center feed | " | 0.145 |
| L-connector | " | 0.145 |
| T-connector | " | 0.145 |
| X-connector | " | 0.200 |
| Cord and plug | " | 0.100 |
| Rigid corner | " | 0.145 |
| Flex connector | " | 0.145 |
| 2 way connector | " | 0.200 |
| Spacer clip | " | 0.050 |
| Grid box | " | 0.145 |
| T-bar clip | " | 0.050 |
| Utility hook | " | 0.145 |
| Fixtures, square | | |

| LIGHTING | UNIT | MAN/ HOURS |
|---|---|---|
| **16510.05 INTERIOR LIGHTING** | | |
| R-20 | EA. | 0.145 |
| R-30 | " | 0.145 |
| 40w flood | " | 0.145 |
| 40w spot | " | 0.145 |
| 100w flood | " | 0.145 |
| 100w spot | " | 0.145 |
| Mini spot | " | 0.145 |
| Mini flood | " | 0.145 |
| Quartz, 500w | " | 0.145 |
| R-20 sphere | " | 0.145 |
| R-30 sphere | " | 0.145 |
| R-20 cylinder | " | 0.145 |
| R-30 cylinder | " | 0.145 |
| R-40 cylinder | " | 0.145 |
| R-30 wall wash | " | 0.145 |
| R-40 wall wash | " | 0.145 |
| Energy saving rapid start fluor. lamps | | |
| F30cw | EA. | 0.100 |
| F40 cw | " | 0.100 |
| F40 cwx | " | 0.100 |
| F30 ww | " | 0.100 |
| F40 ww | " | 0.100 |
| F40 wwx | " | 0.100 |
| Incandescent lamps | | |
| 200w | EA. | 0.100 |
| 300w | " | 0.100 |
| 500w | " | 0.100 |
| 750w | " | 0.145 |
| 1000w | " | 0.200 |
| 1500w | " | 0.200 |
| Energy saving reflector floodlight lamps | | |
| 25w | EA. | 0.100 |
| 30w | " | 0.100 |
| 50w | " | 0.100 |
| 75w | " | 0.100 |
| 120w | " | 0.100 |
| 150w | " | 0.100 |
| 200w | " | 0.100 |
| 300w | " | 0.100 |
| 500w | " | 0.145 |
| 750w | " | 0.145 |
| Reflector spotlight | | |
| 75w | EA. | 0.100 |
| 100w | " | 0.100 |
| 125w | " | 0.100 |
| 150w | " | 0.100 |
| 250w | " | 0.100 |
| 300w | " | 0.100 |
| 400w | " | 0.145 |
| 500w | " | 0.145 |
| 1000w | " | 0.200 |
| Medium par flood lamps | | |
| 75w | EA. | 0.100 |
| 100w | " | 0.100 |

# 16 ELECTRICAL

## LIGHTING

| LIGHTING | UNIT | MAN/ HOURS |
|---|---|---|
| **16510.05 INTERIOR LIGHTING** | | |
| 150w | EA. | 0.100 |
| 200w | " | 0.100 |
| 300w | " | 0.100 |
| 500w | " | 0.145 |
| Medium par spot lamps | | |
| 75w | EA. | 0.100 |
| 120w | " | 0.100 |
| 150w | " | 0.100 |
| Ballast replacements rapid start fluor | | |
| 1f-40-120v | EA. | 0.727 |
| 1f-40-277v | " | 0.727 |
| 1f-96-120v | " | 0.727 |
| 1f-96-277v | " | 0.727 |
| 2f-40-120v | " | 0.727 |
| 2f-40-277v | " | 0.727 |
| 2f-96-120v | " | 0.727 |
| 2f-96-277v | " | 0.727 |
| **16510.10 LIGHTING INDUSTRIAL** | | |
| Surface mounted fluorescent, wrap around lens | | |
| 1 lamp | EA. | 0.800 |
| 2 lamps | " | 0.889 |
| 4 lamps | " | 1.000 |
| Wall mounted fluorescent | | |
| 2-20w lamps | EA. | 0.500 |
| 2-30w lamps | " | 0.500 |
| 2-40w lamps | " | 0.667 |
| Indirect, with wood shielding, 2049w lamps | | |
| 4' | EA. | 1.000 |
| 8' | " | 1.600 |
| Industrial fluorescent, 2 lamp | | |
| 4' | EA. | 0.727 |
| 8' | " | 1.333 |
| Strip fluorescent | | |
| 4' | | |
| 1 lamp | EA. | 0.667 |
| 2 lamps | " | 0.667 |
| 8' | | |
| 1 lamp | EA. | 0.727 |
| 2 lamps | " | 0.889 |
| Wire guard for strip fixture, 4' long | " | 0.348 |
| Strip fluorescent, 8' long, two 4' lamps | " | 1.333 |
| With four 4' lamps | " | 1.600 |
| Wet location fluorescent, plastic housing | | |
| 4' long | | |
| 1 lamp | EA. | 1.000 |
| 2 lamps | " | 1.333 |
| 8' long | | |
| 2 lamps | EA. | 1.600 |
| 4 lamps | " | 1.739 |
| Parabolic troffer, 2'x2' | | |
| With 2 "U" lamps | EA. | 1.000 |
| With 3 "U" lamps | " | 1.143 |

| LIGHTING | UNIT | MAN/ HOURS |
|---|---|---|
| **16510.10 LIGHTING INDUSTRIAL** | | |
| 2'x4' | | |
| With 2 40w lamps | EA. | 1.143 |
| With 3 40w lamps | " | 1.333 |
| With 4 40w lamps | " | 1.333 |
| 1'x4' | | |
| With 1 T-12 lamp, 9 cell | EA. | 0.727 |
| With 2 T-12 lamps | " | 0.889 |
| With 1 T-12 lamp, 20 cell | " | 0.727 |
| With 2 T-12 lamps | " | 0.889 |
| Steel sided surface fluorescent, 2'x4' | | |
| 3 lamps | EA. | 1.333 |
| 4 lamps | " | 1.333 |
| Outdoor sign fluor., 1 lamp, remote ballast | | |
| 4' long | EA. | 6.015 |
| 6' long | " | 8.000 |
| Recess mounted, commercial, 2'x2', 13" high | | |
| 100w | EA. | 4.000 |
| 250w | " | 4.494 |
| High pressure sodium, hi-bay open | | |
| 400w | EA. | 1.739 |
| 1000w | " | 2.424 |
| Enclosed | | |
| 400w | EA. | 2.424 |
| 1000w | " | 2.963 |
| Metal halide hi-bay, open | | |
| 400w | EA. | 1.739 |
| 1000w | " | 2.424 |
| Enclosed | | |
| 400w | EA. | 2.424 |
| 1000w | " | 2.963 |
| High pressure sodium, low bay, surface mounted | | |
| 100w | EA. | 1.000 |
| 150w | " | 1.143 |
| 250w | " | 1.333 |
| 400w | " | 1.600 |
| Metal halide, low bay, pendant mounted | | |
| 175w | EA. | 1.333 |
| 250w | " | 1.600 |
| 400w | " | 2.222 |
| Indirect luminare, square, metal halide, freestanding | | |
| 175w | EA. | 1.000 |
| 250w | " | 1.000 |
| 400w | " | 1.000 |
| High pressure sodium | | |
| 150w | EA. | 1.000 |
| 250w | " | 1.000 |
| 400w | " | 1.000 |
| Round, metal halide | | |
| 175w | EA. | 1.000 |
| 250w | " | 1.000 |
| 400w | " | 1.000 |
| High pressure sodium | | |
| 150w | EA. | 1.000 |
| 250w | " | 1.000 |

| LIGHTING | UNIT | MAN/ HOURS |
|---|---|---|
| **16510.10 LIGHTING INDUSTRIAL** | | |
| 400w | EA. | 1.000 |
| Wall mounted, metal halide | | |
| 175w | EA. | 2.500 |
| 250w | " | 2.500 |
| 400w | " | 3.200 |
| High pressure sodium | | |
| 150w | EA. | 2.500 |
| 250w | " | 2.500 |
| 400w | " | 3.200 |
| Wall pack lithonia, high pressure sodium | | |
| 35w | EA. | 0.889 |
| 55w | " | 1.000 |
| 150w | " | 1.600 |
| 250w | " | 1.739 |
| Low pressure sodium | | |
| 35w | EA. | 1.739 |
| 55w | " | 2.000 |
| Wall pack hubbell, high pressure sodium | | |
| 35w | EA. | 0.889 |
| 150w | " | 1.600 |
| 250w | " | 1.739 |
| Compact fluorescent | | |
| 2-7w | EA. | 1.000 |
| 2-13w | " | 1.333 |
| 1-18w | " | 1.333 |
| Handball & racquet ball court, 2'x2', metal halide | | |
| 250w | EA. | 2.500 |
| 400w | " | 2.759 |
| High pressure sodium | | |
| 250w | EA. | 2.500 |
| 400w | " | 2.759 |
| Bollard light, 42" w/found., high pressure sodium | | |
| 70w | EA. | 2.581 |
| 100w | " | 2.581 |
| 150w | " | 2.581 |
| Light fixture lamps | | |
| Lamp | | |
| 20w med. bipin base, cool white, 24" | EA. | 0.145 |
| 30w cool white, rapid start, 36" | " | 0.145 |
| 40w cool white "U", 3" | " | 0.145 |
| 40w cool white, rapid start, 48" | " | 0.145 |
| 70w high pressure sodium, mogul base | " | 0.200 |
| 75w slimline, 96" | " | 0.200 |
| 100w | | |
| Incandescent, 100a, inside frost | EA. | 0.100 |
| Mercury vapor, clear, mogul base | " | 0.200 |
| High pressure sodium, mogul base | " | 0.200 |
| 150w | | |
| Par 38 flood or spot, incandescent | EA. | 0.100 |
| High pressure sodium, 1/2 mogul base | " | 0.200 |
| 175w | | |
| Mercury vapor, clear, mogul base | EA. | 0.200 |
| Metal halide, clear, mogul base | " | 0.200 |
| 250w | | |

| LIGHTING | UNIT | MAN/ HOURS |
|---|---|---|
| **16510.10 LIGHTING INDUSTRIAL** | | |
| High pressure sodium, mogul base | EA. | 0.200 |
| Mercury vapor, clear, mogul base | " | 0.200 |
| Metal halide, clear, mogul base | " | 0.200 |
| High pressure sodium, mogul base | " | 0.200 |
| 400w | | |
| Mercury vapor, clear, mogul base | EA. | 0.200 |
| Metal halide, clear, mogul base | " | 0.200 |
| High pressure sodium, mogul base | " | 0.200 |
| 1000w | | |
| Mercury vapor, clear, mogul base | EA. | 0.250 |
| High pressure sodium, mogul base | " | 0.250 |
| **16510.30 EXTERIOR LIGHTING** | | |
| Exterior light fixtures | | |
| Rectangle, high pressure sodium | | |
| 70w | EA. | 2.500 |
| 100w | " | 2.581 |
| 150w | " | 2.581 |
| 250w | " | 2.759 |
| 400w | " | 3.478 |
| Flood, rectangular, high pressure sodium | | |
| 70w | EA. | 2.500 |
| 100w | " | 2.581 |
| 150w | " | 2.581 |
| 400w | " | 3.478 |
| 1000w | " | 4.494 |
| Round | | |
| 400w | EA. | 3.478 |
| 1000w | " | 4.494 |
| Round, metal halide | | |
| 400w | EA. | 3.478 |
| 1000w | " | 4.494 |
| Light fixture arms, cobra head, 6', high press. sodium | | |
| 100w | EA. | 2.000 |
| 150w | " | 2.500 |
| 250w | " | 2.500 |
| 400w | " | 2.963 |
| Flood, metal halide | | |
| 400w | EA. | 3.478 |
| 1000w | " | 4.494 |
| 1500w | " | 6.015 |
| Mercury vapor | | |
| 250w | EA. | 2.759 |
| 400w | " | 3.478 |
| Incandescent | | |
| 300w | EA. | 1.739 |
| 500w | " | 2.000 |
| 1000w | " | 3.200 |
| **16510.90 POWER LINE FILTERS** | | |
| Heavy duty power line filter, 240v | | |
| 100a | EA. | 10.000 |

# 16 ELECTRICAL

## LIGHTING

| 16510.90 POWER LINE FILTERS | UNIT | MAN/ HOURS |
|---|---|---|
| 300a | EA. | 16.000 |
| 600a | " | 24.242 |

| 16610.30 UNINTERRUPTIBLE POWER | UNIT | MAN/ HOURS |
|---|---|---|
| Uninterruptible power systems, (U.P.S.), 3kva | EA. | 8.000 |
| 5 kva | " | 11.004 |
| 7.5 kva | " | 16.000 |
| 10 kva | " | 21.978 |
| 15 kva | " | 22.857 |
| 20 kva | " | 24.024 |
| 25 kva | " | 25.000 |
| 30 kva | " | 25.974 |
| 35 kva | " | 27.027 |
| 40 kva | " | 27.972 |
| 45 kva | " | 28.986 |
| 50 kva | " | 29.963 |
| 62.5 kva | " | 32.000 |
| 75 kva | " | 34.934 |
| 100 kva | " | 36.036 |
| 150 kva | " | 50.000 |
| 200 kva | " | 55.172 |
| 300 kva | " | 74.766 |
| 400 kva | " | 89.888 |
| 500 kva | " | 110 |

| 16670.10 LIGHTNING PROTECTION | UNIT | MAN/ HOURS |
|---|---|---|
| Lightning protection | | |
| Copper point, nickel plated, 12' | | |
| 1/2" dia. | EA. | 1.000 |
| 5/8" dia. | " | 1.000 |

## COMMUNICATIONS

| 16720.10 FIRE ALARM SYSTEMS | UNIT | MAN/ HOURS |
|---|---|---|
| Master fire alarm box, pedestal mounted | EA. | 16.000 |
| Master fire alarm box | " | 6.015 |
| Box light | " | 0.500 |
| Ground assembly for box | " | 0.667 |
| Bracket for pole type box | " | 0.727 |
| Pull station | | |
| Waterproof | EA. | 0.500 |
| Manual | " | 0.400 |
| Horn, waterproof | " | 1.000 |
| Interior alarm | " | 0.727 |
| Coded transmitter, automatic | " | 2.000 |
| Control panel, 8 zone | " | 8.000 |
| Battery charger and cabinet | " | 2.000 |

## COMMUNICATIONS

| 16720.10 FIRE ALARM SYSTEMS | UNIT | MAN/ HOURS |
|---|---|---|
| Batteries, nickel cadmium or lead calcium | EA. | 5.000 |
| CO2 pressure switch connection | " | 0.727 |
| Annunciator panels | | |
| Fire detection annunciator, remote type, 8 zone | EA. | 1.818 |
| 12 zone | " | 2.000 |
| 16 zone | " | 2.500 |
| Fire alarm systems | | |
| Bell | EA. | 0.615 |
| Weatherproof bell | " | 0.667 |
| Horn | " | 0.727 |
| Siren | " | 2.000 |
| Chime | " | 0.615 |
| Audio/visual | " | 0.727 |
| Strobe light | " | 0.727 |
| Smoke detector | " | 0.667 |
| Heat detection | " | 0.500 |
| Thermal detector | " | 0.500 |
| Ionization detector | " | 0.533 |
| Duct detector | " | 2.759 |
| Test switch | " | 0.500 |
| Remote indicator | " | 0.571 |
| Door holder | " | 0.727 |
| Telephone jack | " | 0.296 |
| Fireman phone | " | 1.000 |
| Speaker | " | 0.800 |
| Remote fire alarm annunciator panel | | |
| 24 zone | EA. | 6.667 |
| 48 zone | " | 13.008 |
| Control panel | | |
| 12 zone | EA. | 2.963 |
| 16 zone | " | 4.444 |
| 24 zone | " | 6.667 |
| 48 zone | " | 16.000 |
| Power supply | " | 1.509 |
| Status command | " | 5.000 |
| Printer | " | 1.509 |
| Transponder | " | 0.899 |
| Transformer | " | 0.667 |
| Transceiver | " | 0.727 |
| Relays | " | 0.500 |
| Flow switch | " | 2.000 |
| Tamper switch | " | 2.963 |
| End of line resistor | " | 0.348 |
| Printed ckt. card | " | 0.500 |
| Central processing unit | " | 6.154 |
| UPS backup to c.p.u. | " | 8.999 |
| Smoke detector, fixed temp. & rate of rise comb. | " | 1.600 |

| 16720.50 SECURITY SYSTEMS | UNIT | MAN/ HOURS |
|---|---|---|
| Sensors | EA. | |
| Balanced magnetic door switch, surface mounted | " | 0.500 |
| With remote test | " | 1.000 |
| Flush mounted | " | 1.860 |

## COMMUNICATIONS

### 16720.50 SECURITY SYSTEMS

| | UNIT | MAN/ HOURS |
|---|---|---|
| Mounted bracket | EA. | 0.348 |
| Mounted bracket spacer | " | 0.348 |
| Photoelectric sensor, for fence | " | |
| 6 beam | " | 2.759 |
| 9 beam | " | 4.255 |
| Photoelectric sensor, 12 volt dc | " | |
| 500' range | " | 1.600 |
| 800' range | " | 2.000 |
| Capacitance wire grid kit | " | |
| Surface | " | 1.000 |
| Duct | " | 1.600 |
| Tube grid kit | " | 0.500 |
| Vibration sensor, 30 max per zone | " | 0.500 |
| Audio sensor, 30 max per zone | " | 0.500 |
| Inertia sensor | " | |
| Outdoor | " | 0.727 |
| Indoor | " | 0.500 |
| Ultrasonic transmitter, 20 max per zone | | |
| Omni-directional | EA. | 1.600 |
| Directional | " | 1.333 |
| Transceiver | | |
| Omni-directional | EA. | 1.000 |
| Directional | " | 1.000 |
| Passive infra-red sensor, 20 max per zone | " | 1.600 |
| Access/secure control unit, balanced magnetic switch | " | 1.600 |
| Photoelectric sensor | " | 1.600 |
| Photoelectric fence sensor | " | 1.600 |
| Capacitance sensor | " | 1.739 |
| Audio and vibration sensor | " | 1.600 |
| Inertia sensor | " | 1.600 |
| Ultrasonic sensor | " | 1.739 |
| Infra-red sensor | " | 2.000 |
| Monitor panel, with access/secure tone, standard | " | 1.739 |
| High security | " | 2.000 |
| Emergency power indicator | " | 0.500 |
| Monitor rack with 115v power supply | | |
| 1 zone | EA. | 1.000 |
| 10 zone | " | 2.500 |
| Monitor cabinet, wall mounted | | |
| 1 zone | EA. | 1.000 |
| 5 zone | " | 1.600 |
| 10 zone | " | 1.739 |
| 20 zone | " | 2.000 |
| Floor mounted, 50 zone | " | 4.000 |
| Security system accessories | | |
| Tamper assembly for monitor cabinet | EA. | 0.444 |
| Monitor panel blank | " | 0.348 |
| Audible alarm | " | 0.500 |
| Audible alarm control | " | 0.348 |
| Termination screw, terminal cabinet | | |
| 25 pair | EA. | 1.600 |
| 50 pair | " | 2.500 |
| 150 pair | " | 5.000 |
| Universal termination, cabinets & panel | | |

## COMMUNICATIONS

### 16720.50 SECURITY SYSTEMS

| | UNIT | MAN/ HOURS |
|---|---|---|
| Remote test | EA. | 1.739 |
| No remote test | " | 0.727 |
| High security line supervision termination | " | 1.000 |
| Door cord for capacitance sensor, 12" | " | 0.500 |
| Insulation block kit for capacitance sensor | " | 0.348 |
| Termination block for capacitance sensor | " | 0.348 |
| Guard alert display | " | 0.615 |
| Uninterrupted power supply | | |
| Plug-in 40kva transformer | | |
| 12 volt | EA. | 0.348 |
| 18 volt | " | 0.348 |
| 24 volt | " | 0.348 |
| Test relay | " | 0.348 |
| Coaxial cable, 50 ohm | L.F. | 0.006 |
| Door openers | EA. | 0.500 |
| Push buttons | | |
| Standard | EA. | 0.348 |
| Weatherproof | " | 0.444 |
| Bells | " | 0.727 |
| Horns | | |
| Standard | EA. | 1.000 |
| Weatherproof | " | 1.250 |
| Chimes | " | 0.667 |
| Flasher | " | 0.615 |
| Motion detectors | " | 1.509 |
| Intercom units | " | 0.727 |
| Remote annunciator | " | 5.000 |

### 16740.10 TELEPHONE SYSTEMS

| | UNIT | MAN/ HOURS |
|---|---|---|
| Communication cable | | |
| 25 pair | L.F. | 0.026 |
| 100 pair | " | 0.029 |
| 150 pair | " | 0.033 |
| 200 pair | " | 0.040 |
| 300 pair | " | 0.042 |
| 400 pair | " | 0.044 |
| Cable tap in manhole or junction box | | |
| 25 pair cable | EA. | 3.810 |
| 50 pair cable | " | 7.547 |
| 75 pair cable | " | 11.268 |
| 100 pair cable | " | 15.094 |
| 150 pair cable | " | 22.222 |
| 200 pair cable | " | 29.630 |
| 300 pair cable | " | 44.444 |
| 400 pair cable | " | 61.538 |
| Cable terminations, manhole or junction box | | |
| 25 pair cable | EA. | 3.756 |
| 50 pair cable | " | 7.477 |
| 100 pair cable | " | 15.094 |
| 150 pair cable | " | 22.222 |
| 200 pair cable | " | 29.630 |
| 300 pair cable | " | 44.444 |
| 400 pair cable | " | 61.538 |

# 16 ELECTRICAL

## COMMUNICATIONS

### 16780.10 ANTENNAS AND TOWERS

| | UNIT | MAN/ HOURS |
|---|---|---|
| Guy cable, alumaweld | | |
| 1x3, 7/32" | L.F. | 0.050 |
| 1x3, 1/4" | " | 0.050 |
| 1x3, 25/64" | " | 0.059 |
| 1x19, 1/2" | " | 0.070 |
| 1x7, 35/64" | " | 0.080 |
| 1x19, 13/16" | " | 0.100 |
| Preformed alumaweld end grip | | |
| 1/4" cable | EA. | 0.100 |
| 3/8" cable | " | 0.100 |
| 1/2" cable | " | 0.145 |
| 9/16" cable | " | 0.200 |
| 5/8" cable | " | 0.250 |
| Fiberglass guy rod, white epoxy coated | | |
| 1/4" dia. | L.F. | 0.145 |
| 3/8" dia | " | 0.145 |
| 1/2" dia | " | 0.200 |
| 5/8" dia | " | 0.250 |
| Preformed glass grip end grip, guy rod | | |
| 1/4" dia. | EA. | 0.145 |
| 3/8" dia. | " | 0.200 |
| 1/2" dia. | " | 0.250 |
| 5/8" dia. | " | 0.250 |
| Spelter socket end grip, 1/4" dia. guy rod | | |
| Standard strength | EA. | 0.500 |
| High performance | " | 0.500 |
| 3/8" dia. guy rod | | |
| Standard strength | EA. | 0.348 |
| High performance | " | 0.500 |
| Timber pole, Douglas Fir | | |
| 80-85 ft | EA. | 19.512 |
| 90-95 ft | " | 22.222 |
| Southern yellow pine | | |
| 35-45 ft | EA. | 10.959 |
| 50-55 ft | " | 14.035 |

### 16780.50 TELEVISION SYSTEMS

| | UNIT | MAN/ HOURS |
|---|---|---|
| TV outlet, self terminating, w/cover plate | EA. | 0.308 |
| Thru splitter | " | 1.600 |
| End of line | " | 1.333 |
| In line splitter multitap | | |
| 4 way | EA. | 1.818 |
| 2 way | " | 1.702 |
| Equipment cabinet | " | 1.600 |
| Antenna | | |
| Broad band uhf | EA. | 3.478 |
| Lightning arrester | " | 0.727 |
| TV cable | L.F. | 0.005 |
| Coaxial cable rg | " | 0.005 |
| Cable drill, with replacement tip | EA. | 0.500 |
| Cable blocks for in-line taps | " | 0.727 |
| In-line taps ptu-series 36 tv system | " | 1.143 |

## COMMUNICATIONS

### 16850.10 ELECTRIC HEATING

| | UNIT | MAN/ HOURS |
|---|---|---|
| Baseboard heater | | |
| 2', 375w | EA. | 1.000 |
| 3', 500w | " | 1.000 |
| 4', 750w | " | 1.143 |
| 5', 935w | " | 1.333 |
| 6', 1125w | " | 1.600 |
| 7', 1310w | " | 1.818 |
| 8', 1500w | " | 2.000 |
| 9', 1680w | " | 2.222 |
| 10', 1875w | " | 2.286 |
| Unit heater, wall mounted | | |
| 1500w | EA. | 1.667 |
| 2500w | " | 1.818 |
| 4000w | " | 2.286 |
| Thermostat | | |
| Integral | EA. | 0.500 |
| Line voltage | " | 0.500 |
| Electric heater connection | " | 0.250 |

# Supporting Construction Reference Data

This section contains information, text, charts and tables on various aspects of construction. The intent is to provide the user with a better understanding of unfamiliar areas in order to be able to estimate better. This information includes actual takeoff data for some areas and also selected explanations of common construction materials, methods and common practices.

# Additional Costs Involved in Remodeling Construction

## SMALL AMOUNTS OF WORK

### Material

Since remodeling work often entails a smaller scope of work, many of the materials that are bought at a discount in large quantities for a new construction project have to be purchased at a lesser discount or at the regular retail price. To take this factor into account, it is prudent to add an additional 10 percent to the cost of materials for small remodeling projects.

### Labor

In order to obtain maximum labor efficiency and normal production, quantities of work must be sufficient to keep workers busy for long enough periods of time, usually a full day. Even with a full day's worth of a particular task, the start-up and wind-down/cleanup (non-productive time) of a typical day take a significant amount of time. When the task requires less than a full day, the non-productive time increases as a percentage of the work installed. Therefore, the cost of labor per unit increases to a minimum of 2 to 4 hours for the installing function.

## PATCHING TO MATCH

Matching and fitting new materials to the pre-existing construction is usually a costly factor in remodeling work, whether it be the exterior siding of a building or interior paint or plaster. Sometimes creative, and often expensive, methods have to be used to match colors, patterns and materials. Adding 10 to 30 percent to the labor costs will insulate against this.

## WORKING AROUND ONGOING OPERATIONS

One of the most time-consuming parts of a remodeling job is the distractions caused by working around people. Just normal traffic flow of people through or near a work area affects the way the work gets done. For example, when a contractor has to take precautions during demolition work for the safety of passersby, the job will take longer and cost more. These costs increase even more when work is being done in an office or retail store when the business must continue to function during the remodeling. Additng 5 to 50 percent (depending upon severity) to the labor estimate will provide for this.

## DIFFICULT ACCESS

Contractors often find their access to work areas restricted in any number of ways. In remodeling work, the contractor loses the luxury of a logical building process that a new construction project enjoys. Walls have to be cut and patched for the installation of new mechanical or electrical systems; large building materials do not fit through existing doors or windows; even access to the structure's exterior can be limited by added features such as urban locations, landscaping, fences, garages and swimming pools. Delivery and distribution of materials should be approached on a man-hour basis. Add 5 to 20 percent to the labor process for a realistic delivery allowances.

## WORKING AROUND SENSITIVE EQUIPMENT

Doing a remodeling job quickly and efficiently usually means being able to make a mess at first and then cleaning up after. Some job locations, however, only allow a minimum amount of dust and debris. When working in close proximity to sensitive machines, computers especially, a job can become up to 100 percent more expensive for labor. Also to be factored in are the hanging of plastic curtains, longer dumpster rental, and laborers cleaning up on a continual basis.

## WORKING OFF PEAK HOURS

When a remodeling job involves accommodating a functioning business it is often to the contractor's advantage to work odd hours. Sometimes it is even a requirement. Many retail operations that are open late, or even 24 hours, schedule remodeling work around the second or third shift. While this is not overtime, it is more costly for the contractor to hire subcontractors and tradesmen. However, the distractions and accommodations necessitated by building occupants are much less than if the remodeling work was done during the normal business day. The greater efficiency and speed allowed by working in a quiet environment helps to offset the higher labor costs and reduced productivity. Still, it is justifiable to add 10 to 15 percent to the estimated cost of labor scheduled during off peak hours.

## INEFFICIENT WORK METHODS

Working around people and adverse conditions often make the most efficient work methods and the proper tools unusable. The less efficient the method, the more must be spent on additional labor and equipment rental, such as scaffolding and tools a remodeling contractor may not normally have on hand. Depending on the job, another 10 to 40 percent may be added to the labor estimate.

## TYPICAL BUILDING COST BROKEN DOWN BY CSI FORMAT

*(Commercial Construction)*

| Division | | New Construction | Remodeling Construction |
|---|---|---|---|
| 1. | General Requirements | 6 to 8% | Up to 30% |
| 2. | Sitework | 4 to 6% | |
| 3. | Concrete | 15 to 20% | |
| 4. | Masonry | 8 to 12% | |
| 5. | Metals | 5 to 7% | |
| 6. | Wood And Plastics | 1 to 5% | |
| 7. | Thermal And Moisture Protection | 4 to 6% | |
| 8. | Doors And Windows | 5 to 7% | Up to 30% (Divisions 8–9) |
| 9. | Finishes | 8 to 12% | |
| 10. | Specialties | 6 to 10% (Divisions 10–14) | |
| 11. | Architectural Equipment | | |
| 12. | Furnishings | | |
| 13. | Special Construction | | |
| 14. | Conveying Systems | | |
| 15. | Mechanical | 15 to 25% | Up to 40% (Divisions 15–16) |
| 16. | Electrical | 8 to 12% | |
| | **Total Cost** | 100% | |

## CONVERSION FACTORS

| Change | To | Multiply By |
|---|---|---|
| Atmospheres | Pounds per square inch | 14.696 |
| Atmospheres | Inches of mercury | 29.92 |
| Atmospheres | Feet of water | 34 |
| Barrels, oil | Gallons, of oil | 42 |
| Barrels, cement | Pounds of cement | 376 |
| Bags or sacks, cement | Pounds of cement | 94 |
| Btu/min. | Foot-pounds/sec. | 12.96 |
| Btu/min. | Horse-power | 0.02356 |
| Btu/min. | Kilowatts | 0.01757 |
| Btu/min. | Watts | 17.57 |
| Centimeters | Inches | 0.3937 |
| Centimeters of mercury | Atmospheres | 0.01316 |
| Centimeters of mercury | Feet of water | 0.4461 |
| Cubic inches | Cubic feet | 0.00058 |
| Cubic feet | Cubic inches | 1728 |
| cubic feet | Cubic yards | 0.03703 |
| Cubic yards | Cubic feet | 27 |
| Cubic inches | Gallons | 0.00433 |
| Cubic feet | Gallons | 7.48 |
| Feet | Inches | 12 |
| Feet | Yards | 0.3333 |
| Yards | Feet | 3 |
| Feet of water | Atmospheres | 0.02950 |
| Feet of water | Inches of mercury | 0.8826 |
| Gallons | Cubic Inches | 231 |
| Gallons | Cubic feet | 0.1337 |
| Gallons | Pounds of water | 8.33 |
| Gallons | Quarts | 4 |
| Gallons per min. | Cubic feet sec. | 0.002228 |
| Gallons per min. | Cubic feet hour | 8.0208 |
| Gallons water per min. | Tons water/24 hours | 6.0086 |
| Horse-power | Foot-lbs./sec. | 550 |
| Inches | Centimeters | 2.540 |
| Inches | Feet | 0.0833 |
| Inches | Millimeters | 25.4 |
| Inches of water | Pounds per Sq. inch | 0.0361 |
| Inches of water | Inches of mercury | 0.0735 |
| Inches of water | Ounces per square inch | 0.578 |
| Inches of water | Ounces per square foot | 5.2 |
| Inches of mercury | Inches of water | 13.6 |
| Inches of mercury | Feet of water | 1.1333 |
| Inches of mercury | Pounds per square inch | 0.4914 |
| Kilometers | Miles | 0.6214 |
| Meters | Inches | 39.37 |
| Miles | Feet | 5280 |
| Millimeters | Centimeters | 0.1 |
| Millimeters | Inches | 0.03937 |
| Ounces (fluid) | Cubic inches | 1.805 |
| Ounces | Pounds | 0.0625 |
| Pounds | Ounces | 16 |
| Pounds per square inch | Inches of water | 27.72 |
| Pounds per square inch | Feet of water | 2.310 |
| Pounds per square inch | Inches of mercury | 2.04 |
| Pounds per square inch | Atmospheres | 0.0681 |
| Quarts | Cubic Inches | 67.20 |
| Square Inches | Square feet | 0.00694 |
| Square Feet | Square inches | 144 |
| Square Feet | Square yards | 0.11111 |
| Square yards | Square feet | 9 |
| Square miles | Acres | 640 |
| Short tons | Pounds | 2000 |
| Short tons | Long tons | 0.89285 |
| Tons of water/24 hours | Gallons per minute | 0.16643 |
| Yards | Feet | 3 |
| Yards | Centimeters | 91.44 |
| Yards | Inches | 36 |

## CONVERSION CALCULATIONS

### Commercial Measure

| | |
|---|---|
| 16 drams | = 1 ounce |
| 16 ounces | = 1 pound |
| 2,000 pounds | = 1 ton |

### Long Measure

| | |
|---|---|
| 12 inches | = 1 foot |
| 3 feet | = 1 yard |
| 16½ feet | = 1 rod |
| 40 rods | = 1 furlong |
| 8 furlongs (5,280 ft) = | = 1 mile |
| 3 miles | = 1 league |

### Square Measure

| | |
|---|---|
| 144 square inches | = 1 square foot |
| 9 square feet | = 1 square yard |
| 30¼ square yards | = 1 square rod |
| 160 square rods | = 1 acre |
| 4840 square yards | = 1 acre |
| 640 acres | = 1 square mile |
| 36 square miles | = 1 township |

### Surveyors Measure

| | |
|---|---|
| 7.92 inches | = 1 link |
| 25 links | = 1 rod |
| 4 rods (66 ft.) | = 1 chain |
| 10 chains | = 1 furlong |
| 8 furlongs | = 1 mile |
| 1 square mile | = 1 section |

### Cubic Measure

| | |
|---|---|
| 1728 cubic inches | = 1 cubic foot |
| 27 cubic feet | = 1 cubic yard |
| 128 cubic feet | = 1 cord (wood/stone) |
| 231 cubic inches | = 1 U.S. gallon |
| 7.48 U.S. Gallons | = 1 cubic foot |
| 2150.4 cubic inches | = 1 U.S. bushel |

### Liquid Measure

| | |
|---|---|
| 4 fluid ounces | = 1 gill |
| 4 gills | = 1 pint |
| 2 pints | = 1 quart |
| 4 quarts | = 1 gallon |
| 9 gallons | = 1 firkin |
| 31½ gallons | = 1 barrel |
| 2 barrels | = 1 hogshead |

### Dry Measure

| | |
|---|---|
| 2 pints | = 1 quart |
| 8 quarts | = 1 peck |
| 4 pecks | = 1 bushel |
| 2150.42 cubic inches | = 1 bushel |

**SQUARE**

$$A = a^2$$

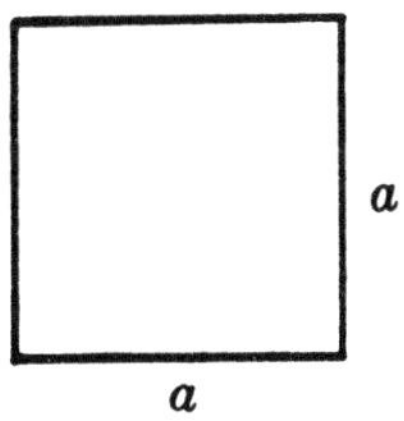


**RECTANGLE**

$$A = bh$$

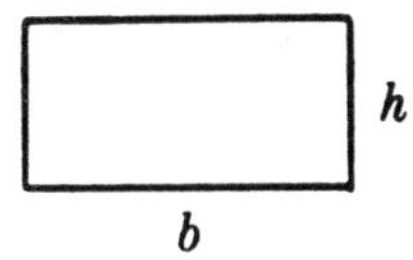


**TRIANGLE**

$$A = \frac{1}{2} bh$$

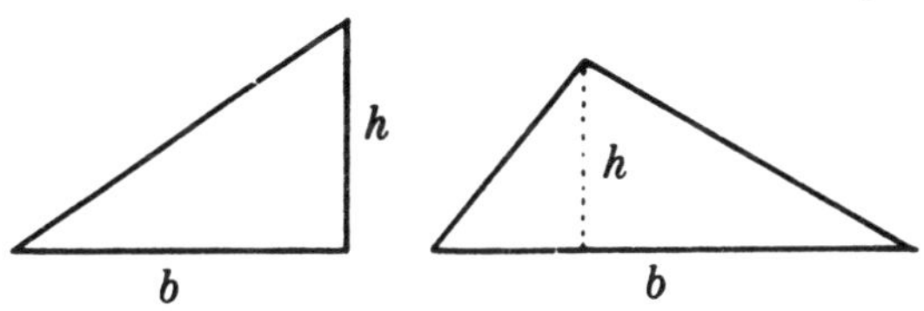


**PARALLELOGRAM**

$$A = bh = ab\ Sin\phi$$

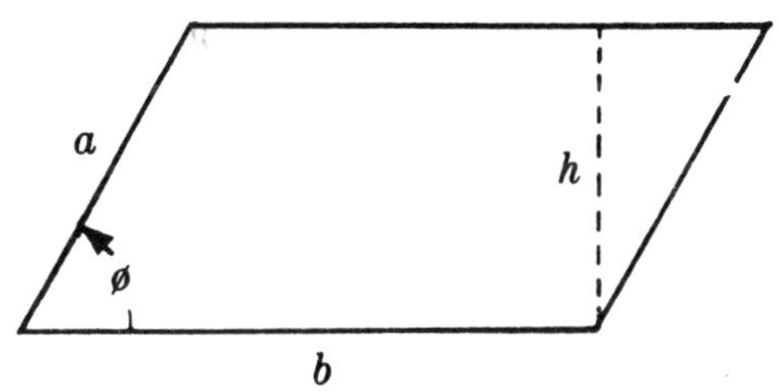


**TRAPEZOID**

$$A = \left(\frac{a+b}{2}\right)h$$

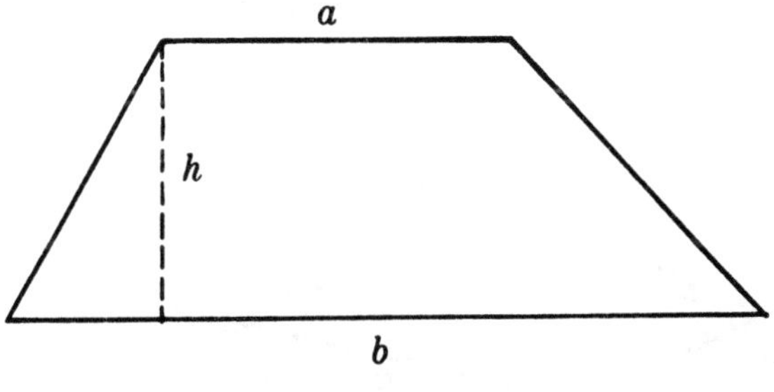


**CIRCLE**

$$A = \pi r^2 = \frac{\pi d^2}{4}$$

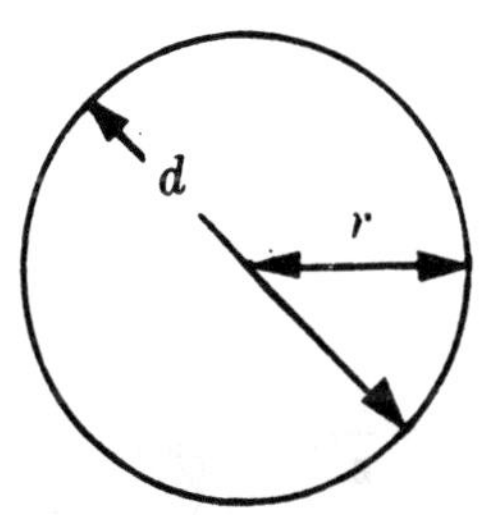


$$Circumference = C = 2\pi r = \pi d$$

**ELLIPSE**

$A = 0.7854\ ab$

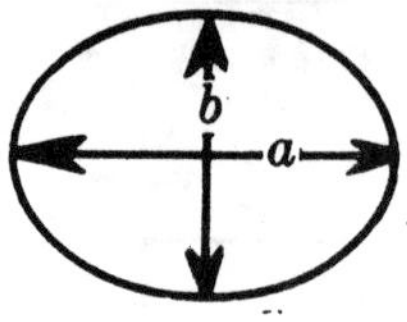


**PARABOLA**

$A = \frac{2}{3} bh$

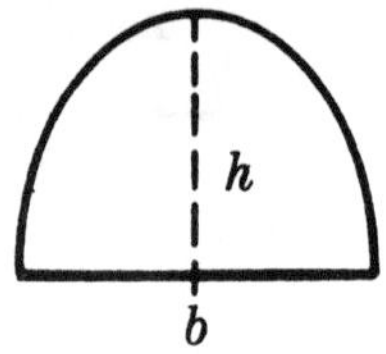


*VOLUMES*

**CUBE**

$V = a^3$

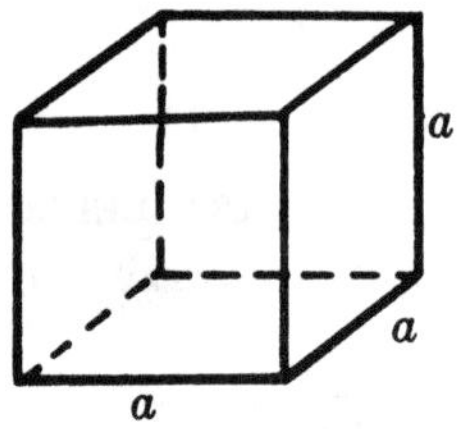


**CYLINDER**

$V = \pi r^2 h$

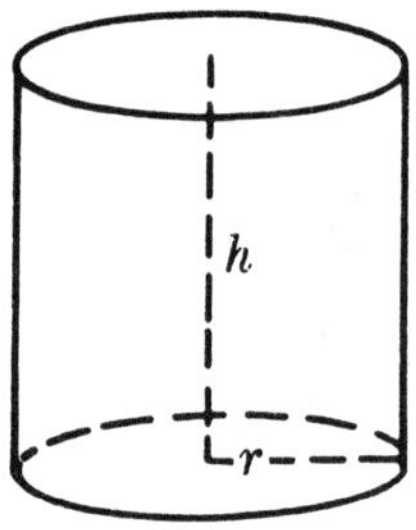


**PYRAMID**

$V = \frac{1}{3}(Base)\ h$

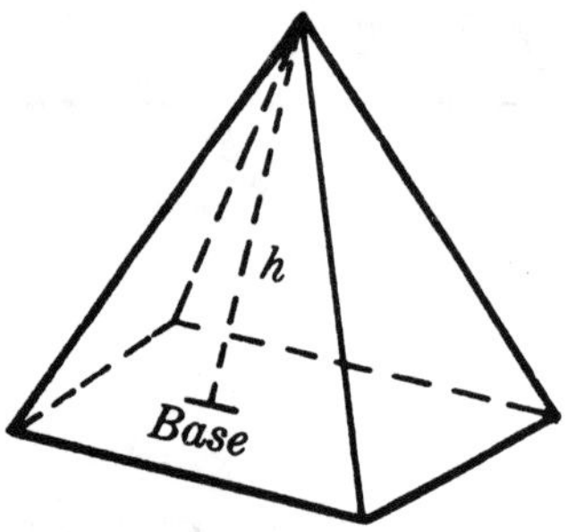


**CONE**

$V = \frac{1}{3} \pi r^2 h$

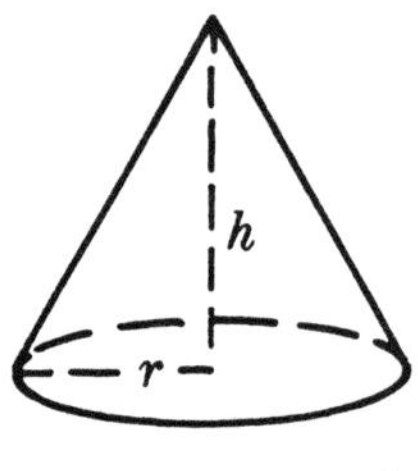


**SPHERE**

$V = \frac{4}{3} \pi r^3 = \frac{1}{6} \pi d^3$

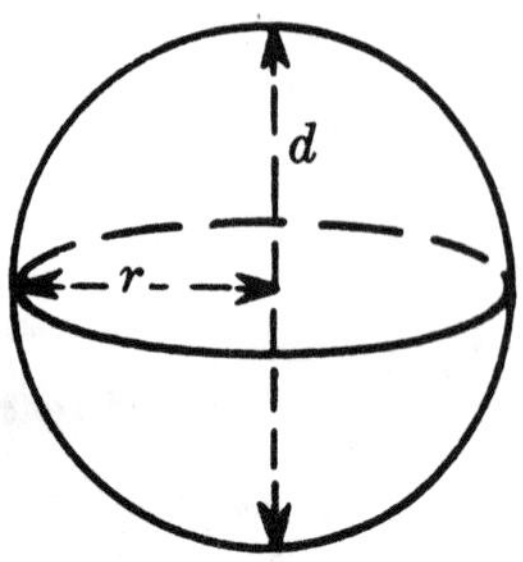


**WEDGE**

$V = \frac{1}{2} abc$

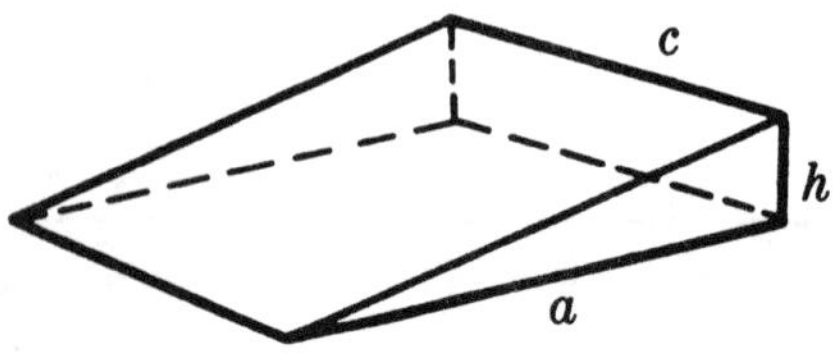

# CONVERSION FACTORS
## ENGLISH TO SI (SYSTEM INTERNATIONAL)

| To Convert from | To | Multiply by |
|---|---|---|
| **LENGTH** | | |
| Inches | Millimetres | 25.4[a] |
| Feet | Metres | 0.3048[a] |
| Yards | Metres | 0.9144[a] |
| Miles (statute) | Kilometres | 1.609 |
| **AREA** | | |
| Square inches | Square millimetres | 645.2 |
| Square feet | Square metres | 0.0929 |
| Square yards | Square metres | 0.8361 |
| **VOLUME** | | |
| Cubic inches | Cubic millimetres | 16.387 |
| Cubic feet | Cubic metres | 0.02832 |
| Cubic yards | Cubic metres | 0.7646 |
| Gallons (U.S. liquid)[b] | Cubic metres[c] | 0.003785 |
| Gallons (Canadian liquid)[b] | Cubic metres[c] | 0.004546 |
| Ounces (U.S. liquid)[b] | Millilitres[c, d] | 29.57 |
| Quarts (U.S. liquid)[b] | Litres[c, d] | 0.9464 |
| Gallons (U.S. liquid)[b] | Litres[c] | 3.785 |
| **FORCE** | | |
| Kilograms force | Newtons | 9.807 |
| Pounds force | Newtons | 4.448 |
| Pounds force | Kilograms force[d] | 0.4536 |
| Kips | Newtons | 4448 |
| Kips | Kilograms force[d] | 453.6 |
| **PRESSURE, STRESS, STRENGTH (FORCE PER UNIT AREA)** | | |
| Kilograms force per sq. centimetre | Megapascals | 0.09807 |
| Pounds force per square inch (psi) | Megapascals | 6895 |
| Kips per square inch | Megapascals | 6.895 |
| Pounds force per square inch (psi) | Kilograms force per square centimetre[d] | 0.07031 |
| Pounds force per square foot | Pascals | 47.88 |
| Pounds force per square foot | Kilograms force per square metre[d] | 4.882 |
| **BENDING MOMENT OR TORQUE** | | |
| Inch-pounds force | Metre-kilog. force[d] | 0.01152 |
| Inch-pounds force | Newton-metres | 0.1130 |
| Foot-pounds force | Metre-kilog. force[d] | 0.1383 |
| Foot-pounds force | Newton-metres | 1.356 |
| Metre-kilograms force | Newton-metres | 9.807 |
| **MASS** | | |
| Ounce (avoirdupois) | Grams | 28.35 |
| Pounds (avoirdupois) | Kilograms | 0.4536 |
| Tons (metric) | Kilograms | 1000[a] |
| Tons, short (2000 pounds) | Kilograms | 907.2 |
| Tons, short (2000 pounds) | Megagrams[e] | 0.9072 |
| **MASS PER UNIT VOLUME** | | |
| Pounds mass per cubic foot | Kilog. per cubic metre | 16.02 |
| Pounds mass per cubic yard | Kilog. per cubic metre | 0.5933 |
| Pds. mass per gallon (U.S. liquid)[b] | Kilog. per cubic metre | 119.8 |
| Pds. mass p/gal. (Canadian liquid)[b] | Kilog. per cubic metre | 99.78 |
| **TEMPERATURE** | | |
| Degrees Fahrenheit | Degrees Celsius | $tK = (1F - 32)/1.8$ |
| Degrees Fahrenheit | Degrees Kelvin | $tK = (1F + 459.67)/1.8$ |
| Degree Celsius | Degree Kelvin | $tK = 1C + 273.15$ |

[a] The factor given is exact.
[b] One U.S. gallon equals 0.8327 Canadian gallon.
[c] 1 litre = 1000 millilitres = 10,000 cubic centimetres = 1 cubic decimetre = 0.001 cubic metre.
[d] Metric but not SI unit.
[e] Called "tonne" in England. Called "metric ton" in other metric systems.

## TRENCH BRACING

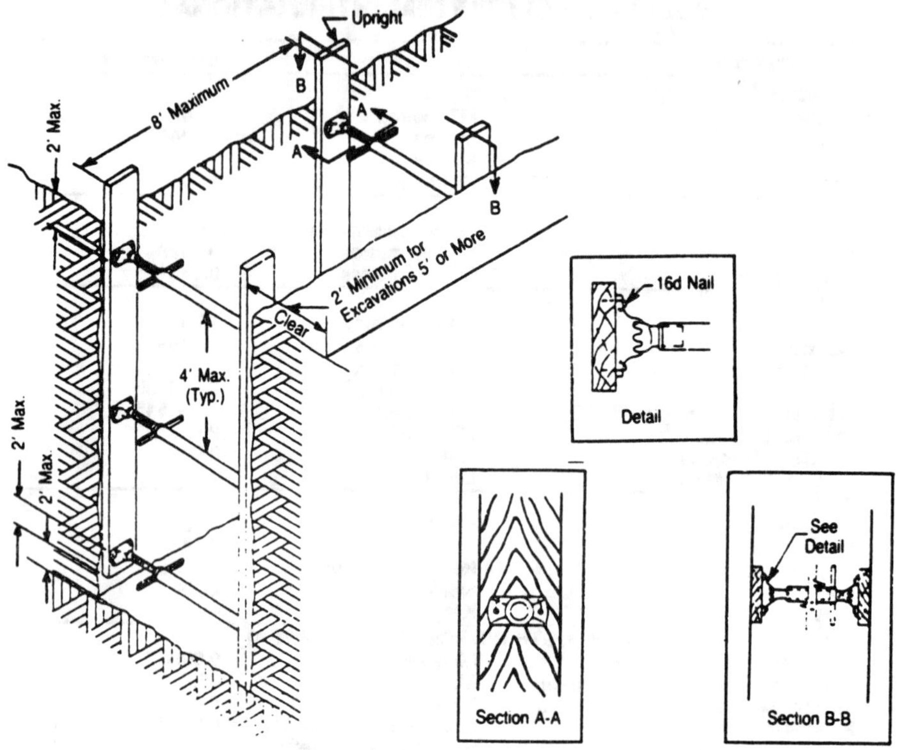


## CLOSED VERTICAL SHEETING

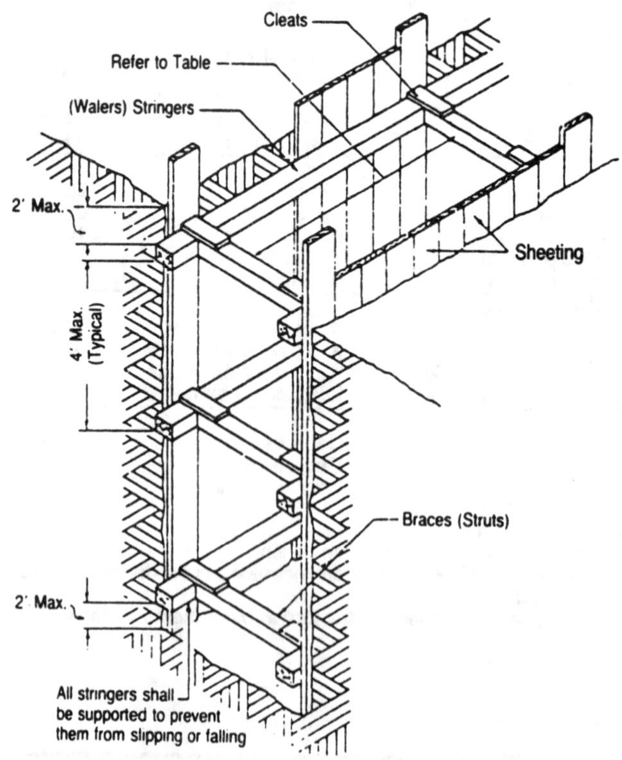

## STANDARD NOMENCLATURE FOR STREET CONSTRUCTION

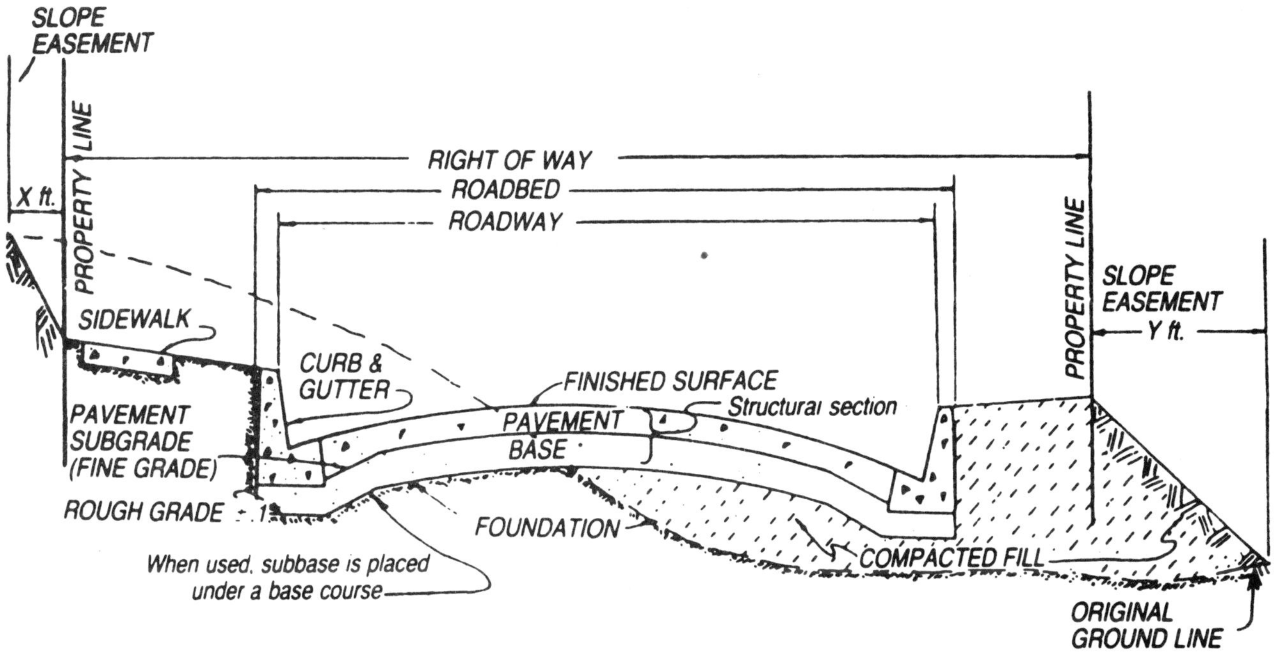

## CONCRETE MASONRY PAVING UNITS

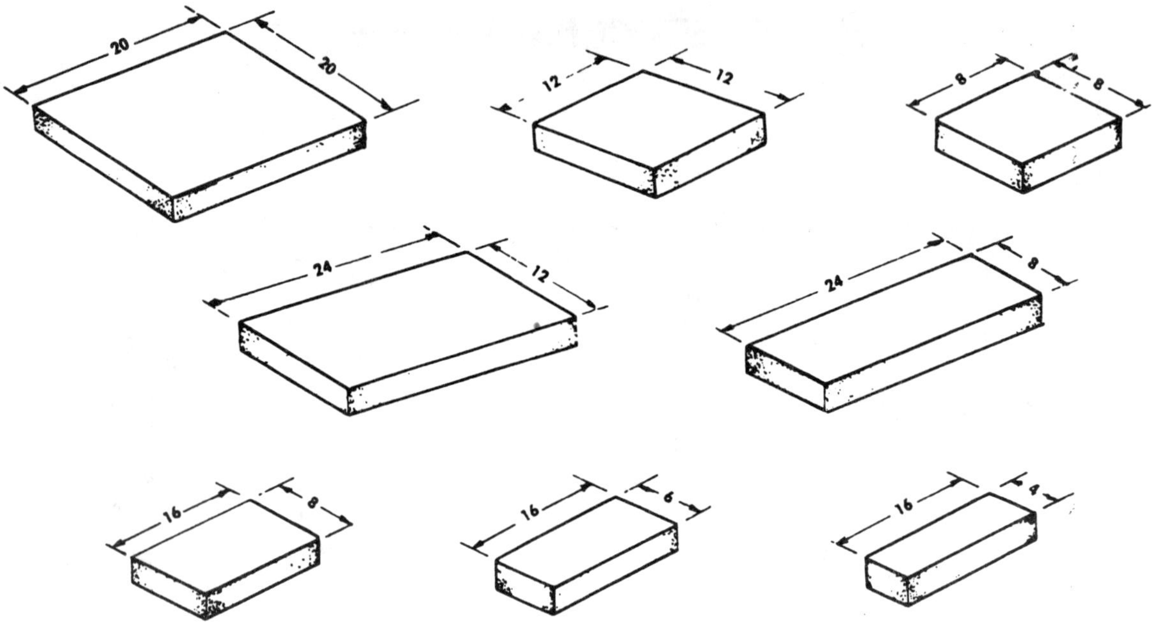


**NOTE:** Sizes are nominal and will vary by manufacturer.

**HEXAGON PAVER UNITS**
**Various Sizes Available**

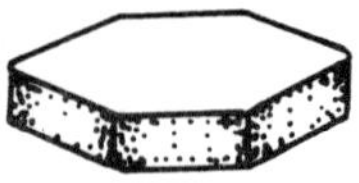

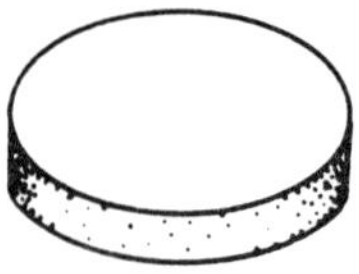

**ROUND PAVING UNITS**
**Various Sizes Available**

## VEHICULAR PAVING UNITS

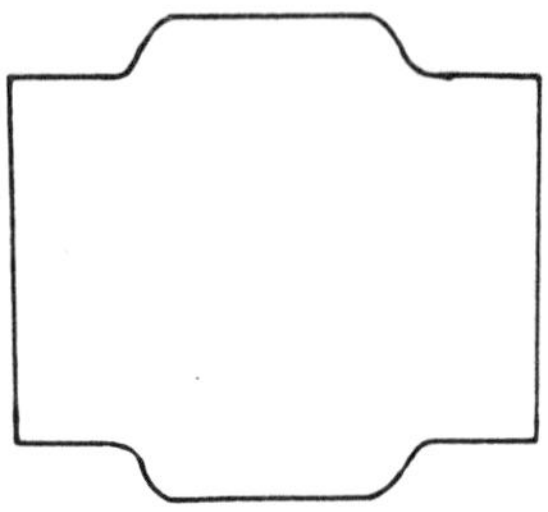

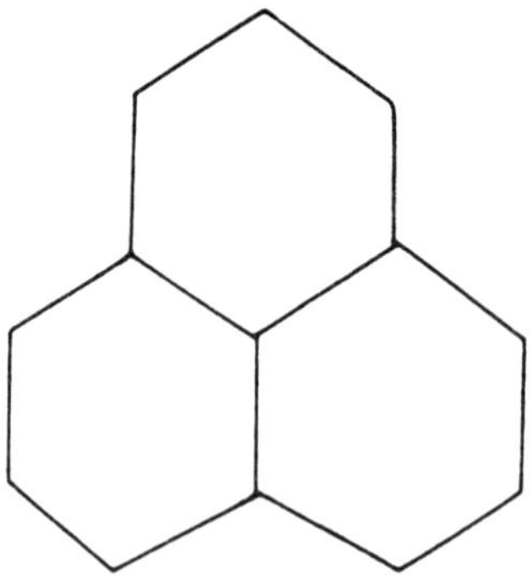

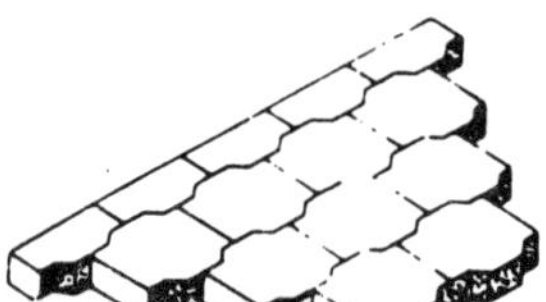

**INTERLOCKING PAVER**
**7¼" x 3" x 8½"**

**INTERLOCKING PAVER**
**12" x 3⅝" x 12"**

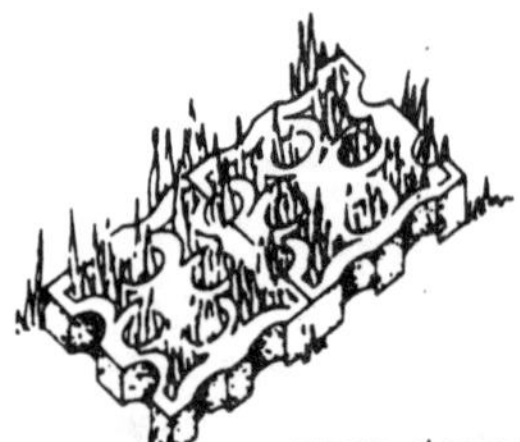

**TURF PAVER**
**24" x 3⅝" x 24"**

## CAPACITIES FOR SEPTIC TANKS SERVING AN INDIVIDUAL DWELLING

| No. of bedrooms | Capacity of tank (gals.) |
|---|---|
| 2 or less | 750 |
| 3 | 900 |
| 4 | 1,000 |

# CONCRETE / FORMWORK 03110

## FORM NOMENCLAURE

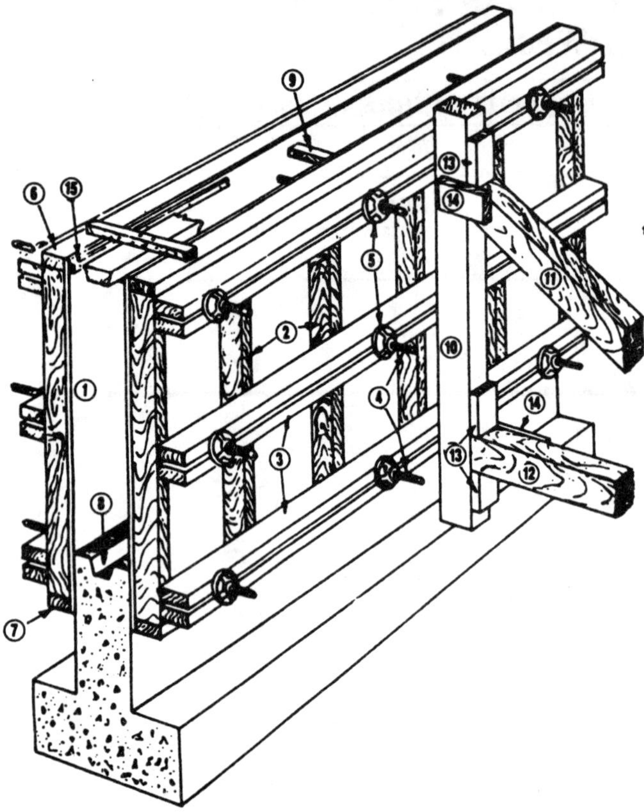

1. SHEATHING
2. STUDS
3. WALES
4. FORM BOLTS
5. NUT WASHER
6. TOP PLATE
7. BOTTOM PLATE
8. KEY-WAY
9. SPREADER
10. STRONGBACK
11. BRACE
12. STRUT
13. CLEATS
14. SCAB
14. POUR STRIP

## FALSEWORK NOMENCLATURE

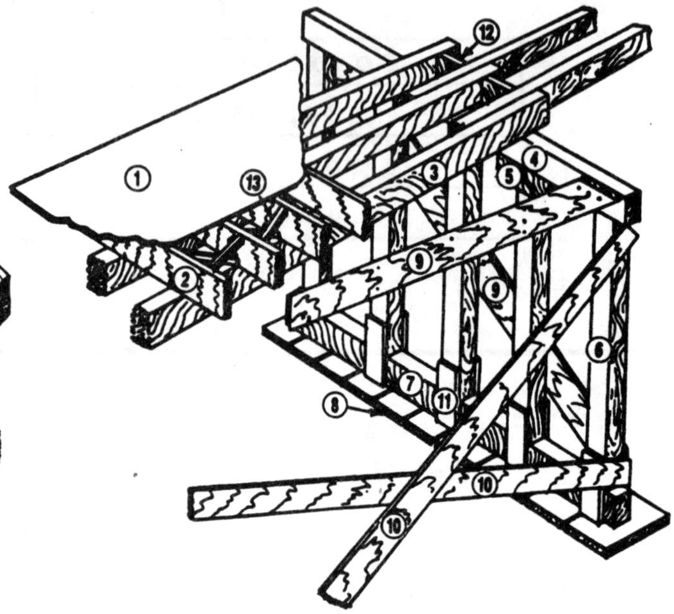

1. SHEATHING
2. JOIST
3. STRINGER
4. CAP
5. CORBEL
6. POST
7. SILL
8. FOOTING
9. SWAY BRACE
10. LONGITUDINAL BRACE
11. SCAB
12. BLOCKING
13. BRIDGING

## TYPICAL PAN-JOIST FORM CONSTRUCTION

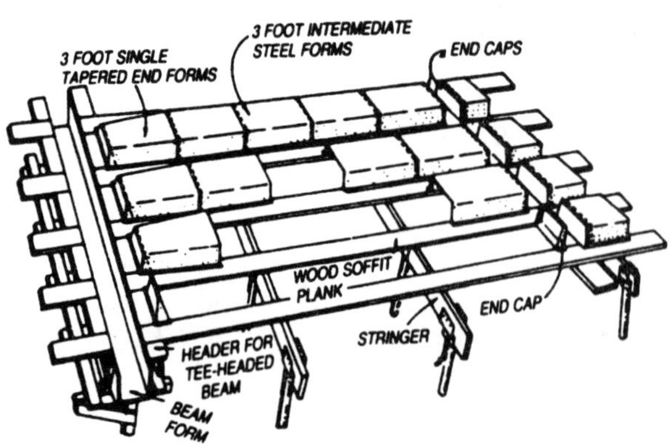


## TYPICAL WAFFLE SLAB FORM CONSTRUCTION

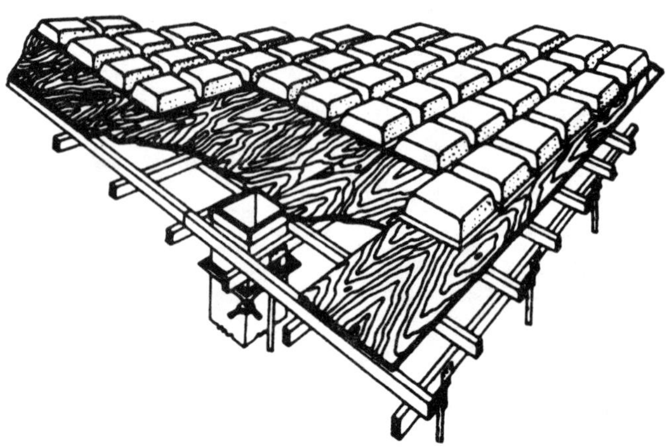

# CONCRETE / REINFORCING STEEL — 03210

## STANDARD SIZES OF STEEL REINFORCEMENT BARS

| STANDARD REINFORCEMENT BARS | | | | |
|---|---|---|---|---|
| | | Nominal Dimensions | | |
| Bar Designation Number* | Nominal Weight, lb. per ft. | Diameter, in. | Cross Sectional Area, sq. in. | Perimeter, in. |
| 3 | 0.376 | 0.375 | 0.11 | 1.178 |
| 4 | 0.668 | 0.500 | 0.20 | 1.571 |
| 5 | 1.043 | 0.625 | 0.31 | 1.963 |
| 6 | 1.502 | 0.750 | 0.44 | 2.356 |
| 7 | 2.044 | 0.875 | 0.60 | 2.749 |
| 8 | 2.670 | 1.000 | 0.79 | 3.142 |
| 9 | 3.400 | 1.128 | 1.00 | 3.544 |
| 10 | 4.303 | 1.270 | 1.27 | 3.990 |
| 11 | 5.313 | 1.410 | 1.56 | 4.430 |
| 14 | 7.65 | 1.693 | 2.25 | 5.32 |
| 18 | 13.60 | 2.257 | 4.00 | 7.09 |

*The bar numbers are based on the number of ⅛ inches included in the nominal diameter of the bar.

| Type of Steel and ASTM Specification No. | Size Nos. Inclusive | Grade | Tensile Strength Min., psi | Yield (a) Min., psi |
|---|---|---|---|---|
| Billet Steel A 615 | 3-11 | 40 | 70,000 | 40,000 |
| | 3-11 14, 18 | 60 | 90,000 | 60,000 |
| | 11, 14, 18 | 75 | 100,000 | 75,000 |

# CONCRETE / WELDED WIRE FABRIC — 03220

## WELDED WIRE FABRIC – COMMON STOCK STYLES OF WELDED WIRE FABRIC

| Style Designation | Steel Area sq. in. per ft. | | Weight Approx. lbs. per 100 sq. ft. |
|---|---|---|---|
| | Longit. | Transv. | |
| **Rolls** | | | |
| 6x6—W1.4xW1.4 | .03 | .03 | 21 |
| 6x6—W2xW2 | .04 | .04 | 29 |
| 6x6—W2.9xW2.9 | .06 | .06 | 42 |
| 6x6—W4xW4 | .08 | .08 | 58 |
| 4x4—W1.4xW1.4 | .04 | .04 | 31 |
| 4x4—W2xW2 | .06 | .06 | 43 |
| 4x4—W2.9xW2.9 | .09 | .09 | 62 |
| 4x4—W4xW4 | .12 | .12 | 86 |
| **Sheets** | | | |
| 6x6—W2.9xW2.9 | .06 | .06 | 42 |
| 6x6—W4xW4 | .08 | .08 | 58 |
| 6x6—W5.5xW5.5 | .11 | .11 | 80 |
| 4x4—W4xW4 | .12 | .12 | 86 |

# COMMON TYPES OF STEEL REINFORCEMENT BARS

ASTM specifications for billet steel reinforcing bars (A 615) require identification marks to be rolled into the surface of one side of the bar to denote the producer's mill designation, bar size and type of steel. For Grade 60 and Grade 75 bars, grade marks indicating yield strength must be show. Grade 40 bars show only three marks (no grade mark) in the following order:

1st — Producing Mill (usually an initial)
2nd — Bar Size Number (#3 through #18)
3rd — Type (N for New Billet)

### NUMBER SYSTEM — GRADE MARKS

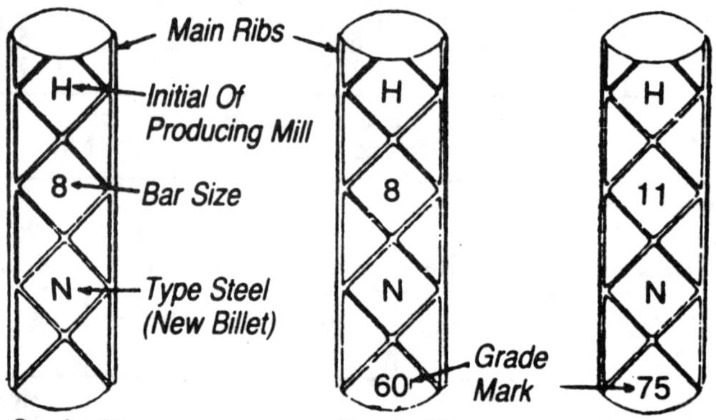


Grade 40
Grade 50

Grade 60

Grade 75

### LINE SYSTEM — GRADE MARKS

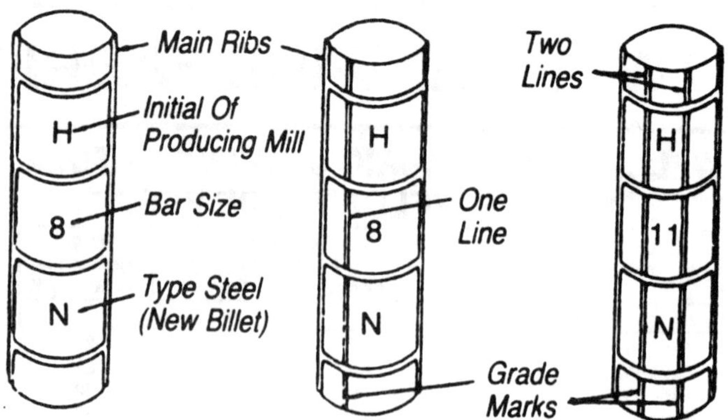


Grade 40
Grade 50

Grade 60

Grade 75

# CONCRETE / PUMPING CONCRETE 03310

Insofar as is possible, the moisture content should be kept uniform to avoid problems in determining the proper amount of water to be added for mixing. Mixing water must be reduced to compensate for moisture in the aggregate in order to control the slump of the concrete and avoid exceeding the specified water-cement ratio.

**Handling Concrete by Pumping Methods.** Transportation and placement of concrete by pumping is another method gaining increased popularity. Pumps have several advantages, the primary one being that a pump will high-lift concrete without the need for an expensive crane and bucket. Since the concrete is delivered through pipe and hoses, concrete can be conveyed to remote locations in buildings, in tunnels, to locations otherwise inaccessible on steep hillside slopes for anchor walls, pipe bedding or encasement, or for placing concrete for chain link fence post bases. Concrete pumps have been found to be economical and expedient in the placement of concrete, and this has promoted the use and acceptance of this development. The essence of proper concrete pumping is the placement of the concrete in its final location without segregation.

Modern concrete pumps, depending on the mix design and size of line, can pump to a height of 200 feet or a horizontal distance of 1,000 feet. They can handle, economically, structural mixes, standard mixes, low slump mixes, mixes with two-inch maximum size aggregate and light weight concrete. When a special pump mix is required for structural concrete in a major structure, the mix design must be approved by the Engineer and checked and confirmed by the Supervisor of the Materials Control Group. The Inspector should obtain the pump manufacturer's printed information and evaluate its characteristics and ability to handle the concrete mixture specified for the project.

If concrete is being placed for a major reinforced structure, it is important that the placement continue without interruption. The Inspector should be sure that the contractor has ready access to a back-up pump to be used in the event of a breakdown. In order to further insure the success of the concrete placement by the pumping method, the user should be aware of the following points:

(a) A protective grating over the receiving hopper of the pump is necessary to exclude large pieces of aggregate or foreign material.

(b) The pump and lines require lubrication with a grout of cement and water. All of the excess grout is to be wasted prior to pumping the concrete.

(c) All changes in direction must be made by a large radius bend with a maximum bend of 90 degrees. Wye connections induce segregation and shall not be used.

(d) Pump lines should be made of a material capable of resisting abrasion and with a smooth interior surface having a low coefficient of friction. Steel is commonly used for pump lines, ance. Aluminum pipe should not be used for pumping concrete and some of the new plastic or rubber tubing is gaining accept- because a chemical reaction occurs between the concrete and the aluminum. Hydrogen is generated which results in a swelling of the concrete, causing a significant reduction in compressive strength. This reaction is aggravated by any of the following: abrasive coarse aggregate, non-uniformly graded sand, low-slump concrete, low sand-aggregate ratio, high-alkali cement or when no air-entraining agent is used.

(e) During temporary interruptions in pumping, the hopper must remain nearly full, with an occasional turning and pumping to avoid developing a hard slug of concrete in the lines.

(f) Excessive line pressures must be avoided. When this occurs, check these points as the probable cause: segregation caused by too low a slump or too high a slump; large particle contamination caused by large pieces of aggregate or frozen lumps not eliminated by the grating; poor gradation of aggregates or particle shape; rich or lean spots caused by improper mixing.

(g) Corrections must be made to correct excessive slump loss as measured at the transit-mixed concrete truck and as measured at the hose outlet. This may be attributable to porous aggregate, high temperature or rapid setting mixes.

(h) Two transit-mix concrete trucks must be used simultaneously to deliver concrete into the pump hopper. These trucks must be discharged alternately to assure a continuous flow of concrete as trucks are replaced.

(i) Samples of concrete for test specimens prepared to determine the acceptance of the concrete quality are to be taken as required for conventional concrete.

Sampling is done before the concrete is deposited in the pump hopper. However, it is suggested that, where possible, the effect of pumping on the compressive strength be checked by taking companion samples, so identified, from the end of the pump line at the same time. The Record of Test must be properly noted as being a special mix used for pumping purposes. This will enable the Materials Control Group to compile a complete history of mix designs and their respective compressive strengths.

The prudent use of pumped concrete can result in economy and improved quality. However, only the control exercised by the operator will assure continued high standards of quality concrete.

Pump lines must be properly fastened to supports to eliminate excessive vibration. Couplings must be easily and securely fastened in a manner that will prevent mortar leakage. It is preferable to use the flexible hose only at the discharge point. This hose must be moved in such a manner as to avoid kinks or sharp bends. The pump line should be protected from excessive heat during hot weather by water sprinkling or shade.

# CONCRETE / STRUCTURAL CONCRETE | 03310

## COMPRESSIVE STRENGTH FOR VARIOUS WATER-CEMENT RATIOS

*(The strengths listed are based on the use of normal portland cement)*

| WATER/CEMENT RATIO | | PROBABLE 28-DAY STRENGTH | |
|---|---|---|---|
| WEIGHT | GALS./100# | PSI | MEGAPASCALS* |
| .40 | 4.8 | 5000 | 34 |
| .45 | 5.4 | 4500 | 31 |
| .50 | 6.0 | 4000 | 28 |
| .55 | 6.6 | 3500 | 24 |
| .60 | 7.2 | 3000 | 21 |
| .65 | 7.8 | 2500 | 17 |
| .70 | 8.4 | 2000 | 14 |

* International system equivalent.

## APPROXIMATE CONTENT OF SAND, CEMENT AND WATER PER CUBIC YARD OF CONCRETE

Based on aggregates of average grading and physical characteristics in concrete mixes having a water-cement ratio (W/C) of about .65 by weight (or 7.8 gallons) per sack of cement; 3-in. slump; and a medium natural sand having a fineness modulus of about 2.75.

| COURSE AGGREGATE MAX. SIZE | WATER | | CEMENT | % SAND |
|---|---|---|---|---|
| | POUNDS | GALLONS | | |
| 3/8 | 385 | 46 | 590 | 57 |
| 1/2 | 365 | 44 | 560 | 50 |
| 3/4 | 340 | 41 | 525 | 43 |
| 1 | 325 | 39 | 500 | 39 |
| 1 1/2 | 300 | 36 | 460 | 37 |

It can be noted from the above chart that, for a given slump, the amount of mixing water increases as the size of the course aggregate decreases. The size of the course aggregate controls the sand content in the same way; that is, the amount of sand required in the mix increases as the size of the course aggregate decreases.

Other typical examples are contained in the pamphlet published by the Portland Cement Association entitled "Design and Control of Concrete Mixtures."

**Effects of Temperature on Concrete.** Concrete mixtures gain strength rapidly in the first few days after placement. While the rate of gain in strength diminishes, concrete continues to become stronger with time over a period of many years, so long as drying of the concrete is prevented. Its strength at 28 days is considered to be the compressive strength upon which the Engineer bases his calculations. The temperature of the atmosphere has a significant effect upon the development of strength in concrete. Lower temperatures retard and higher temperatures accelerate the gain in strength.

Most destructive of the natural forces is freezing and thawing action. While the concrete is still wet or moist, expansion of the water as it is converted into ice results in severe damage to the fresh concrete. In situations where freezing may be encountered, high early strength cement may be used. Also, the mixing water or the aggregate (or both) may be preheated before mixing. Covering the concrete, and using steam or salamanders to heat the concrete under the covering, will help prevent freezing. Air-entraining agents help to diminish the effects of freezing of fresh concrete as well as in subsequent freezing and thawing cycles throughout the life of the concrete.

Hot weather will present problems of a different nature in placing concrete. Concrete will set up faster and tend to shrink and crack at the surface. To minimize this problem, the concrete should be placed without delay after mixing. Avoid the use of accelerators (perhaps even use a retarding agent), dampen all subgrade and forms, protect the freshly placed concrete from hot dry winds, and provide for adequate curing. Crushed ice or chilled water can be used as part of the mixing water to reduce the temperature of the mix in extremely hot areas.

## ARCHITECTURAL WALL PATTERNS (BONDS)

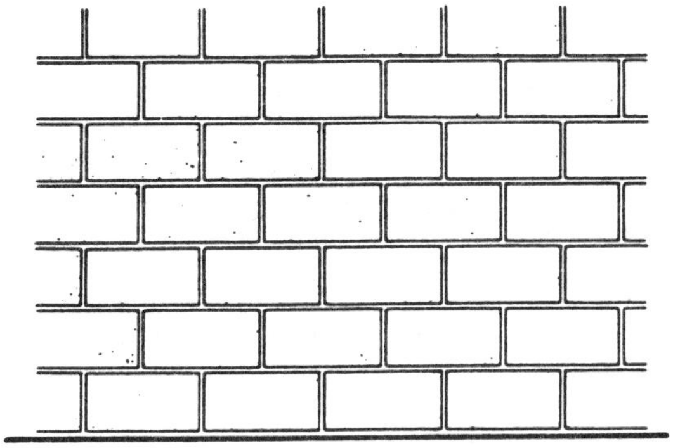

COMMON BOND
8"x 16" UNITS

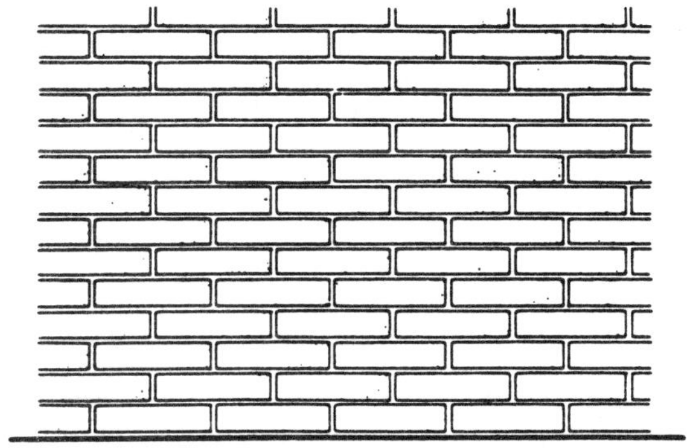

COMMON BOND
4"x 16" UNITS

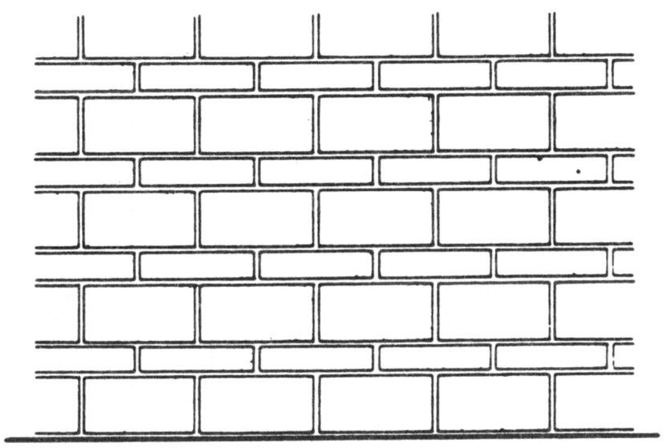

COURSED ASHLAR
8"x 16" & 4"x 16" UNITS

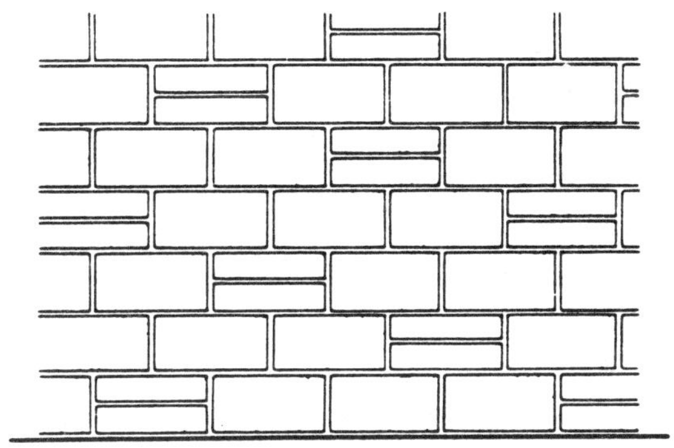

COURSED ASHLAR
8"x 16" & 4"x 16" UNITS

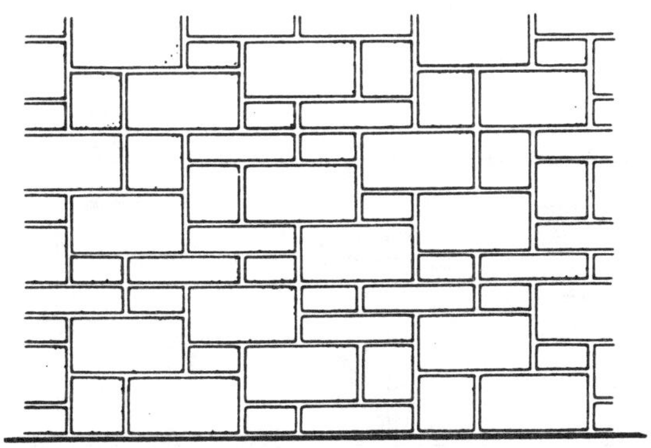

RANDOM ASHLAR
8"x 16", 8"x 8", 4"x 16" AND 4"x 8" UNITS

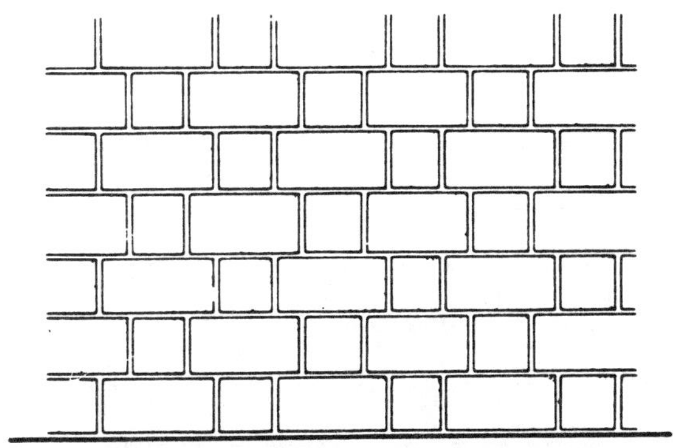

COURSED ASHLAR
8"x 16" & 8"x 8" UNITS

## ARCHITECTURAL WALL PATTERNS (BONDS) — (Continued)

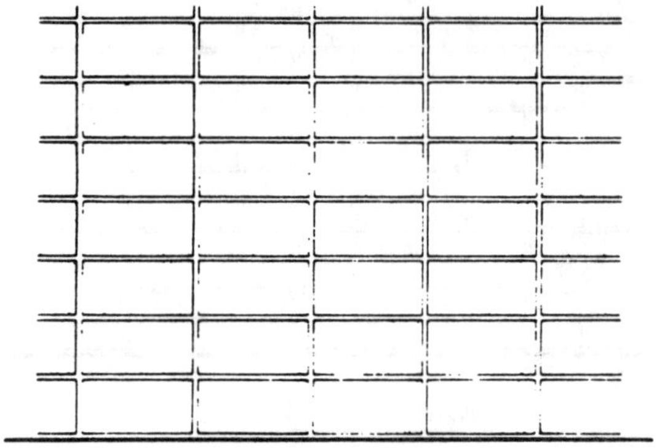

STACKED BOND
8" x 16" UNITS

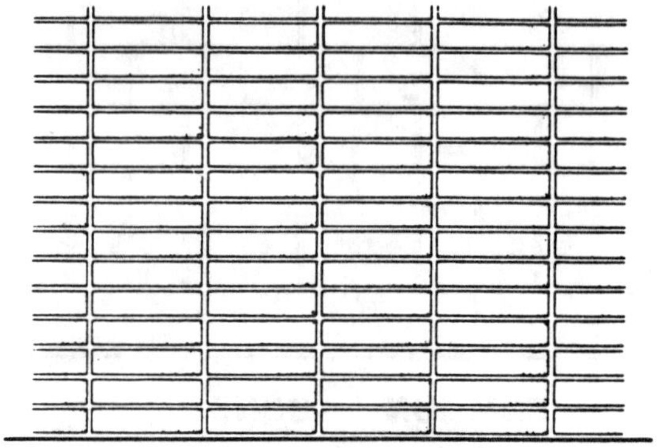

STACKED BOND
4" x 16" UNITS

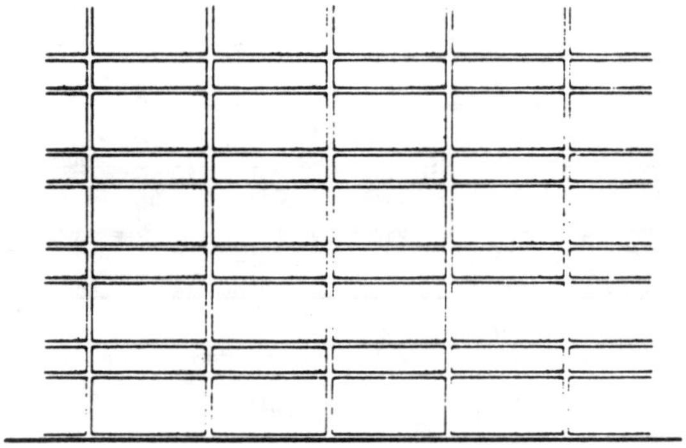

STACKED BOND
8" x 16" & 4" x 16" UNITS

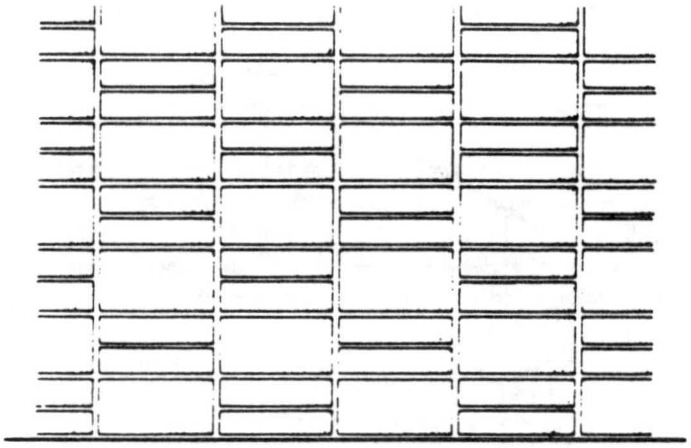

STACKED BOND
8" x 16" & 4" x 16" UNITS

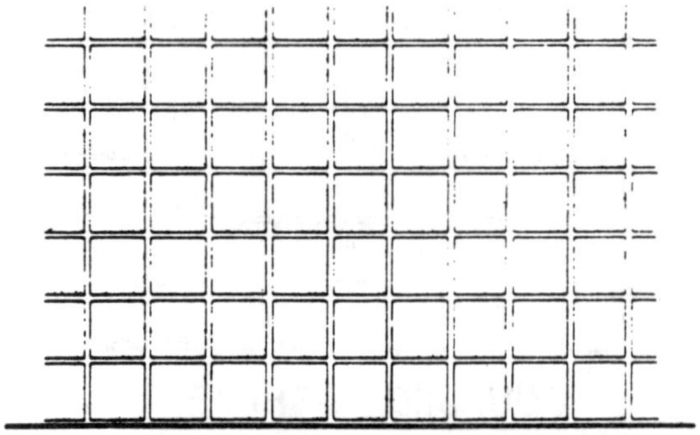

STACKED BOND VERTICAL SCORED UNITS

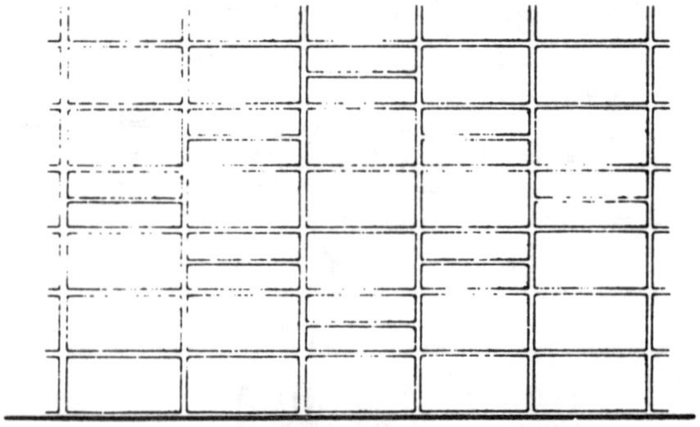

USE OF BLOCK DESIGN IN STACKED BOND

# MASONRY / SPLIT FACE CMU | 04220

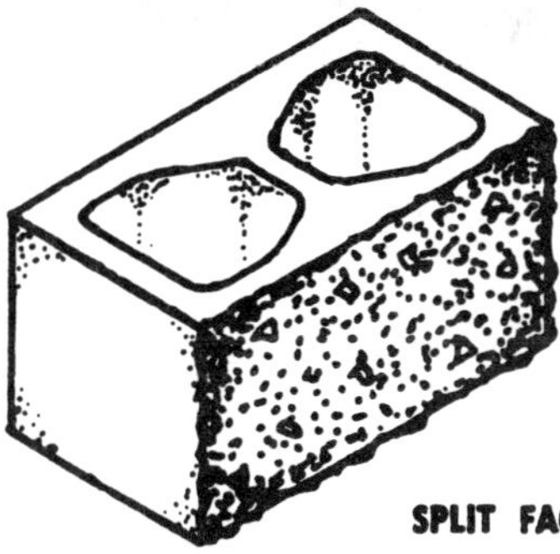

SPLIT FACE

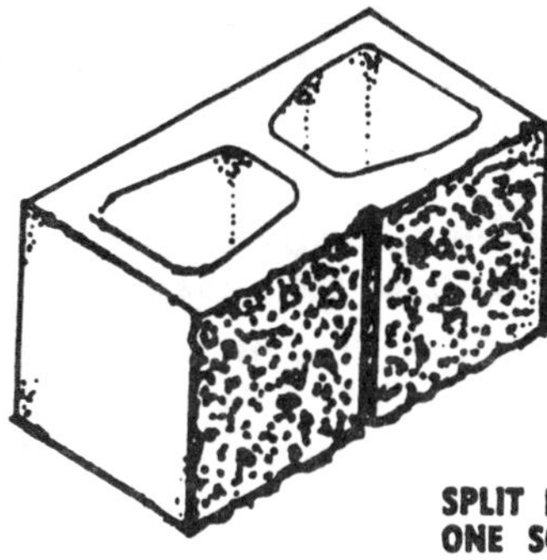

SPLIT FACE
ONE SCORE

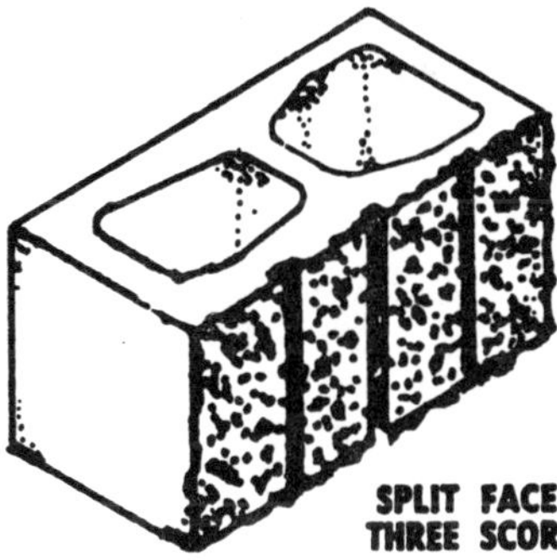

SPLIT FACE
THREE SCORE

SPLIT FACE
FIVE SCORE

## SPLIT FACE BLOCK

Split face block is manufactured as a unit that is normally made double and is literally split apart on a splitter; a machine which resembles a guillotine. The splitter has blades at the top and bottom (and sometimes at the sides) which exert pressure on the blocks, breaking them apart.

Many factors determine the look of the split face, both as to size variances and the amount of aggregate exposure. Split face block is intended to have a rougher texture than precision block. Various configurations of block such as fluted, scored, etc., will split in a different manner than a full split face. The vertical perpendicularity of scored and fluted split face block is subject to variation.

SPLIT FACE
THREE-WIDE SCORE

SPLIT FACE
FOUR-FLUTE

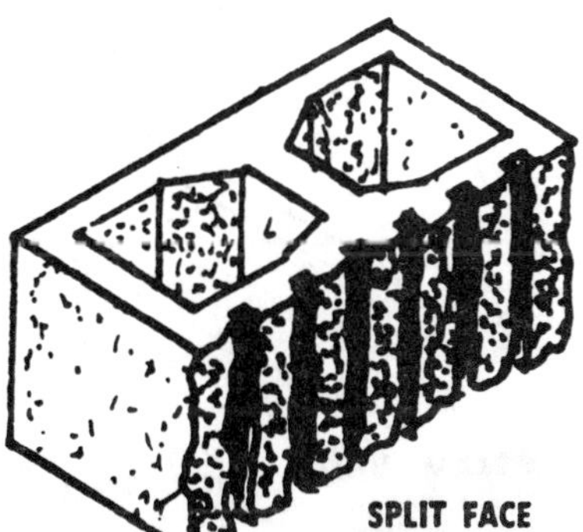

SPLIT FACE
SIX-TAPER SCORE

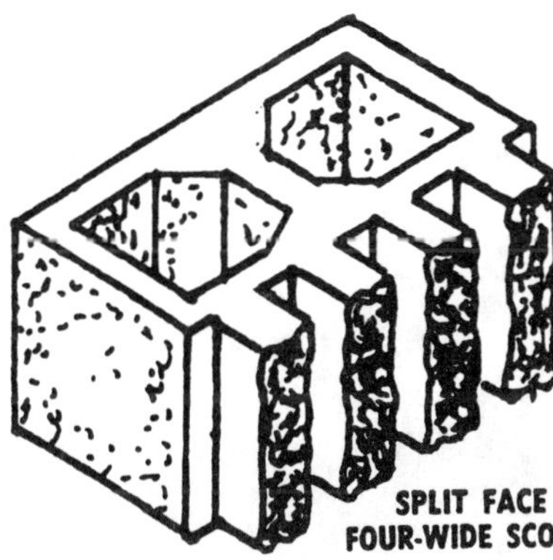

SPLIT FACE
FOUR-WIDE SCORE

**NOTE:** Split face units shown in this manual are a small sampling of the broad range of concrete masonry architectural units available from the industry on special order. Depths and widths of scores vary. Consult a local manufacturer for specific information.

# MASONRY / BOND BEAMS/LINTELS | 04220

## TYPICAL DETAILS—LINTELS AND BOND BEAMS

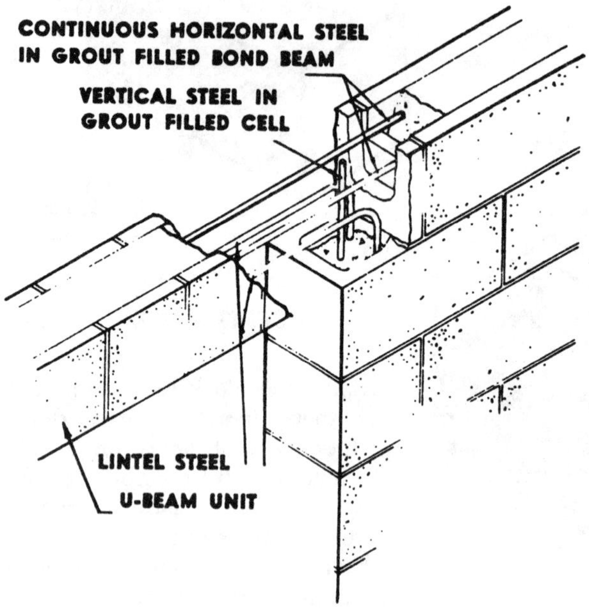


8″ x 8″ LINTEL &
BOND BEAM ON
8″ WALL

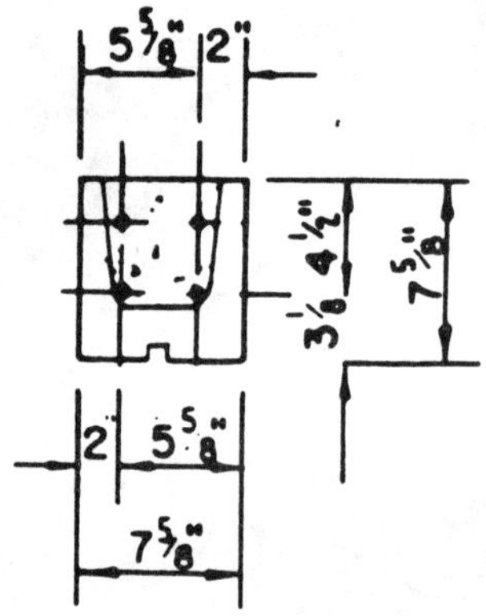


8″ x 16″
BOND BEAM ON
8″ WALL

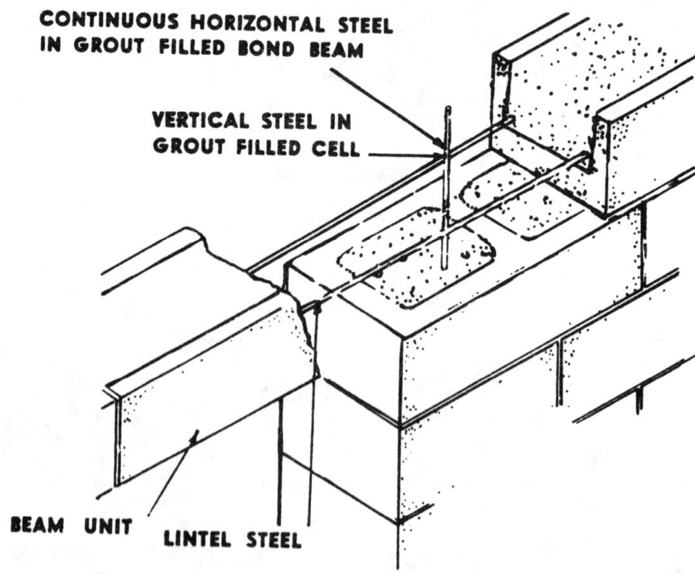

# METALS / WELDING 05050

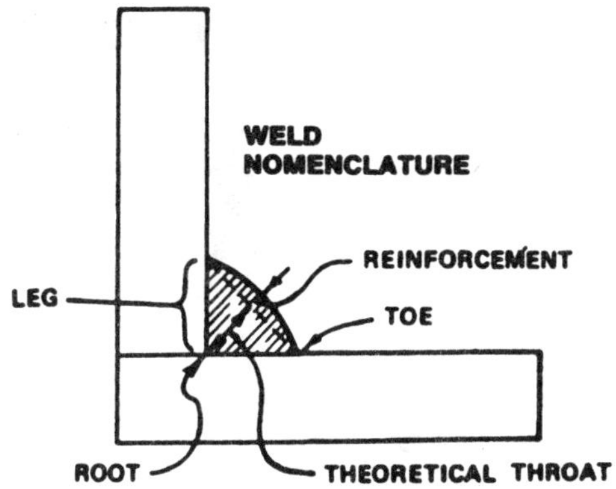


## WELDED JOINTS

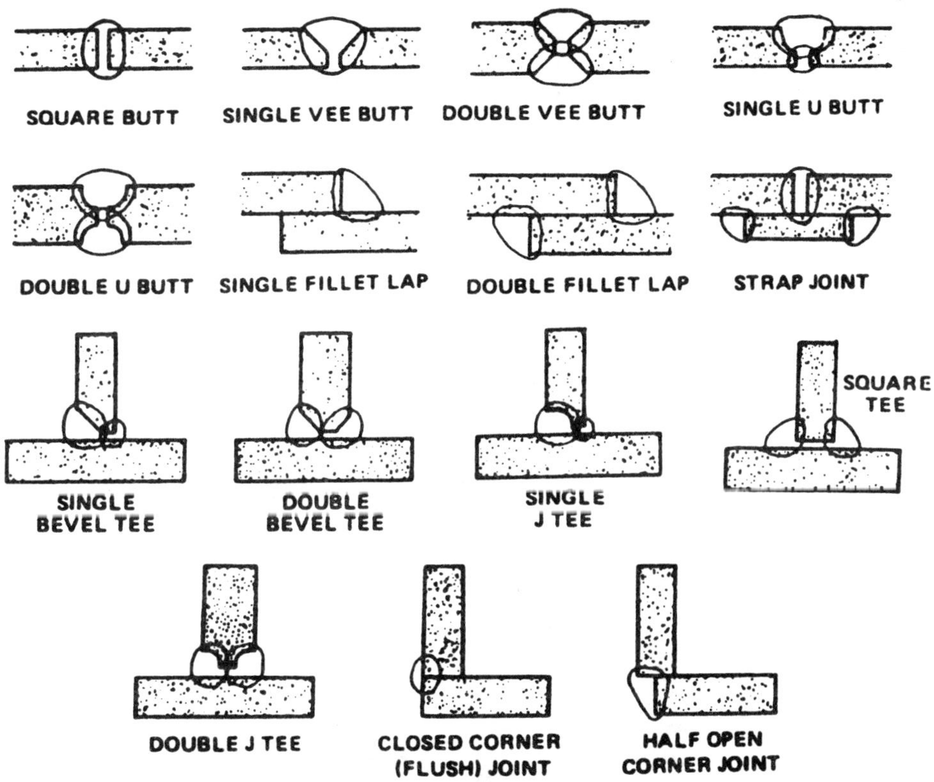


## WELDING POSITIONS

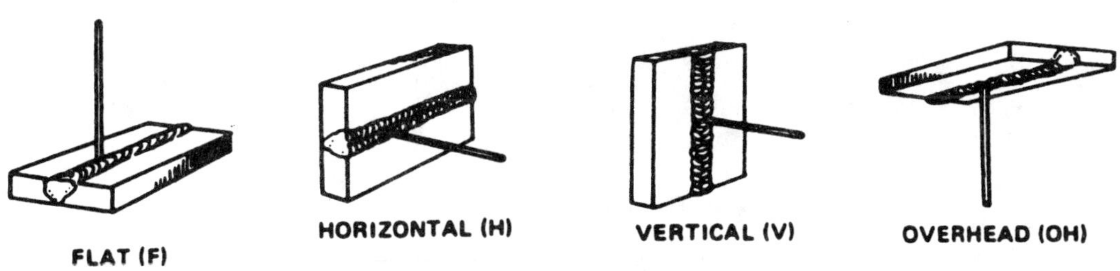

## BOLTS IN COMMON USAGE

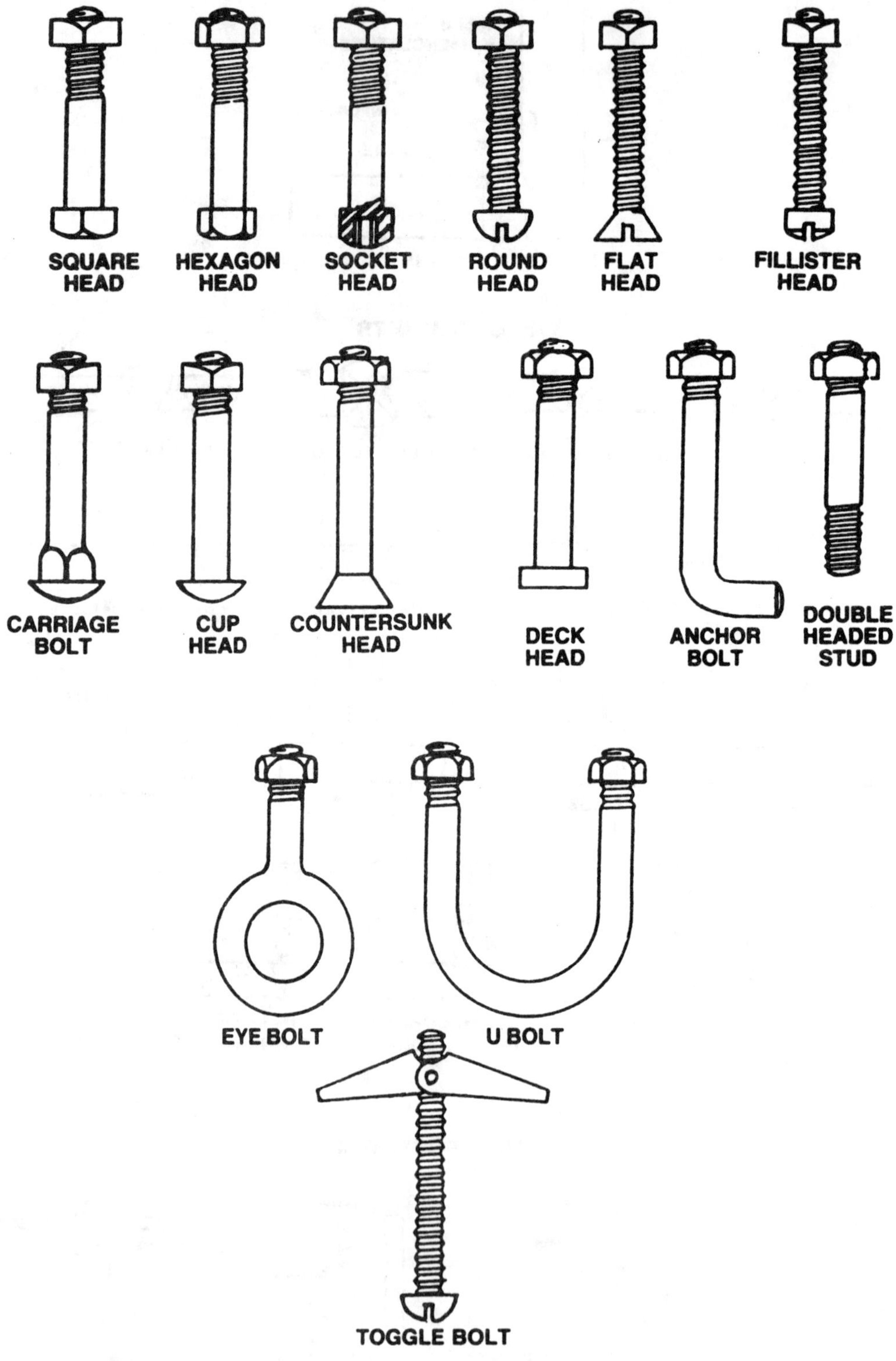

## COMMON WIRE NAILS (ACTUAL SIZE)

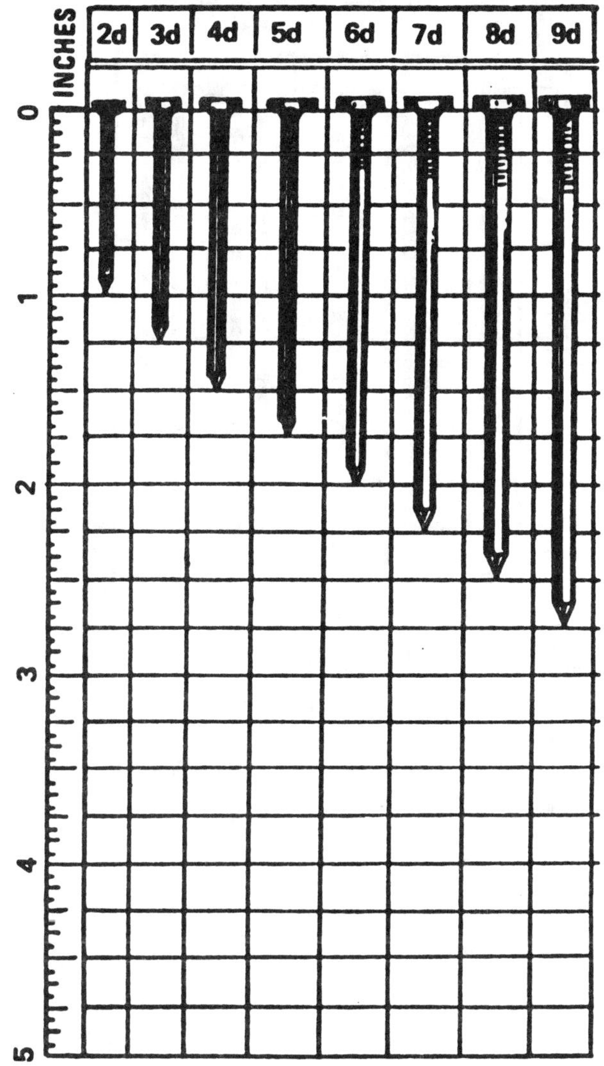


**Cut Nails.** Cut nails are angular-sided, wedge-shaped with a blunt point.

**Wire Nails.** Wire nails are round shafted, straight, pointed nails, and are used more generally than cut nails. They are stronger than cut nails and do not buckle as easily when driven into hard wood, but usually split wood more easily than cut nails. Wire nails are available in a variety of sizes varying from two penny to sixty penny.

**Nail Finishes.** Nails are available with special finishes. Some are galvanized or cadmium plated to resist rust. To increase the resistance to withdrawal, nails are coated with resins or asphalt cement (called cement coated). Nails which are small, sharp-pointed, and often placed in the craftsman's mouth (such as lath or plaster board nails) are generally blued and sterilized.

## COMMON WIRE NAILS (ACTUAL SIZE) (Cont.)

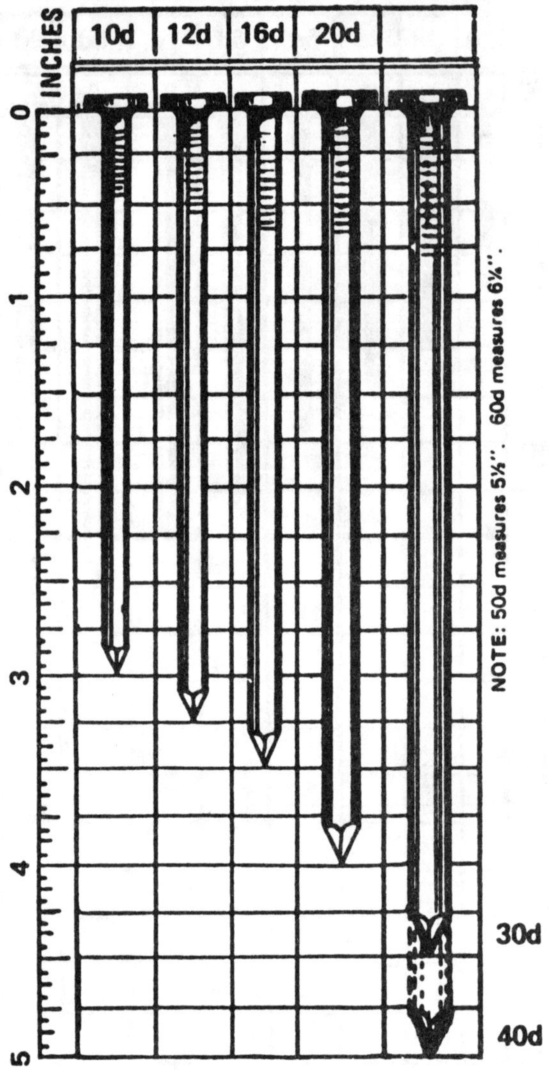


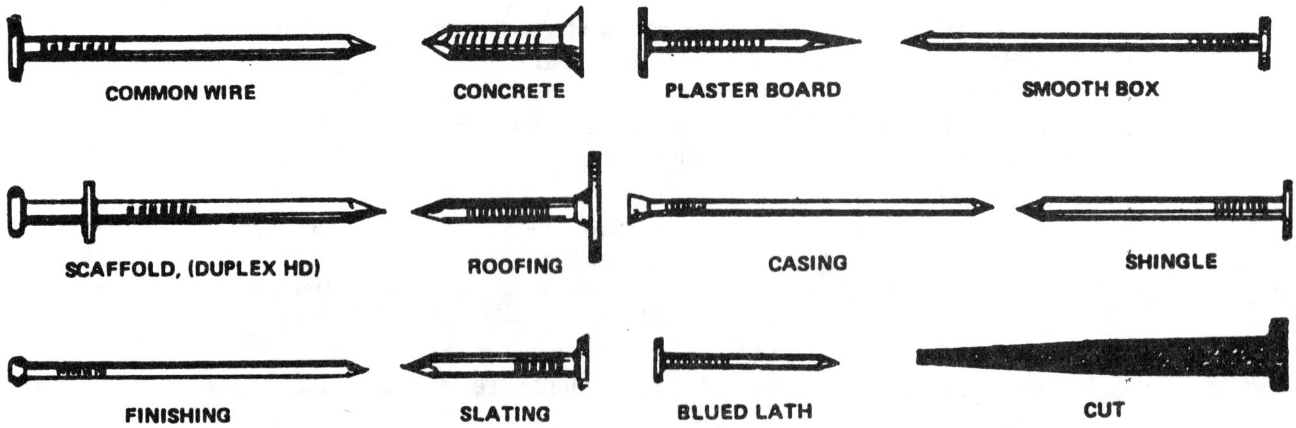

# PLYWOOD – BASIC GRADE MARKS

## AMERICAN PLYWOOD ASSOCIATION (APA)

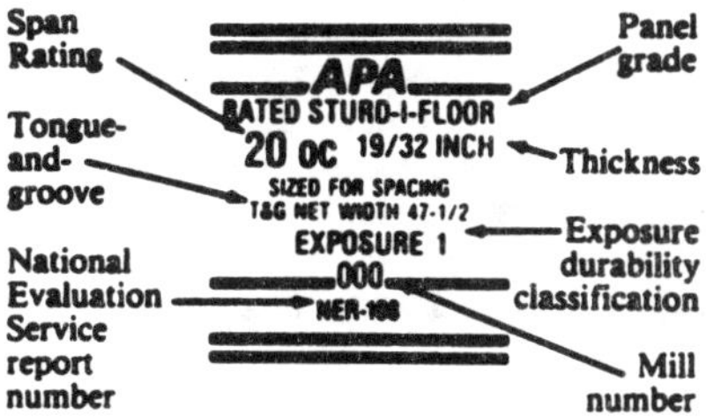


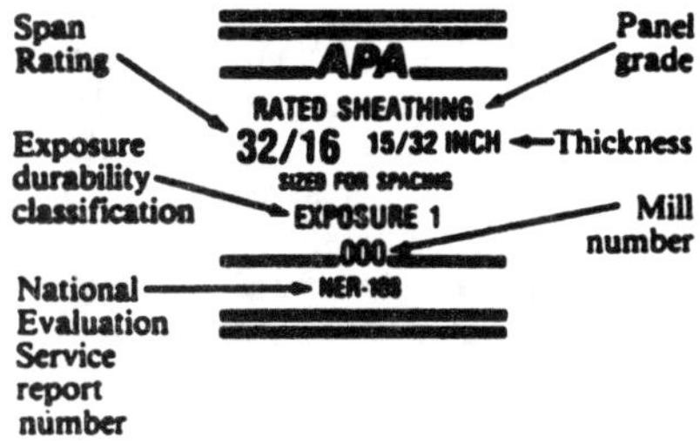


The American Plywood Association's trademarks appear only on products manufactured by APA member mills. The marks signify that the product is manufactured in conformance with APA performance standards and/or U.S. Product Standard PS 1-83 for Construction and Industrial Plywood.

### APA A-C

For use where appearance of one side is important in exterior applications such as soffits, fences, structural uses, boxcar and truck linings, farm buildings, tanks, trays, commercial refrigerators, etc. ***Exposure Durability Classification:*** Exterior. ***Common Thicknesses:*** ¼, 11/32, ⅜, 15/32, ½, 19/32, ⅝, 23/32, ¾.

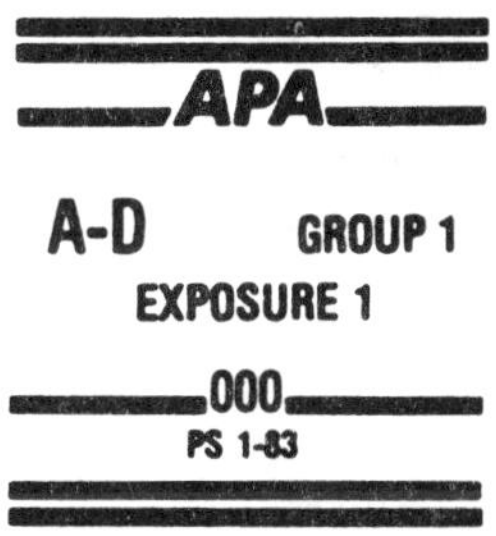


### APA A-D

For use where appearance of only one side is important in interior applications, such as paneling, built-ins, shelving, partitions, flow racks, etc. ***Exposure Durability Classifications:*** Interior, Exposure 1. ***Common Thicknesses:*** ¼, 11/32, ⅜, 15/32, ½, 19/32, ⅝, 23/32, ¾.

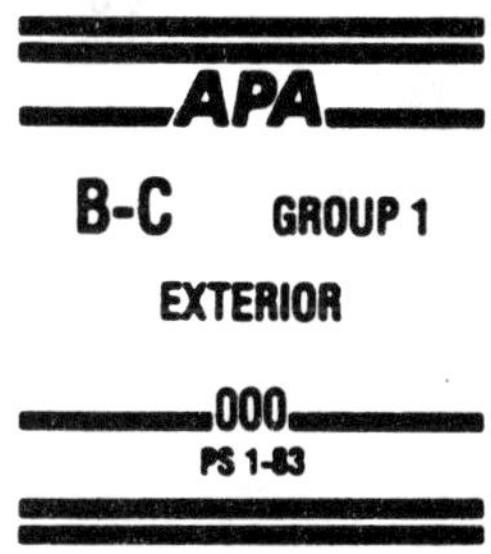


### APA B-C

Utility panel for farm service and work buildings, boxcar and truck linings, containers, tanks, agricultural equipment, as a base for exterior coatings and other exterior uses. ***Exposure Durability Classification:*** Exterior. ***Common Thicknesses:*** ¼, 11/32, ⅜, 15/32, ½, 19/32, ⅝, 23/32, ¾.

### APA B-D

Utility panel for backing, sides or builtins, industry shelving, slip sheets, separator boards, bins and other interior or protected applications. ***Exposure Durability Classifications:*** Interior, Exposure 1. ***Common Thicknesses:*** ¼, 11/32, ⅜, 15/32, ½, 19/32, ⅝, 23/32, ¾.

APA proprietary concrete form panels designed for high reuse. Sanded both sides and mill-oiled unless otherwise specified. Class I, the strongest, stiffest and more commonly available, is limited to Group 1 faces, Group 1 or 2 crossbands, and Group 1, 2, 3 or 4 inner plies. Class II is limited to Group 1 or 2 faces (Group 3 under certain conditions) and Group 1, 2, 3 or 4 inner plies. Also available in HDO for very smooth concrete finish, in Structural I, and with special overlays. ***Exposure Durability Classification:*** Exterior. ***Common Thicknesses:*** 19/32, ⅝, 23/32, ¾.

Plywood panel manufactured with smooth, opaque, resin-treated fiber overlay providing ideal base for paint on one or both sides. Excellent material choice for shelving, factory work surfaces, paneling, built-ins, signs and numerous other construction and industrial applications. Also available as a 303 Siding with texture-embossed or smooth surface on one side only and Structural I. ***Exposure Durability Classification:*** Exterior. ***Common Thicknesses:*** 11/32, ⅜, 15/32, ½, 19/32, ⅝, 23/32, ¾.

## SPECIALTY PANELS

HDO · A-A · G-1 · EXT-APA · 000 · PS1-83

Plywood panel manufactured with a hard, semi-opaque resin-fiber overlay on both sides. Extremely abrasion resistant and ideally suited to scores of punishing construction and industrial applications, such as concrete forms, industrial tanks, work surfaces, signs, agricultural bins, exhaust ducts, etc. Also available with skid-resistant screen-grid surface and in Structural I. ***Exposure Durability Classification:*** Exterior. ***Common Thicknesses:*** 3/8, 1/2, 5/8, 3/4.

MARINE · A-A · EXT-APA · 000 · PS1-83

Specialty designed plywood panel made only with Douglas fir or western larch, solid jointed cores, and highly restrictive limitations on core gaps and faces repairs. Ideal for both hulls and other marine applications. Also available with HDO or MDO faces. ***Exposure Durability Classification:*** Exterior. ***Common Thicknesses:*** 1/4, 3/8, 1/2, 5/8, 3/4.

Unsanded and touch-sanded panels, and panels with "B" or better veneer on one side only, usually carry the APA trademark on the panel back. Panels with both sides of "B" or better veneer, or with special overlaid surfaces (such as Medium Density Overlay), carry the APA trademark on the panel edge, like this:

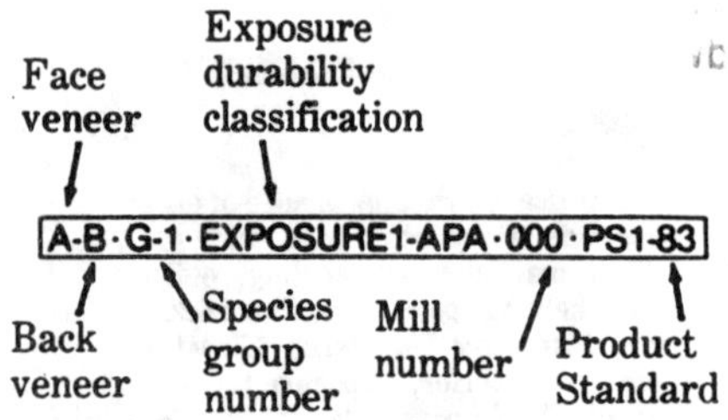


## GLOSSARY OF LUMBER TERMS

Some of the words and terms used in the grading of lumber follow:

***Bow.*** A deviation flatwise from a straight line drawn from end to end of the piece. It is measured at the point of greatest distance from the straight line.

***Checks.*** A separation of the wood which normally occurs across the annual rings and usually as a result of seasoning.

***Crook.*** A deviation edgewise from a straight line drawn from end to end of the piece. It is measured at the point of greatest distance from the straight line.

***Cup.*** A deviation from a straight line drawn across the piece from edge to edge. It is measured at the point of greatest distance from the straight line.

***Flat Grain.*** The annual growth rings pass through the piece at an angle of less than 45 degrees with the flat surface of the piece.

***Mixed Grain.*** The piece may have vertical grain, flat grain or a combination of both vertical and flat grain.

***Pitch.*** An accumulation of resin which occurs in separations in the wood or in the wood cells themselves.

***Shake.*** A separation of the wood which usually occurs between the rings of annual growth.

***Splits.*** A separation of the wood due to tearing apart of the wood cells.

***Vertical Grain.*** The annual growth rings pass through the piece at an angle of 45 degrees or more with the flat surface of the piece.

***Wane.*** Bark or lack of wood from any cause, except eased edges (rounded) on the edge or corner of a piece of lumber.

***Warp.*** Any deviation from a true or plane surface, including crook, cup, bow or any combination thereof.

# LUMBER GRADING

## GRADING-MARK ABBREVIATIONS

### GRADES

*(Listed alphabetically — not by quality)*

| | |
|---|---|
| COM | Common |
| CONST | Construction |
| ECON | Economy |
| No. 1 | Number One |
| SEL-MER | Select Merchantable |
| SEL-STR | Select Structural |
| STAN | Standard |
| UTIL | Utility |

### ALSC TRADEMARKS

| | |
|---|---|
| CLIS | California Lumber Inspection Service |
| NELMA | Northeastern Lumber Manufacturers Association, Inc. |
| NH&PMA | Northern Hardwood & Pine Manufacturers Association, Inc. |
| PLIB | Pacific Lumber Inspection Bureau |
| RIS | Redwood Inspection Service |
| SPIB | Southern Pine Inspection Bureau |
| TP | Timber Products Inspection |
| WCLB | West Coast Lumber Inspection Bureau |
| WWP | Western Wood Products Association |

### SPECIES GROUPINGS

| | |
|---|---|
| AF | Alpine Fir |
| DF | Douglas Fir |
| HF | Hem Fir |
| SP | Sugar Pine |
| PP | Ponderosa Pipe |
| LP | Lodgepole Pine |
| IWP | Idaho White Pine |
| ES | Engelmann Spruce |
| WRC | Western Red Cedar |
| INC CDR | Incense Cedar |
| L | Larch |
| LP | Lodgepole Pine |
| MH | Mountain Hemlock |
| WW | White Wood |

### MOISTURE CONTENT

| | |
|---|---|
| S-GRN | Surfaced at a moisture content of more than 19% |
| S-DRY | Surfaced at a moisture content of 19% or less |
| MC-15 | Surfaced at a moisture content of 15% or less. |

**Framing Estimating Rules of Thumb**

For 16″ O.C. stud partitions figure 1 stud for every L.F. of wall; add for top and bottom plates.

For any type of framing, the quantity of basic framing members (in L.F.) can be determined based on spacing and surface area (S.F.):

| | |
|---|---|
| 12″ O.C. | 1.2 L.F./S.F. |
| 16″ O.C. | 1.0 L.F./S.F. |
| 24″ O.C. | 0.8 L.F./S.F. |

*(Doubled-up members, bands, plates, framed openings, etc., must be added.)*

Framing accessories, nails, joist hangers, connectors, etc., should be estimated as separate material costs. Installation should be included with framing. Rule of thumb allowance is 0.5 to 1.5% of lumber cost for rough hardware. Another is 30 to 40 pounds of nails per M.B.F.

## BOARD FEET/LINEAR FEET FOR LUMBER

| Nominal Size | Actual Size | Board Feet Per Linear Foot | Linear Feet Per 1000 Board Feet |
|---|---|---|---|
| 1 × 2 | ¾ × 1½ | .167 | 6000 |
| 1 × 3 | ¾ × 2½ | .250 | 4000 |
| 1 × 4 | ¾ × 3½ | .333 | 3000 |
| 1 × 6 | ¾ × 5½ | .500 | 2000 |
| 1 × 8 | ¾ × 7¼ | .666 | 1500 |
| 1 × 10 | ¾ × 9¼ | .833 | 1200 |
| 1 × 12 | ¾ × 11¼ | 1.0 | 1000 |
| 2 × 2 | 1½ × 1½ | .333 | 3000 |
| 2 × 3 | 1½ × 2½ | .500 | 2000 |
| 2 × 4 | 1½ × 3½ | .666 | 1500 |
| 2 × 6 | 1½ × 5½ | 1.0 | 1000 |
| 2 × 8 | 1½ × 7¼ | 1.333 | 750 |
| 2 × 10 | 1½ × 9¼ | 1.666 | 600 |
| 2 × 12 | 1½ × 11¼ | 2.0 | 500 |

## BOARD FEET CONVERSION TABLE

| Nominal Size (In.) | ACTUAL LENGTH IN FEET | | | | | | | | |
|---|---|---|---|---|---|---|---|---|---|
| | 8 | 10 | 12 | 14 | 16 | 18 | 20 | 22 | 24 |
| 1 x 2 | | 1⅔ | 2 | 2⅓ | 2⅔ | 3 | 3½ | 3⅔ | 4 |
| 1 x 3 | | 2½ | 3 | 3½ | 4 | 4½ | 5 | 5½ | 6 |
| 1 x 4 | 2¾ | 3⅓ | 4 | 4⅔ | 5⅓ | 6 | 6⅔ | 7⅓ | 8 |
| 1 x 5 | | 4⅙ | 5 | 5⅚ | 6⅔ | 7½ | 8⅓ | 9⅙ | 10 |
| 1 x 6 | 4 | 5 | 6 | 7 | 8 | 9 | 10 | 11 | 12 |
| 1 x 7 | | 5⅝ | 7 | 8⅙ | 9⅓ | 10½ | 11⅔ | 12⅚ | 14 |
| 1 x 8 | 5⅓ | 6⅔ | 8 | 9⅓ | 10⅔ | 12 | 13⅓ | 14⅔ | 16 |
| 1 x 10 | 6⅔ | 8⅓ | 10 | 11⅔ | 13⅓ | 15 | 16⅔ | 18⅓ | 20 |
| 1 x 12 | 8 | 10 | 12 | 14 | 16 | 18 | 20 | 22 | 24 |
| 1¼ x 4 | | 4⅙ | 5 | 5⅚ | 6⅔ | 7½ | 8⅓ | 9⅙ | 10 |
| 1¼ x 6 | | 6¼ | 7½ | 8¾ | 10 | 11¼ | 12½ | 13¾ | 15 |
| 1¼ x 8 | | 8⅓ | 10 | 11⅔ | 13⅓ | 15 | 16⅔ | 18⅓ | 20 |
| 1¼ x 10 | | 10 5/12 | 12½ | 14 7/12 | 16⅔ | 18¾ | 20⅚ | 22 11/12 | 25 |
| 1¼ x 12 | | 12½ | 15 | 17½ | 20 | 22½ | 25 | 27½ | 30 |
| 1½ x 4 | 4 | 5 | 6 | 7 | 8 | 9 | 10 | 11 | 12 |
| 1½ x 6 | 6 | 7½ | 9 | 10½ | 12 | 13½ | 15 | 16½ | 18 |
| 1½ x 8 | 8 | 10 | 12 | 14 | 16 | 18 | 20 | 22 | 24 |
| 1½ x 10 | 10 | 12½ | 15 | 17½ | 20 | 22½ | 25 | 27½ | 30 |
| 1½ x 12 | 12 | 15 | 18 | 21 | 24 | 27 | 30 | 33 | 36 |
| 2 x 4 | 5⅓ | 6⅔ | 8 | 9⅓ | 10⅔ | 12 | 13⅓ | 14⅔ | 16 |
| 2 x 6 | 8 | 10 | 12 | 14 | 16 | 18 | 20 | 22 | 24 |
| 2 x 8 | 10⅔ | 13⅓ | 16 | 18⅔ | 21⅓ | 24 | 26⅔ | 29⅓ | 32 |
| 2 x 10 | 13⅓ | 16⅔ | 20 | 23⅓ | 26⅔ | 30 | 33⅓ | 36⅔ | 40 |
| 2 x 12 | 16 | 20 | 24 | 28 | 32 | 36 | 40 | 44 | 48 |
| 3 x 6 | 12 | 15 | 18 | 21 | 24 | 27 | 30 | 33 | 36 |
| 3 x 8 | 16 | 20 | 24 | 28 | 32 | 36 | 40 | 44 | 48 |
| 3 x 10 | 20 | 25 | 30 | 35 | 40 | 45 | 50 | 55 | 60 |
| 3 x 12 | 24 | 30 | 36 | 42 | 48 | 54 | 60 | 66 | 72 |
| 4 x 4 | 10⅔ | 13⅓ | 16 | 18⅔ | 21⅓ | 24 | 26⅔ | 29⅓ | 32 |
| 4 x 6 | 16 | 20 | 24 | 28 | 32 | 36 | 40 | 44 | 48 |
| 4 x 8 | 21⅓ | 26⅔ | 32 | 37⅓ | 42⅔ | 48 | 53⅓ | 58⅔ | 64 |
| 4 x 10 | 26⅔ | 33⅓ | 40 | 46⅔ | 53⅓ | 60 | 66⅔ | 73⅓ | 80 |
| 4 x 12 | 32 | 40 | 48 | 56 | 64 | 72 | 80 | 88 | 96 |

# MOISTURE PROTECTION / ROOFING — 07300

## SLOPE AREA CALCULATIONS

| Rise and Run | Multiply Flat Area by | LF of Hips or Valleys per LF of Common Run |
|---|---|---|
| 2 in 12 | 1.014 | 1.424 |
| 3 in 12 | 1.031 | 1.436 |
| 4 in 12 | 1.054 | 1.453 |
| 5 in 12 | 1.083 | 1.474 |
| 6 in 12 | 1.118 | 1.500 |
| 7 in 12 | 1.158 | 1.530 |
| 8 in 12 | 1.202 | 1.564 |
| 9 in 12 | 1.250 | 1.600 |
| 10 in 12 | 1.302 | 1.641 |
| 11 in 12 | 1.357 | 1.685 |
| 12 in 12 | 1.413 | 1.732 |

# MOISTURE PROTECTION / DOWNSPOUTS — 07600

## DOWNSPOUT/VERTICAL LEADER CALCULATIONS

| Roof Type | Slope | S.F. Roof/ Sq. In. Leader |
|---|---|---|
| Gravel | Less than ¼″ per foot | 300 |
| Gravel | Greater than ¼″ per foot | 250 |
| Metal or Shingle | Any | 200 |

Alternate calculations:

$$\text{Diameter of downspout/leader} = 1.128\sqrt{\frac{\text{Area of drainage}}{\text{SF Roof/Sq. Inch}}}$$

## TYPICAL MINIMUM SIZE OF VERTICAL CONDUCTORS AND LEADERS

| Size of leader or conductor (Inches) | Maximum projected roof area (Square feet) |
|---|---|
| 2 | 544 |
| 2½ | 987 |
| 3 | 1,610 |
| 4 | 3,460 |
| 5 | 6,280 |
| 6 | 10,200 |
| 8 | 22,000 |

## TYPICAL MINIMUM SIZE OF ROOF GUTTERS

| | Maximum projected roof area for gutters of various slopes | | | |
|---|---|---|---|---|
| Diameter gutter (Inches) | 1/16 in. per Ft. slope (Sq. ft.) | ⅛ in. per Ft. slope (Sq. ft.) | ¼ in. per Ft. slope (Sq. ft.) | ½ in. per Ft. slope (Sq. ft.) |
| 3 | 170 | 240 | 340 | 480 |
| 4 | 360 | 510 | 720 | 1020 |
| 5 | 625 | 880 | 1250 | 1770 |
| 6 | 960 | 1360 | 1920 | 2770 |
| 7 | 1380 | 1950 | 2760 | 3900 |
| 8 | 1990 | 2800 | 3980 | 5600 |
| 10 | 3600 | 5100 | 7200 | 10000 |

# FINISHES / METAL FRAMING SYSTEMS | 09110

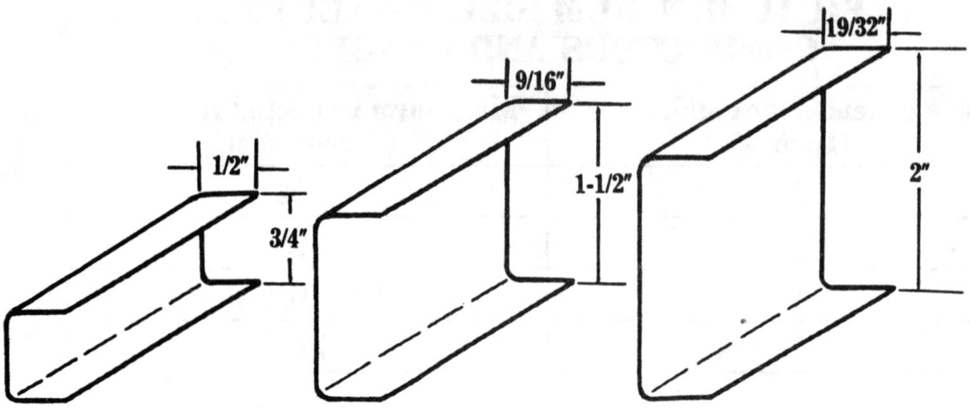


COLD ROLLED CHANNELS

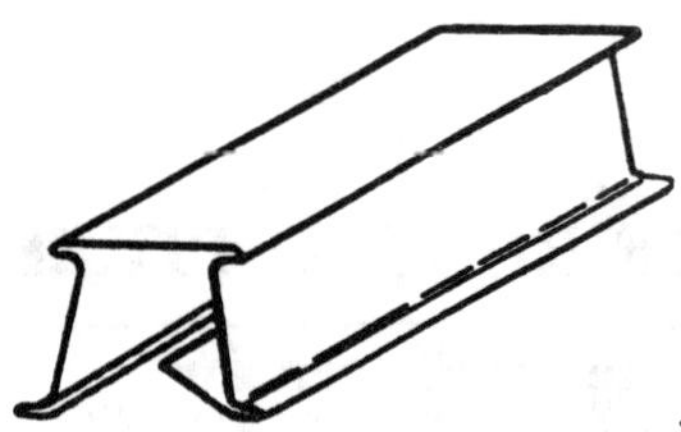

NAILING CHANNEL

SCREW CHANNEL

CHANNEL STUD

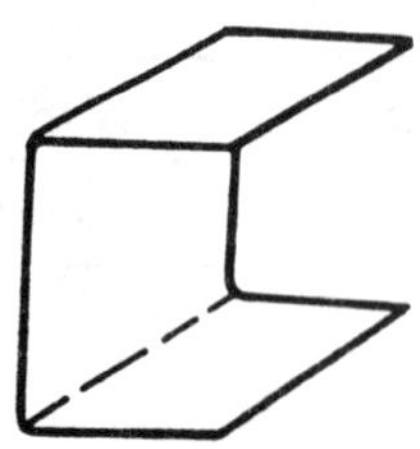

WIDE FLANGE CHANNEL

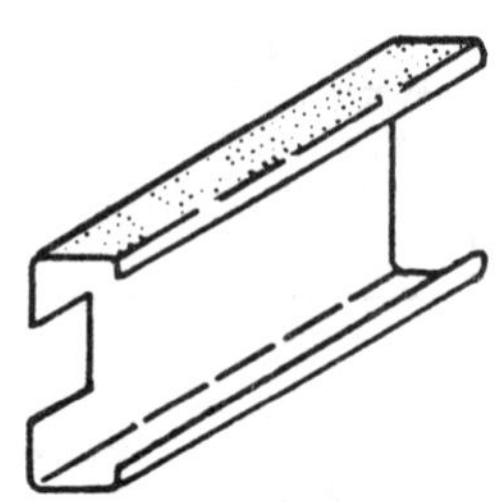

CEE STUD

# STUDLESS SOLID PARTITION

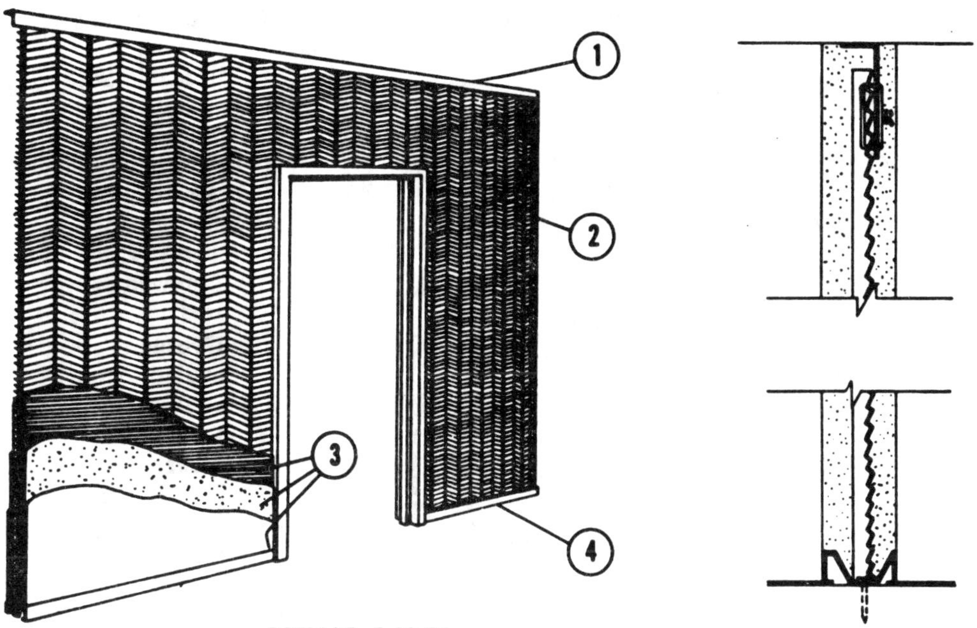


(1) Ceiling Runner
(2) Rib Metal Lath
(3) Plaster
(4) Combination Floor Runner and Screed

METAL LATH

# STUDLESS SOLID PARTITION

(1) Ceiling Runner
(2) Long Length Gypsum Lath
(3) Plaster
(4) Combination Floor Runner and Screed

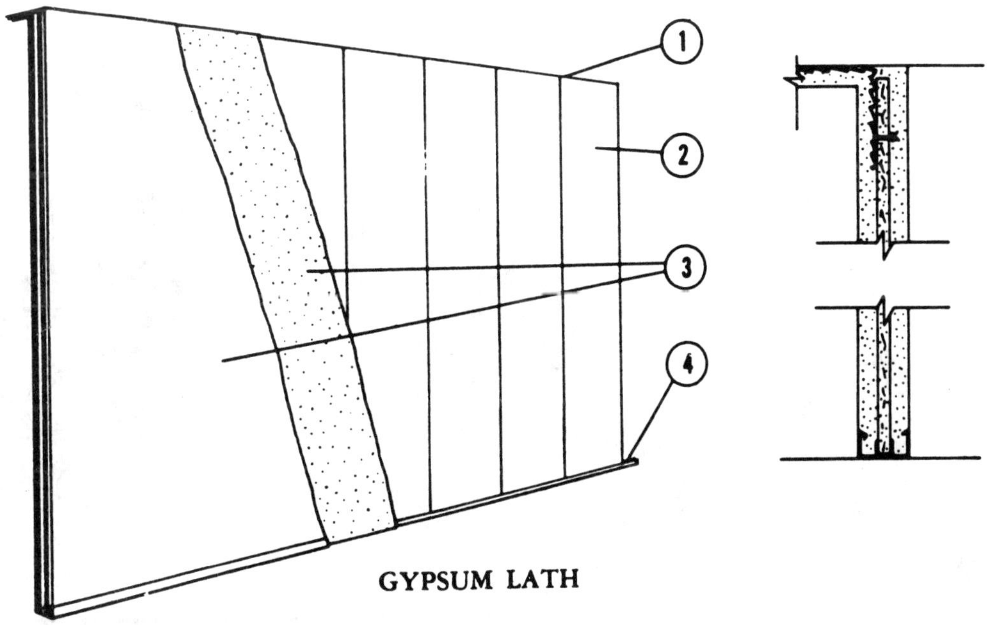


GYPSUM LATH

# STEEL STUD
## Hollow Partition

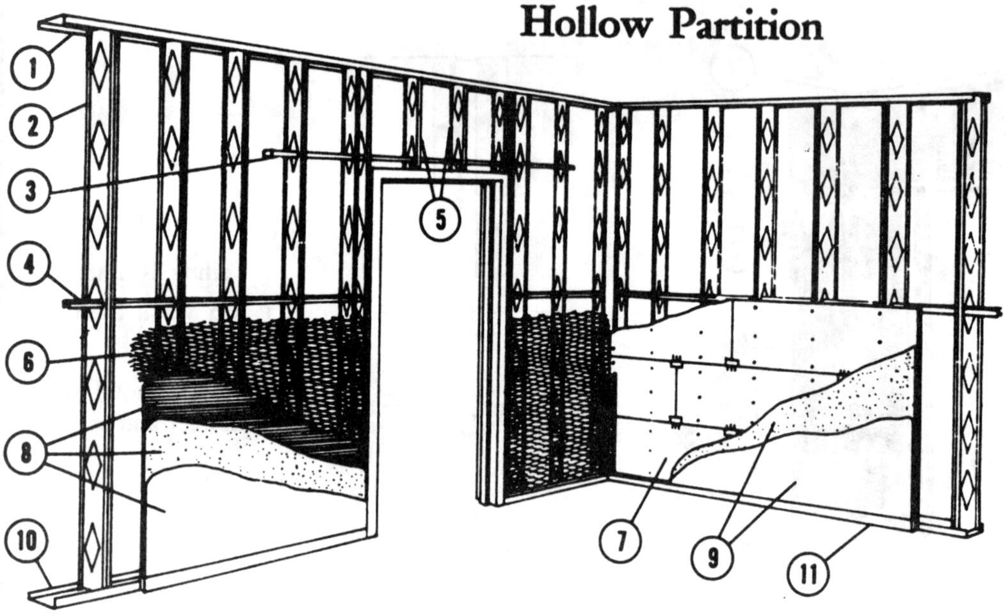


(1) Ceiling Runner Track
(2) Nailable Stud
(3) Door Opening Stiffener
(4) Partition Stiffener
(5) Jack Studs
(6) Metal or Wire Fabric Lath (screwed or wire tied)
(7) Gypsum Lath (nailed, clipped or screwed)
(8) Three Coats of Plaster (Scratch, Brown, Finish)
(9) Two Coats of Plaster (Brown, Finish)
(10) Floor Runner Track
(11) Flush Metal Base

# SCREW STUD
## Hollow Partition

(1) Ceiling Runner Track
(2) Screw Stud
(3) Door Opening Stiffener
(4) Partition Stiffener
(5) Jack Studs
(6) Metal or Wire Fabric Lath (screwed on)
(7) Gypsum Lath (screwed on)
(8) Three Coats of Plaster (Scratch, Brown, Finish)
(9) Two Coats of Plaster (Brown, Finish)
(10) Floor Runner Track
(11) Flush Metal Base

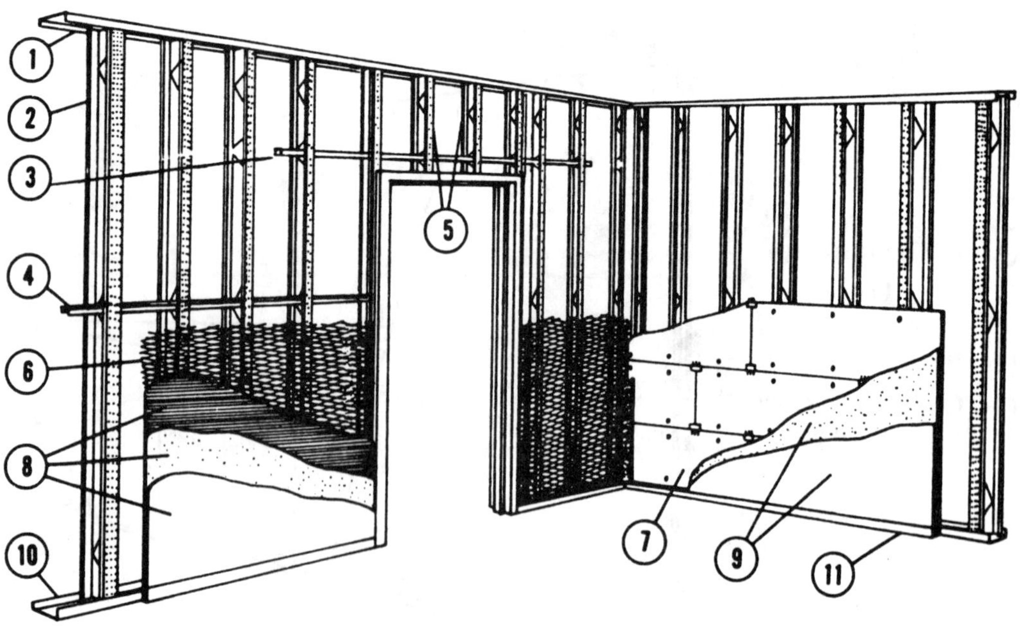

# LOAD-BEARING HOLLOW PARTITION

## Structural Stud

(1) Ceiling Runner Track
(2) Structural Stud (prefabricated)
(3) Structural Stud (nailable)
(4) Jack Studs
(5) Partition Stiffener (bridging)
(6) Metal or Wire Fabric Lath (wire-tied, nailed or stapled)
(7) Gypsum Lath (nailed or stapled)
(8) Three Coats of Plaster (Scratch, Brown, Finish)
(9) Two Coats of Plaster (Brown, Finish)

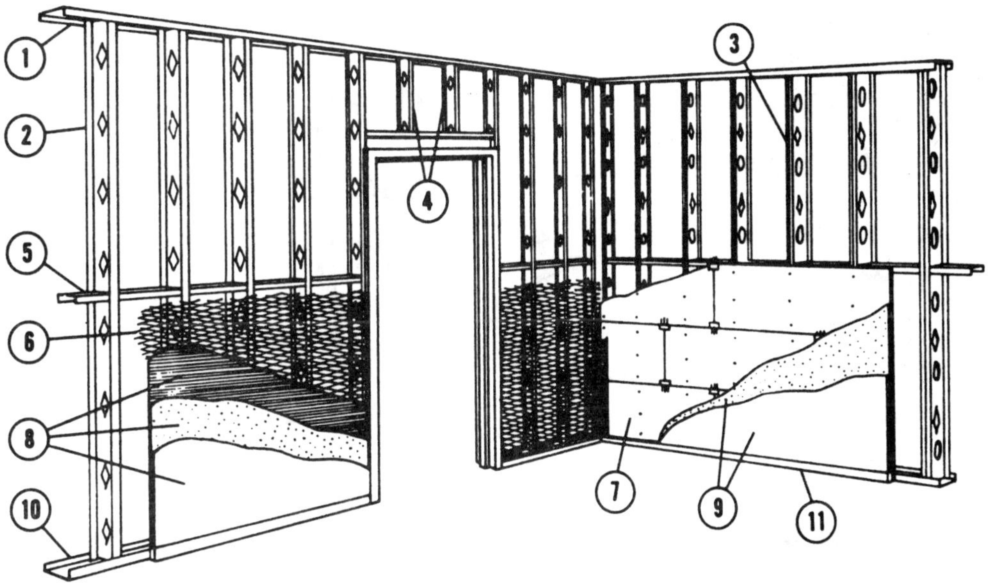

(10) Floor Runner Track
(11) Flush Metal Base

# VERTICAL FURRING

## With Studs

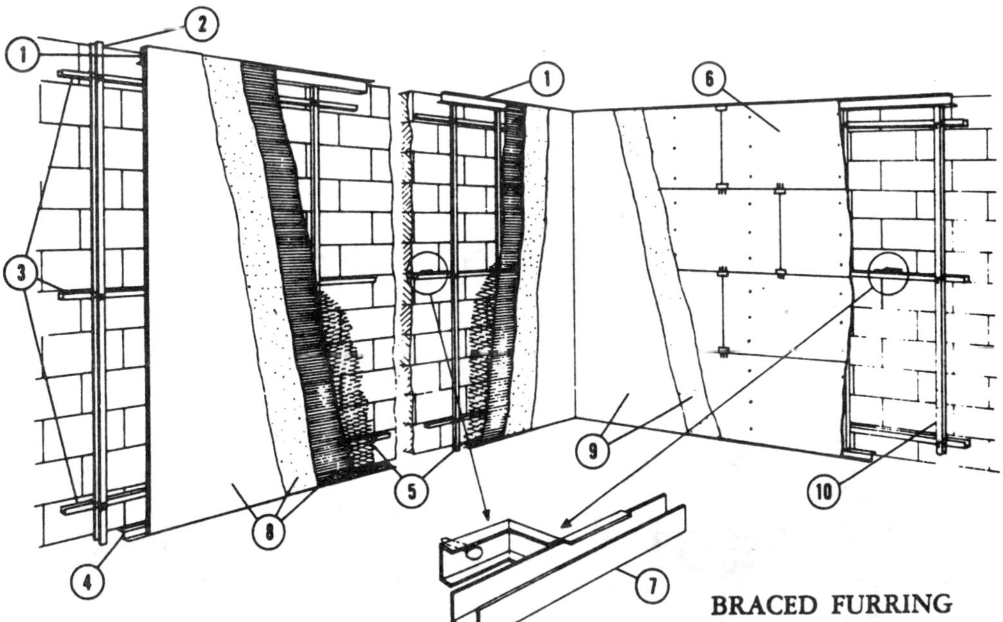

FREE STANDING FURRING

BRACED FURRING

(1) Ceiling Runner
(2) Channel Stud
(3) Horizontal Stiffener
(4) Floor Runner
(5) Metal or Wire Fabric Lath
(6) Gypsum Lath
(7) Bracing
(8) Three Coats of Plaster (Scratch, Brown, Finish)
(9) Two Coats of Plaster (Brown, Finish)
(10) Screw Channel Studs

# DOUBLE CHANNEL STUD

## Hollow Partition

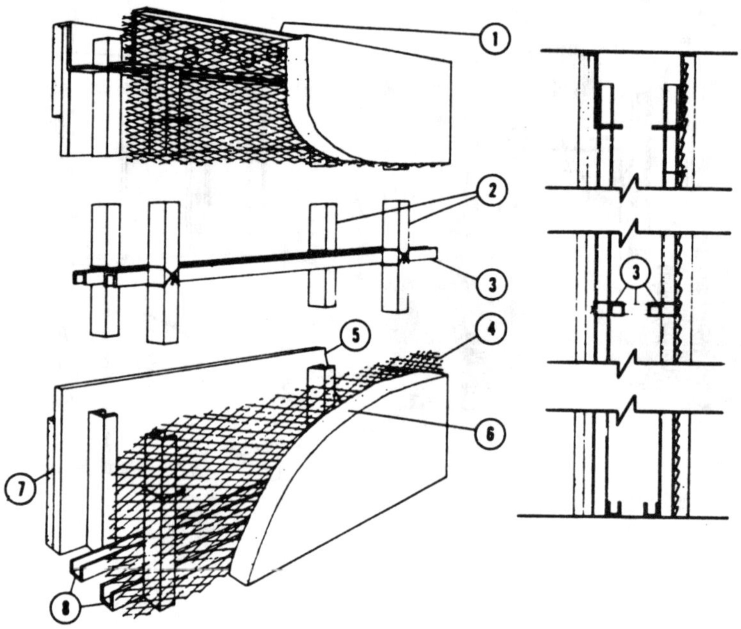

(1) Ceiling Runner
(2) Channel Studs
(3) Partition Stiffener and Channel Spacer
(4) Metal or Wire Fabric Lath (wire-tied)
(5) Gypsum Lath (clipped on)
(6) Three Coats of Plaster (Scratch, Brown, Finish)
(7) Two Coats of Plaster (Brown, Finish)
(8) Floor Runners (channel)

# SINGLE CHANNEL STUD

## Solid Partition

(1) Ceiling Runner
(2) Channel Stud
(3) Metal or Wire Fabric Lath (wire-tied)
(4) Plaster
(5) Combination Floor Runner and Screed

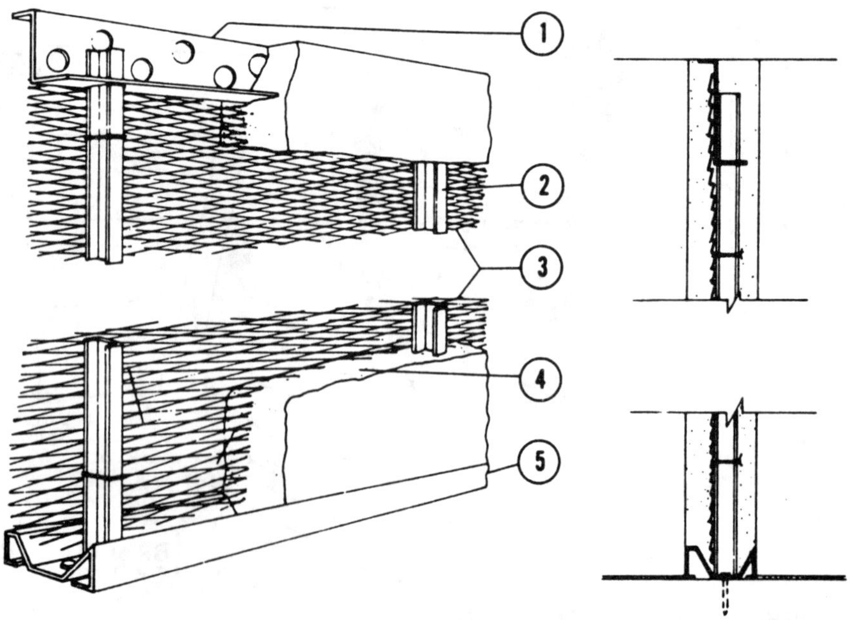

# CEILINGS

## STEEL JOISTS / CONCRETE JOISTS

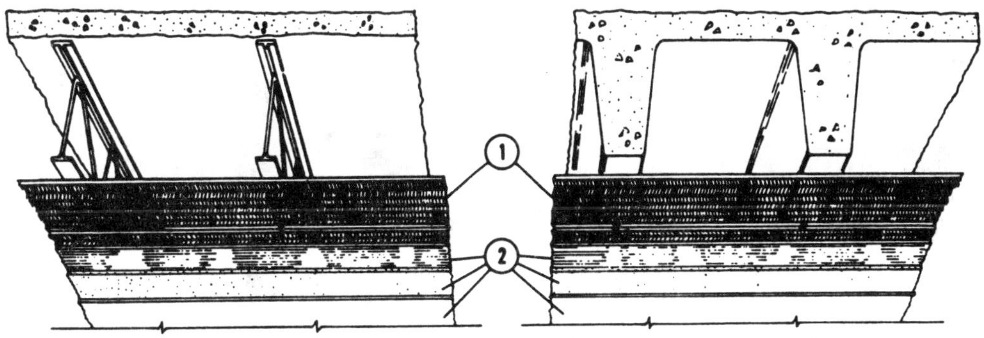


## WOOD JOISTS

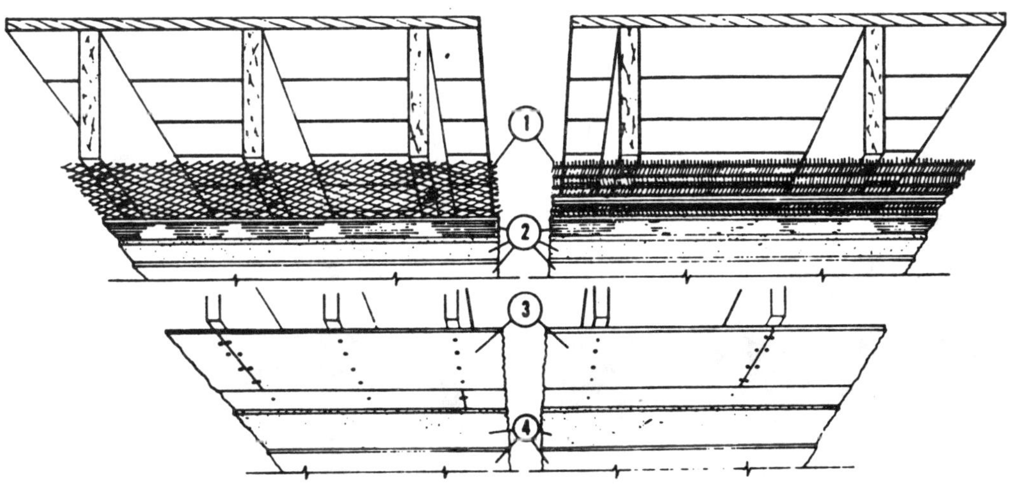


(1) Metal or Wire Fabric Lath

(2) Plaster

(3) Gypsum Lath

(4) Plaster

# SUSPENDED CEILINGS

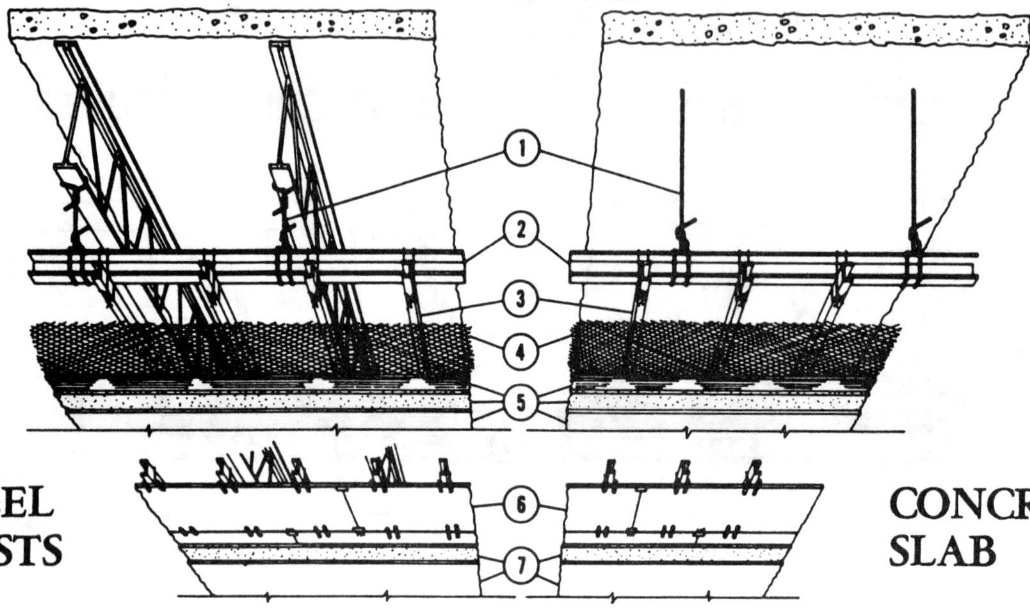


STEEL JOISTS

CONCRETE SLAB

(1) Hanger
(2) Main Runner Channel
(3) Furring Channel
(4) Metal or Wire Fabric Lath
(5) Plaster
(6) Gypsum Lath
(7) Plaster

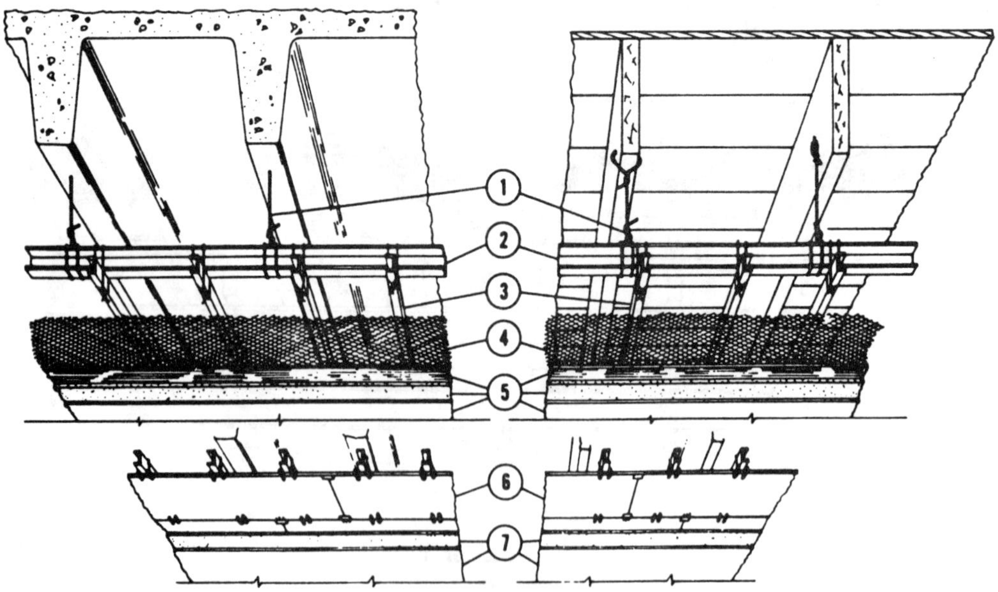


CONCRETE JOISTS

WOOD JOISTS

# FURRED CEILINGS

## STEEL JOISTS CONCRETE JOISTS

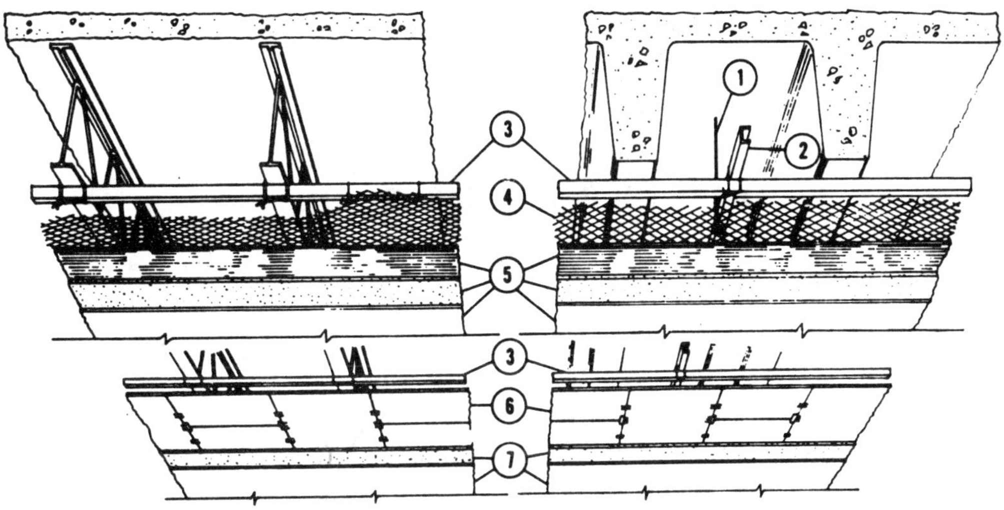

## WOOD JOISTS

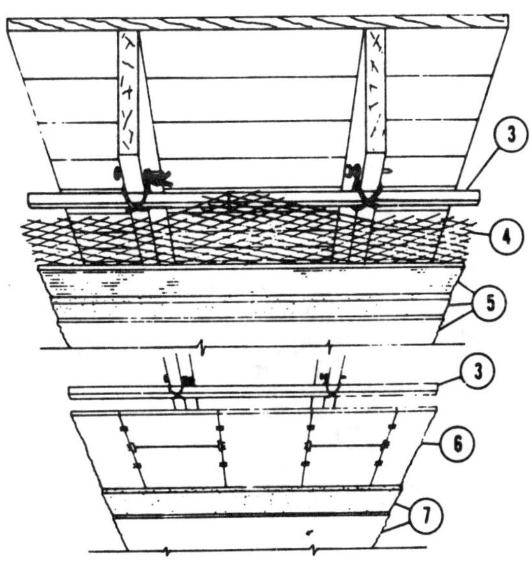

(1) Hanger
(2) ¾-inch Channel
(3) Cross Furring
(4) Metal or Wire Fabric Lath
(5) Plaster
(6) Gypsum Lath
(7) Plaster

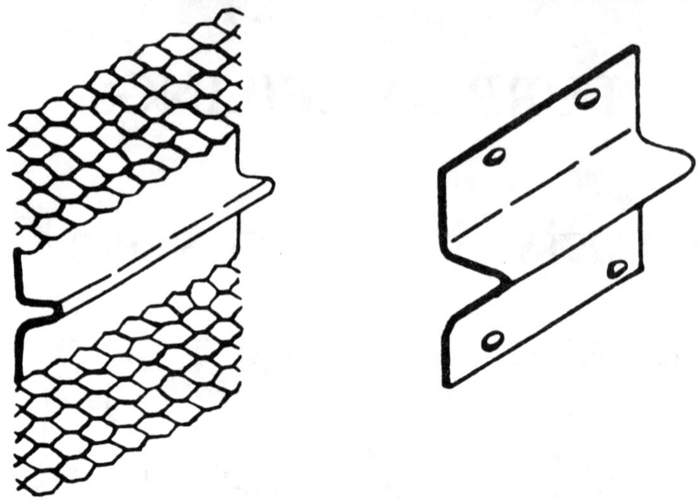

BASE OR PARTING SCREEDS

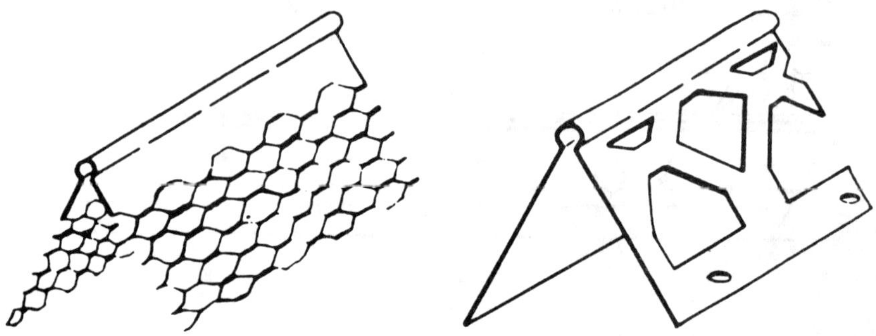

SMALL NOSE CORNER BEADS

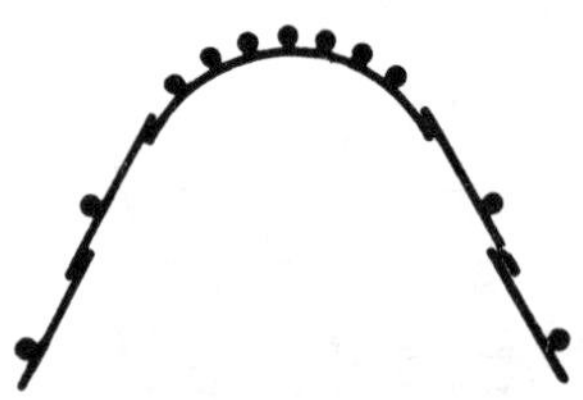

WIRE BULL NOSE CORNER BEADS

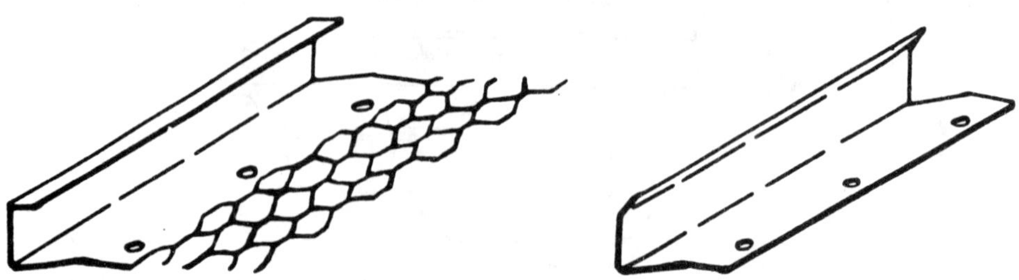

SQUARE CASING BEADS

# FINISHES / METAL LATH — 09205

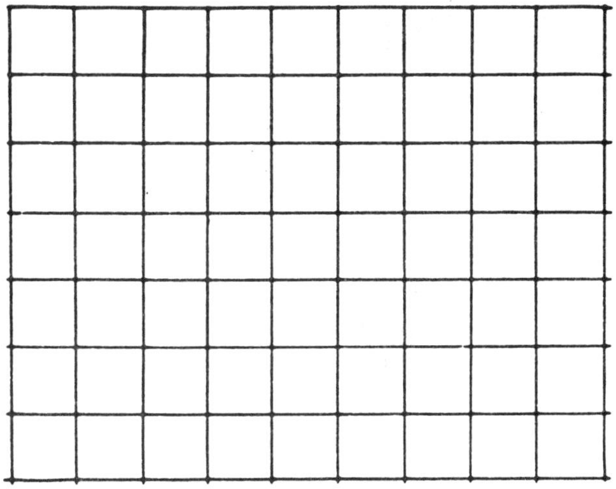

PLAIN WIRE
FABRIC LATH

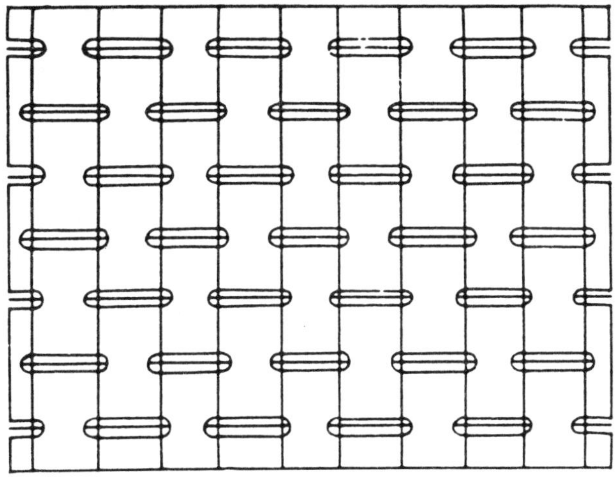

SELF-FURRING
WIRE FABRIC LATH

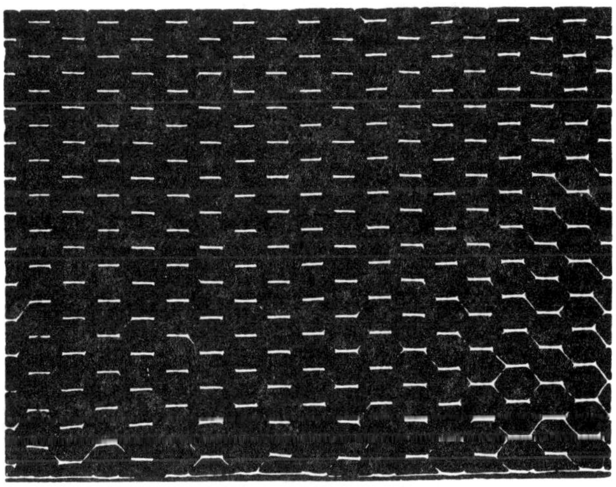

PAPER BACKED
WOVEN WIRE
FABRIC LATH

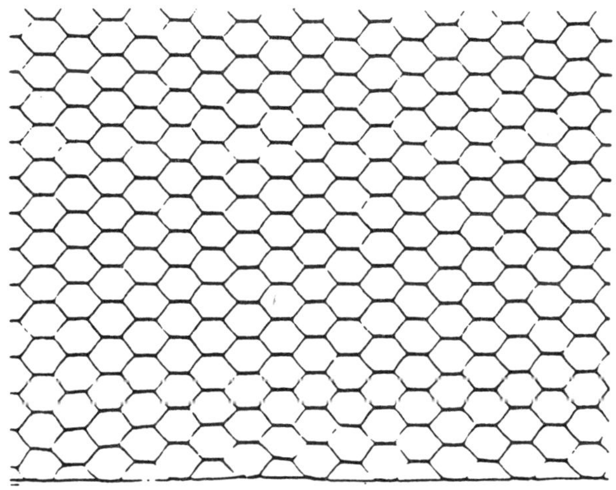

WOVEN WIRE
FABRIC LATH
(Also Available Self-Furred)

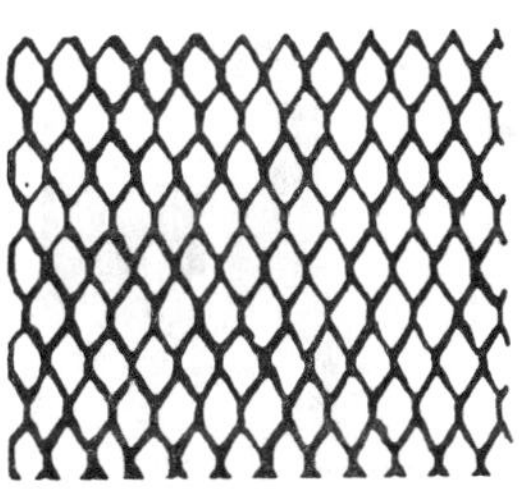

FLAT
DIAMOND MESH
METAL LATH

SELF-
FURRING
METAL LATH

FLAT RIB
METAL LATH

RIB
METAL LATH

RIB
METAL LATH

# VERTICAL FURRING
## Studless

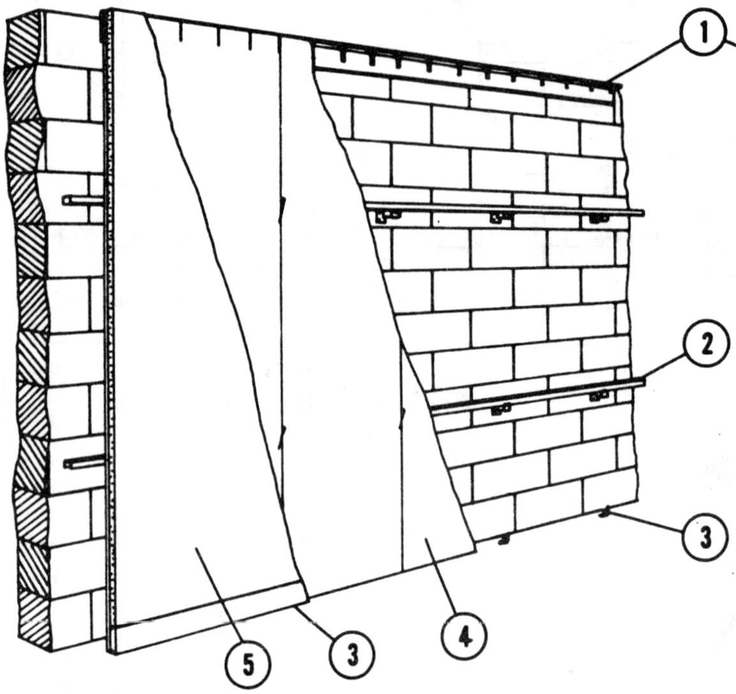


BRACED GYPSUM LATH

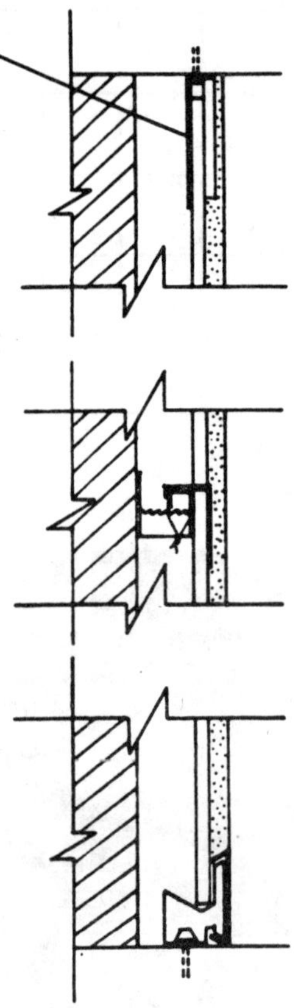

(1) Ceiling Runner

(2) Horizontal Stiffener (secured to bracing attachment)

(3) Metal Base and Clips

(4) Gypsum Lath

(5) Three Coats of Plaster (Scratch, Brown, Finish) (Minimum plaster thickness is ¾ inch.)

# COLUMN FURRING

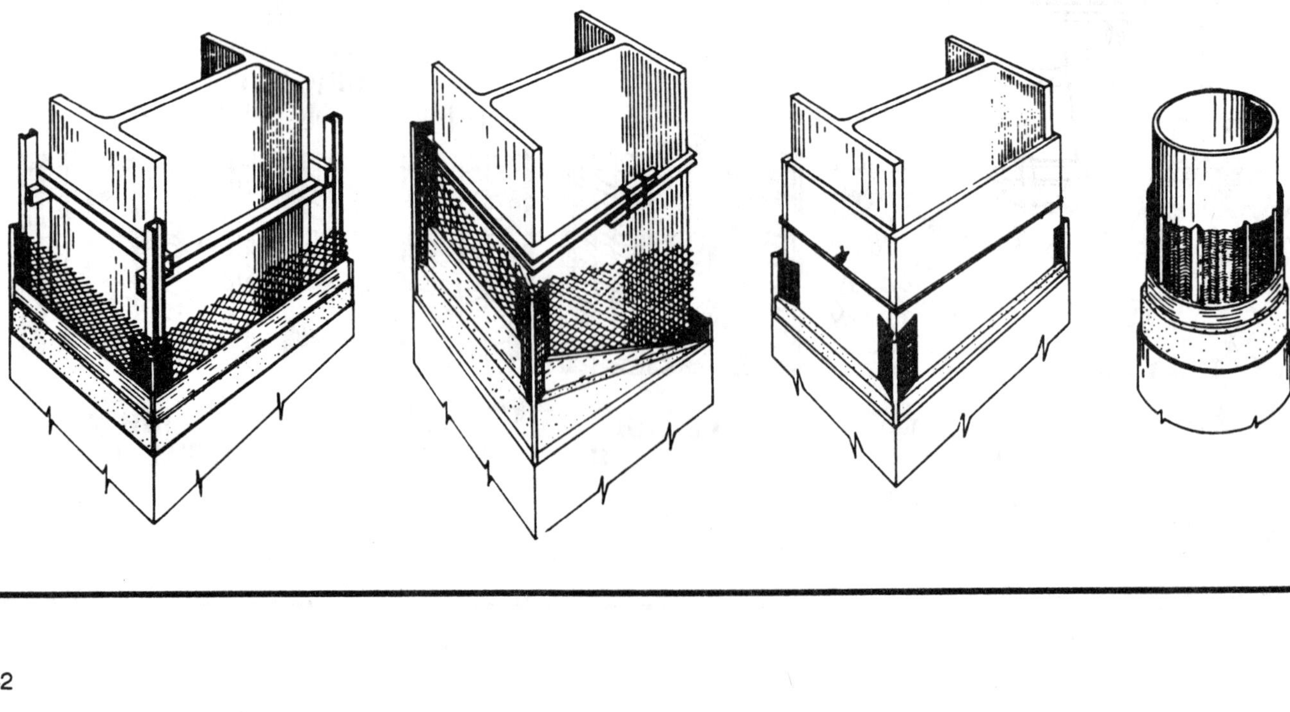

# CONTACT FURRING

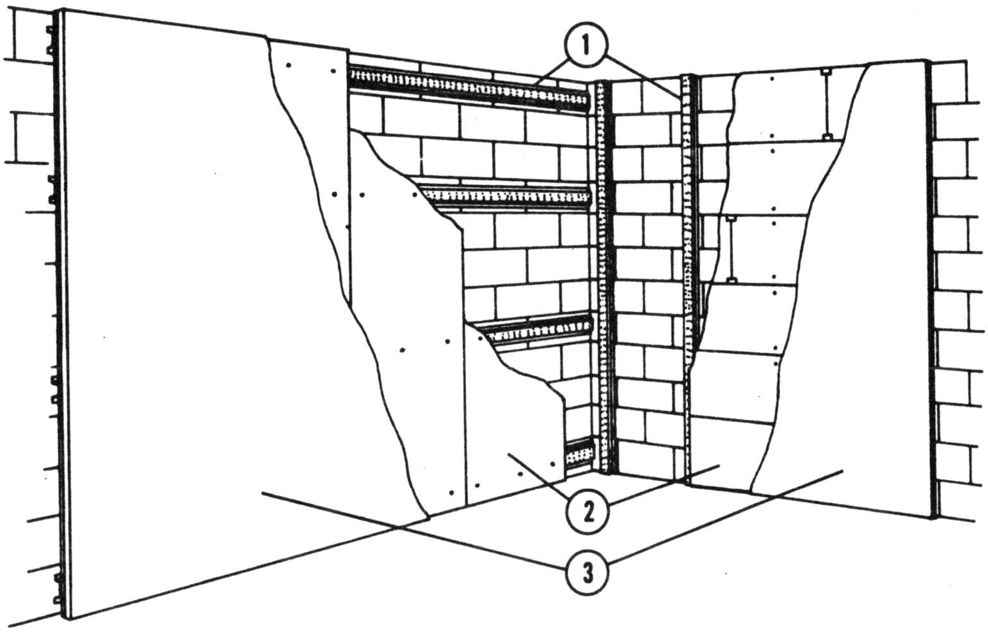


(1) Screw Channel
(2) Gypsum Lath (screwed on)
(3) Two Coats of Plaster (Brown, Finish)

# FALSE BEAMS

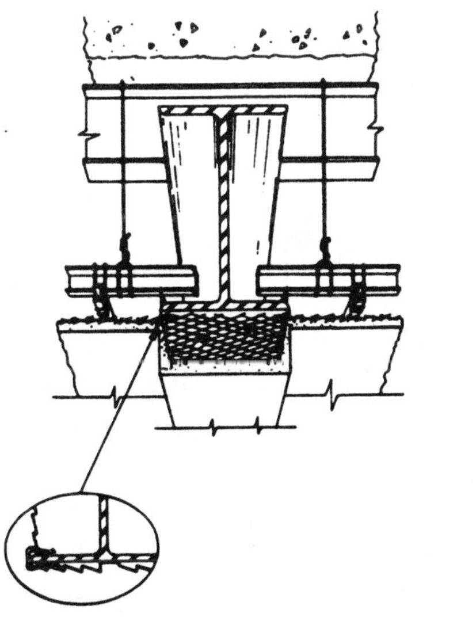

LATH DIRECT TO UNDERSIDE OF BEAM

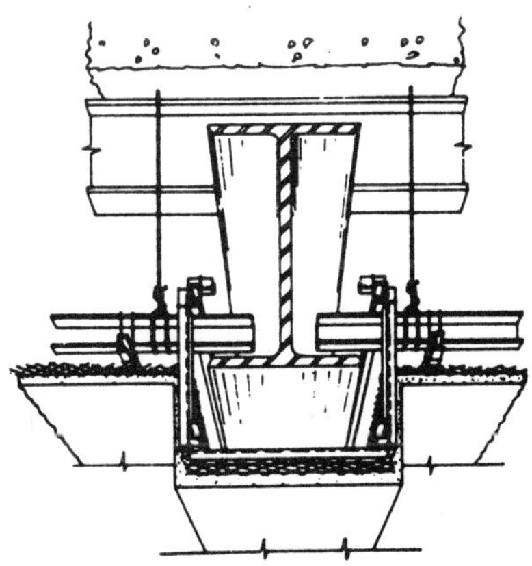

CHANNEL BRACKETS TO MAIN RUNNERS

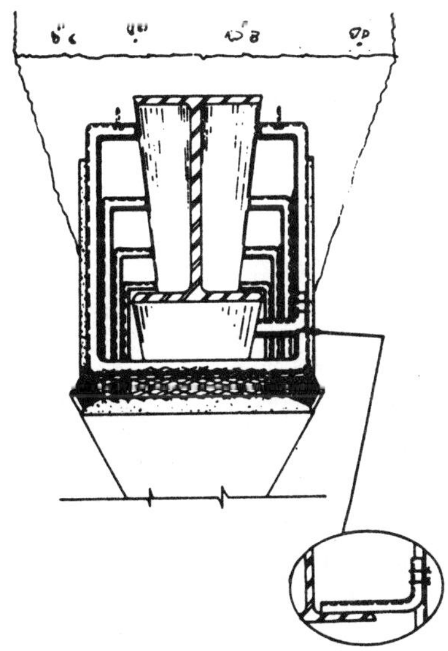

CHANNEL BRACKETS TO SLAB

| MAXIMUM SPACING OF SUPPORTS FOR METAL LATH (Inches) | | | | | | |
|---|---|---|---|---|---|---|
| | | WALLS AND PARTITIONS | | | CEILINGS | |
| Type of Lath | Weight of Lath Lb. Per Sq. Yd. | Wood Studs | Solid Partitions | Steel Studs, Wall Furring, etc. | Wood or Concrete | Metal |
| Diamond Mesh (flat expanded) | 2.5 | 16 | 16 | 13½ | 12 | 12 |
| | 3.4 | 16 | 16 | 16 | 16 | |
| Flat Rib | 2.75 | 16 | 16 | 16 | 16 | 16 |
| | 3.4 | 19 | 24(3) | 19 | 19 | 19 |
| ⅜″ Rib (1) (2) | 3.4 | 24 | (4) | 24 | 24 | 24 |
| | 4.0 | 24 | (4) | 24 | 24 | 24 |
| ¾″ Rib | 5.4 | — | (4) | 24(5) | 36(6) | 36(6) |
| Sheet Lath (2) | 4.5 | 24 | (4) | 24 | 24 | 24 |

**NOTE:** Weights are exclusive of paper, fiber or other backing.

(1) 3.4 lb., ⅜″ Rib Lath is permissible under Concrete Joists at 27″ c.c.

(2) These spacings are based on a narrow bearing surface for the lath. When supports with a relatively wide bearing surface are used, these spacings may be increased accordingly, and still assure satisfactory work.

(3) This spacing permissible for Solid Partitions not exceeding 16′ in height. For greater heights, permanent horizontal stiffener channels or rods must be provided on channel side of partitions, every 6′ vertically, or else spacing shall be reduced 25%.

(4) For studless solid partitions, lath erected vertically.

(5) For interior wall furring or for application over solid surfaces for stucco.

(6) For contact or ceilings only.

# TYPES OF LATH-ATTACHMENT TO WOOD AND METAL SUPPORTS

| TYPE OF LATH | NAILS Type and Size | NAILS MAXIMUM SPACING Vertical (In Inches) | NAILS MAXIMUM SPACING Horizontal (In Inches) | SCREWS MAXIMUM SPACING Vertical (In Inches) | SCREWS MAXIMUM SPACING Horizontal (In Inches) | STAPLES Round or Flattened Wire: Wire Gauge No. | STAPLES Crown (In Inches) | STAPLES Leg (In Inches) | STAPLES MAXIMUM SPACING Vertical (In Inches) | STAPLES MAXIMUM SPACING Horizontal (In Inches) |
|---|---|---|---|---|---|---|---|---|---|---|
| 1. Diamond Mesh Expanded Metal Lath and Flat Rib Metal Lath | 4d blued smooth box 1½ No. 14 gauge 7/32" head (clinched)<br>1" No. 11 gauge 7/16" head, barbed<br>1½" No. 11 gauge 7/16" head, barbed | 6<br>6<br>6 | —<br>—<br>6 | 6 | 6 | 16 | 3/4 | 7/8 | 6 | 6 |
| 2. 3/8" Rib Metal Lath and Sheet Lath | 1½" No. 11 ga. 7/16" head, barbed | 6 | 6 | 6 | 6 | 16 | 3/4 | 1½ | At Ribs | At Ribs |
| 3. 3/4" Rib Metal Lath | 4d common 1½" No. 12½ gauge 1/4" head<br>2" No. 11 gauge 7/16" head, barbed | At Ribs | —<br>At Ribs | At Ribs | At Ribs | 16 | 3/4 | 1⅝ | At Ribs | At Ribs |
| 4. Wire Fabric Lath | 4d blued smooth box (clinched)<br>1" No. 11 gauge 7/16" head, barbed | 6<br>6 | —<br>— | 6 | 6 | 16 | 3/4 | 7/8 | 6 | 6 |
| | 1½" No. 11 gauge 7/16" head, barbed<br>1¼" No. 12 ga. 3/8" head, furring<br>1" no. 12 gauge 3/8" head | 6<br>6<br>6 | 6<br>6 | | | 16 | 7/16 | 7/8 | 6 | 6 |
| 5. 3/8" Gypsum Lath | 1⅛" No. 13 gauge 12/61" head, blued | 8 | 8 | 8 | 8 | 16 | 3/4 | 7/8 | 8 | 8 |
| 6. ½" Gypsum Lath | 1¼" No. 13 gauge 12/61" head, blued | 8 | 8<br>6 | 8 | 8<br>6 | 16 | 3/4 | 1⅛ | 8 | 8<br>6 |

## STRESS RELIEF (CONTROL JOINTS)

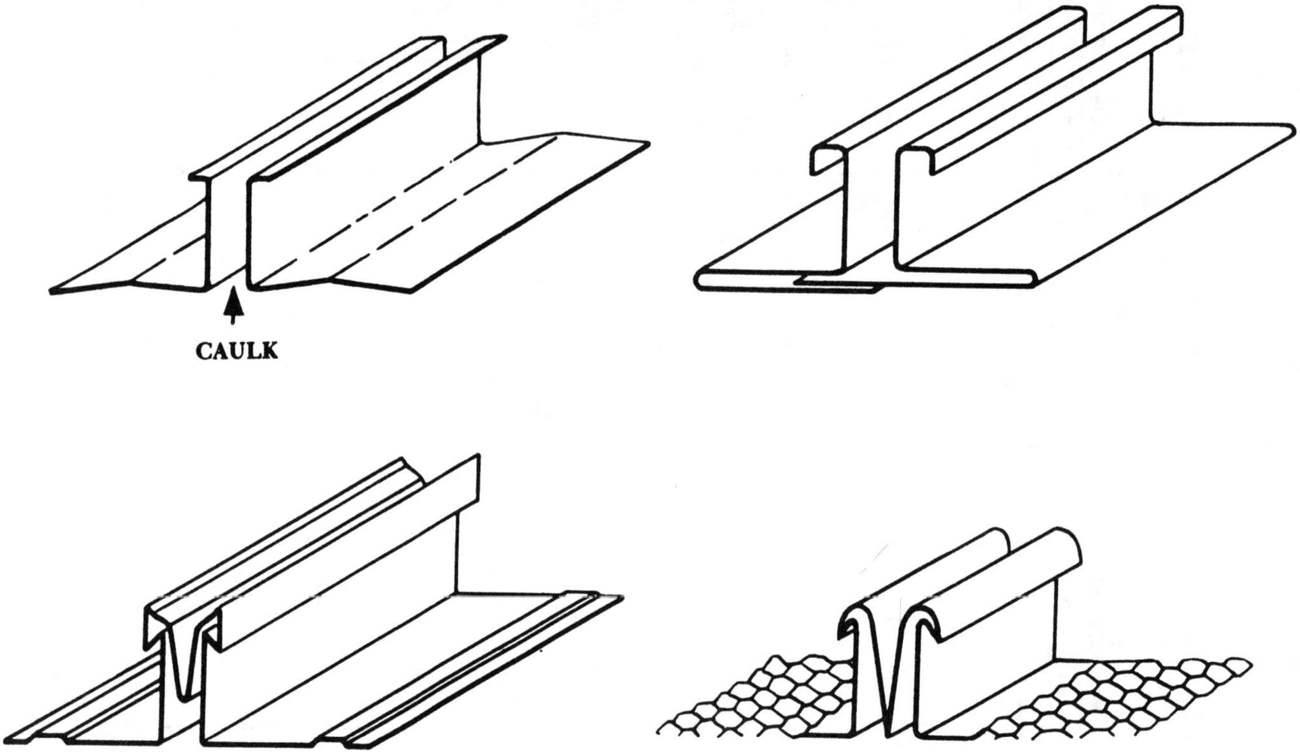


## REVEALS

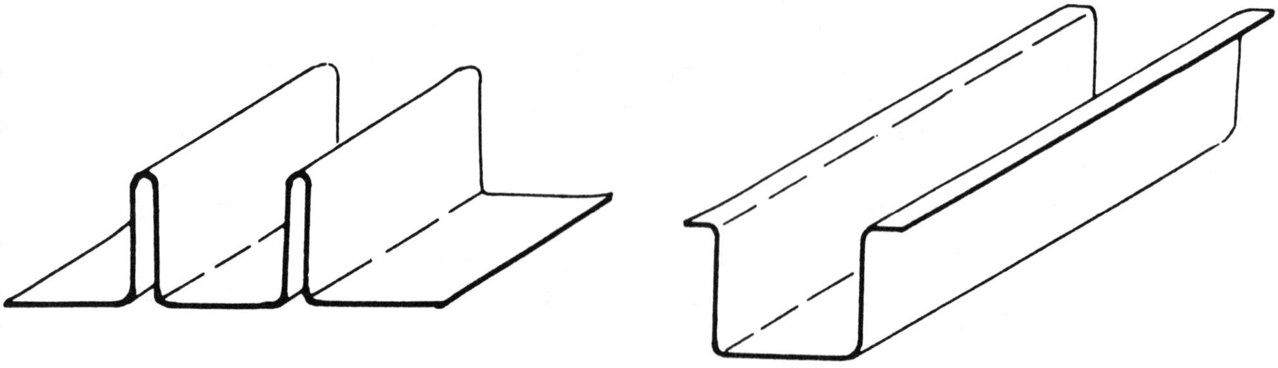

# FINISHES / PLASTER ACCESSORIES | 09205

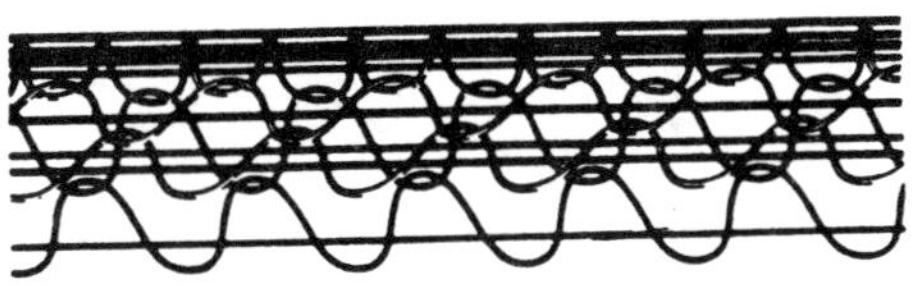

CORNER REINFORCEMENT (EXTERIOR) WIRE

CORNER REINFORCEMENT (EXTERIOR) EXPANDED METAL

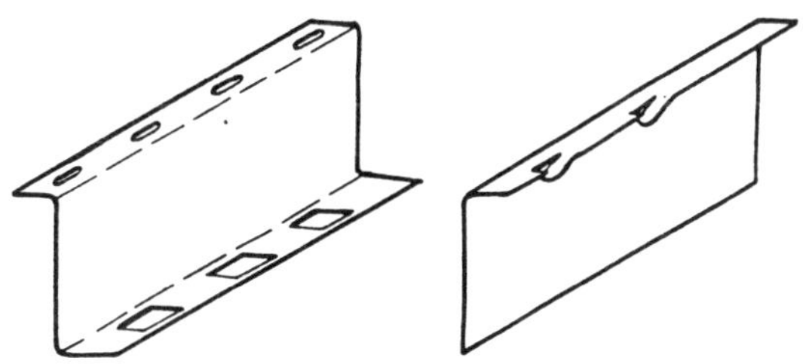

PARTITION RUNNERS (Z AND L SHAPE)

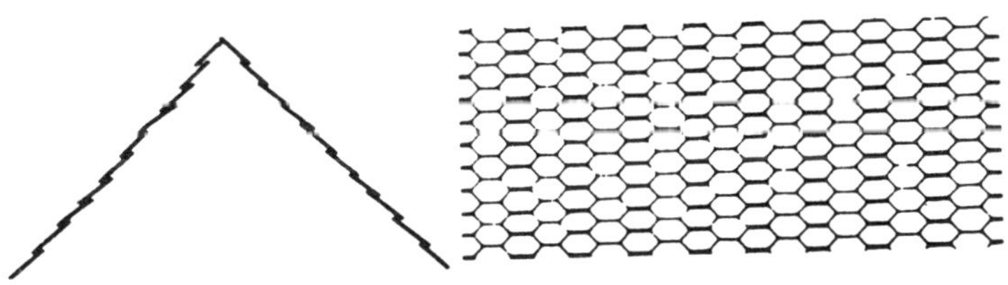

EXPANDED METAL CORNERITE

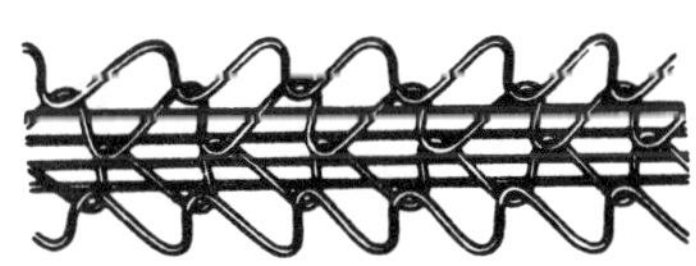

WIRE CORNERITE

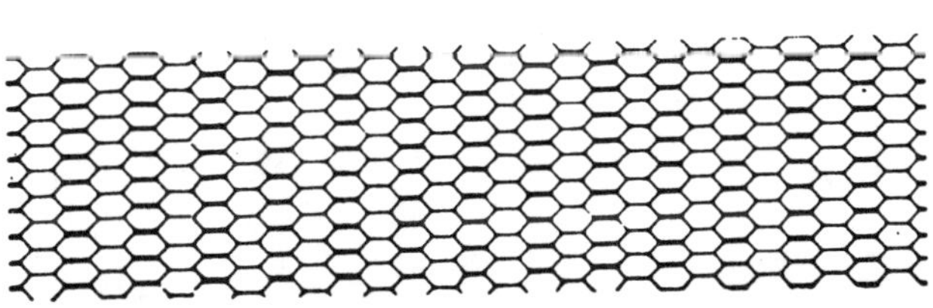

STRIP REINFORCEMENT (EXPANDED METAL)

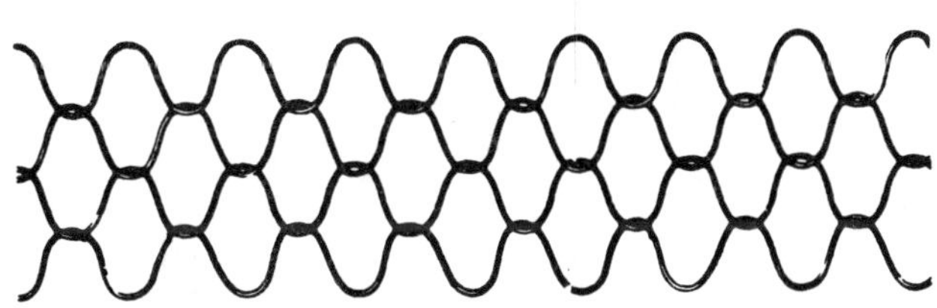

STRIP REINFORCEMENT (WIRE)

# FINISHES / PLASTER ACCESSORIES | 09205

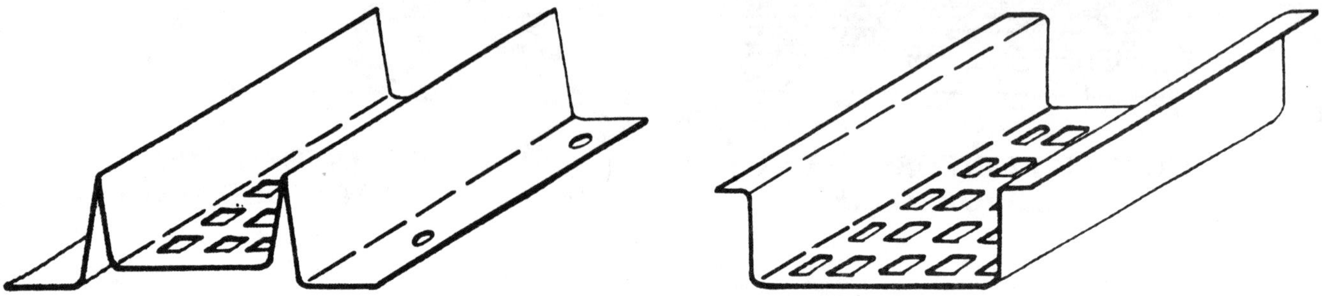

VENTILATING SCREEDS

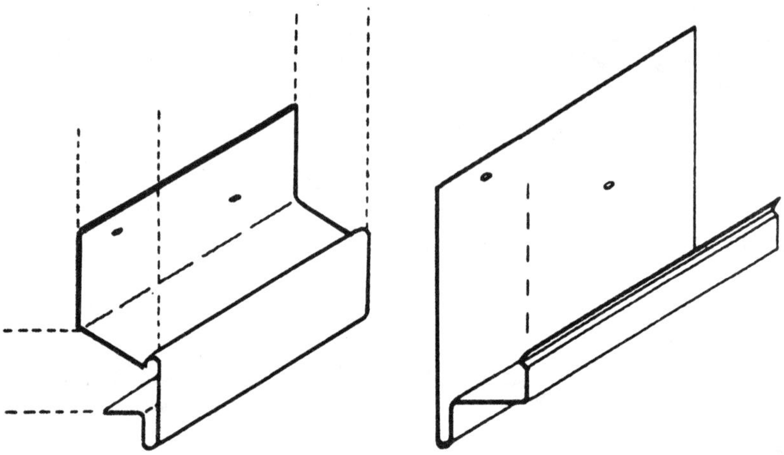

DRIP SCREEDS

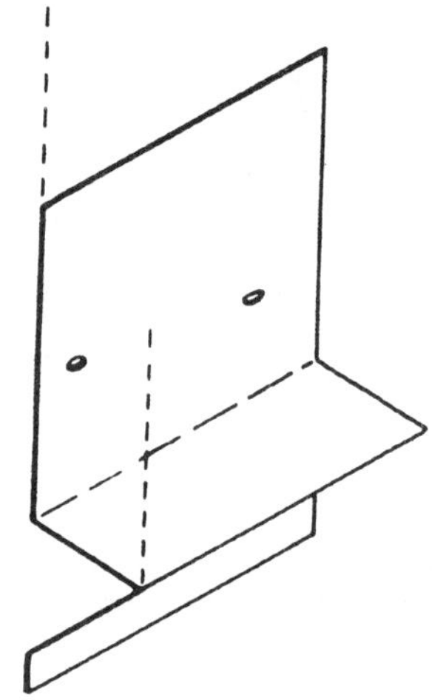

WEEP SCREED
(Also available with perforations)

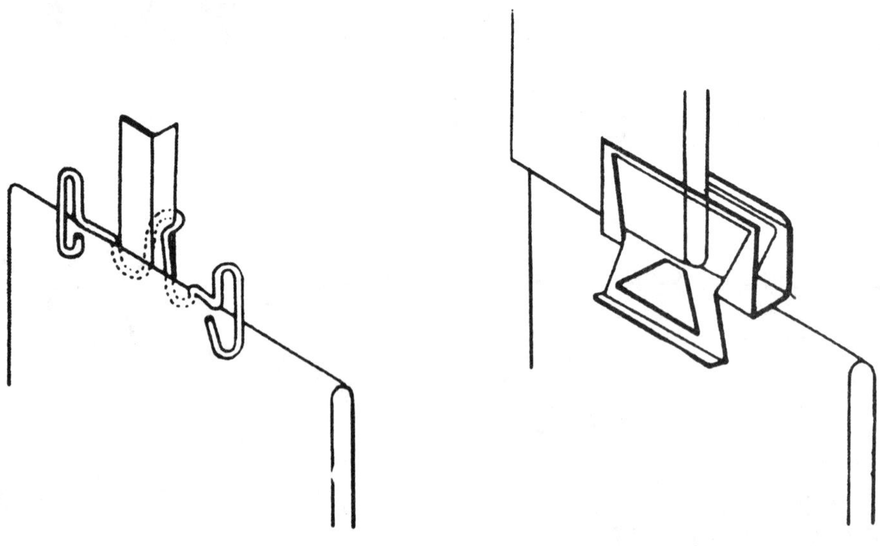

GYPSUM LATH ATTACHMENT CLIPS

## VERTICAL FURRING

| VERTICAL FURRING MEMBER | UNBRACED | | | | BRACED | | | |
|---|---|---|---|---|---|---|---|---|
| | STUD SPACING | | | | STUD SPACING | | | |
| | 24″ | 19″ | 16″ | 12″ | 24″ | 19″ | 16″ | 12″ |
| | Maximum Furring Heights | | | | Maximum Vertical Distance Between Braces | | | |
| ¾″ Channel | 6′ | 7′ | 8′ | 9′ | 5′ | 5′ | 6′ | 7′ |
| 1½″ Channel | 8′ | 9′ | 10′ | 12′ | 6′ | 7′ | 8′ | 9′ |
| 2″ Channel | 9′ | 10′ | 11′ | 13′ | 7′ | 8′ | 9′ | 10′ |
| 2″ Prefab. Stud | 8′ | 9′ | 10′ | 11′ | 6′ | 7′ | 8′ | 9′ |
| 2½″ Prefab. Stud | 10′ | 11′ | 12′ | 14′ | 8′ | 9′ | 10′ | 11′ |
| 3¼″ Prefab. Stud | 14′ | 16′ | 17′ | 20′ | 11′ | 13′ | 14′ | 16′ |

## TYPES OF LATH—MAXIMUM SPACING OF SUPPORTS

| TYPE OF LATH | Minimum Weight (psy), Gauge & Mesh Size | VERTICAL | | | HORIZONTAL | |
|---|---|---|---|---|---|---|
| | | WOOD | METAL | | | |
| | | | Solid Plaster Partitions | Other | Wood or Concrete | Metal |
| Expanded Metal Lath (Diamond Mesh) | 2.5<br>3.4 | 16″<br>16″ | 16″<br>16″ | 12″<br>16″ | 12″<br>16″ | 12″<br>16″ |
| Flat rib Expanded Metal Lath | 2.75<br>3.4 | 16″<br>19″ | 16″<br>24″ | 16″<br>19″ | 16″<br>19″ | 16″<br>19″ |
| Stucco Mesh Expanded Metal Lath | 1.8 and<br>3.6 | 16″ | — | — | — | — |
| ⅜″ Rib Expanded Metal Lath | 3.4<br>4.0 | 24″<br>24″ | | 24″<br>24″ | 24″<br>24″ | 24″<br>24″ |
| Sheet Lath | 4.5 | 24″ | — | 24″ | 24″ | 24″ |
| ¾″ Rib Expanded Metal Lath (Not manufactured in West) | 5.4 | — | — | — | 36″ | 36″ |
| Wire Fabric Lath — Welded | 1.95 lbs., 11ga., 2″ x 2″<br>1.4 lbs., 16 ga., 2″ x 2″<br>1.4 lbs., 18 ga., 1″ x 1″ | 24″<br>16″<br>16″ | 24″<br>16″<br>— | 24″<br>16″<br>— | 24″<br>16″<br>— | 24″<br>16″<br>— |
| Wire Fabric Lath — Woven | 1.4 lbs., 17 ga., 1½″ Hex.<br>1.4 lbs., 18 ga., 1″ Hex. | 24″<br>24″ | 16″<br>16″ | 16″<br>16″ | 24″<br>24″ | 16″<br>16″ |
| ⅜″ Gypsum Lath (plain) | — | 16″ | — | 16″ | 16″ | 16″ |
| (Large Size) | — | 16″ | — | 16″ | 16″ | 16″ |
| ½″ Gypsum Lath (plain) | — | 24″ | — | 24″ | 24″ | 24″ |
| (Large Size) | — | 24″ | No supports; Erected vertically | 24″ | 24″ | 16″ |
| ⅝″ Gypsum Lath (Large Size) | — | 24″ | No supports; Erected vertically | 24″ | 24″ | 16″ |

# PLASTERING TABLES

## THICKNESS OF PLASTER

| PLASTER BASE | FINISHED THICKNESS OF PLASTER FROM FACE OF LATH, MASONRY, CONCRETE | |
|---|---|---|
| | Gypsum Plaster | Portland Cement Plaster |
| Expanded Metal Lath | 5/8" minimum | 5/8" minimum |
| Wire Fabric Lath | 5/8" minimum | 3/4" minimum (interior)<br>7/8" minimum (exterior) |
| Gypsum Lath | 1/2" minimum | |
| Gypsum Veneer Base | 1/16" minimum | |
| Masonry Walls | 1/2" minimum | 1/2" minimum |
| Monolithic Concrete Walls | 5/8" maximum | 7/8" maximum |
| Monolithic Concrete Ceilings | 3/8" maximum | 1/2" maximum |

## GYPSUM PLASTER PROPORTIONS

| NUMBER OF COATS | COAT | PLASTER BASE OR LATH | MAXIMUM VOLUME AGGREGATE PER 100# NEAT PLASTER (CUBIC FEET) | |
|---|---|---|---|---|
| | | | Damp Loose Sand | Perlite or Vermiculite |
| Two-Coat Work | Basecoat | Gypsum Lath | 2½ | 2½ |
| | Basecoat | Masonry | 3 | 3 |
| Three-Coat Work | First Coat | Lath | 2 | 2 |
| | Second Coat | Lath | 3 | 3 |
| | First & Second Coat | Masonry | 3 | 3 |

## PORTLAND CEMENT PLASTER

| COAT | VOLUME CEMENT | MAXIMUM WEIGHT (OR VOLUME) LIME PER VOLUME CEMENT | MAXIMUM VOLUME SAND PER VOLUME CEMENT | APPROXIMATE MINIMUM THICKNESS | MINIMUM PERIOD MOIST CURING | MINIMUM INTERVAL BETWEEN COATS |
|---|---|---|---|---|---|---|
| First | 1 | 20 lbs. | 4 | 3/8" | 48 Hours | 48 Hours |
| Second | 1 | 20 lbs. | 5 | 1st and 2nd Coats total 3/4" | 48 Hours | 7 Days |
| Finish | 1 | 1 | 3 | 1st, 2nd and Finish Coats 7/8" | — | |

## PORTLAND CEMENT-LIME PLASTER

| COAT | VOLUME CEMENT | MAXIMUM VOLUME LIME PER VOLUME CEMENT | MAXIMUM VOLUME SAND PER COMBINED VOLUMES CEMENT AND LIME | APPROXIMATE MINIMUM THICKNESS | MINIMUM PERIOD MOIST CURING | MINIMUM INTERVAL BETWEEN COATS |
|---|---|---|---|---|---|---|
| First | 1 | 1 | 4 | 3/8" | 48 Hours | 48 Hours |
| Second | 1 | 1 | 4½ | 1st and 2nd Coats total 3/4" | 48 Hours | 7 Days |
| Finish | 1 | 1 | 3 | 1st, 2nd and Finish Coats 7/8" | – | |

## METAL STUD CONSTRUCTION

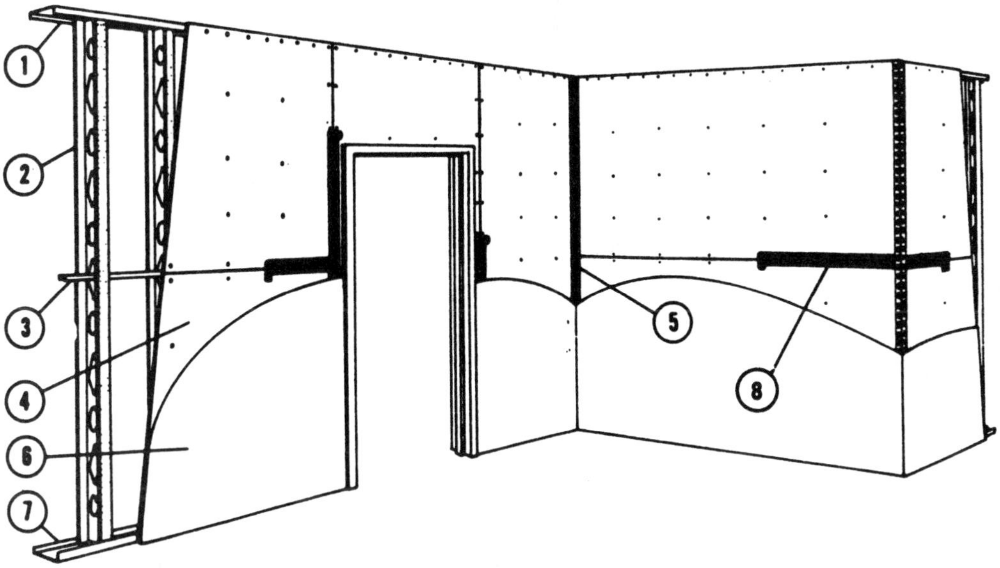


(1) Ceiling Runner Track
(2) Metal Stud (nailable or screw)
(3) Horizontal Stiffener
(4) Large Size Lath
(5) Angle Reinforcement
(6) Veneer Plaster 1/16 to 1/8 inch thick)
(7) Floor Runner Track
(8) Joint Reinforcement

## WOOD STUD CONSTRUCTION

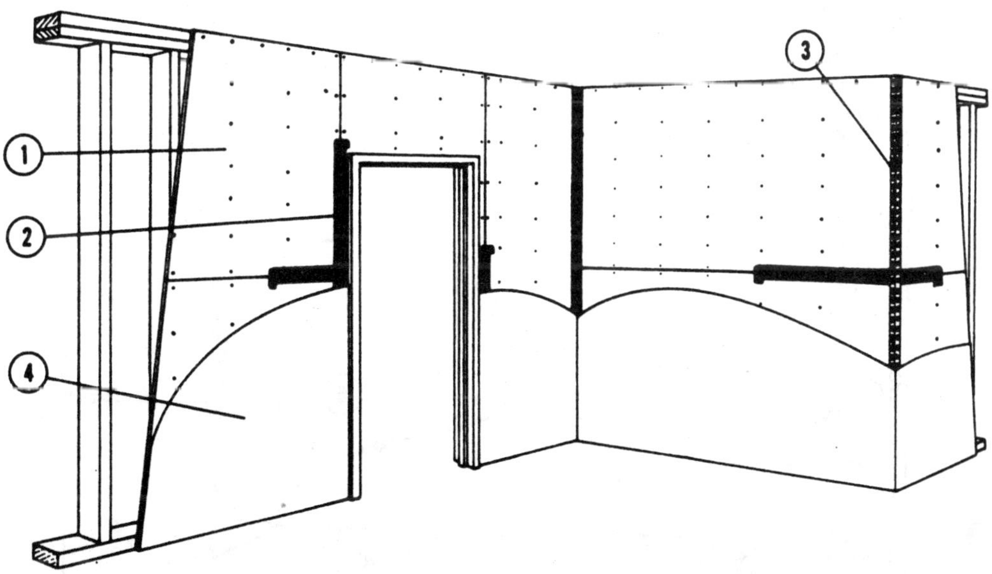


(1) Large Size Lath
(2) Joint Reinforcement
(3) Corner Bead
(4) Veneer Plaster (1/16 to 1/8 inch thick)

# EXTERIOR LATH AND PLASTER

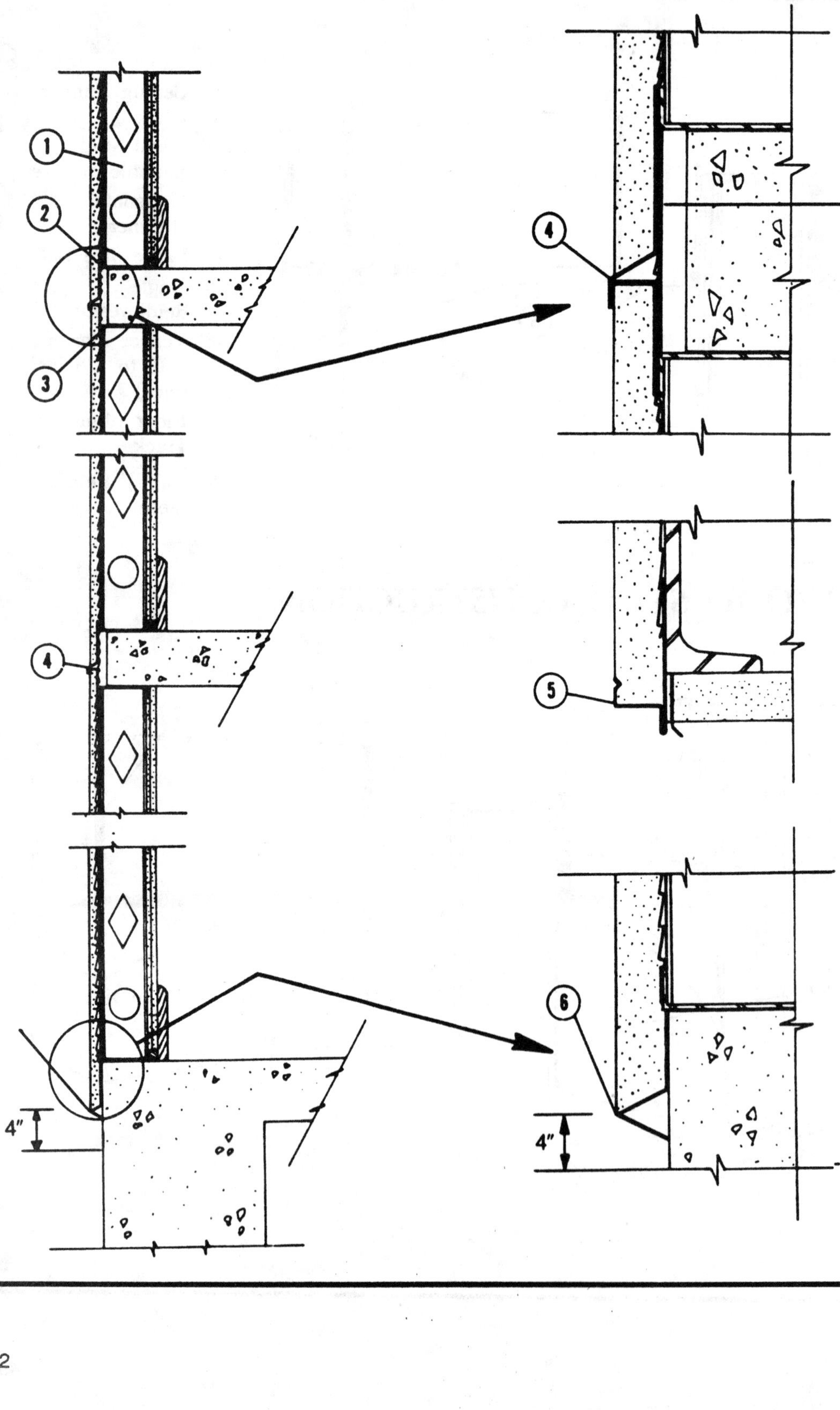


(1) Metal Stud
(2) Floor Runner Track
(3) Ceiling Runner Track
(4) Control Joint
(5) Drip Screed (at soffit)
(6) Weep Screed (at slab or foundation)
(7) Building Paper, Continuous

# EXTERIOR LATH AND PLASTER

## OPEN WOOD FRAME CONSTRUCTION

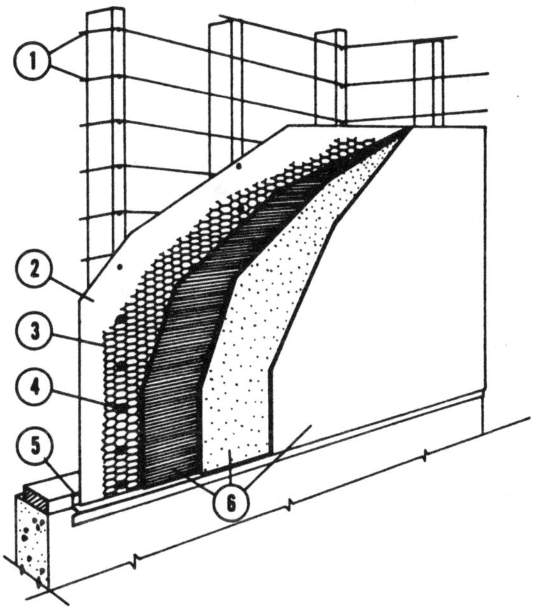


(1) Wire Backing
(2) Building Paper
(3) Wire Fabric Lath
(4) Approved Fasteners
(5) Weep Screed
(6) Three Coats of Plaster (Scratch, Brown, Finish)

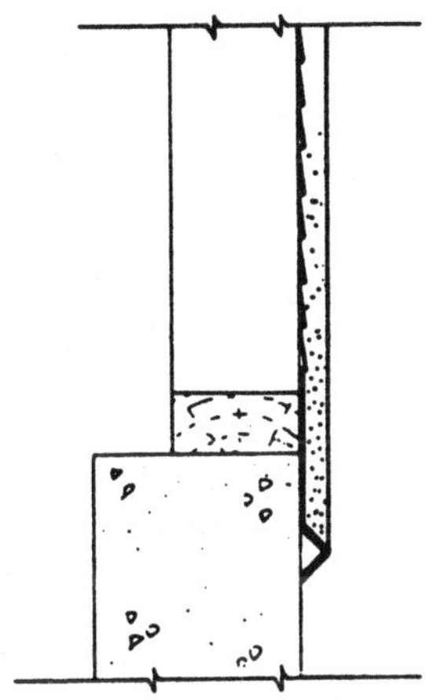

## SHEATHED WOOD FRAME CONSTRUCTION

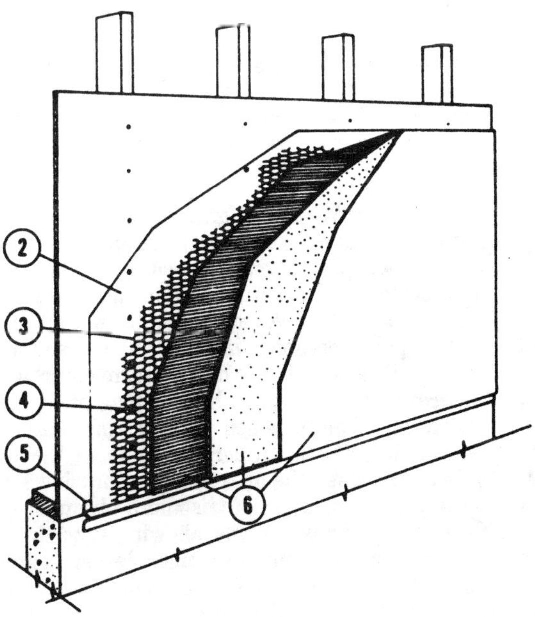


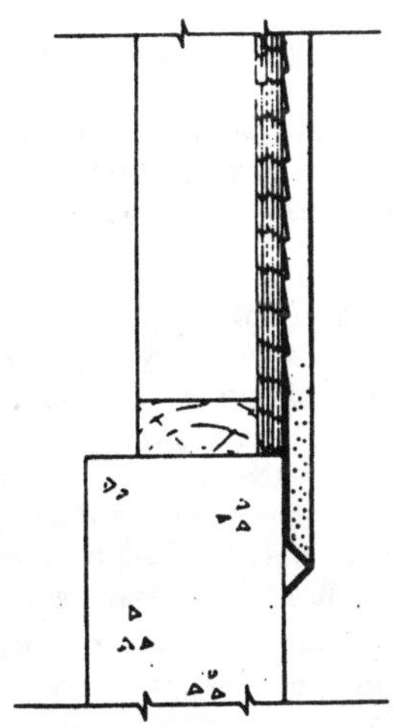

**Paint Failures.** There is a cause for every paint failure, and in most instances the failure can be prevented by observing specific precautions. The most often observed paint failures are: cracking, flaking, scaling, alligatoring, bleeding, excessive chalking, blistering, fading, spotting, washing and discoloration. The probable causes, preventive measures and corrective measures are discussed in the following subsections:

***Alligatoring.*** When a rupturing of the top paint coat causes the surface to break up into irregular areas separated by wide cracks in "alligator hide" fashion, the condition is called alligatoring. Its cause is due to applying paint to unseasoned wood or from applying a heavy coat of paint over a relatively soft undercoat before it has thoroughly dried. To correct, remove the entire paint film and repaint.

***Bleeding.*** When the color of a previous coat is absorbed into the top coat, this condition is called bleeding. It is caused by the partial solubility of the pigment of the undercoat in the vehicle of the top coat. Asphalt or bituminous undercoats are particularly susceptible to bleeding. In order to prevent or correct bleeding, care must be exercised to select a vehicle for the top coats which is not a solvent for the undercoat pigments.

***Blistering.*** Blistering is evidenced by blister-like irregularities on the film of a painted surface. The most common cause of blistering is the application of paint over a damp or wet surface. Under the action of the sun's rays, the moisture is drawn out of the wood (or vaporized on a metallic surface), raising the paint coating with it in the form of blisters. Blistering can also be caused by using too much dryer in the undercoats. To avoid blistering, do not paint damp or wet surfaces, green lumber or greasy spots. Avoid excessive amounts of dryers in undercoats. To correct, remove the paint film in the blistered area, let the surface dry and repaint.

***Blushing.*** A surface in which blushing has occurred is characterized by a white discoloration in the paint film or a precipitation of an ingredient. It is caused by the condensation of moisture on the paint film or by improper composition of the vehicle or solvent. By applying paint under conditions which do not permit moisture to condense on the applied film, blushing can be avoided. Do not paint cold metal under humid conditions. Blushing can be corrected only by removing the paint film in the affected areas and repainting.

***Chalking.*** Chalking of a painted surface can be detected by gentle rubbing which will disclose loose powder on the paint film. Slow, uniform chalking is a desirable quality in the paint, but excessive chalking is undesirable and indicates a paint film failure. It is caused by excessive rain, fog or high humidity during the application of the paint or during the drying period. Paints low in binder content or high in inert pigments have a tendency toward early and excessive chalking.

To prevent chalking, use a paint which is neither deficient in oil content nor high in inert pigment content and apply only under dry conditions. To correct chalked surfaces, remove all loose chalked substance from the surface with a wire brush and repaint the surface with a paint of good quality.

***Checking.*** Checking of a painted surface can be detected by the appearance in the topcoat of small openings or ruptures which divide the surface into small irregular areas, leaving the undercoat visible through the breaks in the topcoat. It is usually caused by too soft an undercoat or by applying a coat over an underlying coat which has not thoroughly dried. To correct, wire brush or sandpaper to remove the loose film and apply a new coat of paint, making sure that it is as elastic as the previous coat.

***Cracking.*** Cracks are breaks in a paint film which extend through to the surface to which the paint was applied. Where cracking is severe, flaking, scaling or peeling usually follows. It is usually caused by improper compounding of the paint, resulting in a hard and brittle paint film lacking in elasticity. Low-grade paints are often inelastic because they are deficient in oils or contain too much inert material. Cracking will also result from painting a surface upon which too many coats of paint already exist. To avoid cracking, use paint of proper consistency and elasticity, and remove heavy coats of paint from the surface to be painted prior to repainting.

***Crawling or Creeping.*** This defect can be distinguished by the little drops or islands which are formed by the paint film. It often occurs when paint is applied on an oily, waxy or greasy surface or on a very smooth surface such as glass and polished metal. To correct, remove all greasy or oily spots from the surface. On glossy surfaces, wash with a mild soda ash solution or sand with fine sandpaper.

***Dulling.*** Dulling is characterized by a loss in gloss which develops in high-gloss paints and enamels and is caused by improper compounding, use of very old stocks of paints and enamels or through the use of too much turpentine as a thinner. To avoid dulling, use fresh paint stocks of good quality. To correct, sand the dulled coat with fine sandpaper and repaint.

***Flaking.*** Flaking is the dropping off of small pieces of a paint coat which has generally started with checking, developing into cracking, and finally the small cracked sections fall away from the surface. The causes and corrective measures are the same as for cracking.

***Mildew.*** Mildew, a form of plant life, is a fungus frequently found on exposed surfaces in damp, warm locations, particularly on soft paint films. It is caused by soft paint films becoming sticky in the warm location and wind-blown spores and decayed and dried vegetation adhering to the surfaces. The oil in the paint becomes infected, and breeding of the mildew spores takes place in the damp, warm environment. Mildew can be prevented by using a hard-drying paint, applied under dry conditions. Paints which contain zinc oxide are resistant to mildew infections. To correct, remove the paint coat with a blowtorch, wash the area with an abrasive soap and water, or with a water solution of trisodium phosphate, rinsing the surface with clear water and allowing it to dry. The addition of one-half ounce of mercuric chloride per gallon of paint or the use of less oils and more turpentine is advisable where mildew is likely to occur. Exercise extreme caution in handling of paints with mercuric chloride or other fungicides to prevent poisoning of the skin.

*Peeling.* Peeling of a paint film is evidenced by large scales of the film curling and peeling off the painted surface. It is usually caused by the application of paint in the presence of moisture or from being applied over a faulty priming coat. It can be caused by moisture getting behind the paint film at a corner, or in the case of wood, at knot-holes which are not properly sealed. Painting unseasoned lumber will also cause peeling. To avoid peeling, seal knot holes and end grains of lumber with shellac before painting and insure that the edges and corners of steel shapes are thoroughly covered. To correct, remove all loose, peeled paint by wire brush or blowtorch, clean the surface of dust film, and repaint on a dry surface only.

*Runs and Sags.* Ripples, runs and sags on a vertical surface are caused by the application of too heavy a coat of paint, or by using a paint which has been thinned excessively. Other contributing causes are the use of stiff, inflexible brushes or incomplete brushing of the film. It can be prevented by applying a uniform coat of paint of the correct consistency, using a flexible brush and properly brushing out the paint. When runs or sags appear on spray-painted surfaces, it is also indicative that too heavy a coat was applied, or that the paint was excessively thinned. To correct, remove the runs or sags with sandpaper and repaint the area.

*Scaling.* Scaling is an aggrevated form of flaking. It is evidenced by large sections of paint coming loose and falling from the painted surface. Scaling is usually preceded by cracking, and like cracking, is usually caused by the paint drying hard and brittle and being unable to expand or contract as the painted object does with changes in temperature and moisture. Scaling frequently occurs when unseasoned lumber is painted. In some cases, previous coats of paint may have lost their elasticity and become "lifeless." This causes poor adhesion in the old coat and, in drying, the new coat shrinks and pulls the old film loose from the painted surface. To avoid, do not paint unseasoned lumber, or over a paint which dries hard and brittle, or over old lifeless paint films. Remove the paint with commercial paint removers, scrapers or blowtorch. Remove dust from the surface and repaint only when the surface is thoroughly dry, using a paint which dries with an elastic film.

*Slow Drying.* The drying time of a paint film varies with the inherent characteristics of the pigment or of the vehicle. However, certain faulty conditions may prolong the drying period and cause the condition to be termed a paint fault. Paints, which under normal drying conditions are tacky or sticky for 12 hours or longer after application, are likely to catch dust and dirt, promote mildew formations or fail by checking or alligatoring. It is usually caused by using paints to which a small amount of mineral oil has been added through negligence or error. Such paint may never dry thoroughly. Old linseed oil that has become fatty by exposure or inferior dryers and thinners frequently contribute to slow drying of paint coats. This condition can be avoided by painting when the temperature is above 50°F or by applying thin coats of paint in colder weather. Use less oil and allow ample time for each coat to dry before applying subsequent coats. Once paint is applied and fails to dry, the condition can be corrected by allowing sufficient additional time for drying, or by removing the paint with commercial paint removers or by scraping, and then repainting with the right type of paint under proper conditions.

*Spotting.* The appearance of discolored spots or craters in a painted surface is called spotting. It is usually caused by too few coats, such as painting on new work using only two coats, or painting old work with one coat. The lack of controlled penetration causes uneven fading in the pigments. Sap in wood may affect the paint and cause spotting. Spotting may be caused by improperly sealed nail heads, or rain or hail on a freshly painted surface. On plastered surfaces, an inferior primer or sealer may cause spotting as the alkali in the plaster burns through. To prevent spotting, apply sufficient coats of paint and avoid painting when rain is imminent. Insure that all old paint, plaster surfaces and nail heads are properly sealed with quality paint films. To correct, apply an additional coat of paint. Paint surfaces spotted from rain or hail must be sanded smooth before repainting.

*Washing.* Washing of paint is evidenced by streaks in the paint surface, the discoloration generally accumulating at the lower edges of boards, panels, girders, etc. Washing is caused by using paints having pigments which form water-soluble compounds or by painting under damp conditions. This defect often occurs when paints of inferior quality are used. To prevent washing, use only good quality paints. To correct, remove the paint film completely and repaint.

*Wrinkling.* Wrinkling of a paint coat is evidenced by the paint film gathering in small wrinkles. Wrinkling may be caused by the application of an excessively thick coat, or by failure to brush out the paint properly. It is frequently caused by too much dryer in the paint or by using paints which have been excessively thinned with oil and applied thick. To avoid wrinkling, do not apply thick coats or use excessive amounts of dryer or oil. To correct, remove the wrinkles with sandpaper and repaint with properly thinned paint which does not have excessive amount of dryer or oil in it. If the coat of paint is excessively wrinkled, strip off the old coat and repaint.

**Painting Checklist.**

(1) Is all paint material tested or approved?

(2) Should surface preparation samples for sandblasting be required?

(3) Are color samples required?

(4) Is the contractor required to submit a list of paint materials for approval?

(5) Are delivered paint materials as specified?

(6) Prior to painting, check each surface with contractor for preparation, type of paint, number of coats and color.

(7) Is pretreatment required?

(8) Is paint properly mixed? Not excessively thinned?

(9) Is coverage satisfactory? Surface appearance and texture? Color?

(10) Is adequate ventilation provided in confined space? Lighting conditions satisfactory?

(11) Are weather conditions satisfactory for painting?

# PIPE WEIGHTS

## CAST IRON PIPE

| SERVICE WEIGHT | | | |
|---|---|---|---|
| Size, Inches | Weight of Pipe | Weight of Water | Total Weight-Lbs. |
| 2″ | 3.8 | 1.45 | 5.3 |
| 3″ | 5.6 | 3.2 | 8.8 |
| 4″ | 7.5 | 5.5 | 13.0 |
| 5″ | 9.8 | 8.7 | 18.5 |
| 6″ | 12.4 | 12.5 | 24.9 |
| 8″ | 18.5 | 21.7 | 40.2 |

| EXTRA HEAVY | | | |
|---|---|---|---|
| Size, Inches | Weight of Pipe | Weight of Water | Total Weight-Lbs. |
| 2″ | 4.3 | 1.45 | 5.8 |
| 3″ | 8.3 | 3.2 | 11.5 |
| 4″ | 10.8 | 5.5 | 16.3 |
| 5″ | 13.3 | 8.7 | 22.0 |
| 6″ | 16.0 | 12.5 | 28.5 |
| 8″ | 26.5 | 21.7 | 48.2 |

## STEEL PIPE

| Pipe Size | W/40 | $H_2O$/Lbs. | Total Lbs./L.F. |
|---|---|---|---|
| 2″ | 3.65 | 1.45 | 5.1 |
| 2½″ | 5.79 | 2.07 | 7.86 |
| 3″ | 7.57 | 3.2 | 10.77 |
| 3½″ | 9.11 | 4.28 | 13.39 |
| 4″ | 10.8 | 5.51 | 16.31 |
| 5″ | 14.6 | 8.66 | 23.26 |
| 6″ | 18.0 | 12.5 | 30.5 |
| 8″ | 28.6 | 21.66 | 50.26 |
| 10″ | 40.5 | 34.15 | 74.65 |
| 12″ | | | |

## SPRINKLER AREA CALCULATIONS

Typical maximum floor area allowed per system riser:

| Light Hazard | Ordinary Hazard | Extra Hazard |
|---|---|---|
| 52,000 S.F. | 40,000-52,000 S.F. | 25,000 S.F. |

Typical maximum floor area coverage allowed per sprinkler head:

| Light Hazard | Ordinary Hazard | Extra Hazard |
|---|---|---|
| 130-200 S.F. | 100-130 S.F. | 90 S.F. |

Typical maximum spacing between lines and sprinkler heads:

| Light Hazard | Ordinary Hazard | Extra Hazard |
|---|---|---|
| 12-15 feet | 12-15 feet | 12 feet |

**Note:** This data is for estimating purposes only. Check all applicable codes and regulations for specific requirements.

## SPRINKLER HEAD CALCULATIONS

***Typical maximum quantity of sprinkler heads allowed by pipe size.***

**Light Hazard:**

For sprinklers below ceiling:

| Steel | | Copper | |
|---|---|---|---|
| 1 in. pipe | 2 sprinklers | 1 in. tube | 2 sprinklers |
| 1¼ in. pipe | 3 sprinklers | 1¼ in. tube | 3 sprinklers |
| 1½ in. pipe | 5 sprinklers | 1½ in. tube | 5 sprinklers |
| 2 in. pipe | 10 sprinklers | 2 in. tube | 12 sprinklers |
| 2½ in. pipe | 30 sprinklers | 2½ in. tube | 40 sprinklers |
| 3 in. pipe | 60 sprinklers | 3 in. tube | 65 sprinklers |
| 3½ in. pipe | 100 sprinklers | 3½ in. tube | 115 sprinklers |

For sprinklers above and below ceiling:

| Steel | | Copper | |
|---|---|---|---|
| 1 in. | 2 sprinklers | 1 in. | 2 sprinklers |
| 1¼ in. | 4 sprinklers | 1¼ in. | 4 sprinklers |
| 1½ in. | 7 sprinklers | 1½ in. | 7 sprinklers |
| 2 in. | 15 sprinklers | 2 in. | 18 sprinklers |
| 2½ in. | 50 sprinklers | 2½ in. | 65 sprinklers |

**Ordinary Hazard:**

For sprinklers above ceiling:

| Steel | | Copper | |
|---|---|---|---|
| 1 in. pipe | 2 sprinklers | 1 in. tube | 2 sprinklers |
| 1¼ in. pipe | 3 sprinklers | 1¼ in. tube | 3 sprinklers |
| 1½ in. pipe | 5 sprinklers | 1½ in. tube | 5 sprinklers |
| 2 in. pipe | 10 sprinklers | 2 in. tube | 12 sprinklers |
| 2½ in. pipe | 20 sprinklers | 2½ in. tube | 25 sprinklers |
| 3 in. pipe | 40 sprinklers | 3 in. tube | 45 sprinklers |
| 3½ in. pipe | 65 sprinklers | 3½ in. tube | 75 sprinklers |
| 4 in. pipe | 100 sprinklers | 4 in. tube | 115 sprinklers |
| 5 in. pipe | 160 sprinklers | 5 in. tube | 180 sprinklers |
| 6 in. pipe | 275 sprinklers | 6 in. tube | 300 sprinklers |

For sprinklers above and below ceiling:

| Steel | | Copper | |
|---|---|---|---|
| 1 in. | 2 sprinklers | 1 in. | 2 sprinklers |
| 1¼ in. | 4 sprinklers | 1¼ in. | 4 sprinklers |
| 1½ in. | 7 sprinklers | 1½ in. | 7 sprinklers |
| 2 in. | 15 sprinklers | 2 in. | 18 sprinklers |
| 2½ in. | 30 sprinklers | 2½ in. | 40 sprinklers |
| 3 in. | 60 sprinklers | 3 in. | 65 sprinklers |

**Extra Hazard:**

For sprinklers below ceiling:

| Steel | | Copper | |
|---|---|---|---|
| 1 in. pipe | 1 sprinkler | 1 in. tube | 1 sprinkler |
| 1¼ in. pipe | 2 sprinklers | 1¼ in. tube | 2 sprinklers |
| 1½ in. pipe | 5 sprinklers | 1½ in. tube | 5 sprinklers |
| 2 in. pipe | 8 sprinklers | 2 in. tube | 8 sprinklers |
| 2½ in. pipe | 15 sprinklers | 2½ in. tube | 20 sprinklers |
| 3 in. pipe | 27 sprinklers | 3 in. tube | 30 sprinklers |
| 3½ in. pipe | 40 sprinklers | 3½ in. tube | 45 sprinklers |
| 4 in. pipe | 55 sprinklers | 4 in. tube | 65 sprinklers |
| 5 in. pipe | 90 sprinklers | 5 in. tube | 100 sprinklers |
| 6 in. pipe | 150 sprinklers | 6 in. tube | 170 sprinklers |

**Note:** This data is for estimating purposes only. Check all applicable codes and regulations for specific requirements.

## SPRINKLER HAZARD OCCUPANCIES

### Typical Light Hazard Occupancies:

- Churches
- Clubs
- Eaves and overhangs, if combustible construction with no combustible beneath
- Educational
- Hospitals
- Institutional
- Libraries, except large stack rooms
- Museums
- Nursing or Convalescent Homes
- Office, including Data Processing
- Residential
- Restaurant seating areas
- Theaters seating areas
- Theaters and Auditoriums excluding stages and prosceniums
- Unused attics

### Typical Ordinary Hazard Occupancies (Group 1):

- Automobile parking garages
- Bakeries
- Beverage manufacturing
- Canneries
- Dairy products manufacturing and processing
- Electronic plants
- Glass and glass products manufacturing
- Laundries
- Restaurant service areas

### Typical Ordinary Hazard Occupancies (Group 2):

- Cereal mills
- Chemical plants — Ordinary
- Cold Storage warehouses
- Confectionery products
- Distilleries
- Leather goods mfg.
- Libraries-large stack room areas
- Mercantiles
- Machine shops
- Metal working
- Printing and publishing
- Textile mfg.
- Tobacco Products mfg.
- Wood product assembly

### Typical Ordinary Hazard Occupancies (Group 3):

- Feed mills
- Paper and pulp mills
- Paper process plants
- Piers and wharves
- Repair garages
- Tire manufacturing
- Warehouses (having moderate to higher combustibility of content, such as paper, household furniture, paint, general storage, whiskey, etc.)
- Wood machining

### Typical Extra Hazard Occupancies (Group 1):

- Combustible Hydraulic Fluid use areas
  - Die Casting
  - Metal Extruding
- Plywood and particle board manufacting
- Printing (using inks with below 100°F [37.8°C] flash points)
- Rubber reclaiming, compounding, drying, milling, vulcanizing
- Saw Mills
- Textile picking, opening, blending, garnetting, carding, combining of cotton, synthetics, wool shoddy or burlap
- Upholstering with plastic foams

### Typical Extra Hazard Occupancies (Group 2):

- Asphalt saturating
- Flammable liquids spraying
- Flow coating
- Mobile Home or Modular Building assemblies (where finished enclosure is present and has combustible interiors)
- Open Oil quenching
- Solvent cleaning
- Varnish and paint dipping

## TYPICAL MINIMUM SIZE OF HORIZONTAL BUILDING STORM DRAINS AND BUILDING STORM SEWERS

| Diameter of drain | Maximum projected area in square feet for various slopes | | |
|---|---|---|---|
| | ⅛ inch per feet slope | ¼ inch per feet slope | ½ inch per feet slope |
| 3 | 822 | 1160 | 1644 |
| 4 | 1880 | 2650 | 3760 |
| 5 | 3340 | 4720 | 6680 |
| 6 | 5350 | 7550 | 10700 |
| 8 | 11500 | 16300 | 23000 |
| 10 | 20700 | 29200 | 41400 |
| 12 | 33300 | 47000 | 66600 |
| 15 | 59500 | 84000 | 119000 |

# MECHANICAL / PLUMBING — 15400

## PLUMBING BUDGETS

Budget estimates for plumbing can be determined as a percentage of total building costs depending on building type. For example:

- Apartments ............................9 to 12 percent
- Assembly ...............................4 to 7 percent
- Banks ....................................3 to 6 percent
- Dormitories ............................7 to 10 percent
- Factories ...............................4 to 8 percent
- Hospitals ...............................8 to 12 percent
- Motels ...................................9 to 12 percent
- Office Buildings .......................4 to 7 percent
- Retail (small) ..........................4 to 7 percent
- Retail (large) ...........................3 to 6 percent
- Schools ..................................3 to 6 percent
- Warehouses ............................3 to 7 percent

# MECHANICAL / PLUMBING VENTS — 15410

## TYPICAL SIZE AND LENGTH OF PLUMBING VENTS

| Diameter of soil or waste stack (in.) | Total fixture units connected to stack (dfu) | | DIAMETER OF VENT PIPE | | | | | | | | | |
|---|---|---|---|---|---|---|---|---|---|---|---|---|
| | | 1¼ | 1 | 2 | 2½ | 3 | 4 | 5 | 6 | 8 | 10 | 12 |
| 1¼ | 2 | 30 | | | | | | | | | | |
| 1½ | 8 | 50 | 150 | | | | | | | | | |
| 1½ | 10 | 30 | 100 | | | | | | | | | |
| 2 | 12 | 30 | 75 | 200 | | | | | | | | |
| 2 | 20 | 26 | 50 | 150 | | | | | | | | |
| 2½ | 42 | | 30 | 100 | 300 | | | | | | | |
| 3 | 10 | | 42 | 150 | 360 | 1040 | | | | | | |
| 3 | 21 | | 32 | 110 | 270 | 810 | | | | | | |
| 3 | 53 | | 27 | 94 | 230 | 680 | | | | | | |
| 3 | 102 | | 25 | 86 | 210 | 620 | | | | | | |
| 4 | 43 | | | 35 | 85 | 250 | 980 | | | | | |
| 4 | 140 | | | 27 | 65 | 200 | 750 | | | | | |
| 4 | 320 | | | 23 | 55 | 170 | 640 | | | | | |
| 4 | 540 | | | 21 | 50 | 150 | 580 | | | | | |
| 5 | 190 | | | | 28 | 82 | 320 | 990 | | | | |
| 5 | 490 | | | | 21 | 63 | 250 | 760 | | | | |
| 5 | 940 | | | | 18 | 53 | 210 | 670 | | | | |
| 5 | 1400 | | | | 16 | 49 | 190 | 590 | | | | |
| 6 | 500 | | | | | 33 | 130 | 400 | 1000 | | | |
| 6 | 1100 | | | | | 26 | 100 | 310 | 780 | | | |
| 6 | 2000 | | | | | 22 | 84 | 260 | 660 | | | |
| 6 | 2900 | | | | | 20 | 77 | 240 | 600 | | | |
| 8 | 1800 | | | | | | 31 | 95 | 240 | 940 | | |
| 8 | 3400 | | | | | | 24 | 73 | 190 | 720 | | |
| 8 | 5600 | | | | | | 20 | 62 | 160 | 610 | | |
| 8 | 7600 | | | | | | 18 | 56 | 140 | 560 | | |

## TYPICAL SIZES OF FIXTURE WATER SUPPLY PIPES

| Fixture | Nominal pipe size (inches) |
|---|---|
| Bath tubs | ½ |
| Combination sink and tray | ½ |
| Drinking fountain | ⅜ |
| Dishwasher (domestic) | ½ |
| Kitchen sink, residential | ½ |
| Kitchen sink, commercial | ¾ |
| Lavatory | ⅜ |
| Laundry tray, 1, 2 or 3 compartments | ½ |
| Shower (single head) | ½ |
| Sinks (service, slop) | ½ |
| Sinks flushing rim | ¾ |
| Urinal (flash tank) | ½ |
| Urinal (direct flush valve) | ¾ |
| Water closet (tank type) | ⅜ |
| Water closet (flush valve type) | 1 |
| Hose bibs | ½ |
| Wall hydrant | ½ |

## TYPICAL VENTILATION AIR REQUIREMENTS FOR SPECIAL USES

| Occupancy Classification | Required ventilation air in cfm per human occupant |
|---|---|
| Special areas | |
| Lockers | 2* (or 30 per locker) |
| Wardrobes | 2* |
| Public bathrooms | 40** |
| Private bathrooms | 25** |
| Swimming pools | 15 (per occupant) |
| Exitways and corridors | 1½* |

*Per square foot floor area.
**Per water closet or urinal.

## TYPICAL VENTILATION AIR REQUIREMENTS FOR RETAIL USES

| Occupancy Classification | Required ventilation air in cfm per human occupant |
|---|---|
| Mercantile | |
| Sales floors and showrooms (basement & grade floors) | 7 |
| Sales and showrooms (upper floors) | 7 |
| Storage areas | 5 |
| Dressing rooms | 7 |
| Malls | 7 |
| Shipping areas | 15 |
| Elevators | 7 |
| Supermarkets | |
| Meat processing rooms | 5 |
| Drugs stores | |
| Pharmacists' work rooms | 20 |
| Specialty shops | |
| Pet shops | 1.0* |
| Florists | 5 |
| Greenhouses | 5 |

*cfm per sq. ft. floor area.

## TYPICAL VENTILATION AIR REQUIREMENTS FOR RESIDENTIAL USES

| Occupancy Classification | Required ventilation air in cfm per human occupant |
|---|---|
| **Residential** | |
| General living areas | 5 |
| Bedrooms | 5 |
| Kitchens | 20 |
| Basements, utility rooms | 5 |
| Mobile homes | 5 |
| **Hotels, motels** | |
| Bedrooms (single, double) | 7 |
| Living rooms (suites) | 10 |
| Corridors | 5 |
| Lobbies | 7 |
| Conference rooms (small) | 20 |
| Assembly rooms (large) | 15 |

## TYPICAL VENTILATION AIR REQUIREMENTS FOR STORAGE USES

| Occupancy Classification | Required ventilation air in cfm per human occupant |
|---|---|
| **Storage** | |
| Garages, service stations, parking garages (enclosed) | 1.5* |
| Auto repair shops | 1.5** |
| **Warehouses** | |
| General | 7 |

*cfm/s.f. floor area.
**Must have positive engine exhaust system.

## TYPICAL VENTILATION AIR REQUIREMENTS FOR FACTORY AND INDUSTRIAL USES

| Occupancy Classification | Required ventilation air in cfm per human occupant |
|---|---|
| Factory and industrial | |
| Metalworking & finishing | 35 |
| Automotive engine test | Require Special Exhaust Systems |
| Paint spray booths | |
| Picking, etching and plating lines | |
| Degreasing booths | |
| Sandblasting booths | |
| Chemicals and pharmaceuticals | |
| Dusty operations | 30 |
| Rooms containing potential gas emitters | 20 |
| Drying oven rooms | 15 |
| Fermentation rooms | 15 |
| Pillmaking booths | 10 |
| Packaging areas | 10 |
| Utility rooms | 7 |
| Computer rooms | 7 |
| Textiles-clothes manufacturer | 15 |
| Electronics and aerospace circuit board and soldering rooms | 20 |
| Wood products, papermaking | 20 |
| Brewing, distilling, wineries, bottling | 20* |
| Food processing | 20 |
| Tobacco processing | 20 |
| Power plants | |
| Control rooms | 10 |
| Boiler rooms | 35 |
| Generator rooms | 20 |
| Sewage treatment plants | |
| Control rooms | 10 |
| Compressor/blower motor rooms | 20 |
| Glass and ceramic manufacturer | 20 |
| Agricultural | 20 |

## TYPICAL VENTILATION AIR REQUIREMENTS FOR BUSINESS USES

| Occupancy Classification | Required ventilation air in cfm per human occupant |
|---|---|
| **Business** | |
| Banks | |
| (see offices) | |
| Vaults | 5 |
| Barber, beauty and health services | |
| Beauty shops (hair dressers) | 25 |
| Reducing salons | 25 |
| Sauna baths, steam rooms | 5 |
| barber shops | 7 |
| Photo studios | |
| Camera rooms, stages | 5 |
| Dark rooms | |
| Shoe repair shops | |
| Workrooms/trade areas | 10 |
| Offices | |
| General office space and showrooms | 15 |
| Conference rooms | 25 |
| Drafting/art rooms | 7 |
| Doctor's consultation rooms | 10 |
| Waiting rooms | 10 |
| Lithographing rooms | 7 |
| Diazo printing rooms | 7 |
| Computer rooms | 5 |
| Keypunch rooms | 7 |
| Communication | |
| TV/radio broadcasting booths, studios | 30 |
| Motion picture and TV stages | 30 |
| Pressrooms | 15 |
| Composing rooms | 7 |
| Engraving rooms | 7 |
| Telephone switchboard rooms (manual) | 7 |
| Telephone switchgear rooms (automatic) | 7 |
| Teletypewriter/facsimile rooms | 5 |
| Research institutes | |
| Laboratories: | |
| Light duty; non-chemical | 15 |
| Chemical | 15 |
| Heavy-duty | 15 |
| Radioisotope, chemical & biologically toxic | 15 |
| Machine shops | 15 |
| Dark rooms, spectroscopy rooms | 10 |
| Animal rooms | 40 |
| Veterinary hospitals | |
| Kennels, stalls | 25 |
| Operating rooms | 25 |
| Reception rooms | 10 |

## TYPICAL VENTILATION AIR REQUIREMENTS FOR INSTITUTIONAL USES

| Occupancy Classification | Required ventilation air in cfm per human occupant |
|---|---|
| Institutional | |
| Prisons | |
| Cell blocks | 7 |
| Eating halls | 15 |
| Guard stations | 7 |

# AIR CONDITIONING

## RECOMMENDED SHEET METAL GAUGES AND CONSTRUCTION FOR RECTANGULAR DUCT

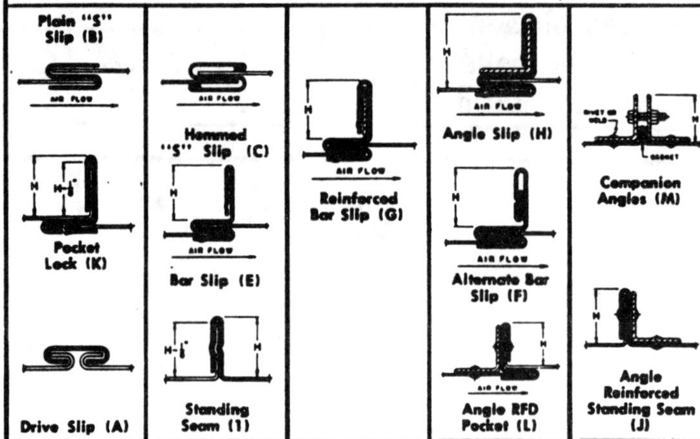

| LOW PRESSURE — LOW VELOCITY = 2" W.G. MAX | | | | | | | | | |
|---|---|---|---|---|---|---|---|---|---|
| | | | | AT JOINTS | | | | | |
| Plate No. | Dimension of Longest Side of Duct | Steel Metal Gauges: Steel | Steel Metal Gauges: Aluminum | Plain "S" Slip (B)<br>Pocket Lock (K)<br>Drive Slip (A) | Hemmed "S" Slip (C)<br>Bar Slip (E)<br>Standing Seam (1) | Reinforced Bar Slip (G) | Angle Slip (H)<br>Alternate Bar Slip (F)<br>Angle RFD Pocket (L) | Companion Angles (M)<br>Angle Reinforced Standing Seam (J) | Reinforcing Between Joints |
| 6 | Thru 12" | 26 | 24 (.020) | A-B-K | — | — | — | — | |
| 6 | 13" thru 18" | 24 | 22 (.025) | A-B-K | — | — | — | — | |
| 7<br>7A | 19" thru 30" | 24 | 22 (.025) | K @ 5' cc<br>A | C-E-@ 5' cc<br>C-E-@ 10' cc | — | — | — | 1"x1"x⅛" @ 5' cc |
| 8 | 31" thru 42" | 22 | 20 (.032) | K @ 5' cc | E-G-K @ 5' cc<br>E-G-K @ 10' cc | — | — | — | 1"x1"x⅛" @ 5' cc |
| 9 | 43" thru 54" | 22 | 20 (.032) | K @ 4' cc<br>K @ 8' cc | E-@ 4' cc<br>E-@ 8' cc | G -@ 4' cc<br>G -@ 8' cc | — | — | 1½"x1½"x⅛" @ 4' cc |
| 9 | 55" thru 60" | 20 | 18 (.040) | K @ 4' cc<br>K @ 8' cc | E-@ 4' cc<br>E-@ 8' cc | G -@ 4' cc<br>G -@ 8' cc | — | — | 1½"x1½"x⅛" @ 4' cc |
| 10 | 61" thru 84" | 20 | 18 (.040) | — | — | G -@ 4' cc<br>G -@ 5' cc | H-@ 4' cc<br>F-@ 4' cc<br>L-@ 4' cc<br>H-@ 5' cc<br>F-@ 5' cc<br>L-@ 5' cc | J-@ 2' cc | 1½"x1½"x⅛" @ 2' cc<br>1½"x1½"x⅛" @ 2'-6" cc |
| 11 | 85" thru 96" | 18 | 16 (.051) | — | — | — | H-@ 4' cc<br>L-@ 4' cc<br>H-@ 5' cc<br>L-@ 5' cc | M-@ 4' cc<br>M-@ 5' cc<br>J-@ 2' cc | 1½"x1½"x3/16" @ 2' cc<br>1½"x1½"x3/16" @ 2'-6" cc<br>1½"x1½"x3/16" @ 2' cc |
| 12 | Over 96" | 18 | 16 (.051) | — | — | — | H-@ 4' cc<br>L-@ 4' cc<br>H-@ 5' cc<br>L-@ 5' cc | M-@ 4' cc<br>M-@ 5' cc<br>J-@ 2' cc | 2"x2"x¼" @ 2' cc<br>2"x2"x¼" @ 2'-6" cc<br>2"x2"x¼" @ 2' cc |

H (height dimension)—up to 42" = 1"
H (height dimension)—43" to 96" = 1½"
H (height dimension)—over 96" = 2"

# AIR CONDITIONING
## TYPICAL DUCT CONNECTIONS
## CROSS JOINTS FOR SHEET METAL DUCTWORK

(NOT TO SCALE)

H=HEIGHT REFERRED TO IN DIMENSIONS

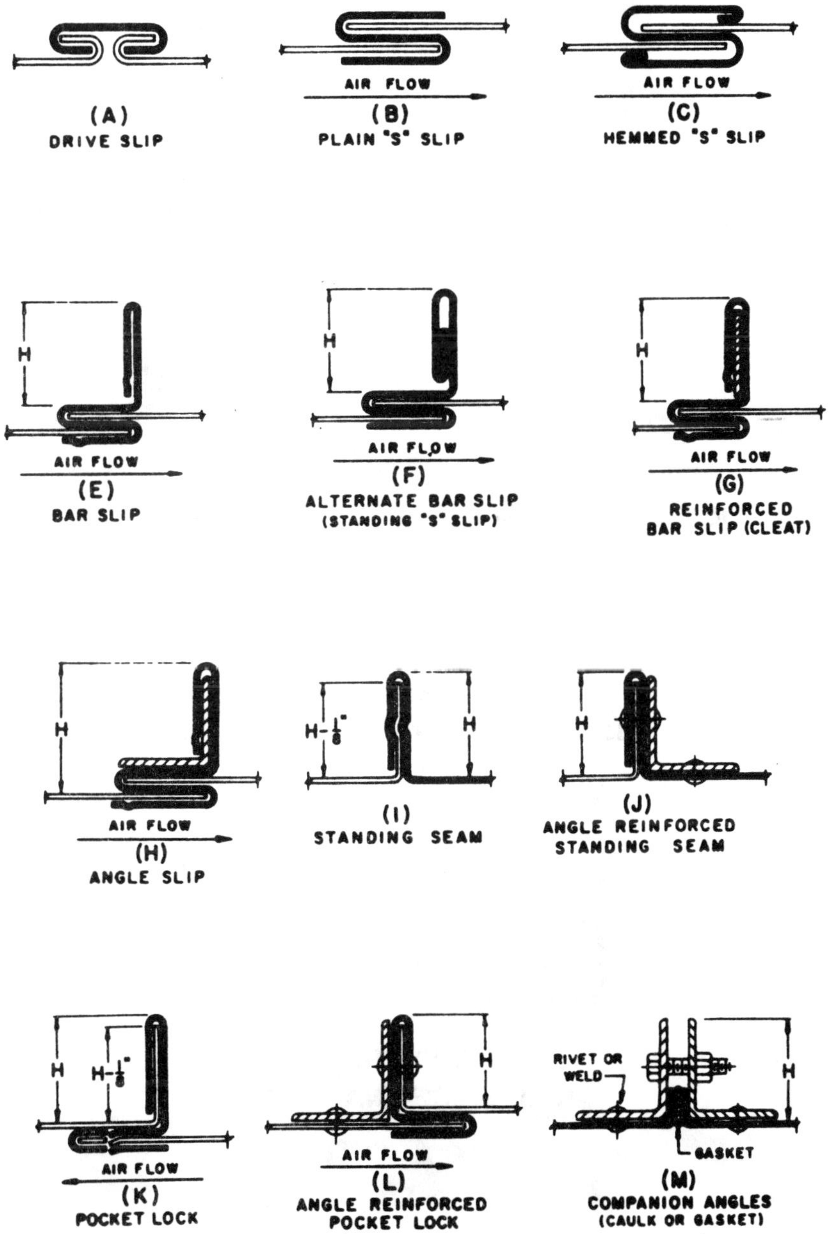

## AIR CONDITIONING (Cont.)

### LONGITUDINAL SEAMS FOR SHEET METAL DUCTWORK

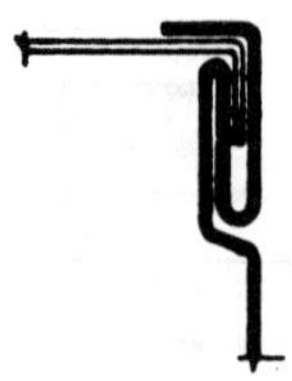

Fig. "N"
PITTSBURGH LOCK

Fig. "Z"
BUTTON PUNCH SNAP LOCK

Fig. "O"
ACME LOCK-GROOVED SEAM

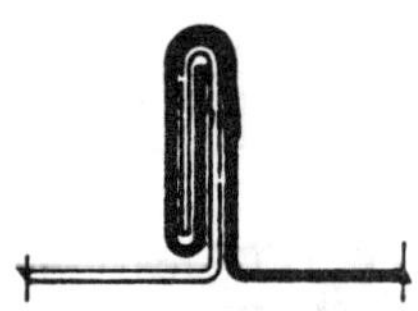

Fig. "T"
DOUBLE SEAM

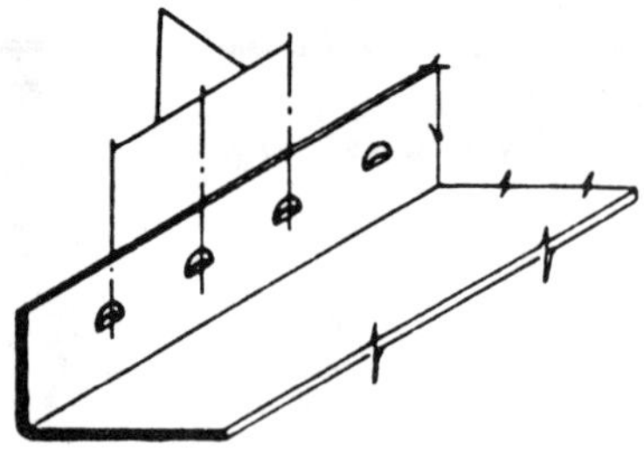


DETAIL NO. 1
MALE PIECE-SNAP LOCK

## TYPICAL DUCT CONSTRUCTION SHEET METAL GAGES IN ONE- AND TWO-FAMILY DWELLINGS

| Metal Gauges (duct not enclosed in partitions) | | |
|---|---|---|
| **ROUND DUCTS** | | |
| Diameter, inches | Minimum thickness galvanized sheet gage | Minimum thickness aluminum B&S gage |
| Less than 12 | 30 | 26 |
| 12-14 | 28 | 26 |
| 15-18 | 26 | 24 |
| Over 18 | 24 | 22 |
| **RECTANGULAR DUCTS** | | |
| Width, inches | Minimum thickness galvanized sheet gage | Minimum thickness aluminum B&S gage |
| Less than 14 | 28 | 24 |
| 14-24 | 26 | 22 |
| 25-30 | 24 | 22 |
| Over 30 | 22 | 20 |
| **Metal Gauges (duct enclosed in partitions)** | | |
| Width, inches | Minimum thickness galvanized sheet gage | Minimum thickness aluminum B&S gage |
| 14 or less | 30 | 26 |
| Over 14 | 28 | 24 |

## TYPICAL DUCT CONSTRUCTION SHEET METAL GAGES

*(All uses except 1- and 2-family dwellings)*

| RECTANGULAR DUCTS | | |
|---|---|---|
| Maximum side inches | Steel min. Galv. Sheet Gage | Aluminum Min. B&S Gage |
| Through 12 | 26 (0.022 in.) | 24 (0.020 in.) |
| 13 through 30 | 24 (0.028 in.) | 22 (0.025 in.) |
| 31 through 54 | 22 (0.034 in.) | 20 (0.032 in.) |
| 55 through 84 | 20 (0.040 in.) | 18 (0.040 in.) |
| Over 84 | 18 (0.052 in.) | 16 (0.051 in.) |

| ROUND DUCTS | | | |
|---|---|---|---|
| Diameter inches | Spiral seam duct | Longitudinal seam duct | Fittings |
| | Steel min. Galv. Sht. Gage | Steel min. Galv. Sht. Gage | Steel min. Galv. Sht. Gage |
| Through 12 | 28.(0.019 in.) | 26 (0.022 in.) | 26 (0.022 in.) |
| 13 through 18 | 26 (0.022 in.) | 24 (0.028 in.) | 24 (0.028 in.) |
| 19 through 28 | 24 (0.028 in.) | 22 (0.034 in.) | 22 (0.034 in.) |
| 29 through 36 | 22 (0.034 in.) | 20 (0.040 in.) | 20 (0.040 in.) |
| 37 through 52 | 20 (0.040 in.) | 18 (0.052 in.) | 18 (0.052 in.) |

## APPLICATIONS FOR CONDUCTORS USED FOR GENERAL WIRING

| | AMBIENT TEMPERATURE | | | | | | | | |
|---|---|---|---|---|---|---|---|---|---|
| | 60°C 140°F | 75°C 167°F | 85°C 185°F | 90°C 194°F | 110°C 230°F | 200°C 392°F | Dry | Dry or Wet | FEATURES |
| R | X | | | | | | X | | Code Rubber |
| RH | | X | | | | | X | | Heat Resistant |
| RHH | | | | X | | | X | | More Heat Resistant |
| RW | X | | | | | | | X | Moisture Resistant |
| RH-RW | X | | | | | | | X | Moisture and Heat Resistant |
| | | X | | | | | X | | Moisture and Heat Resistant |
| RHW | | X | | | | | | X | Moisture and Heat Resistant |
| RU | X | | | | | | X | | Latex Rubber |
| RUH | | X | | | | | X | | Heat Resistant |
| RUW | X | | | | | | | X | Moisture Resistant |
| T | X | | | | | | X | | Thermoplastic |
| TW | X | | | | | | | X | Moisture Resistant |
| THHN | | | | X | | | X | | Heat Resistant |
| THW | | X | | | | | | X | Moisture and Heat Resistant |
| THWN | | X | | | | | | X | Moisture and Heat Resistant |
| MI | | | X | | | | | X | Mineral Insulated Metal Sheathed |
| V | | | X | | | | X | | Varnished Cambric |
| AVA | | | | | X | | X | | With Asbestos |
| AVB | | | | X | | | X | | With Asbestos |
| AVL | | | | | X | | | X | With Asbestos |

This table does not include special condition conductors, thickness of conductor insulation, or reference to all outer protective coverings.

**GENERAL CLASSIFICATION OF INSULATIONS:**

| | |
|---|---|
| A .............. Asbestos | RU ............Latex Rubber |
| H .............. Heat Resistant | V ................Varnished Cambric |
| MI ............ Mineral Insulation | T ................Thermoplastic |
| R .............. Rubber | W ...............(Water) Moisture Resistant |

# WIRE AND SHEET METAL GAGES

(In Decimals of an Inch)

| Name of Gage | American Wire Gage (A.W.G.) (Corresponds to Brown & Sharpe Gage) | Birmingham Iron Wire Gage (B.W.G.) | United States Standard Gage (U.S.S.G.) | |
|---|---|---|---|---|
| Principal Use | Electrical Wire & Non-Ferrous Sheet Metal | Iron or Steel Wire | Ferrous Sheet Metal | |
| Gage No. | | | | Gage No. |
| 00 00000 | | | | 00 00000 |
| 0 00000 | .5800 | | | 0 00000 |
| 00000 | .5165 | .500 | | 00000 |
| 0000 | .4600 | .454 | | 0000 |
| 000 | .4096 | .425 | | 000 |
| 00 | .3648 | .380 | | 00 |
| 0 | .3249 | .340 | | 0 |
| 1 | .2893 | .300 | | 1 |
| 2 | .2576 | .284 | | 2 |
| 3 | .2294 | .259 | 23.91 | 3 |
| 4 | .2043 | .238 | .2242 | 4 |
| 5 | .1819 | .220 | .2092 | 5 |
| 6 | .1620 | .203 | .1943 | 6 |
| 7 | .1443 | .180 | .1793 | 7 |
| 8 | .1285 | .165 | .1644 | 8 |
| 9 | .1144 | .148 | .1495 | 9 |
| 10 | .1019 | .134 | .1345 | 10 |
| 11 | .0907 | .120 | .1196 | 11 |
| 12 | .0808 | .109 | .1046 | 12 |
| 13 | .0720 | .095 | .0897 | 13 |
| 14 | .0641 | .083 | .0747 | 14 |
| 15 | .0571 | .072 | .0673 | 15 |
| 16 | .0508 | .065 | .0598 | 16 |
| 17 | .0453 | .058 | .0538 | 17 |
| 18 | .0403 | .049 | .0478 | 18 |
| 19 | .0359 | .042 | .0418 | 19 |
| 20 | .0320 | .035 | .0359 | 20 |
| 21 | .0285 | .032 | .0329 | 21 |
| 22 | .0253 | .028 | .0299 | 22 |
| 23 | .0226 | .025 | .0269 | 23 |
| 24 | .0201 | .022 | .0239 | 24 |
| 25 | .0179 | .020 | .0209 | 25 |
| 26 | .0159 | .018 | .0179 | 26 |
| 27 | .0142 | .016 | .0164 | 27 |
| 28 | .0126 | .014 | .0149 | 28 |
| 29 | .0113 | .013 | .0135 | 29 |
| 30 | .0100 | .012 | .0120 | 30 |
| 31 | .0089 | .010 | .0105 | 31 |
| 32 | .0080 | .009 | .0097 | 32 |
| 33 | .0071 | .008 | .0090 | 33 |
| 34 | .0063 | .007 | .0082 | 34 |
| 35 | .0056 | .005 | .0075 | 35 |
| 36 | .0050 | .004 | .0067 | 36 |
| 37 | .0045 | | .0064 | 37 |
| 38 | .0040 | | .0060 | 38 |
| 39 | .0035 | | | 39 |
| 40 | .0031 | | | 40 |

# Geographic Cost Modifiers

The costs as presented in this book attempt to represent national averages. Costs, however, vary among regions, states, and even between adjacent localities.

In order to more closely approximate the probable costs for specific locations throughout the U.S., this table of Geographic Cost Modifiers is provided. These adjustment factors are used to modify costs obtained from this book to help account for regional variations of construction costs and to provide a more accurate estimate for specific areas. The factors are formulated by comparing costs in a specific area to the costs as presented in the Costbook pages. An example of how to use these factors is shown below. Whenever local current costs are known, whether material prices or labor rates, they should be used when more accuracy is required.

| Cost Obtained from Costbook Pages | | Location Cost Adjustment Factor | | Adjusted Cost |
|---|---|---|---|---|
| | X | | = | |

For example, a project estimated to cost $125,000 using the Costbook pages can be adjusted to more closely approximate the cost in Los Angeles:

**$125,000 X 1.07 = $133,750**

# Geographic Cost Modifiers

| ALABAMA | |
|---|---|
| BIRMINGHAM | 0.79 |
| HUNTSVILLE | 0.77 |
| MOBILE | 0.81 |
| MONTGOMERY | 0.75 |
| TUSCALOOSA | 0.75 |

| ALASKA | |
|---|---|
| ANCHORAGE | 1.30 |
| JUNEAU | 1.33 |
| FAIRBANKS | 1.37 |
| NOME | 1.43 |

| ARIZONA | |
|---|---|
| FLAGSTAFF | 0.91 |
| PHOENIX | 0.89 |
| PRESCOTT | 0.91 |
| TUCSON | 0.89 |
| YUMA | 0.87 |

| ARKANSAS | |
|---|---|
| FAYETTEVILLE | 0.73 |
| FORTH SMITH | 0.74 |
| LITTLE ROCK | 0.78 |
| PINE BLUFF | 0.75 |

| CALIFORNIA | |
|---|---|
| ANAHEIM | 1.02 |
| BAKERSFIELD | 0.97 |
| LOS ANGELES | 1.07 |
| REDDING | 0.92 |
| RIVERSIDE | 0.97 |
| SACRAMENTO | 1.00 |
| SAN DIEGO | 1.02 |
| SAN JOSE | 1.07 |
| SAN FRANCISCO | 1.12 |
| SANTA BARBARA | 1.07 |

| COLORADO | |
|---|---|
| BOULDER | 0.92 |
| COLORADO SPRINGS | 0.92 |
| DENVER | 0.95 |
| GRAND JUNCTION | 0.90 |
| PUEBLO | 0.90 |

| CONNECTICUT | |
|---|---|
| BRIDGEPORT | 1.03 |
| HARTFORD | 1.01 |
| NEW LONDON | 0.99 |
| STAMFORD | 1.06 |
| WATERBURY | 0.96 |

| DELAWARE | |
|---|---|
| DOVER | 0.89 |
| WILMINGTON | 0.91 |

| FLORIDA | |
|---|---|
| JACKSONVILLE | 0.80 |
| MIAMI | 0.86 |
| ORLANDO | 0.80 |
| TAMPA | 0.82 |
| WEST PALM BEACH | 0.84 |

| GEORGIA | |
|---|---|
| ATLANTA | 0.83 |
| AUGUSTA | 0.75 |
| COLUMBUS | 0.75 |
| MACON | 0.77 |
| SAVANNAH | 0.79 |

| HAWAII | |
|---|---|
| HILO | 1.31 |
| HONOLULU | 1.25 |
| MAUI | 1.28 |

| IDAHO | |
|---|---|
| BOISE | 0.92 |
| LEWISTON | 0.90 |
| POCATELLO | 0.88 |
| TWIN FALLS | 0.88 |

| ILLINOIS | |
|---|---|
| CHICAGO | 1.03 |
| MOLINE | 0.89 |
| PEORIA | 0.92 |
| ROCKFORD | 0.92 |
| SPRINGFIELD | 0.89 |

| INDIANA | |
|---|---|
| FORT WAYNE | 0.88 |
| EVANSVILLE | 0.88 |
| GARY | 0.99 |
| INDIANAPOLIS | 0.95 |
| TERRE HAUTE | 0.86 |

| IOWA | |
|---|---|
| COUNCIL BLUFFS | 0.85 |
| DAVENPORT | 0.87 |
| DES MOINES | 0.89 |
| SIOUX CITY | 0.85 |
| WATERLOO | 0.87 |

| KANSAS | |
|---|---|
| DODGE CITY | 0.77 |
| SALINA | 0.79 |
| TOPEKA | 0.83 |
| WICHITA | 0.81 |

| KENTUCKY | |
|---|---|
| BOWLING GREEN | 0.83 |
| LEXINGTON | 0.85 |
| LOUISVILLE | 0.89 |
| PADUCAH | 0.87 |

| LOUISIANA | |
|---|---|
| BATON ROUGE | 0.84 |
| LAKE CHARLES | 0.82 |
| MONROE | 0.78 |
| NEW ORLEANS | 0.86 |
| SHREVEPORT | 0.78 |

| MAINE | |
|---|---|
| AUGUSTA | 0.85 |
| BANGOR | 0.83 |
| LEWISTON | 0.87 |
| PORTLAND | 0.89 |

| MARYLAND | |
|---|---|
| ANNAPOLIS | 0.92 |
| BALTIMORE | 0.90 |
| HAGERSTOWN | 0.88 |
| ROCKVILLE | 0.95 |

| MASSACHUSETTS | |
|---|---|
| BOSTON | 1.01 |
| LOWELL | 0.94 |
| FALL RIVER | 0.91 |
| SPRINGFIELD | 0.94 |
| WORCESTER | 0.96 |

| MICHIGAN | |
|---|---|
| DETROIT | 0.95 |
| GRAND RAPIDS | 0.88 |
| LANSING | 0.88 |
| MARQUETTE | 0.86 |
| SAGINAW | 0.90 |

| MINNESOTA | |
|---|---|
| DULUTH | 0.93 |
| MINNEAPOLIS | 0.98 |
| ROCHESTER | 0.93 |
| SAINT PAUL | 0.98 |

| MISSISSIPPI | |
|---|---|
| COLUMBUS | 0.77 |
| GULFPORT | 0.79 |
| JACKSON | 0.81 |
| VICKSBURG | 0.79 |

| MISSOURI | |
|---|---|
| JOPLIN | 0.84 |
| KANSAS CITY | 0.86 |
| SAINT LOUIS | 0.90 |
| SAINT JOSEPH | 0.84 |
| SPRINGFIELD | 0.84 |

| MONTANA | |
|---|---|
| BILLINGS | 0.87 |
| BUTTE | 0.85 |
| GREAT FALLS | 0.87 |
| HELENA | 0.85 |
| MISSOULA | 0.85 |

# Geographic Cost Modifiers

| NEBRASKA | |
|---|---|
| GRAND ISLAND | 0.81 |
| LINCOLN | 0.85 |
| NORTH PLATTE | 0.81 |
| OMAHA | 0.87 |

| NEVADA | |
|---|---|
| CARSON CITY | 0.95 |
| LAS VEGAS | 0.97 |
| SPARKS | 0.97 |
| RENO | 1.00 |

| NEW HAMPSHIRE | |
|---|---|
| CONCORD | 0.86 |
| MANCHESTER | 0.88 |
| NASHUA | 0.90 |

| NEW JERSEY | |
|---|---|
| ATLANTIC CITY | 0.97 |
| CAMDEN | 0.95 |
| NEWARK | 1.02 |
| PATERSON | 0.99 |
| TRENTON | 0.97 |

| NEW MEXICO | |
|---|---|
| ALBUQUERQUE | 0.86 |
| CARLSBAD | 0.84 |
| LAS CRUCES | 0.82 |
| SANTA FE | 0.92 |

| NEW YORK | |
|---|---|
| ALBANY | 0.87 |
| BINGHAMTON | 0.83 |
| BUFFALO | 0.90 |
| LONG ISLAND | 1.01 |
| NEW YORK CITY | 1.15 |
| ROCHESTER | 0.90 |
| SYRACUSE | 0.87 |
| WATERTOWN | 0.83 |
| WHITE PLAINS | 1.01 |

| NORTH CAROLINA | |
|---|---|
| ASHEVILLE | 0.75 |
| CHARLOTTE | 0.79 |
| GREENSBORO | 0.75 |
| RALEIGH | 0.77 |
| WILMINGTON | 0.74 |

| NORTH DAKOTA | |
|---|---|
| BISMARK | 0.86 |
| FARGO | 0.90 |
| GRAND FORKS | 0.88 |
| MINOT | 0.84 |

| OHIO | |
|---|---|
| CINCINNATI | 0.83 |
| CLEVELAND | 0.91 |
| COLUMBUS | 0.89 |
| TOLEDO | 0.89 |
| YOUNGSTOWN | 0.85 |

| OKLAHOMA | |
|---|---|
| BARTLESVILLE | 0.77 |
| ENID | 0.77 |
| LAWTON | 0.77 |
| OKLAHOMA CITY | 0.83 |
| TULSA | 0.81 |

| OREGON | |
|---|---|
| EUGENE | 0.90 |
| MEDFORD | 0.90 |
| PORTLAND | 0.95 |
| SALEM | 0.92 |

| PENNSYLVANIA | |
|---|---|
| ALLENTOWN | 0.91 |
| HARRISBURG | 0.85 |
| PHILADELPHIA | 0.98 |
| PITTSBURGH | 0.89 |
| SCRANTON | 0.87 |

| RHODE ISLAND | |
|---|---|
| PAWTUCKET | 0.95 |
| PROVIDENCE | 0.98 |
| NEWPORT | 0.98 |
| WESTERLY | 0.95 |
| WOONSOCKET | 0.95 |

| SOUTH CAROLINA | |
|---|---|
| CHARLESTON | 0.79 |
| COLUMBIA | 0.81 |
| FLORENCE | 0.79 |
| GREENVILLE | 0.81 |

| SOUTH DAKOTA | |
|---|---|
| ABERDEEN | 0.79 |
| PIERRE | 0.79 |
| RAPID CITY | 0.81 |
| SIOUX FALLS | 0.83 |
| WATERTOWN | 0.81 |

| TENNESSEE | |
|---|---|
| CHATTANOOGA | 0.77 |
| JOHNSON CITY | 0.75 |
| KNOXVILLE | 0.79 |
| MEMPHIS | 0.81 |
| NASHVILLE | 0.83 |

| TEXAS | |
|---|---|
| AUSTIN | 0.80 |
| DALLAS | 0.82 |
| HOUSTON | 0.86 |
| LUBBOCK | 0.78 |
| SAN ANTONIO | 0.80 |

| UTAH | |
|---|---|
| LOGAN | 0.83 |
| OGDEN | 0.85 |
| PROVO | 0.83 |
| SALT LAKE | 0.87 |

| VERMONT | |
|---|---|
| BRATTLEBORO | 0.85 |
| BURLINGTON | 0.94 |
| RUTLAND | 0.87 |

| VIRGINIA | |
|---|---|
| ALEXANDRIA | 0.94 |
| LYNCHBURG | 0.81 |
| NORFOLK | 0.83 |
| RICHMOND | 0.85 |
| ROANOKE | 0.81 |

| WASHINGTON | |
|---|---|
| BELLINGHAM | 0.99 |
| SEATLE | 1.06 |
| SPOKANE | 0.99 |
| TACOMA | 1.03 |
| YAKIMA | 0.96 |

| WASHINGTON, D.C. | |
|---|---|
| DISTRICT | 0.95 |

| WEST VIRGINIA | |
|---|---|
| CHARLESTON | 0.87 |
| HUNTINGTON | 0.89 |
| MORGANTOWN | 0.83 |
| PARKERSBURG | 0.83 |

| WISCONSIN | |
|---|---|
| EAU CLAIRE | 0.90 |
| GREEN BAY | 0.92 |
| MADISON | 0.90 |
| MILWAUKEE | 0.97 |
| WAUSAU | 0.90 |

| WYOMING | |
|---|---|
| CASPER | 0.89 |
| CHEYENNE | 0.91 |
| ROCK SPRINGS | 0.87 |
| SHERIDAN | 0.85 |

# Square Foot Tables

The following Square Foot Tables list hundreds of actual projects for dozens of building types, each with associated building size, total square foot building cost and percentage of project costs for total mechanical and electrical components. This data provides an overview of construction costs by building type. These costs are for actual projects. The variations within similar building types may be due, among other factors, to size, location, quality and specified components, materials and processes. Depending upon all such factors, specific building costs can vary significantly and may not necessarily fall within the range of costs as presented. The data has been updated to reflect current construction costs.

| ***BUILDING CATEGORY*** | ***PAGE*** |
|---|---|


*Commercial* ........ *480*
*Residential* ........ *481*
*Educational* ........ *482*
*Hotel/Motel* ........ *483*
*Industrial* ........ *484*
*Medical* ........ *484*
*Public Facilities* ........ *485*
*Offices* ........ *486*
*Recreational* ........ *488*
*Religious* ........ *489*

# SQUARE FOOT TABLES

## COMMERCIAL

### AUTO DEALERSHIP

| Project Size Gross S.F. | Project Cost $/S.F. | % Cost Mechanical | % Cost Electrical |
|---|---|---|---|
| 7,700 | 87.30 | 16.7 | 9.4 |
| 16,100 | 49.50 | 10.2 | 15.9 |
| 20,000 | 52.00 | 12.9 | 23.4 |
| 26,300 | 53.00 | 12.5 | 22.0 |
| 43,600 | 42.00 | 19.4 | 13.2 |
| 53,600 | 70.00 | 12.5 | 11.5 |

### BUSINESS CENTER

| Project Size Gross S.F. | Project Cost $/S.F. | % Cost Mechanical | % Cost Electrical |
|---|---|---|---|
| 3,900 | 52.30 | 12.0 | 9.2 |
| 9,900 | 50.60 | 9.1 | 7.6 |
| 54,400 | 33.80 | 3.4 | 12.2 |
| 135,000 | 36.80 | 8.2 | 1.5 |

### CINEMA

| Project Size Gross S.F. | Project Cost $/S.F. | % Cost Mechanical | % Cost Electrical |
|---|---|---|---|
| 18,000 | 119.50 | 10.9 | 6.7 |
| 22,500 A | 73.50 | 6.6 | 4.2 |

### MALL/PLAZA

| Project Size Gross S.F. | Project Cost $/S.F. | % Cost Mechanical | % Cost Electrical |
|---|---|---|---|
| 9,700 | 36.95 | 15.0 | 13.3 |
| 10,500 | 59.20 | 8.0 | 13.5 |
| 16,300 | 53.90 | 9.4 | 10.6 |
| 26,900 | 56.70 | 15.0 | 7.0 |
| 36,000 | 45.40 | 10.0 | 11.0 |
| 36,300 | 57.60 | 12.4 | 8.2 |
| 44,720 | 53.90 | 18.3 | 12.4 |
| 59,100 | 52.80 | 9.8 | 9.5 |
| 60,000 R | 54.10 | 10.0 | 9.5 |
| 64,100 | 54.40 | 22.4 | 18.4 |
| 66,000 | 62.70 | 13.5 | 11.5 |
| 67,400 | 45.35 | 21.0 | 15.0 |
| 73,500 | 65.45 | 14.9 | 6.9 |

## MALL/PLAZA (Cont.)

| Project Size Gross S.F. | Project Cost $/S.F. | % Cost Mechanical | % Cost Electrical |
|---|---|---|---|
| 142,000 | 41.20 | 7.1 | 8.0 |
| 220,000 | 109.00 | 11.0 | 6.4 |
| 223,700 | 32.00 | 9.0 | 9.3 |
| 321,200 | 38.10 | 7.3 | 7.4 |
| 379,900 | 49.70 | 11.2 | 6.2 |
| 405,100 | 51.20 | 13.9 | 6.0 |
| 482,000 | 85.10 | 10.5 | 9.4 |
| 630,000 | 56.90 | 12.2 | 12.4 |

### RESTAURANT

| Project Size Gross S.F. | Project Cost $/S.F. | % Cost Mechanical | % Cost Electrical |
|---|---|---|---|
| 4,300 R | 94.00 | 6.9 | 8.7 |
| 4,400 R | 138.30 | 14.6 | 8.5 |
| 5,800 | 112.50 | 28.0 | 10.6 |
| 6,800 A | 131.50 | 7.0 | 11.1 |
| 7,360 R | 142.00 | 16.0 | 6.5 |
| 9,600 | 141.00 | 24.7 | 13.1 |
| 10,000 R | 146.30 | 21.0 | 10.0 |
| 10,100 | 126.20 | 28.6 | 18.4 |
| 10,600 | 232.30 | 20.4 | 6.4 |
| 22,900 R | 160.60 | 15.8 | 16.9 |

### RETAIL STORE

| Project Size Gross S.F. | Project Cost $/S.F. | % Cost Mechanical | % Cost Electrical |
|---|---|---|---|
| 1,000 | 141.50 | 12.8 | 6.7 |
| 3,000 R | 126.70 | 14.3 | 10.5 |
| 12,300 | 173.70 | 14.0 | 10.0 |
| 30,000 | 87.80 | 15.6 | 26.2 |
| 61,300 | 45.40 | 13.3 | 13.0 |
| 115,000 | 62.90 | 14.6 | 11.3 |
| 154,700 | 89.00 | 11.2 | 12.4 |
| 314,700 R | 75.10 | 13.8 | 9.4 |

A = Addition R = Remodel

# SQUARE FOOT TABLES

## RESIDENTIAL

### APARTMENTS

| Project Size Gross S.F. | Project Cost $/S.F. | % Cost Mechanical | % Cost Electrical |
|---|---|---|---|
| 3,700 | 52.60 | 14.9 | 4.4 |
| 13,900 | 72.00 | 7.9 | 7.2 |
| 19,200 R | 114.40 | 45.3 | 7.5 |
| 19,700 | 67.10 | 7.4 | 10.4 |
| 23,700 | 73.60 | 10.6 | 4.3 |
| 26,500 | 69.70 | 25.2 | 12.8 |
| 35,100 | 57.10 | 16.4 | 5.6 |
| 54,000 | 98.60 | 23.3 | 13.1 |
| 62,700 | 71.70 | 17.0 | 9.0 |
| 67,300 | 65.50 | 13.8 | 8.4 |
| 70,600 | 36.90 | 18.1 | 7.8 |
| 75,300 | 81.30 | 13.2 | 8.1 |
| 75,600 | 79.20 | 14.5 | 8.9 |
| 72,200 | 68.65 | 18.4 | 10.9 |
| 77,600 | 88.90 | 26.9 | 14.3 |
| 88,100 | 82.40 | 15.3 | 9.3 |
| 89,500 | 77.30 | 10.7 | 11.1 |
| 94,100 | 46.60 | 8.5 | 6.8 |
| 96,000 | 58.10 | 17.0 | 13.3 |
| 102,000 | 100.90 | 17.8 | 12.5 |
| 103,200 | 45.90 | 12.1 | 8.9 |
| 103,600 | 73.80 | 19.1 | 9.0 |
| 105,200 | 94.00 | 14.9 | 8.9 |
| 106,200 | 68.70 | 12.8 | 9.3 |
| 110,900 | 65.10 | 15.6 | 8.4 |
| 111,800 | 92.40 | 17.3 | 7.4 |
| 115,900 | 69.30 | 12.5 | 8.6 |
| 117,200 | 43.30 | 12.9 | 8.0 |
| 119,000 | 59.50 | 15.6 | 8.4 |

### APARTMENTS (Cont.)

| Project Size Gross S.F. | Project Cost $/S.F. | % Cost Mechanical | % Cost Electrical |
|---|---|---|---|
| 119,400 | 39.00 | 18.6 | 8.7 |
| 144,300 | 86.60 | 15.0 | 8.6 |
| 176,300 | 73.60 | 19.1 | 9.0 |
| 192,300 | 43.00 | 10.1 | 6.0 |
| 210,900 | 85.20 | 19.1 | 9.5 |
| 220,200 | 87.70 | 14.8 | 7.5 |
| 253,900 | 114.00 | 20.6 | 7.7 |
| 369,500 | 83.20 | 15.8 | 9.0 |

### CONDOS/TOWNHOUSE

| Project Size Gross S.F. | Project Cost $/S.F. | % Cost Mechanical | % Cost Electrical |
|---|---|---|---|
| 8,600 | 74.70 | 8.5 | 4.4 |
| 16,700 | 60.20 | 13.2 | 6.5 |
| 18,000 | 126.70 | 14.6 | 9.7 |
| 18,400 | 65.00 | 12.9 | 5.8 |
| 74,800 | 59.00 | 13.8 | 5.2 |
| 111,700 | 74.50 | 9.2 | 7.1 |
| 150,300 | 83.10 | 15.9 | 7.9 |
| 278,800 | 129.90 | 14.1 | 7.9 |
| 1,109,900 | 59.80 | 9.8 | 4.7 |

### SINGLE-FAMILY HOMES

| Project Size Gross S.F. | Project Cost $/S.F. | % Cost Mechanical | % Cost Electrical |
|---|---|---|---|
| 600 R | 53.70 | 16.3 | 2.0 |
| 900 | 69.90 | 33.0 | 3.0 |
| 2,100 | 80.90 | 8.8 | 4.0 |
| 2,200 | 173.30 | 23.8 | 3.4 |
| 2,500 | 84.90 | 6.1 | 7.1 |
| 2,900 | 82.00 | 7.7 | 2.8 |
| 3,000 | 57.60 | 9.5 | 6.8 |
| 3,100 | 128.70 | 15.5 | 4.6 |
| 3,600 | 80.60 | 8.5 | 3.0 |
| 3,700 | 112.65 | 9.8 | 3.4 |

A = Addition R = Remodel

# SQUARE FOOT TABLES

## RESIDENTIAL (Cont.)

### SINGLE-FAMILY HOMES (Cont.)

| Project Size Gross S.F. | Project Cost $/S.F. | % Cost Mechanical | % Cost Electrical |
|---|---|---|---|
| 4,200 | 97.90 | 15.5 | 5.7 |
| 4,600 | 155.60 | 9.0 | 4.7 |
| 5,200 | 202.10 | 8.3 | 6.0 |
| 5,700 | 118.40 | 7.4 | 3.7 |
| 5,700 | 80.50 | 8.7 | 12.6 |
| 21,300* | 57.40 | 26.0 | 4.0 |
| 22,700* | 53.30 | 27.0 | 4.6 |
| 45,000* | 81.10 | 7.8 | 2.5 |
| 51,458* | 58.80 | 11.0 | 5.0 |

*TOWNHOUSES

## EDUCATIONAL

### ADMINISTRATION (OFFICES)

| Project Size Gross S.F. | Project Cost $/S.F. | % Cost Mechanical | % Cost Electrical |
|---|---|---|---|
| 53,700 | 138.30 | 15.4 | 9.0 |

### ATHLETIC FACILITY

| Project Size Gross S.F. | Project Cost $/S.F. | % Cost Mechanical | % Cost Electrical |
|---|---|---|---|
| 38,100 | 137.30 | 16.8 | 9.8 |
| 44,100 | 126.70 | 16.8 | 8.4 |
| 100,000 | 86.60 | 19.2 | 5.3 |
| 160,000 | 156.80 | 13.7 | 7.9 |
| 247,500 | 131.60 | 11.2 | 9.0 |
| 271,000 | 123.60 | 13.2 | 7.6 |
| 283,100 | 148.90 | 14.6 | 6.0 |

### AUDITORIUM/PERFORMING ARTS

| Project Size Gross S.F. | Project Cost $/S.F. | % Cost Mechanical | % Cost Electrical |
|---|---|---|---|
| 9,900 | 221.70 | 17.5 | 29.2 |
| 17,800 | 202.70 | 11.4 | 16.1 |
| 29,200 | 200.60 | 16.2 | 10.5 |
| 62,700 | 145.90 | 13.0 | 12.9 |

### CLASSROOM

| Project Size Gross S.F. | Project Cost $/S.F. | % Cost Mechanical | % Cost Electrical |
|---|---|---|---|
| 35,400 | 201.20 | 10.2 | 6.8 |
| 70,000 | 90.80 | 24.3 | 13.1 |
| 78,900 | 169.00 | 13.3 | 14.3 |
| 80,100 | 137.30 | 16.9 | 10.5 |
| 100,000 | 136.80 | 21.1 | 9.8 |
| 166,000 | 94.50 | 16.1 | 11.2 |
| 298,400 | 88.70 | 11.3 | 11.9 |

### COMPLETE COLLEGE FACILITIES

| Project Size Gross S.F. | Project Cost $/S.F. | % Cost Mechanical | % Cost Electrical |
|---|---|---|---|
| 95,300 | 142.60 | 19.4 | 11.7 |
| 450,000 | 170.10 | 18.6 | 10.8 |

### ELEMENTARY SCHOOL

| Project Size Gross S.F. | Project Cost $/S.F. | % Cost Mechanical | % Cost Electrical |
|---|---|---|---|
| 18,000 | 104.25 | 19.1 | 13.8 |
| 30,800 | 95.90 | 16.0 | 10.0 |
| 31,600 | 80.20 | 12.1 | 11.3 |
| 35,700 | 117.25 | 17.3 | 8.7 |
| 40,000 | 101.90 | 18.5 | 13.4 |
| 40,500 | 89.30 | 22.1 | 11.3 |
| 57,000 | 77.20 | 22.1 | 7.5 |
| 69,700 | 120.40 | 20.6 | 9.3 |
| 91,400 | 97.10 | 18.2 | 7.6 |

### HIGH SCHOOL

| Project Size Gross S.F. | Project Cost $/S.F. | % Cost Mechanical | % Cost Electrical |
|---|---|---|---|
| 116,400 | 106.60 | 18.1 | 12.9 |
| 133,000 | 91.70 | 17.8 | 10.7 |
| 184,000 | 161.60 | 23.0 | 10.0 |
| 217,200 R | 87.10 | 26.4 | 10.6 |
| 254,000 R | 67.20 | 17.2 | 14.6 |
| 431,700 | 102.40 | 13.1 | 9.6 |

A = Addition R = Remodel

# SQUARE FOOT TABLES

## EDUCATIONAL (Cont.)

### JUNIOR HIGH SCHOOL

| Project Size Gross S.F. | Project Cost $/S.F. | % Cost Mechanical | % Cost Electrical |
|---|---|---|---|
| 26,000 | 134.20 | 9.5 | 9.3 |
| 28,100 | 83.20 | 11.9 | 9.5 |
| 52,800 | 107.70 | 18.3 | 8.6 |
| 91,600 | 132.40 | 21.2 | 11.0 |
| 123,700 | 108.30 | 29.1 | 9.1 |

### LABORATORY/RESEARCH

| Project Size Gross S.F. | Project Cost $/S.F. | % Cost Mechanical | % Cost Electrical |
|---|---|---|---|
| 9,200 | 232.25 | 25.4 | 4.5 |
| 80,300 | 182.70 | 20.2 | 15.5 |

### LIBRARY

| Project Size Gross S.F. | Project Cost $/S.F. | % Cost Mechanical | % Cost Electrical |
|---|---|---|---|
| 6,900 | 121.90 | 17.9 | 10.3 |
| 8,200 | 111.80 | 18.8 | 10.6 |
| 12,000 | 145.00 | 19.4 | 15.9 |
| 15,000 | 126.70 | 15.7 | 18.7 |
| 16,300 | 107.30 | 11.3 | 7.4 |
| 28,600 | 96.10 | 15.7 | 9.0 |
| 30,100 | 130.40 | 13.0 | 11.0 |
| 37,700 | 102.40 | 14.6 | 8.0 |
| 43,500 | 86.60 | 16.6 | 6.7 |
| 47,900 | 151.70 | 29.7 | 9.0 |
| 51,400 | 155.00 | 13.1 | 12.4 |
| 63,400 A | 116.60 | 13.5 | 8.3 |
| 64,000 | 109.05 | 10.9 | 11.4 |
| 74,000 | 123.60 | 17.2 | 8.0 |
| 75,600 | 164.10 | 12.5 | 16.7 |
| 176,000 | 90.90 | 11.2 | 9.8 |

### SPECIAL NEEDS FUNCTION

| Project Size Gross S.F. | Project Cost $/S.F. | % Cost Mechanical | % Cost Electrical |
|---|---|---|---|
| 15,200 | 103.10 | 16.6 | 8.4 |
| 27,900 | 124.20 | 19.4 | 11.2 |

### STUDENT CENTER/MULTIPURPOSE

| Project Size Gross S.F. | Project Cost $/S.F. | % Cost Mechanical | % Cost Electrical |
|---|---|---|---|
| 90,000 | 156.70 | 16.2 | 10.0 |
| 49,600 | 115.70 | 18.2 | 9.2 |
| 187,700 | 174.20 | 15.3 | 8.8 |
| 194,800 | 92.10 | 17.2 | 8.5 |

## HOTEL/MOTEL

### CONVENTION/CONFERENCE CENTER

| Project Size Gross S.F. | Project Cost $/S.F. | % Cost Mechanical | % Cost Electrical |
|---|---|---|---|
| 8,600 A | 137.60 | 20.7 | 16.3 |
| 71,900 | 152.60 | 12.5 | 13.8 |
| 433,800 | 79.80 | 22.1 | 8.3 |

### HOTEL

| Project Size Gross S.F. | Project Cost $/S.F. | % Cost Mechanical | % Cost Electrical |
|---|---|---|---|
| 19,900 A | 78.70 | 16.8 | 5.8 |
| 25,875 R | 70.60 | 11.8 | 10.5 |
| 48,400 A | 141.10 | 23.6 | 8.2 |
| 64,300 R | 190.10 | 20.3 | 10.1 |
| 104,200 A | 88.30 | 15.0 | 8.8 |
| 108,040 | 75.40 | 13.0 | 7.0 |
| 110,100 | 97.60 | 18.4 | 10.5 |
| 132,000 | 182.50 | 16.4 | 5.4 |
| 135,900 A | 112.80 | 15.2 | 7.7 |
| 144,100 A | 127.70 | 19.3 | 11.5 |
| 231,000 | 137.10 | 15.2 | 8.4 |
| 449,800 A | 79.20 | 13.1 | 7.0 |

### HOTEL/INN

| Project Size Gross S.F. | Project Cost $/S.F. | % Cost Mechanical | % Cost Electrical |
|---|---|---|---|
| 57,400 | 91.90 | 13.0 | 10.0 |
| 73,000 | 59.90 | 24.4 | 18.1 |
| 75,900 | 79.70 | 16.5 | 7.6 |
| 162,000 | 95.90 | 17.5 | 8.0 |
| 197,000 | 91.80 | 15.7 | 7.7 |
| 277,900 | 67.40 | 18.8 | 9.4 |

A = Addition R = Remodel

# SQUARE FOOT TABLES

## INDUSTRIAL

### MANUFACTURING

| Project Size Gross S.F. | Project Cost $/S.F. | % Cost Mechanical | % Cost Electrical |
|---|---|---|---|
| 14,300 | 78.00 | 13.0 | 7.0 |
| 18,500 | 117.50 | 20.5 | 13.5 |
| 26,600 | 42.00 | 6.1 | 12.1 |
| 31,400 | 92.80 | 18.9 | 17.7 |
| 33,400 | 53.70 | 23.1 | 15.8 |
| 37,300 | 40.00 | 3.0 | 24.0 |
| 43,400 | 53.40 | 14.7 | 11.7 |
| 45,400 | 86.60 | 15.9 | 15.0 |
| 79,800 | 90.00 | 16.0 | 14.5 |
| 81,100 | 96.10 | 8.2 | 8.4 |
| 137,400 | 60.30 | 41.6 | 13.6 |
| 179,600 | 69.60 | 26.3 | 13.6 |
| 186,000 | 65.10 | 19.5 | 11.7 |

### RESEARCH AND DEVELOPMENT

| Project Size Gross S.F. | Project Cost $/S.F. | % Cost Mechanical | % Cost Electrical |
|---|---|---|---|
| 89,140 | 124.30 | 20.8 | 9.0 |
| 100,400 | 164.60 | 36.1 | 25.1 |
| 114,200 | 161.30 | 21.4 | 9.8 |
| 125,000 | 109.70 | 20.8 | 8.4 |
| 140,000 | 153.20 | 20.1 | 11.2 |

### WAREHOUSE W/OFFICE

| Project Size Gross S.F. | Project Cost $/S.F. | % Cost Mechanical | % Cost Electrical |
|---|---|---|---|
| 14,000 | 35.80 | 6.5 | 9.2 |
| 19,000 | 32.00 | 2.3 | 1.8 |
| 19,700 | 42.20 | 9.5 | 7.0 |
| 31,200 | 36.30 | 7.0 | 7.0 |
| 40,500 | 46.30 | 6.6 | 10.5 |
| 62,000 | 58.70 | 10.8 | 10.0 |
| 96,200 | 34.00 | 2.2 | 6.3 |
| 105,000 | 33.00 | 5.3 | 11.1 |
| 149,800 | 34.50 | 14.5 | 8.6 |

### WAREHOUSE W/OFFICE (Cont.)

| Project Size Gross S.F. | Project Cost $/S.F. | % Cost Mechanical | % Cost Electrical |
|---|---|---|---|
| 168,600 | 38.00 | 11.4 | 6.3 |
| 209,600 | 35.00 | 4.9 | 10.3 |
| 402,400 | 47.70 | 14.9 | 8.3 |

## MEDICAL

### EDUCATION CENTER

| Project Size Gross S.F. | Project Cost $/S.F. | % Cost Mechanical | % Cost Electrical |
|---|---|---|---|
| 35,400 | 213.30 | 10.2 | 6.8 |

### HOSPITALS

| Project Size Gross S.F. | Project Cost $/S.F. | % Cost Mechanical | % Cost Electrical |
|---|---|---|---|
| 9,300 R | 210.70 | 29.5 | 12.0 |
| 15,900 A | 155.40 | 27.4 | 15.3 |
| 16,600 R | 56.00 | 14.7 | 9.6 |
| 22,000 | 353.70 | 31.6 | 17.5 |
| 39,100 | 176.20 | 23.3 | 7.9 |
| 63,800 | 142.00 | 23.4 | 7.8 |
| 98,000 A | 204.30 | 20.5 | 17.0 |
| 100,200 A | 325.90 | 22.2 | 9.6 |
| 103,900 A | 180.70 | 31.1 | 11.7 |
| 109,300 | 179.00 | 20.7 | 15.7 |
| 148,700 | 140.70 | 25.2 | 11.8 |
| 154,700 | 228.40 | 19.3 | 10.5 |
| 165,484 | 198.50 | 21.5 | 14.8 |
| 165,700 A | 193.80 | 26.2 | 17.7 |
| 179,400 A | 118.10 | 28.9 | 20.3 |
| 182,800 | 175.70 | 19.1 | 14.0 |
| 265,000 | 209.60 | 31.2 | 14.1 |
| 281,100 | 200.70 | 24.1 | 14.6 |
| 435,000 | 208.00 | 21.5 | 13.7 |
| 694,300 | 115.20 | 28.9 | 9.7 |
| 772,300 | 212.20 | 31.5 | 13.4 |

A = Addition  R = Remodel

# SQUARE FOOT TABLES

## MEDICAL (Cont.)

### MEDICAL OFFICES/CENTERS

| Project Size Gross S.F. | Project Cost $/S.F. | % Cost Mechanical | % Cost Electrical |
|---|---|---|---|
| 3,000 | 89.10 | 13.7 | 11.5 |
| 5,500 | 92.70 | 8.5 | 18.7 |
| 10,000 | 112.00 | 13.9 | 9.4 |
| 10,600 | 141.90 | 13.2 | 8.9 |
| 16,300 | 117.40 | 21.8 | 13.2 |
| 18,300 | 64.10 | 7.1 | 13.4 |
| 20,600 | 92.10 | 14.3 | 6.4 |
| 24,900 | 162.10 | 21.0 | 15.1 |
| 27,000 | 163.70 | 19.7 | 10.1 |
| 28,400 | 114.10 | 13.6 | 9.3 |
| 30,500 | 123.60 | 17.2 | 12.7 |
| 32.000 | 98.00 | 18.1 | 10.5 |
| 44,300 | 98.90 | 24.4 | 16.2 |
| 50,200 | 57.10 | 9.8 | 4.9 |
| 51,200 | 144.40 | 21.0 | 12.0 |
| 64,600 | 68.80 | 13.2 | 6.4 |
| 66,000 | 64.10 | 9.8 | 8.8 |
| 80,000 | 55.00 | 8.2 | 8.3 |
| 137,175 | 91.90 | 12.8 | 6.6 |

### NURSING HOMES

| Project Size Gross S.F. | Project Cost $/S.F. | % Cost Mechanical | % Cost Electrical |
|---|---|---|---|
| 11,600 A | 269.20 | 53.2 | 7.9 |
| 16,800 | 175.30 | 33.3 | 7.8 |
| 31,900 A | 132.80 | 22.0 | 11.0 |
| 64,100 | 113.00 | 20.6 | 11.1 |
| 290,000 | 158.40 | 16.1 | 13.6 |

### RESEARCH

| Project Size Gross S.F. | Project Cost $/S.F. | % Cost Mechanical | % Cost Electrical |
|---|---|---|---|
| 34,600 | 150.00 | 18.1 | 3.0 |

## PUBLIC FACILITIES

### ANIMAL CENTER

| Project Size Gross S.F. | Project Cost $/S.F. | % Cost Mechanical | % Cost Electrical |
|---|---|---|---|
| 20,000 | 190.10 | 20.7 | 4.6 |
| 39,100 | 142.00 | 22.9 | 6.8 |
| 44,300 | 135.30 | 8.1 | 5.8 |

### AUTO DEALERSHIP

| Project Size Gross S.F. | Project Cost $/S.F. | % Cost Mechanical | % Cost Electrical |
|---|---|---|---|
| 7,700 | 88.60 | 16.7 | 9.4 |

### BROADCASTING

| Project Size Gross S.F. | Project Cost $/S.F. | % Cost Mechanical | % Cost Electrical |
|---|---|---|---|
| 20,000 | 259.10 | 20.0 | 13.0 |
| 29,500 | 199.90 | 16.6 | 15.0 |
| 45,000 R | 145.00 | 15.0 | 13.0 |

### CIVIC CENTER

| Project Size Gross S.F. | Project Cost $/S.F. | % Cost Mechanical | % Cost Electrical |
|---|---|---|---|
| 6,000 | 142.80 | 9.2 | 2.8 |
| 23,900 | 179.30 | 12.3 | 10.9 |
| 34,400 | 81.80 | 3.5 | 17.8 |
| 69,800 A | 214.10 | 17.7 | 8.8 |
| 206,500 | 115.90 | 15.0 | 11.0 |

### CORRECTION FACILITIES

| Project Size Gross S.F. | Project Cost $/S.F. | % Cost Mechanical | % Cost Electrical |
|---|---|---|---|
| 44,600 | 173.10 | 20.9 | 13.1 |
| 66,000 | 105.60 | 15.0 | 23.0 |
| 257,800 A | 182.80 | 20.9 | 10.7 |
| 360,000 | 123.10 | 32.7 | 13.2 |

### FIRE STATION

| Project Size Gross S.F. | Project Cost $/S.F. | % Cost Mechanical | % Cost Electrical |
|---|---|---|---|
| 6,900 | 154.50 | 12.4 | 9.7 |
| 7,600 | 118.70 | 16.0 | 9.5 |
| 8,430 | 163.30 | 12.8 | 9.6 |
| 9,600 | 142.40 | 13.5 | 11.8 |

A = Addition  R = Remodel

# SQUARE FOOT TABLES

## PUBLIC FACILITIES (Cont.)

### GOVERMENT BUILDINGS

| Project Size Gross S.F. | Project Cost $/S.F. | % Cost Mechanical | % Cost Electrical |
|---|---|---|---|
| 12,300 | 110.80 | 13.5 | 10.7 |
| 23,500 | 181.50 | 11.1 | 15.6 |
| 27,300 | 143.30 | 19.5 | 9.3 |
| 31,600 | 161.40 | 27.1 | 11.3 |
| 46,600 | 151.90 | 17.4 | 13.8 |
| 72,100 | 165.40 | 24.4 | 10.7 |
| 78,200 | 137.20 | 19.7 | 15.2 |
| 332,900 | 133.20 | 16.2 | 14.2 |
| 364,100 | 174.30 | 14.6 | 12.9 |
| 771,000 | 192.60 | 17.6 | 11.3 |

### MUSEUM

| Project Size Gross S.F. | Project Cost $/S.F. | % Cost Mechanical | % Cost Electrical |
|---|---|---|---|
| 27,600 | 133.10 | 17.8 | 14.1 |
| 30,100 | 152.80 | 18.6 | 7.9 |
| 43,264 | 142.40 | 11.1 | 8.3 |
| 63,000 | 154.10 | 8.8 | 18.1 |

### PARKING GARAGE

| Project Size Gross S.F. | Project Cost $/S.F. | % Cost Mechanical | % Cost Electrical |
|---|---|---|---|
| 66,000 | 40.70 | 2.8 | 3.1 |
| 169,000 | 30.00 | 10.3 | 3.7 |
| 562,700 | 27.00 | 2.4 | 6.2 |

### TRANSPORTATION

| Project Size Gross S.F. | Project Cost $/S.F. | % Cost Mechanical | % Cost Electrical |
|---|---|---|---|
| 7,300 | 244.40 | 15.1 | 3.2 |
| 14,300 | 196.40 | 13.3 | 16.6 |
| 23.000 | 135.70 | 9.6 | 13.7 |
| 35,500 | 113.00 | 1.0 | 19.0 |
| 49,100 | 168.50 | 35.5 | 11.6 |
| | | | |

### TRANSPORTATION (Cont.)

| Project Size Gross S.F. | Project Cost $/S.F. | % Cost Mechanical | % Cost Electrical |
|---|---|---|---|
| 288,100 | 97.20 | 8.3 | 13.3 |
| 2,160,000 | 157.30 | 23.3 | 11.5 |

## OFFICES

### BANKS

| Project Size Gross S.F. | Project Cost $/S.F. | % Cost Mechanical | % Cost Electrical |
|---|---|---|---|
| 2,900 | 226.00 | 5.3 | 4.3 |
| 3,100 | 100.90 | 5.9 | 6.7 |
| 3,300 | 146.70 | 10.3 | 16.4 |
| 3,600 | 110.50 | 10.3 | 12.2 |
| 4,000 | 125.90 | 8.0 | 9.0 |
| 4,100 | 124.60 | 21.4 | 13.0 |
| 4,200 | 134.40 | 8.6 | 14.21 |
| 4,400 | 155.20 | 12.2 | 12.7 |
| 4,500 | 96.40 | 12.4 | 13.5 |
| 4,900 | 152.30 | 11.7 | 11.2 |
| 5,900 | 113.20 | 9.3 | 13.8 |
| 6,000 | 127.30 | 11.6 | 7.3 |
| 6,100 | 171.10 | 11.0 | 8.0 |
| 7,000 | 230.20 | 6.0 | 9.0 |
| 7,300 | 140.80 | 11.9 | 11.3 |
| 7,700 | 157.90 | 8.0 | 7.5 |
| 7,800 | 173.00 | 11.0 | 11.7 |
| 8,000 | 87.90 | 10.0 | 14.0 |
| 9,200 | 139.40 | 9.9 | 12.1 |
| 9,400 | 101.00 | 11.7 | 11.8 |
| 10,200 | 210.60 | 12.6 | 12.6 |
| 12,600 | 75.20 | 7.0 | 18.0 |

A = Addition R = Remodel

# SQUARE FOOT TABLES

## OFFICES (Cont.)

### BANKS (Cont.)

| Project Size Gross S.F. | Project Cost $/S.F. | % Cost Mechanical | % Cost Electrical |
|---|---|---|---|
| 13,300 | 132.60 | 9.5 | 8.3 |
| 13,800 | 118.70 | 10.00 | 9.7 |
| 15,000 | 101.70 | 15.4 | 12.4 |
| 15,200 | 81.20 | 9.2 | 12.9 |
| 15,500 | 103.70 | 9.8 | 10.3 |
| 16,000 | 68.00 | 13.4 | 23.1 |
| 20,100 | 64.00 | 13.0 | 11.0 |
| 21,700 | 135.30 | 8.8 | 11.3 |
| 44,800 | 115.60 | 13.0 | 8.2 |
| 53,200 | 208.80 | 14.9 | 7.2 |
| 62,100 | 120.00 | 10.1 | 7.9 |
| 95,100 | 157.80 | 13.5 | 4.3 |

### OFFICE BUILDINGS

| Project Size Gross S.F. | Project Cost $/S.F. | % Cost Mechanical | % Cost Electrical |
|---|---|---|---|
| 2,600 | 141.80 | 17.2 | 9.4 |
| 3,400 | 116.00 | 10.5 | 11.3 |
| 3,800 | 113.90 | 16.4 | 11.8 |
| 4,400 | 112.45 | 12.8 | 8.5 |
| 4,500 | 96.70 | 13.0 | 7.0 |
| 5,100 | 70.20 | 16.6 | 10.2 |
| 5,200 | 92.20 | 8.0 | 5.7 |
| 6,700 | 137.25 | 16.8 | 10.4 |
| 7,500 | 147.80 | 10.1 | 8.0 |
| 7,900 | 115.10 | 17.4 | 9.0 |
| 8,100 | 177.80 | 10.6 | 11.0 |
| 10,600 | 102.90 | 9.9 | 9.7 |
| 10,900 | 59.30 | 13.8 | 10.8 |

### OFFICE BUILDINGS (Cont.)

| Project Size Gross S.F. | Project Cost $/S.F. | % Cost Mechanical | % Cost Electrical |
|---|---|---|---|
| 11,300 | 116.90 | 17.0 | 6.0 |
| 13,000 | 79.50 | 15.0 | 9.0 |
| 14,400 | 104.30 | 19.8 | 12.9 |
| 14,500 | 80.10 | 17.3 | 12.5 |
| 17,000 | 114.25 | 14.7 | 7.9 |
| 17,800 A | 61.80 | 10.4 | 11.0 |
| 18,100 | 119.80 | 22.5 | 11.6 |
| 19,300 | 65.00 | 11.1 | 8.1 |
| 24,600 | 60.00 | 18.5 | 14.1 |
| 27,700 | 72.25 | 19.6 | 5.5 |
| 27,800 | 131.10 | 12.7 | 5.1 |
| 27,800 | 70.60 | 17.8 | 10.3 |
| 32,500 R | 120.70 | 13.9 | 6.4 |
| 35,400 | 70.50 | 15.0 | 12.0 |
| 36.500 | 61.00 | 10.2 | 10.2 |
| 42,300 | 75.80 | 10.3 | 7.7 |
| 44,400 | 103.20 | 23.5 | 14.7 |
| 44,400 | 54.30 | 11.0 | 5.0 |
| 44,500 | 74.70 | 11.5 | 3.1 |
| 45,400 | 64.50 | 19.1 | 13.2 |
| 47,300 | 66.30 | 18.5 | 8.0 |
| 49,700 | 111.50 | 22.1 | 7.2 |
| 50,000 | 106.20 | 19.4 | 15.6 |
| 50,400 | 115.30 | 23.2 | 7.8 |
| 52,200 | 74.90 | 18.3 | 7.8 |
| 52,900 | 87.40 | 4.4 | 3.9 |
| 53,700 | 133.50 | 15.4 | 9.0 |

A = Addition    R = Remodel

# SQUARE FOOT TABLES

## OFFICES (Cont.)

### OFFICE BUILDINGS (Cont.)

| Project Size Gross S.F. | Project Cost $/S.F. | % Cost Mechanical | % Cost Electrical |
|---|---|---|---|
| 54,000 | 54.05 | 14.4 | 2.6 |
| 56,000 | 61.40 | 10.6 | 6.2 |
| 56,500 | 84.40 | 19.1 | 11.0 |
| 72,000 R | 29.70 | 12.9 | 20.2 |
| 74,000 | 58.70 | 13.5 | 6.8 |
| 80,800 | 59.40 | 12.0 | 6.0 |
| 81,800 | 81.80 | 21.1 | 9.5 |
| 81,900 | 68.75 | 19.9 | 9.0 |
| 82,000 | 95.60 | 14.0 | 4.2 |
| 83,100 | 112.70 | 16.0 | 8.6 |
| 85,400 | 98.90 | 18.7 | 8.6 |
| 86,200 | 77.40 | 14.7 | 11.0 |
| 99,900 | 89.60 | 16.6 | 6.6 |
| 100,000 | 100.00 | 10.6 | 12.1 |
| 100,000 | 59.10 | 16.3 | 10.4 |
| 116,400 | 94.50 | 12.6 | 10.0 |
| 134,500 | 192.70 | 14.1 | 12.4 |
| 140,000 | 149.40 | 20.1 | 11.2 |
| 155,700 | 186.10 | 11.1 | 6.1 |
| 171,000 | 87.20 | 24.0 | 7.7 |
| 174,300 | 107.40 | 12.3 | 9.1 |
| 203,300 | 145.20 | 19.0 | 13.2 |
| 265,800 | 200.90 | 11.0 | 10.0 |
| 287,300 | 55.70 | 10.7 | 4.1 |
| 319,800 | 87.20 | 13.4 | 5.6 |
| 350,000 | 128.90 | 20.0 | 13.0 |
| 360,900 | 78.10 | 14.2 | 11.1 |
| 394,000 | 62.00 | 22.5 | 8.7 |

### OFFICE BUILDINGS (Cont.)

| Project Size Gross S.F. | Project Cost $/S.F. | % Cost Mechanical | % Cost Electrical |
|---|---|---|---|
| 430,000 | 92.80 | 21.8 | 9.3 |
| 490,000 | 98.75 | 11.2 | 7.5 |
| 588,400 | 203.00 | 15.0 | 9.0 |
| 606,000 | 81.10 | 9.4 | 8.8 |
| 620,000 | 223.90 | 12.6 | 12.8 |
| 733,500 | 62.00 | 10.8 | 4.2 |

## RECREATIONAL

### ARENA

| Project Size Gross S.F. | Project Cost $/S.F. | % Cost Mechanical | % Cost Electrical |
|---|---|---|---|
| 315,200 | 232.25 | 10.4 | 8.00 |
| 385,800 | 133.10 | 13.2 | 8.80 |
| 727,000 | 120.30 | 14.9 | 7.40 |

### HEALTH CLUB

| Project Size Gross S.F. | Project Cost $/S.F. | % Cost Mechanical | % Cost Electrical |
|---|---|---|---|
| 15,900 | 65.00 | 8.7 | 9.7 |
| 21,800 | 92.00 | 10.3 | 10.3 |
| 30,100 | 142.10 | 11.4 | 19.4 |
| 66,400 | 70.00 | 10.7 | 8.5 |

### RECREATIONAL CENTER

| Project Size Gross S.F. | Project Cost $/S.F. | % Cost Mechanical | % Cost Electrical |
|---|---|---|---|
| 9,900 | 126.30 | 6.6 | 9.8 |
| 14,000 | 82.40 | 11.2 | 5.0 |
| 14,000 | 98.90 | 11.3 | 16.7 |
| 15,700 | 75.20 | 17.8 | 14.3 |
| 20,000 | 170.60 | 19.1 | 8.9 |
| 21,200 | 96.00 | 16.0 | 9.6 |

A = Addition R = Remodel

# SQUARE FOOT TABLES

## RECREATIONAL (Cont.)

### RECREATIONAL CENTER (Cont.)

| Project Size Gross S.F. | Project Cost $/S.F. | % Cost Mechanical | % Cost Electrical |
|---|---|---|---|
| 26,000 | 111.80 | 9.3 | 7.0 |
| 53,400 A | 133.70 | 11.7 | 6.4 |
| 69,800 | 216.90 | 17.7 | 8.8 |

## RELIGIOUS

### CHURCH

| Project Size Gross S.F. | Project Cost $/S.F. | % Cost Mechanical | % Cost Electrical |
|---|---|---|---|
| 4,100 | 152.30 | 8.8 | 13.5 |
| 10,400 R | 207.20 | 12.8 | 8.1 |
| 11,100 | 91.30 | 10.9 | 12.7 |
| 13,400 | 136.80 | 5.5 | 6.0 |
| 14,500 | 93.40 | 12.8 | 7.4 |
| 15,200 | 121.00 | 18.3 | 8.0 |
| 15,700 | 105.40 | 14.4 | 8.4 |

### CHURCH (Cont.)

| Project Size Gross S.F. | Project Cost $/S.F. | % Cost Mechanical | % Cost Electrical |
|---|---|---|---|
| 16,000 | 171.40 | 14.1 | 9.6 |
| 20,900 | 106.90 | 16.0 | 14.0 |
| 21,500 | 117.30 | 7.3 | 8.0 |
| 22,900 | 105.90 | 12.1 | 9.5 |
| 30,600 | 80.20 | 15.5 | 8.0 |
| 42,700 | 90.50 | 18.6 | 7.7 |

### MULTI-PURPOSE

| Project Size Gross S.F. | Project Cost $/S.F. | % Cost Mechanical | % Cost Electrical |
|---|---|---|---|
| 4,400 | 93.60 | 11.5 | 17.5 |
| 5,800 A | 119.20 | 8.9 | 7.5 |
| 6,400 | 162.80 | 15.6 | 15.8 |
| 9,000 | 79.80 | 7.7 | 5.9 |
| 9,000 | 113.20 | 16.0 | 6.6 |
| 10,100 | 75.90 | 11.1 | 12.0 |
| 12,000 | 150.10 | 13.1 | 10.7 |
| 18,400 | 82.25 | 10.8 | 10.1 |
| 19,500 | 145.00 | 16.0 | 17.3 |

A = Addition R = Remodel

For more information subscribe to **Design Cost & Data**

BNi Building News

# INDEX


- A -

A-BAND HANGER 172,352
ABANDON CATCH BASIN 13,275
ABRASIVE SURFACE 58,297
SURFACE TILE 122,123,327
ABS PIPE 192,361
ACCESS CONTROL 139,335
DOOR 95,314
FLOOR 136,334
STAIR 153,342
ACCESS/SECURE
CONTROL UNIT 266,397
ACCESSORY CONCRETE 34,286
ACCORDION FOLDING DOOR 104,319
PARTITION 141,336
ACID-PROOF COUNTER 159,346
ACOUSTICAL BLOCK 48,292
PANEL 125,328
TILE 125,328
TREATMENT 328
ACRYLIC CARPET 129,330
AD PLYWOOD 76,305
ADAPTER TERMINAL 238,383
ADHESIVE 63,299
ROOF 90,312
ADHESIVE-BED TILE 121,327
ADJUSTABLE SHELF 75,305
ADMIXTURE 36
AGGREGATE COATED PANEL 87,310
AIR COMPRESSOR 5,216,372
CONDENSER 218,374
CONDITIONING 218,374
DISTRIBUTION 224,376
DISTRIBUTION PANEL 125,328
ENTRAINING AGENT 36
HANDLING 223,376
MONITORING TEST 12
POWERED HOIST 168,350
TOOL 5
ALARM LOCK 111,321
VALVE 178,354
ALL THREAD ROD 231,380
ALLOWANCES 1
ALUMINUM CONDUIT 231,234,380,381
DOOR 100,116,316,324
DOWNSPOUT 93,313
EXPANSION JOINT 94,314
FACED PANEL 125
FACED TILE 125,329
FLASHING 91,312
GRATING 59,297
GUTTER 93,313
HATCH 95,314
JACKET 176,354
LOUVER 61,298
PLAQUE 138,335
POLE 247,387
RAILING 61,298
ROOF 91,312
ROOF PANEL 86,310
SHINGLE 84,309
SIDING PANEL 87,310
STOREFRONT 106,319
THRESHOLD 113,322
TILE 126,329
TREAD 58,297
VENT 135,333
WINDOW 107,320
AMBIENT DAMPER 218,374
AMP-TRAP FUSE 251,389
ANCHOR 56
BOLT 55
BRICK 44,290
DOVETAIL 44,290
RAFTER 63,299
SLOT 44,290
WOOD 63,299
ANGLE STEEL 50,293
STEEL BRACE 247,388
SUPPORT 172,351
VALVE 206,207,368
ANNUNCIATOR PANEL 265,396
ANODIZED TILE 126,329
ANTENNA 268,269,398
ANTI-SIPHON BREAKER 208,369
APRON 73,304
ARCHITECTURAL EQUIPMENT 146,339
FEE 1
WOODWORK 77,305
ARMOR PLATE 112,322
ARRESTER LIGHTNING 248,388
ASBESTOS REMOVAL 12,275
VACUUM 12
ASH RECEIVER 143,337
ASHLAR VENEER 53,294
ASPHALT DAMPPROOFING 80,307
DEMOLITION 10,274
EXPANSION JOINT 34,286
MEMBRANE 79,307
PRIME COAT 21,279
REPAIR 20,279
ROOF 89,312
SHINGLE 84,309
SURFACE 20,278
ASPHALTIC PAINT 80,307
ASTRAGAL
WEATHERSTRIPPING 112,322
ATHLETIC EQUIPMENT 154,342
ATTENUATION BLANKET 126,329
ATTIC INSULATION 83,309
AUDIO SENSOR 266,397
AUGER FENCE 20,281
AUTO TRANSFER SWITCH 248,388
AUTOMOTIVE EQUIPMENT 216,372
HOIST 166,349
AUTOPSY TABLE 157,344
AWNING WINDOW 109,321

- B -

BACK MOUNTED TILE 122,327
SPLASH 76,305
BACK-UP BLOCK 48,292
BRICK 46,291
BACKBOARD BASKETBALL 154,342
BACKER ROD 97,315
BACKFILL 16,277
HAND 18,277
BACKFLOW PREVENTER 170,351
BACKHOE 6
BACKHOE/LOADER 6
BAKE OVEN 150,341
BALANCE LABORATORY 156,343
BALANCING VALVE 207,368
BALL VALVE 206,207,368
BALL-BEARING HINGE 109
BALLAST REPLACEMENT 260,394
ROOF 90,312
BALUSTER 75,304
BAND SAW 155,343
STAGE 147,339
BANDING GRATING 60,298
BANK COUNTER 146,339
BAR PARALLEL 154,343
REINFORCING 43,290
BARBED WIRE 22, 23,279
BARRICADE 3
BARRIER VAPOR 36, 80,286,307
BASE CABINET 76,159,305,346
COLONIAL 73,304
COLUMN 52, 63,294,299
FLASHING 91,313
GRANITE 53,294
TERRAZZO 124,328
WALL 128,330
BASEBOARD HEATER 269,398
RADIATION 221,375
BASKET LOCKER 139,335
BASKETBALL BACKBOARD 154,342
BATH 210,370
ACCESSORY 143,337
FAUCET 211,370
BATHROOM LOCK 111,321
BATT INSULATION 80,307
BATTEN SIDING 88,311
BEAD MOLDING 74,304
PARTING 75,304
PLASTER 119,121,326
SASH 75,304
BEAM BOND 37, 43, 50,287,290,293
CONCRETE 37,287
FIREPROOFING 83,309
FORMWORK 28,283
FURRING 57,117,296,325
GRADE 38,287
HANGER 63,299
PLASTER 119,326
REINFORCING 31, 33,284,285
SHEETROCK 120,326
BELL END 238,383
BELLOWS COPPER 94,314
BENCH FLOOR MOUNTED 139,335
SAW 155,343
BENTONITE MEMBRANE 79,307
BEVELED SIDING 88,311
BHMA HINGE 109
BI-FOLD DOOR 102,318
BI-PASSING DOOR 102,318
BIBB HOSE 203,366
BICYCLE EXERCISE 156,344
BIN ICE STORAGE 152,342
BINDER COURSE 20,278
BIRCH DOOR 101,317
PANELING 77,305
VENEER 77,305
BIRDSCREEN 162
BITUMINOUS DAMPPROOFING 80,307
MEMBRANE 79,307
SIDEWALK 20,278
BLACK IRON FITTING 198,364
RISER HANGER 172,352
STEEL 198,364
STEEL HANGER 171,351
WIRE HANGER 174,352
BLEACHER GYM 154,342
BLIND 160,346
BLOCK CONCRETE 39, 48,288,292
DAMPPROOFING 80,307
DEMOLITION 8,273
GLASS 51,294
GRANITE 47,292
GROUT 43,290
REMOVAL 10,274
VENT 135,333

# INDEX


BLOCKING WOOD ........................ 64,299
BLOOD ANALYZER .................... 156,343
REFRIGERATOR .................... 156,343
BLOWN-IN INSULATION .............. 83,309
BLUESTONE ................................ 47,292
BOARD CEMENT ......................... 79,307
COMPOSITE ............................ 81,308
DIVING .................................. 162,347
DOCK ..................................... 149,340
INSULATION ............................ 81,308
PROTECTIVE ........................... 79,307
RAFTER .................................. 69,301
RIDGE ..................................... 68,301
SIDING ................................... 88,311
SUB-FLOORING ....................... 70,302
BOILER ...................................... 217,373
BOLLARD PIPE ........................... 26,281
BOLT CARRIAGE ......................... 63,299
DEAD ...................................... 111,321
DOOR ..................................... 112,322
WOOD ..................................... 63,299
BOND ................................................ 7
BEAM ......................... 43, 50,290,293
BEAM PLACEMENT ................. 37,287
ROOF ............................................ 89
BONDERIZED FLAGPOLE .......... 137,334
BOOKKEEPER ................................... 2
BOOTH SPRAY .......................... 154,343
BORED LOCK ............................ 110,321
BOTTLE COOLER ....................... 152,342
BOTTOM BEAM ............................ 28,283
PIVOT HINGE ............................... 109
PLATE ...................................... 70,302
BOWL TOILET ........................... 213,371
BOX BEAM RAIL .......................... 27,281
CABINET ................................. 245,386
DEVICE ................................... 243,386
JUNCTION ............................... 243,385
PLASTER ................................. 244,386
PLASTIC .................................. 244,386
PULL ....................................... 244,386
RECEIVING ............................... 140,336
RIB ROOF ................................. 86,310
SCREW COVER ......................... 245,386
BOX-RIB PANEL ............................ 87,311
BOXING RING ............................ 154,343
BRACKET BOX .................... 243,244,386
BRASS FITTING ......................... 188,359
BREAKER CIRCUIT ..................... 249,388
VACUUM ................................... 208,369
BRICK ANCHOR ........................... 44,290
CHIMNEY ................................... 46,291
FIRE RATED ............................... 46,291
FIREWALL .................................. 46,291
MASONRY .................................. 45,291
PAVER ....................................... 47,292
REMOVAL ......... 8, 10, 54,273,274,295
BRIDGE CRANE ......................... 168,350
BRIDGING ................................... 63,299
BROILER KITCHEN ..................... 150,341
BRONZE PLAQUE ....................... 138,334
RAILING ..................................... 61,298
VALVE ...................................... 203,366
BROOM FINISH ............................ 36,286
BRUSH-OFF BLAST .................... 132,332
BUCKET TRAP ........................... 209,369
BUFF BRICK ................................ 46,291
BUILDING DEMOLITION ................. 8,273
EXCAVATION .............................. 16,277
METAL ....................................... 161,347
PAPER ........................................ 80,307
BUILT-IN OVEN ........................... 153,342
SHELVING .................................. 143,337
BUILT-UP PLAQUE ...................... 138,334
ROOF ......................................... 89,312
ROOF REMOVAL ............................ 9,273
BULK EXCAVATION ...................... 16,276
MAIL SLOT ................................. 140,336
BULKHEAD FORMWORK ............... 30,284
BULLDOZER ........................................ 6
BULLETIN BOARD ....................... 137,334
BULLETPROOF GLASS ............... 115,323
PARTITION ................................. 146,339
BULLNOSE GLAZED ..................... 49,293
BUMPER DOCK ........................... 148,340
DOOR ......................................... 111,321
BURLAP CONCRETE .................... 37,287
RUB ............................................ 36,286
BUSHING CONDUIT .................... 236,382
BUTT HINGE ...................................... 109
BUTTERFLY VALVE .................... 206,368
BUTYL CAULK ............................. 54,295
CAULKING .................................... 96,315
MEMBRANE .................................. 79,307
BX CLAMP ................................... 244,386


- C -


C-CLAMP ..................................... 172,351
CAB ELEVATOR ................................ 163
CABINET BASE ............................. 76,305
EXTINGUISHER ........................... 140,335
HEATER ...................................... 223,376
KITCHEN ..................................... 159,346
LABORATORY .............................. 155,343
MONITOR .................................... 267,397
PAINTING .................................... 131,331
CABLE COMMUNICATION .......... 267,397
GUY ............................................ 268,398
MAN-HOLE .................................. 268,397
THHN-THWN ............................... 241,385
CAGE LADDER .............................. 58,297
CAISSON ...................................... 20,278
CALCIUM SILICATE ...................... 176,353
CANT STRIP .................... 69, 89,301,312
CAP BLACK IRON ......................... 198,364
CONDUIT ..................................... 237,382
COPPER .................. 187,188,358,359
PILE ............................ 29, 38,283,288
PILE REINFORCING ....................... 33,285
STEEL ......................................... 201,365
CAPACITANCE SENSOR ............. 266,397
CARBORUNDUM ............................ 36,286
CARD ACCESS CONTROL .......... 139,335
CARPENTRY ROUGH ..................... 64,299
CARPET ...................................... 129,330
CLEAN ........................................ 129,330
PADDING ..................................... 129,330
TILE ................................ 129,136,330
CARRIAGE BOLT ............................ 63,299
CARRIER CHANNEL ............................ 126
FIXTURE ..................................... 214,372
CART LAUNDRY ........................... 148,340
CARTRIDGE EXTINGUISHER ...... 140,335
CARVED DOOR ........................... 102,317
CASE REFRIGERATED ................ 152,342
CASEMENT WINDOW .......... 107,108,320
CASEWORK ................................. 159,346
CASING ......................................... 74,304
BEAD .......................................... 121,326
RANCH ......................................... 74,304
TRIM ............................................ 75,305
CAST IRON BOILER .................... 217,373
IRON PIPE ................. 179,181,355,356
IRON STRAINER ......................... 209,369
IRON TREAD ................................. 58,297
IRON VALVE ............................... 203,366
CASTING ....................................... 60,298
CAULK MASONRY ......................... 54,295
CAULKING ............................... 96, 97,315
CAVITY WALL .................... 43, 45,290,291
WALL ANCHOR .............................. 44,290
CDX ................................................ 70,302
CEDAR CLOSET ............................. 77,305
DECKING ....................................... 72,303
SHINGLE ............................. 85,309,310
SIDING ............................... 88, 89,311
CEILING ACCESS DOOR ................ 95,314
AND WALL .................................. 125,328
BLOCKING ..................................... 64,299
DIFFUSER .................................... 226,377
EXPANSION ................................... 61,298
FRAMING ....................................... 64,299
FURRING ..................... 66,117,300,325
GRILLE ........................................ 227,378
HEATER ....................................... 153,342
INSULATION ................. 80, 84,307,309
JOIST ............................................ 64,299
LATH ........................... 118,119,325,326
PAINTING ..................................... 131,331
REMOVAL ......................................... 9,273
CELLULAR DECKING ...................... 56,296
CEMENT BOARD ............................ 79,307
FIBER ............................................. 41,289
FIBER BOARD ................................. 40,288
KEENES ........................................ 119,326
PLASTER ...................................... 120,326
CEMENTITIOUS TOPPING ............... 40,288
CENTER BULB ................................. 35,286
CENTRIFUGAL PUMP .................... 215,372
CENTRIFUGE LABORATORY ......... 156,343
CERAMIC TILE ............................... 121,327
CHAIN HOIST DOOR ...................... 105,319
LINK .................................. 13, 22,275,279
CHAIR BAND ................................. 147,339
HYDROTHERAPY .......................... 156,344
LIFEGUARD ................................... 162,347
RAIL ................................................ 74,304
CHALKBOARD ................................ 134,333
CHAMFER STRIP ............................. 30,284
CHANGE TRAILER ................................ 12
CHANNEL DOOR FRAME ..................... 56
FURRING ........................ 57,117,296,325
SLAB ................................................ 40,289
STRAP ............................................. 231,380
CHECK VALVE .............. 178,204,354,367
CHECKROOM EQUIPMENT ........ 147,339
CHEMICAL EXTINGUISHER ......... 140,335
CHERRY VENEER ............................ 77,305
CHESTNUT VENEER ........................ 77,305
CHILLER ......................................... 219,374
CHIME ............................................ 265,396
CHIMNEY BRICK ............................... 46,291
FLUE .................................................. 54,295
MASONRY ......................................... 54,295
CHOPPER FOOD ............................ 151,341
CHUTE ............................................. 166,349
MAIL ................................................ 140,336
TRASH ...................... 149,167,340,349
CIRCUIT BREAKER ........................ 249,388
CL-AY FUSE ................................... 251,390
CLAMP GROUND ROD .................... 252,390
ROD BEAM ...................................... 231,380

CLASSROOM LOCK....................110,321
LOCKSET....................110,321
CLAY BRICK FLOOR....................127,329
TILE....................50,293
TILE FLOOR....................51,293
CLEANING MASONRY....................53,295
CLEANOUT.................... 191,202,361,366
CLEFT NATURAL....................47,292
CLEVIS HANGER....................170,351
CLOCK RECEPTACLE....................245,387
CLOCK-TIMER HOSPITAL....................158,344
CLOSED CELL CAULKING....................96,315
CLOSER....................111,321
CLOSET CEDAR....................77,305
DOOR....................102,317
POLE....................74,304
CLOTH FACED FIBERGLASS....................125,328
CMU....................48,292
GROUT....................43,290
CO2 EXTINGUISHER....................139,335
SPRINKLER SYSTEM....................178,355
COAT RACK....................145,338
COATED ROOF....................90,312
COAX CABLE....................269,398
COAXIAL CABLE....................267,397
COBRA HEAD FIXTURE....................263,395
COFFEE URN....................150,341
COILING DOOR....................103,318
GRILLE....................104,318
COIN DRYER....................148,340
WASHER....................148,340
COLD FORMED FRAMING....................57
COLONIAL BASE....................73,304
MOLDING....................75,304
COLOR GROUP TILE....................122
COLUMN BASE.................... 52, 63,294,299
CONCRETE....................37,287
COVER....................116,323
FIREPROOFING....................83,309
FOOTING....................29,283
FORMWORK....................28,283
FURRING....................57,296
PAINTING....................130,331
PIER....................30,284
PLASTER....................119,326
REINFORCING....................31,284
SHEETROCK....................120,326
TIMBER....................72,303
WOOD....................78,306
COMBINATION RECEPTACLE....................246,387
COMMERCIAL BLAST....................132,332
CARPET....................129,330
DOOR.................... 100,106,316,319
COMMON BRICK....................45,291
COMMUNICATION....................396
COMPACT FLUORESCENT....................262,395
KITCHEN....................149,340
COMPACTOR INDUSTRIAL....................149,340
RESIDENTIAL....................152,342
COMPARTMENT SHOWER....................135,333
COMPOSITE BOARD....................81,308
DECKING....................56,296
COMPRESSION FITTING....................188,359
COMPRESSOR....................5
AUTOMOTIVE....................155,343
COMPUTER FLOOR....................136,334
COMPUTER-ROOM AC....................220,374
CONCEALED Z BAR....................126,329
CONCRETE....................28,283
ACCESSORY....................34,286
ADMIXTURE....................36
BLOCK REMOVAL....................10,274
CONCRETE CUTTING....................11,274
DEMOLITION....................8,273
DOWEL....................32,285
FENCE HOLE....................26,281
FILLED PILE....................19,278
FINISH....................36,286
HARDENER....................36,287
MASONRY UNIT....................48,292
PAINTING....................130,331
PARKING BARRIER....................27,282
PAVING....................20,279
PILING....................18,278
PLANK....................40,288
PLANTER....................27,282
POLE....................247,387
POST....................27,281
PUMP....................37,287
RECEPTOR....................134,333
REINFORCEMENT....................31,284
REMOVAL.................... 10, 14,273,276
REPAIR....................41,289
RESTORATION....................41,289
SLEEPER....................69,302
TESTING....................3
WATERPROOFING....................80,307
CONDENSATE PUMP....................215,372
CONDENSING UNIT....................218,374
CONDUCTIVE FLOOR....................127,128,330
TERRAZZO....................124,328
TILE....................122,327
CONDUCTOR
COPPER.................... 241,242,384,385
STRANDED....................241,385
CONDUIT ALUMINUM.... 231,234,380,381
BODY....................238,383
BUSHING....................236,382
CAP....................237,382
EMT....................232,380
FLEXIBLE....................233,381
GALVANIZED....................234,381
LOCKNUT....................236,382
PENETRATION....................231,380
PLASTIC COATED....................239,383
PVC....................237,383
SPECIALTY....................231,380
STEEL....................233,381
CONNECTOR DIE CAST....................233,381
CONSTRUCTION AID....................4
CONTINUOUS FOOTING....................38,287
CONTROL....................139,335
DAMPER....................229,378
JOINT....................45,291
VALVE....................228,378
CONTROLS....................228,378
CONVECTION OVEN....................150,341
CONVEYING SYSTEM....................163,348
COOLER BOTTLE....................152,342
WATER....................214,372
COOLING TOWER....................219,374
COPING CLAY....................51,294
LIMESTONE....................52,294
COPPER BELLOWS....................94,314
CONDUCTOR....................241,384
DOWNSPOUT....................92,313
FITTING....................183,357
FLASHING....................91,312
GRAVEL STOP....................92,313
HANGER....................172,352
PIPE....................182,356
PIPE FITTING....................187,358
PLAQUE....................138,334
VENT....................94,314
COPPER WIRE HOOK....................174,353
CORD SET....................246,387
CORE DRILLING....................11,274
CORING CONCRETE....................11,274
CORK WALL COVERING....................132,331
CORNER BEAD.................... 119,121,326,327
CABINET....................159,346
GUARD....................56,132,332
POST.................... 22, 23,279,280
CORNICE MOLDING....................75,305
CORRUGATED DOWNSPOUT....................92,313
PANEL....................87,311
ROOF.................... 86, 91,310,312
COUNTER BANK....................146,339
DOOR....................103,318
FLASHING....................91,312
GRIDDLE....................151,341
LABORATORY....................155,343
LAVATORY....................212,370
COUNTER-TOP RANGE....................153,342
COUNTERTOP.................... 76,159,305,346
COUPLING BLACK IRON....................198,364
COPPER.................... 187,188,358,359
ERICKSON....................235,382
NO-HUB....................180
SET SCREW....................233,381
SPLIT....................235,382
WITH STOP....................183,357
COURSE BINDER....................20,278
COURT LIGHTING....................262,395
COVE BASE....................128,330
BASE GLAZED....................49,293
BASE TILE....................123,327
MOLDING....................74,304
COVED STAIR RISER....................128,330
COVER POOL....................162,347
CPVC FITTING....................190,360
PIPE.................... 189,190,360
CRACK PATCH....................120,326
REPAIR....................41,289
SEALING....................21,279
CRANE....................7
JIB....................168,350
MOBILIZATION....................4
TRAVELING....................168,350
CRAWL SPACE INSULATION....................81,307
CRAWLER CRANE....................7
CREOSOTE....................76
CREOSOTED POLE....................246,387
CREW TRAILER....................4
CROSSARM POLE....................247,387
CROWN MOLDING....................74,304
CRUSHED MARBLE....................47,292
CUBICLE HOSPITAL....................135,333
OFFICE....................141,336
CURB & GUTTER....................11,274
FORM....................28,283
FORMWORK.................... 28, 31,283,284
GRANITE....................53,294
ROOF....................225,377
TERRAZZO.................... 123,124,328
CURING CONCRETE....................37,287
PAPER....................37,287
CURTAIN GYM....................154,343
WALL....................115,323
CURVED CURB FORM....................28,283
STAIR TERRAZZO....................124,328
CUSTOM DOOR....................103,318
CUT AND FILL....................16,276
STONE....................52,294

CUT-OUT ........ 9,273
COUNTERTOP ........ 76,305
CUTTING GRATING ........ 60,297

- D -

DAIRY PRODUCT ........ 152,342
DAMPER ........ 225,229,377,378
DAMPPROOFING ........ 80,307
DARBY ........ 36,286
DARKROOM EQUIPMENT ........ 153,342
DB CONDUIT ........ 238,383
DEAD BOLT ........ 111,321
DECK INSULATION ........ 82,308
DECKING CELLULAR ........ 56,296
METAL ........ 56,296
OPEN ........ 56,296
WOOD ........ 72,303
DEHUMIDIFIER ........ 221
DELICATESSEN CASE ........ 152,342
DELIVERY CART ........ 148,340
DELUGE SHOWER ........ 214,372
DEMOLITION ........ 8,273
DENTAL EQUIPMENT ........ 158,344
LIGHT ........ 158,345
DEPOSITORY NIGHT ........ 146,339
DETECTOR SMOKE ........ 112,322
DEVICE COVER ........ 243,386
DIAPHRAGM PUMP ........ 6
DIESEL GENERATOR ........ 248,388
DIFFUSER ........ 226,377
DIMMER SWITCH ........ 256,392
DIRECT DRIVE FAN ........ 224,376
DIRECTIONAL ARROW ........ 21,279
DIRECTORY BOARD ........ 137,334
SHELF ........ 143,337
DISAPPEARING STAIR ........ 153,342
DISH SOAP ........ 123,327
DISHWASHER ........ 151,341
RESIDENTIAL ........ 152,342
DISINFECT PIPE ........ 178,198,355
DISPENSER TISSUE ........ 144,337
DISPOSAL COMMERCIAL ........ 151,341
RESIDENTIAL ........ 153,342
SUIT ........ 12
DISTRIBUTION ........ 252,390
DIVIDER HOSPITAL ........ 135,333
STRIP TERRAZZO ........ 123
DIVING BOARD ........ 162,347
DOCK BOARD ........ 149,340
LEVELER ........ 148,340
DOMESTIC MARBLE ........ 52,294
DOOR ACCESS ........ 95,314
ALUMINUM ........ 116,323
AND WINDOW ........ 98,316
BUMPER ........ 111,321
CHAIN HOIST ........ 105,319
CLOSER ........ 111,321
COMMERCIAL ........ 100,316
COORDINATOR ........ 112,322
FLUSH ........ 102,318
FOLDING ........ 104,319
FRAME ........ 98,102,316,318
FRAME GROUT ........ 43,290
HEADER ........ 70,302
HOLDER ........ 111,321
LOUVER ........ 136,334
MOTOR ........ 104
OPENER ........ 267,397
OVERHEAD ........ 103,318
PAINTING ........ 130,131,331
PLATE ........ 112,322
DOOR REMOVAL ........ 8,273
REVOLVING ........ 106,319
ROLL UP ........ 161
SEAL ........ 149,340
SHOWER ........ 134,333
SLIDING GLASS ........ 105,319
SOLID CORE ........ 101,317
SPECIAL ........ 146,339
SWITCH ........ 266,396
TRIM ........ 111,321
VAULT ........ 146,339
VERTICAL LIFT ........ 105,319
DORMITORY WARDROBE ........ 144,338
DOUBLE HUNG WINDOW ........ 107,108,320
OVEN ........ 150,341
WYE NO-HUB ........ 181,356
DOUGLAS FIR ........ 72,303
DOVETAIL ANCHOR SLOT ........ 44,290
DOWEL REINFORCING ........ 32,285
DOWN DAMPER ........ 225,377
DOWNSPOUT ........ 92,313
ALUMINUM ........ 93,313
DRAIN ........ 209,369
DRAINAGE BOOT ........ 92,313
DRAPERY TRACK ........ 159,346
DRAW OUT BREAKER ........ 249,388
DRIVE-UP WINDOW ........ 146,339
DROP PANEL ........ 28,283
DRY CLEANER ........ 148,340
RUB ........ 36,286
SET TILE ........ 122,327
SPRINKLER SYSTEM ........ 178,355
STONE WALL ........ 52,294
TYPE EXTINGUISHER ........ 140,335
TYPE SPRAY BOOTH ........ 155,343
DRYER COIN ........ 148,340
DARKROOM ........ 153,342
RECEPTACLE ........ 245,387
DRYVIT PANEL ........ 87,310
DRYWALL ........ 120,326
PAINTING ........ 131,331
REMOVAL ........ 8, 9,273
DUCT DETECTOR ........ 265,396
INSULATION ........ 177,354
DUCTWORK METAL ........ 224,376
DUMBELL JOINT ........ 35,286
DUMP FEE ........ 12
TRUCK ........ 6
DUMPSTER ........ 10, 12,274
DUPLEX PUMP ........ 215,372
DUTCH DOOR ........ 98
DWV COPPER ........ 182,356
PIPE ABS ........ 192,361

- E -

EARTHWORK ........ 15,276
EB CONDUIT ........ 238,383
EDGE FORM ........ 29,283
FORMWORK ........ 31,284
REPAIR ........ 41,289
ELASTIC SHEET ROOF ........ 90,312
ELASTOMERIC FLASHING ........ 45,291
ELBOW POLYETHYLENE ........ 192,361
PVC ........ 190,360
ELECTRIC BOILER ........ 217,373
ELEVATOR ........ 163,348
FURNACE ........ 217,373
HEATING ........ 269,398
HOIST ........ 167,349
LOCKSET ........ 111,321
MANHOLE ........ 246,387
ELECTRICAL ........ 231,380
FEE ........ 2
ELEVATED FLOOR ........ 136,334
SLAB ........ 30,284
SLAB FORMWORK ........ 28,283
SLAB REINFORCING ........ 32,284
STAIRS ........ 30,283
ELEVATOR ........ 163,164,165,348
EMT CONDUIT ........ 232,380
ENAMEL PAINT ........ 132,133
ENCASED BURIAL CONDUIT ........ 238,383
ENCLOSURE TELEPHONE ........ 143,337
END BELL ........ 238,383
CAP ........ 255,391
GRIP ........ 268,398
ENGINEER ........ 2
ENGINEERING FEE ........ 1
ENGRAVED PLAQUE ........ 137,138,334
ENTRANCE DOOR ........ 106,319
FITTING ........ 236,382
MAT ........ 160,346
ENTRY LOCK ........ 111,321
LOCKSET ........ 110,321
EPOXY CONCRETE ........ 41,289
FLOORING ........ 130,330
GROUT ........ 41,289
PAINT ........ 130,331
TERRAZZO ........ 124,130,328,331
EQUIPMENT ........ 5
ARCHITECTURAL ........ 146,339
CURB ........ 28,283
INSULATION ........ 177,354
MOBILIZATION ........ 4
PAD ........ 37,287
PAD FORMWORK ........ 29,283
PAD REINFORCING ........ 32,285
ERICKSON COUPLING ........ 235,382
ESCALATOR ........ 165,348
EXERCISE BICYCLE ........ 156,344
EXHAUST FAN ........ 223,376
HOOD ........ 151,341
EXIT LIGHT ........ 258,393
LOCK ........ 110,321
EXPANDED LATH ........ 118,325
EXPANSION CONTROL ........ 61,298
JOINT ........ 34,119,286,326
JOINT COVER ........ 61,298
JOINT ROOF ........ 94,314
SHIELD ........ 55
TANK ........ 175,353
EXPOSED AGGREGATE ........ 47,292
EXTENSION BOX ........ 243,385
EXTERIOR DOOR ........ 102,103,317,318
FIXTURE ........ 263,395
PAINTING ........ 130,331
TILE ........ 122
EXTINGUISHER CABINET ........ 140,335
FIRE ........ 139,335
EXTRACTOR LAUNDRY ........ 148,340
WASHER ........ 148,339
EYE NUT ........ 247,388

- F -

FABRIC CHAIN LINK ........ 22,279
FORM ........ 42
FABRICATION METAL ........ 58, 61,298
FACING PANEL ........ 52,294
PANELS ........ 52,294
FACTORY FINISH FLOOR ........ 127
FAN COIL UNIT ........ 221,375
EXHAUST ........ 223,376

FASCIA BOARD ....69,301
PLASTER....119,326
FASTENER....63,299
INSULATION....81,308
FAUCET....211,370
FEE PROFESSIONAL ....1
FELT CONTROL JOINT....35,286
FEMALE ADAPTER COPPER....185,358
CONNECTOR....188,359
FENCE.... 13, 22,275,279
POST HOLE....26,281
REUSE....13,275
SENSOR....266,397
FIBER BOARD....40,288
CANT ....69,301
FORM CONCRETE....28,283
PANEL....125,328
FIBERBOARD INSULATION....82,308
SHEATHING....71,303
FIBERGLASS DIVING BOARD....162,347
INSULATION.... 80-82,175,307,308,353
PANEL....125,328
PIPE INSULATION ....176,354
PLANTER....27,282
REINFORCED PIPE ....194,362
RIGID....177,354
FIBROUS DAMPPROOFING....80,307
FIELD COAT....130,331
ENGINEER....2
STAFF....2
FILE ROOM DOOR....146,339
FILL GYPSUM ....40,288
FILLED BLOCK.... 39, 48,288,292
FILLER JOINT....45,291
MASONRY....130,331
PAINT....133
FILM VIEWER....157,344
FILTER BARRIER....18,277
LINE....264,395
FIN-TUBE RADIATOR....221,375
FINISH....117,325
CARPENTRY....73,304
PLASTER....119,326
SHOTCRETE....36,287
FINISHING SHEETROCK....121,326
FINK TRUSS....73,303
FIR DECKING....72,303
FLOOR....127,329
SIDING....88,311
FIRE ALARM....265,396
ALARM SYSTEM....265,396
CONNECTION....178,354
DAMPER....226,377
DETECTION ANNUNCIATOR..265,396
DOOR....103,318
EXTINGUISHER....139,335
PROTECTION....139,335
PROTECTION SYSTEM....179,355
RATED BRICK....46,291
RATED DOOR....98
RATED FRAME....99
RATED PANEL....125
RESISTANT SHEETROCK....121
RETARDANT....77
RETARDENT ROOF....85
FIREPROOF VAULT....146,339
FIREPROOFING.... 83,125,231,309,380
FIREWALL BRICK....46,291
FIRING RANGE....147,339
FITTING CPVC....190,360
NO-HUB....179,181,355
PIPE INSULATION....177,354
FITTING POLYETHYLENE....192,361
POLYPROPYLENE....191,360
PVC....190,360
STAINLESS....197,363
WIREMOLD....240,384
FIXED SASH....107,320
TYPE LOUVER....61,298
WINDOW....107,320
FIXTURE CARRIER....214,372
LIGHTING....257,392
PLUMBING....210,370
SQUARE....259,393
FLAGPOLE....137,334
FLAKER ICE....152,342
FLAMEPROOF TILE....126
FLAMEPROOFING....125
FLANGED VALVE....206,367
FLARE FITTING BRASS....188,359
FLASHING....90-92,312,313
MASONRY....45,291
FLAT APRON....73,304
RIB....118,325
FLEXIBLE CONDUIT....233,234,381
DUCTWORK....224,377
FLOAT....36,286
GLASS....114,323
TRAP....209,369
FLOOD LAMP....260,393
FLOODLIGHT....263,395
FLOOR ACCESS....136,334
BASE....128,330
BUMPER....111,321
CLAY TILE....51,293
CLEANOUT....202,366
DRAIN....209,369
FINISH....36,286
FINISHING....127,329
GYM....127,329
HARDENER....36,287
INDUSTRIAL WOOD....127,329
JOIST....65,300
LEVEL....129,330
MASONRY....127,329
MAT....160,346
MOUNTED BENCH....139,335
MOUNTED TRANSFORMER....248,388
PATCH....41
PEDESTAL....136,334
PLATE....59,297
REMOVAL....8,273
SAFE....161,347
SHEATHING....70,302
SLAB....28,283
TERRAZZO....123,327
TILE....121,123,327
FLOORING....329
MARBLE....52,294
FLOURESCENT FIXTURE....260,394
FLOW CHECK VALVE....204,367
FLUE LINER....54,295
FLUID APPLIED FLOOR....128,330
APPLIED MEMBRANE....79,307
APPLIED ROOF....90,312
FLUORESCENT FIXTURE....257,392
FLUSH BOLT....112,322
DOOR....98,102,316,318
PANEL DOOR....116,324
FLUTED COLUMN....52,294
FOAM EXTINGUISHER....140,335
POLYETHYLENE....35,286
FOAMGLASS CANT....89,312
INSULATION....82,308
FOIL BACKED INSULATION....81,308
FOLDING DOOR....104,319
STAIR....153,342
FOOD PREPARATION....151,341
SERVICE EQUIPMENT....149,340
FOOTING CONCRETE....38,287
EXCAVATION....16,277
FORMWORK....29,283
REINFORCING....32,285
FOREMAN....2
FORM FABRIC....42
FORMBOARD....40,288
FORMICA....76,305
PARTITION....134,333
FORMWORK....28,283
OPENINGS....29,283
FOUNDATION BACKFILL....16,277
BLOCK....48,292
REINFORCING....32,285
VENT....43,290
FOUNTAIN WASH....214,372
FRAME DOOR....102,318
FRAMING ANCHOR....63,299
CEILING....64,299
DOOR....70,302
FLOOR....65,300
ROOF....67,301
SOFFIT....69,302
TIMBER....72,303
WALL....69,302
WINDOW....70,302
FREIGHT ELEVATOR....164,348
FRICTION HANGER....172,352
FRONT DOOR LOCK....110,321
FRONT-END LOADER....16,276
FROZEN FOOD CASE....152,342
FRYER KITCHEN....150,341
FULL MORTISE BUTT....110
VISION GLASS....113,322
FURNACE....217,373
FURNISHINGS....159,346
FURRING CEILING....117,325
CHANNEL....57,296
STEEL....117,325
WOOD....66,300
FUSE....250,251,389,390
FUSED SAFETY SWITCH....249,389


- G -


GABLE LOUVER....135,333
TRUSS....73,303
GABLE-END RAFTER....67,301
GALVANIZED CONDUIT....234,381
DUCTWORK....224,376
GRATING....60,297
MESH....33,285
METAL DOOR....98,316
RAILING....59,297
REGLET....31,284
REINFORCEMENT....31,284
SHINGLE....84,309
STEEL....55,296
STEEL PIPE....200,365
WALL TIE....43,290
GARBAGE HANDLING....149,340
GAS FURNACE....218,373
PIPING REMOVAL....13,275
PUMP....216,373
STERILIZER....157,344
GASKET GLASS....115,323

GATE FENCE ..... 24,280
POST ..... 22,279
SWING ..... 23,280
VALVE ..... 203,366
GEL EPOXY ..... 41,289
GENERAL REQUIREMENTS ..... 1
GENERATOR ..... 5,248,388
SET ..... 254,391
GFI BREAKER ..... 249,389
GLASS ..... 113,114,322,323
BACKBOARD ..... 154
BEAD ..... 74,304
BLOCK ..... 51,294
CLOTH TILE ..... 125,328
DOOR ..... 105,319
ENCASED BOARD ..... 137,334
FIBER BOARD ..... 40,288
FRAMING ..... 115,323
PIPE ..... 189,359
WIRE ..... 114,323
GLASSWARE WASHER ..... 156,343
GLAZED BLOCK ..... 49,293
BRICK ..... 46,291
CURTAIN WALL ..... 115,323
FLOOR ..... 127,329
TILE ..... 46,291
WALL TILE ..... 121,327
GLAZING ..... 113,322
ACCESSORY ..... 115,323
GLOBE VALVE ..... 205,367
GLOVE BAG ..... 12
GRAB BAR ..... 143,337
GRADALL ..... 16,277
GRADE BEAM CONCRETE ..... 38,287
BEAM FORMWORK ..... 29,283
BEAM REINFORCING ..... 33,285
GRADING ..... 16,276
GRANITE BLOCK ..... 47,292
VENEER ..... 52,294
GRANULAR WATERPROOFING ..... 79,307
GRASS CLOTH ..... 132,332
GRATE CUTTING ..... 60,297
METAL ..... 59,297
PAINTING ..... 131,331
GRAVEL ROOF ..... 89,312
STOP ..... 92,313
GRAVITY VENTILATOR ..... 95,314
GRAY SPONGE ..... 35,286
GREASE TRAP ..... 202,366
GRIDDLE KITCHEN ..... 151,341
GRILLE DOOR ..... 104,318
METAL ..... 61,298
VENTILATION ..... 227,378
GRINDER ELECTRIC ..... 155,343
GROUND FAULT ..... 249,389
HYDRANT ..... 211,370
ROD ..... 252,390
SLAB ..... 30,284
STRAP ..... 244,386
GROUNDING ..... 252,390
GROUT BEAM ..... 43,290
BLOCK ..... 39, 43, 48,288,290,292
CRACK SEAL ..... 41,289
EPOXY ..... 41
MASONRY ..... 43, 54,290,295
WALL ..... 43,290
GRS CONDUIT ..... 234,381
GUARD ALERT DISPLAY ..... 267,397
GUARDRAIL ..... 26,281
REMOVAL ..... 11,274
GUNITE ..... 36,287
GUTTER ..... 92, 93,313
COPPER ..... 93,313
REMOVAL ..... 11,274
STEEL ..... 93,313
GUY CABLE ..... 268,398
GRIP ..... 247,388
ROD ..... 268,398
GYM CURTAIN ..... 154,343
FLOOR ..... 127,329
GYPSUM BOARD ..... 120,326
FILL ..... 40,288
LATH ..... 118,325
MASONRY ..... 49,293
PLANK ..... 41,289
PLASTER ..... 119,326
SHEATHING ..... 71,303

- H -

H-BEAM FENCE ..... 22,279
H-SECTION PILE ..... 18,278
HALF ROUND GUTTER ..... 93,313
ROUND MOULDING ..... 74,304
SURFACE BUTT ..... 110
HALIDE FIXTURE ..... 261,394
HALON SYSTEM ..... 179
HAMMERED GLASS ..... 114,323
HAND BUGGY ..... 37,287
DRYER ..... 144,337
EXCAVATION ..... 17,277
SPLIT SHAKE ..... 85,310
HANDHOLE ..... 246,387
HANDICAPPED PARTITION ..... 134,333
PLUMBING ..... 211,370
SYMBOL ..... 21,279
HANDLE LOCK DEVICE ..... 253,390
HANDLING AIR ..... 223,376
RUBBISH ..... 10,274
WASTE ..... 149,340
HANDY BOX ..... 243,386
HANGAR DOOR ..... 103,318
HANGER CHANNEL ..... 231,380
JOIST ..... 63,299
ROD ..... 231,380
HARDBOARD DOOR ..... 101,317
UNDERLAYMENT ..... 71,302
HARDENER CONCRETE ..... 36,287
HARDWARE ..... 109,321
HASP ASSEMBLY ..... 112,322
HAT RACK ..... 145,338
HATCH ROOF ..... 94,314
HAULING MATERIAL ..... 15,276
RUBBISH ..... 10,274
HEAD POP-UP ..... 21
HEADER ..... 70,302
CEILING ..... 65,300
HEADWALL REMOVAL ..... 15,276
HEAT DETECTION ..... 265,396
TRANSFER ..... 374
HEAT/SMOKE VENT ..... 94,314
HEATER ..... 5
ELECTRIC ..... 153,269,342,398
UNIT ..... 222,375
WATER ..... 214,371
HEATING & VENTILATING ..... 373
HEAVY CASTING ..... 60,298
DUTY DOOR ..... 98,316
DUTY TERRAZZO ..... 123
TIMBER ..... 72,303
HERRINGBONE FLOOR ..... 127
HEX NUT ..... 172
HIGH PRESSURE SODIUM ..... 261,394
SEAM ROOF ..... 86,310
STRENGTH BLOCK ..... 49,292
HIGHWAY DEMOLITION ..... 10,274
HINGE ..... 109,110
COVER WIREWAY ..... 240,384
HIP RAFTER ..... 68,301
HOIST INDUSTRIAL ..... 167,349
HOLDER DOOR ..... 111,321
HOLE CLIP ..... 173,352
CONDUIT ..... 231,380
PLASTER ..... 120,326
HOLLOW BLOCK ..... 48,292
CLAY TILE ..... 50,293
COLUMN ..... 78,306
CORE DOOR ..... 101,102,317
METAL DOOR ..... 98,316
METAL FRAME ..... 98,316
HOOD EXHAUST ..... 151,341
RANGE ..... 153,342
HOOK WIRE ..... 174,353
HORIZONTAL DAMPER ..... 225,377
REINFORCING ..... 34, 43,285,290
SIDING ..... 87,311
WINDOW ..... 107,320
HORN ..... 265,396
WATERPROOF ..... 265,396
HOSE BIBB ..... 21,203,366
REEL ..... 216,373
HOSPITAL BED ..... 158,344
CUBICLE ..... 135,333
EQUIPMENT ..... 156,344
FAUCET ..... 211,370
RECEPTACLE ..... 246,387
TRACK ..... 135,333
WARDROBE ..... 144,338
HOT LINE CLAMP ..... 248,388
PLATE LABORATORY ..... 156,343
WATER STORAGE TANK ..... 216,372
HUNG SLAB ..... 28,283
HVAC CONTROL ..... 228,378
HYDRANT ..... 211,370
WALL ..... 178,354
HYDRAULIC ELEVATOR ..... 164,348
EXCAVATOR ..... 16, 17,276,277
LIFT ..... 155,343

- I -

ICE CUBE MAKER ..... 152,341
MAKER ..... 153,342
STORAGE BIN ..... 152,342
IDENTIFYING DEVICE ..... 137,334
IMC CONDUIT ..... 239,384
IMPACT BARRIER ..... 27,281
HEAD ..... 21
IN LINE PUMP ..... 215,372
LINE SPLITTER ..... 269,398
INCANDESCENT FIXTURE ..... 258,393
LAMP ..... 259,393
INCINERATOR LABORATORY ..... 156,343
MEDICAL ..... 149,340
INCREASING FEMALE ADAPTER ..... 185,358
MALE ADAPTER ..... 186,358
INCUBATOR ..... 157,344
LABORATORY ..... 156,343
INDIRECT LUMINARE ..... 262,394
INDUSTRIAL COMPACTOR ..... 149,340
EQUIPMENT ..... 154,343
FLOURESCENT ..... 260,394
HOIST ..... 167,349

INDUSTRIAL WINDOW ............... 107,320
WOOD FLOOR ............... 127,329
INERTIA SENSOR ............... 266,397
INFRA-RED SENSOR ............... 266,397
INJECTION GROUT ............... 41,289
INLAID TAPE ............... 21,279
INLET CONNECTION ............... 178,354
INSULATED BLOCK ............... 48,292
GLASS ............... 113,322
GLASS DOOR ............... 105,319
ROOF ............... 90,312
SIDING ............... 88,311
VAULT DOOR ............... 146,339
INSULATION ............... 80,307
BLOWN-IN ............... 83,308
CALCIUM ............... 176,353
DUCT ............... 177,354
EQUIPMENT ............... 177,354
FASTENERS ............... 81,308
MASONRY ............... 83,309
PIPE ............... 175,353
REMOVAL ............... 12,275
RIGID ............... 81,308
SPRAYED ............... 83,309
INTERIOR PAINTING ............... 131,331
SIGN ............... 138,335
WALL FORM ............... 30,284
INTERMEDIATE CONDUIT ............... 239,384
PIVOT HINGE ............... 109
INVERTED ROOF SYSTEM ............... 90,312
TRAP ............... 209,369
IONIZATION DETECTOR ............... 265,396
IRON BLACK ............... 198,364
MISCELLANEOUS ............... 55,296
TREAD ............... 58,297
IRREGULAR STONE ............... 47,292
IRRIGATION LAWN ............... 21
ISOLATION MONITOR ............... 158,344
ISOLATOR VIBRATION ............... 175,353

- J -

JACK RAFTER ............... 68,301
JACKET ALUMINUM ............... 176,354
JALOUSIE WINDOW ............... 107,320
JIB CRANE ............... 168,350
JOB BUILT FORM ............... 28,283
REQUIREMENTS ............... 3
SCHEDULING ............... 2
TRAILER ............... 3
JOINT CLIP ............... 119,326
EXPANSION ............... 34, 61,286,298
FILLER ............... 45,291
REINFORCING ............... 43,290
SEALER ............... 96,315
JOIST BRIDGING ............... 63,299
FLOOR ............... 65,300
HANGER ............... 63,299
PAINTING ............... 130,331
SISTER ............... 65, 66,300
JUMBO BRICK ............... 46,291
JUNCTION
BOX ............... 243,245,268,385,386,397
JUTE PADDING ............... 129,330

- K -

K COPPER ............... 182,356
KEENES CEMENT ............... 119,326
KETTLE STEAM ............... 151,341
KEY LOCK ............... 141,336
KEYED JOINT ............... 35,286
KEYWAY FORM ............... 30,284
KICK PLATE ............... 61,112,298,322
KILN DRIED ............... 77
KING POST TRUSS ............... 73,304
KITCHEN CABINET ............... 76,159,305,346
COMPACT ............... 149,340
FAUCET ............... 211,370
SINK ............... 212,371
KNOTTY PINE ............... 77,305

- L -

L TUBE ............... 183,357
LABORATORY EQUIPMENT ............... 155,343
LADDER METAL ............... 58,297
PAINTING ............... 131,331
POOL ............... 162,347
REINFORCING ............... 43,290
LAMINATED GLASS ............... 114,323
PANEL ............... 141,336
PLASTIC ............... 76,305
RAILING ............... 61,298
RUBBER ............... 148,340
LAMP INCANDESCENT ............... 259,393
LANDING STEEL ............... 58,297
TERRAZZO ............... 124,328
LATCHSET ............... 110,321
LATEX PAINT ............... 133
LATH AND PLASTER ............... 118,325
EXPANDED ............... 118,325
STUCCO ............... 119,326
LATHE SHOP ............... 155,343
LATTICE MOLDING ............... 74,304
LAUAN DOOR ............... 101,317
VENEER ............... 77,305
LAUNDRY EQUIPMENT ............... 148,339
LAVATORY ............... 212,370
CARRIER ............... 214,372
FAUCET ............... 211,370
LAWN IRRIGATION ............... 21
LEAD COATED COPPER ............... 93,313
LINED DOOR ............... 98
LINED FRAME ............... 99
LETTER BOX ............... 141,336
SLOT ............... 141,336
LEVEL FLOOR ............... 129,330
GAUGE ............... 208,369
LEVELER DOCK ............... 148,340
LIFEGUARD CHAIR ............... 162,347
LIFT ............... 166,349
AUTO ............... 166,349
HYDRAULIC ............... 155,343
LIGHT CASTING ............... 60,298
DENTAL ............... 158,345
FIXTURE LAMP ............... 262,395
GAGE FRAMING ............... 57,296
POOL ............... 162,347
SHIELD ............... 27,281
SURGICAL ............... 157,344
TRACK ............... 258,393
LIGHTING ............... 257,258,392,393
LIGHTNING ARRESTER ............... 248,388
PROTECTION ............... 264,396
LIGHTWEIGHT BLOCK ............... 48,292
LIMESTONE AGGREGATE ............... 47,292
PANEL ............... 52,294
LINE FILTER ............... 264,395
MARKING ............... 21,279
POST ............... 22, 23,279,280
LINEN CHUTE ............... 166,167,349
LINER FLUE ............... 54,295
PANEL ............... 86,162,310,347
LINOLEUM REMOVAL ............... 9,273
LINTEL LIMESTONE ............... 52,294
MASONRY ............... 50,293
STEEL ............... 50,293
LIQUID CHALKBOARD ............... 134,333
LEVEL GAUGE ............... 208,369
PENETRATION ............... 170,351
LITHONIA LIGHT ............... 262,395
LOAD BEARING STUD ............... 117,325
CENTER BREAKER ............... 249,388
LOADER ............... 6, 7
LOADING DOCK EQUIPMENT ............... 148,340
LOCKER CLOTHING ............... 139,335
LOCKNUT CONDUIT ............... 236,382
LOCKOUT SWITCH ............... 255,391
LOCKSET ............... 110,321
LONG FORGED NUT ............... 189,359
SWEEP ABS ............... 193,362
LOOSE FILL INSULATION ............... 83,308
LOUVER DOOR ............... 102,136,318,334
METAL ............... 61,298
PENTHOUSE ............... 227,378
VENT ............... 135,333
LUBE EQUIPMENT ............... 216,372
LUMINARE ............... 262,267,394,397

- M -

M TUBE ............... 183,357
MAGNETIC DOOR HOLDER ............... 112,322
STARTER ............... 255,391
MAHOGANY VENEER ............... 77,305
MAIL CHUTE ............... 140,336
MAINTENANCE EQUIPMENT ............... 146,339
MALE ADAPTER COPPER ............... 186,358
CONNECTOR ............... 188,359
CONNECTOR STAINLESS ............... 197,363
MALLEABLE IRON ............... 198,364
MAN HOLE COMMUNICATION ............... 268,397
MANHOLE ELECTRIC ............... 246,387
REMOVAL ............... 13,275
MANUFACTURED ROOF . 86, 94,310,314
WALL ............... 86,310
MAP RAIL CHALKBOARD ............... 134,333
MAPLE FLOOR ............... 127,329
MARBLE IMPORTED ............... 52,294
MASONRY ............... 43,290
ACCESSORY ............... 43,290
BASE ............... 52,294
CHIMNEY ............... 46, 54,291,295
CLEANING ............... 53,295
CONTROL JOINT ............... 45,291
DEMOLITION ............... 8,273
FACING PANEL ............... 52,294
FENCE ............... 13,275
FLASHING ............... 45,291
FLOOR ............... 127,329
FURRING ............... 66,300
GLASS ............... 51,294
GROUT ............... 43,290
GYPSUM ............... 49,293
INSULATION ............... 83,309
LIMESTONE ............... 52,294
MANHOLE ............... 13,275
PAINTING ............... 130,331
PLAQUE ............... 138,335
PLASTER ............... 51,294
REINFORCING ............... 34,285
RESTORATION ............... 53
SANDBLAST ............... 53,295
SLATE ............... 53,294
STONE ............... 52,294

MASONRY WATERPROOFING ....80,307
MAT CONCRETE....39,288
FLOOR....160,346
FORMWORK....29,283
REINFORCING....33, 34,285
MATERIAL HANDLING....166,349
MEAT CASE....152,342
MECHANICAL....170,351
FEE....2
SPECIALTY....170,351
MEDICAL EQUIPMENT....156,343,344
WASTE....149,340
MEDICINE CABINET....144,337
MEMBRANE CURING....37,287
NEOPRENE....79,307
ROOF....89, 90,312
WATERPROOFING....79,307
MERCURY VAPOR....263,395
MESH SLAB....33,285
TIE....45,291
WIRE....39,288
MESSAGE TUBE....167,349
METAL....55,296
ANCHOR....55
BASE....119,326
BUILDING....161,347
CEILING SYSTEM....126,329
CLAD DOOR....103,318
DECKING....56,296
DOOR....98,316
DOOR FRAME....98,316
DOOR REMOVAL....8,273
DUCTWORK....224,376
FABRICATION....58,297
FASTENING....55,296
FENCE....22,279
FLASHING....91,312
FRAMING....57,296
GRATING....59,297
HALIDE....261,262,394,395
LADDER....58,297
LATH....118,325
LINER PANELS....86,310
LINTEL....50, 56,293,296
LOUVER....98
PAN STAIR....58,297
PAN TILE....125,329
PLAQUE....137,334
POLE....247,387
RAILING....59, 61,297,298
ROOF REMOVAL....9,273
SHELVING....143,337
SHINGLE....84,309
SIDING PANEL....87,310
SPECIALTY....61,298
STAIR....58,297
STUD....57,117,296,325
TOILET PARTITION....134,333
WINDOW....106
WINDOW REMOVAL....9,273
METALLIC COLOR TILE....122
HARDENER....36,287
OXIDE MEMBRANE....79,307
METER SERUM....156,343
MICROSCOPE LABORATORY....156,343
MILL FRAMED STRUCTURE....72,303
MILLED WINDOW....109,321
MILLWORK....76,305
PAINTING....131,331
MINERAL FIBER BOARD....40,288
FIBER PANEL....125,328
FIBER TILE....125,328
WOOL....83,309
MINERAL WOOL INSULATION....81,308
MIRROR....144,337
PLATE....114,323
TILE....114,323
MISCELLANEOUS CASTING....60,298
FABRICATION....298
FIXTURE....214,372
FORMWORK....30,284
IRON....55,296
LOCK....111,321
STEEL....130,331
MIXER KITCHEN....152,341
MIXING VALVE....208,369
MOBILIZATION....4
MODIFIER EPOXY....41
MOGUL BASE....263,395
MOISTURE PROTECTION....79,307
RELIEF VENT....94,314
MOLDED CASE BREAKER....249,388
RUBBER BUMPER....149,340
MOLDING....73,304
MONITOR RACK....266,397
MONOLITHIC TERRAZZO....123,328
MOP SINK....213,371
MORTAR GROUT....43,290
MORTISE LOCKSET....110,321
MOSAIC TILE....122
MOTION DETECTOR....267,397
MOTOR CONTROL....254,391
OPERATOR....26,281
STARTER....255,391
MOTORIZED CHECKROOM....147,339
MOVABLE LOUVER....61,298
PARTITION....141,336
MULLION SECTION....115,323
STOREFRONT....100,316
WINDOW....107,320
MUSHROOM STATIONARY....95,314

- N -

NAILER CEILING....65,300
NAMEPLATE PLAQUE....138,334
NAPKIN DISPENSER....144,337
NARROW STILE DOOR. 100,116,316,324
NATURAL CLEFT SLATE....47,292
NEAR-WHITE BLAST....132,332
NEMA SWITCH....250,389
NEOPRENE FLASHING ... 90, 92,312,313
JOINT....34,286
MEMBRANE....79,307
ROOF....90,312
NET SAFETY....4
NIGHT DEPOSITORY....146,339
NIPPLE BLACK IRON....199,364
STEEL....201,366
NM CABLE....242,385
NO-HUB PIPE....179,181,355,356
NON-CELLULAR DECKING....56,296
NON-DESTRUCTIVE TESTING....170,351
NON-DRILLING ANCHOR....56
NON-FUSED SWITCH....250,389
NON-METALLIC HARDENER....36,287
NON-SLIP TERRAZZO....124,328
NORMAN BRICK....46,291
NOSING RUBBER....128,330
NYLON CARPET....129,330

- O -

O RING....96,315
OAK DOOR....103,318
FLOOR....127,329
VENEER....78,305
OBSCURE GLASS....114,323
OFFICE CUBICLE....141,336
PARTITION....141,336
SAFE....146,161,339,347
TRAILER....3
OGEE MOLDING....74,304
OIL BASE CAULKING....97,315
FIRED BOILER....217,373
FURNACE....218,373
SWITCH....255,391
OPEN TYPE DECKING....56,296
OPPOSED BLADE....229,379
ORGANIC BED TILE....121,327
ORNAMENTAL METAL....61,298
OS&Y VALVE....204,367
OUTDOOR FLUORESCENT....261,394
OUTRIGGER FLAGPOLE....137,334
OVEN BAKE....150,341
CONVECTION....150,341
OVERHEAD....1
DOOR....103,104,161,318,319
OVERSIZED BRICK....46,291

- P -

P-TRAP ABS....193,362
NO-HUB....181,356
PACKAGED CHILLER....219,374
HEATER....221,375
PAD EQUIPMENT....29, 37,283,287
REINFORCING....32,285
PADDING CARPET....129,330
PAINT....130,132,331
ASPHALTIC....80,307
DOOR....131,331
LETTER....131,331
MISCELLANEOUS....131,331
PIPE INSULATION....131,331
PREPARATION....132,332
SPRAY BOOTH....154,343
PAN METAL STAIR....58,297
SLAB....28,283
PANEL ACOUSTICAL....125,328
LIMESTONE....52,294
LINER....86,310
SIDING....87,311
WORK....77,305
PANELBOARD....253,390
PANIC DEVICE....111,321
PAPER BACKED INSULATION....81,308
BACKED LATH....119,326
BUILDING....80,307
CURING....37,287
PARABOLIC TROFFER....261,394
PARALLEL BAR....154,343
BLADE....229,379
DAMPER....225,377
PARGING....51,294
PARKING BARRIER....27,282
STALL PAINTING....21,279
PARQUET FLOOR....127,329
PARTICLE BOARD INSULATION....82,308
PARTING BEAD....74,304
PARTITION CLAY TILE....51,293
REMOVAL....8,273
TOILET....134,333
WIRE MESH....142,336

## INDEX


PASS-THROUGH WASHER........148,339
PASSENGER ELEVATOR..........163,348
PATCH CONCRETE......................36,286
HOLE........................................119,326
PATIO BLOCK..............................47,292
PATTERN STRAINER.................209,369
PAVEMENT CUTTING..................11,274
DEMOLITION............................10,274
MARKING.................... 11, 21,274,279
REMOVAL................................10,274
PAVER GRANITE........................47,292
MARBLE..................................52,294
MASONRY...............................47,292
STONE......................... 47, 53,292,294
PAVING CONCRETE....................20,279
SURFACING............................20,278
PECAN VENEER..........................78,305
PEDESTAL FLOOR....................136,334
PEDESTRIAN BARRICADE...................3
PEGBOARD................................77,305
PENETRATION SEAL...................97,315
PENNSYLVANIA SLATE..............85,309
PENTHOUSE LOUVER..............227,378
PERFORATED STRAP...............173,352
PERFORMANCE BOND.........................7
PERLITE.....................................40,288
INSULATION.............................83,309
ROOF INSULATION.................82,308
PERSONNEL LIFT......................166,349
PEST CONTROL........................137,334
PHOTO EQUIPMENT.................153,342
PHOTOELECTRIC SENSOR.......266,397
PHOTOGRAPHS..................................3
PHYSICAL THERAPY.................156,344
PICKUP TRUCK....................................6
PICTURE WINDOW............. 107,108,320
PILASTER........................ 30, 40,284,288
PILE CAP CONCRETE.................38,288
CAP FORMWORK.....................29,283
CAP REINFORCING..................33,285
REPAIR....................................42,289
SPALL......................................42,289
PILING..................................................18
MOBILIZATION..................................4
PINE DECKING..........................72,303
DOOR....................................103,318
RISER......................................78,306
SIDING.....................................89,311
PIPE ABS..................................192,361
BOLLARD................................26,281
CAP........................................237,382
COPPER..................................182,356
CPVC......................................190,360
FIBERGLASS...........................194,362
FLASHING................................90,312
GALVANIZED..........................200,365
GLASS....................................189,359
HANGER................... 170-172,351,352
IDENTIFICATION...........................198
INSULATION............................175,353
INSULATION REMOVAL...........12,275
NO-HUB.................... 179,181,355,356
PAINTING...............................130,331
PILE.........................................19,278
PILE POINT................................19,278
POLYETHYLENE.......................191,361
POLYPROPYLENE....................191,360
PVC.........................................190,360
RAILING....................................59,297
REMOVAL..................................13,275
STAINLESS..............................195,362
PITCH POCKET............................92,313
PIVOT HINGE.......................................109
PLACING CONCRETE...... 36, 37,286,287
PLANTER.........................................27,282
PLAQUE BRONZE.......................137,334
MASONRY...............................138,335
PLASTER..................................119,326
ACCESSORY...........................119,326
MASONRY.................................51,294
PAINTING................................131,331
PATCH....................................120,326
REMOVAL......................................9,273
RING.........................................244,386
PLASTERBOARD........................120,326
PLASTIC COATED CONDUIT.....239,383
CONDUIT................................237,383
CONDUIT BUSHING................236,382
PIPE.........................................194,362
SIDING........................................87,311
SIGN........................................138,335
SKYLIGHT....................................96,315
SQUARE BOX..........................244,386
PLATE.............................................70,302
ARMOR....................................112,322
GLASS......................................113,322
MIRROR....................................114,323
SHEAR........................................63,299
SWITCH.................... 255,256,391,392
TOE...............................................60,297
PLENUM CEILING......................126,329
PLEXIGLASS..............................113,322
PLUG CAST IRON.......................202,366
CONDUIT.................................243,385
NO-HUB...........................................181
PLUMBING.................... 170,179,351,355
FIXTURE..................................210,370
PLUNGER DOOR HOLDER.........111,321
PLYWOOD.......................................70,302
FINISH.......................................76,305
PAINTING.................................131,331
SHEATHING..................................71,303
SHELVING....................................75,305
SIDING..........................................88,311
STRUCTURAL...............................71,303
UNDERLAYMENT.........................71,302
PNEUMATIC CONCRETE.............36,287
HOIST.......................................166,349
SYSTEM....................................167,349
TOOL....................................................5
POCKET PITCH...............................92,313
POINTING MASONRY.......................53,295
POLE CLOSET....................................74,304
ELECTRIC.....................................268,398
LINE HARDWARE...........................247,387
UTILITY..........................................246,387
POLICE EQUIPMENT.........................147,339
POLYACRYLATE FLOOR...........124,328
POLYESTER FLOOR...................124,328
POLYETHYLENE PIPE...............191,361
VAPOR BARRIER.......................80,307
WRAP...........................................42,289
POLYPROPYLENE PIPE............191,360
POLYSTYRENE
INSULATION................ 82, 83,308,309
POLYSULFIDE ACRYLIC.............54,295
POLYURETHANE CAULKING.......97,315
FLOOR......................................127,329
JOINT...........................................34,286
MEMBRANE.................................79,307
POOL EQUIPMENT......................162,347
POP-UP HEAD...................................21
PORCELAIN SHINGLE..................84,309
SINK............................................213,371
POROUS FILL.................................36,286
PORTABLE STAGE........................147,339
PORTLAND CEMENT PLASTER.120,326
POST CORNER...............................23,280
HOLE...........................................26,281
LINE............................................23,280
POSTAL ACCESSORIES............140,336
SPECIALTIES..........................140,336
POSTS TREATED..............................72,303
POURED DECK................................40,288
INSULATION.................................83,309
VERMICULITE...............................83,309
POWER GENERATION..............248,388
LINE FILTER..............................264,395
MANHOLE..................................246,387
TOOL..........................................155,343
PRE-ENGINEERED BUILDING....161,347
PRECAST HANDHOLE...............246,387
LINTEL............................................50,293
TERRAZZO...................................123,328
PREFINISHED SHELVING............75,305
PREFORMED TAPE..........................21,279
PREMOLDED JOINT........................35,286
PREPARATION PAINTING.........132,332
PRESSER CLOTHING...............148,340
PRESSURE GAUGE...................228,378
REDUCING VALVE...................207,368
PRESTRESSED PILING..................18,278
PRIMED DOOR...........................102,318
SIDING.........................................88,311
PRIMER PAINT...................130,133,331
PRIVACY LOCKSET......................110,321
SLAT..............................................26,281
PROCESSOR PHOTO................153,342
PRODUCE CASE.........................152,342
PROFESSIONAL FEE.............................1
PROFIT......................................................1
PROJECTING SASH...................107,320
WINDOW.....................................107,320
PROTECTION SADDLE..............177,354
PROTECTIVE BOARD...................79,307
PULL BOX.............................243,385,386
CABINET......................................244,386
STATION......................................265,396
PUMP.............................................5,215,372
GAS...............................................216,373
SUMP............................................215,372
PUMPED CONCRETE......37, 39,287,288
PUSH PLATE.................................112,322
PVC CAULKING...............................96,315
CEMENT............................................238
CHAMFER STRIP..........................30,284
COATED HANGER........................173,352
CONDUIT.......................................237,383
CONTROL JOINT..............................45,291
FITTING..........................................190,360
FOAM................................................35,286
FORM LINER.....................................30,284
PIPE..................................................190,360
ROOF..................................................90,312
WATERSTOP.......................................35,286
PVC/CPVC PIPE...................................189,359

- Q -

QUARRY TILE.........................................123,327
QUARTER ROUND...................................75,304

- R -

RACEWAY....240,384
RACK-TYPE DISHWASHER....151,341
RADIATION UNIT....221,375
RADIATOR VALVE....207,368
RADIOLOGY EQUIPMENT....157,344
RADIUS CURB....53,294
FORMWORK....30,284
RAFTER....67,301
ANCHOR....63,299
HIP....68,301
RAIL CHAIR....74,304
PAINTING....131,331
RAILING....59, 75,297,304
METAL....61,298
RANCH CASING....74,304
MOLDING....75,304
RANDOM WIDTH FLOOR....127
RANGE ELECTRIC....153,342
FIRING....147,339
HOOD....153,342
KITCHEN....151,341
RAPID START LAMP....259,393
RECEIVING BOX....140,336
RECEPTACLE COMBINATION....246,387
RECEPTOR SHOWER.. 134,212,333,371
RECESS FILLER STRIP....126
RECESSED LIGHT FIXTURE....126
RECIPROCAL CHILLER....219,374
RECTANGULAR WALL TIE....43,290
RED BRICK....45,291
LEAD PAINT....133
REDUCED PRESSURE
ASSEMBLY....170,351
REDUCER COPPER....184,357
NO-HUB....181,356
REDUCING COUPLING COPPER183,357
FEMALE ADAPTER....186,358
TEE....198,364
REDWOOD SIDING....88, 89,311
REEL HOSE....216,373
REFLECTIVE PAINT....21,279
REFLECTOR LAMP....259,393
REFRACTORIES....54,295
REFRIGERATED CASE....152,342
REFRIGERATION....218
TUBING....183,356
REFRIGERATOR BLOOD....156,343
REFUSE HOPPER....149,340
REGISTER....227,378
WALL....227,378
REGLET....31,284
ROOF....92,313
REINFORCEMENT....31,284
REINFORCING
MASONRY....43, 50,290,293
STEEL REPAIRS....41,289
RELIEF VALVE....207,368
VENTILATOR....227,378
REMOVE CATCH BASIN....13,275
DRAINAGE PIPE....13,275
DUCT INSULATION....12,275
FENCING....13,275
GAS PIPING....13,275
HEADWALL....15,276
ROOF....90,312
SEPTIC TANK....14,276
SEWER PIPE....13,275
TANK....14,276
VALVE....14,275
WALL....14,276
REMOVE WATER PIPE....14,275
REPLACEMENT GLASS BLOCK....51,294
SLATE....85,309
REPOINT BLOCK....54,295
RESET MANHOLE....13,275
RESIDENTIAL CARPET....129,330
EQUIPMENT....152,342
LOUVER....61,298
RESILIENT BASE....128
FLOOR....128,330
TILE FLOOR....127,330
RESPIRATOR MASK....12
RESTORATION....53,295
RETARDANT FIRE....77
REVOLVING DOOR....106,319
RIB LATH....118,325
ROOF....86,310
RIBBED BLOCK....49,292
SIDING....88,311
WATERSTOP....35,286
RIDGE BOARD....68,301
ROOF....91,312
VENT....94,314
VENTILATOR....161
RIFLE RANGE....147,339
RIGID STEEL CONDUIT....234,381
URETHANE INSULATION....82,308
RING SPLIT....63,299
TOOTHED....64,299
RISER BAND....147,339
MARBLE....52,294
RESILIENT....128,330
TERRAZZO....124,328
WOOD....78,306
ROCK CAISSON....20,278
DRILL FENCE....26,281
ROCKWOOL INSULATION....83,309
ROD BEAM CLAMP....231,380
GROUND....252,390
GUY....268,398
HANGER....231,380
THREAD....231,380
TRAVERSE....159,346
ROLL ROOF....84, 91,309,312
ROLL-UP DOOR....103,104,161,318
ROOF ASPHALT....89,312
BALLAST....90,312
BOND....89
BUILT-UP....89,312
CURB....225,377
DECKING....56,296
DRAIN....209,369
EXPANSION....61,298
FLUID-APPLIED....90,312
FRAMING....67,301
HATCH....94,314
HIP....68,301
INSULATION....81, 82,308
PANEL....86,162,310,347
SCUPPER....92,313
SHEATHING....71,302
SHINGLE....84,309
SIPHON....95,314
SPECIALTIES....94,314
WALKWAY....89,312
ROOFING....310
REMOVAL....9,273
ROOFTOP AC....223,376
HEATER....221,375
ROSEWOOD VENEER....78,305
ROTARY IMPACT HEAD....21
ROUGH CARPENTRY....64,299
ROUGH-IN PLUMBING....211,370
ROUGH-SAWN SIDING....88,311
ROUND COLUMN....28,283
QUARTER....75,304
RUBBER BUMPER....148,340
LAMINATED....148,340
MAT....112,322
NOSING....128,330
RUBBERIZED ASPHALT....21,279
RUBBISH HANDLING....149,340
REMOVAL....10,274
RUBBLE STONE....52,294
RURAL LETTER BOX....141,336

- S -

S-FORM CONCRETE....31,284
SAFE OFFICE....146,161,339,347
SAFETY GLASS....114,322
NET....4
SWITCH....249,257,389,392
TREAD....128,330
SALAMANDER....5
SALES TAX....2
SALT PRESERVATIVE....77
SAND BED PAVER....47,292
FINISH STUCCO....120,326
SANDBLAST....132,332
MASONRY....53,295
SANDWICH PANEL....86,310
SANITARY CROSS NO-HUB....180,355
PIPING REMOVAL....13,275
T....191,361
T ABS....193,362
T NO-HUB....179,355
T SOIL....182,356
SASH BEAD....75,304
TRANSOM....99,316
SATIN FINISH LETTER....138
SAW BAND....155,343
CUTTING....11,274
SCAFFOLDING....4
SCHEDULING....2
SOFTWARE....3
SCOREBOARD GYM....154,343
SCORED CLAY TILE....50,293
SCRATCH COAT....120,326
SCREED....36,286
SCREEN MOLDING....75,304
VENT....135,333
SCREW COVER BOX....245,386
COVER CABINET....245,386
SCRUB SURGICAL....157,344
SCUPPER ROOF....92,313
SEAL CAULKING....96,315
COAT....20,279
FITTING....235,382
SECTIONAL DOOR....105,319
SECURITY EQUIPMENT....146,339
SASH....107,320
SYSTEM....266,267,396,397
SEDIMENT FENCE....18,277
SELECT COMMON BRICK....46,291
SELF-DRILLING ANCHOR....56
SENSOR PHOTOELECTRIC....266,397
SEPTIC TANK REMOVAL....14,276
SERVICE DOOR....104,318
ELECTRIC....252,390
SINK....212,371
WEIGHT PIPE....182,356
WINDOW....146,339

SET RETARDER ....36
SCREW COUPLING ....233,381
SHAG CARPET ....129,330
SHAKE ROOF....85,310
SHAPER SHOP ....155,343
SHEAR PLATE ....63,299
SHEATHING ROOF ....71,302
WALL ....71,303
SHEET FLASHING ....92,313
FLOOR....128,330
GLASS....113,322
LEAD....126,329
METAL ROOF....91,312
MIRROR ....114,323
PILING ....19,278
ROOF....90,312
SHEETROCK....120,326
FINISHING....121,326
PAINTING....131,331
SHELF ....75,142,305,337
CHECKROOM ....147,339
PAINTING....131,331
UNIT....143,337
SHELTER TRUCK ....149,340
SHIELD EXPANSION ....55
SHINGLE CEDAR....85,310
METAL....84,309
REMOVAL ....9,273
ROOF....84
WOOD ....85,309
SHOE MOLDING ....75,304
SHORT FORGED NUT ....189,359
SHOTCRETE....36,287
SHOWER....212,371
COMPARTMENT ....135,333
DOOR ....134,333
FAUCET....211,370
ROD....144,337
STALL....134,333
SHRUB HEAD....21
SIDE BEAM ....28,283
COILING GRILLE....104,318
SIDELIGHT ....99,100,316
SIDEWALK ....39,288
REMOVAL ....11,274
SIDING.... 86, 87,310,311
PAINTING ....130,331
STEEL....88,311
SIGN ....5
INTERIOR....138,335
PLASTIC ....138,335
SILICON CAULKING....97,315
SILICONE CAULK....54,295
DAMPPROOFING....80,307
ROOF....90,312
SILL ANCHOR ....63,299
LIMESTONE ....52,294
MARBLE ....52,294
SLATE....53,294
WINDOWWALL....116,323
SIMULATED PEG FLOOR....127
SINGLE GATE ....23,280
SINGLE-PLY ROOF....90,312
SINK ....212,371
CARRIER....214,372
DARKROOM ....153,342
SIPHON ROOF ....95,314
SIREN....265,396
SISTER JOIST....65, 66,300
RAFTER....68,301
SITE DEMOLITION ....275
GRADING ....16,276
IMPROVEMENT....21,279
SITEWORK....8,273
SKYLIGHT ....162,347
SLAB CHANNEL....40,289
CONCRETE ....39,288
FORMWORK....28, 29,283
MESH....33,285
REINFORCING .... 32, 33,284,285
SLATE....47,292
PANEL ....53,294
SHINGLE ....85,309
SLEEPER....69,302
SLIDE WATER....162,347
SLIDING FIRE DOOR ....103,318
GLASS DOOR....105,319
WINDOW ....108,320
SLIP COUPLING COPPER....184,357
SLOT ANCHOR ....44,290
SLURRY SEAL....21,279
SMOKE
DETECTOR.... 112,265,266,322,396
VENT ....94,314
SMOKESTACK MASONRY ....54,295
SNAPPED PAVER....47,292
SOAP DISH.... 123,144,327,337
SOCIAL SECURITY ....2
SODIUM LIGHT ....261,394
SOFFIT ....69,302
REPAIR....41,289
TILE ....122
VENT ....94,314
SOFTENER WATER....153,342
SOFTWARE SCHEDULING....3
SOIL PIPE....182,356
STABILIZATION....18,277
TESTING....3
TREATMENT ....18,277
SOLAR SCREEN BLOCK ....49,293
VALVE ....207,368
SOLID BLOCK ....48,292
CORE....101,317
DOOR ....101,317
WALL ....45,291
SOUND ABSORPTION WALL ....126,329
ATTENUATION....126,329
DEADENING BOARD ....119,326
SPALL PILE ....42,289
SPANDREL GLASS....113,322
SPECIAL CONSTRUCTION....161,347
DOOR .... 103,146,318,339
FLOORING....130,330
SYSTEM ....216,372
SPECIALTIES....134,333
SPELTER SOCKET ....268,398
SPINNER ROOF....95,314
SPLINE CEILING TILE....126,329
SPLIT COUPLING....235,382
GROUND FACE....48,292
RIB PROFILE....49,292
RING....63,299
SPONGE NEOPRENE....35,286
RUBBER PADDING....129,330
SPOOL INSULATOR....247,388
SPOT LAMP....260,394
SPOTLIGHT....259,393
SPRAY BOOTH ....154,343
SPRAYED INSULATION....83,309
SPREAD FOOTING ....38,287
TOPSOIL....16,276
SPREADER FEEDER ....148,340
SPRINKLER HEAD ....178,354
PAINTING ....131,331
SQUARE COLUMN....28,283
FACE BRACKET....244,386
FIXTURE....259,393
HEAD COCK....207,368
HEAD PLUG....202,366
TUB....210,370
STAGE BAND ....147,339
THEATER....147,339
STAGING .... 4
STAINLESS COUNTER ....159,346
DOOR.... 104
EXTINGUISHER....140,335
FITTING ....197,363
FLASHING ....91,312
GRAB BAR....144,337
PARTITION ....134,333
PIPE....195,362
RAILING....61,298
REGLET....92,313
SHELVING .... 143
STEEL PIPE....194,362
TILE....126,329
TUBING....196,363
STAIR ACCESS ....153,342
ACCESSORY....128,330
CONCRETE ....39,288
FORMWORK....29,283
LANDING ....58,297
MARBLE....52,294
METAL ....58,297
PAINTING ....131,331
PAN....58,297
RAIL....59,297
REINFORCING ....34,285
TERRAZZO....123,328
TREAD ....124,328
TREAD CERAMIC....123,327
STAIRWORK....78,306
STALL SHOWER ....134,333
STANDING-SEAM ROOF .86, 91,310,312
STARTER....255,391
MAGNETIC ....255,391
STATIONARY SIPHON....95,314
STEAM BOILER....217,373
CLEAN MASONRY ....53,295
GENERATOR MEDICAL....157,344
HEAT....220,374
KETTLE....151,341
STERILIZER....157,344
TRAP....209,210,369
STEAMER ELECTRIC....151,341
STEEL ANGLE....50,293
BOX BEAM....27,281
BRACE....247,388
CONDUIT....239,384
GRATE....59,297
GUTTER....93,313
HATCH....94,314
LINTEL....56,296
PAINTING ....130,331
PILE....18,278
PIPE.... 198,200,364,365
POLE....247,387
REPAIR....41,289
ROOF PANEL ....86,310
SHEET PILING....19,278
SHINGLE....84,309
SIDING....88,311
SIDING PANEL ....87,311
STAIR....58,297

STEEL STUD .......... 117,325
TANK .......... 175,353
TOILET PARTITION .......... 134,333
TRANSOM .......... 99,316
TREAD .......... 58,297
WELDING .......... 55,296
WINDOW .......... 106,320
STEP LIMESTONE .......... 52,294
STERILIZER GAS .......... 157,344
MEDICAL .......... 157,344
STILL WATER .......... 157,344
STIRRUP POLE .......... 247,388
STONE .......... 294
PAVER .......... 47,292
STOOL MARBLE .......... 52,294
MOLDING .......... 75,304
SLATE .......... 53,294
STOP GRAVEL .......... 92,313
MOLDING .......... 75,304
STORAGE SHELVING .......... 143,337
TANK .......... 216,372
TANK REMOVAL .......... 14,276
STOREFRONT .......... 106,319
STOREROOM LOCK .......... 110,321
STORM DOOR .......... 106,319
WINDOW .......... 107,320
STRAIGHT CURB FORM .......... 28,283
STRAINER .......... 208,369
STRANDED WIRE .......... 241,385
STRAP CHANNEL .......... 231,380
TIE .......... 63,299
STRAW BALE .......... 18,277
STREET REMOVAL .......... 10,274
STRETCHER HOSPITAL .......... 158,344
STRINGER TERRAZZO .......... 124,328
STRIP CANT .......... 69,301
CHAMFER .......... 30,284
FLOORING .......... 127,329
FLOURESCENT .......... 260,394
SHINGLE .......... 84,309
STROBE LIGHT .......... 265,396
STRUCTURAL EXCAVATION .......... 16,277
FEE .......... 1
PAINTING .......... 130,331
PLYWOOD .......... 71,303
TILE .......... 46,291
WELDING .......... 55,296
STUCCO .......... 120,326
LATH .......... 119,326
PAINTING .......... 130,331
STUD CLIP .......... 119,326
METAL .......... 57,117,296,325
REMOVAL .......... 8,273
STYROFOAM INSERT .......... 48,292
SUB-FLOORING .......... 70,302
SUBMERSIBLE PUMP .......... 5
SUMP PUMP .......... 215,372
SUPERINTENDENT .......... 2
SUPPORT CONNECTOR .......... 255,391
SYSTEM .......... 117,325
SURFACE BOLT .......... 111,321
COVER .......... 244,386
MOUNTED CLOSER .......... 111,321
MOUNTED RACEWAY .......... 240,384
SURGICAL CHRONOMETER .......... 158,344
LIGHT .......... 157,344
SCRUB .......... 157,344
TABLE .......... 157,344
SURVEILLANCE SYSTEM .......... 146,339
SURVEYING .......... 3
SUSPENDED CEILING .......... 80,307
CEILING REMOVAL .......... 9,273
SUSPENSION SYSTEM .......... 126,329
SWEEP SOIL .......... 182,356
SWIMMING POOL .......... 162,347
SWING CHECK VALVE .......... 178,354
DOOR .......... 142,337
DOWN DOOR .......... 95,314
GATE .......... 23, 25,280
SWITCH BOX .......... 243,386
DIMMER .......... 256,392
OIL .......... 255,391
PLATE .......... 256,392
SAFETY .......... 249,389
TRANSFER .......... 257,392
SWITCHBOARD .......... 252,390
SY-PINE DECKING .......... 72,303
SIDING .......... 88,311
TIMBER .......... 72,303
SYNTHETIC LATEX MASTIC .......... 124,328


- T -


T&G SIDING .......... 89,311
T-BAR SYSTEM .......... 126,329
T-SECTION BUMPER .......... 149,340
WATERSTOP .......... 91,313
T-SERIES TILE .......... 46,291
TABLE LABORATORY .......... 156,343
TACK BOARD .......... 137,334
COAT .......... 21,279
TANDEM BREAKER .......... 249,389
TANK TOILET .......... 213,371
TANK-TYPE EXTINGUISHER .......... 139,335
TARPAULIN .......... 4
TAX .......... 2
TEAK VENEER .......... 78,305
TEE ABS .......... 193,361
BLACK IRON .......... 198,364
BRASS .......... 188,359
CONNECTOR .......... 255,391
COPPER .......... 187,188,358,359
FIBERGLASS .......... 194,362
POLYETHLENE .......... 192,361
PVC .......... 189,190,360
STAINLESS .......... 195,362
TELEPHONE EQUIPMENT .......... 143,337
SYSTEM .......... 267,397
TELESCOPING BLEACHER .......... 154,343
TELEVISION SYSTEM .......... 269,398
TELLER WINDOW .......... 146,339
TEMPERED PEGBOARD .......... 77,305
TEMPERING VALVE .......... 208,368
TEMPORARY FACILITY .......... 3
TERMINAL ADAPTERS .......... 238,383
SCREW .......... 267,397
TERMITE CONTROL .......... 18,137,277,334
TERRA COTTA .......... 51,294
TERRAZZO .......... 123,327
EPOXY .......... 130,331
RECEPTOR .......... 134,333
REMOVAL .......... 8,273
TEST CONNECTION .......... 178,354
TESTING .......... 3
WELD .......... 170,351
TEXTURE 1-11 .......... 88,311
PAINTING .......... 130,331
THEATER EQUIPMENT .......... 147,339
THERMAL AND MOISTURE .......... 79,307
DETECTOR .......... 265,396
THERMOPLASTIC TAPE .......... 21,279
THERMOSTAT .......... 269,398
THERMOSTATIC VALVE .......... 208,369
THHN-THWN CONDUCTOR .......... 241,385
THIN-SET TERRAZZO .......... 123,328
THREAD ROD .......... 231,380
THREADED BOLT .......... 63,299
ROD .......... 171,351
VALVE .......... 203,366
THRESHOLD DOOR .......... 112,322
MARBLE .......... 52,294
THROUGH-WALL FLASHING .......... 45,291
THWN CONDUCTOR .......... 241,385
TIE MESH .......... 45,291
STRAP .......... 63,299
TILE ACOUSTICAL .......... 125,328
CERAMIC .......... 121,327
CLAY .......... 50,293
CONDUCTIVE .......... 122,327
QUARRY .......... 123,327
REMOVAL .......... 8,273
RESILIENT .......... 127,330
STRUCTURAL .......... 46,291
TERRAZZO .......... 123,328
TIMBER FRAMING .......... 66, 72,300,303
GUARD RAIL .......... 27,281
PARKING BARRIER .......... 27,282
PILE .......... 19,278
POLE .......... 268,398
TIME SWITCH .......... 256,392
TIMEKEEPER .......... 2
TINTED GLASS .......... 113,114,322,323
TISSUE DISPENSER .......... 144,337
TOE PLATE .......... 60,297,298
TOGGLE SWITCH .......... 256,392
TOILET .......... 213,371
ACCESSORY .......... 144,337
CARRIER .......... 214,372
PARTITION .......... 134,333
TOOTHBRUSH HOLDER .......... 144,338
TOOTHED RING .......... 64,299
TOP COILING GRILLE .......... 104,318
PLATE .......... 70,302
TOPPING CONCRETE .......... 40,288
TOWEL BAR .......... 144,338
BAR CERAMIC .......... 122,327
DISPENSER .......... 144,337
TOWER .......... 268,398
COOLING .......... 219,374
TRACK LIGHTING .......... 258,393
MOUNTED LOADER .......... 16,276
TRAILER .......... 3
TRANSFER SWITCH .......... 248,257,388,392
TRANSFORMER .......... 247,248,388
TRANSLUCENT PANEL .......... 96,315
TRANSMITTER AUTOMATIC .......... 265,396
ULTRASONIC .......... 266,397
TRANSOM .......... 99
SASH .......... 99,316
TRAP .......... 209,369
FLOAT .......... 209,369
GREASE .......... 202,366
TRASH CHUTE .......... 149,167,340,349
HANDLING .......... 149,340
TRAVELING CRANE .......... 168,350
TRAVERSE ROD .......... 159,346
TRAVERTINE FLOOR .......... 127,330
FLOORING .......... 52,294
TREAD ABRASIVE .......... 58,297
MARBLE .......... 52,294
METAL .......... 58,297
RESILIENT .......... 128,330
SAFETY .......... 128,330
STEEL .......... 58,297
WOOD .......... 78,306

TREATED POLE....246,387
POST....72,303
TREATMENT WOOD....76
TRENCH PAVEMENT....10,274
SHEETING....15,276
TRENCHER....16,277
TRENCHING....16,277
TRIANGULAR PILE CAP....29,283
TRIM ALUMINUM....92,313
CARPENTRY....73,304
CASING....74, 75,304,305
PAINTING....130,331
TRIPLE OVEN....150,341
TROFFER....261,394
TROWEL FINISH....119,326
FINISH STUCCO....120,326
TRUCK....6
CRANE....7
SHELTER....149,340
TRUSS REINFORCING....43,290
WOOD....73,303
TUB BATH....210,370
TUMBLER HOLDER....144,338
TUMBLESTONE AGGREGATE....47,292
TWIST LOCK RECEPTACLE....245,387

- U -

U BOLT....170,351
LAMP....261,394
UF CABLE....243,385
ULTRASONIC TRANSMITTER....266,397
UNDER SLAB SPRAYING....137,334
UNDERGROUND
TANK REMOVAL....14,276
UNDERLAYMENT....71,302
UNDERPINNING....15,276
UNEMPLOYMENT TAX....2
UNFACED INSULATION....81,308
UNGLAZED FLOOR....127,329
FLOOR TILE....121,327
WALL TILE....122,327
UNINTERRUPTIBLE POWER....264,396
UNION BLACK IRON....198,364
COPPER....187,188,358,359
ELBOW BRASS....188,359
STAINLESS....197,363
STEEL....201,365
T STAINLESS....197,363
UNIT HEATER....222,269,375,398
KITCHEN....149,340
MASONRY....45, 49,291,293
UNIVERSAL TERMINATION....267,397
UNRATED DOOR....98,316
UP DAMPER....225,377
UPS BACKUP....266,396
SYSTEM....264,396
URETHANE BOARD....177,354
INSULATION....82, 83,308,309
PADDING....129,330
RIGID....177,354
ROOF....90,312
URINAL....213,371
CARRIER....215,372
SCREEN....134,333
URN COFFEE....150,341
UTENSIL WASHER MEDICAL....157,344
UTILITY EXCAVATION....16,277
POLE....246,387

- V -

V-BEAM ROOF....91,312
VACUUM BREAKER....208,369
CARPET....129,330
SYSTEM....146,339
VALLEY FLASHING....91,312
VALVE....203,366
TRIM....178,354
VACUUM....146,339
VANITY BATH....159,346
VAPOR
BARRIER....36, 80,177,286,307,354
VARNISH PAINT....133
VAULT....161,347
DOOR....103,146,318,339
VAULTED ROOF PANEL....96,315
VEGETATION CONTROL....18,277
VEHICLE LIFT....166,349
VENEER....77,305
ASHLAR....53,294
BRICK....45,291
GRANITE....52,294
VENETIAN BLIND....160,346
TERRAZZO....124
VENT FOUNDATION....43,290
MASONRY....135,333
ROOF....94,314
SMOKE....94,314
SOFFIT....94,314
VENTILATOR GRAVITY....95,314
RELIEF....227,378
VERMICULITE....40,288
INSULATION....83,309
PLASTER....119,326
VERMONT SLATE....85,309
VERTICAL CHECK VALVE....205,367
CUTTER....151,341
GRAIN FLOOR....127,329
LIFT DOOR....105,319
LOUVER....135,333
REINFORCING 34, 43, 50,286,290,293
VIBRATION CONTROL....175,353
VINYL COATED FENCE....22,279
FACED PANEL....125
SHEET....128,330
SHEET FLOOR....128,330
SHEETROCK....121,326
SIDING....87,311
TILE....127,330
WALL COVERING....132,331
VISION GLASS....98
PANEL....146,339
VISUAL AID BOARD....137,334

- W -

WAFERBOARD SHEATHING....71,303
WAINSCOT TERRAZZO....124,328
WALK....39,288
WALKWAY ROOF....89, 90,312
WALL BASE....128,330
BLOCKING....64,299
BRICK....45,291
CABINET....76,159,305,346
CAVITY....40, 43,288,290
CLEANOUT....202,366
CONCRETE....39,288
COVERING....132,331
WALL EXPANSION....61,298
FABRIC....132,332
FINISH....36,286
FOOTING....29,283
FORMWORK....30,284
FRAMING....69,302
FURRING....57, 66,296,300
GRILLE....227,378
HEATER....153,342
HUNG SINK....212,370
HYDRANT....178,211,354,370
INSULATION....81, 82,308
LATH....118,325
LOUVER....135,333
MASONRY REINFORCING....34,285
MOUNTED FLAGPOLE....137,334
MOUNTED HEATER....269,398
PAINTING....131,331
PANEL....116,162,324,347
PENETRATION....170,351
PLASTER....119,326
RAILING....59,297
REINFORCING....34,285
REMOVAL....10, 14,273,276
SHEATHING....71,303
SHEETROCK....120,326
STONE....52,294
TIE....43,290
TILE....121,123,327
WALNUT DOOR....103,318
VENEER....78,305
WARDROBE....159,346
LOCKER....139,335
SPECIALTY....144,338
WASH FOUNTAIN....214,372
WASHER COIN....148,340
EXTRACTOR....148,339
WASHING BRICK....53,295
MACHINE BOX....213,371
WASHROOM FAUCET....211,370
WASTE HANDLING....149,340
RECEPTACLE....144,338
WATCHMAN....2
WATER BATH....156,343
BATH LABORATORY....156,343
CLOSET....213,371
COOLED CHILLER....219,374
COOLER....214,372
HEATER....214,371
MOTOR ALARM....178,354
PIPING REMOVAL....14,275
REDUCING ADMIXTURE....36
SLIDE....162,347
SOFTENER....153,342
STILL....157,344
VALVE....208,368
WATER-RESISTANT SHEETROCK....121
WATERPROOFING....79,307
WATERSTOP....35,286
T-SECTION....91,313
WAX FLOOR....127,329
WEARING SURFACE....20,278
WEATHERHEAD....236,382
WEATHERPROOF BELL....265,396
COVER....243,386
INSULATION....177,354
SWITCH....255,391
WEATHERSTRIPPING....112,322
WEED CONTROL....18,277
WELD TESTING....170,351
X-RAY....170,351
WELDED RAILING....59,297

# INDEX

WELDING ....................................55,296
TEST ..................................................3
WELDMENT CONNECTION ........255,391
WET BED TILE ...........................121,327
LOCATION FLUORESCENT....261,394
PIPE VALVE ...........................178,354
RUB ..........................................36,286
SPRINKLER SYSTEM ............178,354
VALVE .....................................178,354
WHEEL BARROW .........................17,277
CHAIR PARTITION .................134,333
WHEELED EXTINGUISHER........140,335
WHITE CEDAR SHINGLE.............85,310
METAL BLAST........................132,332
WIDE STILE DOOR ....... 100,116,317,324
WIND DRIVEN SPINNER .............95,314
WINDOW ............... 107,108,109,320,321
DRIVE-UP ...............................146,339
FRAME ...................................109,321
HEADER ...................................70,302
PAINTING ........................ 130,131,331
REMOVAL ..................................9,273
SERVICE ................................146,339
SILL ..........................................46,291
SILL MARBLE ............................52,294
STEEL.....................................106,320
STILL CERAMIC ......................123,327
STOOL MARBLE .......................52,294
TREATMENT ..........................159,346
WINDOWWALL............. 115,116,323,324
WIRE GLASS.............................114,323
GUARD ..................................261,394
HOOK HANGER .....................173,352
MESH........................... 33, 39,285,288
MESH REMOVAL .....................10,274
MESH WALL...........................142,336
STRANDED ............................241,385
WIREMOLD .............................240,384
RACEWAY..............................240,384
WIREWAY ...............................240,384
WOOD ANCHOR ..........................63,299
AND PLASTIC...........................63,299
CANT ........................... 69, 89,301,312
COLUMN...................................78,306
DECKING..................................72,303
DOOR ...............................................317
DOOR REMOVAL ........................8,273
FENCE REMOVAL.....................13,275
FIBER INSULATION .................82,308
FIBER PANEL..........................125,328
FIBER TILE.............................125,329
FLOOR REMOVAL ......................9,273
FLOORING .............................127,329
FRAME ...................................102,318
HANDRAIL.................................61,298
INDUSTRIAL FLOOR..............127,329
PAINTING ...............................130,331
PILE ..........................................19,278
PLATE.......................................70,302
POLE ......................................246,387
RISER........................................78,306
SHAKE.......................................85,310
SHELVING ................................75,305
SHINGLE ...................................85,309
SIDING.......................................88,311
SLEEPER ..................................69,302
SOFFIT ......................................69,302
STUD REMOVAL .........................8,273
TREATMENT ..................................76
TRUSS.......................................73,303
WALL .........................................69,302
WINDOW ................................108,320
WOOD WINDOW REMOVAL...........9,273
WOODWORK ARCHITECTURAL..77,305
WOOL CARPET.........................129,330
MINERAL ..................................83,309
WYE.........................................191,361
ABS.......................................193,362
FIBERGLASS..........................194,362
SOIL.......................................182,356

- X -

X-RAY .......................................158,345
EQUIPMENT ...........................157,344
WELD.......................................170,351
XHHW WIRE .............................242,385
XLP WIRE .................................242,385

- Y -

Y STRAINER..............................209,369

- Z -

Z-TIE WALL ................................44,290

# CONSTRUCTION ESTIMATOR(tm)

## for Windows

# System Tutorial

### Construction Cost Estimating System

### by CPR Software

### What is Construction Estimator?

Construction Estimator is a complete, stand-alone cost estimating system running under Windows and Windows 95. It is an abbreviated version of Visual Estimator, Construction Cost Estimating and Bidding System, by CPR Software.

Construction Estimator is distributed free of charge with BNI costbooks. However, for added benefits and productivity, we recommend to upgrade to Visual Estimator, which is described below.

### What is Visual Estimator?

Visual Estimator(tm) is the full-feature version of Construction Estimator which is also developed and published by CPR Software. In order to take advantage from its advanced functions and features, you can upgrade to Visual Estimator without losing your estimate and assembly files created by Construction Estimator. For optimal benefits, we recommend to upgrade to Visual Estimator by contacting BNI Building News at 1-800-873-6397 or CPR Software at 510-654-8338.

*Visual Estimator*

The following are some of the more advanced features and functions provided with Visual Estimator:

- A complete set of on-screen and printed reports.
- Complete cost database in your specific industry.
- Built-in summary and detail reporting. Preview on-screen or print on printer.
- "Adjust Cost Data" module for adjusting material costs for different geographic areas in the USA.
- "Add Cost Data" function used to add custom data items to the Cost Table.
- Printed, easy-to-read User Manual.
- Unlimited, free technical support by phone, fax and e-mail.
- Periodic newsletter and update notifications.

### Technical support for Construction Estimator

CPR Software provides 30-day free technical support for Construction Estimator installation and setup questions at:

Tel: 510-654-8338
Fax: 510-652-6067
E-mail: cpr@cprsoft.com

For usage related questions, there is a per-instance charge of $25.00 payable by VISA, MC or AMEX.

### Construction Estimator Installation

Construction Estimator is provided in a CD-ROM disk. Install Construction Estimator in your computer' s hard disk by running the program **INSTALL.EXE** provided inside the program disks. Construction Estimator can only be run from the hard disk.

The following are minimum system requirements to run Construction Estimator on your computer.

- Personal computer using 386 or higher
- MS-DOS operating system version 3.1 or later
- Microsoft Windows operating system 3.1 or Windows 95
- 8 MB of RAM memory
- Minimum 7 MB hard disk space
- CD-ROM disk drive
- VGA or SVGA monitor
- A compatible printer (such as Laser, InkJet, dot-matrix)
- A Microsoft or compatible mouse

### Running the Install Program

To start the Install program:

1. Insert Construction Estimator CD-ROM in drive.
2. In the File Manager, choose **Run** from the **File** menu. (For Win 95: Click **Start**, then choose **Run** from menu)
3. Type **d:install** (or **e:install** if your CD-ROM drive letter is e)
4. Follow the Install instructions on the screen.

The Install program will install the files in appropriate hard disk locations and also create a Program Group called Construction Estimator. You can start Construction Estimator by double-clicking the Construction Estimator icon.

Construction Estimator default **Installation Directory** is **C:\costest**.Follow screen instructions on the screen. Install program shall create a **Program Group** and **Program Icon** for Construction Estimator.

## What Does Install Program Do?

Construction Estimator's Install program carries out the following:

1. Copies all program files and data files into the **Installation Directory.**

2. Opens a subdirectory entitled '**ESTFILES**' under the Installation Directory and places a sample estimate provided with the distribution disk in this directory. Opens a subdirectory entitled '**ASSFILES**' under the Installation Directory and places a sample assembly provided with the distribution disk in this directory.

3. Writes Installation Directory path and name in the **WIN.INI** file in the Windows directory.

4. Adds "SHARE /F:5100 /L:500" statement to the **AUTOEXEC.BAT** file, if this statement does not already exist. If Install makes changes to the AUTOEXEC.BAT file, the original file is renamed to AUTOEXEC.BAK.

5. Copies necessary 'VBX' files and 'DLL' files to the computer's **WINDOWS/SYSTEM** directory. Install checks the dates on existing VBX and DLL files and only replaces if the existing files are of an earlier date.

## Building an Estimate

At any time, you may have several estimating projects that may be in progress and at different levels of completion. Each file is saved in your hard disk directory.

## Starting a New Estimate

In order to start a new estimate you have to direct Construction Estimator to create a new estimate file and you have to assign an **Estimate Number**. This is done by selecting **New Estimate** from the **File** menu or by clicking on the **New** command button on the **Main** toolbar. This shall open the New Estimate File dialog. Enter the Estimate Number of your choice for you new estimate. Remember, an Estimate Number is a maximum 8 digit alphanumeric, that is you can use numbers or alphabetic characters to assign an Estimate Number.

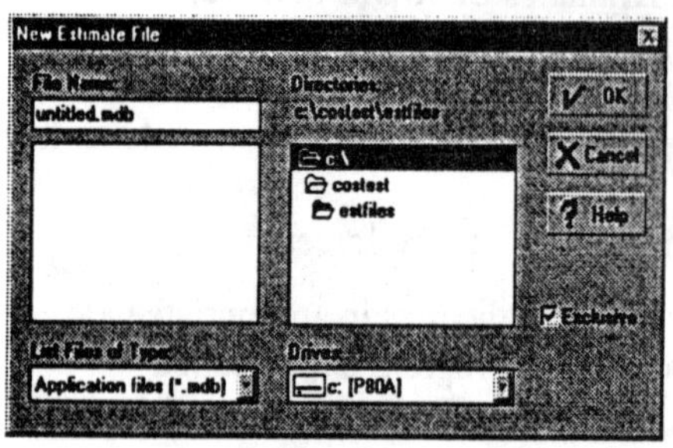


The default installation directory of Construction Estimator is named "**costest**" on the C drive.

Enter a name for the new estimate file. Note that the system indicated the default extension as "**mdb**". <u>Do not change the default extension since Construction Estimator recognizes estimate files by this extension only</u>. After you enter the name of the new estimate file, click your mouse on the **OK** button. The confirmation window shall appear.

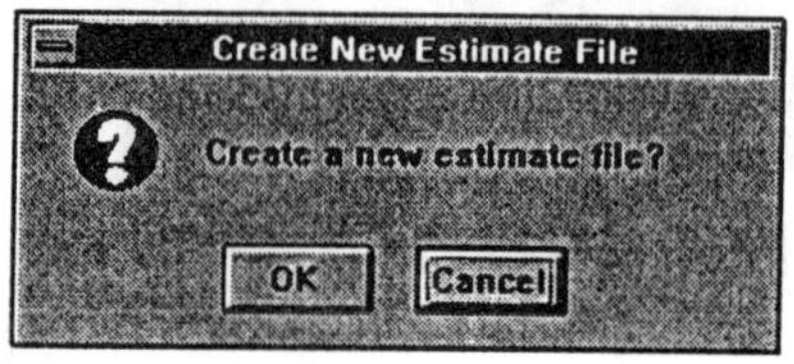


Click on the **OK** button to continue. At this point, clicking on the **Cancel** button shall abort the New estimate procedure.

When you click on the OK button to proceed, Construction Estimator shall quickly create the new estimate file.

At this point your new estimate file is created in the "estfiles" default subdirectory. This file has no records stored in it.

## Opening an Estimate File

To open an estimate file, select **Open Estimate File** menu command from the **File** menu or we click the **Open** button from the **Main** menu. The **Open Estimate File** dialog shall appear.

Select the estimate file to open, then click the **OK** button. When the selection is made, Construction Estimator shall open the estimate screen with the **Worksheet** tab active. The worksheet tab contains the Estimate table which is blank since we have not yet entered any cost data.

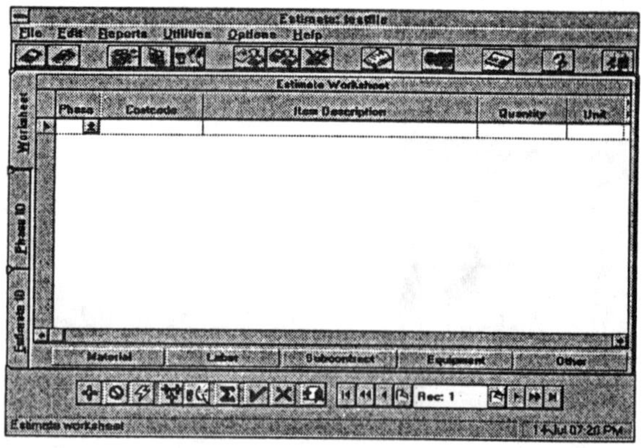


The toolbar below the Estimate table is called **Edit toolbar.** We shall use Edit toolbar buttons to carry out table maintenance functions.

Note that the Estimate screen has 3 tabs identifying pages which are **Worksheet, Phase ID** and **Estimate ID.** Worksheet page contains the Estimate table. Phase ID page contains the Phase table which is used to enter information related to identification of various phases of the project. Estimate ID page is used to enter information related to the estimate as well as estimate markups. The three pages of information entered under the three tabs constitute the estimate file.

## Adding Estimate Identification

Click the Estimate ID tab and Construction Estimator shall switch to that page. The Estimate ID page, in turn, contains two other tabbed pages which logically separate groups of information related to **Estimate Identification** and **Estimate Markups**. You will also see the **Save** button which is used to save entered information.

Enter the information in the appropriate fields. If you do not have the information at this point, you can always come back and edit and add to the data. At any point, click the **Save** button to save information to disk.

The **Estimate Markups** page contains the estimate markups. The markup fields contain default markup data for projects that you enter from the **Options** menu. At this point, you may change the default data and the new data shall be effective for the estimate.

When you complete adding Estimate Identification information, you can switch to any of the other two estimate screen pages.

## Adding Phase Identification

To switch to Phase Identification page, click on the **Phase ID** tab.

At this point, we identify the phases in the project first by a 2-digit **Phase** code, then by description and a brief memo. By using table maintenance functions, we add entries for each phase of the project.

The main section of the estimate file is the **Estimate Worksheet**, where all cost items related to the estimate are stored and maintained.

## Adding a cost item

Click the **Add** button of the **Edit** toolbar and Construction Estimator shall open a new blank record and place the record pointer on this new record. Start adding cell data. Click on the **Phase** code cell and the cell **Lookup** button will show. Click on the cell Lookup button once. The dropdown list shall open showing the contents of the **Phase ID** table. Remember, we entered phase identification data before we started estimate entries. Select the Phase code of your choice that is applicable to this cost item and click on it. The Phase code shall transfer to the Phase code cell and the dropdown list shall disappear.

Next, we click on the **Costcode** cell. When we click, the Costcode **Lookup** button shall appear inside the cell. At this point we can either write a unique costcode or select an existing standard costcode from the **Cost** table. To lookup the Cost table, we click on the Costcode Lookup button. The Lookup window shall open. Note that the "Filter by" option button selection is on **CSI**. This means that our selection in the CSI table (upper table) shall filter the Cost table on the selected CSI section. The filtered costcodes shall be listed in the Costcode table (lower table). We scroll the CSI table and bring the highlight on a CSI section. When we click on the entry, the Costcode table lists all records in the Cost table related to the selected section.

Now, we can scroll and search the Costcode table for the costcode that we want to export into the Estimate table. When we do, we click on the record so that it is selected (highlighted in color).

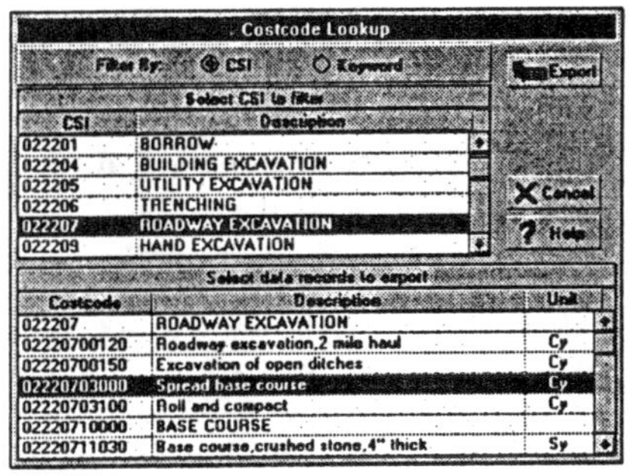


*Lookup*

We click on the **Export** button to indicate we are ready to export the selected record to the Estimate table. After displaying a message box for confirmation, Construction Estimator shall quickly export the selected record to the estimate table.

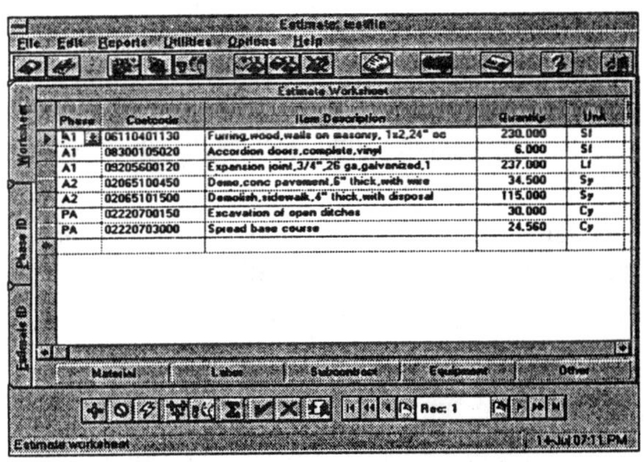

Having imported a cost item from the Cost table to the Estimate table, we proceed to edit the cost item for additional data. We can edit the Phase, Costcode, Description, Quantity, Unit and Note cells by simply clicking on the cell and overwriting the text.

The **Unit Cost** and **Total Cost** cells are displayed in blue which indicates that they cannot be directly modified since they are computed columns. These values are edited by using the **Detail** toolbar buttons at the bottom of the estimate table.

The Detail toolbar contains **Material, Labor, Subcontract, Equipment** and **Other** buttons. These are the cost categories that make up the total unit cost of the cost item. When we click on one of these buttons, the **Edit Detail** table appears. This table contains the user-definable information that are used to calculate the Unit Cost for the selected cost category.

We have entered a cost item in the Estimate table by selecting from the Cost table using the Lookup window. Now let's go one step further and add multiple cost items to the estimate table.

## Deleting cost items

To delete an item from the table, click on the leftmost column of the table. This shall highlight the cost item. You can select one or more items for deletion. After selecting the items for deletion, click on the **Delete** button on the **Edit** toolbar. Construction Estimator shall delete the selected items from the table permanently after user confirmation.

## Adding multiple cost items

To add multiple cost items, click the **Add multiple** button on the **Edit** toolbar. This opens the **Lookup** window. This time, let's view costcodes as filtered by description string. Click the **Keyword** option button of the Filter by window. Then type "window" in the text box. Click the **Done** button to indicate the end of string. Construction Estimator shall quickly filter the Cost table for the typed string and list matching costcodes in the costcode table (lower table). We can now select costcodes by clicking on the record. Remember, since we accessed the Lookup window from the **Add multiple** button, we can now select several items for export. After selecting several costcodes, click the **Export** button. This will cause Construction Estimator to export all selected costcodes to the Estimate table.

As usual, the imported costcodes shall not contain information on fields which are specific to the estimate. Therefore, each transferred costcode should be edited in order to complete all necessary job specific data.

## Appending from Another Estimate

Let's assume that we have an existing estimate which is similar in terms of work scope to the current estimate. In order to save time, we may want to transfer a number of cost items from the existing estimate to the current one. Construction Estimator can do this by using the **Append** function.

When you click the **Append** button from the **Edit** toolbar, Construction Estimator opens the **Append From File** dialog. Select the estimate file that you want to append cost items from. Then click the **OK** button. This shall open the Append window showing cost items from the selected estimate.

Select cost items from the **Append** table by clicking on each record. When finished, click the **Export** button and Construction Estimator shall quickly transfer all selected cost items from the existing estimate to the current estimate. Note that the transferred cost items may need to be edited individually to correctly reflect the current estimate.

## Screen Totals

At any time, you can get running totals for the Estimate table by clicking the **Total** button.

Note that screen totals are for direct costs , that is costs before markup application. You can get fully formatted screen reports with markup applications by making a selection from the list of available reports in the **Report** window. All Construction Estimator reports are available for screen preview prior to printing on printer.

## Adding a Work Assembly to Estimate

To add a work assembly to the estimate worksheet, select the "Add Work Assembly to Estimate" menu command from the **Edit** menu. This shall open the **Open Assembly File** dialog. Select the assembly to be added to the estimate. The **Assembly** window shall open showing the detail as well as total recap for the assembly.

When you click the **Export** button, Construction Estimator shall quickly transfer the recap for the assembly into the current estimate worksheet. The assembly recap shall be displayed as a cost item.

## Applying Crew To Cost item

To apply a crew to a cost item, place your highlight on the cost item. Then select "Apply Crew to Cost Item" menu command from the **Edit** menu. This shall open the **Select Crew** window where you can select the crew of your choice. When you click the **OK** button, Construction Estimator shall incorporate the selected crew data into the current cost item.

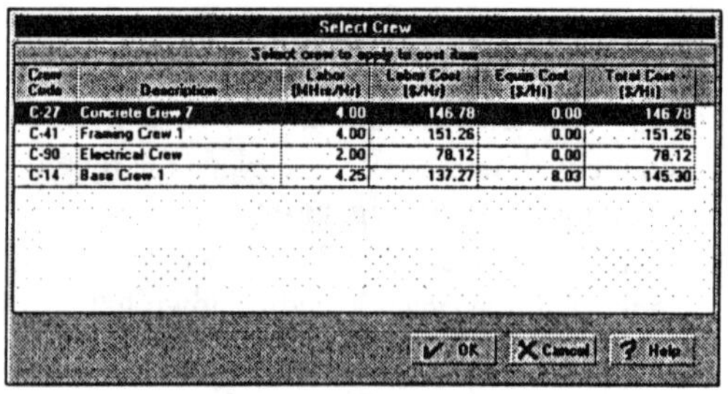


## Creating Estimate Reports

We can create screen and printed reports easily. To create reports, click the **Reports** button from the **Main** menu. This shall open the **Report** window.

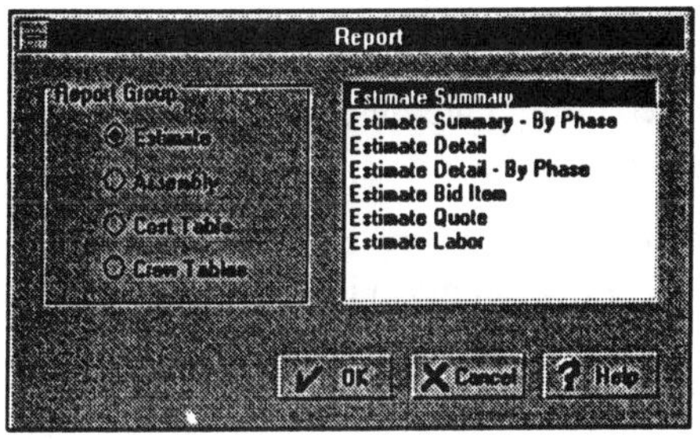


Select Estimate Detail report by clicking on it. Then click the **OK** button and Construction Estimator shall quickly generate and display the screen report for Estimate Detail.

**Important Note: Construction Estimator features Estimate Summary report as a sample for creating reports. Upgrade to Visual Estimator in order to have access to all on-screen and printed reporting functions.**

The **Report** toolbar at the bottom of the screen can be used for navigating the pages of the report as well as sending it to printer.

Construction Estimator offers convenient data export functions that are accessible from the screen report. Click on the **Export** button and you will get the following **Export** window.

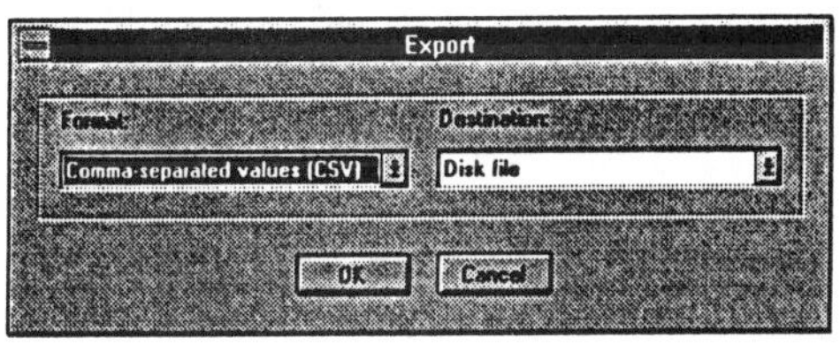


The **Format** list contains a large selection of export formats. Select your choice and Construction Estimator will open a File Dialog where you can indicate the file name and directory. When you make your selection, Construction Estimator shall quickly create the report in the selected format. You can now access and edit the report in your alternate program.

To exit the screen report, click on the **Close** button.

## Exiting Estimate Screen

To exit the estimate screen, click the **Exit** button from the **Edit** toolbar.

## Table Maintenance

**To add a new cost record**

 *Add button*

Click the **Add** button on the **Edit** toolbar. This will open a blank record at the end of the table. Enter your data. When ready, click the **Update** button or **Cancel** button as necessary.

**To edit an existing cost record**

 *Update button*  *Cancel button*

Click at the beginning of the field. Then overwrite edits.Click on **Update** button or **Cancel** button as necessary.

**To delete an existing cost record**

 *Delete button*

Click on the leftmost column of the record. The selected record is highlighted with color. You can select multiple columns for deletion. After selection is complete, click on the **Delete** button.
Construction Estimator will delete selected records after issuing a delete warning.

**To refresh the cost screen**

 *Refresh button*

This button shall read and display an updated copy of the underlying database file. It also generates the sort order of a table.

**To replicate a cost record (Cost table only)**

 *Replicate button*

This button will duplicate the current cost record in the **Cost** table.

**To add multiple cost items to estimate (Estimate worksheet only)**

 *Add Multiple button*

When you click the **Add Multiple** button on the **Edit** toolbar, the **Lookup** window shall open. Click the CSI section to filter, then select one or more records from the Costcode table by clicking on the record. When finished, click the **Export** button.

**To append from another estimate (Estimate worksheet only)**
When you select the corresponding submenu command from the **Edit** menu, the File dialog shall open. Select the source file and click **OK**. The **Append** window shall open. Select one or more records from the table by clicking on the record. When finished, click the **Export** button.

**To add work assembly to estimate (Estimate worksheet only)**

When you select the corresponding submenu command from the **Edit** menu, the File dialog shall open. Select the assembly file and click **OK**. The **Assembly** window shall open. Click the **Export** button to apply the assembly to the estimate.

**To apply crew to a cost item (Estimate and Assembly worksheet only)**

*Crew button*

When you click the **Crew** button on the **Edit** toolbar, the **Crew** screen shall open. Select the crew of your choice. Then click **OK**.

**To get estimate totals (Estimate and Assembly worksheet only)**

*Total button*

When you click the Total button on the Edit toolbar, Construction Estimator shall open the Total window and display totals for direct costs.

**To exit the table view screen**

*Exit button*

This button will exit the table view screen and go back to **Main** menu. The file shall be automatically saved in the hard disk.

**To edit cost category details**

Click on the corresponding button on **Detail** toolbar. The available buttons are **Material, Labor, Subcontract, Equipment** and **Other**. When the detail table opens, carry out entries or edits. When finished, click the **Apply** button to apply the edits to the table or click **Cancel** button to cancel edits.

## Work Assemblies

A work assembly is a work product that is made up of two or more cost items. By building work assemblies for commonly estimated work products and tying them to their measurement units, an estimator can save considerable time in repeated detailed estimating.

A few examples of commonly used assemblies are, cubic yard of concrete, square feet of concrete slab, built-up roofing measured in square feet or square, linear feet of partition wall, etc. You can build work assemblies and incorporate them into your estimates easily with Construction Estimator.

## Utility Functions

The Utilities menu command is used for Cost screen and Estimate screen related utility functions and contains the following submenu commands:

• **CPR Calculator** - A handy on-screen calculator with memory. To activate select menu item or click on **Calculator** button. For more detailed information on the CPR Calculator, see Chapter 3.

• **Adjust Cost Table** - Used to adjust cost data in the **Cost** table for local geographic areas. (This function is available with Visual Estimator only)

• **Compact** - Used to compact **Estimate** files, **Assembly** files and **Cost** table to occupy less space on the hard disk.

• **Repair** - Used to repair damaged **Estimate** files, **Assembly** files and **Cost** table.

• **Update** - Used to update data in **Estimate** files, **Assembly** files and **Cost** table based on current data tables.

## Help System

Construction Estimator's Online Help system is a convenient tool to get immediate context-sensitive help.

For information on Help topics, choose **Contents** from the Help menu or press **F1** and click the Contents button. You can use the Contents screen to jump to topics that describe how to use Construction Estimator and its functions.

• • • • • • • • • • • • • • • • • • • • • •

This tutorial briefly covers the main functions and features of Construction Estimator. Since Construction Estimator is an abbreviated version of Visual Estimator, many of its functions and features are the same as Visual Estimator. However, Visual Estimator also contains several new and advanced features for advanced productivity. It also comes with a printed User Manual and complete BNI cost database of user's choice.

For optimal benefits, we recommend to upgrade to Visual Estimator by contacting BNI Building News at:

**BNi Building News, Inc.**
1612 S. Clementine Street
Anaheim, CA 92802
**Tel: 1-888-BNI-BOOK**